SCHAUM'S OUTLINE OF

THEORY AND PROBLEMS

of

ENGINEERING MECHANICS

Statics and Dynamics

SECOND EDITION

•

BY

W. G. McLEAN, B.S. in E.E., M.S.

Professor of Mechanics
Lafayette College

AND

E. W. NELSON, B.S. in M.E., M.Adm.E.

Engineering Supervisor
Western Electric Co.

•

SCHAUM PUBLISHING CO.

257 Park Avenue South, New York 10

PRINTED IN THE UNITED STATES OF AMERICA

Typography by Signs and Symbols, Inc., New York, N. Y.

Preface

This book is designed to supplement standard texts, primarily to assist students of engineering and science in acquiring a more thorough knowledge and proficiency in analytical and applied mechanics. It is based on the authors' conviction that numerous solved problems constitute one of the best means for clarifying and fixing in mind basic principles. While this book will not mesh precisely with any one text, the authors feel that it can be a very valuable adjunct to all.

The previous edition of this book has been very favorably received. In this second edition most chapters have been revised and enlarged to keep pace with the most recent concepts, methods and terminology. The authors attempt to use the best mathematical tools now available to students at the sophomore level. Thus the vector approach is applied in those chapters where its techniques provide an elegance and simplicity in theory and problems. On the other hand we have not hesitated to use scalar methods elsewhere, since they provide entirely adequate solutions to most of the problems. Chapter 1 is a complete review of the minimum number of vector definitions and operations necessary for the entire book, and applications of this introductory chapter are made throughout the book.

Chapter topics correspond to material usually covered in standard introductory mechanics courses. Each chapter begins with statements of pertinent definitions, principles and theorems. This is followed by graded sets of solved and supplementary problems. The solved problems serve to illustrate and amplify the theory, present methods of analysis, provide practical examples, and bring into sharp focus those fine points which enable the student to apply the basic principles correctly and confidently. Numerous proofs of theorems and derivations of formulas are included among the solved problems. The large number of supplementary problems serve as a complete review of the material of each chapter.

The authors gratefully acknowledge their indebtedness to Professor Paul B. Eaton and Mr. J. Warren Gillon for their helpful advice and encouragement during the preparation of the first edition, and to Professors Charles L. Best and John W. McNabb for constructive criticism and suggestions during the preparation of the second edition. Special appreciation is expressed to Larry Freed and Paul Gary who checked the solutions to the problems. To the staff of Schaum Publishing Company, especially to Henry Hayden and Nicola Miracapillo, we extend particular thanks for valuable suggestions and helpful cooperation. We also wish to thank Mrs. Patricia Henthorn for typing the manuscript.

W. G. McLean
E. W. Nelson

August, 1962

Preface

CONTENTS

CONTENTS

Chapter 1

Vectors

DEFINITIONS

Scalar Quantities possess only magnitude, e.g., time, volume, mass, energy, density, work. Scalars are added by ordinary algebraic methods, e.g., 2 sec + 7 sec = 9 sec.

Vector Quantities possess both magnitude and direction, e.g., force, displacement, velocity, impulse. A vector is represented by an arrow whose head indicates its direction and whose length represents its magnitude. The notation for a vector in print is bold faced type **P**. The magnitude of this vector is represented by $|\mathbf{P}|$ or P. In script it is customary to place an arrow over the letter representing the vector **P** and the magnitude is represented by the letter P without the arrow.

A Free Vector may be moved anywhere in space provided it maintains the same direction and magnitude.

A Sliding Vector may be applied at any point along its line of action. By the *principle of transmissibility* the external effects of a sliding vector remain the same.

A Bound or Fixed Vector must remain at the same point of application.

A Unit Vector is a vector one unit in length.

The Negative of a Vector **P** is the vector −**P** which has the same magnitude but is oppositely directed.

The Resultant of a System of Vectors is the least number of vectors that will replace the given system.

ADDITION of TWO VECTORS

(*a*) **Parallelogram Law** states that the resultant **R** of two vectors **P** and **Q** is the diagonal of the parallelogram for which **P** and **Q** are adjacent sides. All three vectors **P**, **Q**, and **R** are concurrent as shown in Fig. 1-1 below.

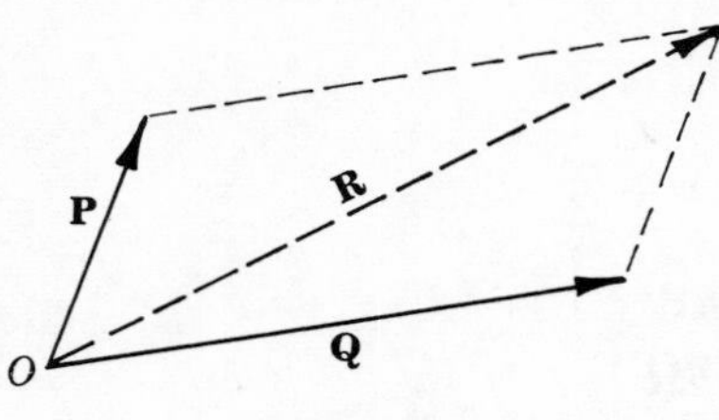

Fig. 1-1

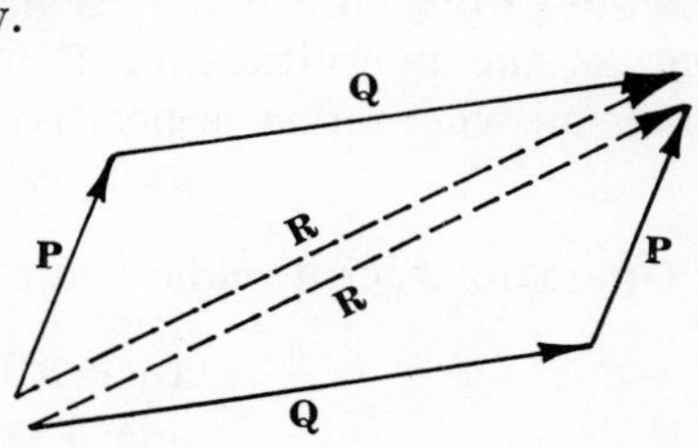

Fig. 1-2

(*b*) **Triangle Law.** Place the tail end of either vector at the arrow end of the other. The resultant is drawn from the tail end of the first vector to the arrow end of the other. The triangle law follows from the parallelogram law because opposite sides of the parallelogram are free vectors as shown in Fig. 1-2 above.

(*c*) **Vector Addition** is commutative, i.e., **P** + **Q** = **Q** + **P**.

SUBTRACTION of a VECTOR is accomplished by adding the negative of the vector, i.e.,

$$\mathbf{P} - \mathbf{Q} = \mathbf{P} + (-\mathbf{Q})$$

Note also that

$$-(\mathbf{P} + \mathbf{Q}) = -\mathbf{P} - \mathbf{Q}$$

ZERO VECTOR

A zero vector is obtained when a vector is subtracted from itself, i.e., $\mathbf{P} - \mathbf{P} = 0$. This is also called a null vector.

COMPOSITION of VECTORS

Composition of vectors is the process of determining the resultant of a system of vectors. A vector polygon is drawn placing the tail end of each vector in turn at the arrow end of the preceding vector as shown in Fig. 1-3. The resultant is drawn from the tail end of the first vector to the arrow end (terminus) of the last vector. As will be shown later, not all vector systems reduce to a single vector. Since the order in which the vectors are drawn is immaterial, it can be seen that for three given vectors **P**, **Q**, and **S**,

$$\begin{aligned} \mathbf{R} &= \mathbf{P} + \mathbf{Q} + \mathbf{S} = (\mathbf{P} + \mathbf{Q}) + \mathbf{S} \\ &= \mathbf{P} + (\mathbf{Q} + \mathbf{S}) = (\mathbf{P} + \mathbf{S}) + \mathbf{Q} \end{aligned}$$

The above equation may be extended to any number of vectors.

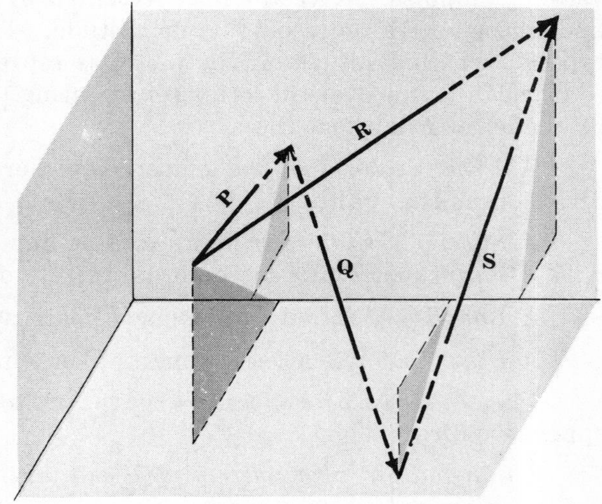

Fig. 1-3

MULTIPLICATION of VECTORS by SCALARS

(*a*) **The Product** of vector **P** and scalar m is a vector $m\mathbf{P}$ whose magnitude is $|m|$ times as great as the magnitude of **P** and which is similarly or oppositely directed to **P** depending on whether m is positive or negative.

(*b*) **Other Operations** with scalars m and n are

$$\begin{aligned} (m+n)\mathbf{P} &= m\mathbf{P} + n\mathbf{P} \\ m(\mathbf{P}+\mathbf{Q}) &= m\mathbf{P} + m\mathbf{Q} \\ m(n\mathbf{P}) &= n(m\mathbf{P}) = (mn)\mathbf{P} \end{aligned}$$

ORTHOGONAL TRIAD of UNIT VECTORS

Orthogonal triad of unit vectors **i**, **j**, and **k** is formed by drawing unit vectors along the x, y, and z axes respectively. A right-handed set of axes is shown in Fig. 1-4 below.

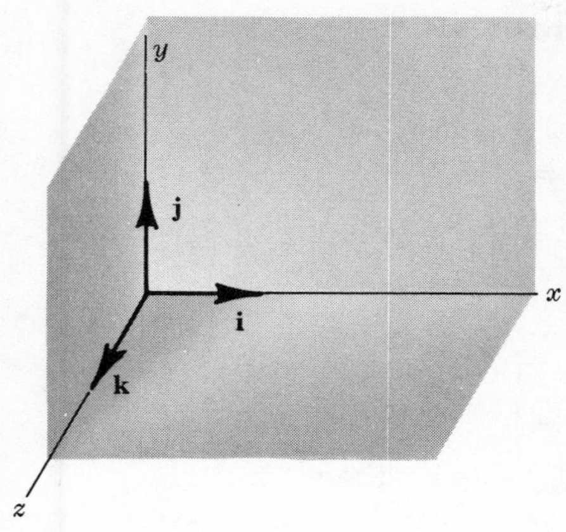

Fig. 1-4

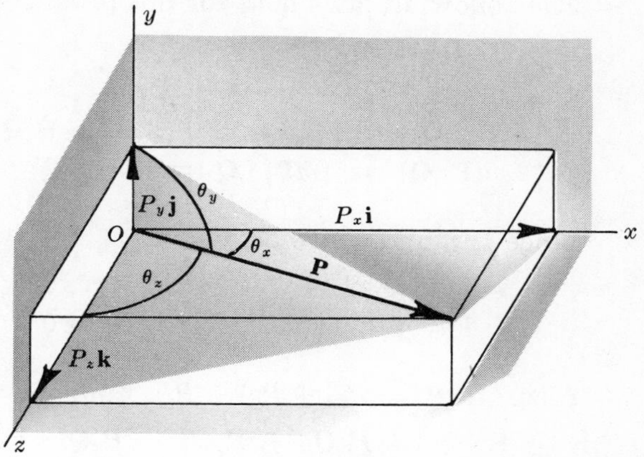

Fig. 1-5

A vector **P** is written as

$$\mathbf{P} = P_x\mathbf{i} + P_y\mathbf{j} + P_z\mathbf{k}$$

where $P_x\mathbf{i}$, $P_y\mathbf{j}$, and $P_z\mathbf{k}$ are the vector components of **P** along the x, y, and z axes respectively as shown in Fig. 1-5 above.

Note that $P_x = P\cos\theta_x$, $P_y = P\cos\theta_y$, and $P_z = P\cos\theta_z$.

The POSITION VECTOR r of a point (x, y, z) in space is written

$$\mathbf{r} = x\mathbf{i} + y\mathbf{j} + z\mathbf{k}$$

where $r = \sqrt{x^2 + y^2 + z^2}$. See Fig. 1-6 below.

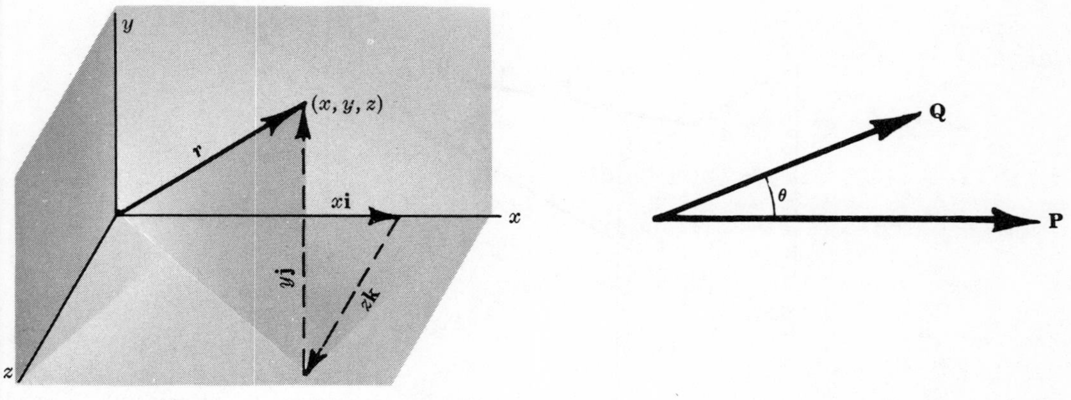

Fig. 1-6

Fig. 1-7

DOT or SCALAR PRODUCT

The dot or scalar product of two vectors **P** and **Q**, written $\mathbf{P}\cdot\mathbf{Q}$, is a scalar quantity and is defined as the product of the magnitudes of the two vectors and the cosine of their included angle θ (see Fig. 1-7 above). Thus

$$\mathbf{P}\cdot\mathbf{Q} = PQ\cos\theta$$

The following laws hold for dot products, where m is a scalar:

$$\mathbf{P}\cdot\mathbf{Q} = \mathbf{Q}\cdot\mathbf{P}$$
$$\mathbf{P}\cdot(\mathbf{Q}+\mathbf{S}) = \mathbf{P}\cdot\mathbf{Q} + \mathbf{P}\cdot\mathbf{S}$$
$$(\mathbf{P}+\mathbf{Q})\cdot(\mathbf{S}+\mathbf{T}) = \mathbf{P}\cdot(\mathbf{S}+\mathbf{T}) + \mathbf{Q}\cdot(\mathbf{S}+\mathbf{T}) = \mathbf{P}\cdot\mathbf{S} + \mathbf{P}\cdot\mathbf{T} + \mathbf{Q}\cdot\mathbf{S} + \mathbf{Q}\cdot\mathbf{T}$$
$$m(\mathbf{P}\cdot\mathbf{Q}) = (m\mathbf{P})\cdot\mathbf{Q} = \mathbf{P}\cdot(m\mathbf{Q})$$

Since **i**, **j**, and **k** are orthogonal,

$$\mathbf{i}\cdot\mathbf{j} = \mathbf{i}\cdot\mathbf{k} = \mathbf{j}\cdot\mathbf{k} = (1)(1)\cos 90^\circ = 0$$
$$\mathbf{i}\cdot\mathbf{i} = \mathbf{j}\cdot\mathbf{j} = \mathbf{k}\cdot\mathbf{k} = (1)(1)\cos 0^\circ = 1$$

Also, if $\mathbf{P} = P_x\mathbf{i} + P_y\mathbf{j} + P_z\mathbf{k}$ and $\mathbf{Q} = Q_x\mathbf{i} + Q_y\mathbf{j} + Q_z\mathbf{k}$, then

$$\mathbf{P}\cdot\mathbf{Q} = P_xQ_x + P_yQ_y + P_zQ_z$$
$$\mathbf{P}\cdot\mathbf{P} = P^2 = P_x^2 + P_y^2 + P_z^2$$

The magnitudes of the vector components of **P** along the rectangular axes can be written

$$P_x = \mathbf{P}\cdot\mathbf{i}, \quad P_y = \mathbf{P}\cdot\mathbf{j}, \quad P_z = \mathbf{P}\cdot\mathbf{k}$$

since, for example, $\mathbf{P}\cdot\mathbf{i} = (P_x\mathbf{i}+P_y\mathbf{j}+P_z\mathbf{k})\cdot\mathbf{i} = P_x + 0 + 0 = P_x$.

Similarly the magnitude of the vector component of **P** along any line L can be written $\mathbf{P}\cdot\mathbf{e}_L$, where $\mathbf{e}_L$ is the unit vector along the line L. (Some authors use **u** as unit vector.) Fig. 1-8 below shows a plane through the tail end A of vector **P** and a plane through the arrow end B, both planes perpendicular to line L. The planes intersect line L in points C and D. The vector **CD** is the component of **P** along L and its magnitude equals $\mathbf{P}\cdot\mathbf{e}_L$.

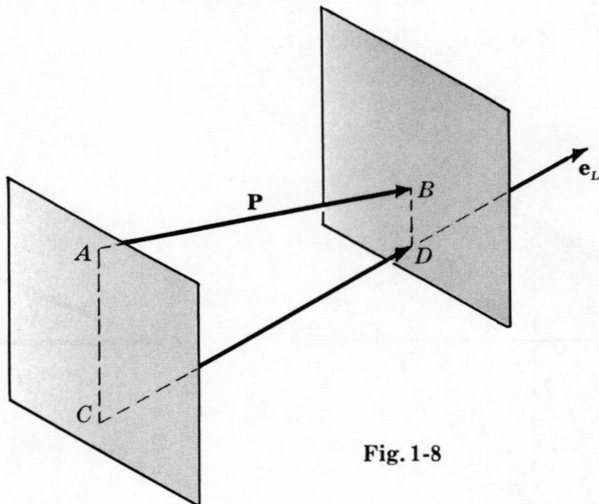

Fig. 1-8

Example. Determine the unit vector $\mathbf{e}_L$ for a line L which originates at point $(2, 3, 0)$ and passes through point $(-2, 4, 6)$. Next determine the projection of the vector $\mathbf{P} = 2\mathbf{i} + 3\mathbf{j} - \mathbf{k}$ along the line L.

The line L changes from $+2$ to -2 in the x direction, or a change of -4. The change in the y direction is $4 - 3 = +1$. The change in the z direction is $6 - 0 = +6$. The unit vector is

$$\mathbf{e}_L = \frac{-4}{\sqrt{(-4)^2 + (+1)^2 + (+6)^2}}\mathbf{i} + \frac{1}{\sqrt{53}}\mathbf{j} + \frac{6}{\sqrt{53}}\mathbf{k} = -0.549\mathbf{i} + 0.137\mathbf{j} + 0.823\mathbf{k}$$

The projection of **P** is thus

$$\mathbf{P}\cdot\mathbf{e}_L = 2(-0.549) + 3(0.137) - 1(0.823) = -1.41$$

The CROSS or VECTOR PRODUCT

The cross or vector product of two vectors **P** and **Q**, written $\mathbf{P}\times\mathbf{Q}$, is a vector **R** whose magnitude is the product of the magnitudes of the two vectors and the sine of their included angle. The vector $\mathbf{R} = \mathbf{P}\times\mathbf{Q}$ is normal to the plane of **P** and **Q** and points in the direction of advance of a right-handed screw when turned in the direction from **P** to **Q** through the smaller included angle θ. Thus if **e** is the unit vector which gives the direction of $\mathbf{R} = \mathbf{P}\times\mathbf{Q}$, the cross product can be written

$$\mathbf{R} = \mathbf{P}\times\mathbf{Q} = (PQ\sin\theta)\mathbf{e}, \quad 0 \leqq \theta \leqq 180^\circ$$

Fig. 1-9 indicates that $\mathbf{P}\times\mathbf{Q} = -\mathbf{Q}\times\mathbf{P}$ (not commutative).

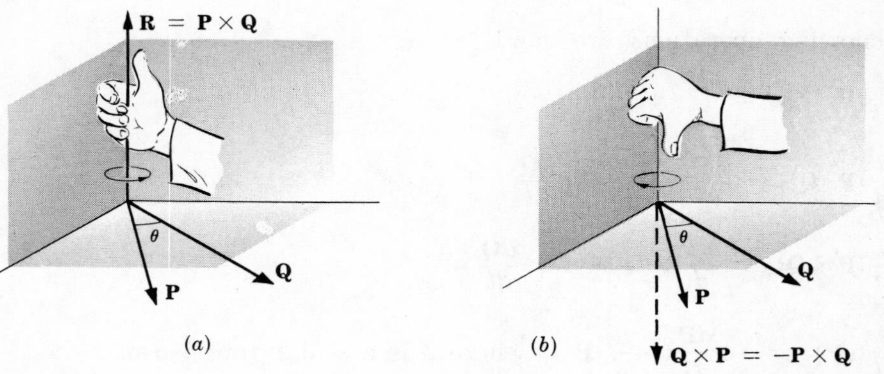

Fig. 1-9

The following laws hold for cross products, where m is a scalar:

$$\mathbf{P}\times(\mathbf{Q}+\mathbf{S}) = \mathbf{P}\times\mathbf{Q} + \mathbf{P}\times\mathbf{S}$$

$$\begin{aligned}(\mathbf{P}+\mathbf{Q})\times(\mathbf{S}+\mathbf{T}) &= \mathbf{P}\times(\mathbf{S}+\mathbf{T}) + \mathbf{Q}\times(\mathbf{S}+\mathbf{T})\\ &= \mathbf{P}\times\mathbf{S} + \mathbf{P}\times\mathbf{T} + \mathbf{Q}\times\mathbf{S} + \mathbf{Q}\times\mathbf{T}\end{aligned}$$

$$m(\mathbf{P}\times\mathbf{Q}) = (m\mathbf{P})\times\mathbf{Q} = \mathbf{P}\times(m\mathbf{Q})$$

Since **i**, **j**, and **k** are orthogonal,

$$\mathbf{i}\times\mathbf{i} = \mathbf{j}\times\mathbf{j} = \mathbf{k}\times\mathbf{k} = 0$$

$$\mathbf{i}\times\mathbf{j} = \mathbf{k}, \quad \mathbf{j}\times\mathbf{k} = \mathbf{i}, \quad \mathbf{k}\times\mathbf{i} = \mathbf{j}$$

Also, if $\mathbf{P} = P_x\mathbf{i} + P_y\mathbf{j} + P_z\mathbf{k}$ and $\mathbf{Q} = Q_x\mathbf{i} + Q_y\mathbf{j} + Q_z\mathbf{k}$, then

$$\mathbf{P}\times\mathbf{Q} = (P_yQ_z - P_zQ_y)\mathbf{i} + (P_zQ_x - P_xQ_z)\mathbf{j} + (P_xQ_y - P_yQ_x)\mathbf{k} = \begin{vmatrix} \mathbf{i} & \mathbf{j} & \mathbf{k} \\ P_x & P_y & P_z \\ Q_x & Q_y & Q_z \end{vmatrix}$$

For proof of this cross product determination see Problem 12.

VECTOR CALCULUS

(*a*) **Differentiation of a Vector P** which varies with respect to a scalar quantity such as time t is performed as follows.

Let $\mathbf{P} = \mathbf{P}(t)$, that is, **P** is a function of time t. A change $\Delta\mathbf{P}$ in **P** as time changes from t to $(t+\Delta t)$ is

$$\Delta\mathbf{P} = \mathbf{P}(t+\Delta t) - \mathbf{P}(t)$$

Then $$\frac{d\mathbf{P}}{dt} = \lim_{\Delta t\to 0}\frac{\Delta\mathbf{P}}{\Delta t} = \lim_{\Delta t\to 0}\frac{\mathbf{P}(t+\Delta t)-\mathbf{P}(t)}{\Delta t}$$

If $\mathbf{P}(t) = P_x\mathbf{i} + P_y\mathbf{j} + P_z\mathbf{k}$, where P_x, P_y and P_z are functions of time t, we have

$$\begin{aligned}\frac{d\mathbf{P}}{dt} &= \lim_{\Delta t\to 0}\frac{(P_x+\Delta P_x)\mathbf{i} + (P_y+\Delta P_y)\mathbf{j} + (P_z+\Delta P_z)\mathbf{k} - P_x\mathbf{i} - P_y\mathbf{j} - P_z\mathbf{k}}{\Delta t}\\ &= \lim_{\Delta t\to 0}\frac{\Delta P_x\mathbf{i} + \Delta P_y\mathbf{j} + \Delta P_z\mathbf{k}}{\Delta t}\\ &= \frac{dP_x}{dt}\mathbf{i} + \frac{dP_y}{dt}\mathbf{j} + \frac{dP_z}{dt}\mathbf{k}\end{aligned}$$

The following operations are valid:

$$\frac{d}{dt}(\mathbf{P}+\mathbf{Q}) = \frac{d\mathbf{P}}{dt} + \frac{d\mathbf{Q}}{dt}$$

$$\frac{d}{dt}(\mathbf{P}\cdot\mathbf{Q}) = \frac{d\mathbf{P}}{dt}\cdot\mathbf{Q} + \mathbf{P}\cdot\frac{d\mathbf{Q}}{dt}$$

$$\frac{d}{dt}(\mathbf{P}\times\mathbf{Q}) = \frac{d\mathbf{P}}{dt}\times\mathbf{Q} + \mathbf{P}\times\frac{d\mathbf{Q}}{dt}$$

$$\frac{d}{dt}(\psi\mathbf{P}) = \psi\frac{d\mathbf{P}}{dt} + \frac{d\psi}{dt}\mathbf{P}$$ where ψ is a scalar function of t

(*b*) **Integration of a Vector P** which varies with respect to a scalar quantity such as time t is performed as follows. Let $\mathbf{P} = \mathbf{P}(t)$, that is, $\mathbf{P}$ is a function of time t. Then

$$\begin{aligned}\int_{t_0}^{t_1}\mathbf{P}(t)\,dt &= \int_{t_0}^{t_1}(P_x\mathbf{i}+P_y\mathbf{j}+P_z\mathbf{k})\,dt\\ &= \mathbf{i}\int_{t_0}^{t_1}P_x\,dt + \mathbf{j}\int_{t_0}^{t_1}P_y\,dt + \mathbf{k}\int_{t_0}^{t_1}P_z\,dt\end{aligned}$$

Solved Problems

1. In a plane, add a 120 lb force at 30° and a −100 lb force at 90° using the Parallelogram Method. Refer to Fig. 1-10(a).

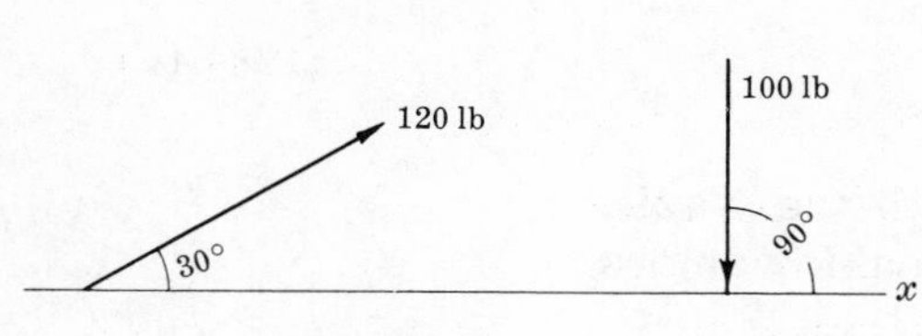

Fig. 1-10(a)

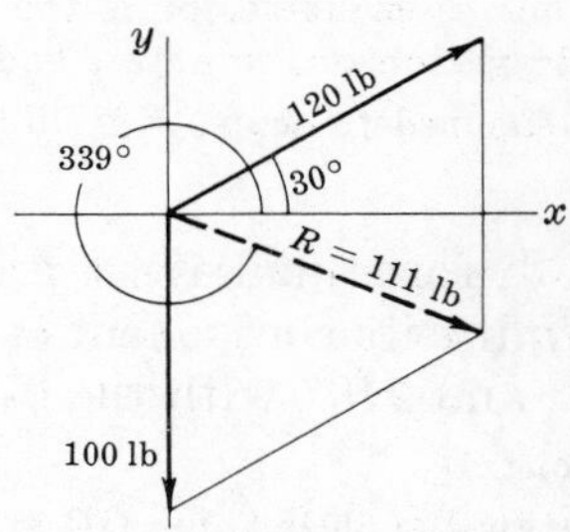

Fig. 1-10(b)

Solution:

Draw a sketch of the problem, not necessarily to scale. The negative sign indicates that the 100 lb force acts along the 90° line downward toward the origin. This is equivalent to a positive 100 lb force along the 270° line, according to the Principle of Transmissibility.

As in Fig. 1-10(b) above, place the tail ends of the two vectors at a common point and draw the vectors to a suitable scale. Complete the parallelogram. The resulting R measures to the chosen scale 111 lb. By protractor, it is at an angle with the x-axis $\theta_x = 339°$.

2. Apply the Triangle Law in Problem 1. See Fig. 1-11.

Solution:

It is immaterial which vector is chosen first. Take the 120 lb force. To the arrow end of this vector attach the tail end of the 100 lb force. Draw the resultant from the tail end of the 120 lb force to the arrow end of the 100 lb force. When measured to the chosen scale and the direction determined, the results are the same as in Problem 1.

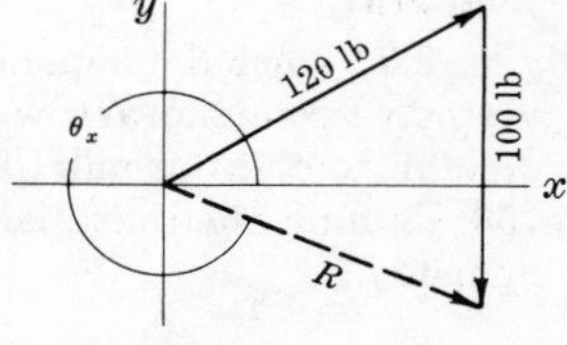

Fig. 1-11

3. The resultant of two forces in a plane is 800 lb at 60°. One of the forces is 160 lb at 30°. Determine the missing force. See Fig. 1-12.

Solution:

Select a point through which to draw the resultant and the given force to a convenient scale.

Draw the line connecting the arrow ends of the resultant and the given force. Place an arrow on the end of this line near the resultant. This line represents the missing force. When measured to scale it is 667 lb with $\theta_x = 67°$.

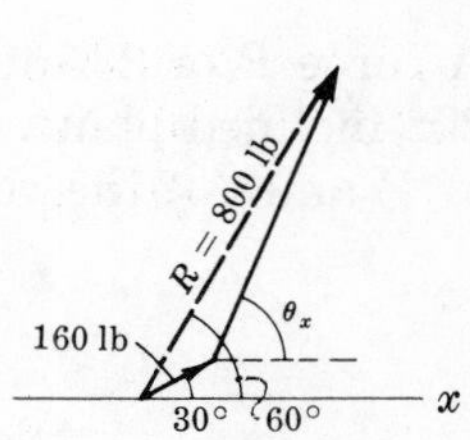

Fig. 1-12

4. In a plane, subtract 130 lb, 60° from 280 lb, 320°. See Fig. 1-13.

Solution:

To the 280 lb, 320° force add the negative of the 130 lb, 60° force obtaining a resultant force of 330 lb, 297°. All angles are measured with respect to the x-axis.

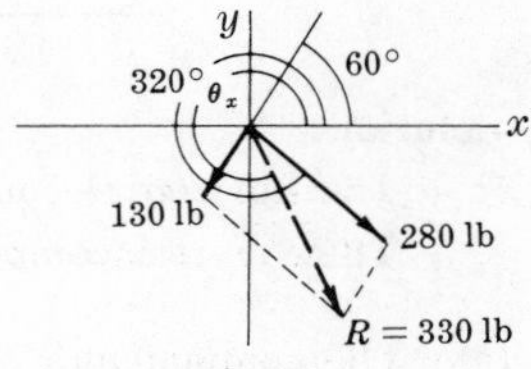

Fig. 1-13

5. Determine the resultant of the following coplanar system of forces: 26 lb, 10°; 39 lb, 114°; 63 lb, 183°; 57 lb, 261°.

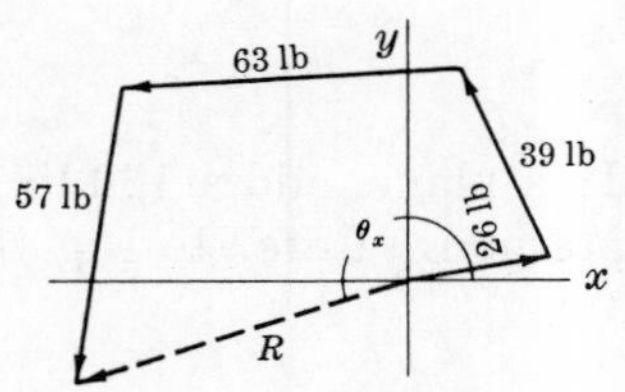

Fig. 1-14

Solution:

Applying the Polygon Method, place the tail end of each vector in turn at the arrow end of the preceding vector.

The resultant vector is the force drawn from the tail end of the first vector to the arrow end of the last vector.

Measured to scale, $R = 65$ lb with $\theta_x = 197°$.

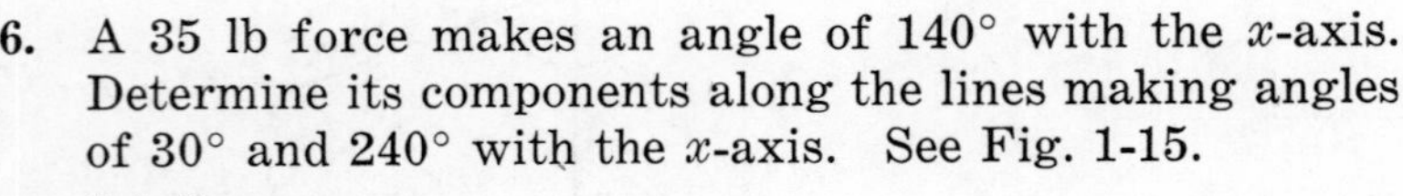

6. A 35 lb force makes an angle of 140° with the x-axis. Determine its components along the lines making angles of 30° and 240° with the x-axis. See Fig. 1-15.

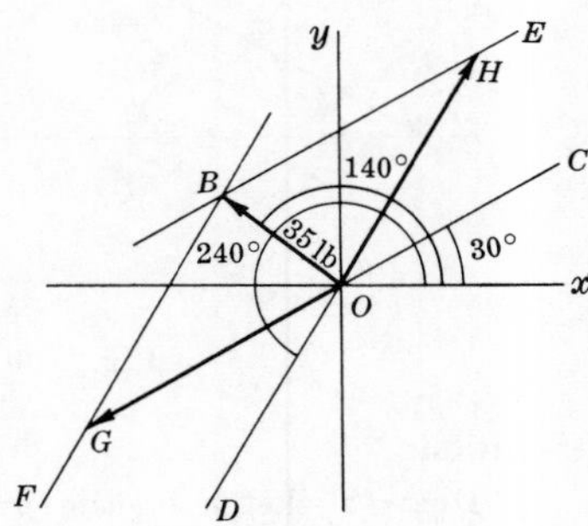

Fig. 1-15

Solution:

Draw the 35 lb force OB at some convenient point O to scale. Through this point O draw the 30° and 240° lines, OC and OD respectively.

Through the point B draw lines BE and BF parallel respectively to OC and OD. Extend OC to meet BF in G, and extend OD to meet BE in H. OG (−68.9 lb) is the desired component along the 30° line, and OH (−65.8 lb) is the result along the 240° line.

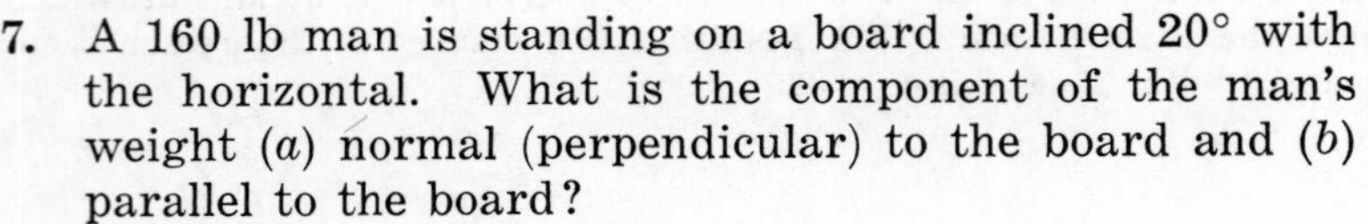

7. A 160 lb man is standing on a board inclined 20° with the horizontal. What is the component of the man's weight (a) normal (perpendicular) to the board and (b) parallel to the board?

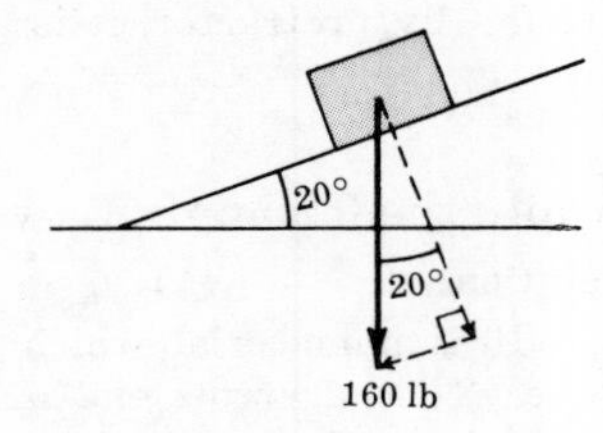

Fig. 1-16

Solution:

The normal component is at an angle of 20° with the weight vector. From the arrow end of the weight vector drop a perpendicular to the normal. To scale, the normal component measures 150 lb and the parallel component measures 55 lb. By trigonometry,

$$\text{normal component} = 160 \cos 20° = 150 \text{ lb}$$
$$\text{parallel component} = 160 \sin 20° = 55 \text{ lb.}$$

8. A force P of 235 lb acts at an angle of 60° with the horizontal on a block resting on a 22° inclined plane. Determine algebraically (a) the horizontal and vertical components of P and (b) the components of P perpendicular to and along the plane.

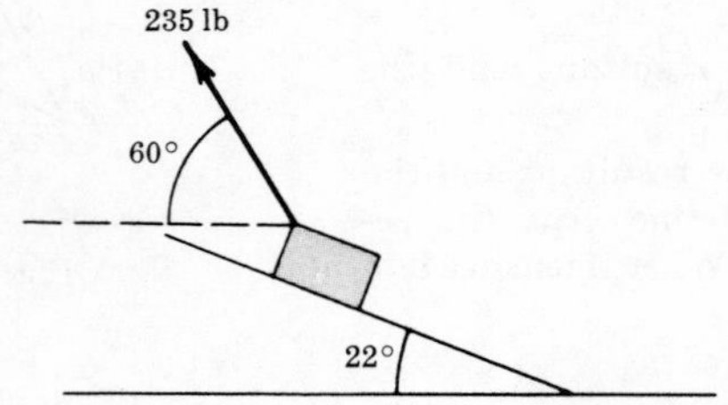

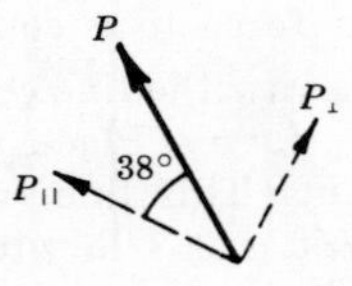

Fig. 1-17

Solution:

(a) The horizontal component P_h acts to the left and is equal to $235 \cos 60° = 118$ lb.
The vertical component P_v acts up and is equal to $235 \sin 60° = 204$ lb.

(b) The component $P_{\parallel}$ parallel to the plane $= 235 \cos(60° - 22°) = 185$ lb acting up the plane.
The component $P_{\perp}$ normal to the plane $= 235 \sin 38° = 145$ lb as shown.

9. Resolve the 20 lb force into two components, one of which will balance the 10 lb horizontal force. See Fig. 1-18.

Solution:

The lower horizontal broken line indicates one of the desired components. Since it is the equilibrant of the given 10 lb force, it is collinear with this given force and equal to it in magnitude but opposite in direction. The other component is concurrent with the given two forces at O and is parallel to the line joining the arrow ends of the 20 lb force and the equilibrant, completing the parallelogram as shown. It scales 26.5 lb at 41°.

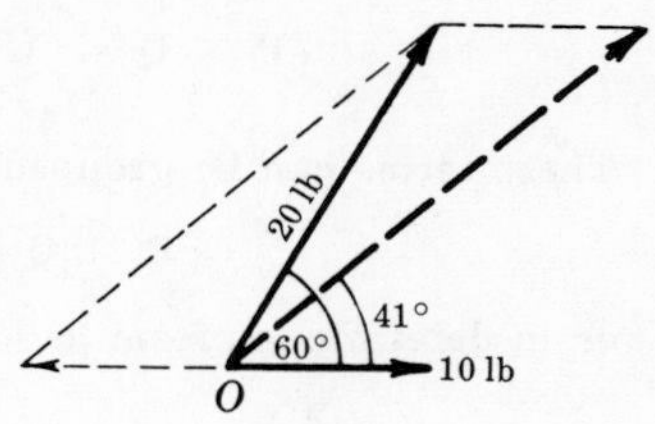

Fig. 1-18

10. Refer to Fig. 1-19. The x, y, z edges of a rectangular parallelepiped are 4, 3, and 2 ft respectively. If the diagonal OP drawn from the origin represents a 50 lb force, determine the x, y, and z components of the force. Express the force as a vector in terms of the unit vectors **i**, **j**, and **k**.

Solution:

Let θ_x, θ_y, θ_z represent respectively the angles between the diagonal OP and the x, y, z axes. Then

$$P_x = P\cos\theta_x, \quad P_y = P\cos\theta_y, \quad P_z = P\cos\theta_z$$

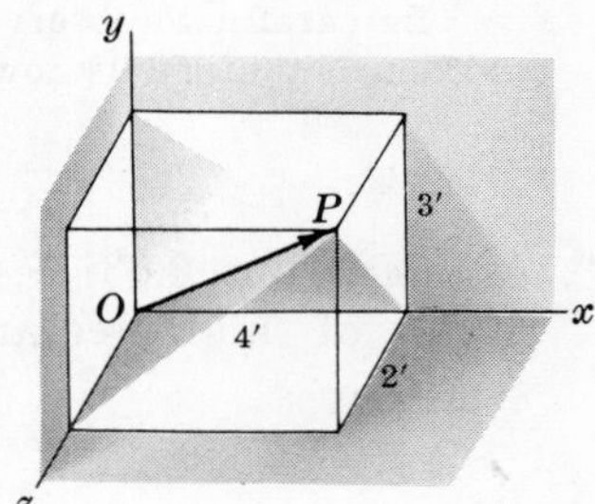

Fig. 1-19

Length of $OP = \sqrt{4^2+3^2+2^2} = 5.38$ ft. Hence,

$$\cos\theta_x = 4/5.38, \quad \cos\theta_y = 3/5.38, \quad \cos\theta_z = 2/5.38.$$

Since each component in the sketch is in the positive direction of the axis along which it acts,

$$P_x = 50\cos\theta_x = 37.2 \text{ lb}, \quad P_y = 50\cos\theta_y = 27.9 \text{ lb}, \quad P_z = 50\cos\theta_z = 18.6 \text{ lb}$$

The vector $\mathbf{P} = P_x\mathbf{i} + P_y\mathbf{j} + P_z\mathbf{k} = 37.2\mathbf{i} + 27.9\mathbf{j} + 18.6\mathbf{k}$ lb.

11. Determine the x, y, and z components of a 100 lb force passing from the origin through point $(2, -4, 1)$. Express the vector in terms of the unit vectors **i**, **j**, and **k**.

Solution:

The direction cosines of the force line are

$$\cos\theta_x = \frac{2}{\sqrt{(2)^2+(-4)^2+(1)^2}} = 0.437, \qquad \cos\theta_y = \frac{-4}{\sqrt{21}} = -0.873, \qquad \cos\theta_z = 0.218$$

Hence $P_x = 43.7$ lb, $P_y = -87.3$ lb, $P_z = 21.8$ lb; and the vector $\mathbf{P} = 43.7\mathbf{i} - 87.3\mathbf{j} + 21.8\mathbf{k}$.

12. Show that the cross product of two vectors **P** and **Q** can be written as

$$\mathbf{P}\times\mathbf{Q} = \begin{vmatrix} \mathbf{i} & \mathbf{j} & \mathbf{k} \\ P_x & P_y & P_z \\ Q_x & Q_y & Q_z \end{vmatrix}$$

Solution:

Write the given vectors in component form and expand the cross product to obtain

$$\begin{aligned}\mathbf{P}\times\mathbf{Q} &= (P_x\mathbf{i}+P_y\mathbf{j}+P_z\mathbf{k})\times(Q_x\mathbf{i}+Q_y\mathbf{j}+Q_z\mathbf{k}) \\ &= (P_xQ_x)\mathbf{i}\times\mathbf{i} + (P_xQ_y)\mathbf{i}\times\mathbf{j} + (P_xQ_z)\mathbf{i}\times\mathbf{k} \\ &\quad + (P_yQ_x)\mathbf{j}\times\mathbf{i} + (P_yQ_y)\mathbf{j}\times\mathbf{j} + (P_yQ_z)\mathbf{j}\times\mathbf{k} \\ &\quad + (P_zQ_x)\mathbf{k}\times\mathbf{i} + (P_zQ_y)\mathbf{k}\times\mathbf{j} + (P_zQ_z)\mathbf{k}\times\mathbf{k}\end{aligned}$$

But $\mathbf{i}\times\mathbf{i} = \mathbf{j}\times\mathbf{j} = \mathbf{k}\times\mathbf{k} = 0$; and $\mathbf{i}\times\mathbf{j} = \mathbf{k}$ and $\mathbf{j}\times\mathbf{i} = -\mathbf{k}$, etc. Hence

$$\mathbf{P}\times\mathbf{Q} = (P_xQ_y)\mathbf{k} - (P_xQ_z)\mathbf{j} - (P_yQ_x)\mathbf{k} + (P_yQ_z)\mathbf{i} + (P_zQ_x)\mathbf{j} - (P_zQ_y)\mathbf{i}$$

These terms can be grouped as

$$\mathbf{P}\times\mathbf{Q} = (P_yQ_z - P_zQ_y)\mathbf{i} + (P_zQ_x - P_xQ_z)\mathbf{j} + (P_xQ_y - P_yQ_x)\mathbf{k}$$

or in determinant form as

$$\mathbf{P}\times\mathbf{Q} = \begin{vmatrix} \mathbf{i} & \mathbf{j} & \mathbf{k} \\ P_x & P_y & P_z \\ Q_x & Q_y & Q_z \end{vmatrix}$$

Be careful to observe that the scalar components of the first vector **P** in the cross product are written in the middle row of the determinant.

13. A force $\mathbf{F} = 2.63\mathbf{i} + 4.28\mathbf{j} - 5.92\mathbf{k}$ newtons acts through the origin. What is the magnitude of this force and what angles does it make with the x, y, and z axes?

Solution:

$$F = \sqrt{(2.63)^2 + (4.28)^2 + (-5.92)^2} = 7.75 \text{ newtons}$$

$$\cos\theta_x = +2.63/7.75,\ \theta_x = 70.2^\circ \qquad \cos\theta_y = +4.28/7.75,\ \theta_y = 56.3^\circ \qquad \cos\theta_z = -5.92/7.75,\ \theta_z = 139.8^\circ$$

14. Find the dot product of $\mathbf{P} = 4.82\mathbf{i} - 2.33\mathbf{j} + 5.47\mathbf{k}$ lb and $\mathbf{Q} = -2.81\mathbf{i} - 6.09\mathbf{j} + 1.12\mathbf{k}$ ft.

Solution:

$$\mathbf{P}\cdot\mathbf{Q} = P_xQ_x + P_yQ_y + P_zQ_z = (4.82)(-2.81) + (-2.33)(-6.09) + (5.47)(1.12) = 6.72 \text{ ft-lb.}$$

15. Determine the projection of the force $\mathbf{P} = 10\mathbf{i} - 8\mathbf{j} + 14\mathbf{k}$ lb on the directed line L which originates at point $(2, -5, 3)$ and passes through point $(5, 2, -4)$.

Solution:

The unit vector along L is

$$\mathbf{e}_L = \frac{5-2}{\sqrt{(5-2)^2 + [2-(-5)]^2 + (-4-3)^2}}\mathbf{i} + \frac{2-(-5)}{\sqrt{107}}\mathbf{j} + \frac{-4-3}{\sqrt{107}}\mathbf{k}$$
$$= 0.290\mathbf{i} + 0.677\mathbf{j} - 0.677\mathbf{k}$$

The projection of **P** on L is

$$\mathbf{P}\cdot\mathbf{e}_L = (10\mathbf{i} - 8\mathbf{j} + 14\mathbf{k}) \cdot (0.29\mathbf{i} + 0.677\mathbf{j} - 0.677\mathbf{k})$$
$$= 2.90 - 5.42 - 9.48 = -12.0 \text{ lb}$$

The minus sign indicates that the projection is directed opposite to the direction of L.

16. Find the cross product of $\mathbf{P} = 2.85\mathbf{i} + 4.67\mathbf{j} - 8.09\mathbf{k}$ ft and $\mathbf{Q} = 28.3\mathbf{i} + 44.6\mathbf{j} + 53.3\mathbf{k}$ lb.

Solution:

$$\mathbf{P}\times\mathbf{Q} = \begin{vmatrix} \mathbf{i} & \mathbf{j} & \mathbf{k} \\ P_x & P_y & P_z \\ Q_x & Q_y & Q_z \end{vmatrix} = \begin{vmatrix} \mathbf{i} & \mathbf{j} & \mathbf{k} \\ 2.85 & 4.67 & -8.09 \\ 28.3 & 44.6 & 53.3 \end{vmatrix}$$

$$= \mathbf{i}[(4.67)(53.3) - (44.6)(-8.09)] - \mathbf{j}[(2.85)(53.3) - (28.3)(-8.09)] + \mathbf{k}[(2.85)(44.6) - (28.3)(4.67)]$$
$$= \mathbf{i}[249 + 361] - \mathbf{j}[152 + 229] + \mathbf{k}[127 - 132] = 610\mathbf{i} - 381\mathbf{j} - 5\mathbf{k} \text{ lb-ft}$$

17. Determine the time derivative of the position vector $\mathbf{r} = 2x\mathbf{i} - 3y\mathbf{j} + z\mathbf{k}$, where **i**, **j**, and **k** are fixed vectors.

Solution:

This is simply $\dfrac{d\mathbf{r}}{dt} = 2\dfrac{dx}{dt}\mathbf{i} - 3\dfrac{dy}{dt}\mathbf{j} + \dfrac{dz}{dt}\mathbf{k}$.

18. Determine the time integral from time $t_1 = 1$ sec to time $t_2 = 3$ sec of the velocity vector

$$\mathbf{v} = t^2\mathbf{i} + 2t\mathbf{j} - \mathbf{k} \text{ ft/sec}$$

where **i**, **j**, and **k** are fixed vectors.

Solution:

$$\int_1^3 (t^2\mathbf{i} + 2t\mathbf{j} - \mathbf{k})\,dt = \mathbf{i}\int_1^3 t^2\,dt + \mathbf{j}\int_1^3 2t\,dt - \mathbf{k}\int_1^3 dt = 8.67\mathbf{i} + 8.00\mathbf{j} - 2.00\mathbf{k}$$

Supplementary Problems

19. Determine the resultant of the coplanar forces 100 lb, 0° and 200 lb, 90°. *Ans.* 224 lb, $\theta_x = 64°$

20. Determine the resultant of the coplanar forces 32 lb, 20° and 64 lb, 190°. *Ans.* 33.0 lb, $\theta_x = 180°$

21. Find the resultant of the coplanar forces −30 lb, 60° and 80 lb, 60°. *Ans.* 50.0 lb, $\theta_x = 60°$

22. Find the resultant of the concurrent coplanar forces 120 newtons, 78° and 70 newtons, 293°. *Ans.* 74.7 nt, $\theta_x = 45.2°$

23. The resultant of two coplanar forces is 18 oz at 30°. If one of the forces is 28 oz at 0°, determine the other. *Ans.* 15.3 oz, 144°

24. The resultant of two coplanar forces is 36 dynes at 45°. If one of the forces is 24 dynes at 0°, find the other force. *Ans.* 25.5 dynes, 87°

25. The resultant of two coplanar forces is 50 newtons at 143°. One of the forces is 120 newtons at 238°. Determine the missing force. *Ans.* 134 nt, $\theta_x = 79.6°$

26. In a plane, subtract 80 lb, 0° from −45 lb, 180°. *Ans.* 35.0 lb, $\theta_x = 180°$ or −35.0 lb, $\theta_x = 0°$

27. In a plane, subtract the force 20 lb, 180° from the force 40 lb, 270°. *Ans.* 44.7 lb, $\theta_x = 296.5°$

28. Determine the resultant of the coplanar forces: 6 oz, 38°; 12 oz, 73°; 18 oz, 67°; 24 oz, 131°. *Ans.* 50.0 oz, $\theta_x = 91°$

29. Determine the resultant of the coplanar forces: 20 lb, 0°; 20 lb, 30°; 20 lb, 60°; 20 lb, 90°; 20 lb, 120°; 20 lb, 150°. *Ans.* 77.2 lb, $\theta_x = 75°$

30. Find the single force to replace the following coplanar forces: 150 dynes, 78°; 320 dynes, 143°; 485 dynes, 249°; 98 dynes, 305°; 251 dynes, 84°. *Ans.* 321 dynes, 171°

31. A sled is being pulled by a force of 25 lb exerted in a rope inclined 30° with the horizontal. What is the effective component of the force pulling the sled? What is the component tending to lift the sled vertically? *Ans.* $P_h = 21.7$ lb, $P_v = 12.5$ lb

32. Determine the resultant of the following coplanar forces: 15 newtons, 30°; 55 newtons, 80°; 90 newtons, 210°; and 130 newtons, 260°. *Ans.* 136 nt, $\theta_x = 235°$

33. A block weighing 19 lb rests on a plane making a 16° angle with the horizontal. Determine the component of the weight normal to the plane. *Ans.* 18.3 lb

34. A telephone pole is supported by a guy wire which exerts a pull of 200 lb on the top of the pole. If the angle between the wire and the pole is 50°, what are the horizontal and vertical components of the pull on the pole? *Ans.* $P_h = 153$ lb, $P_v = 129$ lb

35. A boat is being towed through a canal by a horizontal cable which makes an angle of 10° with the shore. If the pull in the cable is 200 lb, find the force tending to move the boat along the canal. *Ans.* 197 lb

36. Express in terms of the unit vectors **i**, **j**, and **k** the force of 200 lb that starts at the point (2, 5, −3) and passes through the point (−3, 2, 1). *Ans.* $\mathbf{F} = -141\mathbf{i} - 84.9\mathbf{j} + 113\mathbf{k}$

37. Determine the resultant of the three forces $\mathbf{F}_1 = 2.0\mathbf{i} + 3.3\mathbf{j} - 2.6\mathbf{k}$, $\mathbf{F}_2 = -\mathbf{i} + 5.2\mathbf{j} - 2.9\mathbf{k}$, and $F_3 = 8.3\mathbf{i} - 6.6\mathbf{j} + 5.8\mathbf{k}$, which are concurrent at the point (2, 2, −5). *Ans.* $\mathbf{R} = 9.3\mathbf{i} + 1.9\mathbf{j} + 0.3\mathbf{k}$ lb at (2, 2, −5)

38. The pulley shown in Fig. 1-20 below is free to ride on the supporting guide wire. If the pulley supports a 160 lb man, what is the tension in the wire? *Ans.* $T = 234$ lb

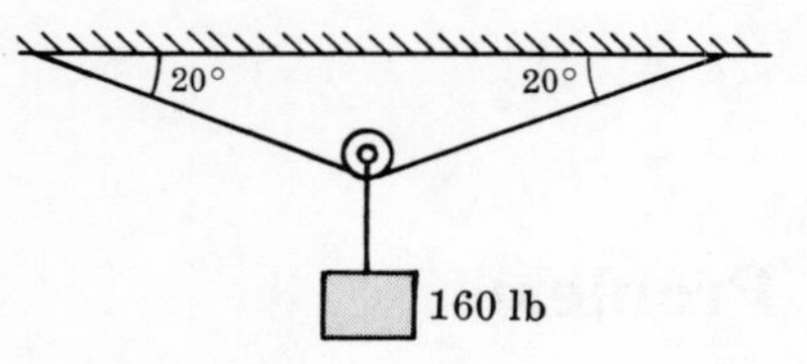

Fig. 1-20

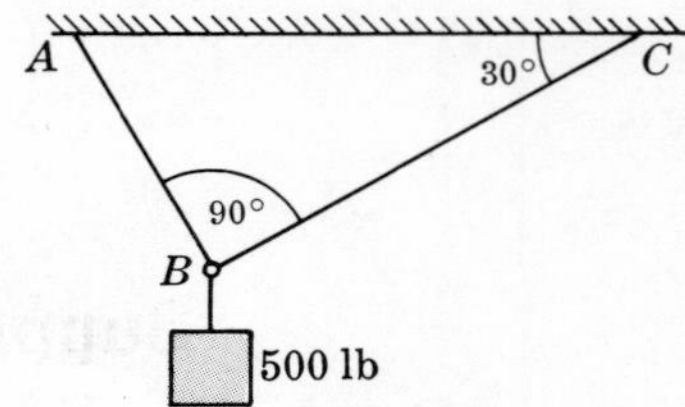

Fig. 1-21

39. The two cables support the 500 lb weight as shown in Fig. 1-21 above. Determine the tension in each cable. *Ans.* $T_{AB} = 433$ lb, $T_{BC} = 250$ lb

40. What force P is required to hold the 10 lb weight W in the position shown in Fig. 1-22 below? *Ans.* $P = 3.25$ lb

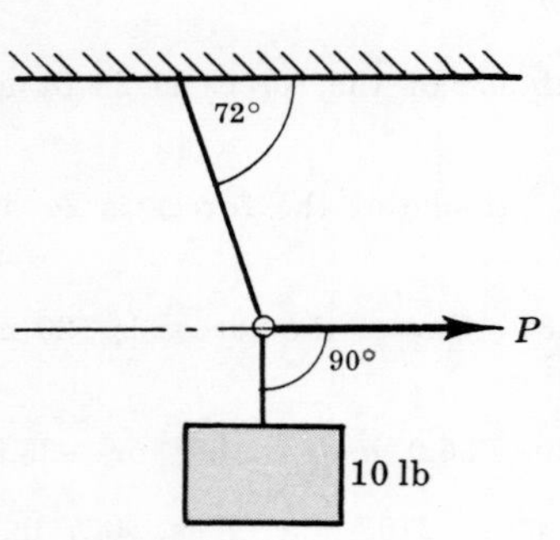

Fig. 1-22

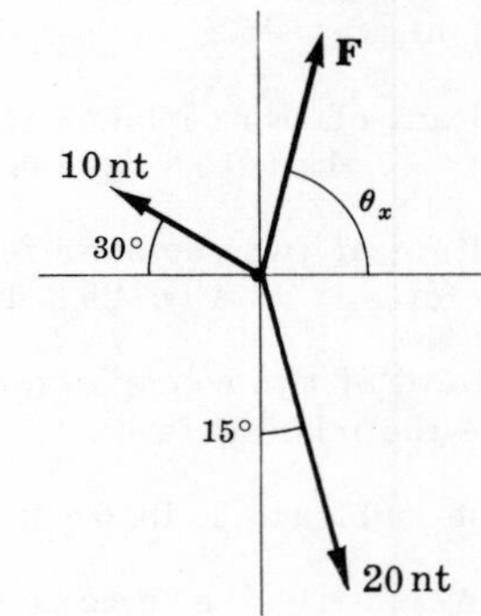

Fig. 1-23

41. A charged particle is at rest under the action of three other charged particles. The forces exerted by two of the particles are shown in Fig. 1-23 above. Determine the magnitude and direction of the third force. *Ans.* $F = 14.7$ nt, $\theta_x = 76.8°$

42. Determine the resultant of the coplanar forces 200 lb, 0° and 400 lb, 90°. Since each force in Problem 19 has been multiplied by the scalar 2, the magnitude of the resultant in this problem should be double that of Problem 19. The angle should be the same.

43. What vector must be added to the vector $\mathbf{F} = 30$ lb, 60° to yield the zero vector? *Ans.* 30 lb, $\theta_x = 240°$

44. At time $t = 2$ sec, a point moving on a curve has coordinates $(3, -5, 2)$. At time $t = 3$ sec, the coordinates of the point are $(1, -2, 0)$. What is the change in the position vector?
Ans. $\Delta\mathbf{r} = -2\mathbf{i} + 3\mathbf{j} - 2\mathbf{k}$

45. Determine the dot product of $\mathbf{P} = 4\mathbf{i} + 2\mathbf{j} - \mathbf{k}$ and $\mathbf{Q} = -3\mathbf{i} + 6\mathbf{j} - 2\mathbf{k}$. *Ans.* $+2$

46. Determine the cross product of the vectors in Problem 45. *Ans.* $\mathbf{P} \times \mathbf{Q} = 2\mathbf{i} + 11\mathbf{j} + 30\mathbf{k}$

47. Determine the derivative with respect to time of $\mathbf{P} = x\mathbf{i} + 2y\mathbf{j} - z^2\mathbf{k}$.

Ans. $\dfrac{d\mathbf{P}}{dt} = \dfrac{dx}{dt}\mathbf{i} + 2\dfrac{dy}{dt}\mathbf{j} - 2z\dfrac{dz}{dt}\mathbf{k}$

48. If $\mathbf{P} = 2t\mathbf{i} + 3t^2\mathbf{j} - t\mathbf{k}$ and $\mathbf{Q} = t\mathbf{i} + t^2\mathbf{j} + t^3\mathbf{k}$, show that $\dfrac{d}{dt}(\mathbf{P}\cdot\mathbf{Q}) = 4t + 8t^3$. Check the result by using $\dfrac{d\mathbf{P}}{dt}\cdot\mathbf{Q} + \mathbf{P}\cdot\dfrac{d\mathbf{Q}}{dt} = \dfrac{d}{dt}(\mathbf{P}\cdot\mathbf{Q})$.

49. In the preceding problem show that $\dfrac{d}{dt}(\mathbf{P}\times\mathbf{Q}) = (15t^4 + 3t^2)\mathbf{i} - (8t^3 + 2t)\mathbf{j} - 3t^2\mathbf{k}$.
Check the result by using $\dfrac{d\mathbf{P}}{dt}\times\mathbf{Q} + \mathbf{P}\times\dfrac{d\mathbf{Q}}{dt} = \dfrac{d}{dt}(\mathbf{P}\times\mathbf{Q})$.

50. Determine the dot products for the following vectors.

	$\mathbf{P}$	$\mathbf{Q}$	$\mathbf{P}\cdot\mathbf{Q}$
(*a*)	$3\mathbf{i} - 2\mathbf{j} + 8\mathbf{k}$	$-\mathbf{i} - 2\mathbf{j} - 3\mathbf{k}$	-23
(*b*)	$0.86\mathbf{i} + 0.29\mathbf{j} - 0.37\mathbf{k}$	$1.29\mathbf{i} - 8.26\mathbf{j} + 4.0\mathbf{k}$	-2.77
(*c*)	$a\mathbf{i} + b\mathbf{j} - c\mathbf{k}$	$d\mathbf{i} - e\mathbf{j} + f\mathbf{k}$	$ad - be - cf$

51. Determine the cross products for the following vectors.

	$\mathbf{P}$	$\mathbf{Q}$	$\mathbf{P}\times\mathbf{Q}$
(*a*)	$3\mathbf{i} - 2\mathbf{j} + 8\mathbf{k}$	$-\mathbf{i} - 2\mathbf{j} - 3\mathbf{k}$	$22\mathbf{i} + \mathbf{j} - 8\mathbf{k}$
(*b*)	$0.86\mathbf{i} + 0.29\mathbf{j} - 0.37\mathbf{k}$	$1.29\mathbf{i} - 8.26\mathbf{j} + 4.0\mathbf{k}$	$-1.90\mathbf{i} - 3.92\mathbf{j} - 7.49\mathbf{k}$
(*c*)	$a\mathbf{i} + b\mathbf{j} - c\mathbf{k}$	$d\mathbf{i} - e\mathbf{j} + f\mathbf{k}$	$(bf - ec)\mathbf{i} - (af + cd)\mathbf{j} - (ae + bd)\mathbf{k}$

52. Determine the unit vector along the line which originates at the point $(2, 3, -2)$ and passes through the point $(1, 0, 5)$. *Ans.* $\mathbf{e}_L = -0.13\mathbf{i} - 0.391\mathbf{j} + 0.912\mathbf{k}$

53. Determine the component of the vector $\mathbf{P} = 1.52\mathbf{i} - 2.63\mathbf{j} + 0.83\mathbf{k}$ on the line which originates at the point $(2, 3, -2)$ and passes through the point $(1, 0, 5)$. *Ans.* $P_L = +1.59$

Chapter 2

Operations with Forces

The MOMENT M of a FORCE F

The moment **M** of a force **F** with respect to a point O is the cross product $\mathbf{M} = \mathbf{r} \times \mathbf{F}$, where **r** is the position vector relative to O of any point P on force **F**.

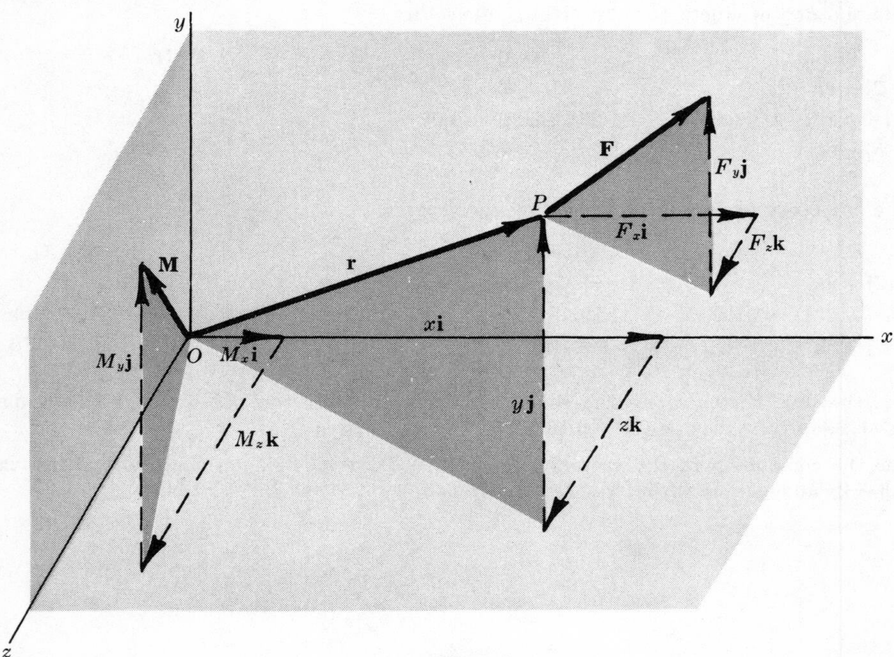

Fig. 2-1

If a set of x, y, and z axes is drawn through O as shown in Fig. 2-1 above,

$$\mathbf{r} = x\mathbf{i} + y\mathbf{j} + z\mathbf{k}, \qquad \mathbf{F} = F_x\mathbf{i} + F_y\mathbf{j} + F_z\mathbf{k}, \qquad \mathbf{M} = M_x\mathbf{i} + M_y\mathbf{j} + M_z\mathbf{k}$$

and, by definition

$$\mathbf{M} = \mathbf{r} \times \mathbf{F} = \begin{vmatrix} \mathbf{i} & \mathbf{j} & \mathbf{k} \\ x & y & z \\ F_x & F_y & F_z \end{vmatrix}$$

Expanding the determinant,

$$\mathbf{M} = \mathbf{i}(F_z y - F_y z) + \mathbf{j}(F_x z - F_z x) + \mathbf{k}(F_y x - F_x y)$$

Comparing this expression for **M** with the one listed above, it can be seen that

$$M_x = F_z y - F_y z, \qquad M_y = F_x z - F_z x, \qquad M_z = F_y x - F_x y$$

The scalar quantities M_x, M_y, and M_z are the respective moments of the force **F** about the x, y, and z axes through O. See Problems 3 and 4.

Note that the scalar component M_x of the moment **M** can be obtained by the dot product of the moment **M** and the unit vector **i** along the x-axis. Thus

$$\mathbf{M}\cdot\mathbf{i} = (M_x\mathbf{i}+M_y\mathbf{j}+M_z\mathbf{k})\cdot\mathbf{i} = M_x\mathbf{i}\cdot\mathbf{i} + M_y\mathbf{j}\cdot\mathbf{i} + M_z\mathbf{k}\cdot\mathbf{i} = M_x(1) + 0 + 0 = M_x$$

Similarly the scalar moment M_L (along any line L) of the moment **M** can be obtained by the dot product of the moment **M** and the unit vector $\mathbf{e}_L$ along the line L. Thus

$$M_L = \mathbf{M}\cdot\mathbf{e}_L$$

A COUPLE consists of two forces equal in magnitude, parallel, but oppositely directed.

The MOMENT C of a COUPLE

The moment **C** of a couple with respect to any point O is the sum of the moments with respect to O of the two forces that constitute the couple.

The moment **C** of the couple shown in Fig. 2-2 is

$$\mathbf{C} = \Sigma\mathbf{M}_O = \mathbf{r}_1\times\mathbf{F} + \mathbf{r}_2\times(-\mathbf{F})$$
$$= (\mathbf{r}_1-\mathbf{r}_2)\times\mathbf{F}$$
$$= \mathbf{a}\times\mathbf{F}$$

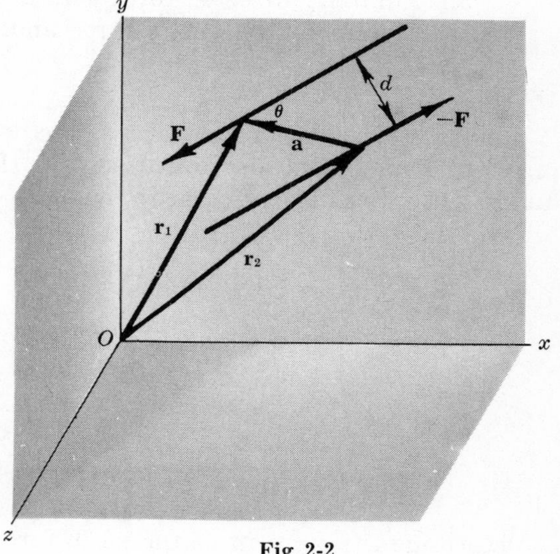

Fig. 2-2

Thus **C** is a vector perpendicular to the plane containing the two forces (**a** is in the same plane). By definition of the cross product, the magnitude of **C** is $|\mathbf{a}\times\mathbf{F}| = aF\sin\theta$. Since d, the perpendicular distance between the two forces of the couple, is equal to $a\sin\theta$, the magnitude of **C** is $C = Fd$.

Couples obey the laws of vectors. Any couple **C** can be written $\mathbf{C} = C_x\mathbf{i} + C_y\mathbf{j} + C_z\mathbf{k}$, where C_x, C_y, and C_z are the scalar components.

A SINGLE FORCE F

A single force **F** acting at point P may be replaced by (a) an equal and similarly directed force acting through any point O and (b) a couple $\mathbf{C} = \mathbf{r}\times\mathbf{F}$, where **r** is the vector from O to P. See Problems 10, 11.

COPLANAR FORCE SYSTEMS

Coplanar Force Systems occur in many problems of Mechanics. The following scalar treatment is useful in dealing with these two-dimensional problems.

1. **The Moment** M_O of a force about a point O in a plane containing the force is the scalar moment of the force about an axis through the point and perpendicular to the plane. As such, the moment is the product of (1) the force and (2) the perpendicular distance from the point to the line of action of the force. It is customary to assign a positive sign to the moment if the force tends to turn in a counterclockwise direction about the point.

2. **Varignon's Theorem** states that the moment of a force about any point is equal to the algebraic sum of the moments of the components of the force about that point.

3. **The External Effects of a Couple** will not be changed if
   (*a*) the couple is rotated or translated in its plane,
   (*b*) the couple is transferred to a parallel plane,
   (*c*) the size of its forces is changed provided the moment arm is also altered to keep the moment the same.

4. **A Couple and a Single Force** in the same plane or parallel planes may be combined into one force of the same magnitude and sense as the given force and parallel to it. See Problem 8.

5. **Conversely, a Single Force** as indicated above may be replaced by (*a*) an equal and similarly directed force acting through any point and (*b*) a couple lying in the same plane as the single force and the chosen point. See Problem 10.

NOTE

In some of the solved problems, vector equations are used, but in other problems the equivalent scalar equations are used. In the figures, vectors are identified by their magnitudes when the directions are obvious.

Solved Problems

1. Determine the moment of the 20 lb force about the point O. See Fig. 2-3.

Solution:

Drop the perpendicular OD from O to the action line of the 20 lb force. Its length to scale is 4.33 ft. The moment of the force about O (actually about an axis through O perpendicular to the xy-plane) is therefore $-(20 \times 4.33) = -86.6$ lb-ft.

The minus sign is used because the direction of rotation viewed from the positive end of the z-axis (not shown) is clockwise.

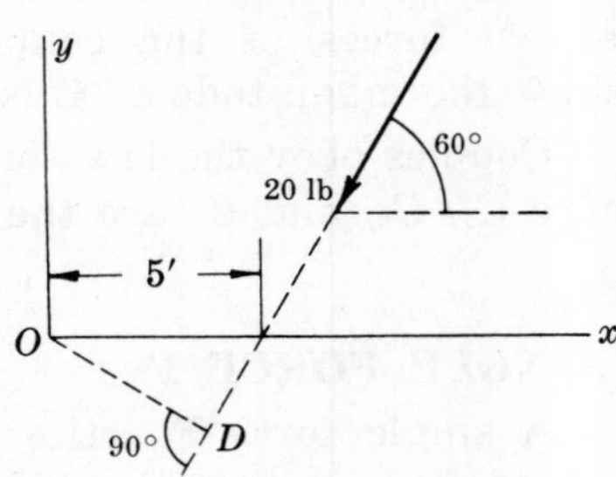

Fig. 2-3

2. Solve Problem 1 using Varignon's Theorem. See Fig. 2-4.

Solution:

In using this theorem the 20 lb force is replaced with its rectangular components parallel to the x and y axes and acting at *any* convenient point along the line of action. If the point is chosen on the x-axis, then it is apparent the x component has no moment about O.

The moment of the 20 lb force about O is then only the moment of the y component about O, or $-(17.3 \times 5) = -86.5$ lb-ft.

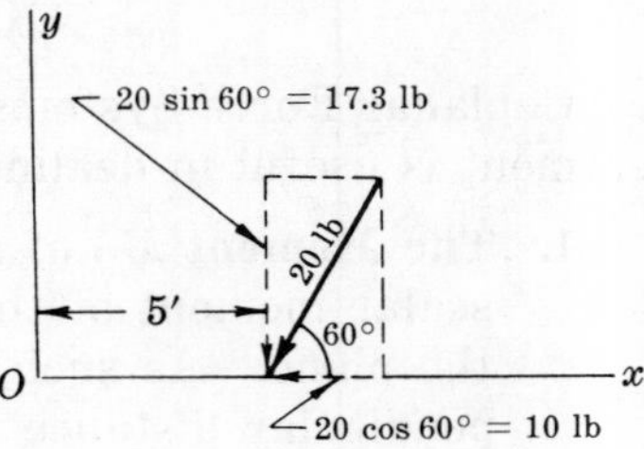

Fig. 2-4

3. A 100 lb force is directed along the line drawn from the point whose x, y, z coordinates are (2, 0, 4) ft to the point whose coordinates are (5, 1, 1) ft. What are the moments of this force about the x, y, and z axes?

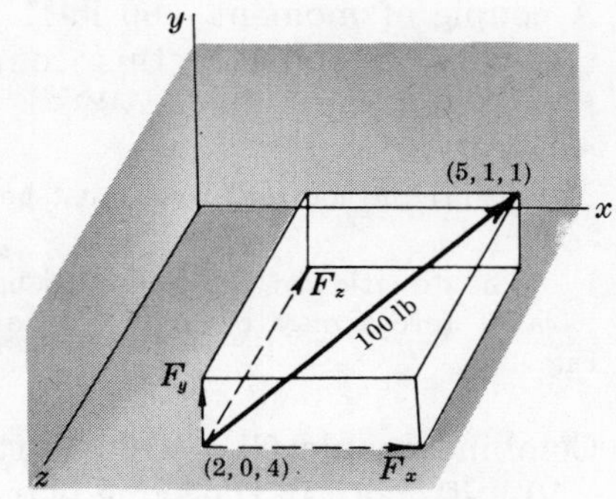

Fig. 2-5

Solution:

In Fig. 2-5, assume the scale is such that the 100 lb force is measured by the diagonal of the parallelepiped whose sides are parallel to the axes. The sides represent to the same scale the components of the force.

The x side is $5 - 2 = 3$ ft long.

The y side is $1 - 0 = 1$ ft long.

The z side is $1 - 4 = -3$ ft long. This means that the component F_z is directed toward the back or negative direction of the z-axis.

$$F_x = \frac{\text{length of } x \text{ side}}{\text{length of diagonal}} \times 100 \text{ lb} = \frac{3}{\sqrt{3^2 + 1^2 + 3^2}} \times 100 = \frac{3}{\sqrt{19}} \times 100 = 68.7 \text{ lb.}$$

$$\text{Similarly,} \quad F_y = \frac{1}{\sqrt{19}} \times 100 = 22.9 \text{ lb} \quad \text{and} \quad F_z = \frac{-3}{\sqrt{19}} \times 100 = -68.7 \text{ lb.}$$

To find the moment of the 100 lb force about the x-axis, determine the moments of its components about the x-axis. By inspection the only component that has such a moment is F_y. Therefore M_x for the 100 lb force is the moment of F_y about the x-axis and equals $-22.9 \times 4 = -91.6$ lb-ft. The minus sign indicates that the rotation of F_y is clockwise about the x-axis when viewed from the positive end of the x-axis.

In finding the moment about the y-axis, note that F_y is parallel to the y-axis and has no moment about it. Now, however, both F_z and F_x must be considered. It is better to determine the sign of the moment by inspection rather than by writing signs for the component and its arm. Accordingly, $M_y = +(68.7 \times 2) + (68.7 \times 4) = +412$ lb-ft.

By similar reasoning using F_y only (since F_z is parallel to the z-axis and F_x perpendicular to it), $M_z = +(22.9 \times 2) = +45.8$ lb-ft.

Be sure to affix signs for moments and to understand the significance thereof.

4. Repeat Problem 3 using the cross product definition of moment.

Solution:

From Problem 3, $\mathbf{F} = 68.7\mathbf{i} + 22.9\mathbf{j} - 68.7\mathbf{k}$

The vector $\mathbf{r}$ is the position vector of any point on the action line of $\mathbf{F}$ with respect to the origin. If we use point (2, 0, 4), $\mathbf{r} = 2\mathbf{i} + 0\mathbf{j} + 4\mathbf{k}$. Then

$$\begin{aligned} \mathbf{M} = \mathbf{r} \times \mathbf{F} &= \begin{vmatrix} \mathbf{i} & \mathbf{j} & \mathbf{k} \\ 2 & 0 & 4 \\ 68.7 & 22.9 & -68.7 \end{vmatrix} \\ &= \mathbf{i}[0 - 4(22.9)] - \mathbf{j}[2(-68.7) - 4(68.7)] + \mathbf{k}[2(22.9) - 0] \\ &= -91.6\mathbf{i} + 412\mathbf{j} + 45.8\mathbf{k} \end{aligned}$$

Next using point (5, 1, 1) on the action line of $\mathbf{F}$, $\mathbf{r} = 5\mathbf{i} + \mathbf{j} + \mathbf{k}$. Then

$$\begin{aligned} \mathbf{M} &= \begin{vmatrix} \mathbf{i} & \mathbf{j} & \mathbf{k} \\ 5 & 1 & 1 \\ 68.7 & 22.9 & -68.7 \end{vmatrix} \\ &= \mathbf{i}[-1(68.7) - 22.9(1)] - \mathbf{j}[5(-68.7) - 1(68.7)] + \mathbf{k}[5(22.9) - 68.7(1)] \\ &= -91.6\mathbf{i} + 412\mathbf{j} + 45.8\mathbf{k} \end{aligned}$$

The scalar moments about the x, y, and z axes are the coefficients of the unit vectors $\mathbf{i}$, $\mathbf{j}$, and $\mathbf{k}$.

5. A couple of moment +60 lb-ft acts in the plane of the paper. Indicate this couple with (*a*) 10 lb forces and (*b*) 30 lb forces.

Solution:

In (*a*) the moment arm must be 6 ft, while in (*b*) it is 2 ft.

The direction of rotation must be counterclockwise. The parallel forces may be drawn at any angle as shown in Fig. 2-6.

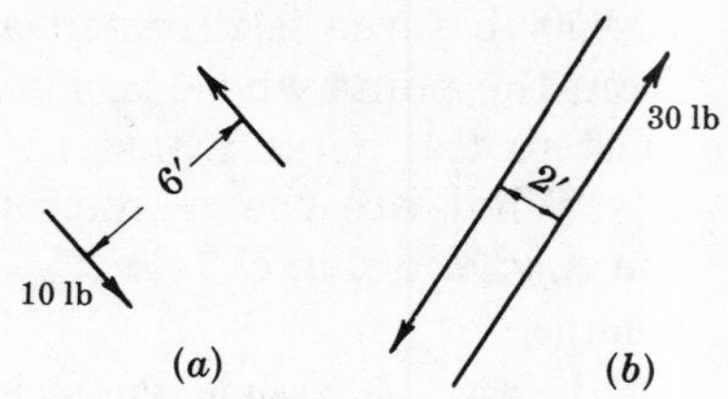

Fig. 2-6

6. Combine couple $C_1 = +20$ lb-ft with couple $C_2 = -50$ lb-ft both in the same plane. See Fig. 2-7.

Solution:

To combine graphically show both couples with forces of the same magnitude, say 10 lb, and drawn in such a way that two of the forces, one from each couple, are collinear but oppositely directed.

It is evident that the collinear forces cancel, leaving two 10 lb forces with an arm of 3 ft. The resultant couple is −30 lb-ft, a result which can also be obtained by algebraic addition.

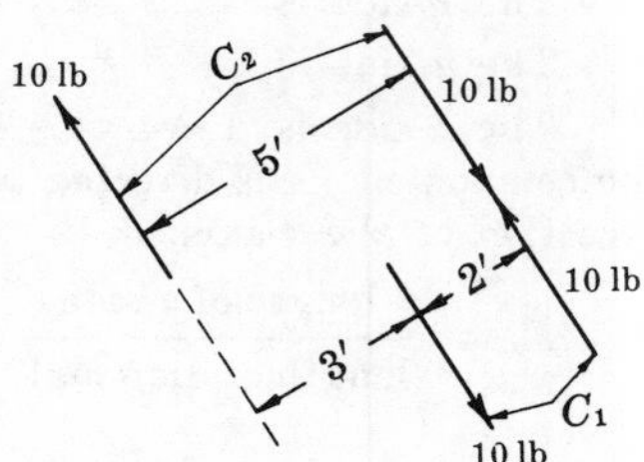

Fig. 2-7

7. A couple of −30 lb-ft acts in a horizontal plane and a couple of 60 lb-ft acts in a 45° plane, as shown in Fig. 2-8(*a*). Determine their resultant graphically.

Solution:

Draw each couple with forces of the same magnitude, say 30 lb, and in such a way that two of the forces, one from each couple, are collinear along the line of intersection of the planes but oppositely directed. Note that the broken line represents a 30 lb force in the horizontal plane. See Fig. 2-8(*b*).

The collinear forces cancel, leaving two oppositely directed parallel and equal forces forming a couple in a plane shown with angle θ_x with the horizontal plane, as shown in Fig. 2-8(*c*).

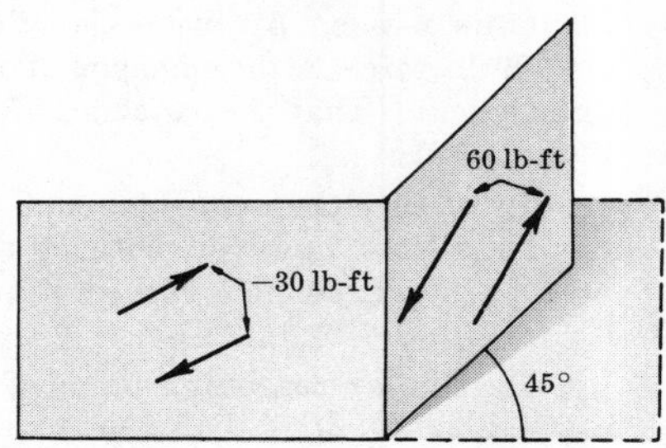

Fig. 2-8(*a*)

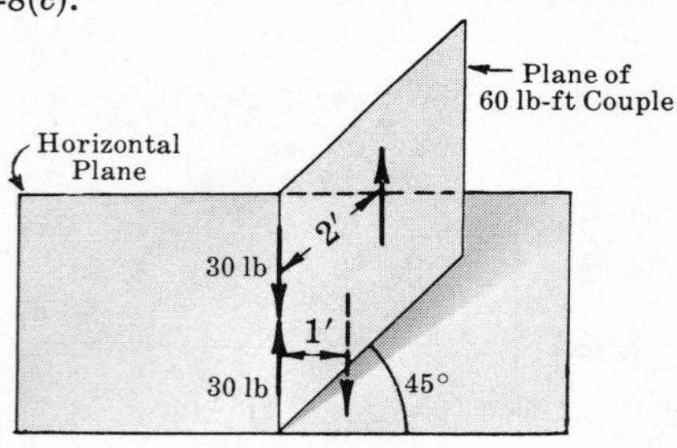

Fig. 2-8(*b*)

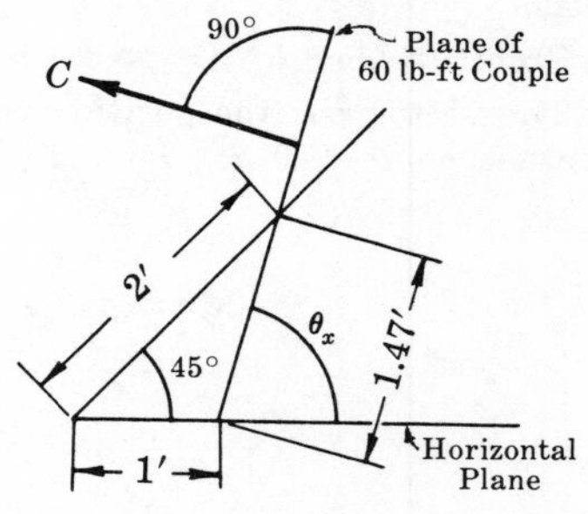

Fig. 2-8(*c*)

The arm of the resultant couple when measured to scale is 1.47 ft.

The couple may be represented as shown by a vector **C** perpendicular to the plane in which it acts and of magnitude (1.47 × 30) or 44.1 lb-ft. The angle θ_x is 74°.

Another graphical solution is obtained by using vector representation of the given couples. The vectors are shown at any point on the line of intersection and at right angles to their respective planes. See Fig. 2-8(*d*).

Their resultant is **C**. Its length to scale is 44.2 lb-ft. The vector **C** represents a couple in the plane perpendicular to itself, i.e., a plane with $\theta_x = 74°$.

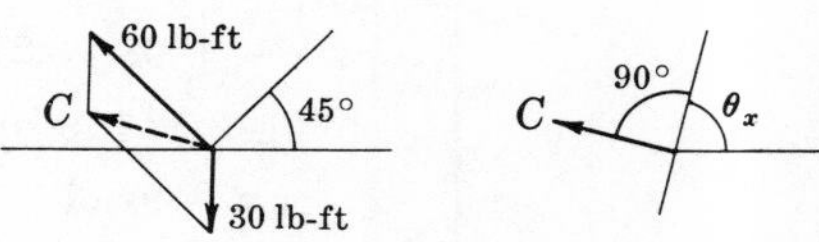

Fig. 2-8(*d*)

8. Combine a force of 25 lb, 80° with a +60 lb-ft couple in the same plane.

Solution:

A couple as such cannot be reduced to a simpler system but it can be combined with another force.

Draw the given couple with 25 lb forces and in such a way that one of its forces is collinear with the given single 25 lb force but oppositely directed.

By inspection the collinear forces cancel, leaving only a single force of 25 lb parallel to and in the same direction as the original force but at a distance 2.4 ft from it.

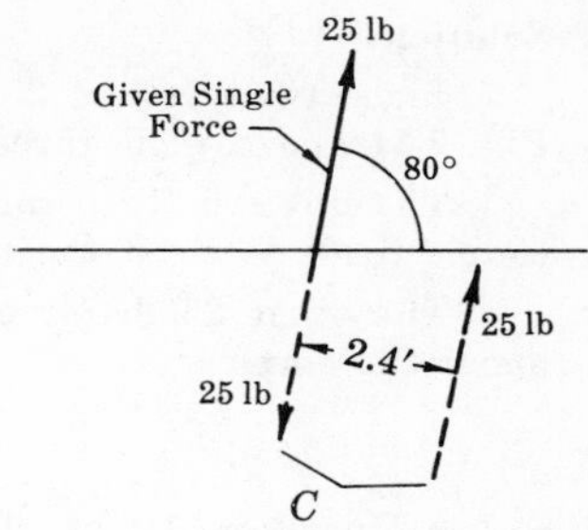

Fig. 2-9

9. A couple C_1 of 20 lb-ft acts in the xy-plane, a couple C_2 of 40 lb-ft acts in the yz-plane, and a couple C_3 of −55 lb-ft acts in the xz-plane. Determine the resultant couple.

Solution:

The couple C_1 is positive and acting in the xy-plane. When viewed from the positive end of the z-axis it tends to turn in a counterclockwise direction about the z-axis. By the *right hand rule* it is represented by a vector along the z-axis drawn towards the positive end. Using this type of reasoning, all three couples are drawn in the figure. Adding vectorially,

$$C = \sqrt{C_1^2 + C_2^2 + C_3^2} = \sqrt{(20)^2 + (40)^2 + (-55)^2} = 70.9 \text{ lb-ft}$$

$$\cos\phi_x = C_2/C = +0.564, \qquad \cos\phi_y = C_3/C = -0.777,$$

$$\cos\phi_z = C_1/C = +0.282$$

These are the direction cosines of the couple **C**. The couple acts in a plane perpendicular to this vector.

The couple **C** may be written in vector notation,

$$\mathbf{C} = +40\mathbf{i} - 55\mathbf{j} + 20\mathbf{k},$$

from which the value of **C** is derived as above.

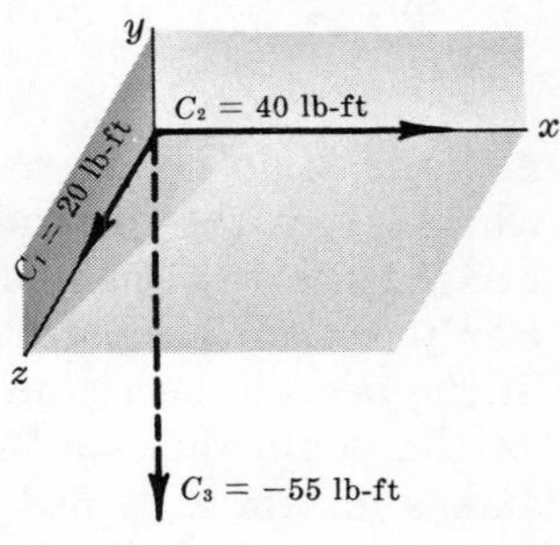

Fig. 2-10

10. A pipe 2 inches in diameter is subjected to a force of 25 lb applied vertically downward to the horizontal rod at an arm of 14 inches. Replace the 25 lb force with (1) a force at the end of the pipe which causes bending and (2) a couple which twists the shaft placing it in torsion. What are the moments of the force and the couple? See Fig. 2-11 below.

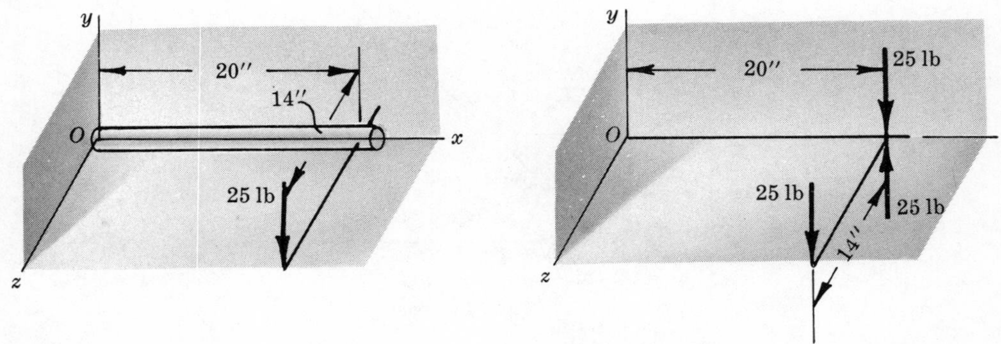

Fig. 2-11

Solution:

Place two vertical 25 lb forces oppositely directed through the center of the pipe as shown in Fig. 2-11 above. The three forces are still equivalent to the original force.

The upward force combines with the original to form a couple $C = 25 \times 14 = 350$ lb-in. This couple tends to twist the pipe counterclockwise when viewed from the right.

The other 25 lb force down on the pipe causes a bending moment $M = -25 \times 20 = -500$ lb-in about the z-axis.

11. Solve Problem 10 by determining the moment of the 25 lb force about the point O.

Solution:

The position vector of the point of application of the 25 lb force with respect to the origin is $\mathbf{r} = 20\mathbf{i} + 14\mathbf{k}$. The force $\mathbf{F} = -25\mathbf{j}$. Thus the moment of the 25 lb force with respect to the origin is

$$\mathbf{M} = \mathbf{r} \times \mathbf{F} = \begin{vmatrix} \mathbf{i} & \mathbf{j} & \mathbf{k} \\ 20 & 0 & 14 \\ 0 & -25 & 0 \end{vmatrix} = \mathbf{i}[0 - 14(-25)] - \mathbf{j}[0 - 0] + \mathbf{k}[-20(25) - 0] = 350\mathbf{i} - 500\mathbf{k}$$

This agrees with the results of Problem 10.

12. A crane is on level ground. The x-axis is through the contact points of the rearmost wheels with the ground, the y-axis is parallel to the front-to-back centerline, and the z-axis is vertical as shown. The bed (platform) of the crane is 3 ft above the ground. For practical purposes the pivot point of the bottom of the boom can be considered in the bed of the crane and 6 ft from the center of the cab. The center of the cab is on the centerline 15 ft forward (to the left) of the rearmost axle. The 50 ft boom makes an angle of 60° with the bed of the crane in a vertical plane and the cab and boom are swivelled 45° horizontally from the fore and aft centerline of the truck bed. The distance between the contact points of the rear wheels is considered to be 8 ft. Determine the turning moment of the 4000 lb load about the x-axis.

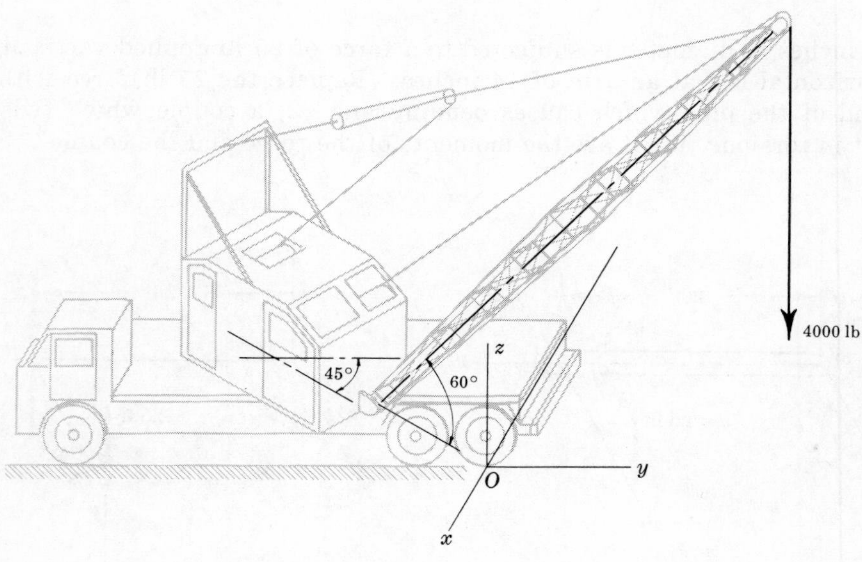

Fig. 2-12

Solution:

Relative to the origin O of the axes, the coordinates of the cab center are $(-4, -15, +3)$. The coordinates of the bottom of the boom are $(-4 + 6 \sin 45°, -15 + 6 \cos 45°, +3)$ or $(+0.24, -10.8, +3)$. The coordinates of the top of the boom are $(+0.24 + 50 \cos 60° \sin 45°, -10.8 + 50 \cos 60° \cos 45°, +3 + 50 \sin 60°)$ or $(+17.9, +6.91, +46.3)$.

The moment of the 4000 lb weight about O is $\mathbf{M} = \mathbf{r} \times \mathbf{F} = \begin{vmatrix} \mathbf{i} & \mathbf{j} & \mathbf{k} \\ 17.9 & 6.91 & 46.3 \\ 0 & 0 & -4000 \end{vmatrix}$

The scalar coefficient of the $\mathbf{i}$ term is the moment about the x axis. Hence $M_x = -27{,}600$ lb-ft. Thus the moment is clockwise about the x axis when viewed from the front.

Supplementary Problems

13. In each case, find the moment of the force **F** about the origin.

Magnitude of **F**	Angle of **F** with horizontal	Coordinates of point of application of **F**	*Answer*
20 lb	30°	(5, −4) ft	+119 lb-ft
64 lb	140°	(−3, 4) ft	+72.9 lb-ft
15 lb	337°	(8, −2) ft	−19.3 lb-ft
8 oz	45°	(6, 1) in.	+28.3 oz-in.
4 dynes	90°	(0, −20) cm	0
96 newtons	60°	(4, 2) meters	+236 nt-m

14. A 50 lb force is directed along the line drawn from the point whose x, y, z coordinates are $(8, 2, 3)$ ft to the point whose coordinates are $(2, -6, 5)$ ft. What are the scalar moments of the force about the x, y, and z axes? *Ans.* $M_x = +137$ lb-ft, $M_y = -167$ lb-ft, $M_z = -255$ lb-ft

15. Combine $C_1 = +20$ lb-ft, $C_2 = -80$ lb-ft, and $C_3 = -18$ lb-ft, all acting in the same plane.
Ans. $C = -78$ lb-ft acting in the same or parallel plane

16. Add couples $C_1 = -20$ oz-in. and $C_2 = -48$ oz-in. acting in parallel planes 6 inches apart.
Ans. $C = -68$ oz-in. in any parallel plane

17. Determine the resultant vector of the three couples +16 lb-ft, −45 lb-ft, and +120 lb-ft acting respectively in the xy, yz, and xz planes.
Ans. $C = +129$ lb-ft, $\cos \phi_x = -0.349$, $\cos \phi_y = 0.931$, $\cos \phi_z = 0.124$

18. The 24 lb forces are applied at the corners A and B of the parallelepiped, shown in Fig. 2-13, and act along AE and BF respectively. Show the given couple may be replaced by a set of vertical forces consisting of a 16 lb force up at point C and a 16 lb force down at point D.

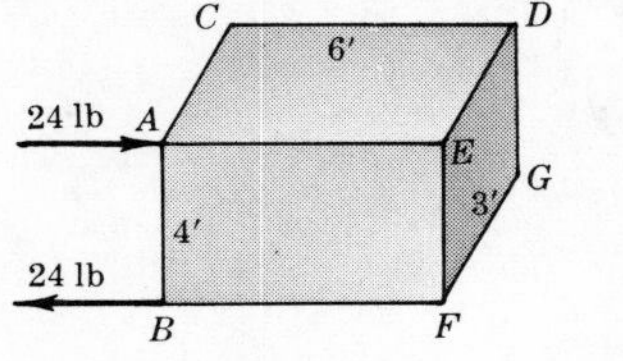

Fig. 2-13

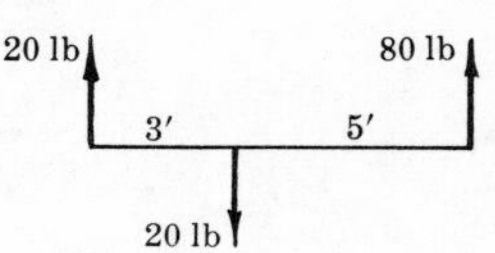

Fig. 2-14

19. Show that the given set of three vertical forces, shown in Fig. 2-14, may be replaced by a single 80 lb force acting vertically up and 0.75 ft to the left of the given 80 lb force.

20. A horizontal bar 8 ft long is acted upon by a downward vertical 12 lb force at the right end as shown in Fig. 2-15 below. Show this is equivalent to a 12 lb vertical force acting down at the left end *and* a clockwise couple of 96 lb-ft.

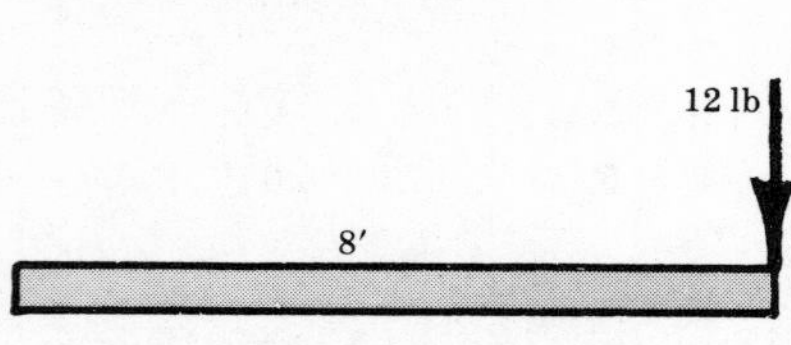

Fig. 2-15

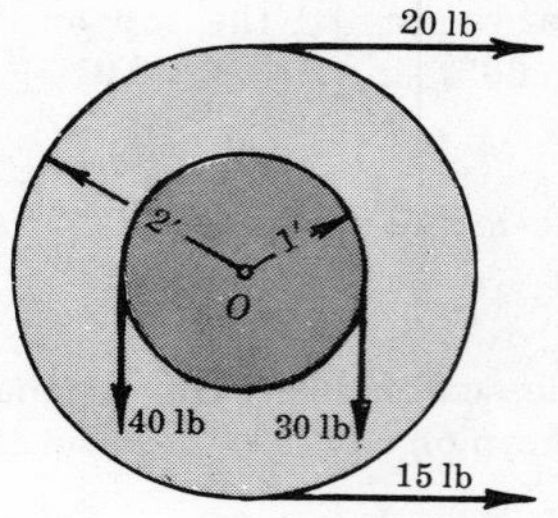

Fig. 2-16

21. Reduce the system of forces in the belts, shown in Fig. 2-16 above, to a single force at O and a couple. Forces are either vertical or horizontal. *Ans.* 78.3 lb, $\theta_x = 296.5°$, $C = 0$

22. Reduce the system of forces acting on the beam, shown in Fig. 2-17 below, into a force at A and a couple. *Ans.* R = 100 lb up at A, C = 6000 lb-ft

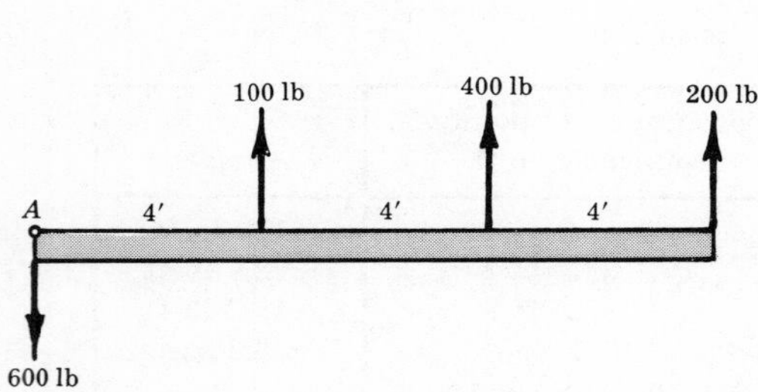

Fig. 2-17

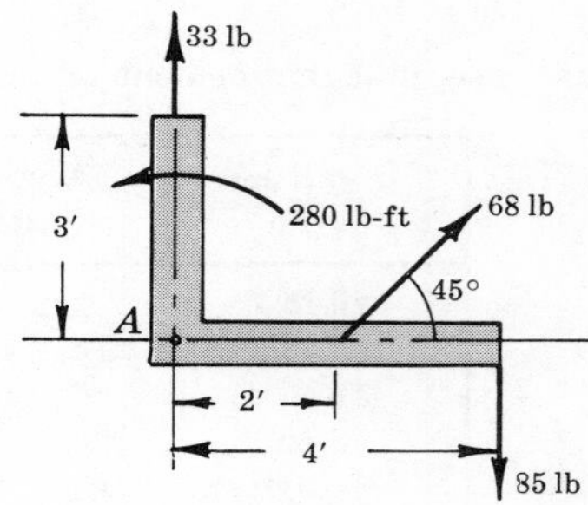

Fig. 2-18

23. Reduce the system of forces and couples to the simplest system using point A. Refer to Fig. 2-18 above. *Ans.* R_x = +48.1 lb, R_y = −3.9 lb, C = +36.2 lb-ft

24. Determine the moments of the two forces about the x, y, and z axes shown in Fig. 2-19 below. *Ans.* $\mathbf{M} = 488\mathbf{i} + 732\mathbf{k}$ lb-ft or M_x = +488 lb-ft, $M_y = O$, M_z = +732 lb-ft

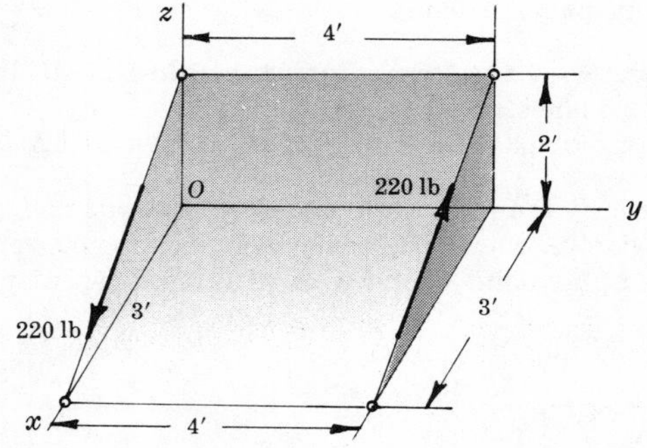

Fig. 2-19

Chapter 3

Resultants of Coplanar Force Systems

COPLANAR FORCES

Coplanar forces lie in the one plane. A concurrent system consists of forces which intersect at a point called the concurrence. A parallel system consists of forces which intersect at infinity. A non-concurrent, non-parallel system consists of forces which are not all concurrent and not all parallel.

Vector equations may be applied to the above systems to determine resultants, but the following derived scalar equations will be more useful for a given system.

CONCURRENT SYSTEM

The resultant **R** may be (*a*) a single force through the concurrence or (*b*) zero. Algebraically,

$$R = \sqrt{(\Sigma F_x)^2 + (\Sigma F_y)^2} \qquad \text{and} \qquad \tan\theta_x = \frac{\Sigma F_y}{\Sigma F_x}$$

where $\Sigma F_x, \Sigma F_y$ = algebraic sums of the x and y components, respectively, of the forces of the system,

θ_x = angle which the resultant **R** makes with the x axis.

PARALLEL SYSTEM

The resultant may be (*a*) a single force **R** parallel to the system, (*b*) a couple in the plane of the system or in a parallel plane, or (*c*) zero. Algebraically,

$$R = \Sigma F \qquad \text{and} \qquad R\bar{a} = \Sigma M_O$$

where ΣF = algebraic sum of the forces of the system,

O = any moment center in the plane,

$\bar{a}$ = perpendicular distance from the moment center O to the resultant R,

$R\bar{a}$ = moment of R with respect to O,

ΣM_O = algebraic sum of the moments of the forces of the system with respect to O.

If ΣF is not zero, apply the equation $R\bar{a} = \Sigma M_O$ to determine $\bar{a}$ and hence the action line of R. If $\Sigma F = O$, the resultant couple, if there is one, has a magnitude of ΣM_O.

NON-CONCURRENT, NON-PARALLEL SYSTEM

The resultant may be (*a*) a single force **R**, (*b*) a couple in the plane of the system or in a parallel plane, or (*c*) zero. Algebraically,

$$R = \sqrt{(\Sigma F_x)^2 + (\Sigma F_y)^2} \qquad \text{and} \qquad \tan\theta_x = \frac{\Sigma F_y}{\Sigma F_x}$$

where $\Sigma F_x, \Sigma F_y$ = algebraic sums of the x and y components, respectively, of the forces of the system,

θ_x = angle which the resultant **R** makes with the x axis.

To determine the action line of the resultant force, employ the equation

$$R\bar{a} = \Sigma M_O$$

where O = any moment center in the plane,
$\bar{a}$ = perpendicular distance from the moment center O to the resultant R,
$R\bar{a}$ = moment of R with respect to O,
ΣM_O = algebraic sum of the moments of the forces of the system with respect to O.

Note that even if $R = 0$, a couple may exist with magnitude equal to ΣM_O.

GRAPHICAL SOLUTIONS

Graphical solutions are illustrated in the solved problems which follow.

In the following problems, the forces in the diagrams are represented by their magnitudes if the directions are obvious by inspection.

Solved Problems

1. Determine the resultant of the concurrent force system shown in Fig. 3-1.

Solution:

Find the x and y components of each of the four given forces. Add the x components algebraically to determine ΣF_x. Find ΣF_y for the y components. A tabular form may present the information more clearly.

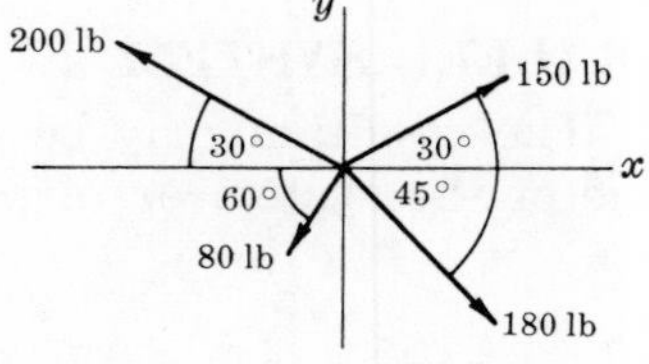

Fig. 3-1

Force	Cos θ_x	Sin θ_x	F_x	F_y
150	+0.866	+0.500	+129.9	+75.0
200	−0.866	+0.500	−173.2	+100.0
80	−0.500	−0.866	−40.0	−69.2
180	+0.707	−0.707	+127.3	−127.3

$\Sigma F_x = +44.0$, $\Sigma F_y = -21.5$ and $R = \sqrt{(\Sigma F_x)^2 + (\Sigma F_y)^2} = \sqrt{(44.0)^2 + (-21.5)^2} = 49.0$ lb.

$\text{Tan}\,\theta_x = \dfrac{\Sigma F_y}{\Sigma F_x} = \dfrac{-21.5}{+44.0} = -0.489$, from which $\theta_x = 360° - 26° = 334°$.

2. Solve Problem 1 graphically. See Fig. 3-2.

Solution:

Starting with point O apply the Polygon Law, adding vectorially each force in turn. The resultant OK of the system is drawn from O to the arrow end of the last force.

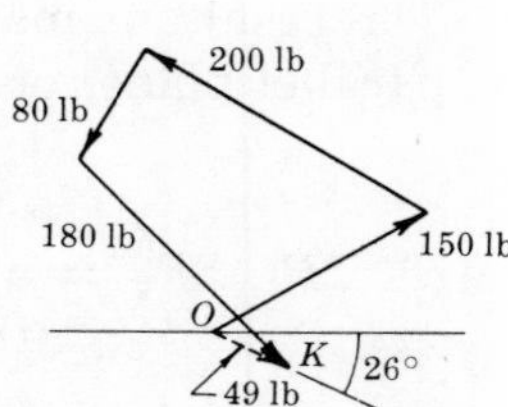

Fig. 3-2

3. Determine algebraically the resultant of the force system shown in the figure. Note that the slope of the action line of each force is indicated in Fig. 3-3.

Solution:

Force	F_x	F_y
50	$+50 \times \dfrac{1}{\sqrt{1^2+1^2}}$	$+50 \times \dfrac{1}{\sqrt{2}}$
100	$-100 \times \dfrac{3}{\sqrt{1^2+3^2}}$	$-100 \times \dfrac{1}{\sqrt{10}}$
30	$+30 \times \dfrac{1}{\sqrt{1^2+2^2}}$	$-30 \times \dfrac{2}{\sqrt{5}}$

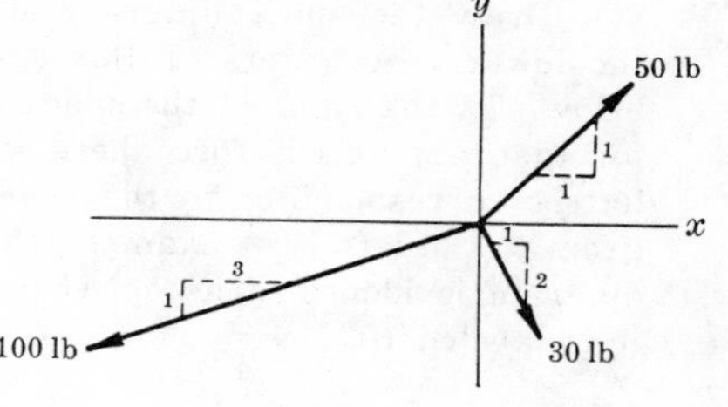

Fig. 3-3

$\Sigma F_x = -46.1$, $\Sigma F_y = -23.0$, and $R = \sqrt{(-46.1)^2+(-23.0)^2} = 51.6$ lb with $\theta_x = 207°$.

4. Determine the resultant of the parallel system.

Solution:

In Fig. 3-4 the action lines of the forces are vertical as shown.

$R = -20+30+5-40 = -25$ lb, i.e., down.

To determine the action line of this 25 lb force choose any moment center O. Since the moment of a force about a point on its own action line is zero, it is advisable but not necessary to choose O on one of the given forces. Let O be on the 30 lb force.

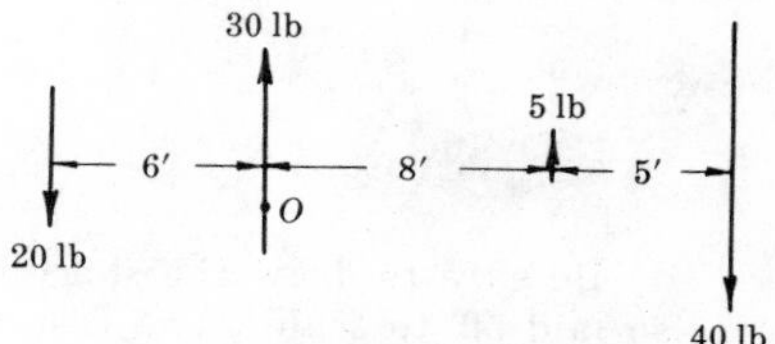

Fig. 3-4

$$\Sigma M_O = +(20 \times 6) + (30 \times 0) + (5 \times 8) - (40 \times 13) = -360 \text{ lb-ft}$$

Then the moment of R must equal -360 lb-ft. This means that R, which is down $(-)$, must be placed to the right of O because only then will its moment be clockwise $(-)$.

Apply $R\bar{a} = \Sigma M_O$ to obtain $\bar{a} = \dfrac{360 \text{ lb-ft}}{25 \text{ lb}} = 14.4$ ft to the right of O.

Note the determination of $\bar{a}$ without regard to the signs of R or ΣM_O but by using reasoning.

5. Determine the resultant of the parallel force system in Fig. 3-5 below.

Solution:

$R = -100+200-200+400-300 = 0$. This means that the resultant is not a single force.

Next find ΣM_O. Choose O, as shown, on the 100 lb force.

$$\Sigma M_O = +(100 \times 0) + (200 \times 2) - (200 \times 5) + (400 \times 9) - (300 \times 11) = -300 \text{ lb-ft}$$

The resultant is therefore a couple $C = -300$ lb-ft which can be represented in the plane of the paper in accordance with the laws governing couples.

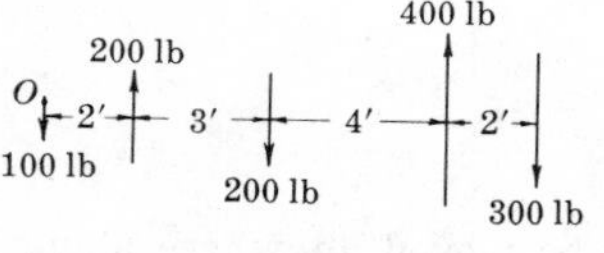

Fig. 3-5

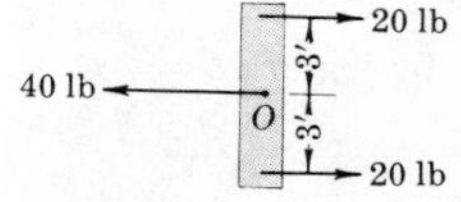

Fig. 3-6

6. Determine the resultant of the horizontal force system acting on the bar shown in Fig. 3-6 above.

Solution:

$R = \Sigma F_h = +20+20-40 = 0$. This means that the resultant is not a single force, but it may be a couple.

$\Sigma M_O = -(20 \times 3) + (20 \times 3) = 0$.

Hence in this system the resultant force is zero and the resultant couple is also zero.

7. Determine graphically the resultant of the parallel force system shown in Fig. 3-7(a).

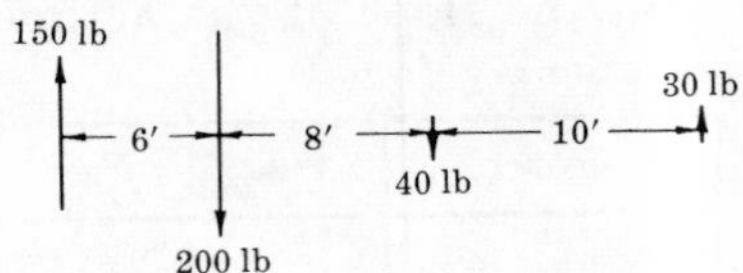

Fig. 3-7(a)

Solution:

Draw the space diagram identifying each force with the lower case letters of Bow's notation as in Fig. 3-7(b) below. To the right of the space diagram is drawn the vector diagram. Each force here is labeled with the capital letters corresponding to the lower case letters in the diagram to the left. For example, the 200 lb force in the space diagram is identified as bc whereas in the vector diagram it is labeled BC.

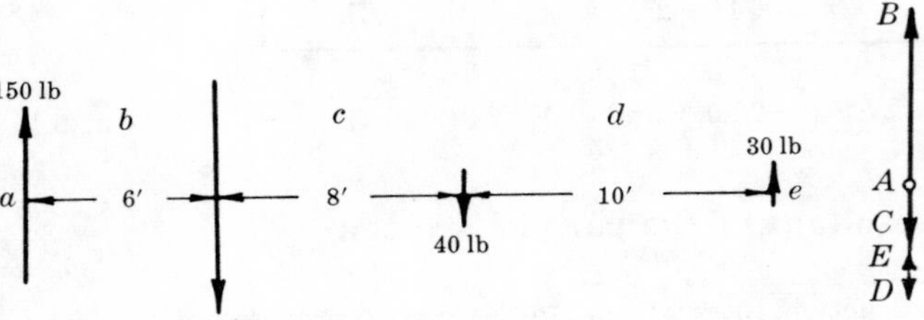

Fig. 3-7(b)

Be sure to draw the space diagram with all distances to scale. The forces AB, BC, CD, DE are also laid off to scale. The resultant is from A to E, or a force of 60 lb down.

The next step is to choose a pole point O as shown in Fig. 3-7(c) below. Draw rays OA, OB, OC, OD, OE. Actually, AO and OB are components which make up AB as shown.

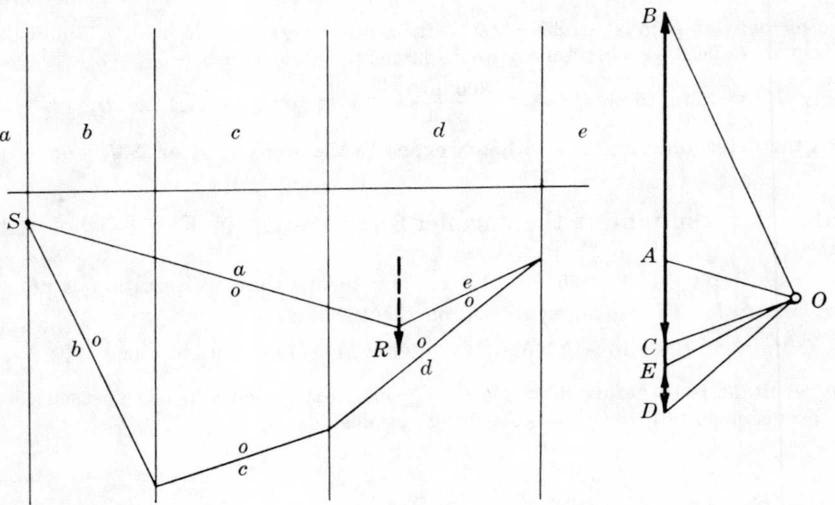

Fig. 3-7(c)

Since by the principle of transmissibility a force may be moved anywhere along its line of action, choose any point S on the action line of the first force ab as shown. Through it draw strings oa and ob parallel to the rays OA and OB. Next draw through bc the action lines of its components ob and oc. Since ob is already drawn, use the point on bc where ob cuts it and through this point draw oc parallel to OC.

Continue in this fashion until all strings have been drawn.

At this stage notice strings oa and oe in the space diagram. Continue these until they intersect in point R. Since they are the action lines of the components of the resultant, they meet on the action line of the resultant. Draw a 60 lb force down through this point which when measured to the space scale is 17.3 ft to the right of the first line ab.

8. Determine the resultant of the vertical force system shown in Fig. 3-8(a).

Solution:

Label the fields as before a, b, c, d, e. Draw the space diagram, shown in Fig. 3-8(b) below, to scale. Next draw the free vectors on a line to the right of the figure.

Now note that A and E coincide, merely indicating that the resultant is not a single force. Choose pole point O and draw rays as before. Again choose any point on the action line of ab. Draw ob and oa. Through point of intersection of ob with bc draw oc. Through point of intersection of oc with cd draw do. Finally through point of intersection of do with de draw oe, which of course is parallel to oa and does not intersect it. This fact can be used to determine the moment of the couple.

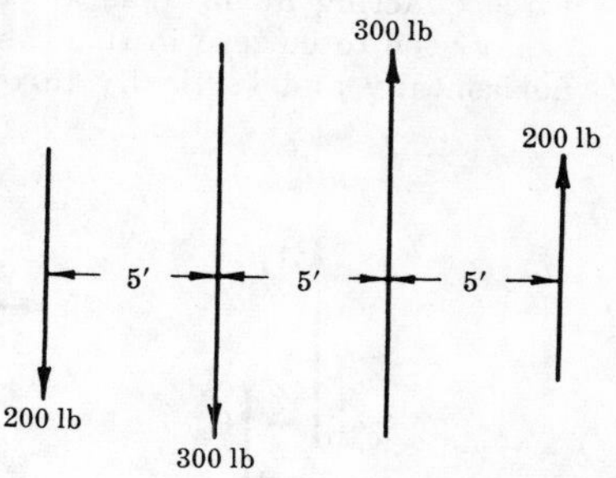

Fig. 3-8(a)

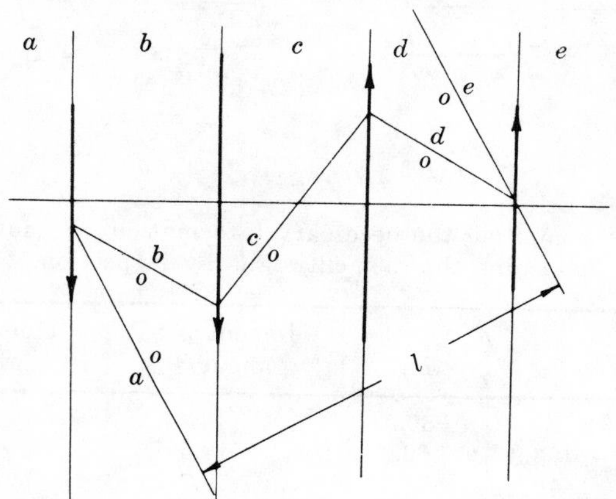

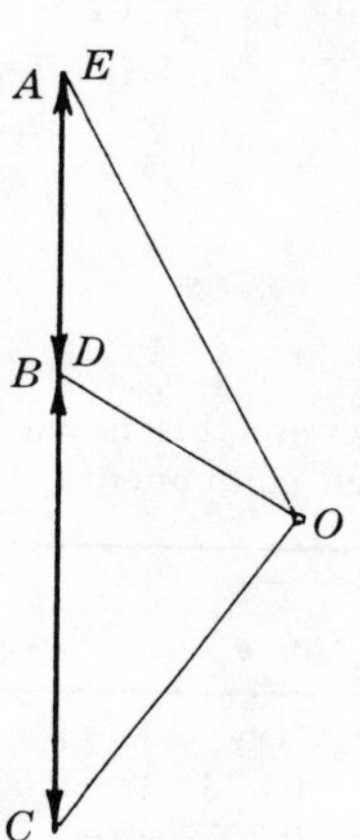

Fig. 3-8(b)

AB is represented in the diagram to the right by its components which by the triangle law are A to O and O to B. Therefore in the space diagram the string ao is the action line of component AO acting down and to the right. By similar reasoning, the string oe is the action line of component OE acting up and to the left. The size in lb of these components as measured to the force scale is 329 lb. The perpendicular distance l between those two strings is measured to the distance scale and equals 13.7 ft. Thus the resultant is a couple with forces of 329 lb and a moment arm of 13.7 ft, making $C = +4510$ lb-ft.

Had the strings oe and ao coincided the moment-arm would be zero, indicating no moment. The system would then be in equilibrium.

A different starting point on ab or a different pole point will affect the shape of the string polygon. This is often disturbing to students. But one should remember that a couple depends on the *product* of the arm and the size of the forces which constitute it, and is also independent of the slope of its component forces in the plane.

9. Determine the resultant of the non-concurrent, non-parallel system shown in Fig. 3-9(a). Assume that the coordinates are in feet.

Solution:

For convenience the forces are lettered A, B, C, D. The simplest method of attack is to use a tabular form listing x and y components for each force and also the moment of each component about some moment center – for this example O. The forces are now replaced with their components at the same points in the action lines as given as indicated in Fig. 3-9(b) below. It may be convenient at times to use components at a different point in the action line than that which is given, e.g.,

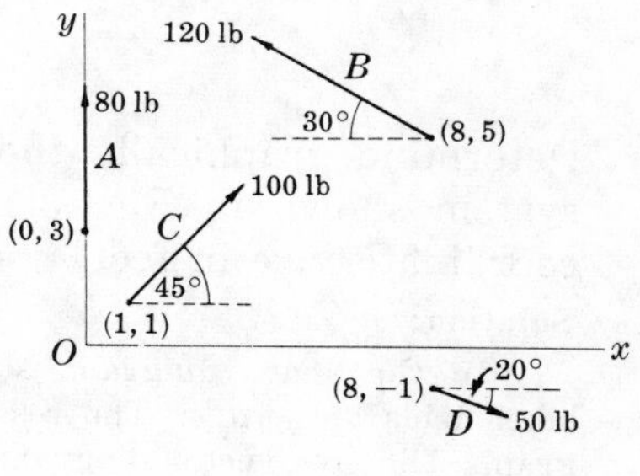

Fig. 3-9(a)

force C acting at 45° has an action line that passes through the origin O. The moment about O is easily seen to be zero in this case. However, the components used in this example will be shown acting horizontally and vertically through the given points of application.

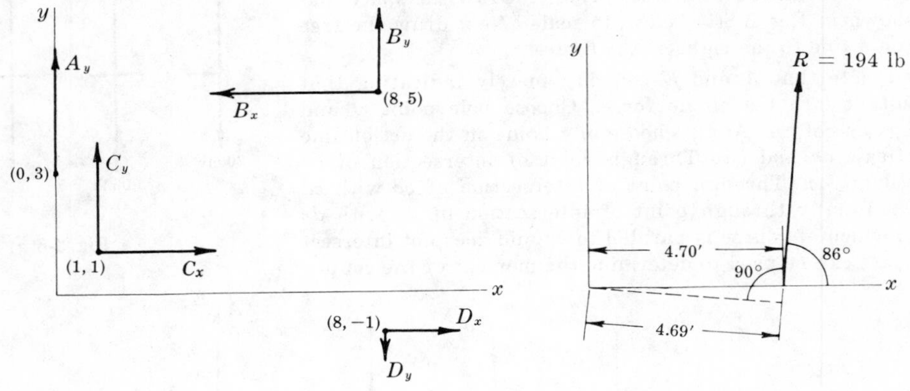

Fig. 3-9(b)

The following table is useful in compiling the necessary information. Be sure to place the proper sign before each component and to determine the moment signs by inspection.

Force	Cos θ_x	Sin θ_x	F_x	F_y	Moment of F_x about O	Moment of F_y about O	M_O
A	0	+1	0	+80.0	0	0	0
B	−0.866	+0.500	−103.9	+60.0	+519.5	+480.0	+999.5
C	+0.707	+0.707	+70.7	+70.7	−70.7	+70.7	0
D	+0.940	−0.342	+47.0	−17.1	+47.0	−136.8	−89.8

$$\Sigma F_x = +13.8 \text{ lb}, \quad \Sigma F_y = +193.6 \text{ lb}, \quad \Sigma M_O = +910 \text{ lb-ft}$$

$$R = \sqrt{(\Sigma F_x)^2 + (\Sigma F_y)^2} = \sqrt{(+13.8)^2 + (+193.6)^2} = 194 \text{ lb}, \quad \theta_x = \tan^{-1}\frac{+193.6}{+13.8} = 86°$$

To find the moment arm of the resultant, divide 910 by 194 to obtain 4.69 ft.

Since R acts upward and slightly to the right it must be placed as shown because ΣM_O is positive, i.e., R must have a counterclockwise moment.

Another method to locate the action line of the resultant is to determine its intercept with say the x-axis. If the components of the resultant are drawn through the intercept on the x-axis, the x component would have no moment about O. The moment would be determined solely by the y component and would be equal to the product of the y component and the x distance to the intercept (x coordinate of the intercept).

$\bar{x} = \dfrac{\Sigma M_O}{\Sigma F_y} = \dfrac{910}{193.6} = 4.70$ ft. Draw the resultant as shown with the intercept +4.70 ft to the right because ΣF_y is positive and ΣM_O is positive.

10. Determine graphically the resultant of the force system shown in Fig. 3-10(a). Assume that the coordinates are in feet.

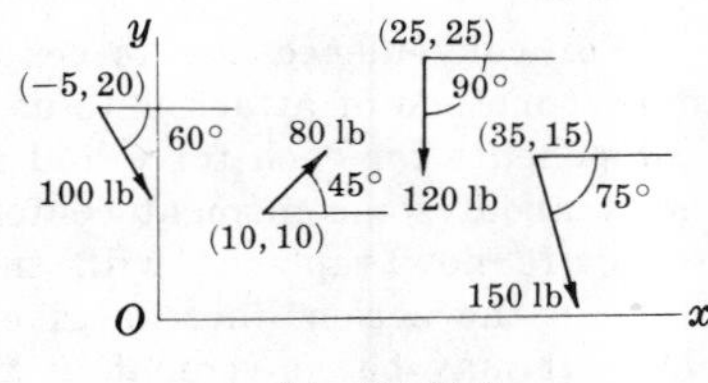

Fig. 3-10(a)

Solution:

In the space diagram, shown in Fig. 3-10(b) below, label fields a, b, c, d, e. The resultant will be ae in this diagram. The free vector diagram is drawn to the right of the space diagram.

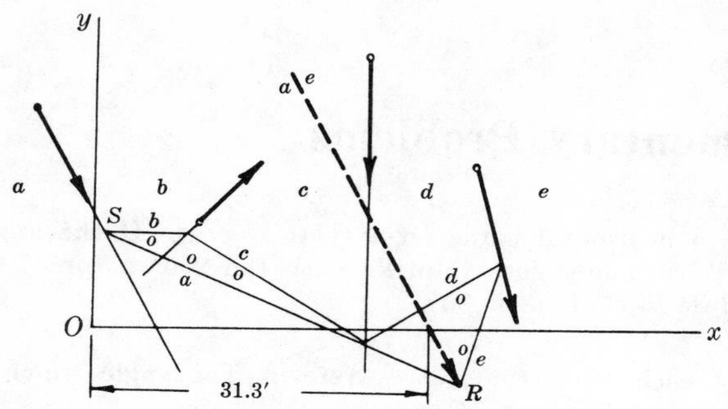

Fig. 3-10(*b*)

Starting with point S on the action line of ab, draw strings oa and ob parallel to the rays OA and OB. Continue as shown until the intersection of ao and oe is determined. Through this point R draw the action line of the resultant. This intersects the x-axis in a point 31.3 ft from the origin O. The resultant measures 330 lb with $\theta_x = 298°$.

11. Determine the resultant of the four forces tangent to the circle of radius 3 ft shown in Fig. 3-11(*a*). What will be its location with respect to the center of the circle?

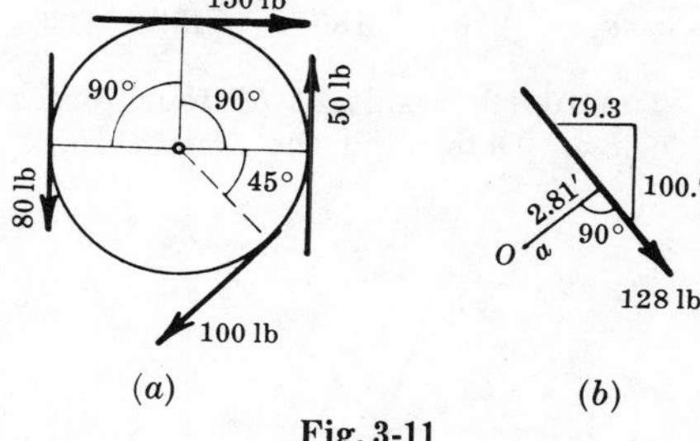

Fig. 3-11

Solution:

Note that the horizontal and vertical components of the 100 lb force are both −70.7 lb. Hence $\Sigma F_h = +150 - 70.7 = +79.3$ lb, i.e., to the right; and $\Sigma F_v = +50 - 80 - 70.7 = -100.7$ lb, i.e., down. The resultant $R = \sqrt{(\Sigma F_h)^2 + (\Sigma F_v)^2} = 128$ lb.

The moment of R with respect to O is $R \times a$, and this equals the sum of the moments of all the given forces about O. Hence $128a = +50 \times 3 - 150 \times 3 + 80 \times 3 - 100 \times 3 = -360$. The resultant is shown, in Fig. 3-11(*b*), at a distance of 2.81 ft from the center O of the circle, causing a negative moment.

12. In triangle ABC, two forces are drawn from vertex B along the sides as shown in Fig. 3-12 below. If the two forces are proportional to the lengths of the two sides, show that their resultant bisects the third side.

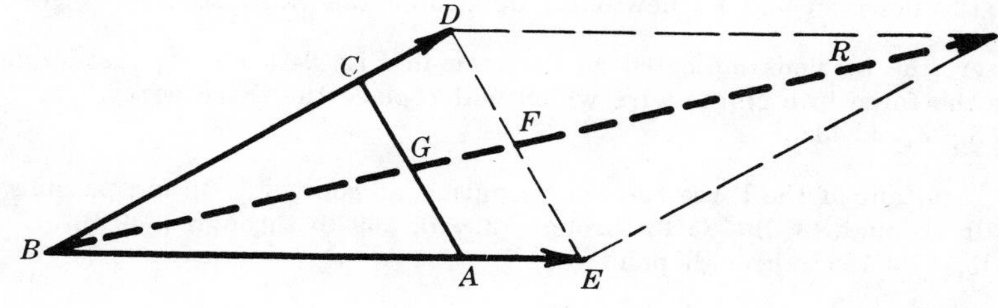

Fig. 3-12

Solution:

The force $BD = n \times BC$ and the force $BE = n \times BA$. The diagonal R of the parallelogram bisects the other diagonal DE in F. Also, $BD/BC = BE/BA = n$. Hence CA is parallel to DE.

From similar triangles BCG and BDF, $CG/DF = BC/BD = 1/n$. From similar triangles BGA and BFE, $GA/FE = BA/BE = 1/n$. Therefore $CG/GA = DF/FE$. This means that G is the midpoint of CA and the resultant bisects CA.

Supplementary Problems

13. Two forces of 200 lb and 300 lb pull in a horizontal plane on a vertical post. If the angle between them is 85°, what is their resultant? What angle does it make with the 200 lb force? Solve both graphically and algebraically. *Ans.* $R = 375$ lb, $\theta = 53°$

In Problems 14-17, find the resultant of each concurrent force system. The angle which each force makes with the x-axis (measured counterclockwise) is given. Forces are in lb units.

14.	Force	85	126	65	223		
	θ_x	38°	142°	169°	295°		*Ans.* $R = 59.8$ lb, $\theta_x = 268°$
15.	Force	22	13	19	8		
	θ_x	135°	220°	270°	358°		*Ans.* $R = 21.3$ lb, $\theta_x = 214°$
16.	Force	1250	1830	855	2300		
	θ_x	62°	125°	340°	196°		*Ans.* $R = 2520$ lb, $\theta_x = 138°$
17.	Force	285	860	673	495	241	
	θ_x	270°	180°	45°	330°	100°	*Ans.* $R = 181$ lb, $\theta_x = 89°$

18. The 100 lb resultant of four forces together with three of those four forces is shown in Fig. 3-13 below. Determine the fourth force. *Ans.* $F = 203$ lb, $\theta_x = 49°$

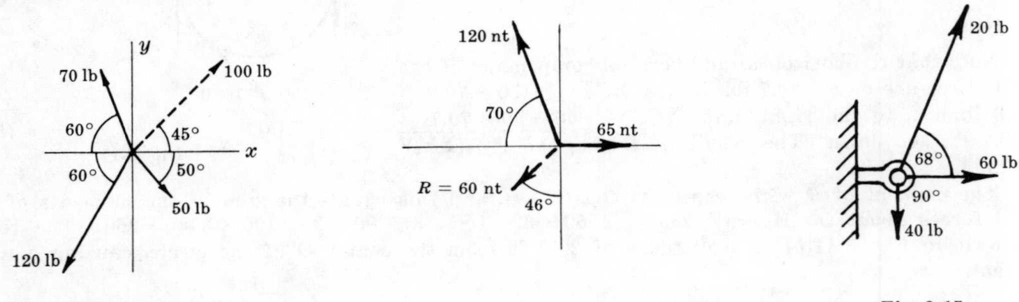

Fig. 3-13 Fig. 3-14 Fig. 3-15

19. Three coplanar forces of 80 lb each pull on a small ring (diameter negligible). Assuming that their lines of action make equal angles with each other (120°), determine the resultant. This system is said to be in equilibrium. *Ans.* $R = 0$

20. The resultant of three forces is 60 newtons as shown in Fig. 3-14 above. Two of the three forces are also shown as 120 newtons and 65 newtons. Determine the third force. *Ans.* 169 nt, $\theta_x = 246°$

21. Three wires exert the tensions indicated on the ring in Fig. 3-15 above. Assuming a concurrent system, determine the force in a single wire which will replace the three wires.
Ans. $T = 70.8$ lb, $\theta_x = 343°$

22. Determine the resultant of the three forces originating at point $(3, -3)$ and passing through the points indicated: 126 lb through $(8, 6)$, 183 lb through $(2, -5)$, 269 lb through $(-6, 3)$.
Ans. $R = 263$ lb, $\theta_x = 159°$ through point $(3, -3)$

Find the resultant in each of Problems 23-25. The forces are horizontal and expressed in lb. The y distances to the action lines are expressed in ft.

23.	Force	+50	+20	−10			
	y	+3	−5	+6			*Ans.* $R = +60$ lb, $\bar{y} = -0.167$ ft
24.	Force	+800	−300	+1000	−600		
	y	−6	−5	−4	0		*Ans.* $R = +900$ lb, $\bar{y} = -8.11$ ft
25.	Force	+160	−220	+80	−180	+160	
	y	+3	−7	−3	+10	0	*Ans.* $C = +20$ lb-ft

26. See Fig. 3-16 below. Find the resultant of the three loads shown acting on the beam.
Ans. $R = 38$ T down, distance from left support = 8.37 ft

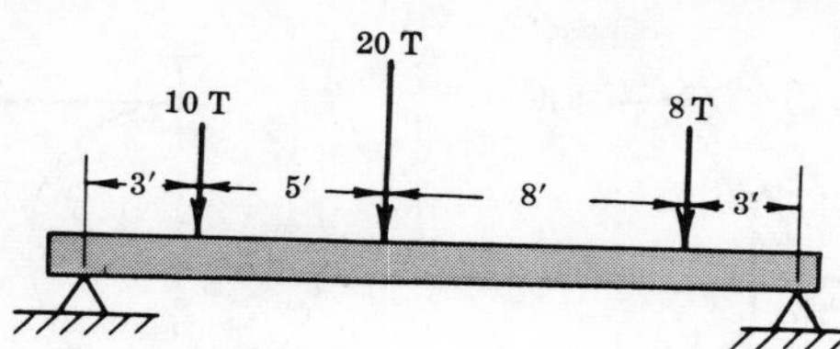

Fig. 3-16

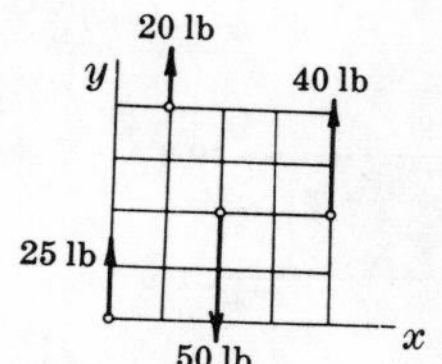

Fig. 3-17

27. Determine the resultant of the four forces shown in Fig. 3-17 above. The side of each small square is one ft. *Ans.* $R = +35$ lb, $\bar{x} = 2.29$ ft

28. Six weights of 30, 20, 40, 25, 10, and 35 lb hang in one plane from a horizontal support at distances 2, 3, 5, 7, 10, and 12 ft respectively from a wall. What single force would replace these six? *Ans.* $R = -160$ lb, 6.34 ft from wall

29. Three forces act on the beam, two of which are shown in Fig. 3-18 together with the resultant of all three. What is the third force?
Ans. $F = 20$ T down, $\bar{x} = 10$ ft from left support

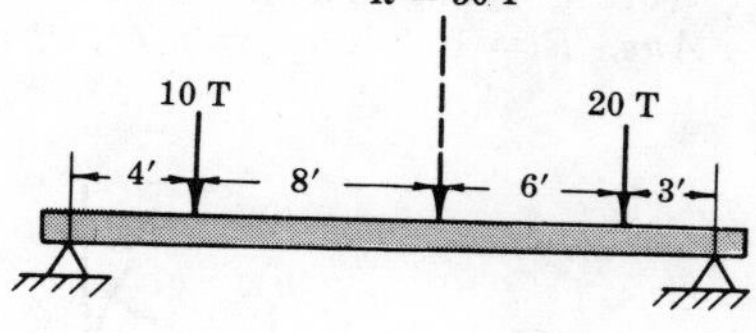

Fig. 3-18

In Problems 30-32, determine the resultant of the non-concurrent, non-parallel systems of forces. F is in lb, and the coordinates are in ft.

30.

F	20	30	50	10	*Ans.* $R = 54.7$ lb
θ_x	45°	120°	190°	270°	$\theta_x = 157°$
Coord. of point of application	(1, 3)	(4, −5)	(5, 2)	(−2, −4)	x int. = 3.52 ft

31.

F	50	100	200	90	*Ans.* $R = 303$ lb
θ_x	90°	150°	30°	45°	$\theta_x = 60.3°$
Coord. of point of application	(2, 2)	(4, 6)	(3, −2)	(7, 2)	x int. = 6.77 ft

32.

F	2	4	5	8	*Ans.* $R = 7.12$ lb
θ_x	45°	290°	183°	347°	$\theta_x = 322°$
Coord. of point of application	(0, 5)	(4, 3)	(9, −4)	(2, −6)	x int. = 1.20 ft

33. Determine completely the resultant of the 5 forces shown in Fig. 3-19 below. The forces are in oz, and the squares are 1 inch by 1 inch. *Ans.* $C = -268$ oz-in.

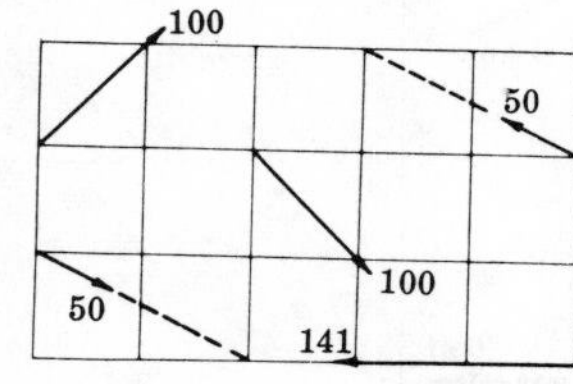

Fig. 3-19

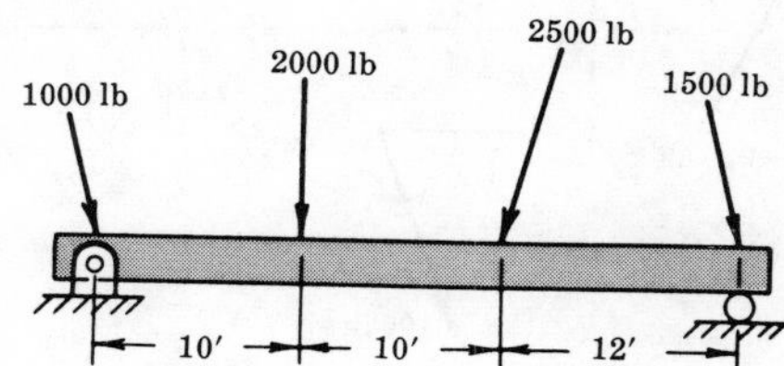

Fig. 3-20

34. Determine completely the resultant of the four forces shown in Fig. 3-20 above. Each force makes a 15° angle with the vertical, except the 2000 lb force which is vertical.
Ans. $R = 6830$ lb down, horizontal distance from hinge = 16.8 ft

35. A thin steel plate is subjected to the three forces shown in Fig. 3-21 below. What single force would have an equivalent effect on the plate?
Ans. $R = 18.7$ lb, $\theta_x = 285°$, bottom intercept = 4.23 ft to left of O

36. Determine the resultant of the forces acting on the bell crank. See Fig. 3-22 below.
Ans. $R = 247$ lb, $\theta_x = 259°$, horizontal intercept = −6.3 in. from O

37. Determine the resultant of the three forces acting on the pulley shown in Fig. 3-23 below.
Ans. $R = 742$ lb, $\theta_x = 357°$, R cuts vertical diameter 2.27 ft above O

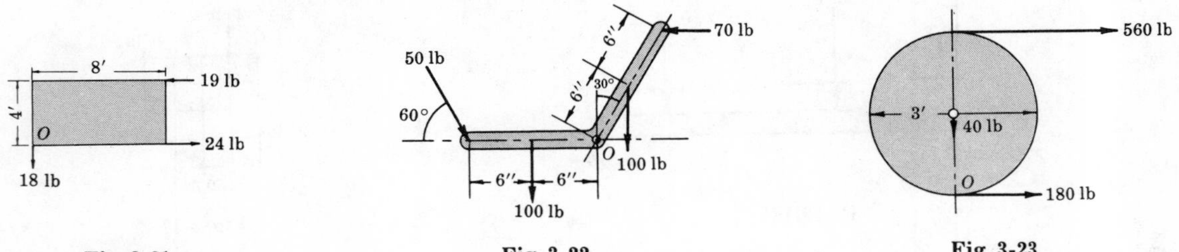

Fig. 3-21 Fig. 3-22 Fig. 3-23

38. Solve for the resultant of the six loads on the truss shown in Fig. 3-24 below. The loads are given in kips (1 kip = 1000 lb). Three loads are vertical. The wind loads are perpendicular to the side. The truss is symmetrical.
Ans. $R = 10.7$ K, $\theta_x = 281°$, R cuts lower member of truss +15.3 ft from left support

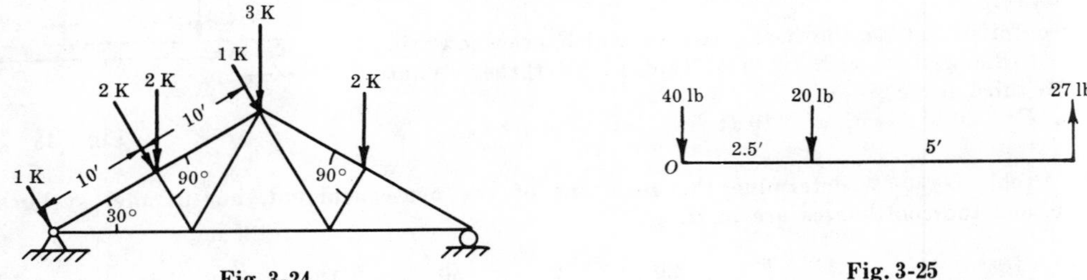

Fig. 3-24 Fig. 3-25

39. The resultant of four vertical forces is a couple of 300 lb-ft counterclockwise. Three of the four forces are shown in Fig. 3-25 above. Determine the fourth. *Ans.* 33 lb up at 4.46 ft to right of O

40. Determine the resultant of the forces shown in Fig. 3-26 below. The coordinates are in feet.
Ans. 73.4 lb, $\theta_x = 107°$, x intercept = 8.38 ft to left of O

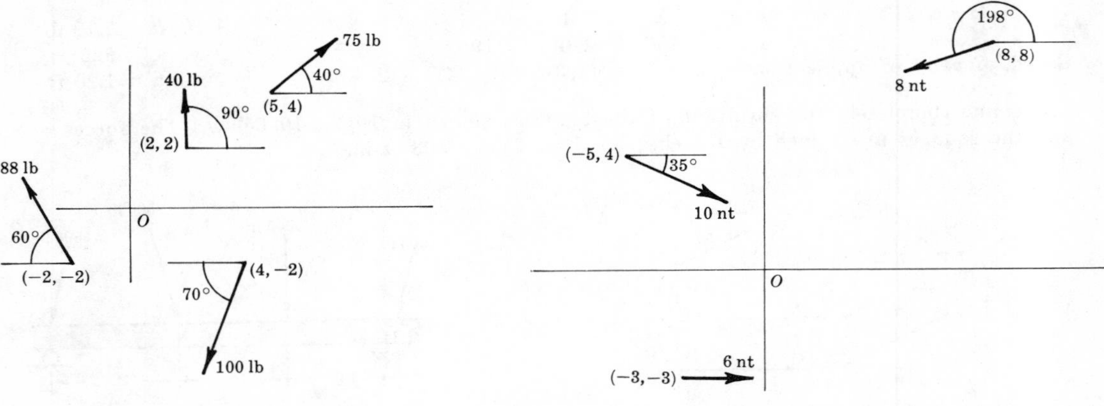

Fig. 3-26 Fig. 3-27

41. Determine the resultant of the force system shown in Fig. 3-27 above. Coordinates are in centimeters.
Ans. 10.5 nt, $\theta_x = 309°$, x intercept = 6.69 cm to left of O

Chapter 4

Resultants of Spatial Force Systems

SPATIAL FORCE SYSTEMS

Spatial force systems are categorized as follows. A concurrent system consists of forces which intersect at a point called the concurrence. A parallel system consists of forces which intersect at infinity. The most general system is called non-concurrent, non-parallel (or skew) and, as the name implies, the forces are not all concurrent and not all parallel.

RESULTANT of a SPATIAL FORCE SYSTEM

The resultant of a spatial force system is a force **R** and a couple **C** where $\mathbf{R} = \Sigma\mathbf{F}$, the vector sum of all the forces of the system and $\mathbf{C} = \Sigma\mathbf{M}$, the vector sum of the moments (relative to a selected base point) of all the forces of the system. The value of **R** is independent of the choice of the base point but the value of **C** depends on the base point. For any force system it is possible to select a unique base point so that the vector **C** representing the couple is parallel to **R**. This special combination is called a *wrench* or *screw*.

The vector equations in the preceding paragraph may be applied directly to spatial systems to determine the resultant, or the following derived scalar equations may be used.

CONCURRENT SYSTEM

The resultant **R** may be (*a*) a single force through the concurrence or (*b*) zero. Algebraically,

$$R = \sqrt{(\Sigma F_x)^2 + (\Sigma F_y)^2 + (\Sigma F_z)^2}$$

with direction cosines

$$\cos\theta_x = \frac{\Sigma F_x}{R}, \quad \cos\theta_y = \frac{\Sigma F_y}{R}, \quad \cos\theta_z = \frac{\Sigma F_z}{R}$$

where $\Sigma F_x, \Sigma F_y, \Sigma F_z$ = algebraic sums respectively of the x, y, and z components of the forces of the system,

$\theta_x, \theta_y, \theta_z$ = angles which the resultant **R** makes with the x, y, and z axes respectively.

PARALLEL SYSTEM

The resultant may be (*a*) a single force **R** parallel to the system, (*b*) a couple, or (*c*) zero. Assume the y axis is parallel to the system. Then algebraically,

$$R = \Sigma F, \quad R\bar{x} = \Sigma M_z, \quad R\bar{z} = \Sigma M_x$$

where ΣF = algebraic sum of the forces of the system,

$\bar{x}$ = perpendicular distance from the yz plane to the resultant,

$\bar{z}$ = perpendicular distance from the xy plane to the resultant,

$\Sigma M_x, \Sigma M_z$ = algebraic sums of the moments of the forces of the system about the x and z axes respectively.

If $\Sigma F = 0$, the resultant couple **C**, if there is one, will be determined by the following equation:

$$C = \sqrt{(\Sigma M_x)^2 + (\Sigma M_z)^2} \qquad \text{with} \qquad \tan\phi = \frac{\Sigma M_z}{\Sigma M_x}$$

where ϕ = angle which the vector representing the resultant couple makes with the x axis.

NON-CONCURRENT, NON-PARALLEL SYSTEM

As already indicated, the resultant is a force and a couple where the couple varies with the choice of a base point. In the following discussion a set of x, y, and z axes is placed with their origin at the base point.

Replace each force of the given system by the following setup: (1) an equal parallel force but acting through any chosen origin and (2) a couple acting in the plane containing the given force and the origin.

The magnitude of the resultant **R** of the concurrent system at the origin is given by the equation

$$R = \sqrt{(\Sigma F_x)^2 + (\Sigma F_y)^2 + (\Sigma F_z)^2}$$

with direction cosines

$$\cos\theta_x = \frac{\Sigma F_x}{R}, \qquad \cos\theta_y = \frac{\Sigma F_y}{R}, \qquad \cos\theta_z = \frac{\Sigma F_z}{R}$$

where the above quantities have the same meaning as listed under *Concurrent Spatial Systems.*

The magnitude of the resultant couple **C** is given by

$$C = \sqrt{(\Sigma M_x)^2 + (\Sigma M_y)^2 + (\Sigma M_z)^2}$$

with direction cosines

$$\cos\phi_x = \frac{\Sigma M_x}{C}, \qquad \cos\phi_y = \frac{\Sigma M_y}{C}, \qquad \cos\phi_z = \frac{\Sigma M_z}{C}$$

where $\Sigma M_x, \Sigma M_y, \Sigma M_z$ = algebraic sums of the moments of the forces of the system about the x, y, and z axes respectively,

ϕ_x, ϕ_y, ϕ_z = angles which the vector representing the couple **C** makes with the x, y, and z axes respectively.

Solved Problems

In the following problems equivalent scalar equations are used when more convenient than the vector equations. Similarly, in the diagrams the forces are indicated by their magnitudes if the directions are clearly indicated.

1. Forces of 20 lb, 15 lb, 30 lb, and 50 lb are concurrent at the origin and are respectively directed through the points whose coordinates are (2, 1, 6), (4, −2, 5), (−3, −2, 1), and 5, 1, −2). Determine the resultant of the system.

Solution:

F	Coordinates	Cos θ_x	Cos θ_y	Cos θ_z	F_x	F_y	F_z
20	(2, 1, 6)	$\frac{+2}{\sqrt{41}} = +0.313$	$\frac{+1}{\sqrt{41}} = +0.156$	$\frac{+6}{\sqrt{41}} = +0.938$	+6.26	+3.12	+18.8
15	(4, −2, 5)	$\frac{+4}{\sqrt{45}} = +0.597$	$\frac{-2}{\sqrt{45}} = -0.298$	$\frac{+5}{\sqrt{45}} = +0.745$	+8.96	−4.47	+11.2
30	(−3, −2, 1)	$\frac{-3}{\sqrt{14}} = -0.803$	$\frac{-2}{\sqrt{14}} = -0.535$	$\frac{+1}{\sqrt{14}} = +0.268$	−24.1	−16.1	+8.04
50	(5, 1, −2)	$\frac{+5}{\sqrt{30}} = +0.912$	$\frac{+1}{\sqrt{30}} = +0.183$	$\frac{-2}{\sqrt{30}} = -0.365$	+45.6	+9.15	−18.3

The denominator in each case is determined by taking the square root of the sum of the squares of the x, y, and z intervals or differences. For the 30 lb force this is

$$\sqrt{(-3)^2 + (-2)^2 + (1)^2} = \sqrt{14}$$

F_x is the product of F and $\cos\theta_x$. Watch signs. It is advisable to write signs before inserting any values.

$\Sigma F_x = +6.26 + 8.96 - 24.1 + 45.6 = +36.7$. Similarly, $\Sigma F_y = -8.30$ and $\Sigma F_z = +19.7$. Then

$$R = \sqrt{(\Sigma F_x)^2 + (\Sigma F_y)^2 + (\Sigma F_z)^2} = 42.5 \text{ lb}$$

$$\cos\theta_x = \frac{\Sigma F_x}{R} = \frac{+36.7}{42.5} = +0.864, \quad \theta_x = 30.2°$$

$$\cos\theta_y = \frac{\Sigma F_y}{R} = \frac{-8.30}{42.5} = -0.192, \quad \theta_y = 79.0°$$

$$\cos\theta_z = \frac{\Sigma F_z}{R} = \frac{+19.7}{42.5} = +0.463, \quad \theta_z = 62.4°$$

The negative value of $\cos\theta_y$ signifies that the resultant has a negative component in the y direction. This is illustrated in Fig. 4-1.

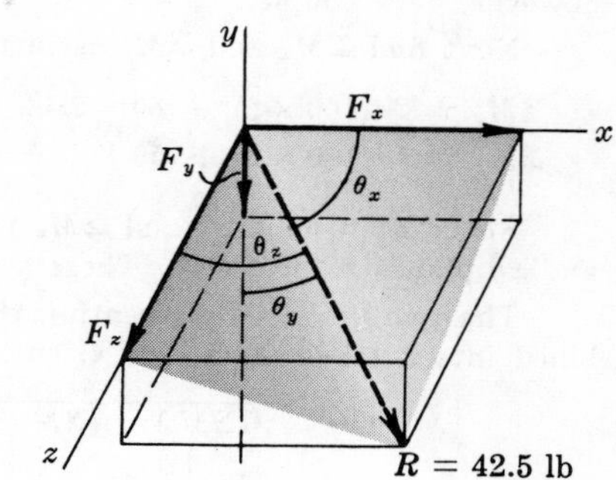

Fig. 4-1

2. Three forces of +20 lb, −10 lb, and +30 lb are shown in Fig. 4-2. The y-axis is chosen parallel to the action lines of the forces. These lines pierce the xz-plane in the points whose x and z coordinates in feet are respectively (2, 3), (4, 2), and (7, 4). Locate the resultant.

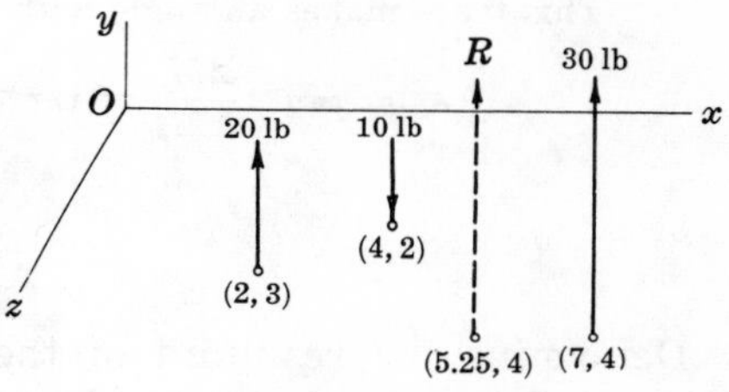

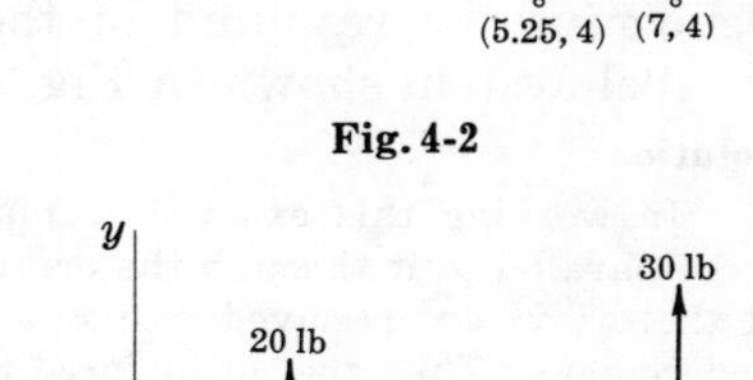

Fig. 4-2

Solution:

$$R = \Sigma F = +20 - 10 + 30 = +40 \text{ lb}$$

To determine the x coordinate of the resultant (i.e., of the point where the action line of the resultant pierces xz-plane), use the projected system in the xy-plane as shown in Fig. 4-3. Apply the equation $R\bar{x} = \Sigma M_z$.

$$\Sigma M_z = \Sigma M_O = +(20 \times 2) - (10 \times 4) + (30 \times 7) = +210 \text{ lb-ft}$$

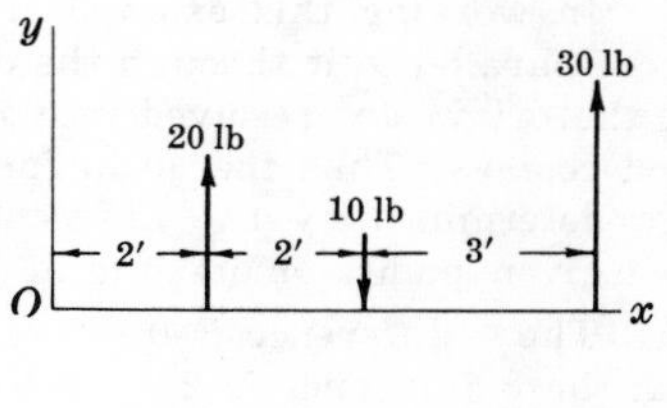

Fig. 4-3

The x coordinate must be such that a force of +40 lb (acting up) will have a positive or counterclockwise moment. Therefore R must be to the right of O.

$$\bar{x} = +210/40 = +5.25 \text{ ft}$$

Be sure to determine the sign by inspection as indicated in the previous paragraph, and not by combining the signs of the moment and the force.

Fig. 4-4 shows the projection of the system into the yz-plane.

$$\Sigma M_x = \Sigma M_O = -(30 \times 4) - (20 \times 3) + (10 \times 2) = -160 \text{ lb-ft}$$

The z coordinate must be such that a force of 40 lb (acting up) will have a negative or clockwise moment of 160 lb-ft. Therefore R must be to the left of O. In this case the z coordinate is positive when it is to the left of O (refer to the space diagram).

$$\bar{z} = +160/40 = +4.00 \text{ ft}$$

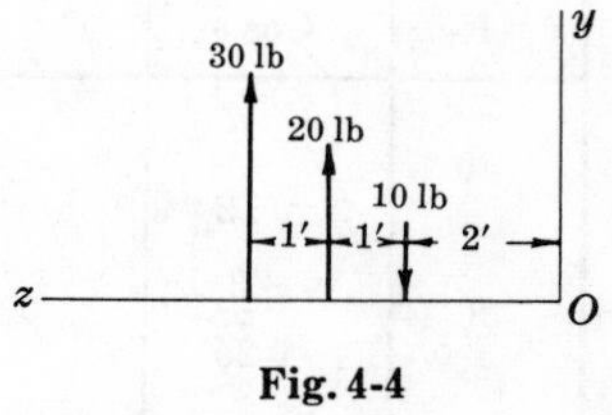

Fig. 4-4

The problem may now be summarized by saying that the resultant is a 40 lb force acting up. Its action line is parallel to the y-axis and pierces the xz-plane in the point whose x, z coordinates are (+5.25, +4.00) ft. This is shown in the original figure of this problem.

3. Find the resultant of the system of forces shown in Fig. 4-5. The coordinates are in feet.

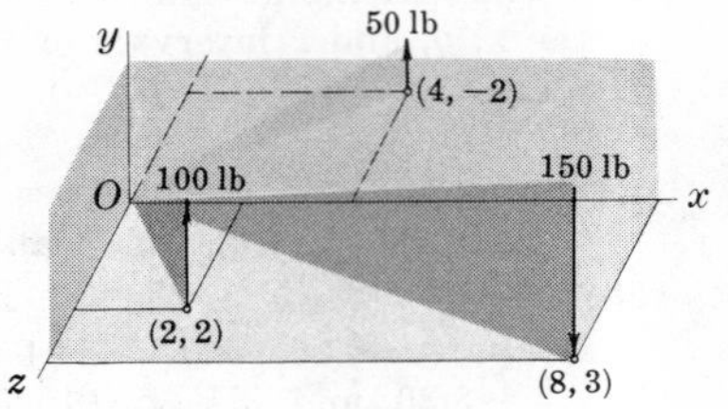

Fig. 4-5

Solution:

$$R = \Sigma F = +100 + 50 - 150 = 0$$

This indicates that the resultant is not a single force. It may, however, be a couple.

Next find ΣM_x and ΣM_z as in the preceding problem.

$$\Sigma M_x = -(100 \times 2) + (50 \times 2) + (150 \times 3) = +350 \text{ lb-ft}$$
$$\Sigma M_z = +(100 \times 2) + (50 \times 4) - (150 \times 8) = -800 \text{ lb-ft}$$

Since $\Sigma F = 0$, ΣM_x and ΣM_z represent couples in the yz and xy planes respectively. These are shown in Fig. 4-6.

The two vectors representing the couples are shown combined into a resultant couple **C** with magnitude

$$C = \sqrt{(\Sigma M_x)^2 + (\Sigma M_z)^2} = 874 \text{ lb-ft}$$

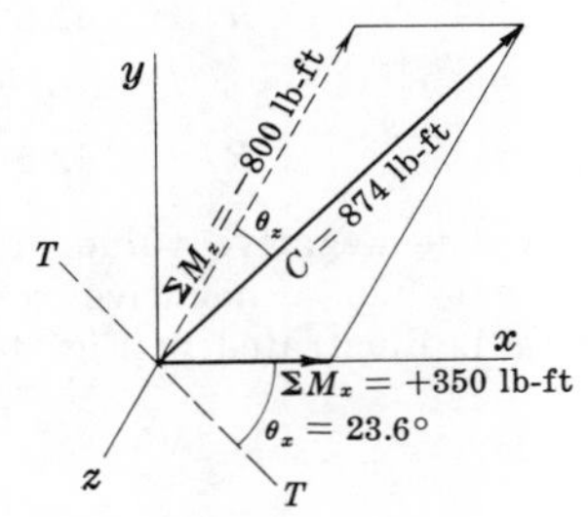

Fig. 4-6

The vector **C** in the xz-plane makes an angle θ_z with the z-axis as shown in the figure, where $\theta_z = \theta_x$.

According to the convention about couples, the resultant couple acts in a plane perpendicular to the vector **C** representing it. In the figure this could be in a plane containing the y-axis with trace TT in the xz-plane.

This trace makes an angle with the x-axis of

$$\theta_x = \tan^{-1}\frac{\Sigma M_x}{\Sigma M_z} = \tan^{-1}\frac{350}{800} = 23.6°.$$

4. Determine the resultant of the non-concurrent, non-parallel system shown in Fig. 4-7.

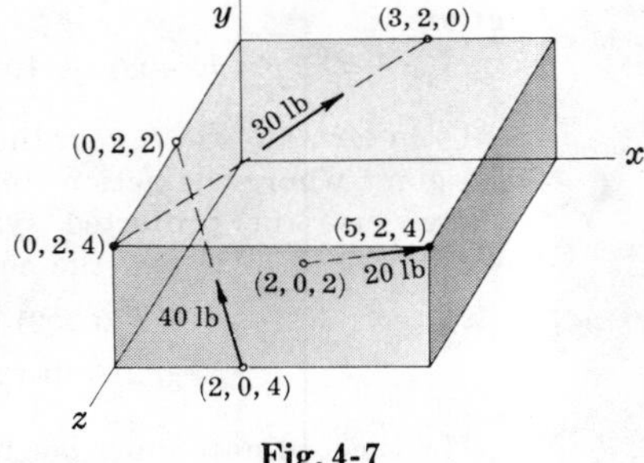

Fig. 4-7

Solution:

In working this example, replace each force by an equal force parallel to it through the origin and a couple. The forces at the origin are resolved into x, y, z components using direction cosines. Thus the 40 lb force has direction cosines which are determined by the differences in the coordinates of the two given points on its line of action.

The x difference is $0 - 2 = -2$; the y difference is $2 - 0 = +2$; the z difference is $2 - 4 = -2$. The cosine of the angle the 40 lb force makes with the x-axis is $\cos\theta_x = \dfrac{-2}{\sqrt{(-2)^2 + (+2)^2 + (-2)^2}} = \dfrac{-2}{\sqrt{12}}$. Similarly, $\cos\theta_y = \dfrac{+2}{\sqrt{12}}$ and $\cos\theta_z = \dfrac{-2}{\sqrt{12}}$.

The results are most conveniently set down in tabular form.

F	Cos θ_x	Cos θ_y	Cos θ_z	F_x	F_y	F_z
40	$\frac{-2}{\sqrt{12}}$	$\frac{+2}{\sqrt{12}}$	$\frac{-2}{\sqrt{12}}$	−23.1	+23.1	−23.1
30	$\frac{+3}{\sqrt{25}}$	0	$\frac{-4}{\sqrt{25}}$	+18.0	0	−24.0
20	$\frac{+3}{\sqrt{17}}$	$\frac{+2}{\sqrt{17}}$	$\frac{+2}{\sqrt{17}}$	+14.6	+9.71	+9.71

$$\Sigma F_x = +9.5, \quad \Sigma F_y = +32.8, \quad \Sigma F_z = -37.4$$

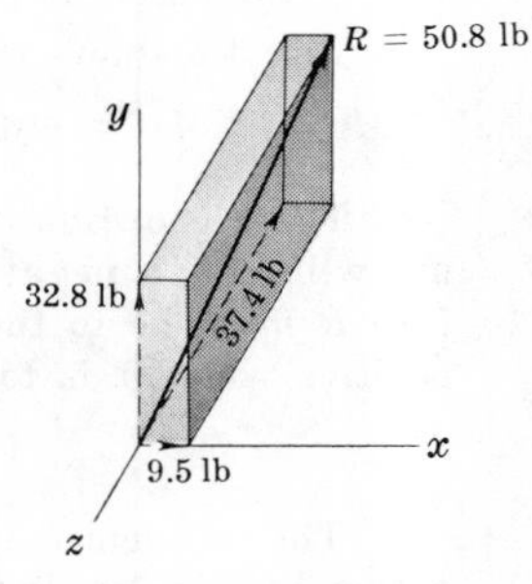

Fig. 4-8

From the above table determine the resultant of this system of transferred forces concurrent at the origin.

$$R = \sqrt{(\Sigma F_x)^2 + (\Sigma F_y)^2 + (\Sigma F_z)^2} = \sqrt{(+9.5)^2 + (+32.8)^2 + (-37.4)^2} = 50.8 \text{ lb}$$

$$\cos\theta_x = \frac{\Sigma F_x}{R} = \frac{+9.5}{50.8} = +0.187, \quad \cos\theta_y = \frac{\Sigma F_y}{R} = +0.645, \quad \cos\theta_z = \frac{\Sigma F_z}{R} = -0.737$$

The resultant of the transferred forces is shown graphically in Fig. 4-8 above.

The foregoing did not use the couple associated with each transferred force. Determine the moments of each of the three forces about the three axes to establish the magnitude and direction of these couples. Referring to the first figure, consider the 40 lb force acting at the point (2, 0, 4). Its moment about the x-axis is the algebraic sum of the moments of its three components about the x-axis. However, its only component possessing a moment about the x-axis is the y component. The moment of the 40 lb force about the x-axis is therefore $-(23.1 \times 4) = -92.4$ lb-ft. In finding the moment about the y-axis (i.e., M_y) consider the moments of both the x and z components. The moment of the x component about the y-axis is $-(23.1 \times 4) = -92.4$ lb-ft. The moment of the z component about the y-axis is $+(23.1 \times 2) = +46.2$ lb-ft. Therefore M_y is equal to $-92.4 + 46.2 = -46.2$ lb-ft. The moment of the 40 lb force about the z-axis is the same as the moment of its y component about the z-axis. Hence $M_z = +(23.1 \times 2) = +46.2$ lb-ft.

A tabular solution is given for the moments of the forces.

F	F_x	F_y	F_z	M_x	M_y	M_z
40	−23.1	+23.1	−23.1	−92.4	−46.2	+46.2
30	+18.0	0	−24.0	−48.0	+72.0	−36.0
20	+14.6	+9.71	+9.71	−19.4	+9.8	+19.4

$$\Sigma M_x = -159.8, \quad \Sigma M_y = +35.6, \quad \Sigma M_z = +29.6$$

The magnitude of the resultant couple is

$$C = \sqrt{(\Sigma M_x)^2 + (\Sigma M_y)^2 + (\Sigma M_z)^2} = \sqrt{(-159.8)^2 + (+35.6)^2 + (+29.6)^2} = 166 \text{ lb-ft}$$

with direction cosines

$$\cos\phi_x = \frac{\Sigma M_x}{C} = \frac{-159.8}{166} = -0.963,$$

$$\cos\phi_y = \frac{\Sigma M_y}{C} = +0.214,$$

$$\cos\phi_z = \frac{\Sigma M_z}{C} = +0.178$$

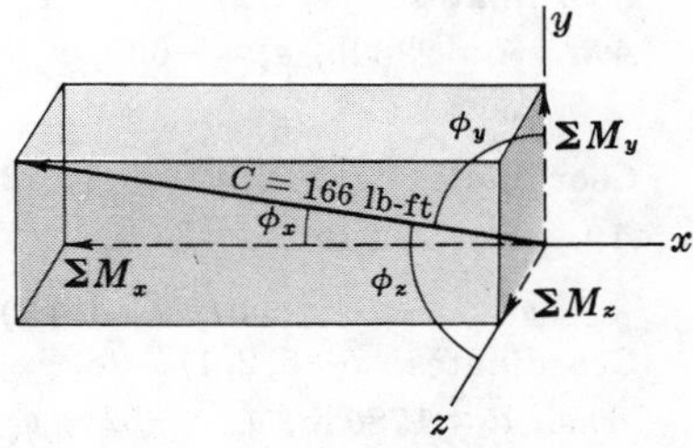

Fig. 4-9

Vector **C** is shown in Fig. 4-9. The resultant couple acts in plane perpendicular to the vector **C**.

The resultant of the system is the combination of the force **R** and the couple **C**.

5. In Problem 4, find the couple **C** by determining the sum of the moments of the three forces about the origin using vector cross products.

Solution:

The components of the forces are listed in the first table in Problem 4.

The moment $\mathbf{M}_1$ of the 30 lb force [using point (0, 2, 4) on its line of action to determine $\mathbf{r} = 2\mathbf{j} + 4\mathbf{k}$] is

$$\mathbf{M}_1 = \begin{vmatrix} \mathbf{i} & \mathbf{j} & \mathbf{k} \\ 0 & 2 & 4 \\ 18.0 & 0 & -24.0 \end{vmatrix} = -48.0\mathbf{i} + 72.0\mathbf{j} - 36.0\mathbf{k}$$

Similarly using point (2, 0, 4), the moment $\mathbf{M}_2$ of the 40 lb force about the origin is

$$\mathbf{M}_2 = \begin{vmatrix} \mathbf{i} & \mathbf{j} & \mathbf{k} \\ 2 & 0 & 4 \\ -23.1 & +23.1 & -23.1 \end{vmatrix} = -92.4\mathbf{i} - 46.2\mathbf{j} + 46.2\mathbf{k}$$

Then using point (2, 0, 2), the moment $\mathbf{M}_3$ of the 20 lb force about the origin is

$$\mathbf{M}_3 = \begin{vmatrix} \mathbf{i} & \mathbf{j} & \mathbf{k} \\ 2 & 0 & 2 \\ +14.6 & +9.71 & +9.71 \end{vmatrix} = -19.4\mathbf{i} + 9.8\mathbf{j} + 19.4\mathbf{k}$$

The resultant is

$$\mathbf{C} = \mathbf{M}_1 + \mathbf{M}_2 + \mathbf{M}_3 = -159.8\mathbf{i} + 35.6\mathbf{j} + 29.6\mathbf{k}$$

which agrees with the values of M_x, M_y, and M_z given in Problem 4.

Supplementary Problems

Determine the resultants in Problems 6-9, which involve concurrent systems of forces. The forces are in lb and the coordinates of the points on the lines of action are in ft. All forces in each problem are concurrent at the origin. (Negative answers involve negative components.)

6. F 100 200 500 300
Coordinates (1, 1, 1) (2, 3, 1) (−2, −3, 4) (−1, 1, −2)
Ans. $R = 286$ lb, $\theta_x = -60°$, $\theta_y = 78°$, $\theta_z = 33°$

7. F 5 2 3 4 8
Coordinates (2, 2, 3) (5, 1, −2) (−3, −4, 5) (2, 1, −4) (5, 2, 3)
Ans. $R = 13.3$ lb, $\theta_x = 33°$, $\theta_y = 70°$, $\theta_z = 66°$

8. F 1000 1500 1800
Coordinates (−5, 2, 1) (6, −3, −2) (−2, −1, −1)
Ans. $R = 1780$ lb, $\theta_x = -52°$, $\theta_y = -55°$, $\theta_z = -57°$

9. F 40 80 30 20
Coordinates (6, 5, 4) (1, −3, −2) (8, 10, −7) (−10, −9, −10)
Ans. $R = 80.1$ lb, $\theta_x = 49°$, $\theta_y = -67°$, $\theta_z = -51°$

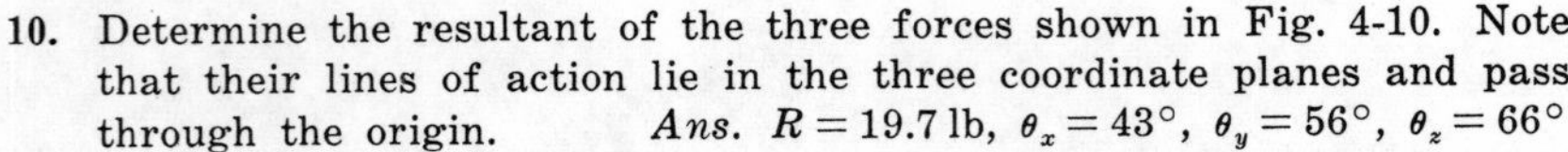

10. Determine the resultant of the three forces shown in Fig. 4-10. Note that their lines of action lie in the three coordinate planes and pass through the origin. *Ans.* $R = 19.7$ lb, $\theta_x = 43°$, $\theta_y = 56°$, $\theta_z = 66°$

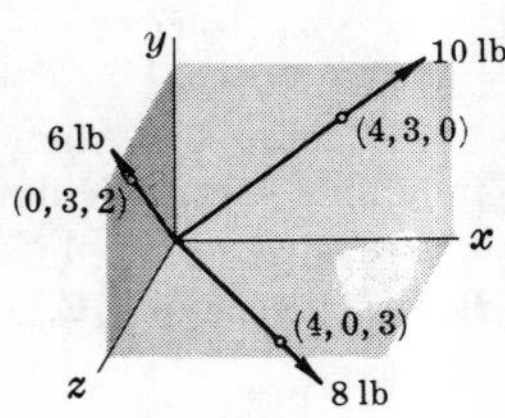

Fig. 4-10

In each of Problems 11-14, find the resultant and the coordinates of the intersection of its line of action with the *xz*-plane. The given forces are in lb and parallel to the *y*-axis, and the coordinates of the intersection of each action line with the *xz*-plane are in ft.

11. F 100 150 200 300
(x, z) (3, −2) (1, 6) (2, −3) (−1, −1)
Ans. $R = 750$ lb, $\bar{x} = 0.733$ ft, $\bar{z} = -0.267$ ft

12. F −25 18 −12 −30 36
(x, z) (1, 2) (2, −1) (0, 0) (−6, −2) (3, 2)
Ans. $R = -13$ lb, $\bar{x} = -23.0$ ft, $\bar{z} = -4.92$ ft

13. F 3 -4 2 -5
(x, z) $(2, 5)$ $(1, -5)$ $(3, 3)$ $(-4, -4)$
Ans. $R = -4$ lb, $\bar{x} = -7.00$ ft, $\bar{z} = -15.3$ ft

14. F $+10$ $+20$ -30
(x, z) $(1, 1)$ $(2, -5)$ $(3, -4)$
Ans. $C_x = -30$ lb-ft, $C_z = -40$ lb-ft

15. Five weights of 20 lb, 15 lb, 12 lb, 6 lb, and 10 lb rest on a table with the following coordinates respectively: $(0.5, 15°)$, $(1.5, 90°)$, $(0.8, 185°)$, $(0.7, 262°)$, and $(1.2, 340°)$, where the first number in parentheses represents the radial distance in ft from the center of the table and the angle is measured counterclockwise from a reference radius as the table is viewed from above. Determine the resultant weight. *Ans.* $R = 63$ lb (down), $r = 0.3$, $\theta = 56°$

16. Find the resultant of the system shown in Fig. 4-11. The forces are in lb and the distances in ft.
Ans. $R = 40.3$ lb, $\cos\theta_x = +0.594$, $\cos\theta_y = +0.673$, $\cos\theta_z = -0.428$ at origin
$C = 251$ lb-ft, $\cos\phi_x = -0.660$, $\cos\phi_y = +0.633$, $\cos\phi_z = +0.396$

17. In determining the moment of a force about a moment center, the position vector **r** may be drawn from the moment center to *any* point on the action line of the force. To illustrate this independence the reader should rework Problem 5 using points $(0, 2, 2)$, $(3, 2, 0)$ and $(5, 2, 4)$ on the 40 lb, 30 lb and 20 lb forces respectively.

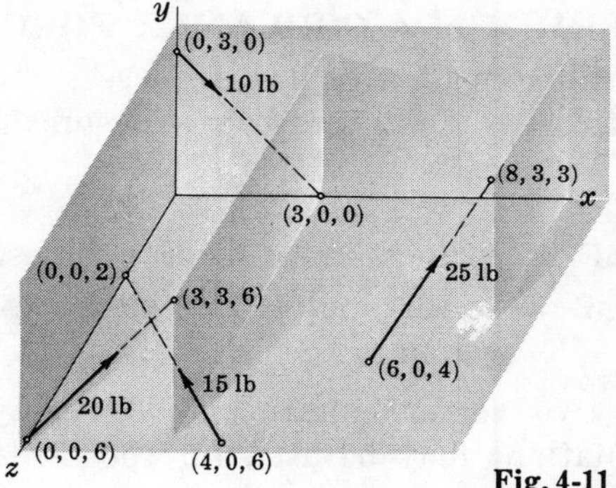

Fig. 4-11

18. Determine the resultant of the three forces shown in Fig. 4-12 below. Coordinates are in ft. Use the origin as the base point.
Ans. $\mathbf{R} = 3530\mathbf{i} + 267\mathbf{j} + 1200\mathbf{k}$ lb through the origin, and $\mathbf{C} = -3200\mathbf{i} + 4810\mathbf{j} - 534\mathbf{k}$ lb-ft

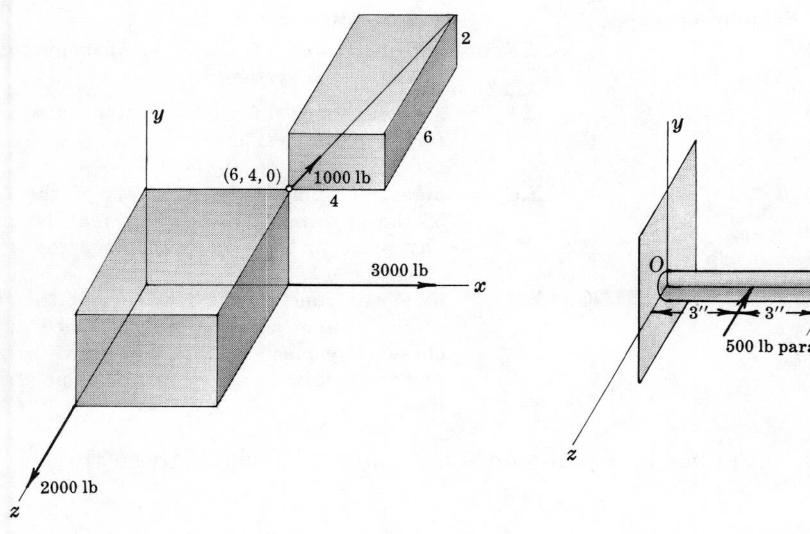

Fig. 4-12

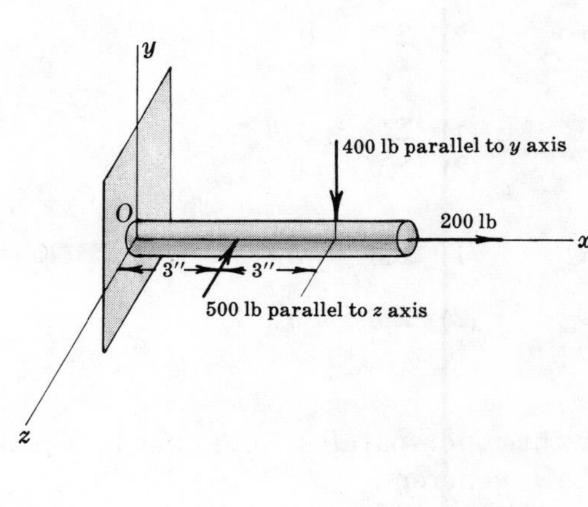

Fig. 4-13

19. Replace the three forces shown in Fig. 4-13 above by a resultant force **R** through O and a couple.
Ans. $\mathbf{R} = 200\mathbf{i} - 400\mathbf{j} - 500\mathbf{k}$
or $R = 671$ lb with $\cos\theta_x = 0.298$, $\cos\theta_y = -0.597$, $\cos\theta_z = -0.745$
$\mathbf{C} = 1500\mathbf{j} - 2400\mathbf{k}$ lb-in
or $C = 2830$ lb-in with $\cos\phi_x = 0$, $\cos\phi_y = 0.530$, $\cos\phi_z = -0.848$

Chapter 5

Equilibrium of Coplanar Force Systems

EQUILIBRIUM of a COPLANAR FORCE SYSTEM

Equilibrium of a coplanar force system occurs if the resultant is neither a force **R** nor a couple **C**. The necessary and sufficient conditions that **R** and **C** be zero vectors are

$$\mathbf{R} = \Sigma\mathbf{F} = 0 \qquad \text{and} \qquad \mathbf{C} = \Sigma\mathbf{M} = 0$$

where $\Sigma\mathbf{F}$ = vector sum of all the forces of the system,

$\Sigma\mathbf{M}$ = vector sum of the moments (relative to any point) of all the forces of the system.

The two vector equations above may be applied directly or the following derived scalar equations may be used for the three types of coplanar force systems.

CONCURRENT SYSTEM

Any of the following sets of equations insures equilibrium of a concurrent system, i.e., the resultant is zero. Assume concurrency at origin.

Set	Equations of Equilibrium	Remarks
A	(1) $\Sigma F_x = 0$	ΣF_x = algebraic sum of the x components of the forces of the system.
	(2) $\Sigma F_y = 0$	ΣF_y = algebraic sum of the y components of the forces of the system.
B	(1) $\Sigma F_x = 0$ (2) $\Sigma M_A = 0$	ΣM_A = algebraic sum of the moments of the forces of the system about A which may be chosen any place in the plane except on the y axis.
C	(1) $\Sigma M_A = 0$ (2) $\Sigma M_B = 0$	ΣM_A and ΣM_B = algebraic sums of the moments of the forces of the system about A and B which may be chosen any place in the plane provided A, B, and the origin do not lie on the same straight line.

If only three non-parallel forces act in a plane on a body in equilibrium, these three forces must be concurrent.

PARALLEL SYSTEM

Any of the following sets of equations insures equilibrium of a parallel system, i.e., the resultant is neither a force nor a couple.

Set	Equations of Equilibrium	Remarks
A	(1) $\Sigma F = 0$	ΣF = algebraic sum of the forces of the system parallel to the action lines of the forces.
	(2) $\Sigma M_A = 0$	ΣM_A = algebraic sum of the moments of the forces of the system about any point A in the plane.

	Equations		Remarks
B	(1) $\Sigma M_A = 0$ (2) $\Sigma M_B = 0$	ΣM_A and ΣM_B =	algebraic sums of the moments of the forces of the system about A and B which may be chosen any place in the plane provided the line joining A and B is not parallel to the forces of the system.

NON-CONCURRENT, NON-PARALLEL SYSTEM

Any of the following sets of equations insures equilibrium of a non-concurrent, non-parallel system, i.e., the resultant is neither a force nor a couple.

Set	Equations		Remarks
A	(1) $\Sigma F_x = 0$ (2) $\Sigma F_y = 0$ (3) $\Sigma M_A = 0$	ΣF_x = ΣF_y = ΣM_A =	algebraic sum of x components of the forces. algebraic sum of y components of the forces. algebraic sum of the moments of the forces of the system about any point A in the plane.
B	(1) $\Sigma F_x = 0$ (2) $\Sigma M_A = 0$ (3) $\Sigma M_B = 0$	ΣF_x = ΣM_A and ΣM_B =	algebraic sum of x components of the forces. algebraic sums of the moments of the forces of the system about any two points A and B in the plane, provided that the line joining A and B is not perpendicular to the x-axis.
C	(1) $\Sigma M_A = 0$ (2) $\Sigma M_B = 0$ (3) $\Sigma M_C = 0$	ΣM_A, ΣM_B and ΣM_C =	algebraic sums of the moments of the forces of the system about any three points A, B and C in the plane, provided that A, B and C do not lie on the same straight line.

GRAPHICAL ANALYSIS

In any of the above systems, the closing of the force polygon insures that the resultant is not a single force. The closing of the funicular polygon in the parallel system and in the non-concurrent, non-parallel system indicates further that the resultant is not a couple and therefore the system is in equilibrium.

REMARKS

In the solution of problems the following comments may be of assistance.

(1) Draw free body diagrams. A roller or knife edge support means that the reaction is shown perpendicular to the member, whereas a pin connection means that the reaction may be at an angle other than ninety degrees with the member which is supported.

(2) Unless the direction of a reaction is apparent by inspection, assume a positive direction. A negative sign simply indicates that the direction of the reaction is actually opposite to the one assumed.

(3) It may not be necessary to use all three equations of a set to obtain a solution. The proper choice of a moment center, for example, may yield an equation containing only one unknown.

(4) The x and y axes in the above equations need not necessarily be chosen horizontally and vertically respectively. Actually, if a system is in equilibrium the algebraic sum of the components of the forces of the system along any axis must be zero.

(5) In the diagrams, forces are identified by their magnitudes if the directions are readily apparent.

Solved Problems

1. Fig. 5-1(a) shows a 25 lb lamp supported by two cables AB and AC. Find the tension in each cable.

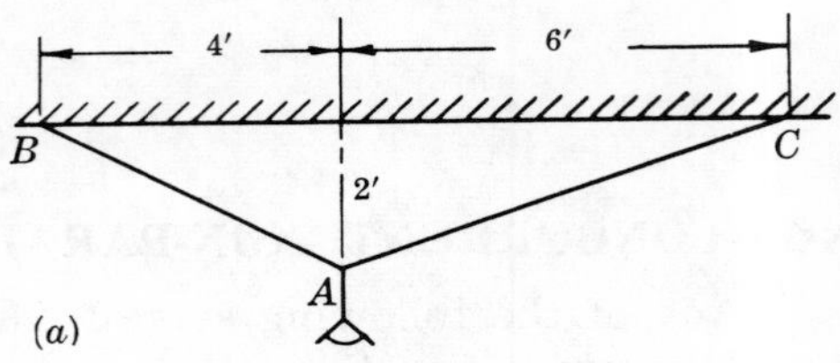

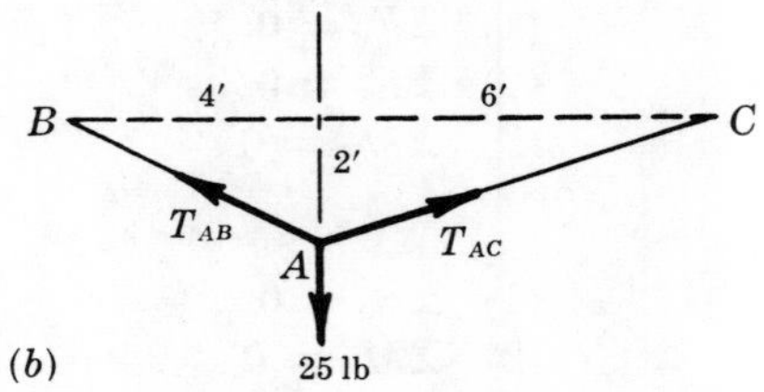

Fig. 5-1

Solution:

A free body diagram of the ring at A is shown in Fig. 5-1(b) with the 25 lb force (weight of lamp) acting vertically down and the tensions in AC and AB.

Using Set A of the equations for a *concurrent system*, we have

$$(1)\quad \Sigma F_x = 0 = +T_{AC}\frac{6}{\sqrt{40}} - T_{AB}\frac{4}{\sqrt{20}},$$

$$(2)\quad \Sigma F_y = 0 = T_{AC}\frac{2}{\sqrt{40}} + T_{AB}\frac{2}{\sqrt{20}} - 25$$

Note that in each equation the zero is inserted immediately after the sigma sign. This prevents the common mistake of not placing the zero in these equations.

There are two equations in two unknowns. The problem is therefore statically determinate, i.e., can be solved.

From equation (1), $T_{AC} = \frac{2}{3}\sqrt{2}\,T_{AB} = 0.942\,T_{AB}$. Substituting into (2),

$$0.942\,T_{AB}\frac{2}{\sqrt{40}} + T_{AB}\frac{2}{\sqrt{20}} - 25 = 0$$

from which $T_{AB} = 33.6$ lb, and $T_{AC} = 0.942\,T_{AB} = 31.7$ lb.

The solution could be obtained by using Sets B or C for a *concurrent system*. By choosing a moment center on one of the unknown forces, an equation is obtained which yields one unknown. Suppose, for example, point B is chosen as a moment center. Then,

$$\Sigma M_B = 0 = -25\times 4 + T_{AC}\times\frac{2}{\sqrt{40}}\times 4 + T_{AC}\times\frac{6}{\sqrt{40}}\times 2 \quad \text{or} \quad T_{AC} = 31.7 \text{ lb}$$

The moment of the force T_{AC} is equal to the moment of its components taken about point B. Another moment center, say at C, will yield an equation involving only the unknown T_{AB}.

2. Solve Problem 1 graphically.

Solution:

For equilibrium of this concurrent system, the force polygon must close. Draw to scale the force of 25 lb vertically down as shown in Fig. 5-2. From the top of this force draw a line parallel to AB. From the bottom of the 25 lb force draw a line parallel to AC. Measure T_{AC} and T_{AB} to scale to obtain the values found in Problem 1.

Note the directions as indicated by the arrow heads. These directions must be followed if the force polygon is to close. The arrows then indicate the directions in which the forces act on the ring A. Since AC pulls away from A, then A pulls away from AC (equal and opposite reaction). AC is therefore in tension. By similar reasoning, AB is in tension.

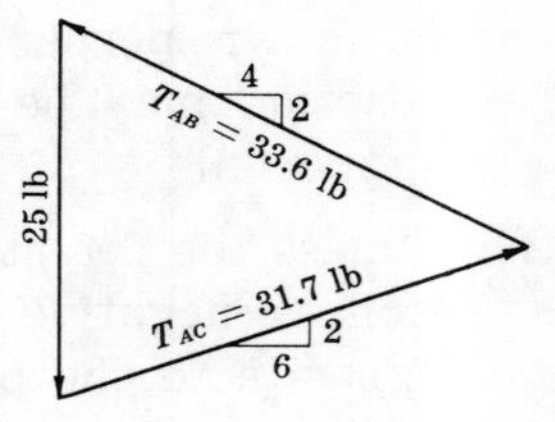

Fig. 5-2

3. A boom 20 ft long supports a load of 1200 lb as shown in Fig. 5-3(a) below. The cable BC is horizontal and 10 ft long. Solve for the stresses in the cable and the boom.

Solution:

F_2 is the force in the boom, and F_1 is the force in the cable. By inspection the arrows are placed as shown in Fig. 5-3(b), indicating tension in the cable and compression in the boom.

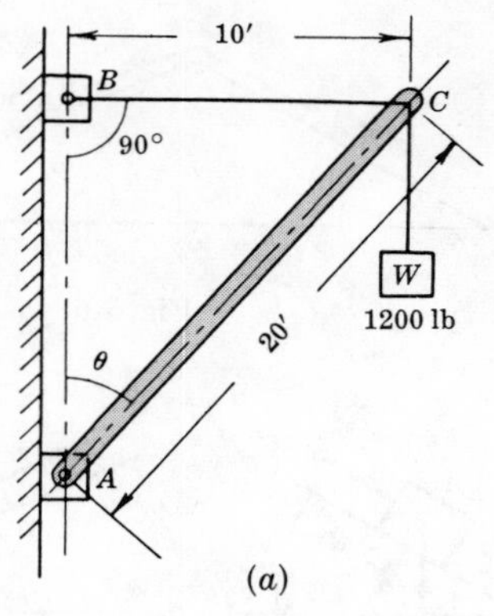

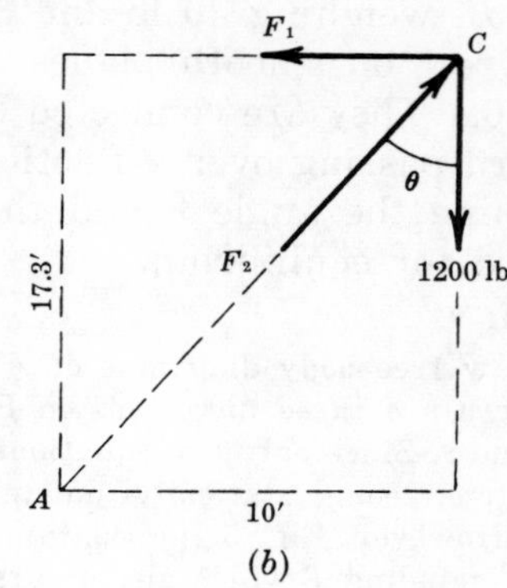

Fig. 5-3

$$AB = \sqrt{(20)^2 - (10)^2} = 17.3 \text{ ft}, \qquad \cos\theta = 17.3/20 = 0.866$$

By taking moments about A, only one unknown enters into the equation.

$$\Sigma M_A = 0 = +(F_1 \times AB) - (1200 \times 10), \qquad 0 = F_1(17.3) - 12{,}000, \qquad F_1 = 693 \text{ lb}$$

Summing forces vertically: $\Sigma F_v = 0 = -1200 + F_2 \cos\theta, \quad 0 = -1200 + 0.866F_2, \quad F_2 = 1390$ lb.

4. Solve Problem 3 graphically.

Solution:

As usual, start the force polygon with a force that is known both in magnitude and direction, in this case the 1200 lb weight. See Fig. 5-4(a). Draw lines parallel to the action lines of F_1 and F_2 through the tail end and arrow end of this vector until they intersect. The same results are obtained whether F_1 is drawn parallel to its action line through the tail end or the arrow end of the 1200 lb force as shown in Fig. 5-4(b).

Measure F_1 and F_2 to obtain 700 and 1400 lb respectively. F_1 is tension and F_2 is compression.

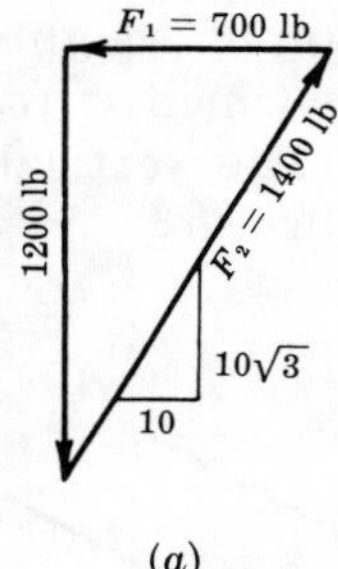

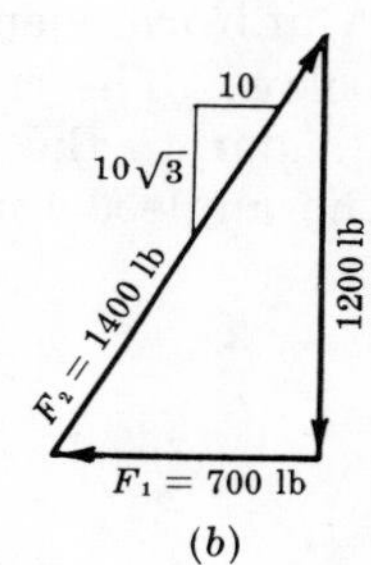

Fig. 5-4

5. The bar AB weighs 10 lb per ft and is supported by cable AC and a pin at B. See Fig. 5-5(a) below. Determine graphically the reaction at B and the tension in the cable.

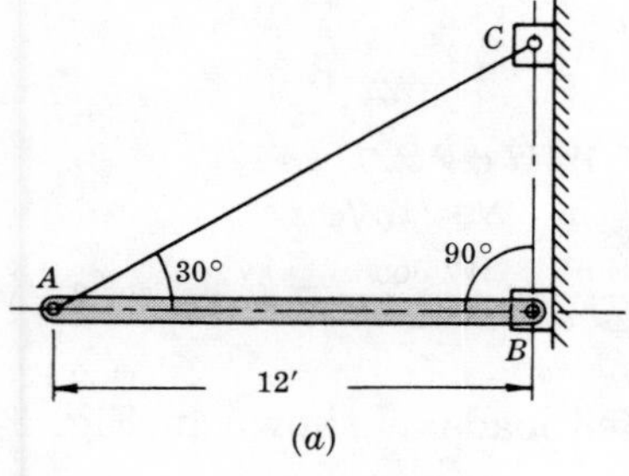

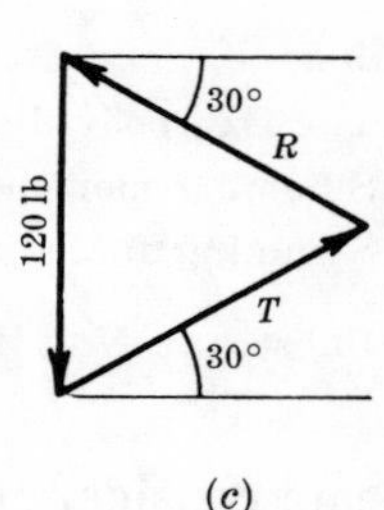

Fig. 5-5

Solution:

In the free body diagram of the bar (Fig. 5-5(b) above), drawn only with the given information, R is unknown both in direction and magnitude. However, when only three forces act on a body in equilibrium they must be concurrent. This means that R will pass through the intersection D of the force T and the weight 120 lb, thereby determining its direction. The force polygon is now drawn as described previously. See Fig. 5-5(c) above.

By measurement, $T = R = 120$ lb. R makes an angle of 30° with AB.

6. A and B, weighing 40 lb and 30 lb respectively, rest on smooth planes as shown in Fig. 5-6. They are connected by a weightless cord passing over a frictionless pulley. Determine the angle θ and the tension in the cord for equilibrium.

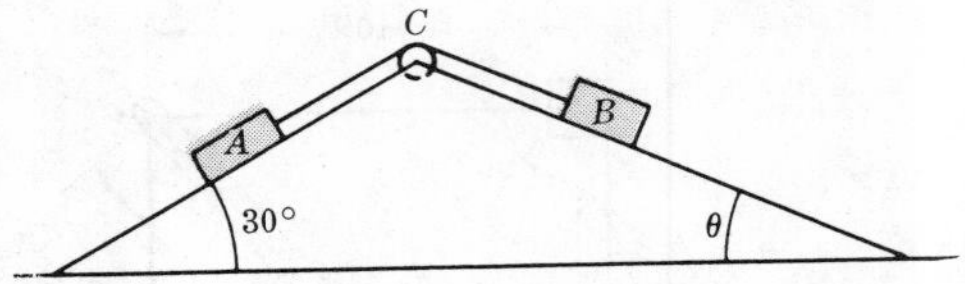

Fig. 5-6

Solution:

Draw free body diagrams of A and B.

There are three unknowns in Fig. 5-7(b), i.e., T, N_B and θ. Since only two equations are available, the system seems statically indeterminate as it stands. However, Fig. 5-7(a) contains only two unknowns including T which also occurs in Figure (b), thereby making that system determinate when T has been found.

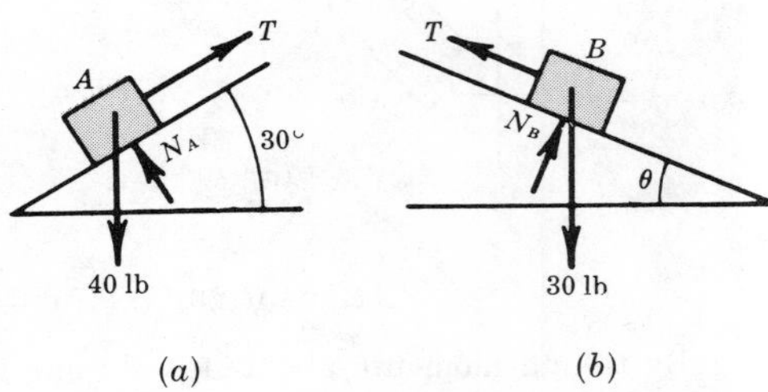

Fig. 5-7

Summing forces parallel to the 30° plane, the equation of equilibrium obtained is

$$\Sigma F_{\parallel} = 0 = +T - 40 \sin 30° \qquad \text{or} \qquad T = 20 \text{ lb}$$

Returning to Fig. 5-7(b) and summing forces parallel to the plane, we have

$$\Sigma F_{\parallel} = 0 = +T - 30 \sin \theta, \qquad \sin \theta = T/30 = 2/3, \qquad \theta = 41.8°$$

7. A uniform slender beam of weight W has its center of gravity as shown in Fig. 5-8(a) below. The corner on which it rests is a knife-edge; hence the reaction N is perpendicular to the beam. The vertical wall on the left is smooth. What is the value of the angle θ for equilibrium?

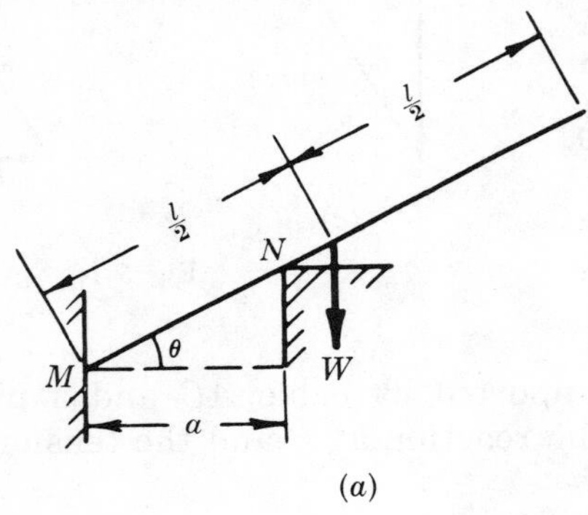

(b)

Fig. 5-8

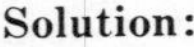

Solution:

The free body diagram is shown in Fig. 5-8(b) above.

Summing moments about M, $\quad \Sigma M_M = 0 = +N(a/\cos\theta) - W(\tfrac{1}{2}l \cos\theta)$.

Summing forces vertically, $\quad \Sigma F_v = 0 = N\cos\theta - W \quad$ or $\quad N = W/\cos\theta$.

Substitute $N = W/\cos\theta$ into the first equation to obtain $\dfrac{Wa}{\cos^2\theta} - \dfrac{Wl\cos\theta}{2} = 0$ or $\cos\theta = \sqrt[3]{2a/l}$.

8. A beam considered weightless is loaded with concentrated loads as shown in Fig. 5-9(a) below. Determine the reactions at A and B.

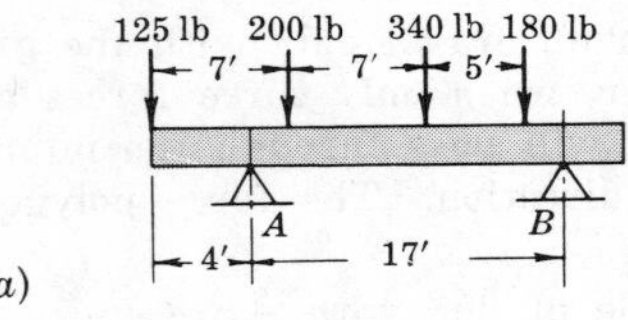

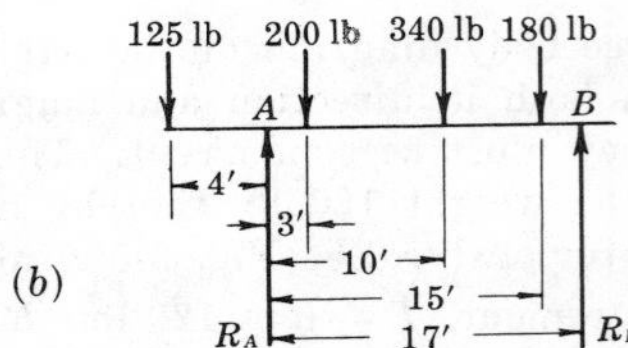

Fig. 5-9

Solution:

To find beam reactions R_A and R_B shown in Fig. 5-9(*b*) above, it is advisable to take moments about A and then about B. In this way each equation yields only one unknown. Each reaction is thus found independently of the other. Then the summation of the forces should equal zero, providing an excellent check. Many readers may prefer to determine one reaction by a moment equation and then determine the other by the sum of the forces. This is, of course, a correct procedure. But the authors prefer the two summations of moments, thereby reserving the summation of forces as a check equation. Using this procedure, the two equations are:

(*1*) $\Sigma M_A = 0 = +(125 \times 4) - (200 \times 3) - (340 \times 10) - (180 \times 15) + R_B \times 17$ or $R_B = 365$ lb

(*2*) $\Sigma M_B = 0 = +(125 \times 21) - R_A \times 17 + (200 \times 14) + (340 \times 7) + (180 \times 2)$ or $R_A = 480$ lb

Checking, $\Sigma F = -125 + 480 - 200 - 340 - 180 + 365 = 0$. This sum should, within the limits of accuracy, equal zero. Since it does, the beam reactions are correct.

9. Determine graphically the reactions for the beam with both concentrated and distributed loads, as shown in Fig. 5-10(*a*) below.

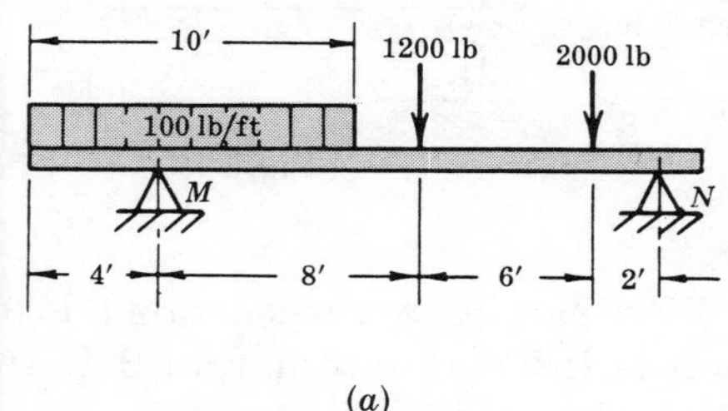

(*a*)

1′ 1000 lb 1200 lb 2000 lb a b c d e R_M R_N

(*b*)

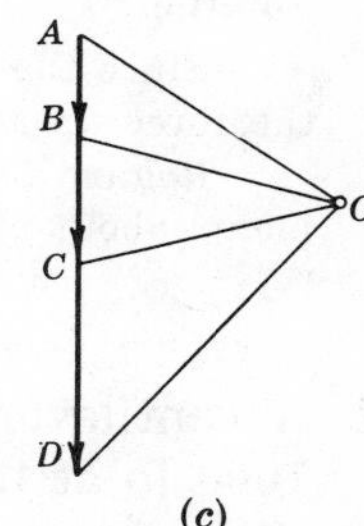

(*c*)

Fig. 5-10

Solution:

Note the distributed load of 100 lb per ft is replaced in the space diagram with a concentrated load of 1000 lb at its midpoint. This is permissible only in determining beam reactions and not in analyzing stresses within the beam.

The fields are labeled in the space diagram (Fig. 5-10(*b*) above) with lower case letters. Letter *e* represents the field between the beam reactions. The force polygon is drawn in Fig. 5-10(*c*) above, starting with the known forces *AB*, *BC*, and *CD*. Since the force polygon must close, the sum of the two beam reactions must be included between *D* and *A*. To determine the point *E* dividing the line *DA* into the two distances proportional to the reactions, first choose a pole point *O* from which to draw rays *OA*, *OB*, *OC*, and *OD*.

The additional Fig. 5-11(*a*) is now drawn to show the next stage of development.

Choose a convenient point *T* on the action line *ab*. Through this point draw strings *ao* and *ob* parallel to the rays *AO* and *OB* respectively. Next draw *oc* through the point where *bo* crosses *bc*. Draw *od* through the point where *co* crosses *cd*. Through the point *S* where *do* crosses *de*, the string *oe* should be drawn. This string must also be drawn through the point *R* where *oa* crosses *ea*. Hence *RS* is the closing string of the polygon, thereby insuring that a couple does not exist.

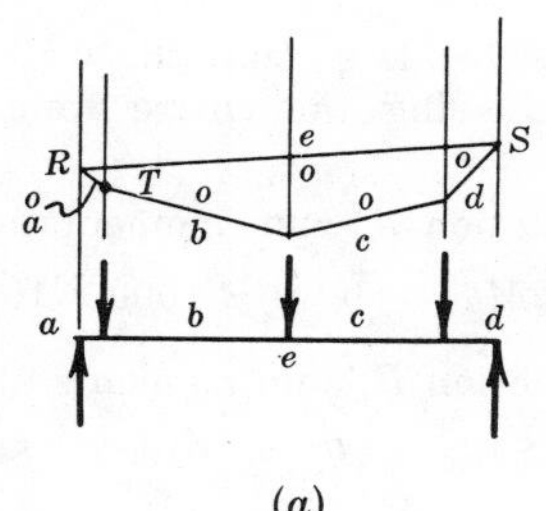

(*a*)

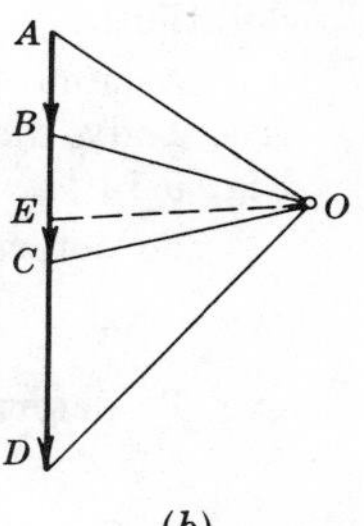

(*b*)

Fig. 5-11

Returning to the force polygon [Fig. 5-11(*b*)], draw ray *EO* parallel to the string *eo*. *DE* equals 2410 lb and *EA* equals 1790 lb. This means that $R_M = 1790$ lb and $R_N = 2410$ lb, both acting up.

10. Determine the force *P* required to hold a weight of 100 lb in equilibrium utilizing the system of pulleys shown in Fig. 5 12(*a*) below. Assume that all pulleys are the same size.

Solution:

Fig. 5-12(*b*) is a free body diagram of the lowest pulley. The 100 lb force (weight) acts down. Acting up is the pull in the rope on each side of the pulley. Since the rope is continuous and since frictionless pulleys are assumed, the tension in the rope leaving one side is the same as the tension in the rope entering the other side. The tension T_1 is therefore equal to 50 lb, since a vertical summation of forces yields $\Sigma F_v = 0 = +2T_1 - 100$.

Next draw a free body diagram of the middle pulley, Fig. 5-12(c). From the reasoning just explained the tension in the rope around this pulley is T_2. Summing forces vertically, the equation obtained is $\Sigma F_v = 0 = +2T_2 - T_1 = 2T_2 - 50$, whence $T_2 = 25$ lb.

Finally, draw a free body diagram of the top pulley, Fig. 5-12(d). Since the rope is continuous, $P = T_2$, or $P = 25$ lb.

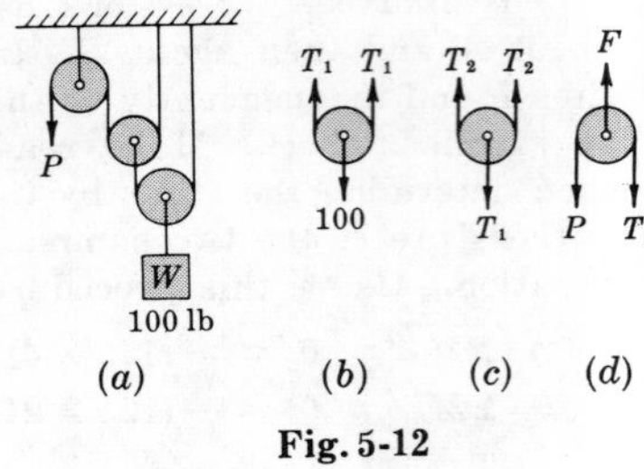

Fig. 5-12

11. How far along the beam should the vertical force F be placed so as to hold the beam in a horizontal position?

Solution:

Since the rope over the pulley is continuous and inextensible, the force up at the end of the beam is W lb.

Regard only the beam as a free body diagram and take moments about pin O to obtain $\Sigma M_O = 0 = -Fx + Wl$, or $x = Wl/F$.

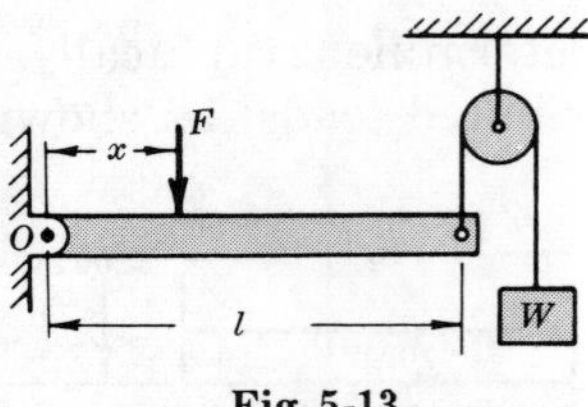

Fig. 5-13

12. A cantilever beam 10 ft long and weighing 100 lb/ft carries a concentrated load of 1000 lb at its free end. The other end of the beam is inserted into a wall 2 ft thick. What are the reactions on the beam at A and B? Refer to Fig. 5-14(a).

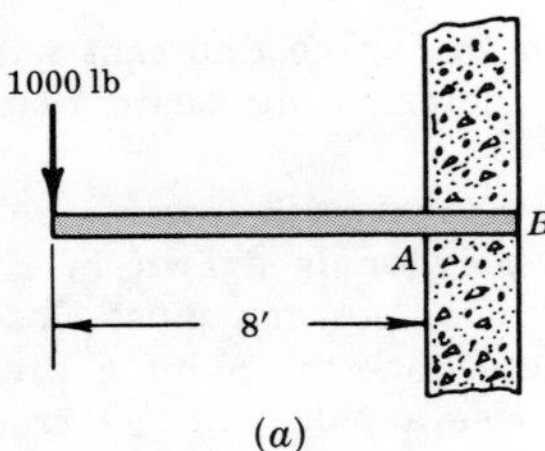

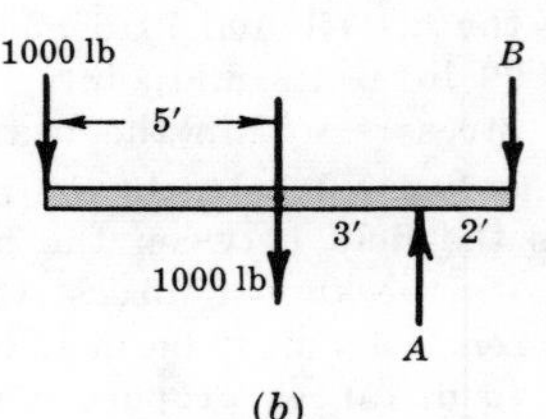

Fig. 5-14

Solution:

Assume the beam bends so that the wall pushes up at A and down at B on the beam. Draw the free body diagram showing the entire weight of 1000 lb (10 ft × 100 lb/ft) at the midpoint. See Fig. 5-14(b).

To determine reaction A, sum moments about B to obtain

$$\Sigma M_B = 0 = +1000 \times 10 + 1000 \times 5 - 2A, \qquad A = 7500 \text{ lb}$$

To determine reaction B, sum moments about A to obtain

$$\Sigma M_A = 0 = +1000 \times 8 + 1000 \times 3 - 2B, \qquad B = 5500 \text{ lb}$$

13. Blocks A and B weigh 400 and 200 lb respectively. They rest on a 30° inclined plane and are attached to the post which is held perpendicular to the plane by force P parallel to the plane (see Fig. 5-15(a) below). Assume that all surfaces are smooth and that the cords are parallel to the plane. Determine the value of P.

Solution:

Draw free body diagrams of A, B, and the post as shown in Fig. 5-15(b) below. It is seen by inspection that T_A and T_B may be found by a summation of forces parallel to plane. Thus, $T_A = 400 \sin 30° = 200$ lb. Similarly, $T_B = 100$ lb.

In the free body diagram of the post, sum moments about O to obtain $-P \times 24 + T_B \times 12 + T_A \times 6 = 0$. Substitute the values of T_A and T_B to obtain $P = 100$ lb.

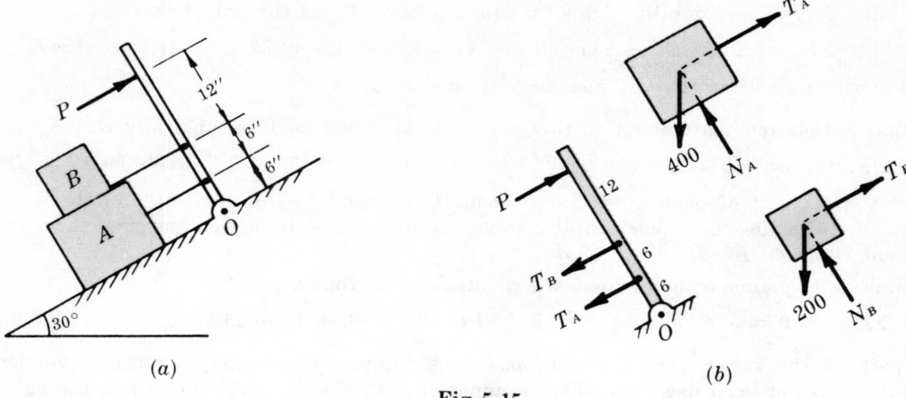

Fig. 5-15

14. A vertical force F of 50 lb is applied to the bell crank at point A. A force P applied at B as shown in Fig. 5-16(a) below prevents rotation of the crank about point O. Determine the force P and the bearing reaction R at O.

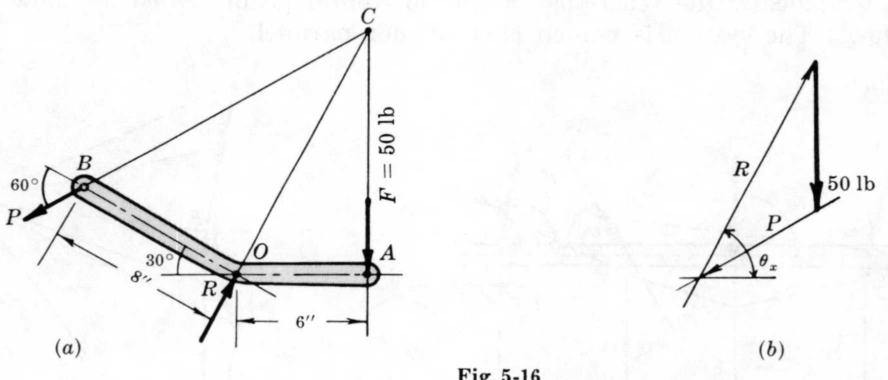

Fig. 5-16

Solution:

Since only three forces act on the bell crank which is in equilibrium, they must be concurrent. This means that the reaction R must pass through point C where F and P intersect.

The force triangle is shown in Fig. 5-16(b) above. The reaction R is parallel to the line OC. Measurement of the force triangle yields $P = 43.0$ lb, and $R = 80.8$ lb with $\theta_x = 62.5°$.

15. Determine algebraically the reactions on the beam loaded as shown in Fig. 5-17(a). The loads are in kips. Neglect the thickness and weight of the beam.

Solution:

In the free body diagram, Fig. 5-17(b), the horizontal and vertical components C_h and C_v of pin reaction at C are assumed positive. The roller support at D is normal to the beam, as shown.

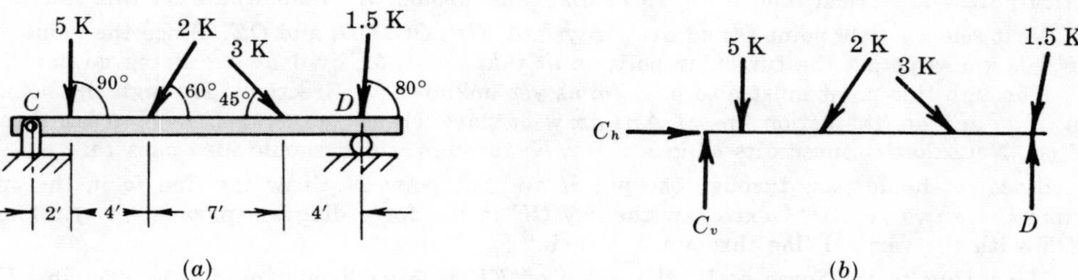

Fig. 5-17

Summing forces horizontally yields an equation with C_h as the only unknown.

$$\Sigma F_h = 0 = C_h - 2\cos 60° + 3\cos 45° - 1.5\cos 80°, \qquad C_h = -0.86K$$

This means that C_h acts to the left instead of as assumed.

To find D take moments about C; this will yield an equation involving only D.

$$\Sigma M_C = 0 = -5\times 2 - (2\sin 60°)\times 6 - (3\sin 45°)\times 13 - (1.5\sin 80°)\times 17 + D\times 17$$

Note that the moment of each given force about C is equal to the moment of only its vertical component about C because each horizontal component has a line of action passing through C. Solving this moment equation for D, $D = 4.3\,K$.

To find C_v, the summation of moments about D is convenient.

$$\Sigma M_D = 0 = -C_v\times 17 + 5\times 15 + 2\sin 60°\times 11 + 3\sin 45°\times 4, \qquad C_v = 6.03K$$

A check on the values for C_v and D may be obtained by summing the forces vertically, because this equation has not been used yet. This summation ΣF_v should equal zero when the values of C_v and D are inserted.

$$\Sigma F_v = 6.03 - 5 - 2\times 0.866 - 3\times 0.707 - 1.5 + 4.30 = -0.02$$

Since this is within the limits of accuracy of the problem, the values check closely enough.

16. Determine graphically the reactions on the horizontal beam loaded as shown in Fig. 5-18(a) below. The system is non-concurrent, non-parallel.

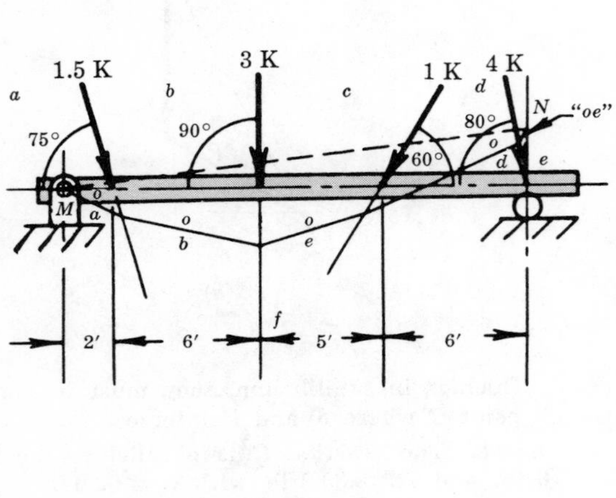

(a)

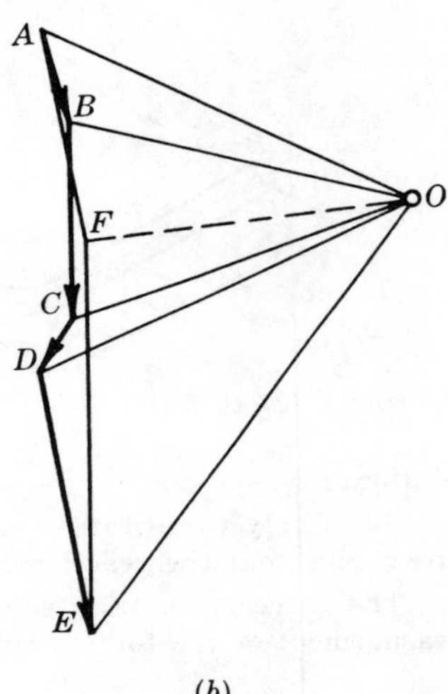

(b)

Fig. 5-18

Solution:

The space diagram, Fig. 5-18(a) above, is labeled with the small letters of Bow's notation. The right reaction EF is vertical. The only known characteristic of the left reaction FA is its point of application at the pin M.

Draw AB, BC, CD, and DE in the force diagram, Fig. 5-18(b) above. Since the right reaction EF is vertical, draw a vertical line in the force diagram through E. Somewhere on this line is the point F.

Next select a pole point O and draw rays OA, OB, OC, OD, and OE. Since the point of application of FA is known, start the funicular polygon at that point, M, by drawing string ao parallel to the ray AO. Through this point must also pass fo, as yet unknown in direction. Through the point of intersection of ao and ab (the action line of AB) draw bo parallel, of course, to BO. In order then draw co, do and eo. Note that eo intersects ef in a point, N, through which should also pass fo.

Since fo should pass through the pin M and the point N, draw the line fo in the space diagram joining these two points. Next draw the ray OF in the force diagram parallel to of. The intersection of OF with the vertical line through E is point F.

Measuring to the force scale, the value of EF is 5950 lb and of FA is 3350 lb. FA makes an angle θ of 80° with the horizontal.

17. Determine the tension in the cable AB which holds a post BC from sliding. Fig. 5-19(a) shows the essential data. Weight of the post is 18 lb. Assume all surfaces are smooth.

Solution:

Fig. 5-19(b) is the free body diagram. Note that R_D is normal to the post and R_B is normal to the floor, because frictionless surfaces are assumed.

One procedure is to take moments about B to find R_D, and then to sum the forces horizontally to find T. The following equations ensue.

$$\Sigma M_B = 0 = -18(7\cos 60°) + R_D \frac{10}{\cos 30°}$$

$$\Sigma F_h = 0 = T - R_D \cos 30° = T - 5.46(0.866)$$

from which $R_D = 5.46$ lb, $T = 4.72$ lb.

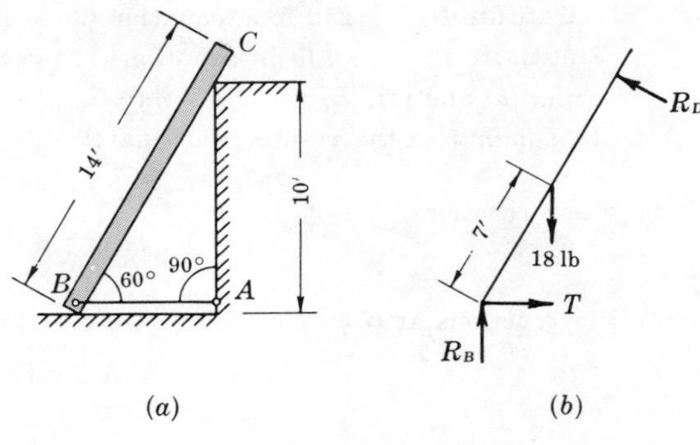

Fig. 5-19

18. Determine the following forces for the A-frame shown in Fig. 5-20(a): (1) the floor reactions at A and E, (2) the pin reactions at C on CE. (3) the pin reactions at B on AC. The floor is assumed to be smooth. Neglect the weight of the members.

Solution:

To determine the floor reactions at A and E, consider the entire frame as a solid free body as shown in Fig. 5-20(b). The manner in which the frame distributes the 400 lb load within itself has no bearing in determining the external reactions at A and E. Taking moments about A and E, the following solutions result.

$$\Sigma M_A = 0 = R_E \times 8.00 - 400 \times (8.00 - 3.54), \qquad R_E = 223 \text{ lb}$$

$$\Sigma M_E = 0 = -R_A \times 8.00 + 400 \times 3.54, \qquad R_A = 177 \text{ lb}$$

As a check, the vertical summation of forces does equal zero.

In working parts (2) and (3) of the problem, draw a free body diagram of the member CE in Fig. 5-20(c). Assume that the reactions of the pins *on* CE are as shown. There are four unknowns in the figure with only three equations available. Another free body must now be drawn involving some of the same unknowns. Draw a free body diagram of member BD in Fig. 5-20(d) showing the pin reactions acting at D on BD opposite in direction to those assumed in Fig. (c).

The 2.13′ dimension in Fig. (d) is obtained by subtracting the horizontal projection of the 2.00′ dimension from 3.54′. By similar reasoning the other dimensions are obtained.

The following equations may be written for Fig. (c).

(*1*) $\Sigma F_h = 0 = +C_h + D_h$

(*2*) $\Sigma F_v = 0 = +C_v + D_v + 223$

(*3*) $\Sigma M_D = 0 = +223 \times 2\cos 45° - C_v \times 5.18 \cos 45° - C_h \times 5.18 \sin 45°$

For Fig. (d) write the following equations

(*4*) $\Sigma M_B = 0 = -400 \times 3.64 - D_v \times 5.77$

(*5*) $\Sigma M_D = 0 = -B_v \times 5.77 + 400 \times 2.13$

(*6*) $\Sigma F_h = 0 = +B_h - D_h$

In solving the above equations, look for equations each with only one unknown.

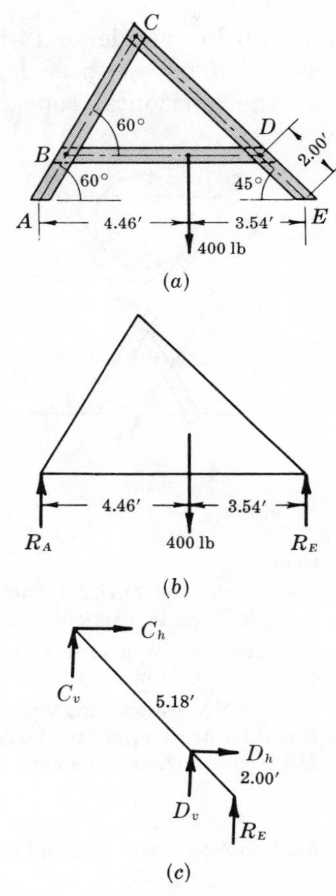

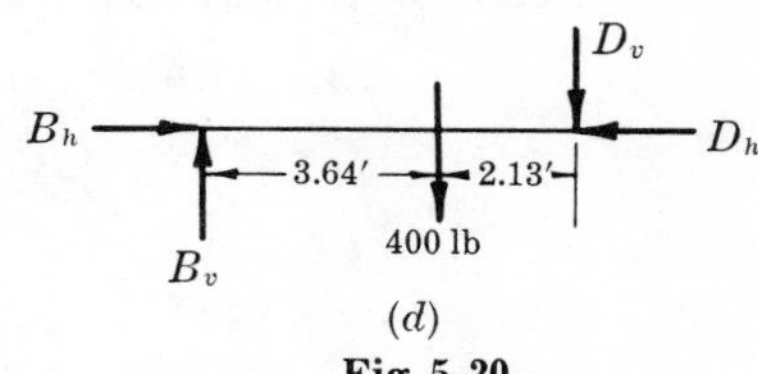

Fig. 5-20

$$\text{From } (4),\ D_v = \frac{-400 \times 3.64}{5.77} = -252 \text{ lb.} \quad \text{From } (5),\ B_v = \frac{400 \times 2.13}{5.77} = 148 \text{ lb.}$$

Substitute $D_v = -252$ lb in equation (*2*) to obtain $C_v = -223 - (-252) = 29.0$ lb.
Substitute $C_v = 29.0$ lb in equation (*3*) to obtain $C_h = 57.1$ lb.
From (*6*) and (*1*), $B_h = +D_h = +(-C_h) = +(-57.1) = -57.1$ lb.
To summarize the results, with particular emphasis on signs:

(1) Floor reactions.

$$R_A = 177 \text{ lb up}, \qquad R_E = 223 \text{ lb up}$$

(2) Pin reactions at C on CE. These were assumed to act on CE in positive directions.

$$C_h = 57.1 \text{ lb to right}, \qquad C_v = 29.0 \text{ lb up}$$

(3) Pin reactions at B on AC. In Figure (*d*) the pin reactions at B were shown acting on BD. Therefore they act in the opposite direction on AC. In the solution for the reaction *on* BD it was found that: $B_h = -57.1$ lb, i.e. to left; $B_v = 148$ lb, i.e. up. Therefore, *on* AC the pin reactions are

$$B_h = 57.1 \text{ lb to right}, \qquad B_v = 148 \text{ lb down}$$

19. A 100 lb cylinder 4 ft in diameter is lodged between the cross pieces which make an angle of 60° with each other as shown in Fig. 5-21(*a*) below. Determine the tension in the horizontal rope DE assuming a smooth floor.

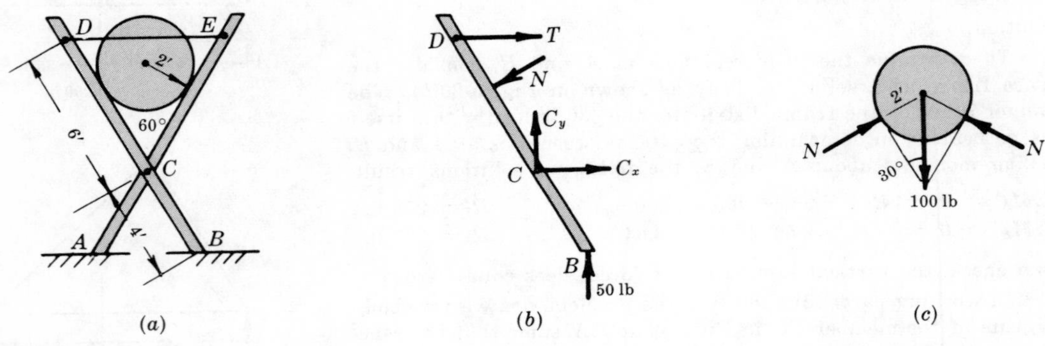

Fig. 5-21

Solution:

Treat the entire structure as a free body diagram. Because of symmetry it is apparent that $A = B = 50$ lb vertically upward.

Next draw a free body diagram of the arm DB showing the rope tension T and the reactions C_x and C_y of pin C (Fig. 5-21(*b*) above). The reaction N of the cylinder is perpendicular to the arm.

If N were known, then a summation of moments about C would yield the tension T. But N can be found by drawing a free body diagram of the cylinder (Fig. 5-21(*c*) above). From the geometry involved, N extends through the center of the cylinder. Hence a vertical summation of forces yields

$$\Sigma F_v = 0 = 2N \sin 30° - 100, \qquad N = 100 \text{ lb}$$

Also note that the perpendicular distance from N to C is $2/(\tan 30°) = 3.47$ ft.

Return to the free body diagram of the arm BD and sum moments about C to obtain

$$\Sigma M_C = 0 = -T \times 6 \cos 30° + 100 \times 3.47 + 50 \times 4 \sin 30°, \qquad T = 86 \text{ lb}$$

Supplementary Problems

20. A weight of 100 lb is suspended by a rope from a ceiling. The weight is pulled by a horizontal force until the rope makes an angle of 70° with the ceiling. Find the horizontal force and the tension in the rope. Use graphical analysis. *Ans.* $H = 36.4$ lb, $T = 106$ lb

21. Solve Problem 20 by algebraic analysis.

22. A rubber band has an unstretched length of eight inches. It is pulled until its length is ten inches as shown in Fig. 5-22 below. The horizontal force P is 6 oz. What is the tension in the band?
Ans. $T = 5$ oz

23. A small piece of round rod is welded to the pinch bar at A to serve as a fulcrum. A force P equal to 120 lb is required to lift the left side of a box B as shown in Fig. 5-23 below. To lift the right side of the box the same pinch bar is used and P is then 100 lb. What does the box weigh?
Ans. $W = 1980$ lb

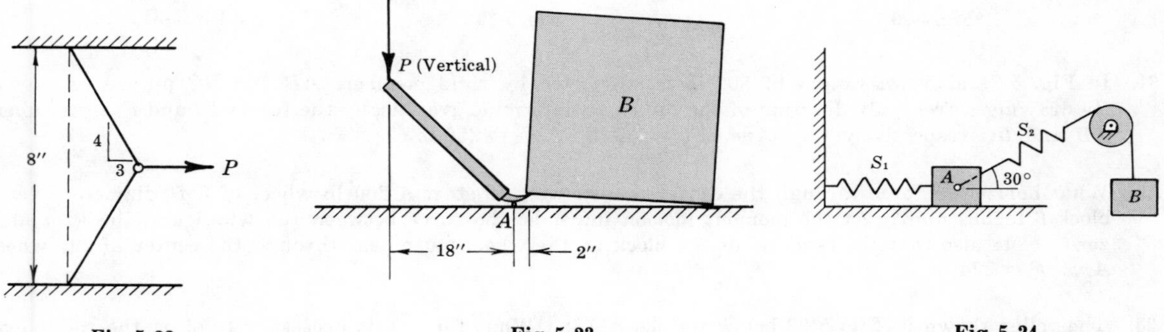

Fig. 5-22 Fig. 5-23 Fig. 5-24

24. Body A weighs 32.8 lb and rests on a smooth surface. Body B weighs 14.3 lb. Determine the tensions in S_1 and S_2 and the normal reaction of the horizontal surface on A. Refer to Fig. 5-24 above.
Ans. $S_1 = 12.4$ lb, $S_2 = 14.3$ lb, $N = 25.7$ lb

25. Two guy wires are fastened to an anchor bolt in a foundation as shown in Fig. 5-25 below. What pull does the bolt exert on the foundation? *Ans.* $P = 1030$ lb, $\theta_x = 135°$

26. A pulley to which is attached W_1 rides on a wire which is attached to a support at the left and which passes over a pulley on the right to weight W (see Fig. 5-26 below). The horizontal distance between the left support and the pulley (neglect the dimensions of the pulleys) is L. Express the sag d at the center in terms of W, W_1, and L. *Ans.* $d = \frac{1}{2}L/\sqrt{(2W/W_1)^2 - 1}$

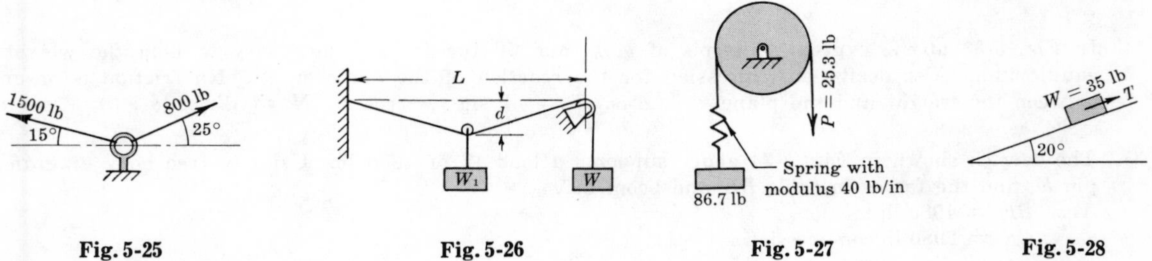

Fig. 5-25 Fig. 5-26 Fig. 5-27 Fig. 5-28

27. What is the normal reaction between the weight shown in Fig. 5-27 above and the ground? The pulley is assumed weightless and in frictionless bearings. *Ans.* $N = 61.4$ lb

28. Refer to Fig. 5-28 above. What force T parallel to the smooth plane is necessary to hold the 35 lb weight, W, in equilibrium? *Ans.* $T = 12.0$ lb

29. Refer to Fig. 5-29 below. A weight of 80 lb is suspended from a weightless bar AB which is supported by cable CB and a pin at A. Determine the tension in the cable and the pin reaction at A on the bar AB. *Ans.* $T = 197$ lb, $A_x = 180$ lb, $A_y = 0$ lb

30. Refer to Fig. 5-30 below. In the figure, three spheres each weighing 20.5 lb and each 14 inches in diameter rest in a box 30 inches wide. Find (1) the reaction of B on A, (2) the reaction of the wall on C, and (3) the reaction of the floor on B.
Ans. (1) 12.5 lb along the line joining their centers, (2) 7.14 lb to left, (3) 30.8 lb up

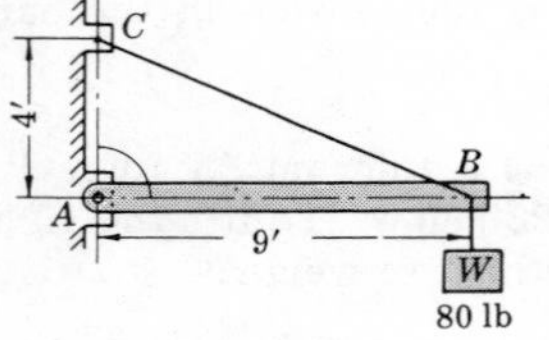

Fig. 5-29

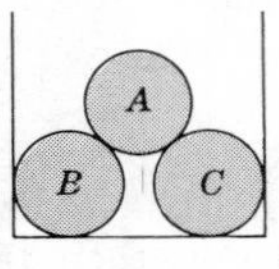

Fig. 5-30

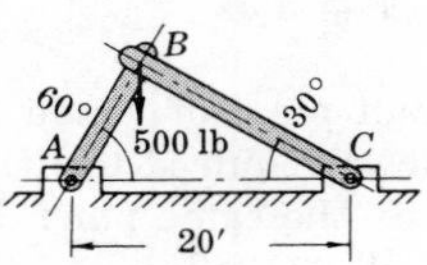

Fig. 5-31

31. In Fig. 5-31 above, a weight of 500 lb is supported by rigid members AB and BC pinned as shown. By drawing a free body diagram of the pin at B, determine graphically the forces F_1 and F_2 in members AB and BC respectively. *Ans.* $F_1 = 433$ lb C, $F_2 = 250$ lb C

32. What horizontal force through the center is necessary to start a 200 lb wheel of 3 ft diameter over a block 6 inches high? At the moment motion impends, the force between the wheel and the ground is zero. Note also that the reaction of the block on the wheel must pass through the center of the wheel. *Ans.* $F = 224$ lb

33. The roller shown in Fig. 5-32 below weighs 339 lb. What force T is necessary to start the roller over the block A? *Ans.* $T = 403$ lb

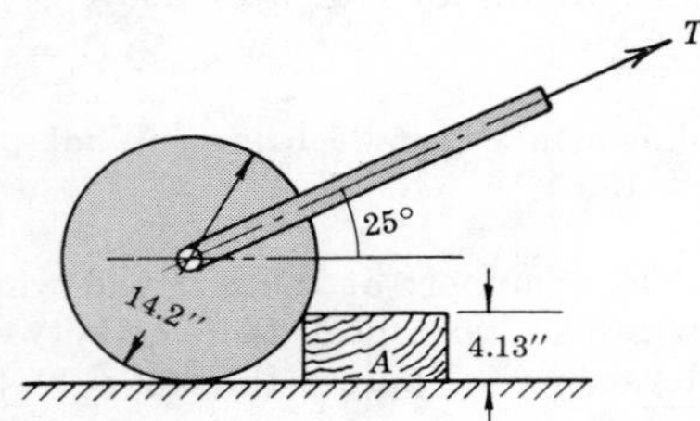

Fig. 5-32

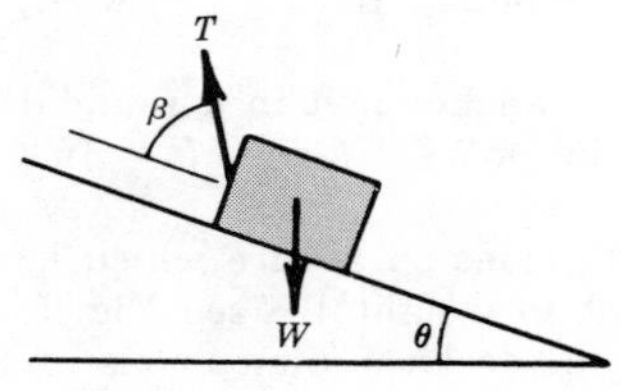

Fig. 5-33

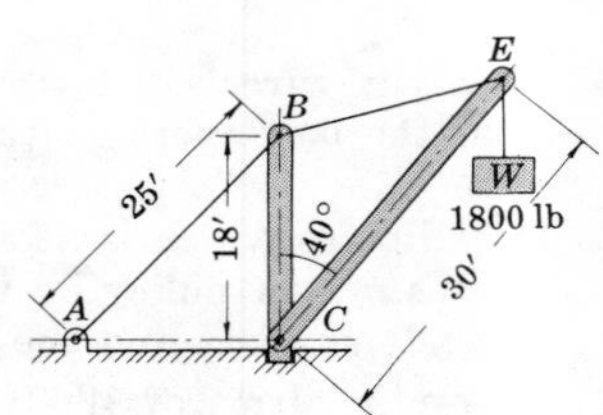

Fig. 5-34

34. In Fig. 5-33 above, express in terms of θ, β, and W the force T necessary to hold the weight in equilibrium. Also derive an expression for the reaction of the plane on W. No friction is assumed between the weight and the plane. *Ans.* $T = W \sin\theta \div \cos\beta$, $N = W\cos(\theta+\beta) \div \cos\beta$

35. The derrick shown in Fig. 5-34 above supports a load W of 1800 lb. Using a free body diagram of pin E, find the forces in cable BE and boom CE.
Ans. BE = 1980 lb tension
CE = 2980 lb compression

36. If the structure in Problem 31 rested at A and C on the smooth floor but a cable joined pins A and C, determine the tension in the cable AC. *Ans.* $T = 217$ lb

37. Solve Problem 8 graphically.

38. Solve Problem 9 algebraically.

39. Determine the reactions on the beam loaded as shown in Fig. 5-35 below. The uniformly distributed load is 200 lb per ft. Neglect the weight of the beam. *Ans.* $R_A = 792$ lb, $R_B = 208$ lb

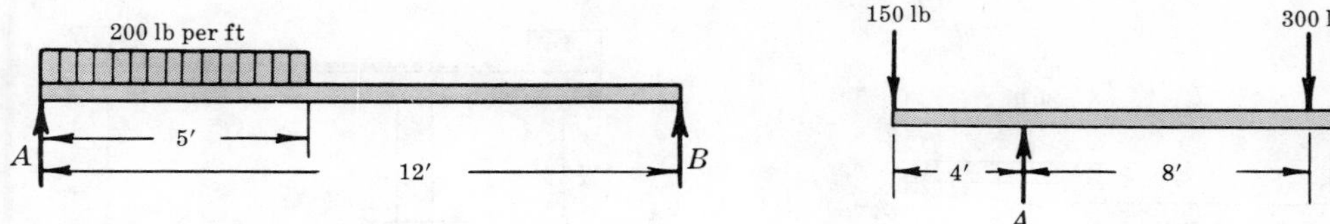

Fig. 5-35

Fig. 5-36

40. Determine beam reactions in Fig. 5-36 above. Consider only the two concentrated loads. *Ans.* $R_A = 286$ lb, $R_B = 164$ lb

41. The bar A shown in Fig. 5-37 weighs 30 lb per ft and is 12 ft long. The left end is inserted into a wall 14 inches thick. The 2 ft diameter pulley weighs 40 lb. The tension T in the rope is 80 lb. Determine the reactions at points B and C on the bar which fits loosely in the wall. *Ans.* $R_C = 3910$ lb up, $R_B = 3350$ lb down

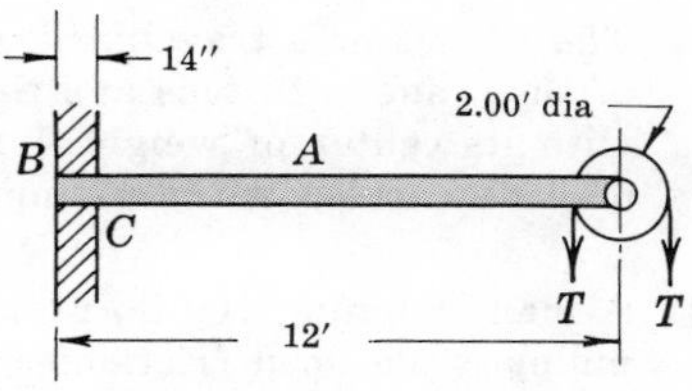

Fig. 5-37

42. A beam 8 ft long and weighing 15 lb per ft rests on supports at its ends. How far from the left support should a concentrated load of 100 lb be placed so that the left reaction will be 85 lb? *Ans.* $d = 6$ ft

43. In Fig. 5-38 below, what force P is required to raise a weight of 200 lb at constant speed? *Ans.* $P = 100$ lb

44. In Fig. 5-39 below, what force P is required to hold a weight of 600 lb in equilibrium? *Ans.* $P = 200$ lb

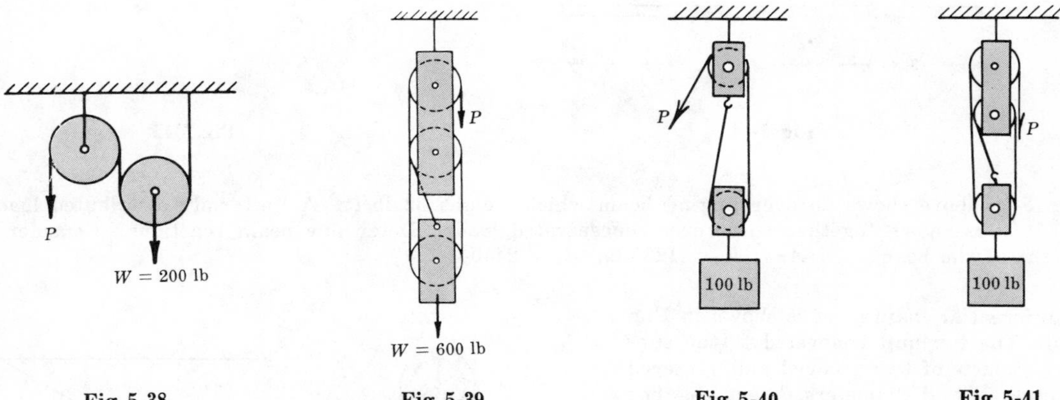

Fig. 5-38 Fig. 5-39 Fig. 5-40 Fig. 5-41

45. The upper block in Fig. 5-40 above is suspended from a fixed support. The rope is secured to the lower end of the casing of the upper block and then passes around the sheave in the lower block. Then it passes around the sheave in the upper block and is acted upon by force P. Show that for equilibrium the force P is 50 lb when a 100 lb weight is suspended from the bottom of the casing of the lower block.

46. The upper block in Fig. 5-41 above contains two sheaves and the lower block has one sheave. The rope is attached to the upper end of the casing of the lower block and then passes around one sheave in the upper block. It then returns to the sheave in the lower block and finally passes around the second sheave in the upper block where force P is exerted upon it. Show that for equilibrium the force P is 33.3 lb when a 100 lb weight is suspended from the bottom of the casing of the lower block.

47. In Fig. 5-42 below, a system of levers is shown supporting a load of 80 lb. Determine the reactions at A and B on the lever. *Ans.* $R_A = -71.1$ lb, $R_B = +124$ lb

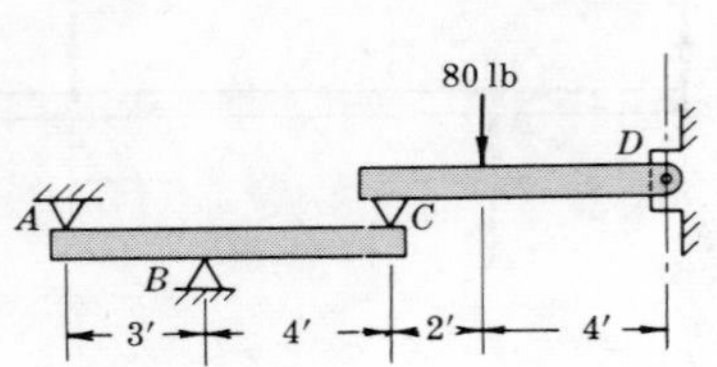

Fig. 5-42

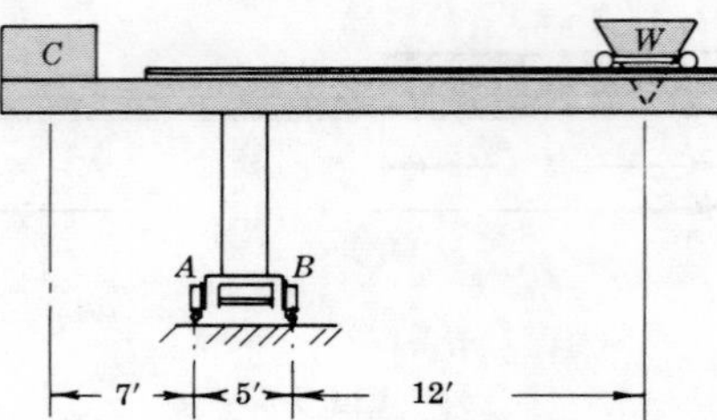

Fig. 5-43

48. The wheels of a traveling crane move on tracks at A and B as shown in Fig. 5-43 above. The weight of the crane is 10 tons and the center of weight is 3 ft to the right of A. The counterweight C is 4 tons with its center of weight 7 ft to the left of A. What maximum weight W, 12 ft to the right of B, may be carried without tipping? *Ans.* $W = 5.67$ tons

49. A man weighing 160 lb, represented by W, holds the 56.3 lb weight as shown in Fig. 5-44 below. The pulley is assumed frictionless. The platform on which the man is standing is suspended by two ropes at A and two ropes at B. What is the tension in one of the ropes at A? *Ans.* $A = 34.6$ lb

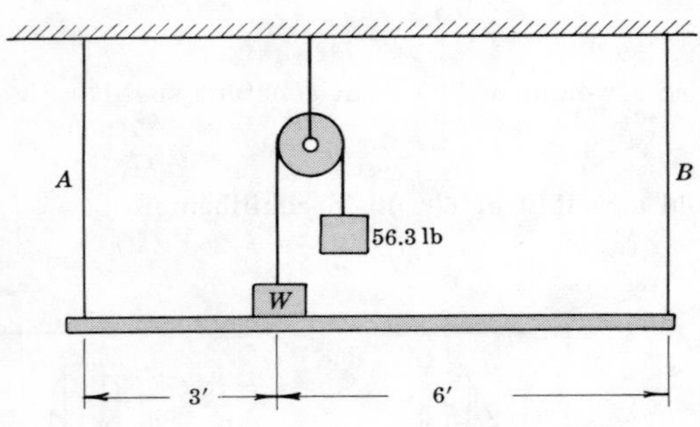

Fig. 5-44

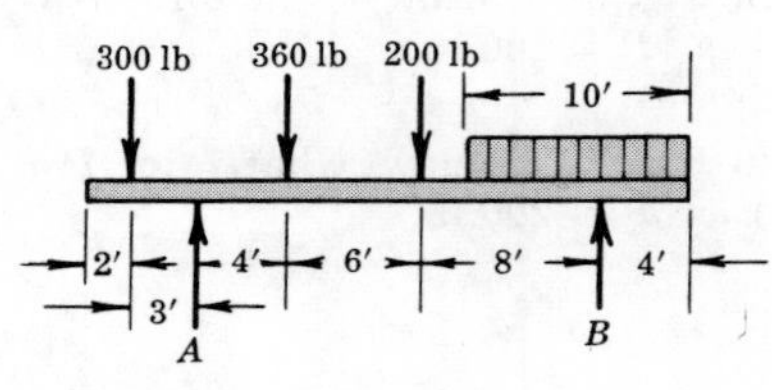

Fig. 5-45

50. Fig. 5-45 above shows an overhanging beam which weighs 32 lb/ft. A uniformly distributed load of 200 lb/ft is shown together with three concentrated loads. Determine beam reactions. Consider the weight of the beam. *Ans.* $R_A = 1280$ lb, $R_B = 2440$ lb

51. A differential chain hoist is shown in Fig. 5-46. The top unit connected to the support consists of two grooved pulleys keyed together but of diameters d_1 and d_2. The lower pulley is of diameter $\frac{1}{2}(d_1 + d_2)$. The weight is attached to the lower pulley. A continuous chain passes over the pulleys as shown. Assume no tension in the slack side (that part of the chain shown to the right of the smaller pulley at the top). What force P is necessary to just start the weight W in the upward direction? *Ans.* $P = \frac{1}{2}W(d_2 - d_1)/d_2$

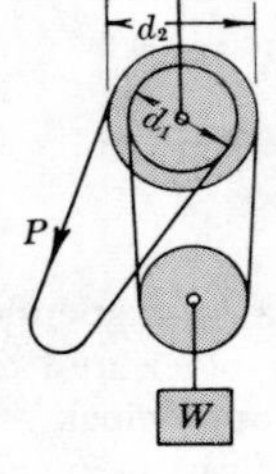

Fig. 5-46

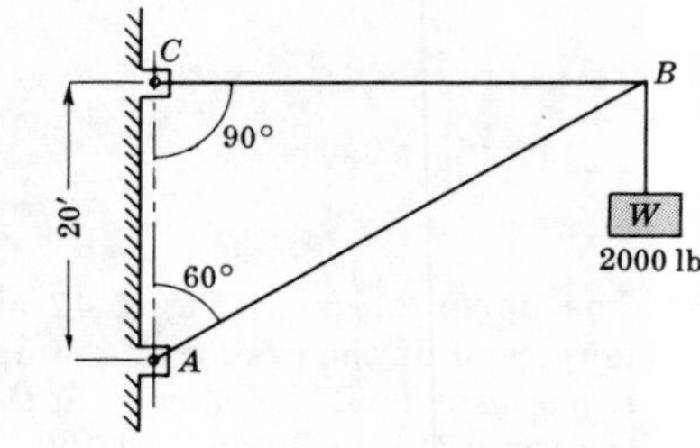

Fig. 5-47

52. In Fig. 5-47 above, AB is a rigid rod and CB is a cable. If W is 2000 lb, what is the pin reaction at A on the rod AB? What is the tension in the cable? *Ans.* $A_h = 3460$ lb, $A_v = 2000$ lb, $T = 3460$ lb

53. A horizontal force F of 5 lb is applied to the hammer shown in Fig. 5-48 below. Assuming the hammer to pivot about point A, what force is exerted on the vertical nail that is being pulled from the horizontal floor? *Ans.* $P = 15.8$ lb

54. Determine the force P to maintain the bell crank shown in Fig. 5-49 below in equilibrium. Neglect friction at the pivot point O. *Ans.* $P = 52.2$ lb

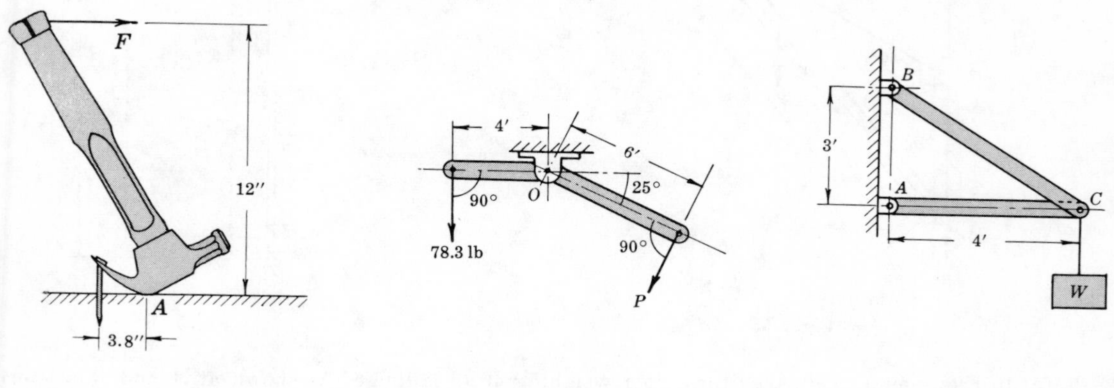

Fig. 5-48 Fig. 5-49 Fig. 5-50

55. The 1000 lb weight W is attached to the pin at C as shown in Fig. 5-50 above. Determine the forces acting in members AC and BC. *Ans.* $AC = 1330$ lb, $BC = 1670$ lb

56. In Fig. 5-51 below, the forces are shown acting on the beam divided into one foot intervals. The loads are in kips. Determine the reactions at A and B. *Ans.* $A_h = 1.65$ K, $A_v = 4.60$ K, $B = 8.71$ K

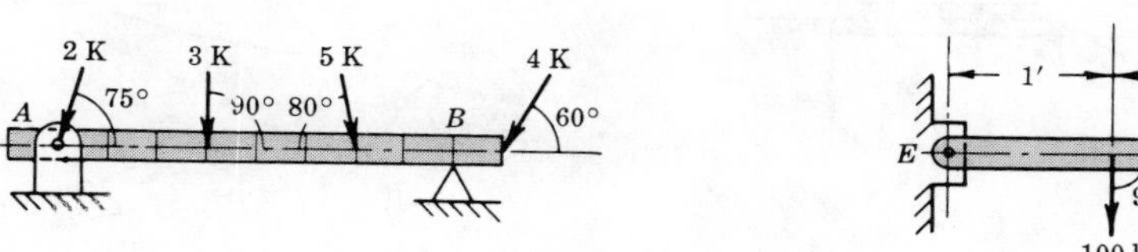

Fig. 5-51 Fig. 5-52

57. Beam ED is loaded as shown in Fig. 5-52 above. The beam is pin connected to the wall at E. At D an 8 inch diameter pulley is attached through frictionless bearings to the beam. A rope passes around the pulley and is connected to A and B vertically above the extremities of the horizontal diameter of the pulley. Determine the pin reaction at E, and the tension in the rope.
Ans. $E = 160$ lb, $\theta_x = 51°$, $T = 74.4$ lb

58. Find the pin reaction at A and the knife edge reaction at B using graphical means. Refer to Fig. 5-53.
Ans. $A = 82.0$ lb, $\theta = 94°$, $B = 15.0$ lb

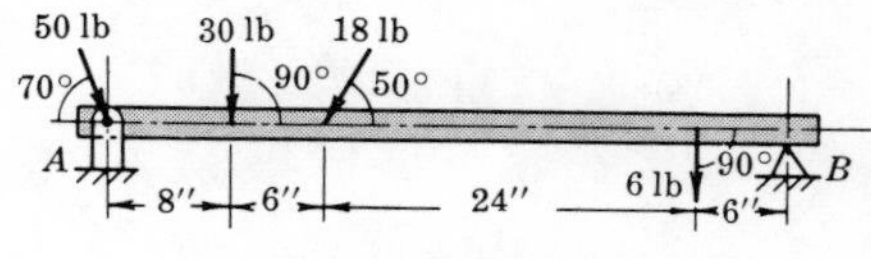

Fig. 5-53

59. An A-frame supports a load of 250 lb as shown in Fig. 5-54. The weights of the side members AB and BC are 60 lb and 90 lb respectively. Determine the reactions of pins D and E on DE.
Ans. $D_v = 57$ lb up,
$E_v = 193$ lb up,
$D_h = 135$ lb to left,
$E_h = 135$ lb to right

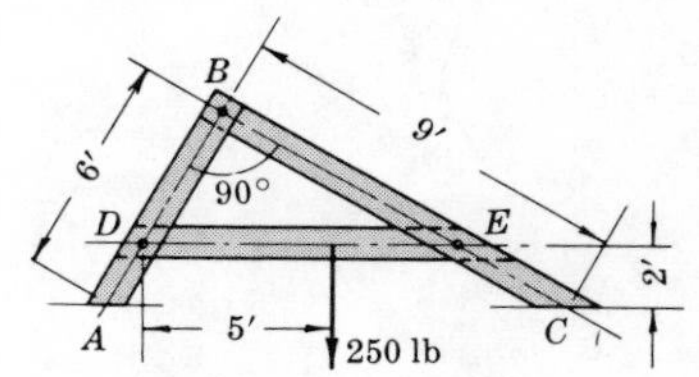

Fig. 5-54

60. Refer to Fig. 5-55 below. Determine the tension in cable BC by graphical analysis. Neglect the weight of AB. *Ans.* $T = 1000$ lb

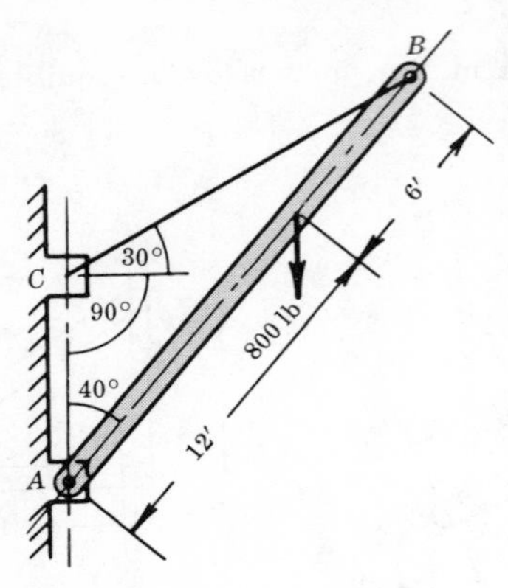

Fig. 5-55

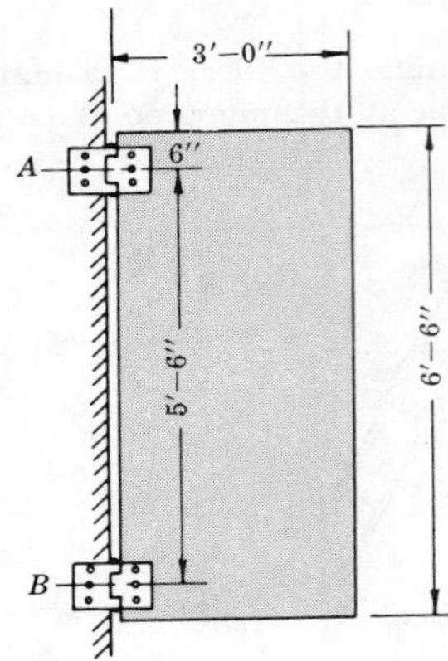

Fig. 5-56

61. Refer to Fig. 5-56 above. A uniform door weighing 40 lb is hinged as shown at A and B. Determine hinge reactions at A and B on the door. Assume that the vertical components of the reactions at A and B are equal. *Ans.* $A = 22.8$ lb, $\theta_x = 118.6°$; $B = 22.8$ lb, $\theta_x = 61.4°$

62. Refer to Fig. 5-57 below. The frame illustrated is used to support a weight of 450 lb at F. Determine (1) the pin reaction of E on DE, (2) the pin reaction of C on CF, and (3) the floor reaction at B on AB. *Ans.* $E = 1510$ lb, $\theta_x = 206.6°$; $C = 1370$ lb, $\theta_x = 190°$; $B_x = 540$ lb, $B_y = 450$ lb

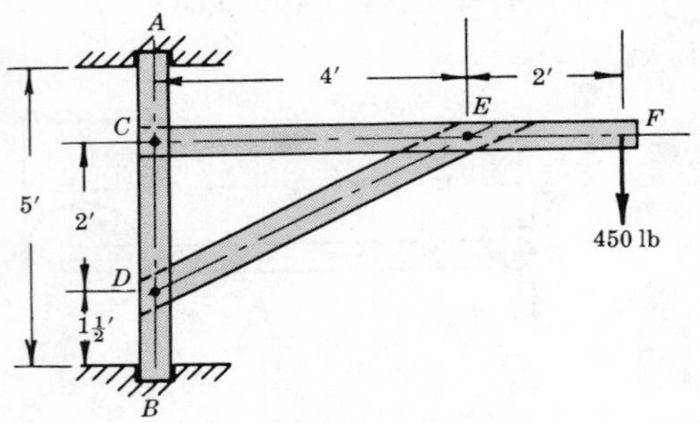

Fig. 5-57

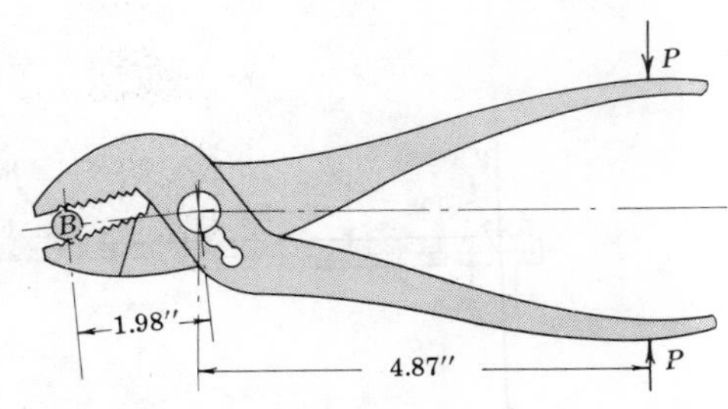

Fig. 5-58

63. The bolt B is held by a gripping force of 10 lb perpendicular to the jaws of the pliers as shown in Fig. 5-58 above. What forces P must be applied perpendicular to the handles to supply the gripping force? *Ans.* $P = 4.07$ lb

Chapter 6

Trusses and Cables

TRUSSES and CABLES

These are examples of coplanar force systems in equilibrium (see Chapter 5).

TRUSSES

A. **Assumptions**

(*a*) Each truss is assumed to be composed of rigid members all lying in one plane. This means that coplanar force systems are involved.

(*b*) Forces are transmitted from one member to another through smooth pins fitting perfectly in the members. These are called two-force members.

(*c*) The weights of the members are neglected because they are small in comparison with the loads.

B. **Solution by the Method of Joints**

To use this technique draw a free body diagram of any pin in the truss, provided no more than two unknown forces act on that pin. This limitation is imposed because the system of forces is a concurrent one for which, of course, only two equations are available for a solution. Proceed from one pin to another until all unknowns have been determined.

C. **Solution by the Method of Sections**

In the method of joints, forces in various members are determined by using free body diagrams of the pins. In the method of sections, a section of the truss is taken as a free body diagram. This involves cutting through a number of members, including those members whose forces are unknown, in order to isolate one part of the truss. The forces in the members cut act as external forces helping to hold that part of the truss in equilibrium. Since the system is non-concurrent, non-parallel, three equations are available. Therefore in any one sectioning no more than three unknown forces can be found. Be sure to isolate the free body completely and at the same time have no more than three unknown forces.

D. **Graphical Solution of Trusses**

As in the algebraic solution of trusses by the method of joints, the graphical solution is applied at a joint on which no more than two unknown forces act. Since each member acts on two pins, one on each end, it is evident that the force in that member is drawn in the free body diagrams of both pins. Instead of redrawing the force in a separate free body diagram of the second pin, it is customary to use that force as a common link between the two free body diagrams. In this way it is possible to group all the free body diagrams into one figure called a Maxwell diagram or a Maxwell-Cremona diagram.

CABLES

A. Parabolic

This is a cable loaded with w lb per horizontal foot. Fig. 6-1 illustrates such a suspension from supports on the same level. Temperature variations which change the tension are neglected.

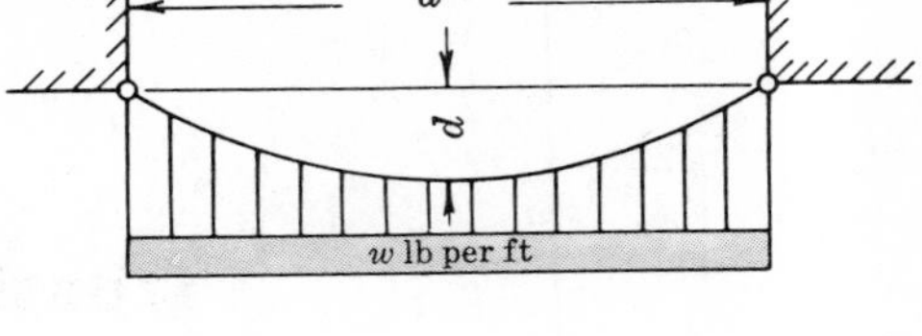

Fig. 6-1

The following equations apply to this coplanar system:

(1) $d = \dfrac{wa^2}{8H}$

(2) $T = \frac{1}{2}wa\sqrt{1 + \dfrac{a^2}{16d^2}}$

(3) $l = a\left[1 + \frac{8}{3}\left(\frac{d}{a}\right)^2 - \frac{32}{5}\left(\frac{d}{a}\right)^4 + \frac{256}{7}\left(\frac{d}{a}\right)^6 - \cdots\right]$

where d = sag in ft, w = load in lb per horizontal ft, a = span in ft, H = tension at midpoint in lb, T = tension at supports in lb, l = length of cable in ft.

B. Catenary

This is the curve assumed by a cable carrying a load of w lb per foot along the cable rather than horizontally as in the parabolic case. Fig. 6-2 illustrates such a suspension from supports at the same level. Neglect temperature changes.

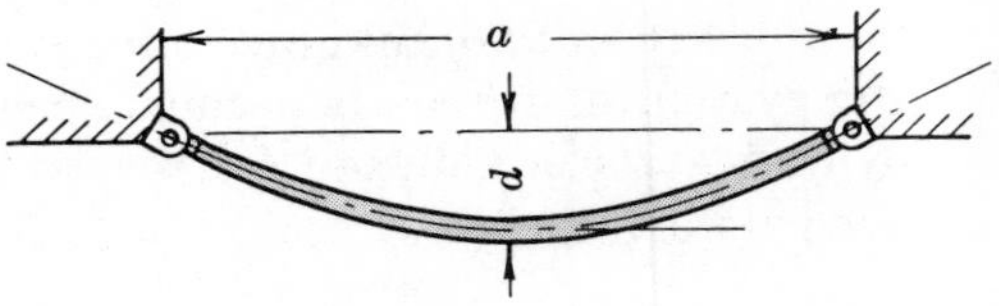

Fig. 6-2

To solve this type of problem, let:

T = tension at distance x from midpoint,
s = length along the cable from midpoint to point where tension is T,
w = load in lb per ft along the cable, e.g., its weight per ft,
a = span in ft, d = sag in ft, l = total length in ft of the cable,
H = tension at midpoint in lb, $T_{\max}$ = tension at support in lb.

Referring to the free body diagram of a portion of the cable to the right of center (Fig. 6-3), note the x axis is at the distance c below the center of the cable. This simplifies the derivation. Equation (1) gives the value of c.

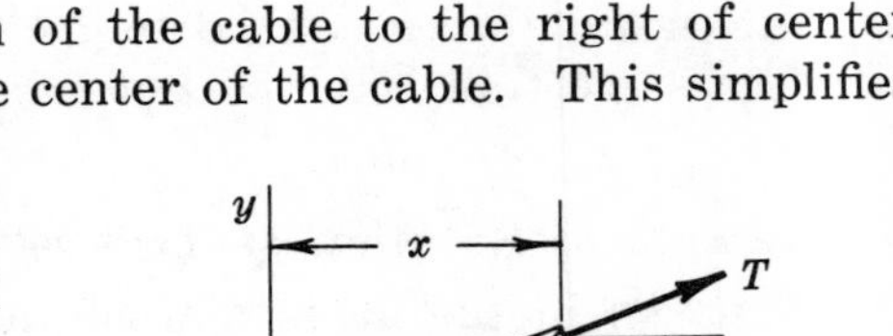

Fig. 6-3

The following equations apply to the catenary. Note that T becomes $T_{\max}$ when $x = a/2$ and $y = c + d$.

(1) $c = H/w$

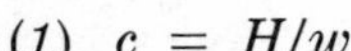

(2) $y = c\cosh x/c$ and $c + d = c\cosh a/2c$

(3) $T = wy$ and $T_{\max} = w(c + d)$

(4) $s = c\sinh x/c$ and $l/2 = c\sinh a/2c$

(5) $y^2 = c^2 + s^2$ and $(c + d)^2 = c^2 + l^2/4$

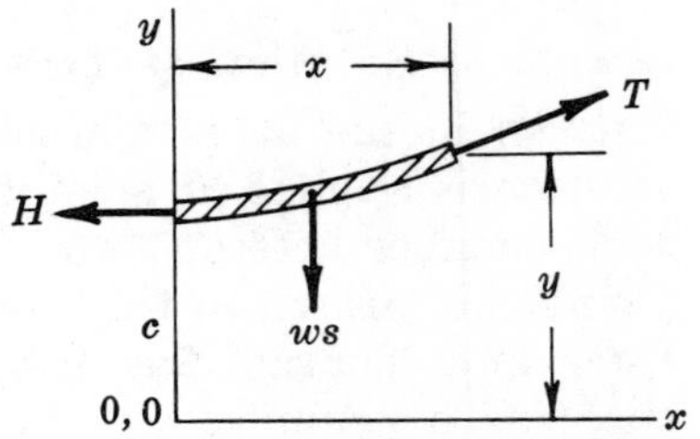

The following types of problems involving cables of weight w lb per ft may present themselves for solution:

(a) given the span and sag, i.e., a and d
(b) given the span and length, i.e., a and l
(c) given the sag and length, i.e., d and l

In case (a), solve equation (2) by trial to obtain c. Then equation (3) yields T_{max}, and equation (4) or (5) determines l.

In case (b), solve equation (4) by trial to obtain c. Then equation (5) yields d, and equation (3) T_{max}.

In case (c), solve equation (5) for c. Then T_{max} may be obtained from equation (3). To find a, solve either equation (2) or (4).

Solved Problems

1. The simple triangle truss in Fig. 6-4(a) supports two loads as shown. Determine reactions and the forces in each member.

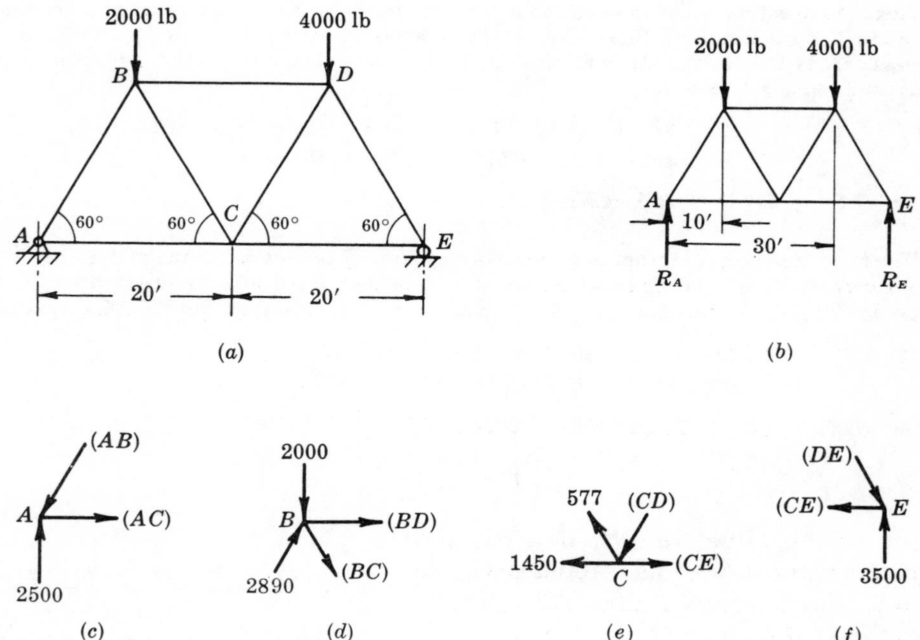

Fig. 6-4

Solution:

Fig. 6-4(b) is a free body diagram of the entire truss from which to determine R_A and R_E. Since the two loads are vertical, only one component of the pin reaction at A is shown.

(1) $\Sigma M_A = 0 = R_E \times 40 - 4000 \times 30 - 2000 \times 10, \qquad R_E = 3500$ lb
(2) $\Sigma M_E = 0 = -R_A \times 40 + 2000 \times 30 + 4000 \times 10, \qquad R_A = 2500$ lb

Of course a vertical summation of the two given forces and the two reactions just determined equals zero, thereby checking the results.

Fig. 6-4(*c*) above is a free body diagram of pin *A*. The 2500 lb reaction is drawn up. The only force which can have a downward component to balance R_A is the force in the member *AB*. This is shown acting towards the pin, which means that the member *AB* is in compression. Since force (*AB*) acts to the left as well as down, some force must act to the right to balance it. Therefore force (*AC*) is shown to the right pulling on the pin. The pin pulls to the left on the member *AC*, which means that (*AC*) is a tensile force.

Writing the equations of the concurrent system of Fig. 6-4(*c*),

$$(3)\quad \Sigma F_h = 0 = +(AC) - (AB)\cos 60^\circ$$

$$(4)\quad \Sigma F_v = 0 = +2500 - (AB)\sin 60^\circ$$

Solving, $(AB) = +2500/0.866 = +2890$ lb, $(AC) = (AB)\cos 60^\circ = +1450$ lb. The + signs indicate that the directions chosen are correct. Hence $(AB) = 2890$ lb *C*, $(AC) = 1450$ lb *T*.

Next draw a free body diagram for pin *B*. See Fig. 6-4(*d*) above. Some might have chosen pin *C*, but there are three unknown forces there, i.e., (*BC*), (*CD*) and (*CE*). In this figure, member *AB* is in compression and must be shown pushing on the pin. The 2000 lb load is shown acting directly down on the pin. The directions of forces (*BD*) and (*BC*) are unknown. Instead of spending time trying to decide the direction of each of these, assume they are in tension. A plus sign in the result indicates that tension is correct, whereas a minus sign indicates compression. The equations for this system are:

$$(5)\quad \Sigma F_h = 0 = (BD) + 2890\cos 60^\circ + (BC)\cos 60^\circ$$

$$(6)\quad \Sigma F_v = 0 = 2890\sin 60^\circ - 2000 - (BC)\sin 60^\circ$$

Solving equation (*6*), $(BC) = 577$ lb *T*. Substituting in equation (*5*), $(BD) = -1730$ lb. Since the sign is minus, this force is actually a compression.

Next draw a free body diagram of pin *C*, as shown in Fig. 6-4(*e*) above. The two known values (*AC*) and (*BC*) are inserted. Since (*BC*) has a component acting vertically up, (*CD*) must be shown as compression. If this is not clear assume it in tension and a minus sign will result, indicating compression. The equations are:

$$(7)\quad \Sigma F_h = 0 = (CE) - 1450 - 577\cos 60^\circ - (CD)\cos 60^\circ$$

$$(8)\quad \Sigma F_v = 0 = +577\sin 60^\circ - (CD)\sin 60^\circ$$

Solving, $(CD) = 577$ lb *C* and $(CE) = 2020$ lb *T*.

The next free body diagram may be either at *D* or *E* to determine the last force (*DE*). Fig. 6-4(*f*) above shows the free body diagram for pin *E*. Note that force (*CE*) is inserted as an unknown. This is done deliberately to provide a check on this value as obtained at pin *C*. The equations are:

$$(9)\quad \Sigma F_v = 0 = 3500 - (DE)\sin 60^\circ$$

$$(10)\quad \Sigma F_h = 0 = (DE)\cos 60^\circ - (CE)$$

Solving, $(DE) = 4030$ lb *C* and $(CE) = 2020$ lb *T*.

2. Determine the forces in *FH, HG, IG,* and *IK* in the truss shown diagrammatically in Fig. 6-5. Each load is 2 kips. All triangles are equilateral with sides of 12 ft.

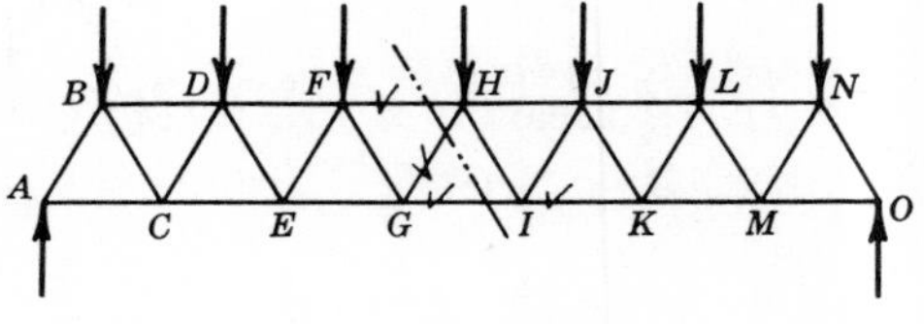

Fig. 6-5

Solution:

As a first step check the members in which the forces are to be found. Cut through as many members as possible, but no more than three in which the forces are unknown. The first cut should be through *FH, HG,* and *GI*. A free body diagram of either the left or right portion may now be drawn. Choose the one involving as few external forces as possible – the left part in this case. Draw a free body diagram of this portion as shown in Fig. 6-6. It is usually wise to assume the

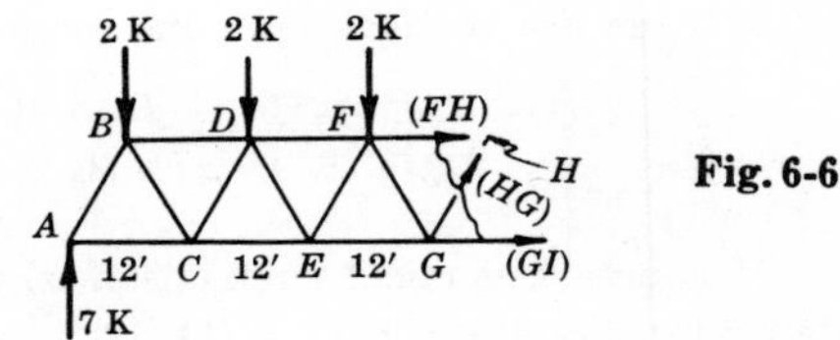

Fig. 6-6

forces in the members as tension, realizing that a minus sign in the result indicates compression. An arrow pointing away from the free body means that the member pulls on the body and is therefore in tension.

The left reaction of 7 kips (i.e., 7K) is determined by inspection of the *entire* truss which is symmetrical and symmetrically loaded.

Any three equations of equilibrium may be applied to the free body diagram. The summation of moments about G yields an equation with only one unknown force (FH). A summation of moments about H (external to the figure) involves one unknown force (GI), since members FH and HG intersect in H. Finally, a vertical summation of forces will result in a solution for force (HG). Using this procedure, the results are:

(1) $\Sigma M_G = 0 = -(FH) \times 6 \tan 60° - 7 \times 36 + 2 \times 30 + 2 \times 18 + 2 \times 6, \qquad (FH) = -13.9 \text{ K}, C$

(2) $\Sigma M_H = 0 = +(GI) \times 6 \tan 30° - 7 \times 42 + 2 \times 36 + 2 \times 24 + 2 \times 12, \qquad (GI) = 14.4 \text{ K}, T$

(3) $\Sigma F_v = 0 = +7 - 2 - 2 - 2 + (HG) \sin 60°, \qquad (HG) = -1.15 \text{ K}, C$

As a check on this particular free body diagram and its solution, sum the forces horizontally – an equation not used in the solution – to determine whether the result is zero or not.

(4) $$\Sigma F_h = -13.9 + 14.4 - 1.15 \cos 60° = 0$$

To determine the force in member IK, make a cut as shown in the adjacent Fig. 6-7. Take moments about point J, yielding equation (5).

(5) $$\Sigma M_J = 0 = -(IK) \times 6 \tan 60° - 2 \times 12 - 2 \times 24 + 7 \times 30, \qquad (IK) = 13.3 \text{ K}, T$$

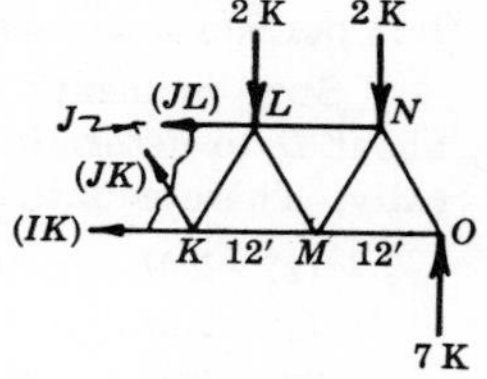

Fig. 6-7

3. The steel cable is hoisting a 2000 lb weight. Determine the forces in AC and BD assuming the cable is parallel to the sides as shown.

Solution:

Tension T is equal to 2000 lb on the assumption that the weight is not accelerating. To determine the forces in AC and BD, use a section as shown in Fig. 6-8(*a*) and draw a free body diagram of the top portion [Fig. 6-8(*b*)].

Through the center of the pulley draw two 2000 lb forces vertically but in opposite directions. Thus the 2000 lb vertical force is replaced by a new 2000 lb vertical force down at the center plus a couple $C = -2000 \times$ radius.

Similarly replace the 2000 lb diagonal tension with a 2000 lb force acting down along GE plus a couple $C = +2000 \times$ radius. The two couples then cancel each other, leaving only the two 2000 lb forces along GF and GE.

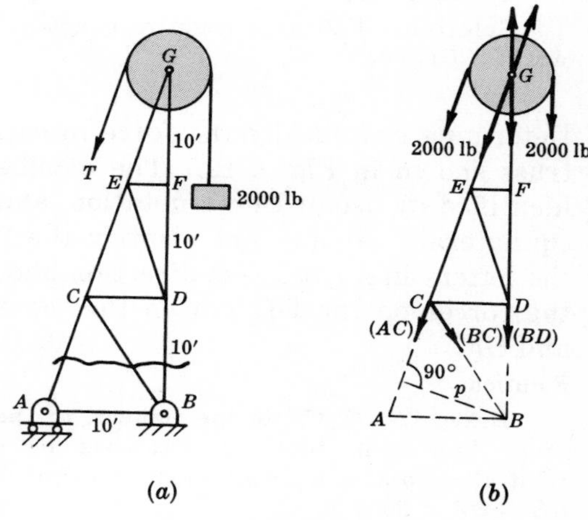

Fig. 6-8

Next sum moments about point C to obtain $\Sigma M_C = 0 = -(BD) \times 6.67 - 2000 \times 6.67$, or $(BD) = -2000$ lb. The minus sign indicates the force is compression instead of tension as assumed.

Finally sum moments about point B to obtain $\Sigma M_B = 0 = +(AC) \times p + 2000 \times p$, where p is the perpendicular distance from B to the line of action of both forces. Therefore $(AC) = -2000$ lb, where again the minus sign indicates compression instead of tension as assumed.

4. Determine the forces in members BD, CD, and CE of the Fink truss shown in Fig. 6-9 below.

Solution:

Use the method of sections to solve this problem. First determine the vertical reaction at A and the pin reaction at G by treating the entire truss as a free body as shown in Fig. 6-10 below.

Distance $FG = 12 \cos 30° = 10.4$ ft, and distance $DG = 18/(\cos 30°) = 20.8$ ft. To find force A, sum moments about G to obtain $\Sigma M_G = 0 = +2000 \times 12 + 4000 \times 10.4 + 2000 \times 20.8 - 36A$. Hence $A = 2980$ lb.

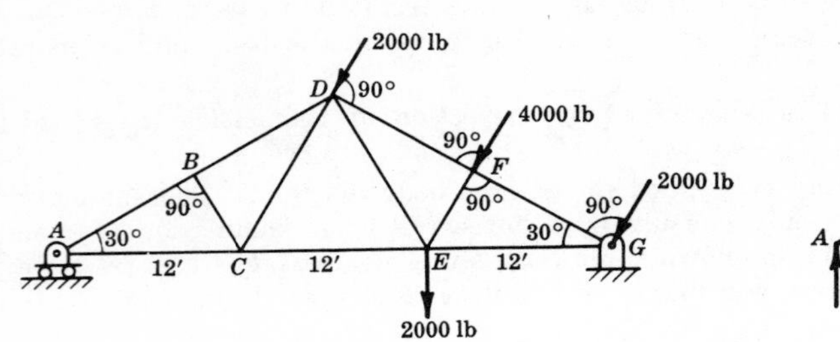

Fig. 6-9

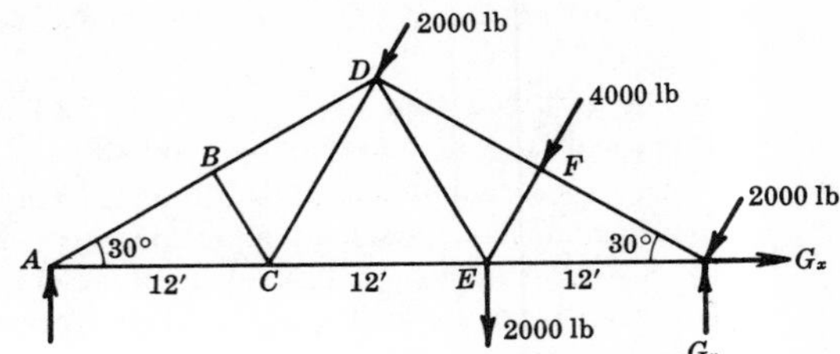

Fig. 6-10

To determine G_x, sum forces horizontally to obtain $\Sigma F_h = 0 = G_x - 8000 \sin 30°$. Hence $G_x =$ 4000 lb. A vertical summation yields $G_y = 5950$ lb.

To determine the forces in the members, select the section as shown in Fig. 6-11 below. The left portion is chosen since only one known force A is acting together with the unknowns.

Sum moments about C to determine (BD). Sum moments about D to determine (CE). To obtain (CD) sum the forces vertically. These equations in the order given are as follows:

(1) $\Sigma M_C = 0 = -(BD) \times 6 - 2980 \times 12$

(2) $\Sigma M_D = 0 = -2980 \times 18 + (CE) \times 10.4$

(3) $\Sigma F_v = 0 = +2980 + (BD) \cos 60° + (CD) \cos \theta$

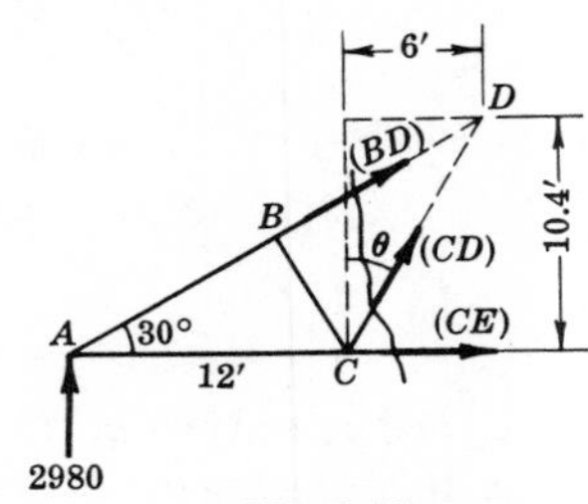

Fig. 6-11

Note that $\tan \theta = 6/10.4$; hence, $\cos \theta = 0.866$. From (1), $(BD) = -5960$ lb, i.e., compression. From (2), $(CE) = +5170$ lb, i.e., tension as assumed.

Substitute (BD) as a negative quantity into equation (3) to obtain $(CD) = 0$.

5. Determine graphically the force in each member of the truss shown in Fig. 6-12. The members and loads are identified by use of Bow's notation, and all triangles are equilateral. At any pin identify the forces by reading the letters in a clockwise direction about the pin. Thus the forces on the left pin on the top are FA, AB, BG, and GF.

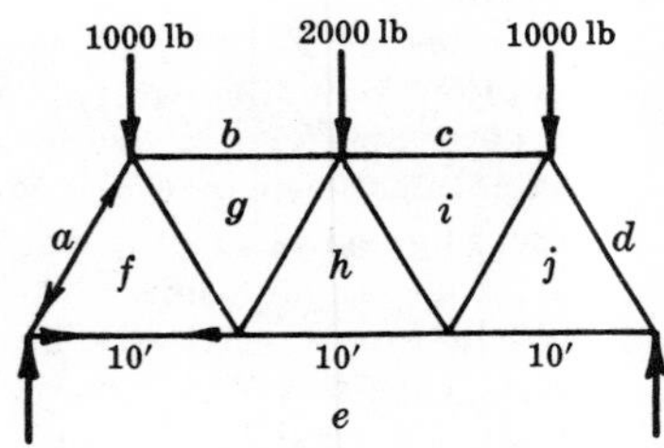

Fig. 6-12

Solution:

Draw AB, BC, CD to the right of the figure in a force diagram (Fig. 6-13). Since the reactions are equal by symmetry, point E is located midway on the vertical line of forces; hence $DE = EA = 2000$ lb.

At the lower pin the force EA is known. It is balanced by the forces AF and FE (note the clockwise reading). In the force diagram, through A must pass AF parallel to the member af in the space diagram. Draw a line through A, therefore, parallel to af. Similarly, through E must pass FE parallel to fe in the space diagram. Draw a line through E parallel to fe. The two lines just drawn intersect in the point F. Measure AF and FE to the force scale.

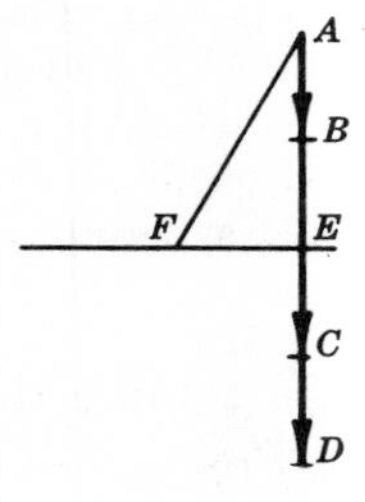

Fig. 6-13

To determine the direction of the force, note that EA is read first and is up. AF is read next and acts from A to F. Insert an arrow in the space diagram acting down and to the left. This indicates that the member pushes on the pin, or conversely that the pin pushes on the member placing it in compression. An arrow should also be inserted at this time at the other end of the member AF showing the member pushing on the pin at that end. Continuing around the triangle EAF, the force FE acts to the right. Place an arrow in the space diagram near the lower left pin showing the member pulling the pin to the right. Of course the pin pulls the member to the left, thereby placing it in tension. Show an arrow at the right end of this member acting to the left on the second pin.

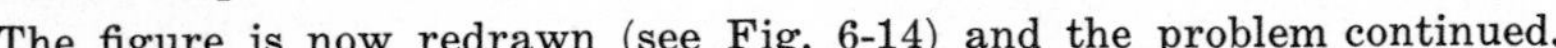
The figure is now redrawn (see Fig. 6-14) and the problem continued.

Continuing with the left pin on top, the forces known are FA and AB. The unknowns are BG and GF. In the force diagram (Fig. 6-15) the polygon needed is $FABGF$. FA and AB are already drawn. Through B draw a line parallel to bg of the space diagram and through F draw a line parallel to gf of the space diagram. These intersect in point G. As before, the order of travel around the polygon (clockwise at the pin) determines directions of the forces at the pin. $FABGF$ therefore means that BG is to the left and GF down to the right. Hence on the top left pin, BG acts to the left and GF down to the right. The arrows indicate these directions.

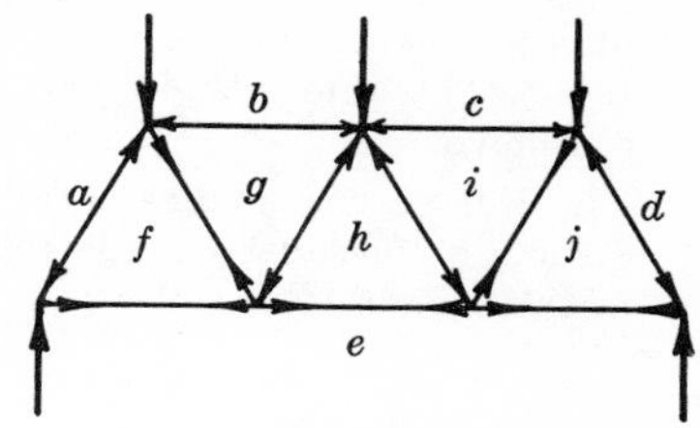

Fig. 6-14

The next pin is the bottom one, second from the left at which EF and FG are known and GH and HE unknown. Through G draw a line parallel to gh of the space diagram. Through E draw a line parallel to he of the space diagram. These intersect in point H. As before, follow around the force polygon $EFGHE$. Since GH is down to the left, insert an arrow toward the pin and an arrow at the other end of the member toward the top pin. Also, HE is toward the right.

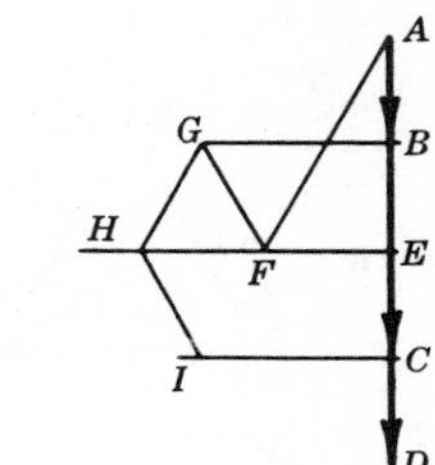

Fig. 6-15

Choosing the middle pin on the top, the force polygon is $HGBCIH$. Determine CI and IH by drawing lines through C and H parallel to ci and ih respectively. These intersect in point I. Force CI is to the left and IH up to the left, as indicated by the arrows.

Any one of the three remaining pins may be chosen, since each involves no more than two unknowns. However, by symmetry the magnitudes of the forces IJ, JE, and DJ are equal respectively to FG, EF, and FA.

The values obtained above are:

$AF = DJ = 2310$ lb C $\quad$ $GF = JI = 1160$ lb T
$FE = EJ = 1160$ lb T $\quad$ $GH = IH = 1160$ lb C
$BG = IC = 1740$ lb C $\quad$ $HE = 2310$ lb T.

6. Determine the forces in the members of the truss shown in Fig. 6-16 below. All triangles are equilateral.

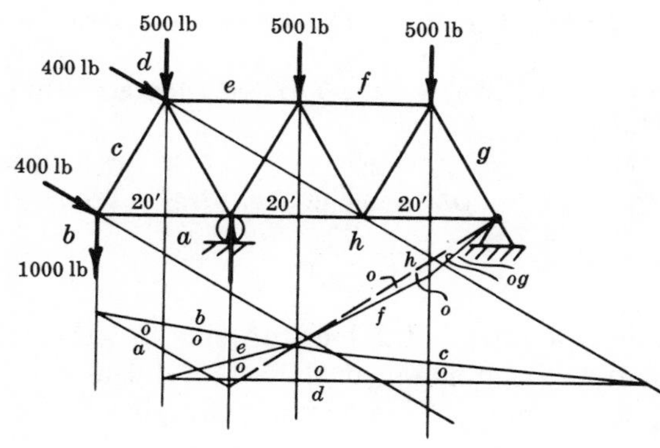

Fig. 6-16

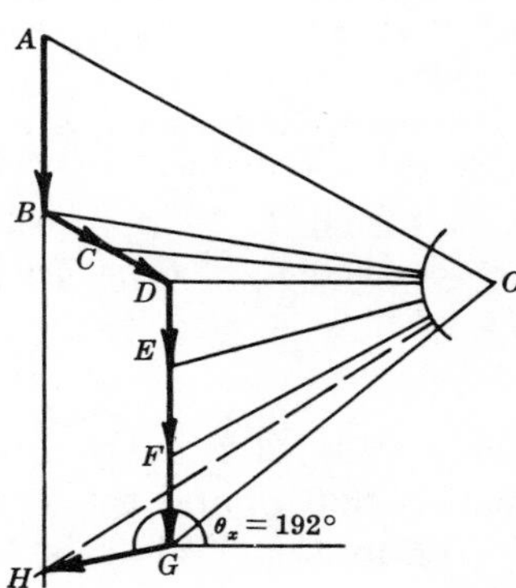

Fig. 6-17

Solution:

As shown, first determine the reactions HA and GH by the graphical analysis previously described in Problem 16 of Chapter 5. To review this technique in a few words, label the fields surrounding the forces with lower case letters. Draw the known forces AB, BC, CD, DE, EF and FG to the right. To complete the force polygon (Fig. 6-17 above), reactions GH and HA are needed. Since HA is a roller reaction, its line of action is vertical. This means in the force polygon that H is on a vertical line through A – in this case AB extended. Select pole O and draw rays to all known points (H is still unknown). Start the string or funicular polygon at the pin, the only known characteristic of reaction GH. Through this point draw go parallel to GO, keeping in mind that through this point must also pass ho. Continue the string polygon until the string ao cuts the action line ha. Through this point of intersection and the starting pin, draw the broken line ho. In the force diagram draw OH parallel to ho. This line cuts AB extended in H. Measure $HA = 3050$ lb and $GH = 710$ lb at an angle $\theta_x = 192°$ as shown.

For clarity in presentation, only the truss is shown again in Fig. 6-18 below without the construction lines used in determining reactions. Construct a Maxwell Diagram to determine forces in the individual members.

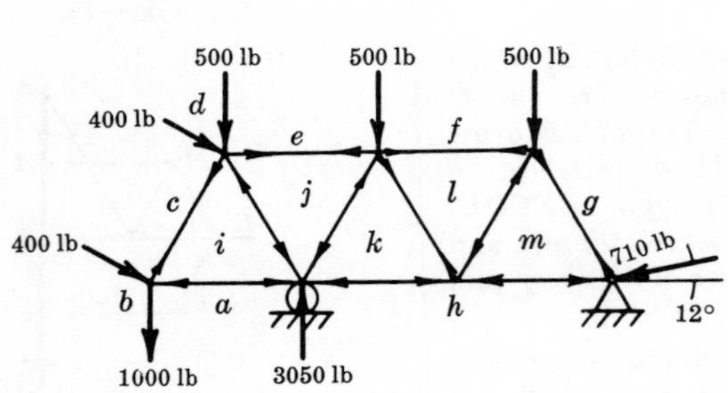

Fig. 6-18

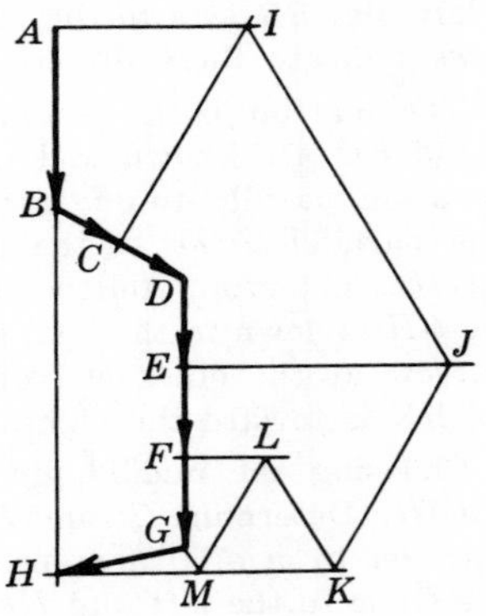

Fig. 6-19

Starting with the lower pin on the left, the force polygon is $ABCIA$ with I to be determined (see Fig. 6-19 above). Through A draw a line parallel to ia. Through C draw a line parallel to ci. The lines intersect in I. The arrows are shown at the pin as obtained in going around the polygon in the order above, $ABCIA$.

Use the left pin on the top as the next free body diagram and draw lines through E and I parallel respectively to ej and ji. These determine J. Again the direction is clockwise around the pin $ICDEJI$. In traversing the force polygon in this direction, EJ is to the right and JI up to the left. The arrows at the top left pin are drawn to correspond. Of course, EJ is to the left when considering the top middle pin, and IJ is down to the right when considering the pin directly above the left support.

Next consider the pin above the left support. The polygon is $HAIJKH$. Draw lines through J and H parallel to jk and kh respectively and intersecting in K. JK is down to the left and KH to the left.

Consider the remaining pins in similar fashion and complete the Maxwell Diagram. By measuring to the force scale the following values are obtained:

$CI = 1390$ lb T, $IA = 1050$ lb C, $EJ = 1450$ lb T, $JI = 2200$ lb C, $JK = 1300$ lb C, $KH = 1500$ lb C, $FL = 440$ lb T, $LK = 730$ lb T, $GM = 150$ lb T, $HM = 770$ lb C, and $ML = 740$ lb C.

7. Each cable of a suspension bridge carries a horizontal load of 800 lb per ft. If the span is 600 ft and the sag 40 ft, determine the tension at the ends of the cable and at the midpoint. What is the length of the cable? This is an example of a parabolic cable.

Solution:

The tension at either end is the same.

$$T = \tfrac{1}{2}wa\sqrt{1 + \frac{a^2}{16d^2}} = \tfrac{1}{2} \times 800 \times 600\sqrt{1 + \frac{(600)^2}{16(40)^2}} = 932{,}000 \text{ lb}$$

$$H = \frac{wa^2}{8d} = \frac{800 \times (600)^2}{8 \times 40} = 900{,}000 \text{ lb}$$

$$l = a\left[1 + \frac{8}{3}\left(\frac{d}{a}\right)^2 - \frac{32}{5}\left(\frac{d}{a}\right)^4 + \frac{256}{7}\left(\frac{d}{a}\right)^6 - \cdots\right]$$

Usually, two or three terms of a series which converges rapidly are sufficient for the accuracy desired. It is well to check as shown below.

$$\frac{d}{a} = \frac{40}{600} = 0.0667 \quad \text{and} \quad \left(\frac{d}{a}\right)^2 = 0.0045$$

Using two terms, $l = 600[1 + \frac{8}{3}(0.0045)] = 607$ ft.

Using three terms, $l = 600[1 + \frac{8}{3}(0.0045) - \frac{32}{5}(0.0667)^4] = 607$ ft.

If four terms are used the value will increase slightly, but the magnitude of each term beyond the second is negligible.

8. Solve for the desired tensions in Problem 7 without recourse to formulas.

Solution:

A free body diagram of the right half of the cable contains the horizontal tension H at the low point, the maximum T_{max} at the support, and the load $800(300) = 240{,}000$ lb (see Fig. 6-20). Since only three forces act on the cable, they must be concurrent as shown. Summing the forces horizontally and then vertically,

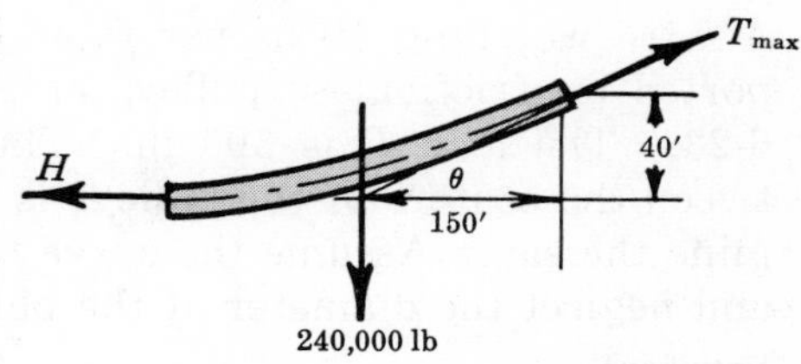

Fig. 6-20

$$\Sigma F_h = 0 = T_{max} \cos\theta - H$$
$$\Sigma F_v = 0 = T_{max} \sin\theta - 240{,}000$$

from which (1) $T_{max} \sin\theta = 240{,}000$ and (2) $T_{max}\cos\theta = H$. Dividing (1) by (2), $\tan\theta = 240{,}000/H$. But $\tan\theta = 40/150$; hence $40/150 = 240{,}000/H$ and $H = 900{,}000$ lb.

Squaring equations (1) and (2) and adding,

$$T^2_{max}(\sin^2\theta + \cos^2\theta) = (240{,}000)^2 + H^2, \quad T^2_{max} = (240{,}000)^2 + (900{,}000)^2, \quad T_{max} = 932{,}000 \text{ lb}$$

9. A cable weighing 2 lb per ft is stretched between two supports not at the same level as shown in Fig. 6-21. Determine the maximum tension.

Solution:

In Fig. 6-22, a free body diagram is shown of a piece of the cable to the right of the lowest point P whose location is unknown. Let H be the unknown tension in the cable at point P, and let T be the tension at a point a distance x to the right of P. The weight of the cable for this distance x is approximately $wx = 2x$. Summing forces horizontally and vertically,

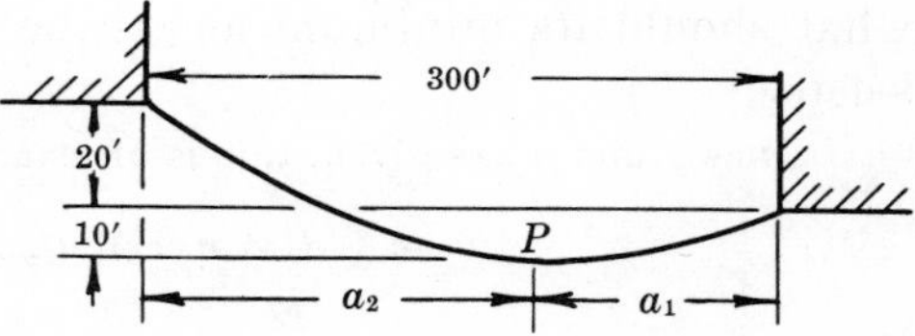

Fig. 6-21

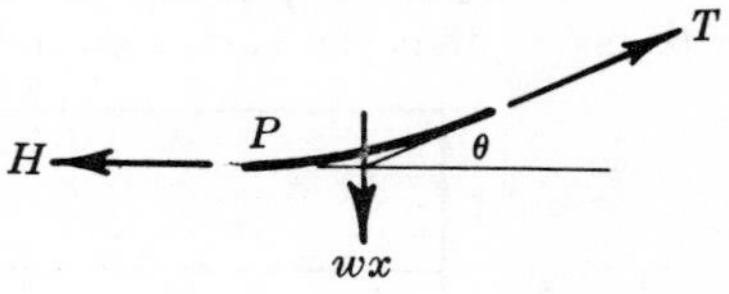

Fig. 6-22

$$\Sigma F_x = 0 = T\cos\theta - H \quad \text{or} \quad (1)\ T\cos\theta = H$$
$$\Sigma F_y = 0 = T\sin\theta - 2x \quad \text{or} \quad (2)\ T\sin\theta = 2x$$

Dividing (2) by (1), $\tan\theta = 2x/H$. But from the free body diagram, $\tan\theta = y/(x/2) = 2y/x$; hence, $2x/H = 2y/x$ or $Hy = x^2$. Since $x = a_1$ when $y = 10$ ft, $10H = a_1^2$. Similarly for the section to the left of P, $30H = a_2^2$. Now

$$a_1 + a_2 = 300, \quad \sqrt{10H} + \sqrt{30H} = 300, \quad H = 1210 \text{ lb}$$

and $a_1 = \sqrt{10H} = 110$ ft, $\quad a_2 = \sqrt{30H} = 190$ ft.

Squaring equations (1) and (2) and adding, $T^2 = 4x^2 + H^2$. Since the maximum tension occurs at the left support where $x = -190$ ft, $T^2_{max} = 4(-190)^2 + (1210)^2$ or $T_{max} = 1270$ lb.

10. A horizontal load of 120 lb per ft is to be carried by a cable suspended between supports on the same level and 60 ft apart. The maximum permissible tension is 36,000 lb. Determine the required length l of the cable and its sag d.

Solution:

$$T = \tfrac{1}{2}wa\sqrt{1+\frac{a^2}{16d^2}}, \qquad 36{,}000 = \tfrac{1}{2}\times 120\times 60\sqrt{1+\frac{(60)^2}{16d^2}}, \qquad d = 1.51 \text{ ft}$$

Using $l = a\left[1+\frac{8}{3}\left(\frac{d}{a}\right)^2 - \frac{32}{5}\left(\frac{d}{a}\right)^4 + \frac{256}{7}\left(\frac{d}{a}\right)^6 - \cdots\right]$, where $\frac{d}{a} = \frac{1.51}{60} = 0.0251$, it is evident that terms with powers greater than 2 will not affect the result. The length is 60.1 ft.

11. A wire weighing 10 oz per ft of length is supported on frictionless pulleys as shown in Fig. 6-23. The load P is 500 lb. The distance between the centers of the pulleys is 80 ft. Determine the sag. Assume the curve to be parabolic and neglect the diameter of the pulleys.

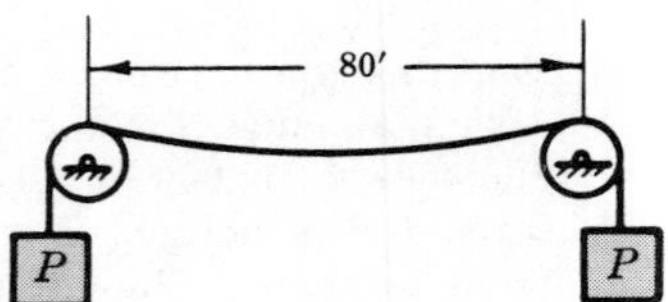

Fig. 6-23

Solution:

The tension at the pulleys is the weight P, 500 lb. Then

$$500 = \frac{1}{2}\times\frac{10}{16}\times 80\sqrt{1+\frac{(80)^2}{16d^2}} \qquad \text{or} \qquad d = 1 \text{ ft}$$

12. A wire weighing 0.518 lb per ft is suspended between two towers at the same level and 500 ft apart. If the sag is 50 ft, what is the maximum tension in the wire and what should its minimum length be? This is an example of a catenary.

Solution:

Since a and d are given, this is an example of case (a) under the catenary. Equation (*2*) yields

$$c + d = c\cosh a/2c \qquad \text{or} \qquad c + 50 = c\cosh 500/2c$$

A graphical solution is perhaps easier than substitution of values. Plot the line $c+50$ against c, and the curve $c\cosh 500/2c$ against c. The value of c at which these two curves intersect is the solution. The values for plotting are shown in the table below. The extra points at 150 and 250 were chosen to insure that the curves do not meet at a lower value of c.

c	$c+50$	$\frac{500}{2c}$	$\cosh\frac{500}{2c}$	$c\cosh\frac{500}{2c}$
0	50	∞	∞	——
100	150	2.5	6.1323	613.2
200	250	1.25	1.8884	377.7
300	350	.833	1.3678	410.3
400	450	.625	1.2018	480.7
500	550	.500	1.1276	563.8
600	650	.417	1.0882	652.9
700	750	.357	1.0644	745.1
800	850	.313	1.0494	839.5
150	200	1.667	2.7427	411.4
250	300	1.000	1.5431	385.8

The curves are plotted against c and intersect at approximately $c = 650$. See Fig. 6-24.

For greater accuracy an enlarged plot could be made of that portion of the graph where the curves cross. However, try $c = 650$ in the equation to determine a more accurate balance:

$$c + 50 = 650 + 50 = 700$$

But $650 \cosh \dfrac{500}{2 \times 650} = 698.8.$

Now try $c = 640$: $640 \cosh \dfrac{500}{2 \times 640} =$ 689.5, instead of $640 + 50$ or 690.

A check on $c = 635$ indicates this to be sufficiently accurate.

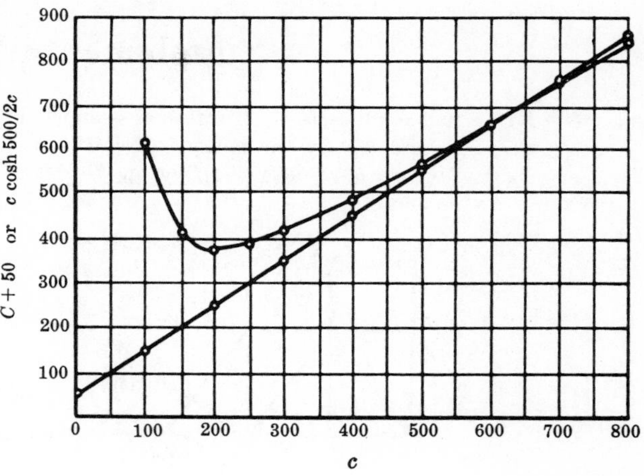

Fig. 6-24

Using equation (*3*): $T_{max} = w(c + d) = 0.518(635 + 50) = 355$ lb

Using equation (*5*): $(c + d)^2 = c^2 + \frac{1}{4}l^2$, $(635 + 50)^2 = (635)^2 + \frac{1}{4}l^2$, $l = 514$ ft

13. A cable weighing 0.4 lb per ft and 800 ft long is to be suspended with a sag of 80 ft. Determine the maximum tension and maximum span.

Solution:

This is an example of case (*c*) under the catenary.

Using equation (*5*),

$$(c + d)^2 = c^2 + \tfrac{1}{4}l^2, \quad (c + 80)^2 = c^2 + \tfrac{1}{4}(800)^2, \quad c = 960$$

Using equation (*3*):

$$T_{max} = w(c + d) = 0.4(960 + 80) = 416 \text{ lb}$$

Using equation (*4*):

$$\tfrac{1}{2}l = c \sinh a/2c, \quad \tfrac{1}{2}(800) = 960 \sinh a/1920, \quad a = 778 \text{ ft}$$

Supplementary Problems

14. The Howe truss supports the three loads shown in Fig. 6-25 below. Determine the forces in members *AB, BD, CD,* and *EF* by the method of joints.
Ans. AB = 25,000 lb C, BD = 15,000 lb C, CD = 12,500 lb C, EF = 22,500 lb T

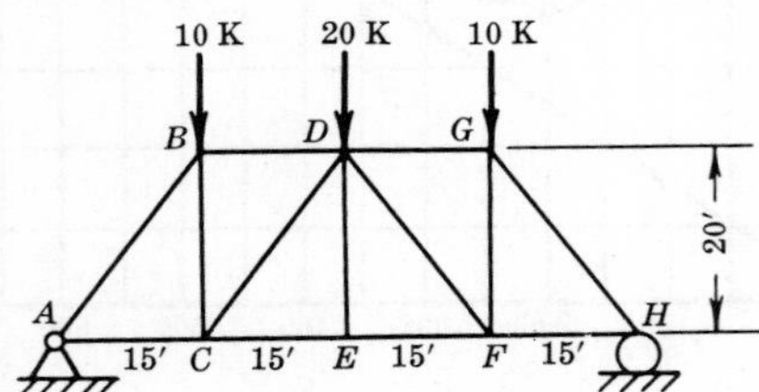

Fig. 6-25

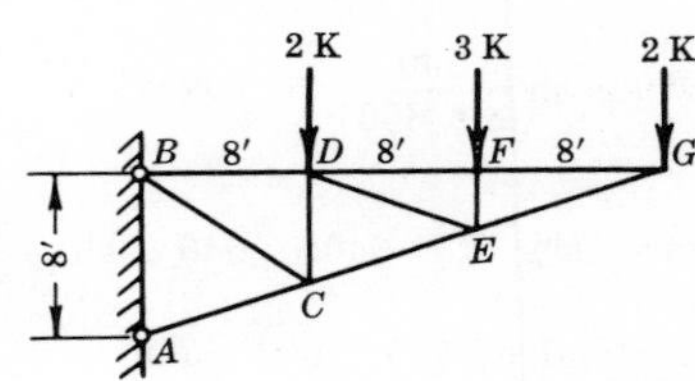

Fig. 6-26

15. Determine the forces in all members of the cantilever truss shown diagrammatically in Fig. 6-26 above. Loads are in kips. Solution may be started with pin *G*.
Ans. BD = 10,500 lb T, BC = 4200 lb T, AC = 14,800 lb C, DF = 6000 lb T, DE = 4730 lb T, CE = 11,100 lb C, FG = 6000 lb T, EG = 6330 lb C, CD = 3490 lb C, EF = 3000 lb C

16. In the truss shown in Fig. 6-27 below, determine the forces in members *AC* and *BD* by the method of joints. *Ans.* AC = 35,400 lb C, BD = 48,700 lb T

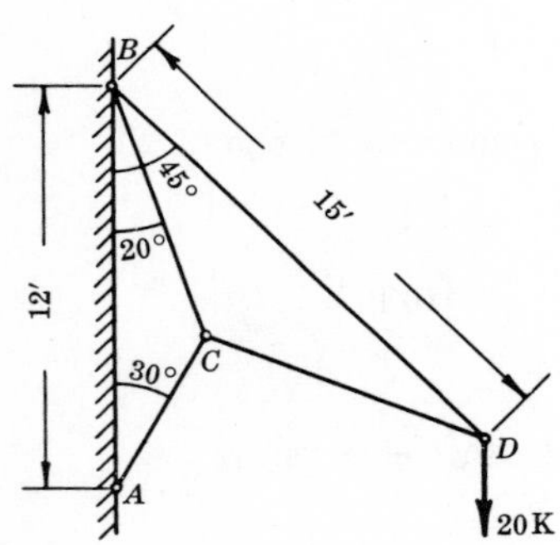

Fig. 6-27

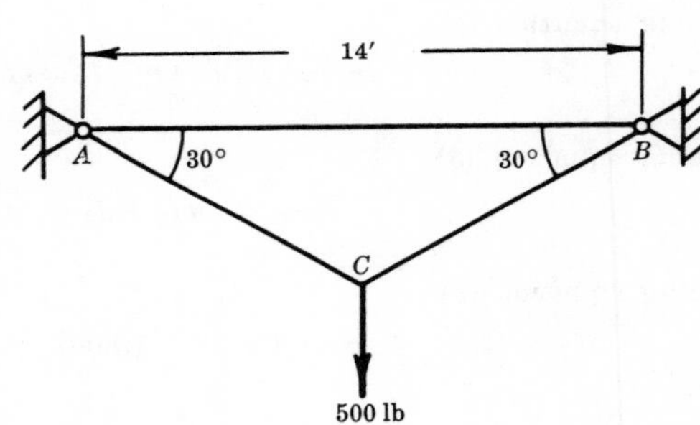

Fig. 6-28

17. See Fig. 6-28 above. The truss is supported at *A* and *B* and supports 500 lb at its midpoint. Determine the forces in *AC* and *BC*. *Ans.* AC = 500 lb T, BC = 500 lb T

18. The Fink truss is subjected to the loading shown in the adjacent Fig. 6-29. Determine the forces in all members by the method of joints.
Ans. AB = 10.04 K T, AC = 9.87 K C, CD = 8.87 K C, BD = 3.72 K T, BE = 6.33 K T, DF = 8.30 K C, FG = 9.30 K C, EG = 8.06 K T

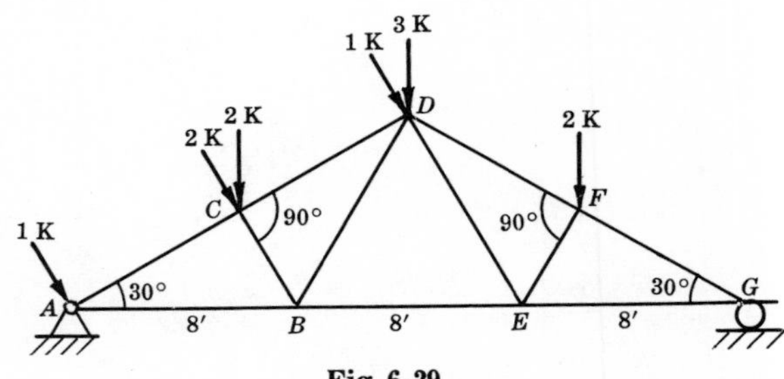

Fig. 6-29

19. Determine forces in *AB, AC, EG,* and *FG* in Fig. 6-30 below.
Ans. AB = 1730 lb T, AC = 866 lb C, EG = 2890 lb C, FG = 5770 lb T

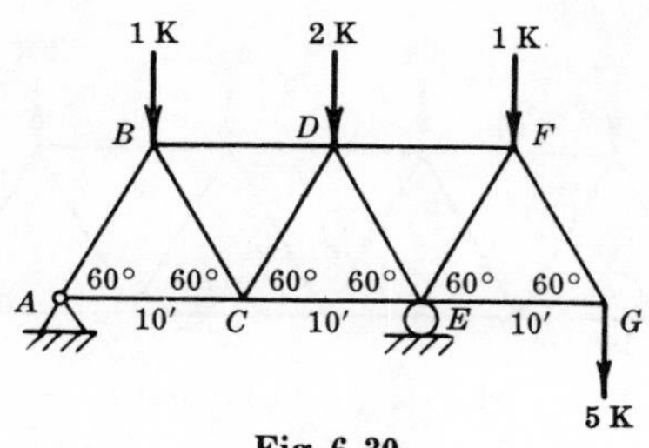

Fig. 6-30

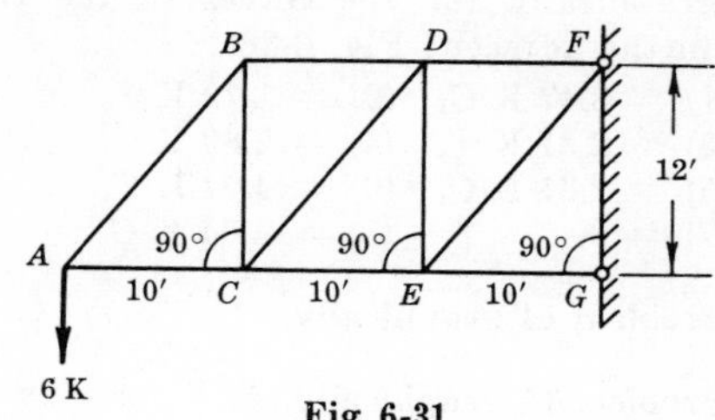

Fig. 6-31

20. Determine the forces in AB and CD in the cantilever truss using the method of joints. See Fig. 6-31 above. *Ans.* $AB = 7.83$ K T, $CD = 7.83$ K T

21. Determine the forces in AC and AB in the Howe truss using the method of joints. See Fig. 6-32 below. *Ans.* $AC = 4.3$ K C, $AB = 2.8$ K T

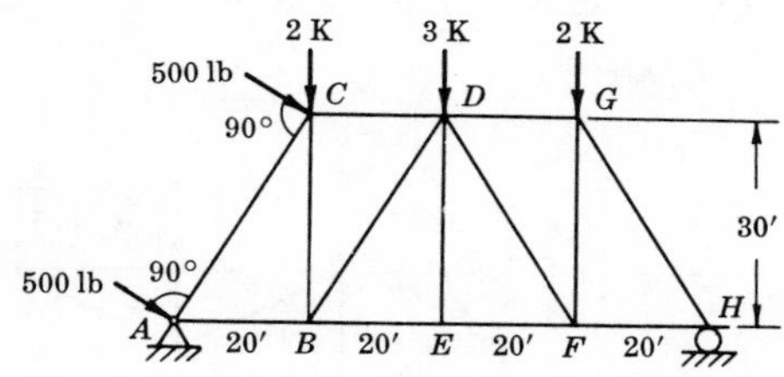

Fig. 6-32

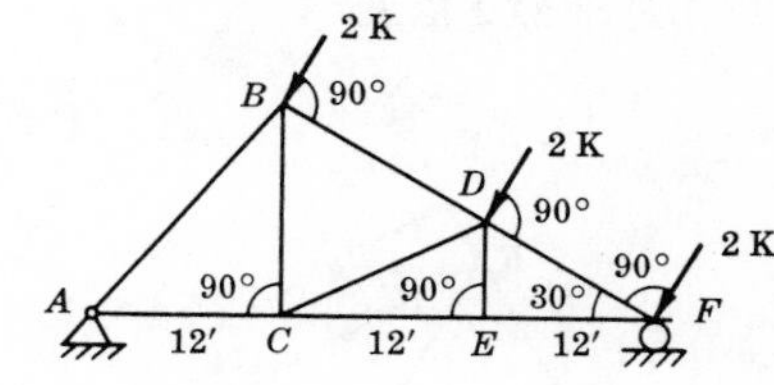

Fig. 6-33

22. Find the forces in members AB and CD in Fig. 6-33 above. *Ans.* $AB = 3060$ lb C, $CD = 2310$ lb C

Solve Problems 23-26 by the method of sections for the forces in the members checked in the individual figures.

23. Fig. 6-34 below. *Ans.* $DF = 3.46$ K C, $DE = 0$, $CE = 3.46$ K T

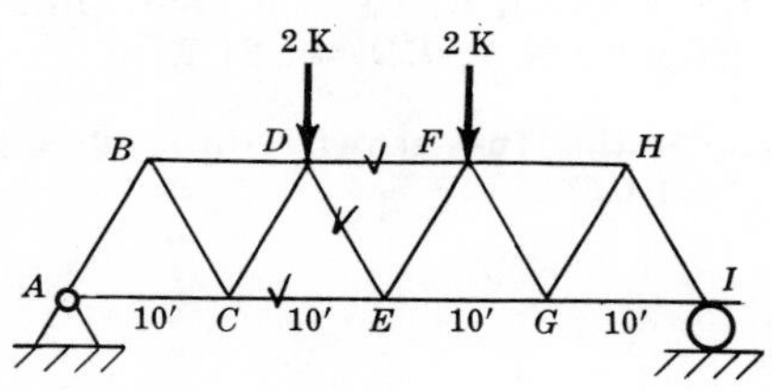

Fig. 6-34

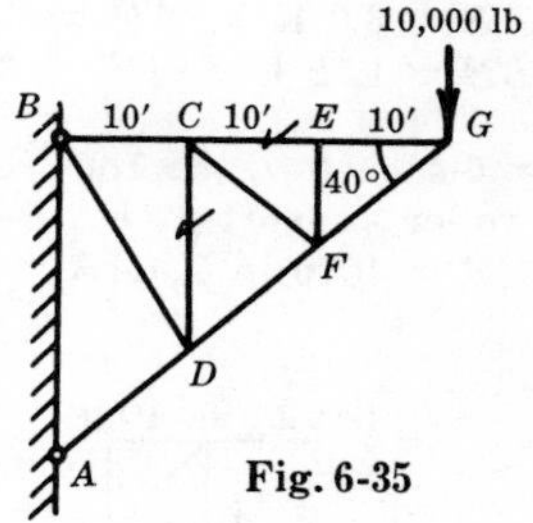

Fig. 6-35

24. Fig. 6-35 above. *Ans.* $CE = 11.9$ K T, $CD = 0$

25. Fig. 6-36 below. *Ans.* $DF = 830$ lb C, $DE = 320$ lb T

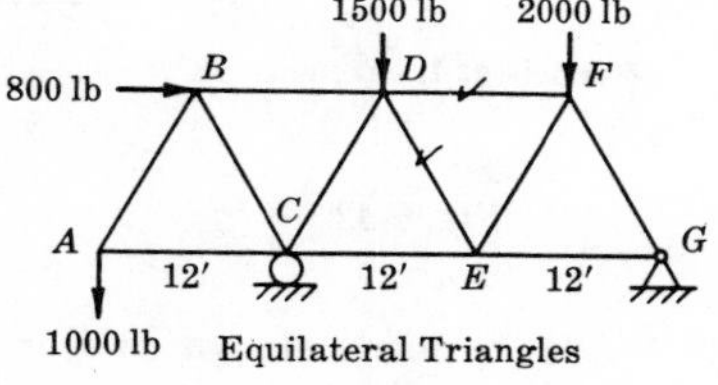

Fig. 6-36

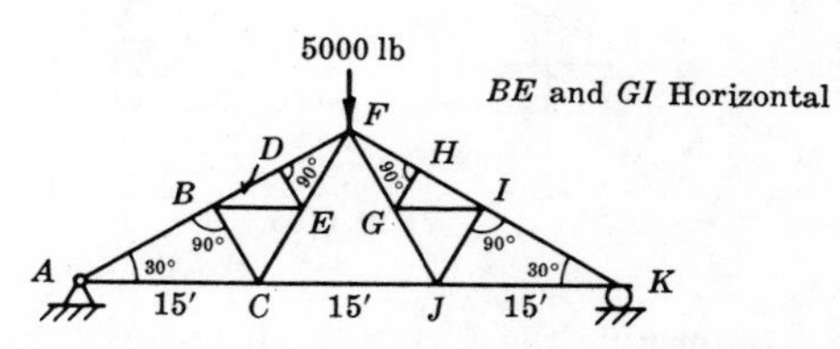

Fig. 6-37

26. Fig. 6-37 above. *Ans.* $BD = 5000$ lb C, $EF = 0$

27. Solve graphically for the forces in the truss shown in the adjacent Fig. 6-38.
Ans. (1) = 3.47 K *C*, (2) = 1.73 K *T*, (3) = 2.31 K *T*, (4) = 2.89 K *C*, (5) = 2.31 K *C*, (6) = 4.04 K *T*, (7) = 0, (8) = 4.04 K *C*

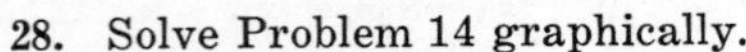

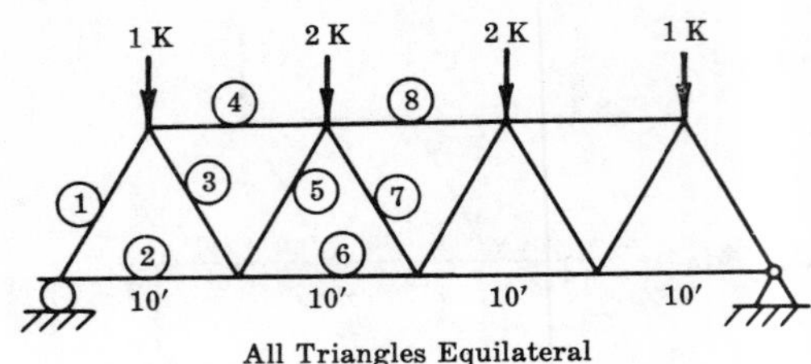

Fig. 6-38

28. Solve Problem 14 graphically.

29. Solve Problem 15 graphically.

30. Solve Problem 17 graphically.

31. Solve Problem 18 graphically.

32. Solve Problem 25 graphically.

33. Determine the forces in all members in Fig. 6-39 below.
Ans. AB = 4.00 K *C*, AD = 2.00 K *C*, BD = 4.00 K *T*, BC = 3.46 K *C*, CE = 3.46 K *C*, CD = 4.00 K *C*, DE = 2.00 K *T*

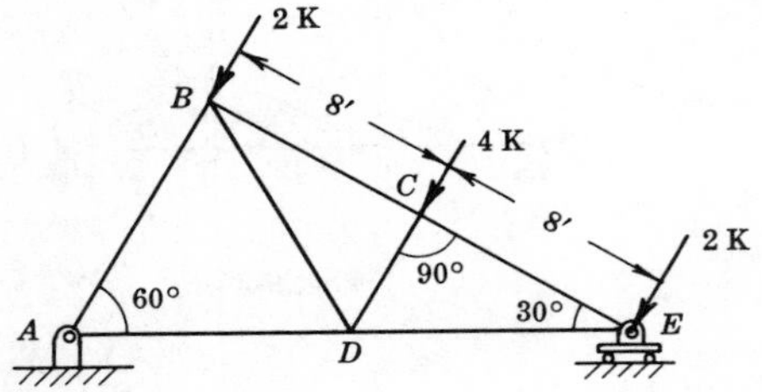

Fig. 6-39

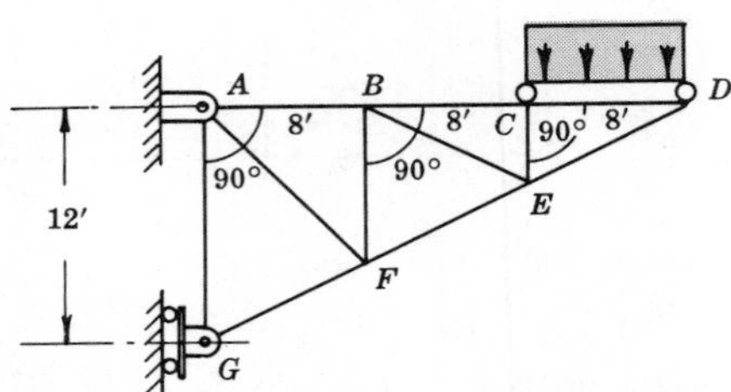

Fig. 6-40

34. The car is located on the span CD as shown in Fig. 6-40 above. It impresses a load of 4000 lb which is equally distributed on all four wheels. Since there are two identical trusses to support the rails, the load is 1000 lb at C and 1000 lb at D. What are the forces in the members of the truss?
Ans. AB = 3.0 K *T*, AF = 0.47 K *T*, AG = 1.67 K *T*, GF = 3.72 K *C*, BF = 0.5 K *C*, BC = 2.0 K *T*, BE = 1.12 K *T*, FE = 3.35 K *C*, CE = 1.0 K *C*, CD = 2.0 K *T*, ED = 2.24 K *C*

35. In Fig. 6-41 below, the 1800 lb load is applied horizontally to the truss shown with a pin support at A and a roller support at E. Determine the forces in AB and CE.
Ans. CE = 4070 lb *T*, AB = 1300 lb *C*

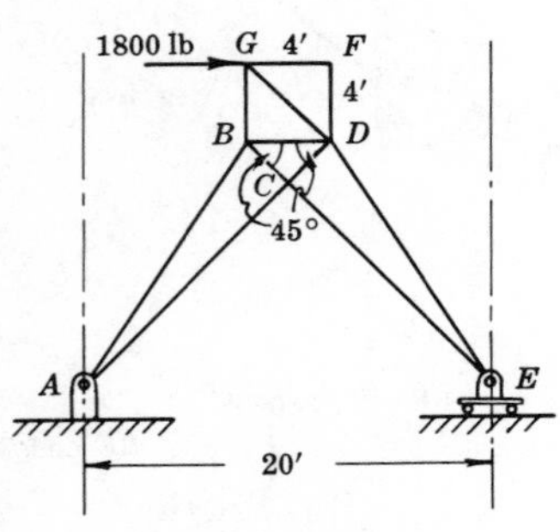

Fig. 6-41

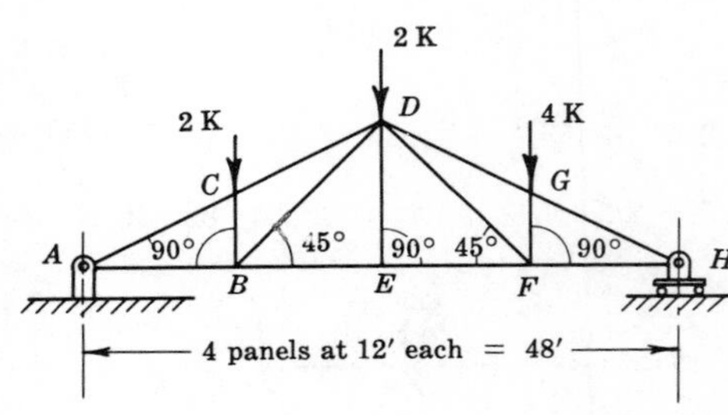

Fig. 6-42

36. Determine the forces in all members of the truss subjected to the three loads shown in Fig. 6-42.
Ans. AC = 7.82 K *C*, AB = 7.00 K *T*, BC = 2.00 K *C*, CD = 7.82 K *C*, BD = 2.83 K *T*, DE = 0, DG = 10.1 K *C*, DF = 5.67 K *T*, BE = 5.00 K *T*, EF = 5.00 K *T*, FG = 3.98 K *C*, FH = 9.00 K *T*, GH = 10.1 K *C*

37. For the truss shown (Fig. 6-43 below) pin supported at A and roller supported at E, determine the forces in CE and DE when a horizontal load of 3000 lb is applied at D.
Ans. CE = 943 lb T, DE = 2150 lb C

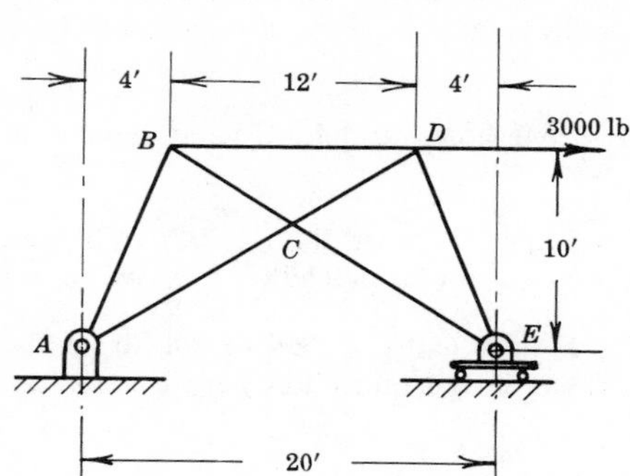

Fig. 6-43

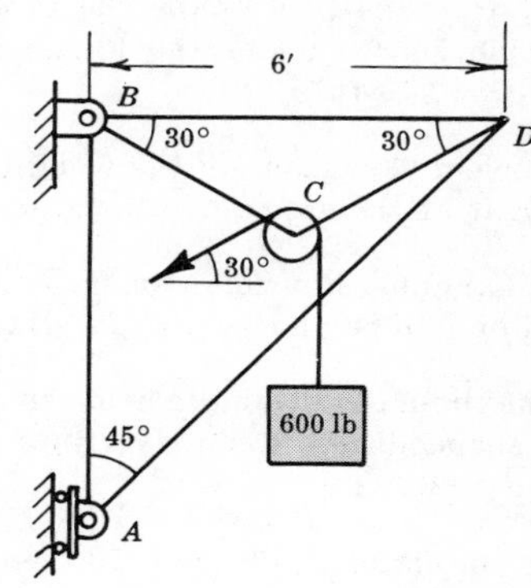

Fig. 6-44

38. For the truss shown in Fig. 6-44 above determine the forces in members BC, AD, and CD. Assume the pulley is weightless and frictionless. Its diameter is 2 ft. The load of 600 lb is held by a rope inclined 30° with the horizontal. *Ans.* BC = 600 lb T, CD = 1200 lb T, AD = 850 lb C

39. The truss which is pin supported at A and roller supported at D is inclined 40° with the vertical as shown in Fig. 6-45 below. The members AC and CD are cables which are designed for a maximum load of 2000 lb (in tension obviously). What is the maximum value which W may have?
Ans. 1770 lb

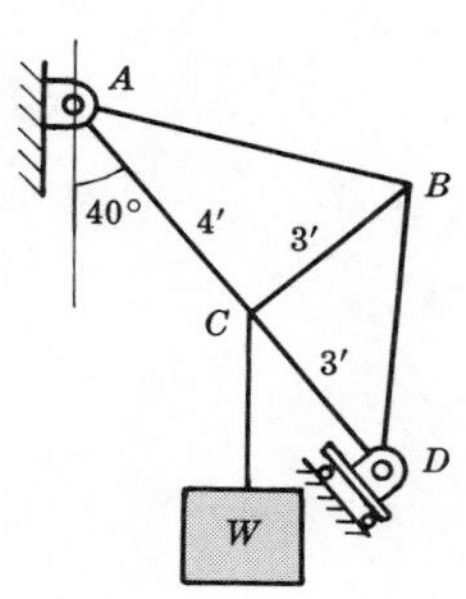

Fig. 6-45

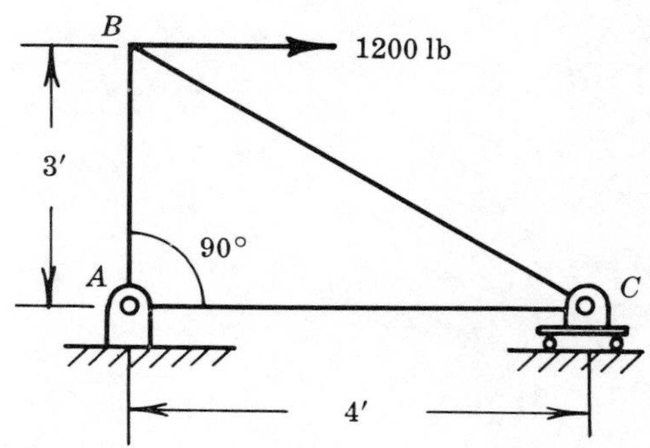

Fig. 6-46

40. In Fig. 6-46 above, the truss is pin supported at A and roller supported at C. Determine the forces in each member caused by the horizontal 1200 lb load.
Ans. AC = 1200 lb T, AB = 900 lb T, BC = 1500 lb C

41. The truss is pin supported at A and roller supported at D as shown in Fig. 6-47. Determine the force in each member caused by the 88.5 lb load.
Ans. $AB = BD$ = 73.8 lb C,
$AC = CD$ = 59.0 lb T,
BC = 88.5 lb T

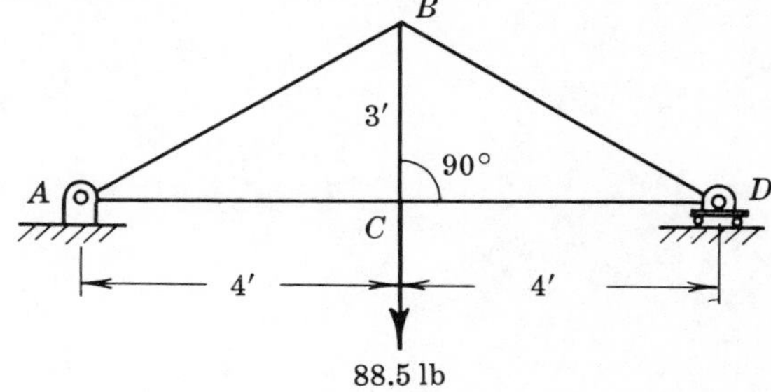

Fig. 6-47

42. A cable is suspended between two supports at the same elevation and 600 ft apart. The load is 400 lb per horizontal ft. The sag is 40 ft. Find the length of the cable and the tension at the support.
Ans. $l = 607$ ft, $T = 466{,}000$ lb

43. Calculate the sag in a tape 100 ft long held between two supports 99.8 ft apart and at the same level. Use equation (*3*) under the Parabolic Cable section, neglecting powers of d/a beyond the second.
Ans. $d = 2.73$ ft

44. In Problem 43, assuming the weight of the tape is 0.01 lb per horizontal foot, what tension is necessary to hold it at the support? *Ans.* $T = 4.59$ lb

45. The maximum allowable tension in a cable is 1000 lb. A sag of no more than 4 ft is to be allowed under a load of 1 lb per horizontal ft. How far apart may the supports be placed? *Ans.* $a = 178$ ft

46. The maximum allowable tension in a cable is 90,000 lb. It is to carry a load of 200 lb per horizontal ft when suspended between level supports 300 ft apart. What is the minimum permissible sag?
Ans. $d = 26.6$ ft

47. Using the data of Problem 43, compute the maximum tension in the tape and its sag assuming the weight is 0.01 lb per lineal ft along the tape. *Ans.* $T = 4.4$ lb, $d = 2.9$ ft

48. A wire weighing 0.5 lb per ft is stretched between two level supports 160 ft apart. If the sag is 40 ft, determine the length of the wire and the maximum tension. *Ans.* $l = 184$ ft, $T = 63$ lb

49. A transmission wire 753 ft long weighing 0.653 lb per lineal ft is suspended between towers of equal elevation 750 ft apart. Find the maximum tension and the sag. *Ans.* $T = 1580$ lb, $d = 30$ ft

50. A rope 50 ft long weighs 0.1 lb per lineal ft. How far apart should two supports be placed so that a maximum tension of 10 lb is induced? Use equations (*3*) and (*5*) in the section on the catenary.
Ans. $a = 49.4$ ft

Chapter 7

Equilibrium of Spatial Force Systems

EQUILIBRIUM of a SPATIAL FORCE SYSTEM

Equilibrium of a spatial force system occurs if the resultant is neither a force **R** nor a couple **C**. The necessary and sufficient conditions that **R** and **C** be zero vectors are

$$\mathbf{R} = \Sigma\mathbf{F} = 0 \quad \text{and} \quad \mathbf{C} = \Sigma\mathbf{M} = 0$$

where $\Sigma\mathbf{F}$ = vector sum of all the forces of the system,

$\Sigma\mathbf{M}$ = vector sum of the moments (relative to any point) of all the forces of the system.

The two vector equations above may be applied directly or, in the simpler problems, the following derived scalar equations may be used for the three types of spatial force systems.

CONCURRENT SYSTEM

The following set of equations insures equilibrium of a concurrent, spatial system of forces.

(1) $\Sigma F_x = 0$

(2) $\Sigma F_y = 0$

(3) $\Sigma F_z = 0$

where ΣF_x, ΣF_y and ΣF_z = algebraic sums of the $x, y,$ and z components respectively of the forces of the system.

$\Sigma M = 0$ may be used as an alternate to one of the above equations. For example, if it replaces equation *(3)* then ΣM must be the algebraic sum of the moments of the forces of the system about an axis neither parallel to nor intersecting the z-axis.

PARALLEL SYSTEM

The following set of equations insures equilibrium of a parallel, spatial system of forces.

(1) $\Sigma F_y = 0$

(2) $\Sigma M_x = 0$

(3) $\Sigma M_z = 0$

where ΣF_y = algebraic sum of the forces of the system along the y-axis which is chosen parallel to the system,

ΣM_x and ΣM_z = algebraic sums of the moments of the forces of the system about the x and z axes respectively.

NON-CONCURRENT, NON-PARALLEL SYSTEM

The following six equations are necessary and sufficient conditions for equilibrium of the most general force system in three dimensional space.

(1) $\Sigma F_x = 0$

(2) $\Sigma F_y = 0$

(3) $\Sigma F_z = 0$

(4) $\Sigma M_x = 0$

(5) $\Sigma M_y = 0$

(6) $\Sigma M_z = 0$

where ΣF_x, ΣF_y, and ΣF_z = algebraic sums of the $x, y,$ and z components respectively of the forces of the system,

ΣM_x ΣM_y, and ΣM_z = algebraic sum of the moments of the forces of the system about x, y, and z axes respectively.

All systems previously studied are special cases of this system. Not all of the six equations are necessary in these special cases.

Solved Problems

Note that in the diagrams, forces are identified by their magnitudes if the directions are readily apparent.

1. In Fig. 7-1 below, a pole 30 ft high is shown supporting a wire in the xy-plane, exerting a force of 150 lb on the top at an angle of 10° below the horizontal. Two guy wires are affixed as shown. Determine the tension in each guy wire and the compression in the pole.

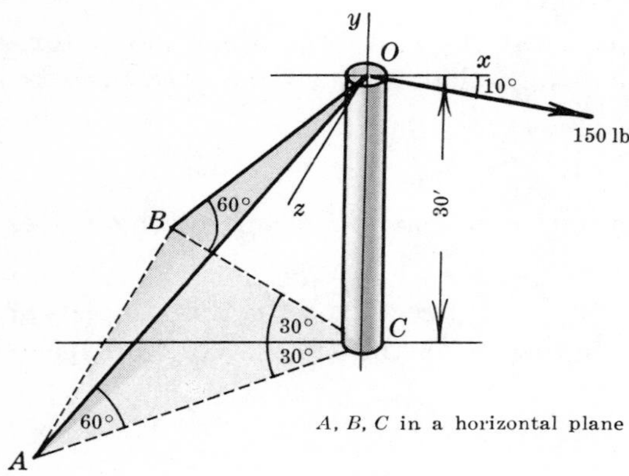

Fig. 7-1

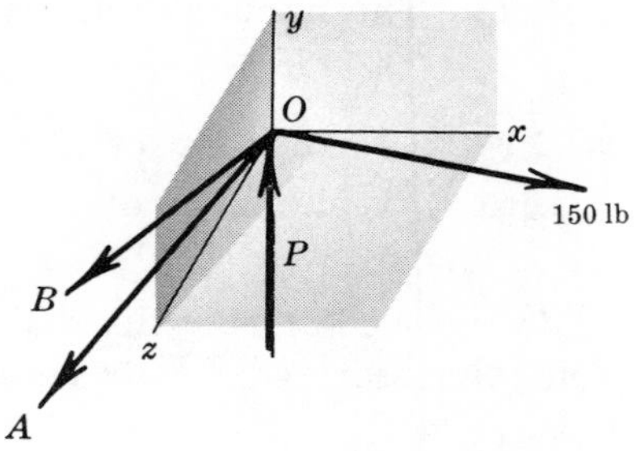

Fig. 7-2

Solution:

Since the pole is subjected only to end loads, it is a two force member carrying an axial compressive load P. For the concurrent system shown in the free body diagram (see Fig. 7-2 above),

$$\Sigma F_z = 0 = +A\cos 60° \sin 30° - B\cos 60°\sin 30° \qquad \text{or} \qquad A = B$$

and

$$\Sigma F_x = 0 = +150\cos 10° - B\cos 60°\cos 30° - A\cos 60° \cos 30°$$

Substituting A for B and solving, we obtain $A = 171$ lb T.

To determine P, sum the forces vertically along the y-axis.

$$\Sigma F_y = P - 150 \sin 10° - 2A\sin 60° = 0 \qquad \text{or} \qquad P = 322 \text{ lb C}$$

2. Rework Problem 1 using $\Sigma \mathbf{M}_C = 0$.

Solution:

The pole is in equilibrium under the action of the following four forces as shown in Fig. 7-3.

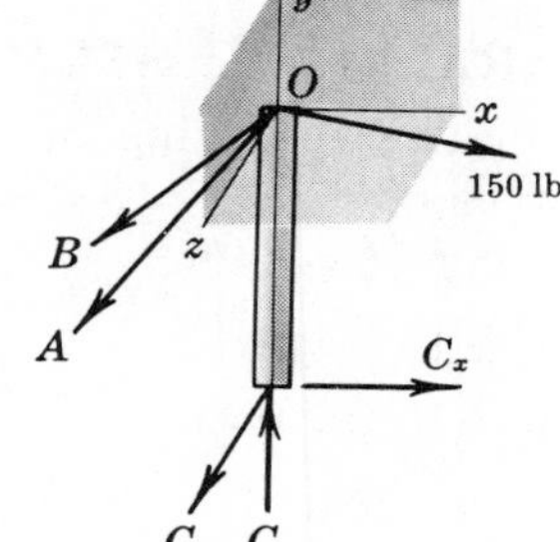

Fig. 7-3

(1) $\mathbf{A} = (-A\cos 60° \cos 30°)\mathbf{i} + (-A\sin 60°)\mathbf{j} + (A\cos 60° \sin 30°)\mathbf{k}$
$= -0.433A\mathbf{i} - 0.866A\mathbf{j} + 0.25A\mathbf{k}$

(2) $\mathbf{B} = (-B\cos 60°\cos 30°)\mathbf{i} + (-B\sin 60°)\mathbf{j} + (-B\cos 60° \sin 30°)\mathbf{k}$
$= -0.433B\mathbf{i} - 0.866B\mathbf{j} - 0.25B\mathbf{k}$

(3) $\mathbf{C} = C_x\mathbf{i} + C_y\mathbf{j} + C_z\mathbf{k}$

(4) The 150 lb force: $+(150\cos 10°)\mathbf{i} - (150\sin 10°)\mathbf{j} = +149\mathbf{i} - 25.9\mathbf{j}$

The position vector of O relative to C is $\mathbf{r} = 30\mathbf{j}$. Then using $\Sigma \mathbf{M}_C = 0$,

$$\Sigma(\mathbf{r}\times\mathbf{F}) = A\begin{vmatrix} \mathbf{i} & \mathbf{j} & \mathbf{k} \\ 0 & 30 & 0 \\ -0.433 & -0.866 & +0.25 \end{vmatrix} + B\begin{vmatrix} \mathbf{i} & \mathbf{j} & \mathbf{k} \\ 0 & 30 & 0 \\ -0.433 & -0.866 & -0.25 \end{vmatrix} + \begin{vmatrix} \mathbf{i} & \mathbf{j} & \mathbf{k} \\ 0 & 30 & 0 \\ +149 & -25.9 & 0 \end{vmatrix} = 0$$

Expanding the determinants and combining,

$$(7.5A - 7.5B)\mathbf{i} + (0)\mathbf{j} + (13A + 13B - 4470)\mathbf{k} = 0$$

or $7.5A - 7.5B = 0$ and $13A + 13B - 4470 = 0$, from which $A = 171$ lb T, $B = 171$ lb T.

Summing forces in the y direction yields C_y which is also the compression in the pole:

$$\Sigma F_y = C_y - 0.866A - 0.866B - 25.9 = 0 \qquad \text{or} \qquad C_y = 322 \text{ lb}$$

The sum of the moments about O indicates $C_x = C_z = 0$.

3. A weight of 60 lb is supported by the three wires as shown in Fig. 7-4. AB and AC are in the xz-plane. Determine tensions T_1, T_2, and T_3.

Solution:

$$AD = \sqrt{3^2+2^2+6^2} = 7, \quad AC = \sqrt{4^2+2^2} = 4.47, \quad AB = 5.$$

First sum the forces in the y direction because this equation involves only one unknown, T_1.

$$\Sigma F_y = 0 = -60 + T_1(6/7), \qquad T_1 = 70 \text{ lb}$$

Now sum forces in the x and z directions.

$$\Sigma F_x = 0 = T_2(4/4.47) - T_3(3/5) - 70(3/7)$$
$$\Sigma F_z = 0 = T_3(4/5) - T_2(2/4.47) - 70(2/7)$$

Multiply the ΣF_z equation by 2 and add to the ΣF_x equation to obtain $T_3 = 70$ lb. Substitute this value into the ΣF_x or ΣF_z equation to obtain $T_2 = 80.5$ lb.

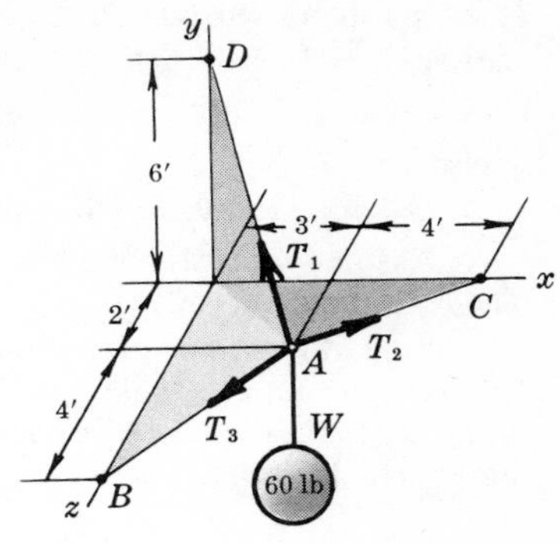

Fig. 7-4

4. Solve Problem 3 by expressing each force in terms of its **i**, **j**, and **k** components.

Solution:

Point A is in equilibrium under the action of the following four concurrent forces:

$$(1) \quad \mathbf{T}_1 = -(3/7)T_1\mathbf{i} + (6/7)T_1\mathbf{j} - (2/7)T_1\mathbf{k}$$
$$(2) \quad \mathbf{T}_2 = +(4/4.47)T_2\mathbf{i} + 0 - (2/4.47)T_2\mathbf{k}$$
$$(3) \quad \mathbf{T}_3 = -(3/5)T_3\mathbf{i} + 0 + (4/5)T_3\mathbf{k}$$
$$(4) \quad \mathbf{W} = -60\mathbf{j}$$

Then $\quad \Sigma \mathbf{F}_y = 0 = (6/7)T_1\mathbf{j} - 60\mathbf{j} \quad$ or $\quad T_1 = 70$ lb magnitude

In vector form, $\mathbf{T}_1 = -30\mathbf{i} + 60\mathbf{j} - 20\mathbf{k}$.

Also,

$$\Sigma \mathbf{F}_x = 0 = -(3/7)T_1\mathbf{i} + (4/4.47)T_2\mathbf{i} - (3/5)T_3\mathbf{i}$$
$$\Sigma \mathbf{F}_z = 0 = -(2/7)T_1\mathbf{k} - (2/4.47)T_2\mathbf{k} + (4/5)T_3\mathbf{k}$$

or $\quad (4/4.47)T_2 - (3/5)T_3 = (3/7)T_1 = 30 \quad$ and $\quad -(2/4.47)T_2 + (4/5)T_3 = (2/7)T_1 = 20$

whose simultaneous solution is $T_2 = 80.5$ lb and $T_3 = 70.0$ lb.

5. A vertical pole AB is 24 ft high and is guyed by four wires AC, AD, AE and AF each 30 ft long (see Fig. 7-5). What is the compressive force in the pole if each wire is stressed 1000 lb? $CDEF$ is a square with B at its center.

Solution:

Point A is the concurrency of the compressive force P and the given four equal tensile forces.

A vertical summation yields: $+P - 4(1000)(24/30) = 0$. Hence the force P in the pole = 3200 lb.

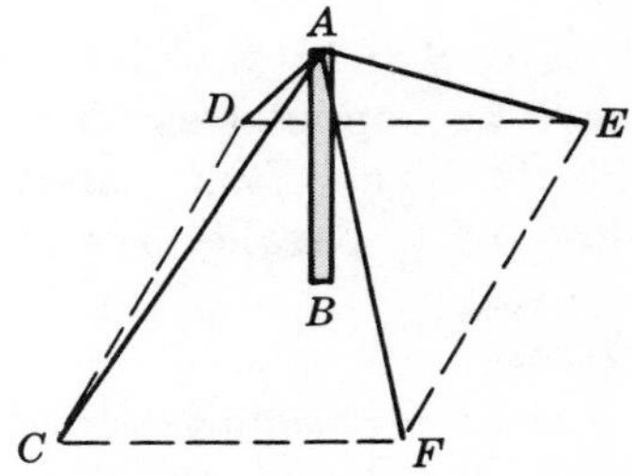

Fig. 7-5

6. A 400 lb weight is being lifted out of a hole 10 ft in diameter. Three ropes attached to it are held by three men equally spaced around the edge of the hole. What is the pull in each rope when the weight is 5 ft from the top and in the center?

Solution:

By symmetry, each rope is subjected to the same force. Hence the vertical components of the three forces must equal the 400 lb weight, i.e., $3T\cos\theta = 400$ lb, where θ is the angle between each rope and the vertical line through the weight. Then $\theta = \tan^{-1} 5/5 = 45°$, and $T = 189$ lb.

7. The system shown in Fig. 7-6 is subjected to a horizontal load P of 100 lb lying in the xy-plane. Determine the force in each leg.

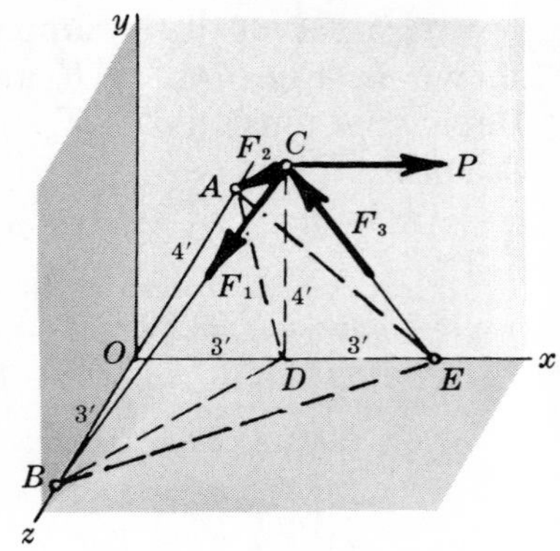

Fig. 7-6

Solution:

Calculations yield $CE = 5$, $BC = \sqrt{34}$ and $AC = \sqrt{41}$ ft.

Assume that the *three non-coplanar forces* F_1, F_2, F_3 are in the directions shown.

Sum forces parallel to the z-axis to obtain a relationship between F_1 and F_2. Take moments about the line AB to determine F_3. Then sum forces parallel to the x-axis to obtain another relationship between F_1 and F_2. These equations are:

(1) $\Sigma F_z = 0 = +(3/\sqrt{34})F_1 - (4/\sqrt{41})F_2$

(2) $\Sigma M_z = 0 = +(4/5)F_3 \times 6 - 100 \times 4$

(3) $\Sigma F_x = 0 = +100 - (3/5)F_3 - (3/\sqrt{34})F_1 - (3/\sqrt{41})F_2$

The results are: $F_1 = 55.6$ lb tension; $F_2 = 45.7$ lb tension; $F_3 = 83.3$ lb compression.

8. Solve Problem 7 using vector notation.

Solution:

The four forces may be expressed for the assumed directions as follows:

$$\mathbf{F}_1 = -(3/\sqrt{34})F_1\mathbf{i} - (4/\sqrt{34})F_1\mathbf{j} + (3/\sqrt{34})F_1\mathbf{k}$$
$$\mathbf{F}_2 = -(3/\sqrt{41})F_2\mathbf{i} - (4/\sqrt{41})F_2\mathbf{j} - (4/\sqrt{41})F_2\mathbf{k}$$
$$\mathbf{F}_3 = -(3/5)F_3\mathbf{i} + (4/5)F_3\mathbf{j} + 0$$
$$\mathbf{P} = 100\mathbf{i}$$

Since the system is in equilibrium, the sums of the **i**, **j**, and **k** components must each equal zero:

$$-(3/\sqrt{34})F_1 - (3/\sqrt{41})F_2 - (3/5)F_3 + 100 = 0$$
$$-(4/\sqrt{34})F_1 - (4/\sqrt{41})F_2 + (4/5)F_3 = 0$$
$$+(3/\sqrt{34})F_1 - (4/\sqrt{41})F_2 = 0$$

The simultaneous solution of these three equations is: $F_1 = 55.6$ lb (tension as assumed), $F_2 = 45.7$ lb (tension as assumed), $F_3 = 83.3$ lb (compression as assumed).

9. A table 6 ft by 6 ft is mounted on three legs. Four loads are applied as shown in Fig. 7-7. Determine the three reactions. Since three equations are available for a parallel system, only three supports are necessary.

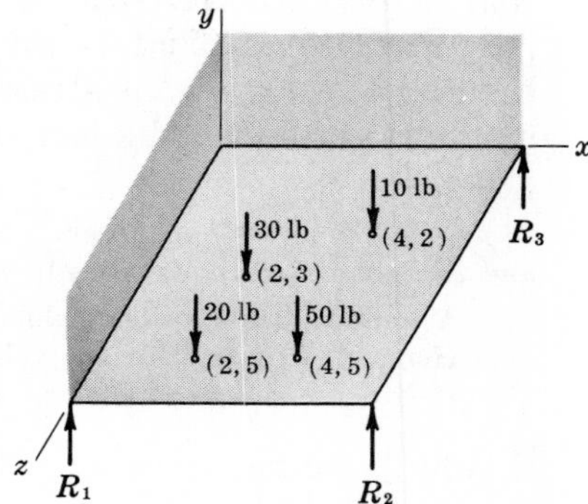

Fig. 7-7

Solution:

Applying the three equations of the parallel system there result the following equations:

(1) $\Sigma F_y = 0 = R_1 + R_2 + R_3 - 20 - 30 - 10 - 50$

(2) $\Sigma M_x = 0 = -R_1 \times 6 - R_2 \times 6 + 20 \times 5 + 30 \times 3 + 50 \times 5 + 10 \times 2$

(3) $\Sigma M_z = 0 = +R_2 \times 6 + R_3 \times 6 - 20 \times 2 - 50 \times 4 - 10 \times 4 - 30 \times 2$

Upon simplification, these become:

(1′) $R_1 + R_2 + R_3 = 110$

(2′) $R_1 + R_2 = 76.7$

(3′) $R_2 + R_3 = 56.7$

Substitute $R_1 + R_2 = 76.7$ into equation (1′) to obtain $76.7 + R_3 = 110$, or $R_3 = 33.3$ lb.
Substitute $R_2 + R_3 = 56.7$ into equation (1′) to obtain $R_1 + 56.7 = 110$, or $R_1 = 53.3$ lb.
Finally, from equation (1′), $R_2 = 110 - R_1 - R_3 = 23.4$ lb.

Another method. Sum moments about edges R_1R_2 and R_2R_3 to obtain R_3 and R_1 respectively.

10. A crankshaft is subjected to pulls F_1 and F_3 parallel to the z-axis, and F_2 and F_4 parallel to the y-axis. See Fig. 7-8. What are the bearing reactions at A and B if the pulls are each equal to F?

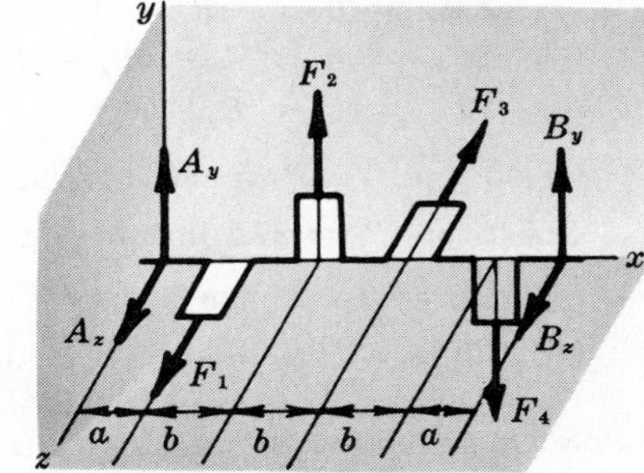

Fig. 7-8

Solution:

The bearing reactions are assumed to act in the positive directions of the y and z axes. The equations of equilibrium are:

(1) $\Sigma M_z = 0 = +F_2(a+b) - F_4(a+3b) + B_y(2a+3b)$

(2) $\Sigma M_y = 0 = -F_1(a) + F_3(a+2b) - B_z(2a+3b)$

(3) $\Sigma M_{B_z} = 0 = -A_y(2a+3b) - F_2(a+2b) + F_4(a)$

(4) $\Sigma M_{B_y} = 0 = +A_z(2a+3b) + F_1(a+3b) - F_3(a+b)$

In an actual engine the F's would not be equal, and each of the above equations would be solved for the unknown it contains. If the F's are assumed equal, then

$$B_y = \frac{2b}{2a+3b}F, \qquad B_z = \frac{2b}{2a+3b}F, \qquad A_y = \frac{-2b}{2a+3b}F, \qquad A_z = \frac{-2b}{2a+3b}F$$

The minus signs indicate that the components A_z and A_y actually act respectively to the rear and down. The total reaction at B is parallel to the total reaction at A; it is equal to it in magnitude but opposite in direction. The two form a couple as one would expect because F_1, F_3 and F_2, F_4 form couples when the forces are assumed to be of equal magnitudes.

11. Assume that an automobile door weighing 60 lb is of rectangular shape 3 ft by 4 ft high with its center of gravity at the geometric center. The door is opened 45°. A wind load of 50 lb is applied perpendicular to the door and is assumed concentrated at the geometric center. A door handle is 28 inches from the bottom and 3 inches from the right edge. What force P, applied in a horizontal plane at the handle but at an angle of 20° with the perpendicular to the door, is necessary to keep the door open? What are the components of the hinge reactions at A and B? Choose the x-axis along the door of the automobile. Assume that the lower hinge B carries all vertical loads, i.e., $A_y = 0$.

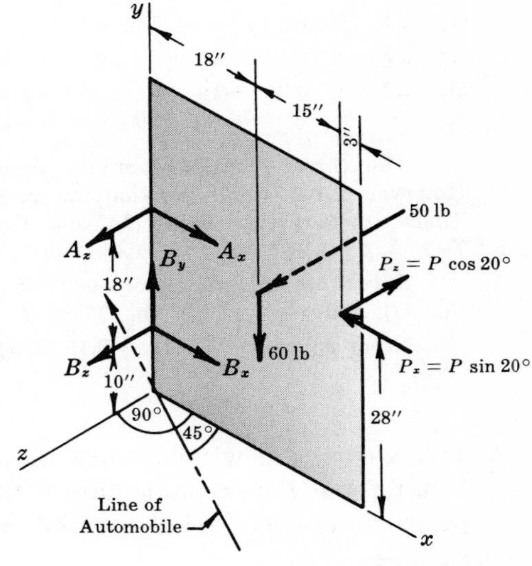

Fig. 7-9

Solution:

In Fig. 7-9 the components (assumed positive) of each hinge reaction are shown. Note that the two components of the force P are shown, one perpendicular to the door and one parallel to the door.

Taking moments about the y-axis yields an equation with only one unknown P_z from which P can be found.

Summing forces along the z-axis yields an equation in the unknowns A_z and B_z. Taking moments about the x-axis produces another equation involving A_z and B_z. Solve simultaneously.

Taking moments about the z-axis and summing forces along the x-axis yield two equations in A_x and B_x.

Summing forces vertically yields an equation involving B_y.

The above paragraphs are written to indicate the type of analysis that can be made before any equations are written. The equations follow:

(1) $\Sigma M_y = 0 = -50 \times 18 + P_z \times 33$

(2) $\Sigma F_z = 0 = A_z + B_z + 50 - P_z$

(3) $\Sigma M_x = 0 = +A_z \times 28 + B_z \times 10 + 50 \times 24 - P_z \times 28$

(4) $\Sigma M_z = 0 = -B_x \times 10 + P_x \times 28 - 60 \times 18 - A_x \times 28$

(5) $\Sigma F_x = 0 = A_x + B_x - P_x$

(6) $\Sigma F_y = 0 = B_y - 60$

From (1), $P_z = (50 \times 18)/33 = 27.3$ lb. But $P \cos 20° = P_z = 27.3$ lb; then $P = 29.1$ lb.

Substitute $P_z = 27.3$ lb into equations (2) and (3) and regroup terms as indicated below.

(2′) $A_z + B_z = -50 + 27.3$

(3′) $28A_z + 10B_z = -1200 + 765$

Solve simultaneously equations (2′) and (3′) to obtain $A_z = -11.6$ lb, $B_z = -11.1$ lb.
Next, substituting in equations (4) and (5) we obtain:

(4′) $-10B_x + 28(29.1 \times 0.342) - 28A_x = 1080$

(5′) $A_x + B_x - 29.1 \times 0.342 = 0$

Solve simultaneously equations (4′) and (5′) to obtain $A_x = -50$ lb, $B_x = +60$ lb.
From equation (6), $B_y = 60$ lb.

12. A 10 ft homogeneous rod BC weighing 10 lb rests against a smooth wall at B and on a smooth floor at C. Determine the tensions in AB and DC. Note in Fig. 7-10 that BD is perpendicular to the z-axis, and that AB is in the yz-plane. AB and DC are strings which hold the rod in equilibrium.

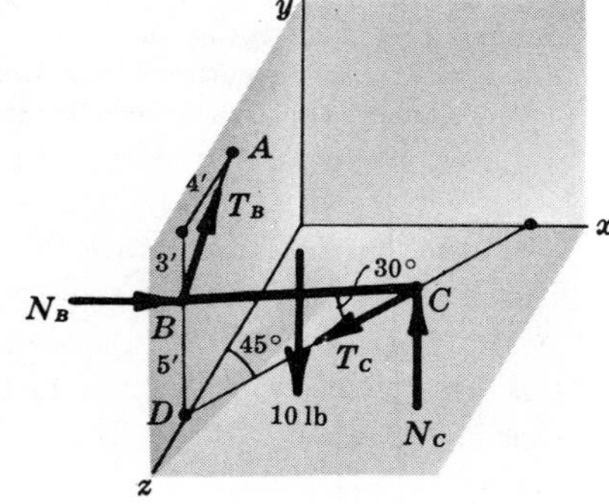

Fig. 7-10

Solution:

Add the wall and floor normal reactions N_B and N_C to complete the free body diagram of the rod. The following equations of equilibrium apply:

(1) $\Sigma F_x = 0 = N_B - T_C \cos 45°$

(2) $\Sigma F_y = 0 = T_B \times \frac{3}{5} - 10 + N_C$

(3) $\Sigma M_z = 0 = -10 \times 5 \cos 30° \cos 45° + N_C \times 10 \cos 30° \cos 45° - N_B \times 5$

The three equations contain four unknowns and appear at first glance impossible to solve. However, for a stable position the sum of the forces perpendicular to the plane BCD must be zero. There are only two forces (N_B and T_B) which have components perpendicular to this plane. Hence, $T_B \times \frac{4}{5} \times \cos 45° = N_B \cos 45°$, or $T_B \times \frac{4}{5} = N_B$.

Substitute this value into equation (2) and obtain $N_C = 10 - \frac{3}{4}N_B$. This, substituted into equation (3), yields $N_B = 3.19$ lb. Then $T_B = \frac{5}{4} \times 3.19 \text{ lb} = 3.99$ lb.

From equation (1), $T_C = 3.19/0.707 = 4.51$ lb.

13. Two views of a windlass are shown in Fig. 7-11 below. The bearings are frictionless. What force P perpendicular to the crank is necessary to hold a 200 lb weight in the position shown? What are the bearing reactions at A and B?

Solution:

A free body diagram is drawn in Fig. 7-12 below showing all forces acting on the windlass. Since no forces act along the axis of the windlass, no x components of the bearing reactions are shown.

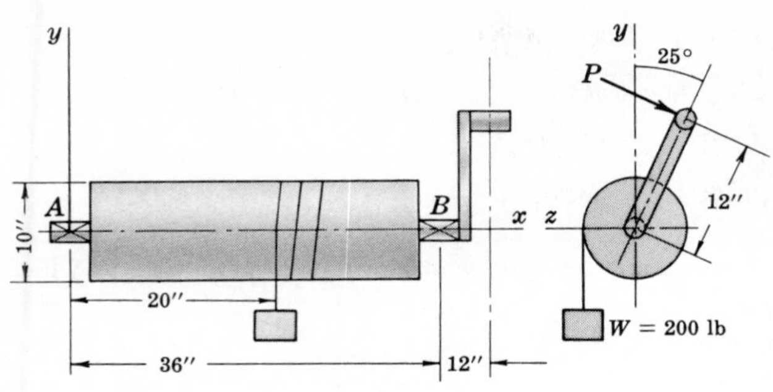

Fig. 7-11

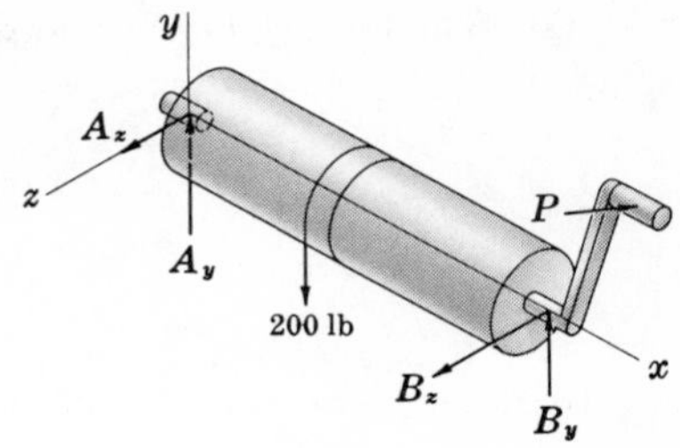

Fig. 7-12

Sum moments about the x-axis: $\Sigma M_x = 0 = -P \times 12 + 200 \times 5$. Hence, $P = 83.3$ lb.

To determine the bearing reactions, the following four equations may be used:

(1) $\Sigma M_y = 0 = -B_z \times 36 + P \cos 25° \times 48$

(2) $\Sigma M_z = 0 = -200 \times 20 + B_y \times 36 - P \sin 25° \times 48$

(3) $\Sigma F_y = 0 = A_y - 200 + B_y - P \sin 25°$

(4) $\Sigma F_z = 0 = A_z + B_z - P \cos 25°$

Note that the force P (83.3 lb) is resolved into its components $P \cos 25°$ and $P \sin 25°$ along the z and y axes respectively. The moment, for example, of P about the y-axis is then only the moment of its z component about the y-axis because the y component is parallel to the y-axis and therefore has no moment about it.

From equation (1), $B_z = 101$ lb; from equation (2), $B_y = 158$ lb.

Substituting for B_y, B_z and P in (3) and (4), we obtain $A_y = 77.2$ lb and $A_z = -25.5$ lb.

14. The beam EF weighs 10 lb/ft and carries the 138 lb weight at its end. It is supported by a ball and socket joint at E and the cables AB and CD. Determine the tensions in AB and CD. Find the pin reactions at E.

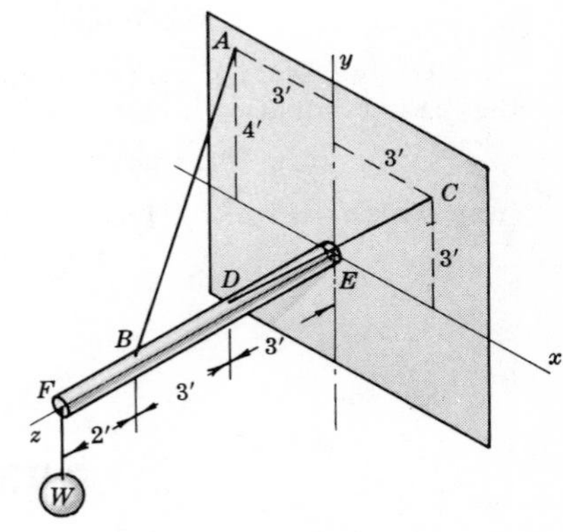

Fig. 7-13

Solution:

Select x, y, and z axes as shown in Fig. 7-13. For equilibrium of the beam EF, choose $\Sigma \mathbf{M}_E = 0$ and $\Sigma \mathbf{F} = 0$.

The forces acting on the system are:

(1) The weight W of 138 lb acts vertically down and may be written $-138\mathbf{j}$.

(2) The weight of the beam is 8 times 10 lb and may be written $-80\mathbf{j}$.

(3) The ball and socket reaction is $E_x\mathbf{i} + E_y\mathbf{j} + E_z\mathbf{k}$.

(4) The tension in AB may be written $A_x\mathbf{i} + A_y\mathbf{j} + A_z\mathbf{k}$, where

$$A_x = A \cos \theta_x = -(3/\sqrt{3^2+4^2+6^2})A = -(3/\sqrt{61})A = -0.384A$$

$$A_y = A \cos \theta_y = +(4/\sqrt{61})A = +0.512A$$

$$A_z = A \cos \theta_z = -(6/\sqrt{61})A = -0.768A$$

The sign of each component is defined once we assume tension in AB which must then pull on the beam EF in the direction from B to A which is in the negative x direction, the positive y direction, and the negative z direction.

(5) The tension in CD may be written $C_x\mathbf{i} + C_y\mathbf{j} + C_z\mathbf{k}$, where

$$C_x = C\cos\theta_x = +(3/\sqrt{3^2+3^2+3^2})C = +(3/\sqrt{27})C = +0.577C$$
$$C_y = C\cos\theta_y = +(3/\sqrt{27})C = +0.577C$$
$$C_z = C\cos\theta_z = -(3/\sqrt{27})C = -0.577C$$

It is advisable to summarize the five forces and the points on their lines of action to which the position vectors from point E will be drawn.

(1′) $-138\mathbf{j}$ at $(0, 0, 8)$
(2′) $-80\mathbf{j}$ at $(0, 0, 4)$
(3′) $E_x\mathbf{i} + E_y\mathbf{j} + E_z\mathbf{k}$ at $(0, 0, 0)$
(4′) $-0.384A\mathbf{i} + 0.512A\mathbf{j} - 0.768A\mathbf{k}$ at $(0, 0, 6)$
(5′) $+0.577C\mathbf{i} + 0.577C\mathbf{j} - 0.577C\mathbf{k}$ at $(0, 0, 3)$

The sum of the moments of these five forces about E will be set equal to zero. Using each position vector from point $(0, 0, 0)$ to the point on the force vector listed above, we have

$$\begin{vmatrix} \mathbf{i} & \mathbf{j} & \mathbf{k} \\ 0 & 0 & 8 \\ 0 & -138 & 0 \end{vmatrix} + \begin{vmatrix} \mathbf{i} & \mathbf{j} & \mathbf{k} \\ 0 & 0 & 4 \\ 0 & -80 & 0 \end{vmatrix} + \begin{vmatrix} \mathbf{i} & \mathbf{j} & \mathbf{k} \\ 0 & 0 & 0 \\ E_x & E_y & E_z \end{vmatrix} + \begin{vmatrix} \mathbf{i} & \mathbf{j} & \mathbf{k} \\ 0 & 0 & 6 \\ -0.384A & +0.512A & -0.768A \end{vmatrix}$$
$$+ \begin{vmatrix} \mathbf{i} & \mathbf{j} & \mathbf{k} \\ 0 & 0 & 3 \\ 0.577C & 0.577C & -0.577C \end{vmatrix} = 0$$

or $$8(138)\mathbf{i} + 4(80)\mathbf{i} + 0 + [-6(0.512A)\mathbf{i} - 6(0.384A)\mathbf{j}] + [-3(0.577C)\mathbf{i} + 3(0.577C)\mathbf{j}] = 0$$

Equating the coefficients of **i** to zero, and then the coefficients of **j** to zero,

$$1424 - 3.072A - 1.731C = 0 \quad \text{and} \quad -2.304A + 1.731C = 0$$

from which $A = 265$ lb and $C = 353$ lb. Hence

$$\mathbf{A} = -0.384(265)\mathbf{i} + 0.512(265)\mathbf{j} - 0.768(265)\mathbf{k} = -102\mathbf{i} + 136\mathbf{j} - 204\mathbf{k}$$
$$\mathbf{C} = 0.577(353)\mathbf{i} + 0.577(353)\mathbf{j} - 0.577(353)\mathbf{k} = 204\mathbf{i} + 204\mathbf{j} - 204\mathbf{k}$$

To find the pin reaction, equate successively to zero the coefficients of the **i**, **j**, and **k** terms in the $\Sigma\mathbf{F} = 0$ equation:

$$E_x - 102 + 204 = 0, \quad -138 - 80 + E_y + 136 + 204 = 0, \quad E_z - 204 - 204 = 0$$

from which $E_x = -102$ lb, $E_y = -122$ lb, $E_z = +408$ lb.

Supplementary Problems

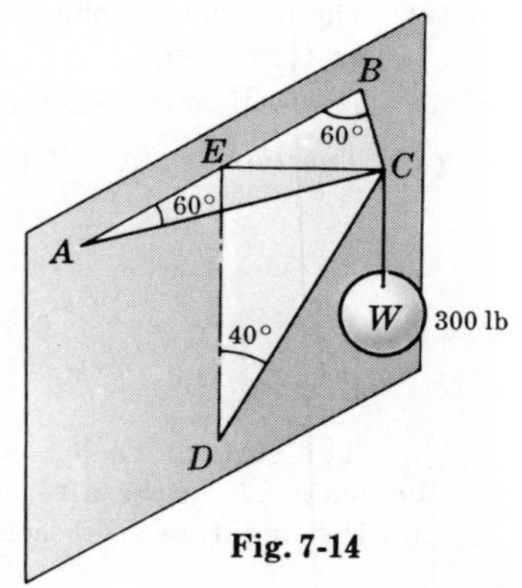

Fig. 7-14

15. In Fig. 7-14, a weight of 300 lb is supported by a compression member CD and two tension members AC and BC. CD makes an angle of 40° with the wall. A, B, and C are in a horizontal plane. $AE = EB = 3$ ft. Find the forces in AC, BC, and CD.
Ans. $AC = BC = 146$ lb T, $CD = 392$ lb C

16. In Fig. 7-15 below, the crane consists of a boom BE, column BD (vertical), and three cables AD, CD, DE. A, B, and C are in a horizontal plane. AC is bisected by the plane containing BD, BE, and DE. Determine the forces in AD, CD, and BD. (Hint: First consider the coplanar, concurrent system at E to obtain force in DE. Then consider spatial, concurrent system at D.)
Ans. $AD = CD = 2910$ lb T, $BD = 2970$ lb C

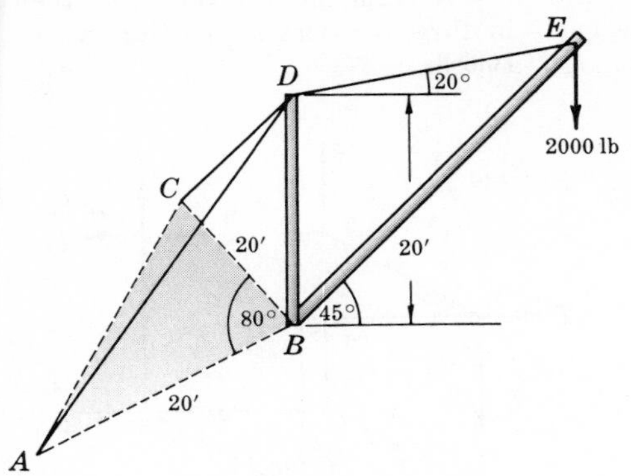

Fig. 7-15

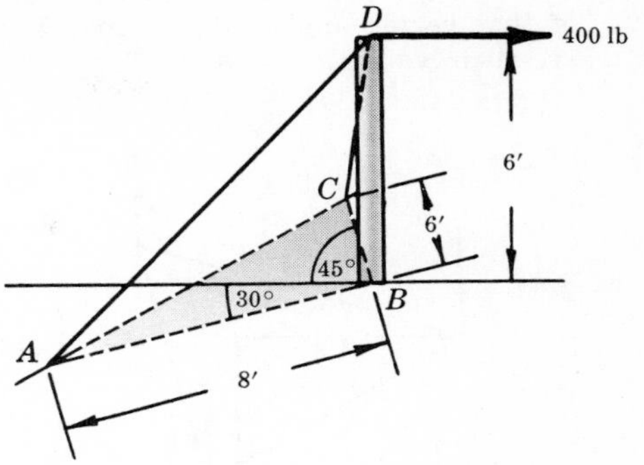

Fig. 7-16

17. In Fig. 7-16 above, a horizontal pull of 400 lb is shown acting on the top of post DB. The post is held in equilibrium by the two guy wires AD and CD. A, B, and C are on level ground. Find the forces in AD and CD. *Ans.* $AD = 366$ lb T, $CD = 293$ lb T

18. A camera weighing 4 lb rests on a tripod with legs equally spaced and each making 20° with the vertical. Assuming that the system is concurrent at a point 4 ft above level ground, determine the force in each leg. *Ans.* $C = 1.42$ lb

19. A 500 lb weight is hung on a rope in a tripod with legs of equal length, as shown in Fig. 7-17 below. Each leg makes an angle of 30° with the rope. A, B, and C are in the horizontal plane and form an equilateral triangle. Determine the force in each leg. *Ans.* $AD = BD = CD = 192$ lb C

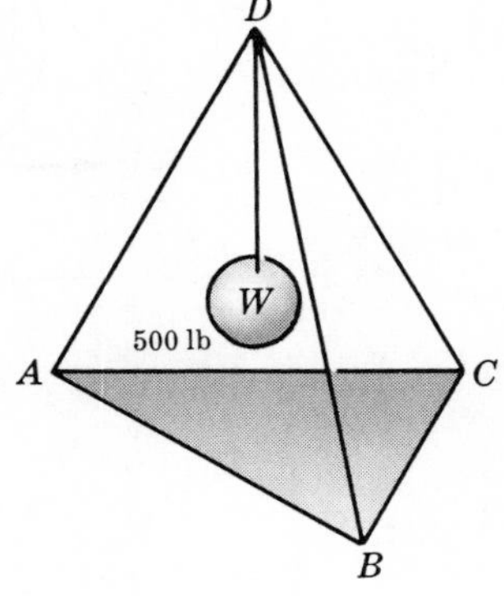

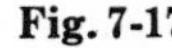
Fig. 7-17

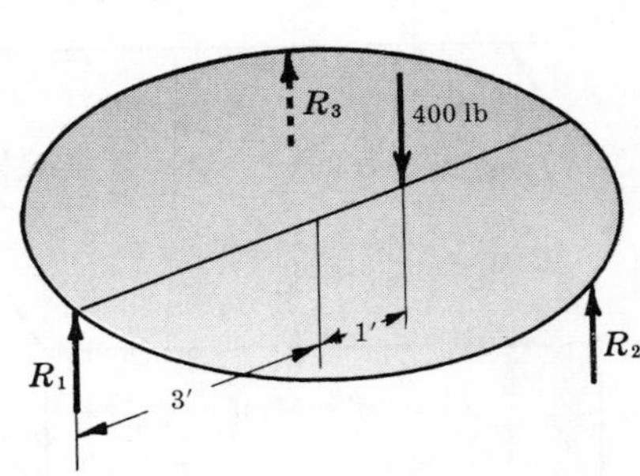

Fig. 7-18

20. The circular table 6 ft in diameter, shown in Fig. 7-18 above, supports a load of 400 lb located on a diameter through the support R_1 and 1 ft from the center on the opposite side from R_1. R_1, R_2, and R_3 are equally spaced. Determine their magnitudes. *Ans.* $R_1 = 44$ lb, $R_2 = 178$ lb, $R_3 = 178$ lb

21. A homogeneous circular plate weighing 60 lb is supported by three vertical wires, as shown in the adjacent Fig. 7-19. Note the unequal spacing of the wires. The plate is 2 ft in diameter. Determine the tensile forces in the wires.

Ans. $T_1 = 12.6$ lb
$T_2 = 25.4$ lb
$T_3 = 22.0$ lb

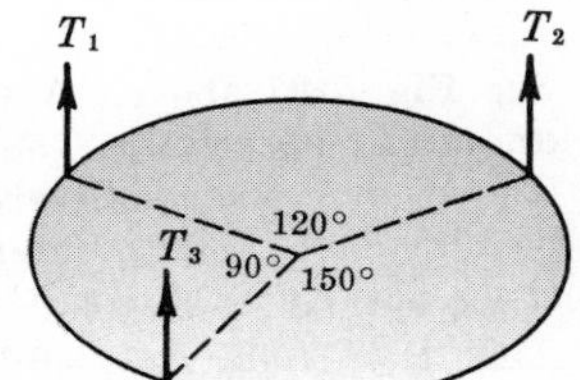

Fig. 7-19

22. The triangular plate, in Fig. 7-20 below, carries a 30 lb load 4 ft from the left vertex on the bisector of that vertex angle. T_1, T_2, and T_3 are the tensile loads in three vertical supporting wires. What are their values? *Ans.* $T_1 = 8.8$ lb, $T_2 = 10.6$ lb, $T_3 = 10.6$ lb

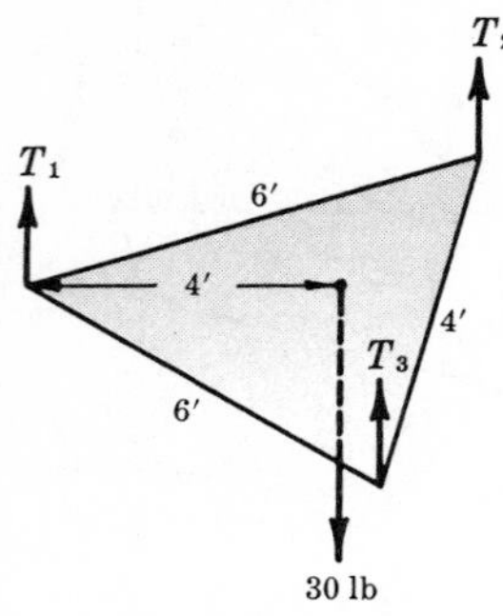

Fig. 7-20

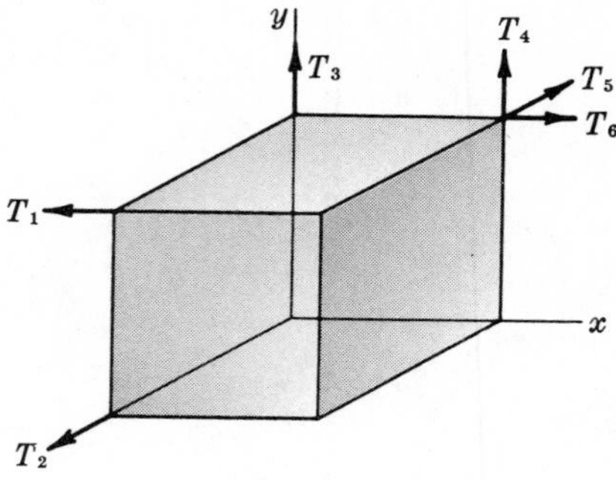

Fig. 7-21

23. A uniform cube of weight W is supported by six strings attached to corners, as shown in Fig. 7-21 above. Each string is perpendicular to a face, i.e., a continuation of an edge of the cube. Determine the tension in each string to hold the cube in equilibrium. *Ans.* $T = W/2$

24. See Fig. 7-22 below. Assume that a motor weighing 500 lb has its center of gravity $\frac{5}{8}$ of the overall length from the front on the longitudinal center line. If the base is 22 inches wide and 34 inches long, what is the magnitude of the reactions of the supports assuming one at each front corner and one at the middle in the rear of the motor. A block diagram is shown in the figure.
Ans. $R_F = 94$ lb, $R_R = 312$ lb

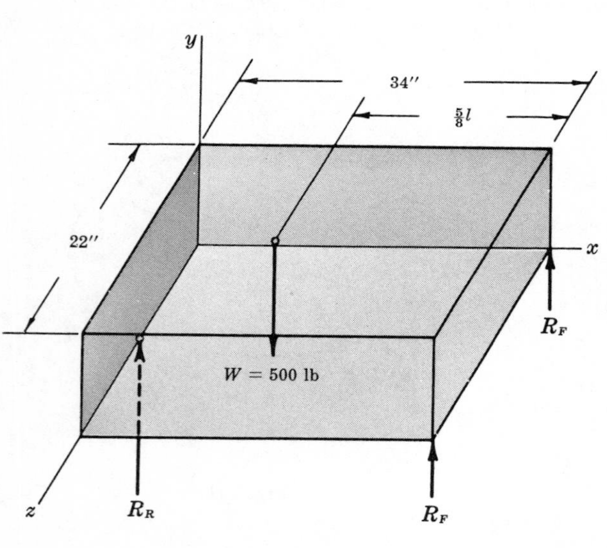

Fig. 7-22

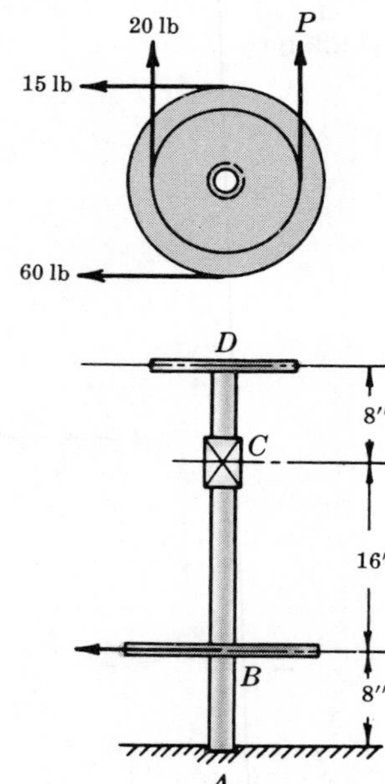

Fig. 7-23

25. See Fig. 7-23 above. A vertical shaft weighing 40 lb carries two pulleys at B and D weighing 12 lb and 9 lb respectively. The pulley at B is 16 inches in diameter and the one at D is 12 inches in diameter. The 15 and 60 lb pulls are parallel to the x-axis. The 20 lb pull and P are parallel to the z-axis. The bearing at C and the step bearing at A are to be considered frictionless. Determine the force P and the reactions at A and C.
Ans. $P = 80$ lb, $A_x = 50$ lb, $A_y = 61$ lb, $A_z = -33$ lb, $C_x = 25$ lb, $C_z = 133$ lb, $C_y = 0$

26. In the simple crane shown in Fig. 7-24 below, CH is vertical, GD is horizontal, and AC and BC are guy wires. Points A and B are equidistant from the plane which contains CH, DG, and EF. Weight $W = 4000$ lb. Determine the tension in AC and the reactions at H.

Ans. $T = 2590$ lb, $H_x = 2800$ lb, $H_y = 8040$ lb, $H_z = 0$

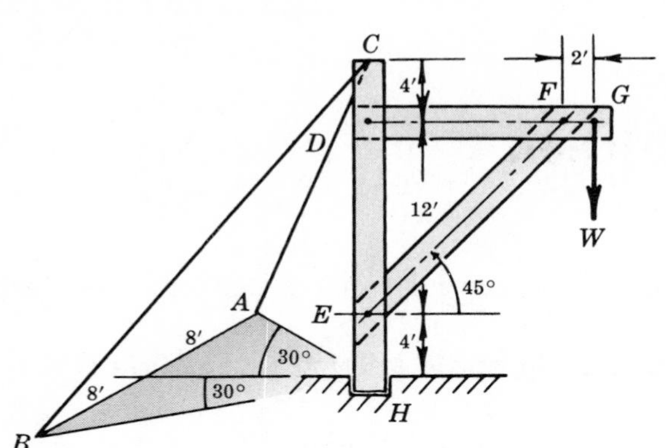

Fig. 7-24

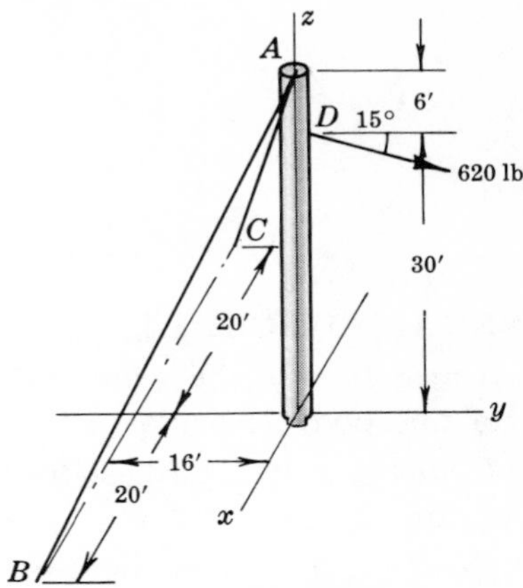

Fig. 7-25

27. A vertical pole is subjected to a pull of 620 lb in the yz-plane and 15° below the horizontal. The guy wires AB and AC are attached to supports in the xy-plane. The pole rests in a socket. See Fig. 7-25 above. What is the tension in each cable? *Ans.* $T_{AB} = T_{AC} = 688$ lb

28. The homogeneous trapdoor weighs 96 lb. What rope tension T is needed to hold the door in the 26° position shown in Fig. 7-26 below? What are the hinge reactions at A and B? Assume the pulley D is on the vertical z-axis.

Ans. $T = 52.4$ lb, $A_x + B_x = -28.8$ lb, $A_y = +30.0$ lb, $A_z = +30.7$ lb, $B_y = -4.3$ lb, $B_z = +29.9$ lb

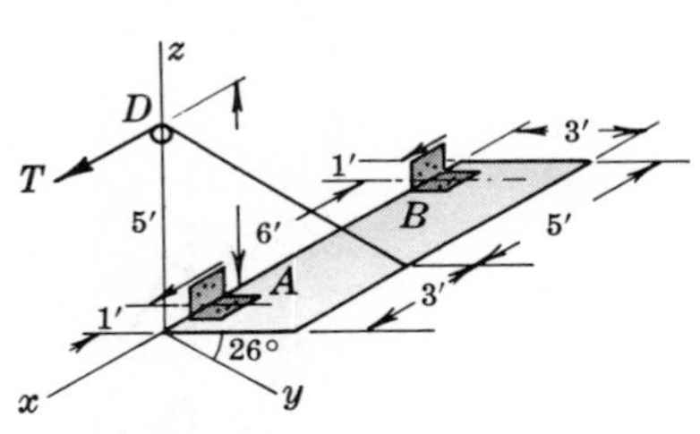

Fig. 7-26

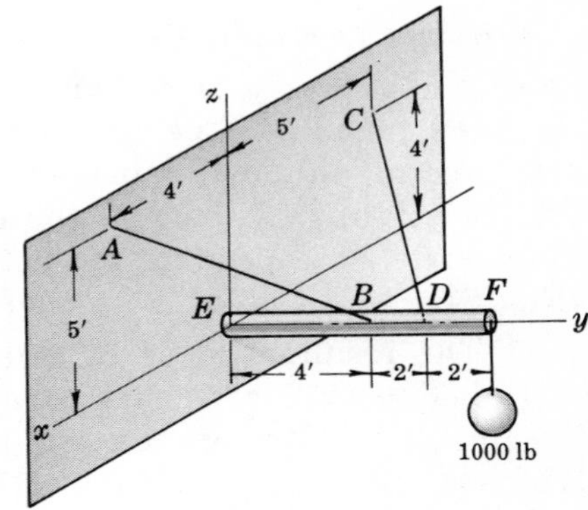

Fig. 7-27

29. The boom EF shown in Fig. 7-27 above may be considered weightless. It is supported by the cables AB and CD and a socket at E. Determine the tensions in the two cables and the pin reactions at E.

Ans. Tension in $AB = 1840$ lb;
tension in $CD = 1140$ lb;
$E_x = -326$ lb, $E_y = 1760$ lb, $E_z = -740$ lb.

30. The door shown in Fig. 7-28 is attached to straps which are welded to the rod EB. The rod EB is carried in bearings at A and B and carries a gear E at its end. A pinion (not shown) exerts a horizontal force F at the bottom of gear E. Assuming the homogeneous door weighs 30 lb, determine the force F and the bearing reactions when angle α is 58°.

Ans. $F = 71.5$ lb, $A_x = -89.5$ lb, $A_y = 15$ lb, $A_z = 0$, $B_x = 17.9$ lb, $B_y = 15$ lb, $B_z = 0$

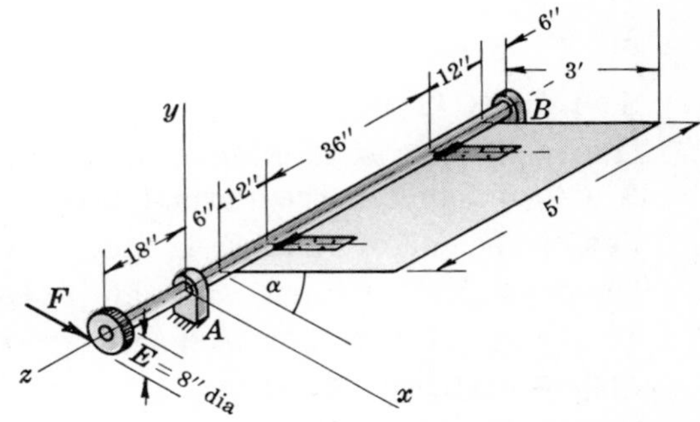

Fig. 7-28

Chapter 8

Friction

GENERAL CONCEPTS

(1) *Static friction* between two bodies is the tangential force which opposes the sliding of one body relative to the other.

(2) *Limiting friction* F' is the maximum value of static friction that occurs when motion is impending.

(3) *Kinetic friction* is the tangential force between two bodies after motion begins. It is less than static friction.

(4) *Angle of friction* is the angle between the action line of the total reaction of one body on another and the normal to the common tangent between the bodies when motion is impending.

(5) *Coefficient of static friction* is the ratio of the limiting friction F' to the normal force N.

$$\mu = \frac{F'}{N}$$

(6) *Coefficient of kinetic friction* is the ratio of the kinetic friction to the normal force.

(7) *Angle of repose* α is the angle to which an inclined plane may be raised before an object resting on it will move under the action of its weight and the reaction of the plane. This state of impending motion is shown in Fig. 8-1.

The resultant R of F' and N is shown acting opposite but equal in magnitude to weight W. Although motion impends, the body is still in equilibrium. The following relations hold by the trigonometry of the figure.

$$\mu = \frac{F'}{N} = \tan\phi = \tan\alpha$$

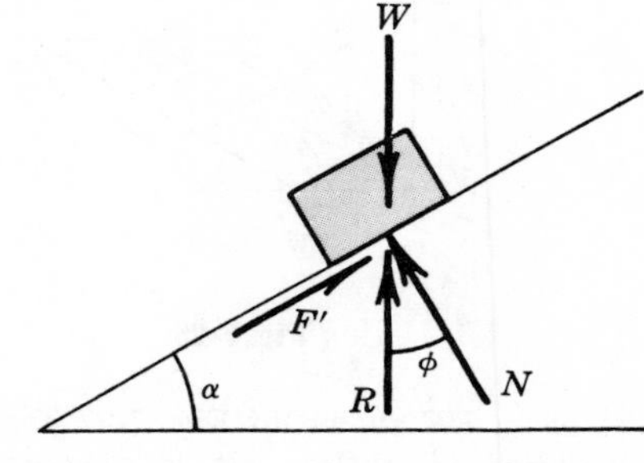

Fig. 8-1

LAWS of FRICTION

(a) The coefficient of friction is independent of the normal force; however, the friction is proportional to the normal force.

(b) The coefficient of friction is independent of the area of contact.

(c) The coefficient of kinetic friction is less than that of static friction.

(d) At low speeds, friction is independent of the speed. At higher speeds a decrease in friction has been noticed.

(e) The static frictional force is never greater than that which is necessary to hold the body in equilibrium.

JACK-SCREW

The jack-screw is an example of a frictional device. For the square-threaded screw shown in Fig. 8-2, there are essentially two problems: (*a*) The moment of the force P necessary to raise the load and (*b*) the moment of the force P necessary to lower the load.

In each case the turning moment is about the longitudinal (vertical in the figure) axis of the screw.

In case (*a*) the turning moment must overcome friction and raise the load W, whereas in case (*b*) the load W helps to overcome the friction. To raise the load W, the screw must turn counterclockwise when viewed from the top.

Let β be the lead angle, i.e., the angle whose tangent is equal to the lead divided by the mean circumference. Let ϕ be the angle of friction. The formulas for the two cases, where r is the mean radius of the thread, are:

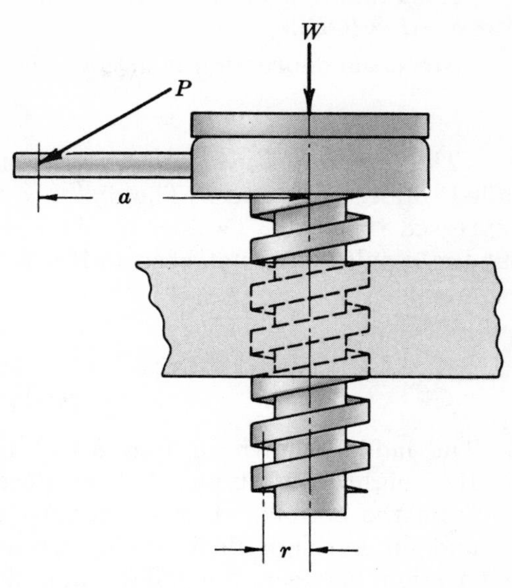

Fig. 8-2

$$(a) \quad M = W\,r\tan(\phi+\beta)$$
$$(b) \quad M = W\,r\tan(\phi-\beta)$$

These formulas hold also if the screw is turning at constant speed. Of course ϕ is then the angle of kinetic friction. They also apply to a screw jack in which a cap is added to act as a bearing for a load. To be accurate, another term should be added to each formula on the right side representing the moment necessary to overcome the friction between the cap and the screw. This extra term is of the form $W\mu r_c$, where W is the load, μ is the coefficient of friction between the cap and the screw, and r_c is the mean radius of the bearing surface between the cap and screw. See Problem 13 for an application of this formula. This is an approximation for the more accurate expression for collar frictional moment.

BELT FRICTION and BRAKE BANDS

Belt friction and brake bands also illustrate the use of friction. When a belt or band passes over a rough pulley, the tensions in the belt or band on the two sides of the pulley will differ. When slip is about to occur, the following formula applies.

$$T_1 = T_2 e^{\mu\alpha}$$

where T_1 = larger tension, T_2 = smaller tension, μ = coefficient of friction, α = angle of wrap in radians, e = 2.718 (base of natural logarithms).

ROLLING FRICTION (RESISTANCE)

Rolling friction (resistance) occurs because of the deformation of the surface under a rolling load. Fig. 8-3 exaggerates this effect. A wheel of weight W and radius r is being pulled out of the depression and over the point A by the horizontal force P. Naturally this is a continuous process as the wheel rolls.

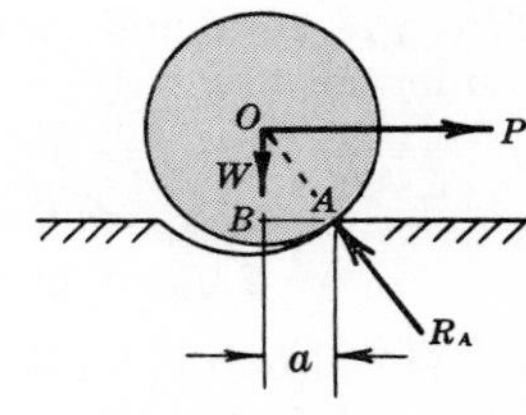

Fig. 8-3

A summation of moments about point A yields the following equation: $\Sigma M_A = 0 = W \times a - P \times (OB)$.

Since the depression is actually very small, distance (OB) may be replaced by r; hence,

$$P \times r = W \times a$$

The horizontal component of the surface reaction R is equal to P by inspection and is called rolling resistance. The distance a is called the coefficient of rolling resistance and is expressed in inches. Values of the coefficients for various materials have been tabulated, but the results have not been uniform.

Solved Problems

1. The ladder shown in Fig. 8-4(a) is non-homogeneous. Its weight of 90 lb may be considered concentrated 6 ft from the bottom. It rests against a smooth wall at A and on a rough floor at B. The coefficient of static friction between the ladder and floor is $\frac{1}{3}$. Will the ladder stand in the 60° position shown?

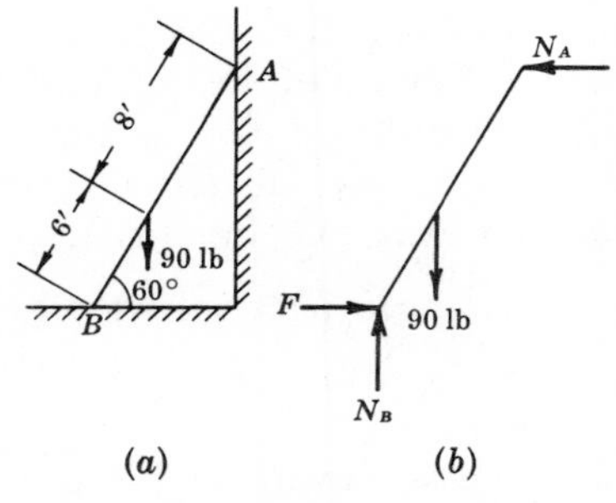

Fig. 8-4

Solution:

This is an example in which it is unknown whether or not motion impends.

The free body diagram is shown in Fig. 8-4(b). Since there is no friction at A, the wall being smooth, the only reaction is the normal pressure N_A. At B the normal pressure is N_B. The friction is unknown and is represented as F acting to the right. It acts to the right since it tends to prevent the ladder from slipping to the left.

Determine the value of F and N_B. If this value of F which is necessary to hold the ladder from slipping is less than the limiting value obtainable, i.e., $F' = \mu N_B = N_B/3$, then the ladder will stand. The following equations of equilibrium are used.

$$(1) \quad \Sigma F_v = 0 = N_B - 90$$

$$(2) \quad \Sigma M_A = 0 = +F \times 14 \sin 60° + 90 \times 8 \cos 60° - N_B \times 14 \cos 60°$$

From equation (1), $N_B = 90$. Substituting this value into (2) gives $F = 22.3$ lb.

The limiting value is $F' = \frac{1}{3} \times N_B = \frac{1}{3} \times 90 = 30$ lb.

Therefore the ladder will stand because only 22.3 lb is needed, whereas 30 lb maximum friction is available. Some may prefer to determine the ratio $22.3/90 = 0.248$. Since this is less than $\frac{1}{3}$, the ladder will stand.

2. Determine the smallest angle θ for equilibrium of the homogeneous ladder of length l. The coefficient of friction for all surfaces is μ.

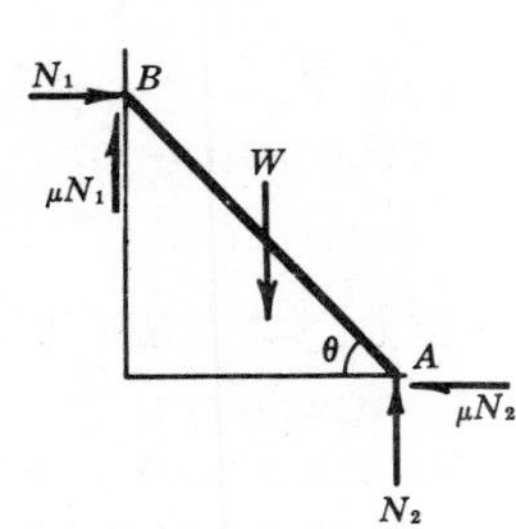

Fig. 8-5

Solution:

Assuming slipping to impend, draw a free body diagram of the ladder. There are three unknowns: N_1, N_2, θ (see Fig. 8-5). The three equations of equilibrium are:

$$(1) \quad \Sigma F_h = 0 = N_1 - \mu N_2$$

$$(2) \quad \Sigma F_v = 0 = N_2 - W + \mu N_1$$

$$(3) \quad \Sigma M_B = 0 = -W \tfrac{1}{2} l \cos \theta + N_2 \, l \cos \theta - \mu N_2 \, l \sin \theta$$

Substitute $N_1 = \mu N_2$ into (2) to get $N_2 = W/(1 + \mu^2)$. Substitute this value of N_2 into (3) to get $\theta = \tan^{-1} (1 - \mu^2)/2\mu$. This is the critical value of θ below which slip will occur.

3. Determine the value of P which will cause the 150 lb block in Fig. 8-6(*a*) below to move. The coefficient of static friction between the block and the horizontal surface is $\frac{1}{4}$.

Solution:

Motion may ensue in two ways, assuming that P is applied gradually. The block may slide to the left or it may tip about the forward edge O.

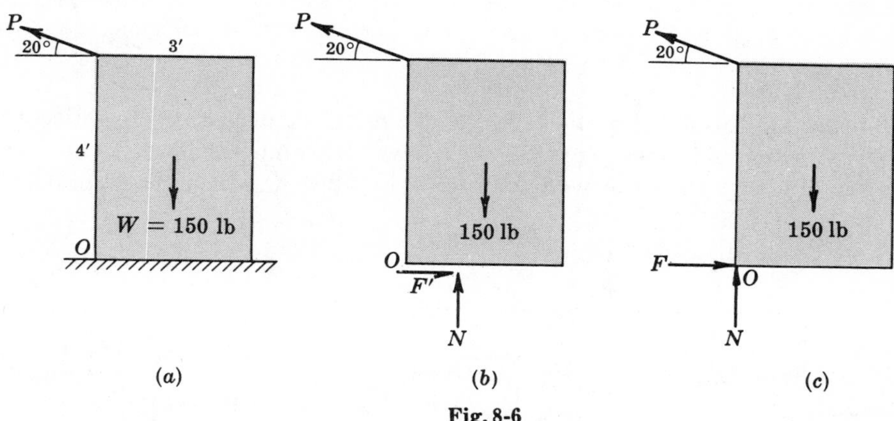

Fig. 8-6

First determine the value of P to cause sliding to the left. In this case limiting friction is used as shown in Fig. 8-6(*b*) above. The equations which apply are:

$$(1)\quad \Sigma F_v = 0 = P\sin 20^\circ - 150 + N$$
$$(2)\quad \Sigma F_h = 0 = -P\cos 20^\circ + F'$$

Since $F' = \mu N = N/4$, substitute into equation (*2*). The equations are then

$$(3)\quad P\sin 20^\circ + N = 150 \qquad\qquad (5)\quad P\sin 20^\circ + N = 150$$
$$(4)\quad -P\cos 20^\circ + N/4 = 0 \qquad \text{or} \qquad (6)\quad 4P\cos 20^\circ - N = 0$$

Add equations (*5*) and (*6*) to obtain $P = 36.6$ lb.

Refer to Fig. 8-6(*c*) above. Next assume that the block will tip about the forward edge at O. Determine the value of P to do this. Note that in this determination no indication of the size of the frictional force is given. It must be labeled F, as shown in Fig. 8-6(*c*). Since the block is assumed to tip, the normal pressure is at O. The equations of equilibrium are:

$$(7)\quad \Sigma M_O = 0 = P\cos 20^\circ \times 4 - 150 \times 3/2$$
$$(8)\quad \Sigma F_h = 0 = -P\cos 20^\circ + F$$
$$(9)\quad \Sigma F_v = 0 = +P\sin 20^\circ + N - 150$$

Equation (*7*) yields directly $P = 59.8$ lb. The investigation can be stopped here, since to cause sliding P must equal 36.6 lb whereas to cause tipping P must go to 59.8 lb. It is thus seen that sliding will be the first to occur as P increases steadily from zero to a maximum.

4. Will the 200 lb block shown in Fig. 8-7(*a*) be held in equilibrium by the horizontal force of 300 lb? The coefficient of static friction is 0.3.

Solution:

In this case it is unlikely that the force of friction is exactly the limiting value. Assume that the 300 lb force is more than sufficient to hold the block from sliding down the plane. Then it may be large enough to cause motion up the plane. To check this condition assume that F acts down the plane as shown in the free body diagram in Fig. 8-7(*b*).

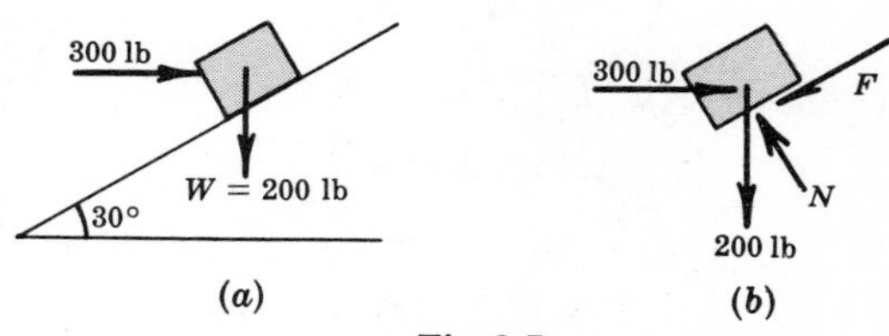

Fig. 8-7

The equations summing the forces parallel and perpendicular to the plane are:

$$(1)\quad \Sigma F_{\parallel} = 0 = -F - 200 \sin 30° + 300 \cos 30°$$
$$(2)\quad \Sigma F_{\perp} = 0 = +N - 200 \cos 30° - 300 \sin 30°$$

Solving these equations, $F = 160$ lb, $N = 323$ lb.

This indicates that the value of F necessary to hold the block from moving up the plane is 160 lb. However, the maximum value obtainable is $F' = 0.3N = 0.3 \times 323 = 97$ lb. This means that the block will move up the plane. What happens after motion starts is the subject of later chapters.

5. What horizontal force P on the wedges B and C is necessary to raise the weight of 20 tons resting on A? See Fig. 8-8(*a*) below. Assume μ between the wedges and the ground is $\frac{1}{4}$ and between the wedges and A is 0.2. Also assume symmetry of loading.

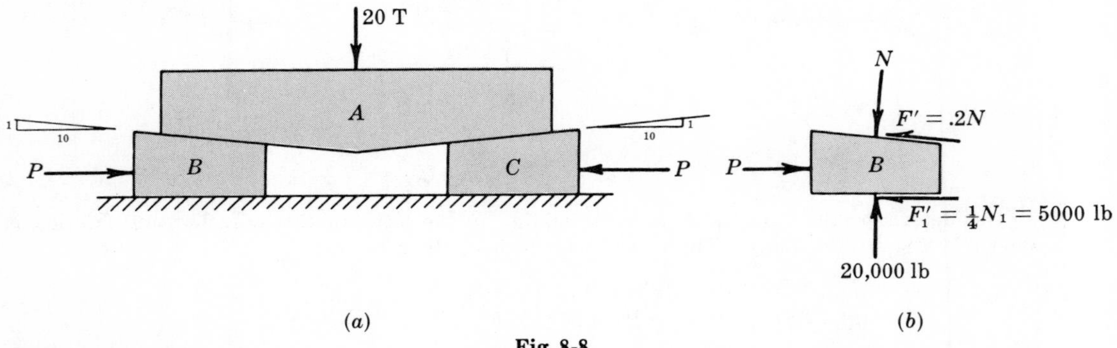

Fig. 8-8

Solution:

By considering the entire setup as a free body, it is evident that the normal force between the ground and each wedge is 10 tons or 20,000 lb.

Draw a free body diagram of wedge B (see Fig. 8-8(*b*) above). Here the normal force of 20,000 lb is shown acting vertically up. The force of friction between the wedge and ground opposes motion, as shown. Since motion impends, its value is $1/4 \times 20{,}000$ lb $= 5000$ lb. The normal force of A on B is of course perpendicular to their tangent surface, while F' is the limiting value of friction drawn so as to oppose motion. This completes the free body diagram of B, showing only forces acting on B. The equations of equilibrium are then:

$$(1)\quad \Sigma F_h = 0 = P - 5000 - N\frac{1}{\sqrt{101}} - 0.2N\frac{10}{\sqrt{101}}, \qquad (2)\quad \Sigma F_v = 0 = 20{,}000 - N\frac{10}{\sqrt{101}} + 0.2N\frac{1}{\sqrt{101}}$$

From equation (*2*), $N = 20{,}500$ lb. Substituting into equation (*1*), $P = 11{,}100$ lb.

6. Block B rests on block A and is attached by a horizontal rope BC to the wall as shown in Fig. 8-9(*a*) below. What force P is necessary to cause motion of A to impend? The coefficient of friction between A and B is $\frac{1}{4}$, and between A and the floor is $\frac{1}{3}$. A weighs 30 lb and B weighs 20 lb.

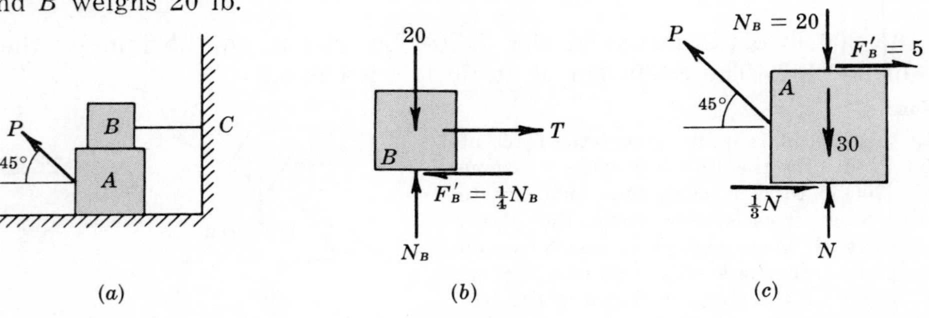

Fig. 8-9

Solution:

Since motion of A impends to the left, it will tend to drag B with it because of friction. A free body diagram of B shows F'_B acting to the left (see Fig. 8-9(*b*) above).

A summation of forces vertically indicates that $N_B = 20$ lb. Therefore $F'_B = 5$ lb and, since B does not move, the tension T is also 5 lb.

Draw a free body diagram of body A as shown in Fig. 8-9(*c*) above. In this figure F'_B is shown acting to the right, opposing motion. The equations of equilibrium are:

$$(1)\quad \Sigma F_h = 0 = -P\cos 45° + N/3 + 5$$
$$(2)\quad \Sigma F_v = 0 = +P\sin 45° + \text{N} - 30 - 20$$

Since $\sin 45° = \cos 45°$, add the two equations to obtain $N = 33.8$ lb. Substitute for N in equation (*2*) to obtain $P = 22.9$ lb.

7. What should be the value of the angle θ so that motion of the 90 lb block impends down the plane? The coefficient of friction μ for all surfaces is $\frac{1}{3}$. Refer to Fig. 8-10(*a*) below.

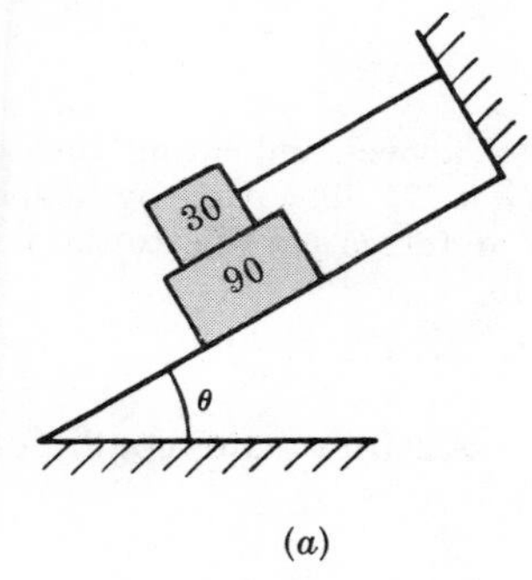

(*a*)

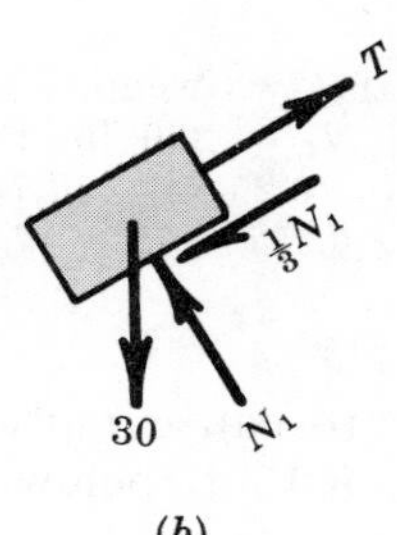

(*b*)

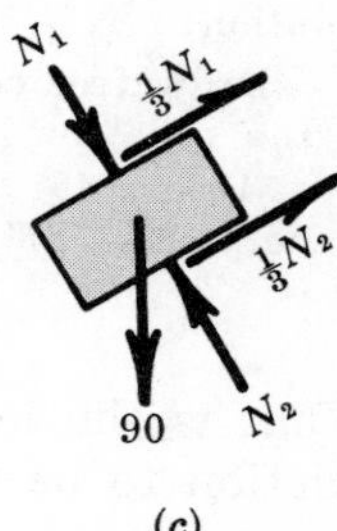

(*c*)

Fig. 8-10

Solution:

Draw free body diagrams of the two weights. By inspection, $N_1 = 30\cos\theta$ in the free body diagram of the 30 lb weight. The equations for the 90 lb weight are:

$$(1)\quad \Sigma F_{\parallel} = 0 = -90\sin\theta + \tfrac{1}{3}N_1 + \tfrac{1}{3}N_2$$
$$(2)\quad \Sigma F_{\perp} = 0 = N_2 - 90\cos\theta - N_1$$

Substitute $N_1 = 30\cos\theta$ into (*1*) and (*2*). Then eliminate N_2 to find $\theta = \tan^{-1}\frac{5}{9} = 29°3'$.

8. A body weighing 350 lb rests on a plane inclined 30° with the horizontal as shown in Fig. 8-11. The angle of static friction between the body and the plane is 15°. What horizontal force P is necessary to hold the body from sliding down the plane?

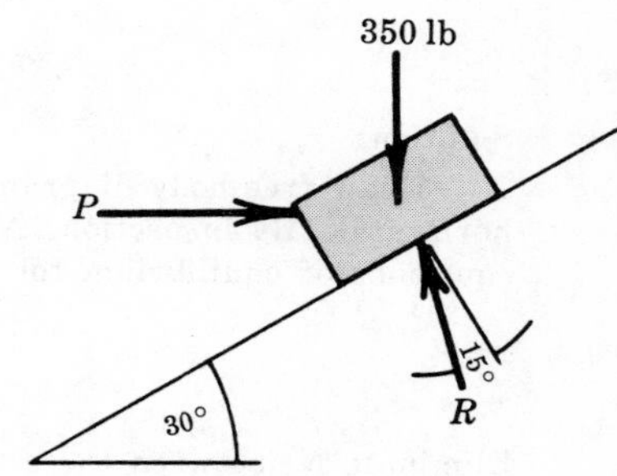

Fig. 8-11

Solution:

Since a slight decrease in P would cause motion down the plane, the limiting value of friction F' is used. The reaction R is shown acting at the angle of friction ϕ of 15° with the normal and in such a way as to assist the force P.

It should be noted that the angle of static friction can be used only if motion impends.

The equations of motion are obtained by summing forces parallel and perpendicular to the plane.

$$\Sigma F_{\parallel} = 0 = P\cos 30° - 350\sin 30° + R\sin 15°$$
$$\Sigma F_{\perp} = 0 = R\cos 15° - 350\cos 30° - P\sin 30°$$

Solving simultaneously by multiplying the first equation by cos 15° and the second equation by sin 15°, we obtain $P = 93.4$ lb.

9. Determine the necessary force P acting parallel to the plane to cause motion to impend. Assume the coefficient of friction is 0.25 and the pulley smooth. Refer to Fig. 8-12(*a*) below.

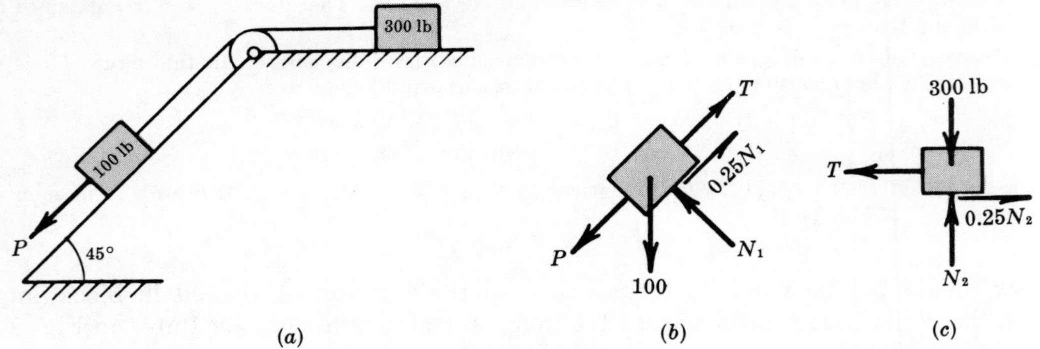

Fig. 8-12

Solution:

Draw free body diagrams of the two weights (Fig. 8-12(*b*) and (*c*) above), indicating the tension in the rope by T. By inspection, $N_2 = 300$ lb; hence $T = 0.25\,N_2 = 75$ lb. Also by inspection, $N_1 = 100\cos 45° = 70.7$ lb; thus $F_1' = 0.25\,N_1 = 17.7$ lb. A summation of forces parallel to the inclined plane yields $\Sigma F = 0 = P + 100 \times 0.707 - 75 - 17.7$, or $P = 22.0$ lb.

10. What is the least value of P to cause motion to impend? Assume the coefficient of friction to be 0.20. See Fig. 8-13(*a*) below.

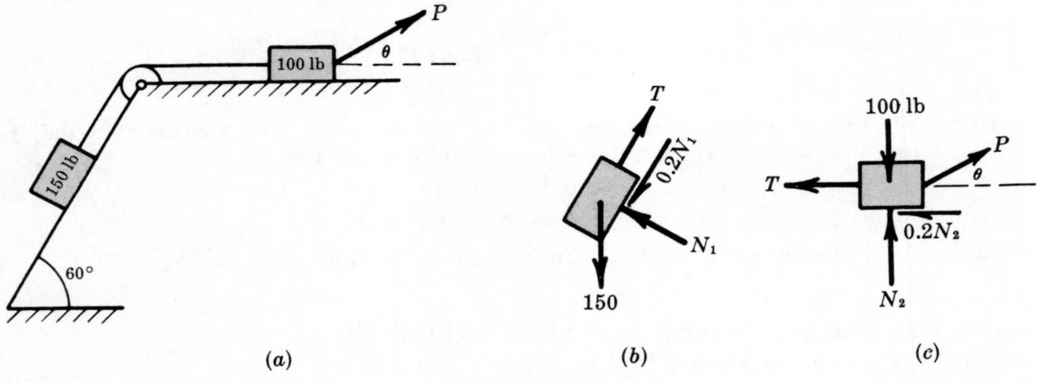

Fig. 8-13

Solution:

Draw free body diagrams of both weights. Note that P is shown at an unknown angle θ with the horizontal. By inspection, $N_1 = 150\cos 60° = 75$ lb. Hence $T = 0.20\,N_1 + 150\cos 30° = 145$ lb. The equations of equilibrium for the 100 lb weight are:

$$(1)\quad \Sigma F_h = 0 = P\cos\theta - 0.20N_2 - 145$$

$$(2)\quad \Sigma F_v = 0 = P\sin\theta + N_2 - 100$$

Eliminate N_2 between these two equations to obtain $P = \dfrac{165}{\cos\theta + 0.20\sin\theta}$.

The value of P will be a minimum when the denominator $(\cos\theta + 0.20\sin\theta)$ is maximum. Take the derivative of the denominator with respect to θ and set this equal to zero to determine the value of θ which will make P a minimum.

$$\frac{d}{d\theta}(\cos\theta + 0.20\sin\theta) = -\sin\theta + 0.20\cos\theta = 0 \quad \text{or} \quad \theta = \tan^{-1} 0.20 = 11°20'$$

Hence the minimum value of P is $P = \dfrac{165}{\cos 11°20' + 0.20\sin 11°20'} = 162$ lb.

11. Will the 40 lb force cause the 200 lb cylinder to slip? The coefficient of friction is 0.25. See Fig. 8-14.

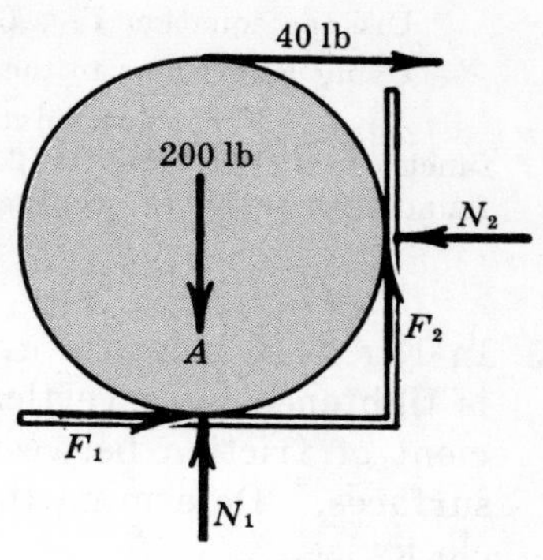

Fig. 8-14

Solution:

Since it is unknown whether the cylinder slips or not, it is not possible to say $F_1 = \mu N_1$ and $F_2 = \mu N_2$. Therefore consider F_1 and F_2 as unknowns together with N_1 and N_2. The equations of equilibrium are:

$$(1)\quad \Sigma F_h = 0 = F_1 - N_2 + 40$$
$$(2)\quad \Sigma F_v = 0 = N_1 + F_2 - 200$$
$$(3)\quad \Sigma M_A = 0 = -40 \times 2r + F_2 \times r + N_2 \times r$$

Solve for N_1, N_2, and F_1 in terms of F_2. These values are $N_1 = 200 - F_2$, $N_2 = 80 - F_2$, and $F_1 = 40 - F_2$.

Let us assume F_2 is its maximum value, i.e., $0.25\,N_2$, and solve for N_2, N_1, and F_1 in the equations in the preceding paragraph. Then $N_2 = 64$ lb, $N_1 = 184$ lb, and $F_1 = 24$ lb.

This means that if F_2 assumes its maximum static value, then F_1 must be 24 lb to hold the system in equilibrium. Since the maximum value of F_1 obtainable is $0.25\,N_1 = 46$ lb, the cylinder will not rotate.

12. The mean diameter of the threads of a square-threaded screw is 2 inches. The pitch of the thread is $\frac{1}{4}$ inch. The coefficient of friction $\mu = 0.15$. What force must be applied at the end of a 2 ft lever which is perpendicular to the longitudinal axis of the screw to raise a load of 4000 lb? To lower the load?

Solution:

To raise the load apply formula (a) under The Jack-screw: $M = Wr \tan(\phi + \beta)$.

The turning moment M is equal to the product of the force and the length of the lever. ϕ is the angle whose tangent is 0.15, or $\phi = 8.53°$.

$$\beta = \tan^{-1}\frac{\text{lead}}{\text{mean circumference}} = \tan^{-1}\frac{0.25}{\pi 2} = \tan^{-1} 0.0397 = 2.27°.$$

$M = P \times 24 = 4000 \times 1 \tan(8.53° + 2.27°)$. Hence $P = 31.8$ lb, to raise the load.

To determine the force to lower the load apply the formula: $M = Wr \tan(\phi - \beta)$.

$M = P \times 24 = 4000 \times 1 \tan(8.53° - 2.27°)$. Hence $P = 18.3$ lb, to lower the load.

13. A screw-jack has 4 threads per inch. The mean radius of the threads is 2.338 inches. The mean diameter of the bearing surface under the cap is 3.25 inches. The coefficient of friction for all surfaces is 0.06. What turning moment is necessary to raise 1500 lb?

Solution:

As indicated in the theory, a term must be added to account for the additional moment necessary to overcome the friction between the cap and the screw.

$M = Wr \tan(\phi + \beta) + \mu W r_c$, where r_c is the mean radius of the bearing surfaces between cap and screw. $\phi = \tan^{-1} 0.06 = 3.43°$. $\beta = \tan^{-1} 0.25/(2\pi \times 2.338) = 1.00°$.

$M = 1500 \times 2.338 \tan(3.43° + 1.00°) + 0.06 \times 1500 \times 3.25/2 = 418$ lb-in.

14. Two equal pulleys, of diameter 30 inches, are connected by a belt. The tension in the tight side of the belt is 200 lb. If the coefficient of friction is 0.25, determine the tension in the slack side when the belt is about to slip.

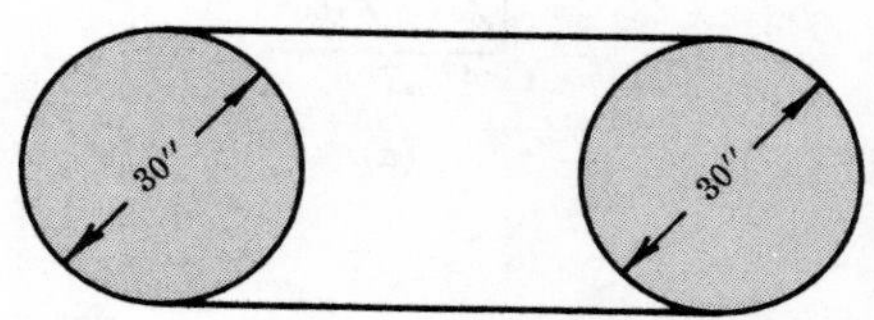

Fig. 8-15

Solution:

From Fig. 8-15 it is evident that the angle of wrap of the belt on either pulley is 180° or π radians.

Use the equation $T_1 = T_2 e^{\alpha\mu}$. Then $200 = T_2 \times 2.718^{0.25\pi}$.

Using logarithms to the base 10, $\log T_2 = 2.3010 - 0.3410 = 1.9600$. Hence $T_2 = 91.2$ lb.

As an alternative solution, find the value of $e^{0.25\pi}$ by interpolation from tables of exponential functions: $e^{0.25\pi} = e^{0.7854} = 2.1933$. Hence $T_2 = 200/2.1933 = 91.2$ lb. Of course, these values can be found with sufficient accuracy using a slide rule.

15. In Fig. 8-16 below, a drum 25 inches in diameter is encircled by a brake band which is tightened by a vertical force P of 40 lb on the lever AC. Assume that the coefficient of friction between the drum and the band is $\frac{1}{3}$. Neglect friction on all other surfaces. Determine the net braking moment on the drum if rotation is impending clockwise.

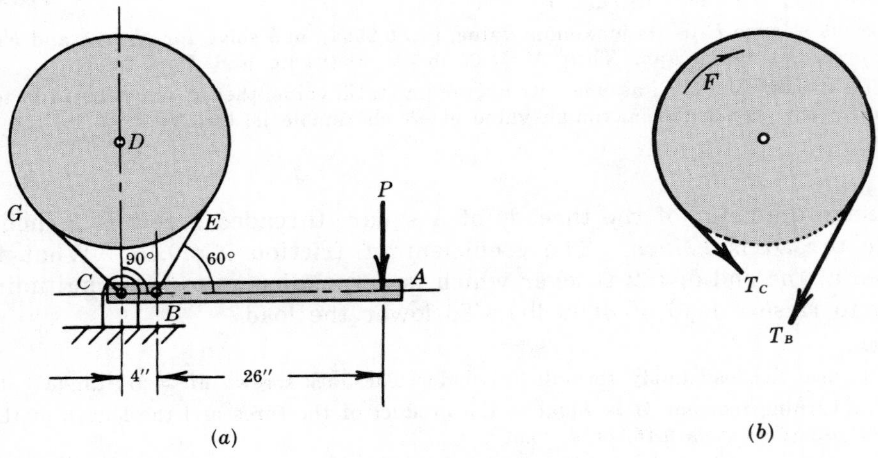

Fig. 8-16

Solution:

In solving with rotation of the drum impending clockwise, note that friction F *on* the drum acts to oppose motion, i.e., counterclockwise. This same friction acts *on* the belt in an opposite or clockwise direction. In considering the free body diagram of the belt, the band tension T_C must be greater than T_B since it holds both T_B and F in equilibrium.

Taking moments about C of the forces acting on the lever, we obtain

$$\Sigma M_C = 0 = +T_B \sin 60^\circ \times 4 - 40 \times 30 \quad \text{or} \quad T_B = 347 \text{ lb}$$

The angle of wrap must be found before the formula for tensions can be applied. Fig. 8-17 is drawn to show the trigonometry of the problem. First we determine the value of θ.

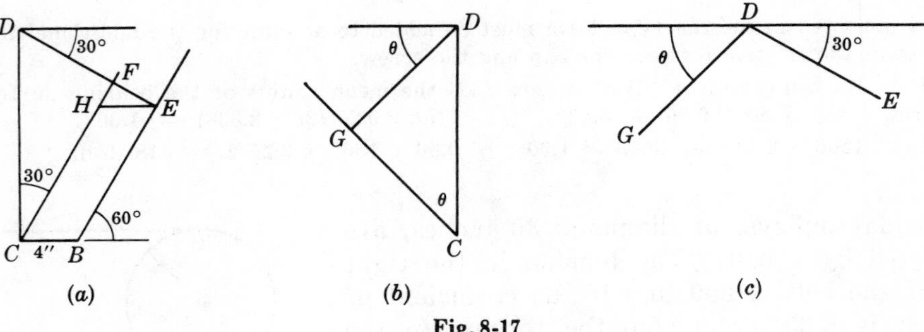

Fig. 8-17

In Fig. 8-17(a), $CD = \dfrac{DF}{\sin 30^\circ} = \dfrac{DE - EF}{\sin 30^\circ} = \dfrac{12.5 - HE\cos 30^\circ}{\sin 30^\circ} = \dfrac{12.5 - 4\cos 30^\circ}{\sin 30^\circ} = 18.1''$.

From Fig. 8-17(b), $\theta = \sin^{-1} GD/CD = \sin^{-1} 12.5/18.1 = \sin^{-1} 0.691 = 43.7^\circ$.

From Fig. 8-17(c) above, angle of wrap $= 180° + 30° + 43.7° = 253.7° = 4.4279$ radians.
T_C is greater than T_B. Hence $T_C = T_B e^{\alpha\mu} = 347e^{4.4279/3} = 1520$ lb.
Hence the braking moment is $(T_C - T_B) \times 12.5 = (1520 - 347) \times 12.5 = 14{,}700$ lb-in.

16. Four turns of rope around a horizontal post will hold a 1000 lb weight with a pull of 10 lb. Determine the coefficient of friction between the rope and the post.

Solution:

Use the equation $T_1 = T_2 e^{\alpha\mu}$, where $T_1 = 1000$ lb and $T_2 = 10$ lb. Hence $e^{\alpha\mu} = 100$.
But α = the angle of wrap in radians $= 4 \times 2\pi$ radians.
From exponential tables, $e^{4.605} = 100$. Hence $\alpha\mu = 8\pi\mu = 4.605$, or $\mu = 0.18$.

17. What force is necessary to hold a weight of 2000 lb suspended on a rope wrapped twice around a post? Assume the coefficient of friction $\mu = 0.20$.

Solution:

Use the equation $T_1 = T_2 e^{\alpha\mu}$, where T_1 is the larger force (2000 lb). Hence the holding force $T_2 = 2000/e^{\alpha\mu}$, where $\alpha = 2 \times 2\pi$ radians and $\mu = 0.20$. The solution is $T_2 = 162$ lb.

18. A steel wheel 30 inches in diameter rolls on a horizontal steel rail. It carries a weight of 500 lb. The coefficient of rolling resistance is 0.012 in. What is the force P necessary to roll the wheel along the rail?

Solution:

$$P = \frac{Wa}{r} = \frac{500 \times 0.012}{15} = 0.4 \text{ lb}$$

19. A circular shaft D inches in diameter is resting in a support as shown in Fig. 8-18. The shaft supports a weight W lb. Assuming the coefficient of friction is μ, determine the twisting moment M necessary to cause rotary motion to impend.

Solution:

The resisting friction dF' acting on the differential area $dA = \rho\, d\rho\, d\theta$ is shown. The moment of this friction about the centerline of the shaft is $\rho\, dF'$.

But the normal force on this differential area is the product of the area and the unit load which is the total weight divided by the total area. Thus we can write

$$dF' = \mu\, dN = \mu \frac{W}{\frac{1}{4}\pi D^2} \rho\, d\rho\, d\theta$$

and
$$M = \int_0^{\frac{1}{2}D} \int_0^{2\pi} \rho \frac{4\mu W}{\pi D^2} \rho\, d\rho\, d\theta$$

$$= \int_0^{\frac{1}{2}D} \frac{4\mu W}{\pi D^2} \rho^2\, d\rho (2\pi) = \frac{8\mu W}{D^2}\left[\frac{\rho^3}{3}\right]_0^{\frac{1}{2}D} = \frac{\mu W D}{3}$$

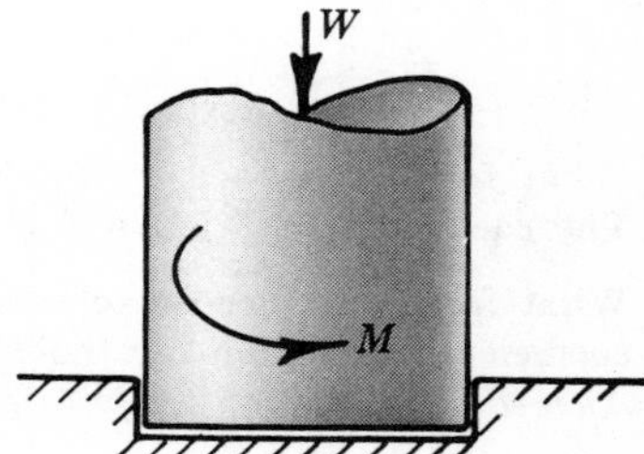

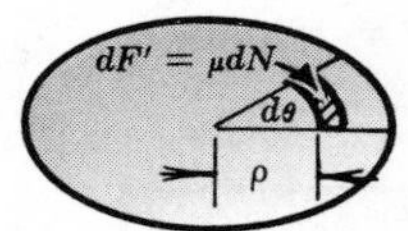

Fig. 8-18

Supplementary Problems

20. The clamp exerts a normal force of 100 lb on the three pieces held together as shown in Fig. 8-19. What force P may be exerted before motion impends? The coefficient of friction between the pieces is 0.30. *Ans.* $P = 60$ lb

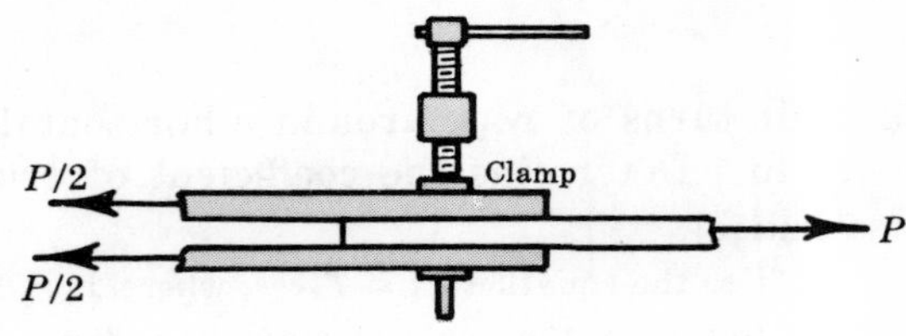

Fig. 8-19

21. A homogeneous ladder 18 ft long and weighing 120 lb rests against a smooth wall. The angle between it and the floor is 70°. The coefficient of friction between the floor and the ladder is $\frac{1}{4}$. How far up the ladder can a 180 lb man walk before the ladder slips? *Ans.* 14.6 ft

22. A man can pull horizontally with a force of 100 lb. An 800 lb weight is resting on a horizontal surface for which the coefficient of friction is 0.20. The vertical cable of a crane is attached to the top of the block as shown in Fig. 8-20 below. What will be the tension in the cable if the man is just able to start the block to the right? *Ans.* $T = 300$ lb

23. Refer to Fig. 8-21 below. The wedge B is used to raise the weight of 2000 lb resting on block A. What horizontal force P is required to do this if the coefficient of friction for all surfaces is 0.2? *Ans.* $P = 1540$ lb

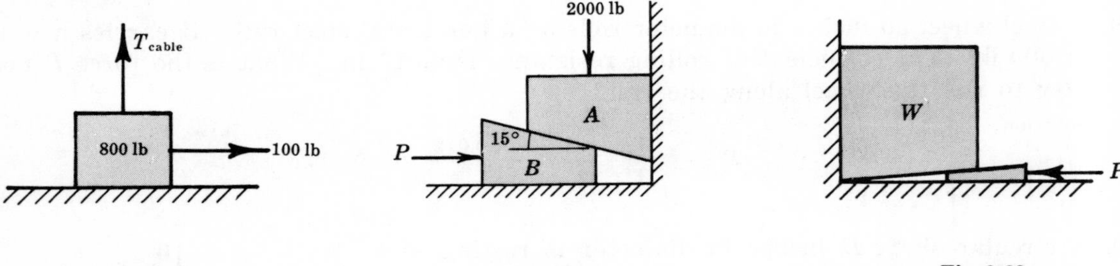

Fig. 8-20 Fig. 8-21 Fig. 8-22

24. What force P is needed to drive the 5° wedge to raise the 1000 lb weight shown in Fig. 8-22 above? The coefficient of friction for all surfaces is 0.25. *Ans.* $P = 596$ lb

25. What force P is needed to raise the 1200 lb weight with the 3° wedge shown in Fig. 8-23 below? The coefficient of friction for the wedge and the weight and for the wedge and the horizontal plane is 0.25. *Ans.* $P = 332$ lb

26. Refer to Fig. 8-24 below. Block A weighing 100 lb rests on block B and is tied with a horizontal string to the wall at C. Block B weighs 200 lb. If the coefficient of friction between A and B is $\frac{1}{4}$ and between B and the surface is $\frac{1}{3}$, what horizontal force P is necessary to move block B? *Ans.* $P = 125$ lb

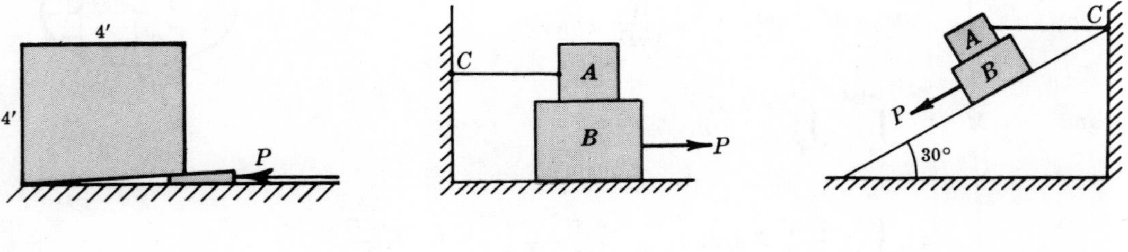

Fig. 8-23 Fig. 8-24 Fig. 8-25

27. Refer to Fig. 8-25 above. Block A weighing 60 lb rests on block B which weighs 80 lb. Block A is restrained from moving by a horizontal rope tied to the wall at C. What force P parallel to the plane inclined 30° with the horizontal is necessary to start B down the plane? Assume μ for all surfaces to be $\frac{1}{3}$. *Ans.* $P = 40.3$ lb

28. Refer to Fig. 8-26 below. Cube A weighing 16 lb is 4 inches on a side. Angle $\theta = 15°$. If the coefficient of friction is $\frac{1}{4}$, will the cube slide or tip as the force P is gradually increased?
Ans. Slide. $P_{\text{tipping}} = 9.80$ lb, $P_{\text{sliding}} = 8.00$ lb

29. Refer to Fig. 8-27 below. The weight of A is 50 lb; the weight of B is 80 lb. The coefficients of friction are 0.60 between A and B, 0.20 between B and the plane, and 0.30 between the rope and the fixed drum. Determine the maximum weight of W before motion impends. *Ans.* $W = 41.7$ lb

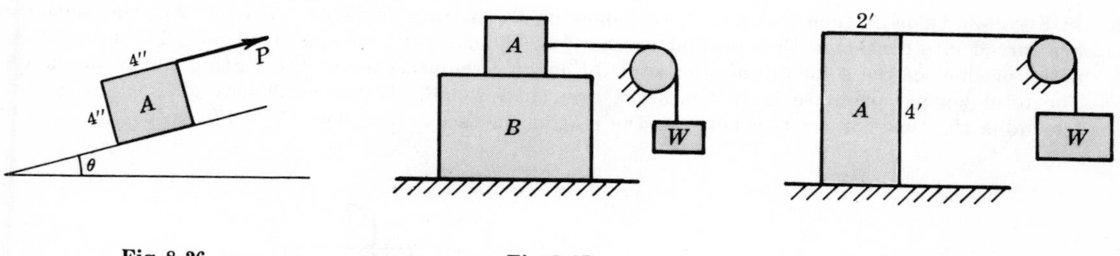

Fig. 8-26 Fig. 8-27 Fig. 8-28

30. Refer to Fig. 8-28 above. The homogeneous body A weighs 120 lb. The coefficients of friction are 0.30 between A and the plane and $2/\pi$ between the rope and the drum. What value of W will cause motion of A to impend? *Ans.* $W = 81.5$ lb

31. The rope holding the 50 lb weight E passes over the pulley and is attached to the frame at A as shown in Fig. 8-29 below. The weight of C is 60 lb. What is the minimum coefficient of friction μ between the outer rope and the pulley for equilibrium? *Ans.* $\mu = 0.291$

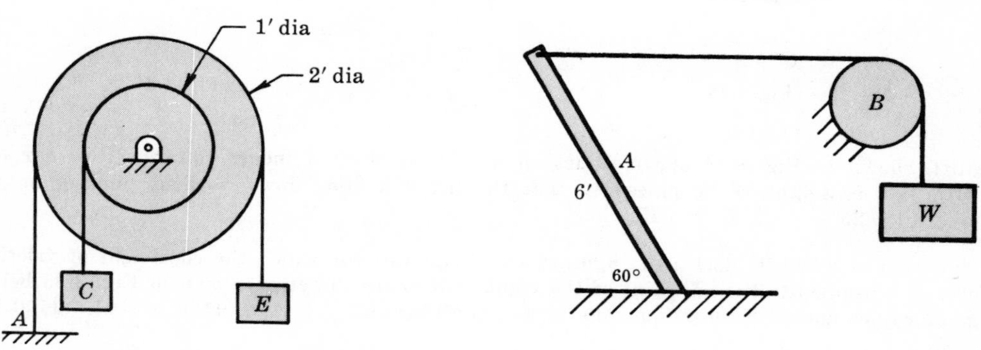

Fig. 8-29 Fig. 8-30

32. The homogeneous bar A weighs 40 lb, is 6 ft long and is inclined 60° with the horizontal plane. The weight W of 15.8 lb is connected by the rope to the bar as shown in Fig. 8-30 above. The rope leaving the bar is horizontal. The coefficient of friction between the rope and the drum B is 0.20. What is the minimum coefficient of friction between the bar and the plane for equilibrium? *Ans.* $\mu = 0.288$

33. The angle of static friction between the block of weight W and the plane is ϕ. For the angles shown in the adjacent Fig. 8-31, what is the expression for the force P to move the block up the plane?
Ans. $P = \dfrac{W \sin(\theta + \phi)}{\cos(\beta - \phi)}$

34. In Problem 33, show that the minimum value of P for a given angle θ is $W \sin(\theta + \phi)$.

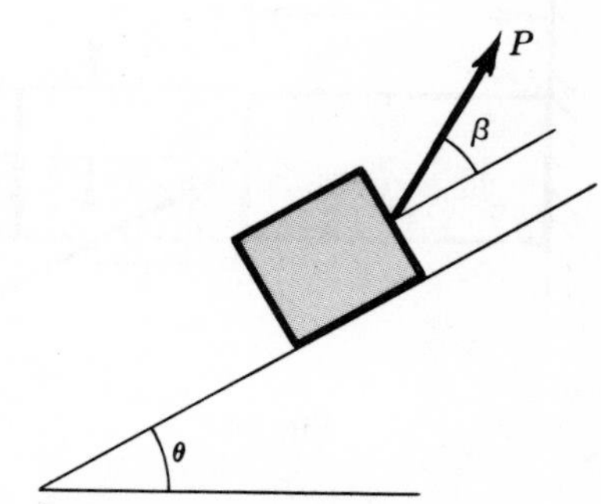

Fig. 8-31

35. A homogeneous bar of length l and weight w lb rests horizontally as shown in Fig. 8-32 with its free end on a block of weight W. The block W is at rest on a plane inclined at the angle α with the horizontal. Determine the necessary coefficient of friction μ between the block and the plane for equilibrium. Assume no friction between the bar and the block.
Ans. $\mu = \dfrac{\sin 2\alpha}{w/W + 2\cos^2\alpha}$

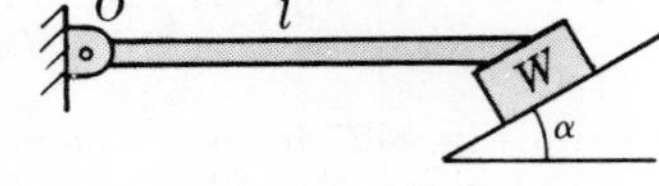

Fig. 8-32

36. In Fig. 8-33 below, a cone clutch is drawn showing the mating surfaces. Assume that the mating parts are forced into contact with a normal pressure on the mating surfaces of 10 psi. The area of contact is the product of the 2 inch dimension and the mean circumference of the mating parts, i.e., $2 \times \pi \times 8$. The total normal pressure is 10 times the area just found. If the coefficient of friction is $\mu = 0.35$, determine the force of friction between the mating surfaces. *Ans.* $F = 176$ lb

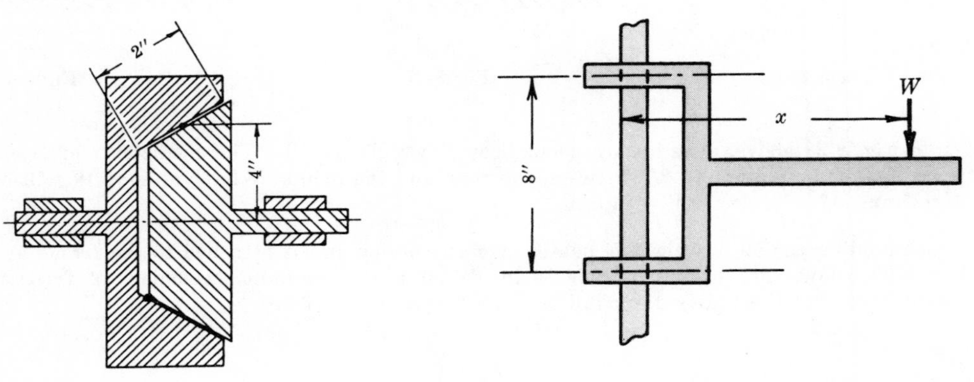

Fig. 8-33 Fig. 8-34

37. The lift, shown in Fig. 8-34 above, slides on a vertical shaft 3 inches square. How far out on the platform can a weight W be placed so that the lift will slide down without binding on the shaft? Assume $\mu = 0.25$. *Ans.* $x = 17.5$ in.

38. The homogeneous 800 lb block rests against a vertical wall for which the coefficient of friction is 0.25. A force P is applied to the midpoint of the right face in the direction shown in Fig. 8-35 below. What range of values may P have without disturbing equilibrium? *Ans.* 1190 lb $< P <$ 1790 lb

39. What couple M is needed to cause the wheel of weight W and radius r shown in Fig. 8-36 below to have impending motion? The coefficient of friction for all surfaces is μ. *Ans.* $M = \mu Wr(1+\mu)/(1+\mu^2)$

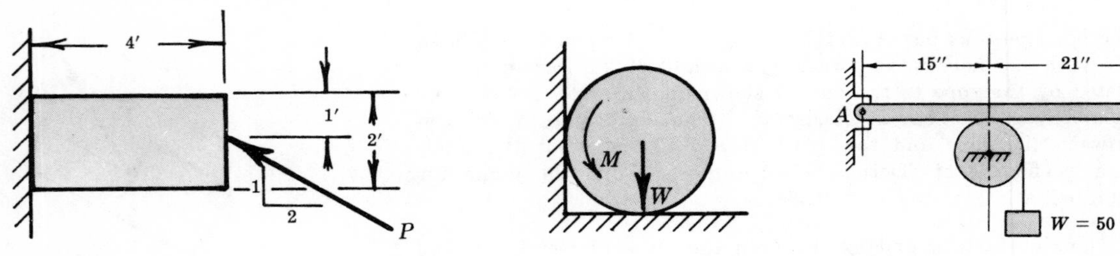

Fig. 8-35 Fig. 8-36 Fig. 8-37

40. A vertical force P exerted on lever AB holds the weight of 50 lb from falling, as shown in Fig. 8-37 above. The coefficient of friction between the lever and the 1 ft drum is $\frac{1}{4}$. Neglect the weight of the drum and lever, and determine the force P to hold the weight. *Ans.* $P = 83.3$ lb

41. In Fig. 8-38, W_1 weighs 50 lb and W_2 weighs 30 lb. They are tied together by a rope parallel to the plane. The coefficient of friction between W_1 and the plane is $\frac{1}{4}$, and between W_2 and the plane it is $\frac{1}{2}$. Determine the value of the angle θ at which sliding will occur. What is the tension in the rope? *Ans.* $\theta = 19.0°$, $T = 4.4$ lb

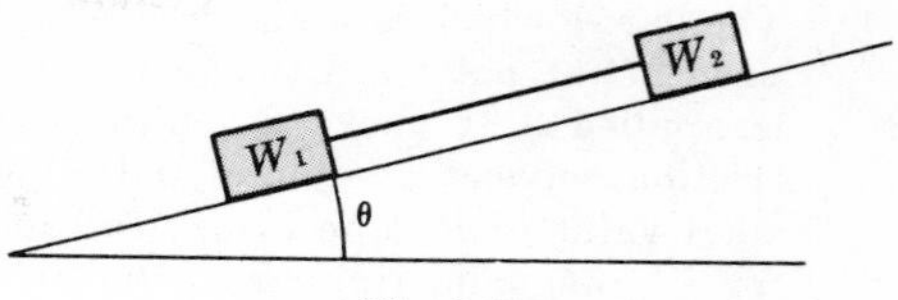

Fig. 8-38

42. Determine the force P to cause motion to impend if the coefficient of friction for both blocks and the plane shown in Fig. 8-39 below is 0.25. The force P and the ropes are parallel to the plane. The pulley is frictionless. *Ans.* $P = 1.5$ lb

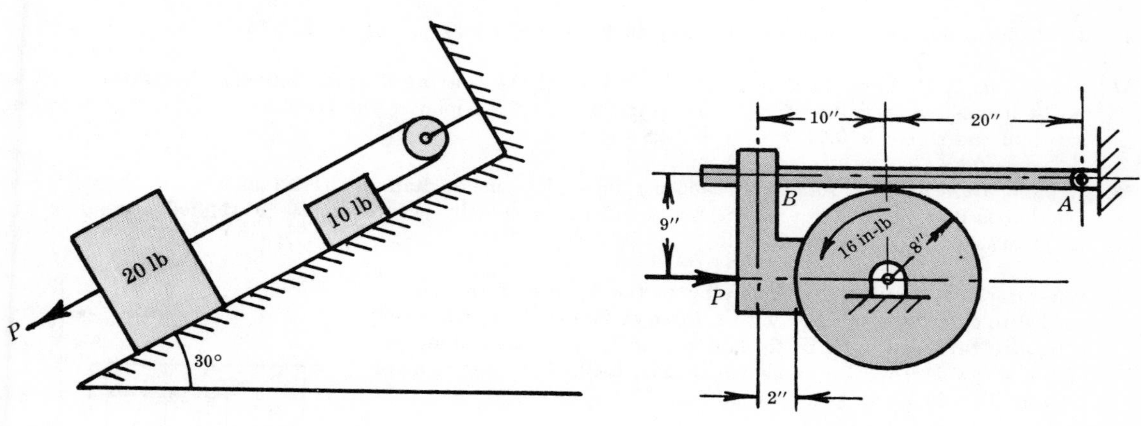

Fig. 8-39

Fig. 8-40

43. The drum shown in Fig. 8-40 above is subjected to a counterclockwise torque of 16 in-lb. What horizontal force P is necessary to resist motion? The coefficient of friction between both braking members (pinned together at B) and the drum is 0.40. Neglect weight of braking members. *Ans.* $P = 3.43$ lb

44. A body weighing 50 lb rests on a plane inclined 60° with the horizontal. The coefficient of friction between the two surfaces is $\frac{1}{3}$. The body is acted upon by a horizontal force P. What is the value of P so that the body will not slide down the plane? What is the value of P so that the body will slide up the plane? In between these values the body will be at rest. *Ans.* $P = 44.3$ lb, $P = 244$ lb

45. A plane is inclined at an angle θ with the horizontal. A body can just rest on the plane. Determine the least force P to draw the body up the plane. (Hint: $\theta = \alpha$, since θ is the angle of repose; also assume that P acts at angle β with the plane.) *Ans.* $P = W \sin 2\theta$

46. The uniform bar weighs 80 lb. What rightward force P is needed to start the bar moving? The coefficient of friction for all surfaces is 0.30. See Fig. 8-41. *Ans.* $P = 57.5$ lb

47. Two blocks of equal weight W can slide on a horizontal rod. The coefficient of friction between the blocks and the rod is μ. A string of length l is suspended between the blocks and carries a weight M at its midpoint. How far apart will the blocks be for equilibrium?

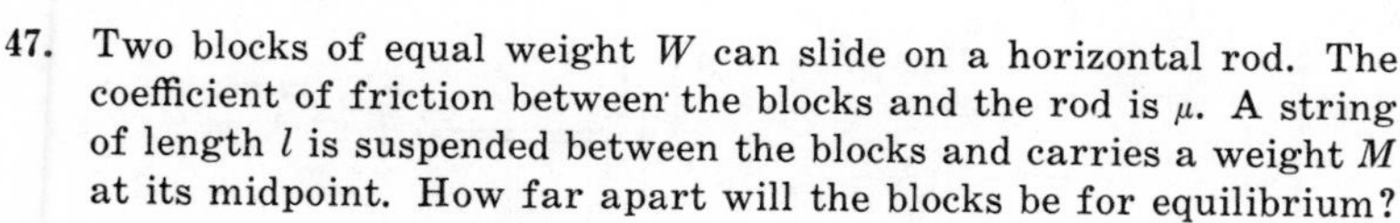

Ans. $x = \dfrac{l\mu(W + M/2)}{\sqrt{(M/2)^2 + \mu^2(W + M/2)^2}}$

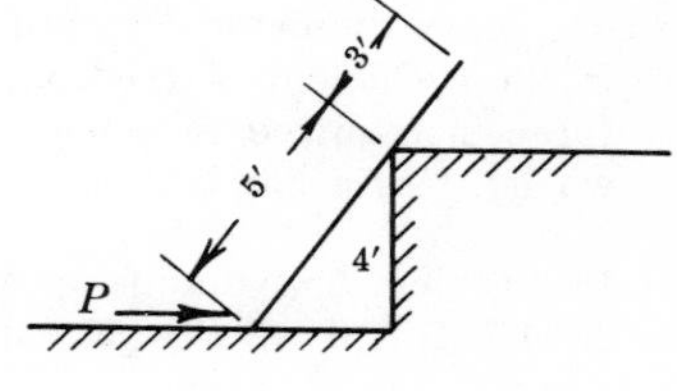

Fig. 8-41

48. A prism whose cross section is a regular polygon of n sides rests on a horizontal face. An insect is crawling up the inside. The coefficient of friction between the insect and the inner faces is μ. Show that the highest face which the insect can climb (counting the horizontal bottom face as one) is given by the expression: $n/360 \times (\tan^{-1} \mu + 360/n)$.

49. The bar A which is weightless is pinned to the homogeneous 600 lb prism B at point C as shown in Fig. 8-42. A horizontal force P is applied 2 ft above the horizontal plane. If the coefficient of friction between the plane and either the bar or the prism is 0.4, what value of P is necessary to cause motion to impend? Analyze for slipping and tipping of the prism. *Ans.* $P = 209$ lb

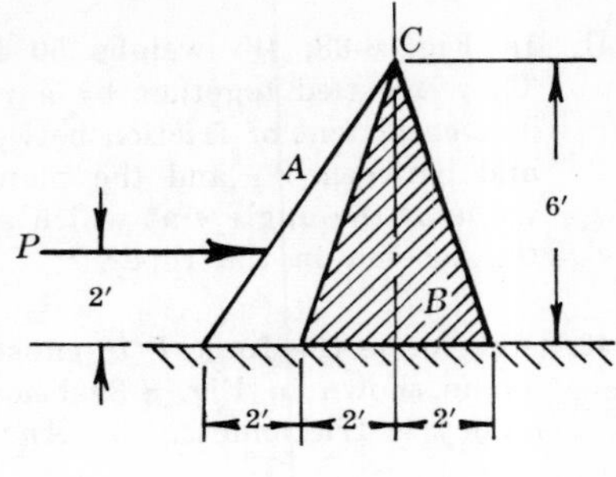

Fig. 8-42

50. A jack-screw has 3 threads per inch with a mean radius of thread equal to 0.648 inch. A lever 18 inches long is used to raise a load of 2400 lb. If the coefficient of friction is 0.10, what force is necessary when applied normal to the lever at its free end? *Ans.* 15.9 lb

51. In Problem 50, what force is necessary to lower the load? *Ans.* 1.55 lb

52. Supposing in Problem 50 that the mean radius of the bearing surface between the cap and the screw is 1.4 inches, what value of P is necessary to raise the load if the coefficient of friction between the cap and the screw is 0.07? *Ans.* 28.9 lb

53. A hand press has five square threads per inch. The mean diameter is 1.2 inches. If the coefficient of friction is 0.08, what force in the press can be exerted by a force of 20 lb applied normal to a lever 20 inches long? *Ans.* 5000 lb

54. What force P applied perpendicular to the handle of the vise is needed to clamp the piece A with a force of 20 lb? The jack-screw is square threaded with 10 threads per inch. The mean diameter of the screw is 0.438 in. as shown in Fig. 8-43. The coefficient of friction is 0.20. *Ans.* $P = 0.3$ lb

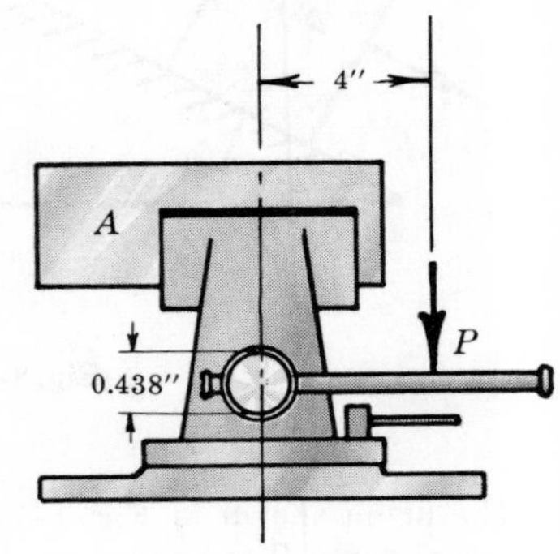

Fig. 8-43

55. A square-threaded jack-screw has a pitch of 0.4 in. and a mean diameter of 3.0 in. The mean diameter of the bearing surface between the cap and the screw is 3.5 in. The coefficient of friction between all surfaces is 0.10. What force is required at the end of a lever 36 inches long to raise 4000 lb? *Ans.* $P = 43.3$ lb

56. Determine the load that can be raised with a single-threaded jack-screw, having $2\frac{1}{2}$ threads per inch and a mean diameter of 3 inches, when a force of 200 lb is exerted on a bar 30 inches long. Use $\mu = 0.05$. *Ans.* $W = 43{,}100$ lb

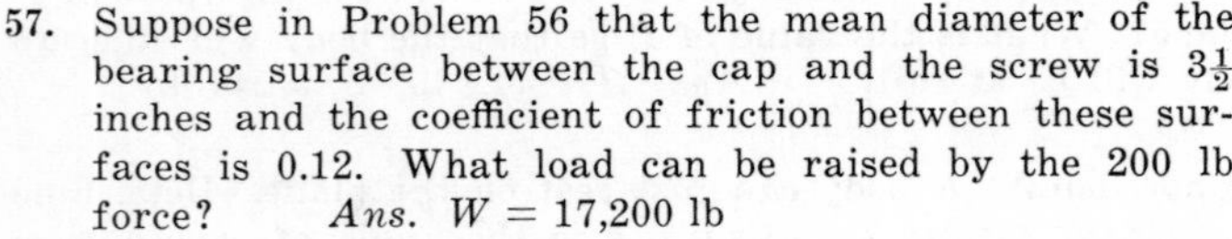

57. Suppose in Problem 56 that the mean diameter of the bearing surface between the cap and the screw is $3\frac{1}{2}$ inches and the coefficient of friction between these surfaces is 0.12. What load can be raised by the 200 lb force? *Ans.* $W = 17{,}200$ lb

58. A jack-screw has a square thread with a pitch of 0.3 inch. The mean diameter of thread is 2 inches. The cap has inner diameter 2 inches and outer diameter 3 inches. If the coefficient of friction for all surfaces is 0.15, what force is required to cause motion of a 5000 lb load upward? Use a bar 3 ft long. *Ans.* $P = 53.7$ lb

59. In Problem 58, what force is necessary to start the load down? *Ans.* $P = 40.1$ lb

60. Refer to Fig. 8-44. The weight of A is 500 lb and the weight of B is 100 lb. What force P must be exerted perpendicular to the arm of the jack and 20 in. from the jack centerline to raise A? The jack-screw has a lead of 0.32 in. and a mean diameter of 2.00 in. The coefficient of friction between the jack and the screw is 0.15. *Ans.* $P = 3.04$ lb

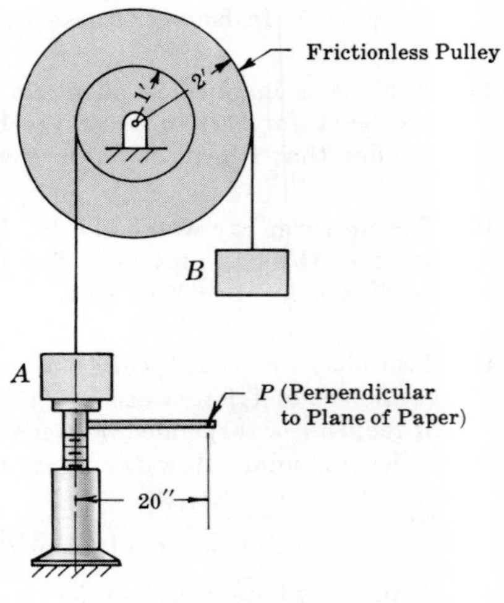

Fig. 8-44

61. What force is necessary to hold a weight of 100 lb suspended on a rope wrapped three times around a post? Assume $\mu = 0.3$. *Ans.* $F = 0.35$ lb

62. In Problem 61, what force is needed to raise the weight? *Ans.* $F = 28{,}500$ lb

63. In the adjacent Fig. 8-45, a weight of 50 lb is held from falling by the force in the rope passing over the fixed drum. What is the value of F if $\mu = 0.25$? *Ans.* $F = 33.7$ lb

64. In Problem 63, what must be the value of F to start the weight up? *Ans.* $F = 74$ lb

65. A laborer lowers a 400 lb boiler section into a pit by means of a rope snubbed $1\frac{1}{4}$ turns around a horizontal pole. If the coefficient of friction is 0.35, what force must he exert? *Ans.* $F = 25.6$ lb

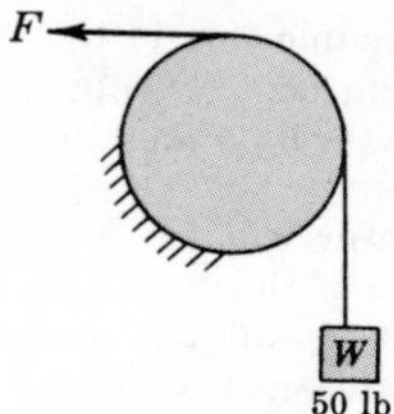

Fig. 8-45

66. Three turns of a rope around a horizontal post with a pull of 8 lb will hold a 250 lb weight. Determine the coefficient of friction between the rope and post. *Ans.* $\mu = 0.18$

67. Refer to Fig. 8-46. The two pulleys are held apart by a spring with compressive force S. The diameter of each pulley is d. The coefficient of friction between the belt and the pulley is μ. Determine the maximum torque which may be transmitted.
Ans. Torque $= \frac{1}{2}Sd(e^{\mu\pi} - 1)/(e^{\mu\pi} + 1)$

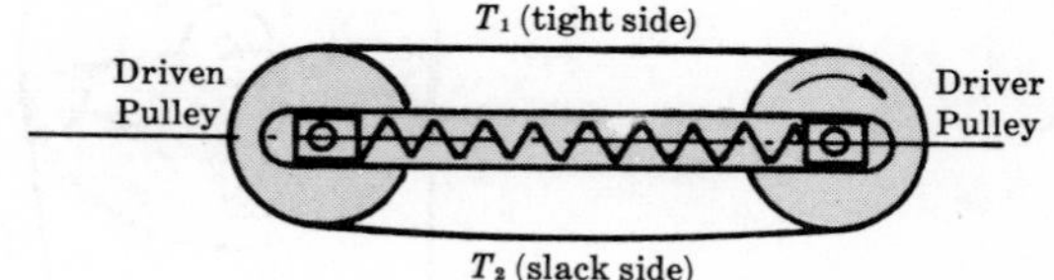

Fig. 8-46

68. Refer to Fig. 8-47. A weight of 120 lb is prevented from falling by a rope wrapped $\frac{1}{4}$ turn around drum B and $1\frac{1}{4}$ turns around drum A. Assuming that drum B is smooth and the coefficient of friction between the rope and A is $\frac{1}{4}$, what is the value of the holding force F? *Ans.* $F = 16.9$ lb

69. In Problem 68, assume that the coefficient of friction between the rope and the drum B is not zero but $\frac{1}{3}$. What is the value of the holding force F? (Hint: Use free body diagram of B to determine tension in the rope between A and B.) *Ans.* $F = 10.0$ lb

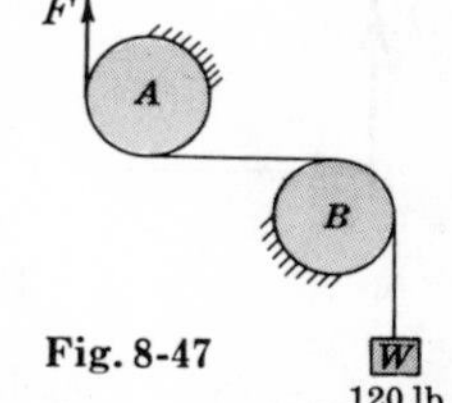

Fig. 8-47

70. A weight of 500 lb is held by passing a rope around a horizontal post and exerting a pull of 50 lb. If the coefficient of friction between the post and the rope is $\frac{1}{4}$, how many turns of the rope around the post are necessary? *Ans.* 1.47 turns

71. Refer to Fig. 8-48. Determine the range of values of tension T for equilibrium. The coefficient of friction $\mu = 1/\pi$ between the belt and each fixed drum. The tension at the other end of the belt is 15 lb.
Ans. $1.23 \text{ lb} < T < 183 \text{ lb}$

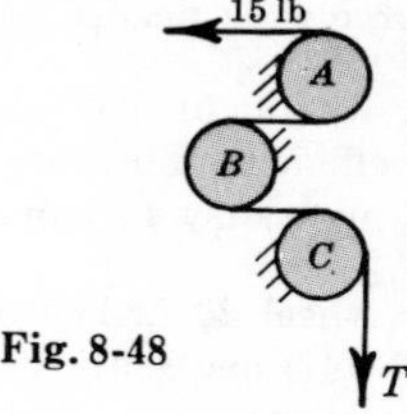

Fig. 8-48

72. In the adjacent Fig. 8-49, body A weighs 100 lb and body B weighs 300 lb. The coefficient of friction between A and the plane is 0.20. The coefficient of friction between A and B is 0.20. The coefficient of friction between the rope and the drum is 0.25. Determine the minimum weight W to cause motion of B to impend.
Ans. $W = 167$ lb

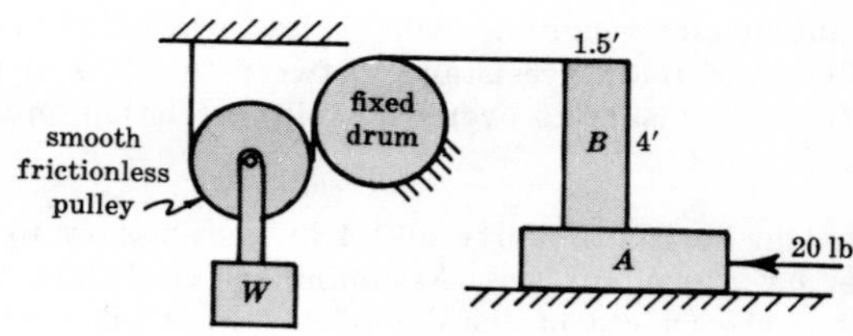

Fig. 8-49

73. A "weightless" beam of length $2l$ is supported by a cord which passes over two fixed drums as shown in Fig. 8-50. How far from the center can a weight W be placed without disturbing equilibrium?
Ans. $x = l(e^{\mu\pi} - 1)/(e^{\mu\pi} + 1)$

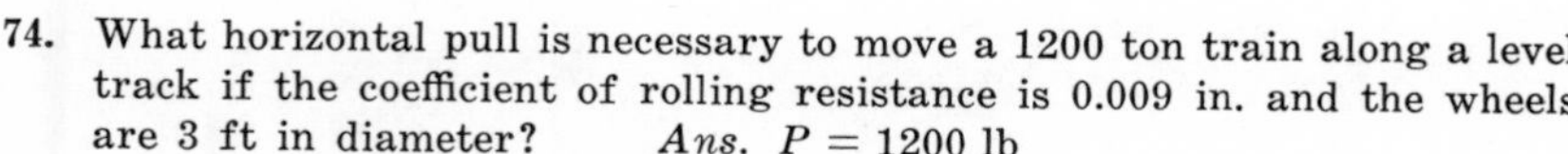

74. What horizontal pull is necessary to move a 1200 ton train along a level track if the coefficient of rolling resistance is 0.009 in. and the wheels are 3 ft in diameter? *Ans.* $P = 1200$ lb

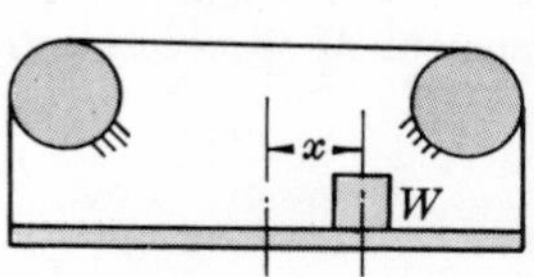

Fig. 8-50

75. A brake band encircles drum D and is fastened to the horizontal lever at B and C, as shown in Fig. 8-51. The diameter of the drum is 18 inches. The coefficient of friction between the brake band and the drum is $\frac{1}{3}$. Force P is 30 lb. What braking moment on the drum is possible (a) if the drum is rotating clockwise, (b) if the drum is rotating counterclockwise?
Ans. $M_a = 409$ lb-in, $M_b = 1170$ lb-in

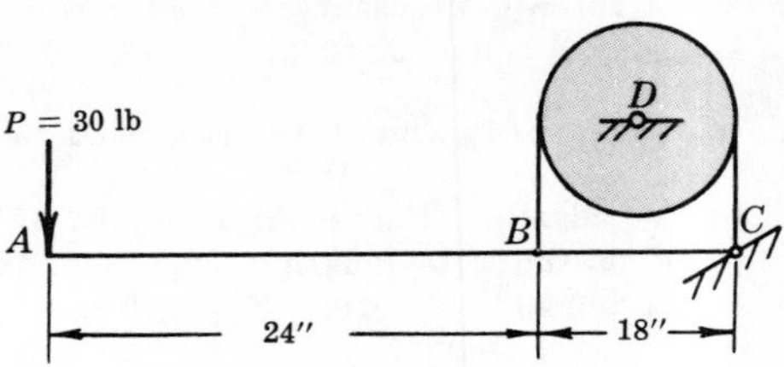

Fig. 8-51

76. Refer to Fig. 8-52 below. What force P is necessary to cause the brake to resist motion of the drum under the action of the moment M? The coefficient of friction between the brake and the drum is μ. The angle of wrap is α.

Ans. $P = \frac{M}{rc}\left(\frac{ae^{\mu\alpha} - b}{e^{\mu\alpha} - 1}\right)$

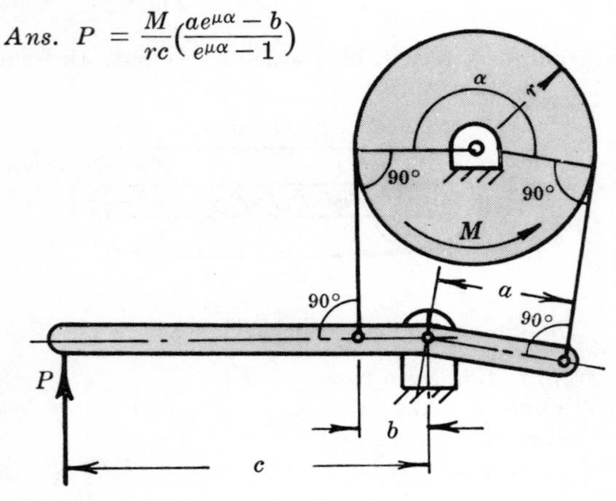

Fig. 8-52

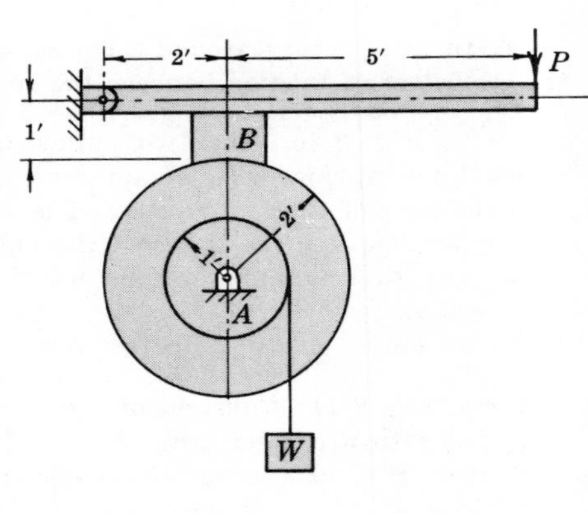

Fig. 8-53

77. The weight $W = 200$ lb hangs on a pulley A free to rotate in frictionless bearings as shown in Fig. 8-53 above. The coefficient of friction between the facing of brake B and the pulley is 0.25. What minimum force P is needed to prevent rotation? *Ans.* $P = 129$ lb

78. A weight of 3000 lb rests on an oak beam. The beam rests on 8 inch diameter rollers. Assuming the coefficient of rolling resistance between the beam and the rollers to be 0.035 in., what horizontal force is necessary to move the load on a level surface? *Ans.* $P = 26.3$ lb

79. A wheel 20 inches in diameter carries a load of 20,000 lb. If a horizontal force of 20 lb is necessary to move it over a level surface, determine the coefficient of rolling resistance. *Ans.* $a = 0.01$ in.

80. An automobile weighing 3900 lb has wheels 29 inches in diameter. Assuming a coefficient of rolling resistance between the tires and the road of 0.02 in., determine the force necessary to overcome rolling friction on a level road.
Ans. $P = 5.4$ lb

81. A central horizontal force of 1.4 lb is necessary to move a drum 36 inches in diameter on a level surface. Assuming the coefficient of rolling resistance is 0.025 in., what is the weight of the drum? *Ans.* $W = 1010$ lb

82. The collar bearing shown in Fig. 8-54 supports a load of 680 lb. If the coefficient of friction is 0.20 and if uniform distribution of pressure is assumed, what turning moment M is necessary? *Ans.* $M = 212$ lb-in

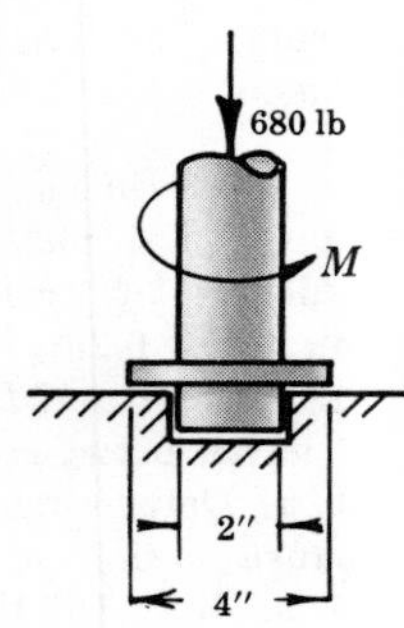

Fig. 8-54

Chapter 9

First Moments and Centroids

CENTROID of an ASSEMBLAGE

The centroid of an assemblage of n similar quantities, $\Delta_1, \Delta_2, \Delta_3, \ldots, \Delta_n$ situated at points $P_1, P_2, P_3, \ldots, P_n$ for which the position vectors relative to a selected point O are $\mathbf{r}_1, \mathbf{r}_2, \mathbf{r}_3, \ldots, \mathbf{r}_n$, has a position vector $\bar{\mathbf{r}}$ defined as

$$\bar{\mathbf{r}} = \frac{\sum_{i=1}^{n} \mathbf{r}_i \Delta_i}{\sum_{i=1}^{n} \Delta_i} \qquad (1)$$

where Δ_i = ith quantity. For example this could be an element of length, area, volume, or mass.

$\mathbf{r}_i$ = position vector of ith element

$\sum_{i=1}^{n} \Delta_i$ = sum of all n elements

$\sum_{i=1}^{n} \mathbf{r}_i \Delta_i$ = first moment of all elements relative to the selected point O.

The centroid has coordinates

$$\bar{x} = \frac{\sum_{i=1}^{n} x_i \Delta_i}{\sum_{i=1}^{n} \Delta_i}, \qquad \bar{y} = \frac{\sum_{i=1}^{n} y_i \Delta_i}{\sum_{i=1}^{n} \Delta_i}, \qquad \bar{z} = \frac{\sum_{i=1}^{n} z_i \Delta_i}{\sum_{i=1}^{n} \Delta_i} \qquad (2)$$

where Δ_i = magnitude of the ith quantity (element)

$\bar{x}, \bar{y}, \bar{z}$ = coordinates of centroid of the assemblage

x_i, y_i, z_i = coordinates of P_i at which Δ_i is concentrated.

CENTROID of a CONTINUOUS QUANTITY

The centroid of a continuous quantity may be located by calculus using infinitesimal elements of the quantity (such as dL of a line, dA of an area, dV of a volume, or dm of a mass). Thus for a mass m we can write

$$\bar{\mathbf{r}} = \frac{\int \mathbf{r}\, dm}{\int dm} \qquad (3)$$

In terms of x, y, and z coordinates the centroid of the continuous quantity has coordinates

$$\bar{x} = \frac{\int x\, dm}{\int dm} = \frac{Q_{yz}}{m}, \qquad \bar{y} = \frac{\int y\, dm}{\int dm} = \frac{Q_{xz}}{m}, \qquad \bar{z} = \frac{\int z\, dm}{\int dm} = \frac{Q_{xy}}{m} \qquad (4)$$

where Q_{xy}, Q_{yz}, Q_{xz} = first moments with respect to the xy, yz, and xz planes.

The centroid of a homogeneous mass coincides with the centroid of its volume.

The following table indicates the first moments Q of various quantities Δ about the coordinate planes.

Δ	Q_{xy}	Q_{yz}	Q_{xz}	Units
Line	$\int z\,dL$	$\int x\,dL$	$\int y\,dL$	L^2
Area	$\int z\,dA$	$\int x\,dA$	$\int y\,dA$	L^3
Volume	$\int z\,dV$	$\int x\,dV$	$\int y\,dV$	L^4
Mass	$\int z\,dm$	$\int x\,dm$	$\int y\,dm$	mL

where Q_{xy}, Q_{yz}, Q_{xz} = first moments with respect to the xy, yz, xz planes,

L = unit of length,

m = unit of mass,

dL, dA, dV, dm = differential elements respectively of line, area, volume, mass.

Note that in two dimensional work, e.g., in the xy plane, Q_{xz} becomes Q_x and Q_{yz} becomes Q_y.

THEOREMS of PAPPUS and GULDINUS

The theorems of Pappus and Guldinus may be stated as follows:

The area of a surface generated by revolving an arc of a plane curve about a non-intersecting axis in its plane is equal to the product of the length of the arc and the distance travelled by the centroid of the arc during the generation.

The volume of the solid generated by revolving a plane area about a non-intersecting axis in its plane is equal to the product of the area and the distance travelled by the centroid of the area during the generation.

CENTER of PRESSURE

The center of pressure is explained as follows:

When an area is subjected to a pressure, a point in the area exists through which the entire force could be concentrated with the same external effect. This point is called the *center of pressure*. If the pressure is uniformly distributed over an area, the center of pressure coincides with the centroid of the area.

Solved Problems

1. Determine Q_y for the area bounded by the parabola $y^2 = 4ax$ and the lines $y = 0$, $x = b$.

Solution:

The problem is set up in Fig. 9-1. Choose a differential element dA, as shown, parallel to the y-axis. It is at a constant distance x from the y-axis. Hence

$$Q_y = \int x\,dA$$

But dA is the product of y, the distance from the x-axis to the parabola, and dx, the width of the element. Also, x must vary from 0 to b to include the given area.

$$Q_y = \int_0^b xy\,dx = \int_0^b x\sqrt{4ax}\,dx$$

$$= 2\sqrt{a}\int_0^b x^{3/2}\,dx = \tfrac{2}{5}(2\sqrt{a})x^{5/2}\Big]_0^b = \tfrac{4}{5}b^2\sqrt{ab}$$

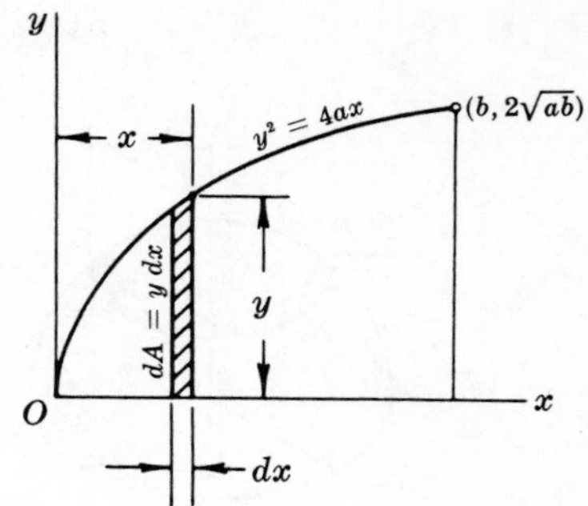

Fig. 9-1

2. In Problem 1, determine Q_x for the same area. Choose the differential strip parallel to the x-axis, as shown in Fig. 9-2 below.

Solution:

The height of the strip is dy and the width is $b - x$.

$$Q_x = \int y\,dA = \int_0^{2\sqrt{ab}} y(b-x)\,dy$$

Note that the upper limit of y is determined by allowing x to equal b, whence $y = \pm\sqrt{4ab}$. Choose the positive value.

$$Q_x = \int_0^{2\sqrt{ab}} y\Big(b - \frac{y^2}{4a}\Big)dy = \frac{b(2\sqrt{ab})^2}{2} - \frac{(2\sqrt{ab})^4}{16a} = ab^2$$

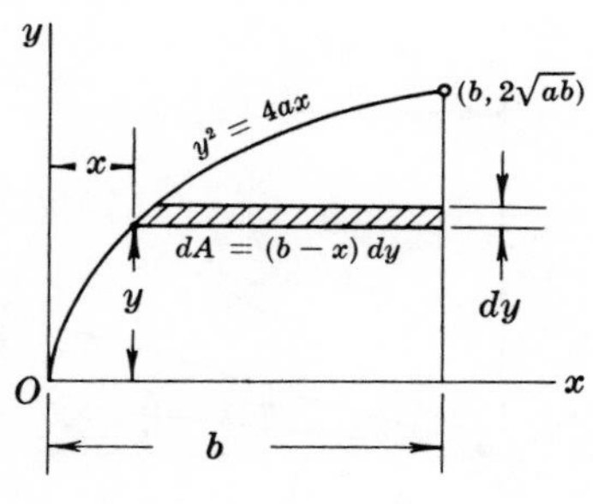

Fig. 9-2

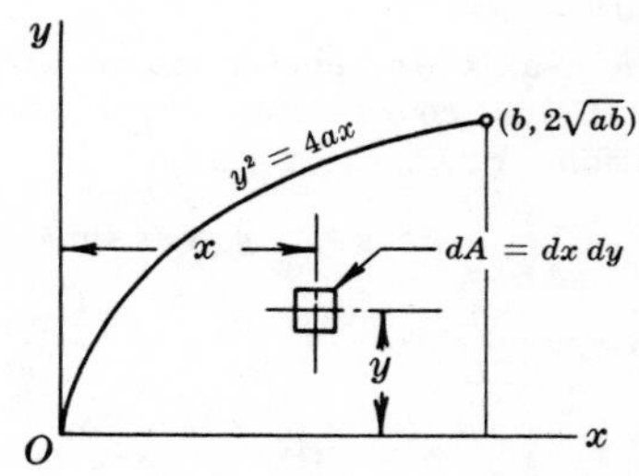

Fig. 9-3

3. Determine Q_x and Q_y in Problem 2 using the differential element shown in Fig. 9-3 above.

Solution:

In this case a double integration is involved, as shown in the following.

$$Q_x = \int y\,dA = \int_0^b\int_0^{2\sqrt{ax}} y\,dy\,dx = \int_0^b \Big[\tfrac{1}{2}y^2\Big]_0^{2\sqrt{ax}} dx = \int_0^b 2ax\,dx = ab^2.$$

$$Q_y = \int x\,dA = \int_0^b\int_0^{2\sqrt{ax}} dy\,x\,dx = \int_0^b [y]_0^{2\sqrt{ax}}\,x\,dx = \int_0^b 2\sqrt{ax}\,x\,dx = \tfrac{4}{5}b^2\sqrt{ab}.$$

The upper limits of the variable y must be expressed in terms of x as shown, because the summation vertically is limited by the curve which is of varying height.

4. Determine the moment of the volume of a hemisphere with respect to its base. Refer to Fig. 9-4 below.

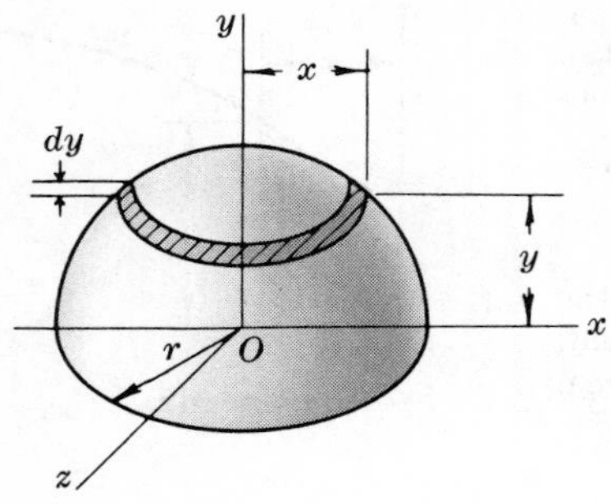

Fig. 9-4

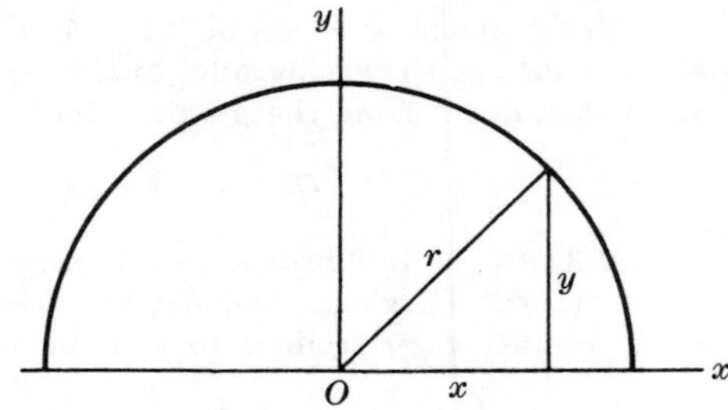

Fig. 9-5

Solution:

Choose a differential volume parallel to the base. It is at a distance y from the base which is shown in the xz-plane.

A cross section in the xy-plane yields a right triangle with sides x, y, and radius r (see Fig. 9-5 above). Hence $x^2 + y^2 = r^2$.

This could of course also be deduced from the equation of the hemisphere by substituting $z = 0$, i.e., $x^2 + y^2 + z^2 = r^2$ reduces then to $x^2 + y^2 = r^2$.

$$Q_{xz} = \int y\,dV = \int_0^r y\pi x^2\,dy = \pi \int_0^r y(r^2 - y^2)dy = \tfrac{1}{4}\pi r^4$$

5. Locate the centroid of the arc of the circle in Fig. 9-6.

Solution:

The x-axis was chosen as an axis of symmetry. Polar coordinates often simplify the integration. From the figure,

$$x = r\cos\theta, \qquad y = r\sin\theta$$

$$\bar{x} = \frac{Q_y}{L} = \frac{\int x\,dL}{\int dL} = \frac{\int_{-\frac{1}{2}\alpha}^{\frac{1}{2}\alpha} xr\,d\theta}{\int_{-\frac{1}{2}\alpha}^{\frac{1}{2}\alpha} r\,d\theta}$$

$$= \frac{\int_{-\frac{1}{2}\alpha}^{\frac{1}{2}\alpha} r^2\cos\theta\,d\theta}{\int_{-\frac{1}{2}\alpha}^{\frac{1}{2}\alpha} r\,d\theta}$$

$$= \frac{r^2[\sin\frac{1}{2}\alpha - \sin(-\frac{1}{2}\alpha)]}{r[\frac{1}{2}\alpha - (-\frac{1}{2}\alpha)]}$$

$$= \frac{r(\sin\frac{1}{2}\alpha + \sin\frac{1}{2}\alpha)}{\alpha}$$

$$= \frac{2r\sin\frac{1}{2}\alpha}{\alpha}$$

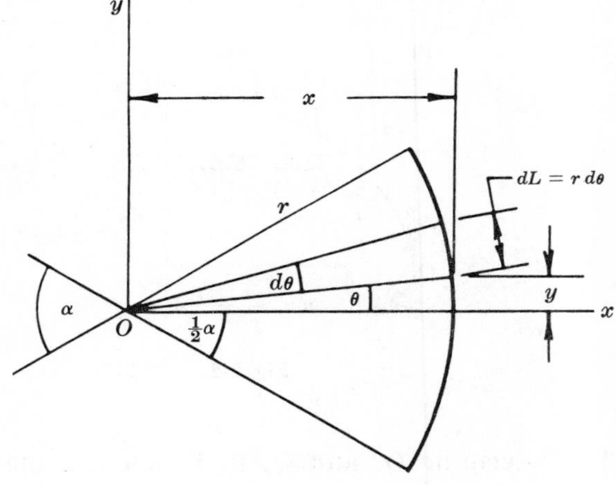

Fig. 9-6

Note that α is the subtended angle for the entire arc of the circle.

If $\bar{y}$ were determined by the same method, the integration would yield a cosine term which when the limits were substituted, would disappear. Hence $\bar{y} = 0$. This can be observed directly, however, because the centroid always lies on the axis of symmetry.

If the arc is a semicircle, α equals 180° or π radians. Then $\bar{x} = \dfrac{2r\sin\frac{1}{2}\pi}{\pi} = \dfrac{2r}{\pi}$.

6. Locate the centroid of the bent wire shown in the adjacent Fig. 9-7.

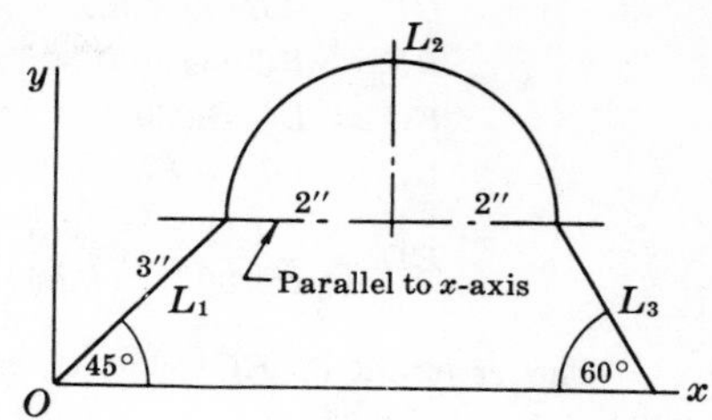

Fig. 9-7

Solution:

Let L_1 = 3-inch piece at 45° with the x-axis,
L_2 = semicircular piece,
L_3 = piece at 60° with the x-axis.

$$\text{Length of } L_3 = \frac{3 \cos 45°}{\sin 60°} = 2.45 \text{ in.}$$

The following table indicates the centroidal distances for each component.

Component	Length	$\bar{x}$	$\bar{y}$
L_1	3	$3/2 \cos 45° = 1.06$	$3/2 \sin 45° = 1.06$
L_2	$\pi r = 6.28$	$3 \cos 45° + 2 = 4.12$	$3 \sin 45° + 2r/\pi = 3.39$
L_3	2.45	$3 \cos 45° + 4 + 2.45/2 \cos 60° = 6.73$	$2.45/2 \sin 60° = 1.06$

$$\bar{x} = \frac{L_1\bar{x}_1 + L_2\bar{x}_2 + L_3\bar{x}_3}{L_1 + L_2 + L_3} = \frac{3(1.06) + 6.28(4.12) + 2.45(6.73)}{3 + 6.28 + 2.45} = 3.88 \text{ in.}$$

$$\bar{y} = \frac{L_1\bar{y}_1 + L_2\bar{y}_2 + L_3\bar{y}_3}{L_1 + L_2 + L_3} = \frac{3(1.06) + 6.28(3.39) + 2.45(1.06)}{11.73} = 2.31 \text{ in.}$$

Note that the value $2r/\pi$ in determining $\bar{y}_2$ is taken from Problem 5.

7. Locate the centroid of the bar built up as shown in Fig. 9-8 below. Assume the diameter of the bar is negligible compared with the dimensions of the figure.

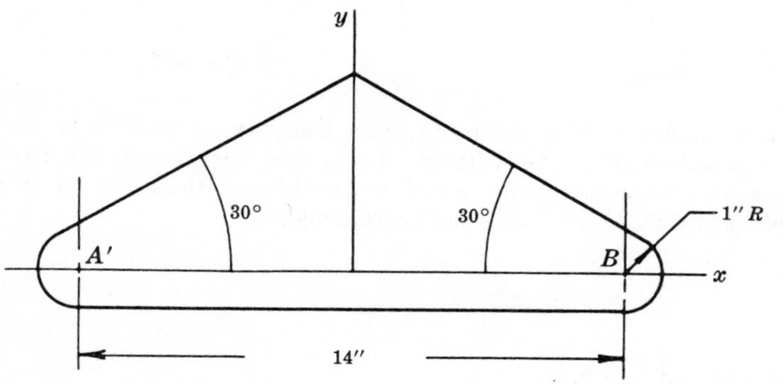

Fig. 9-8

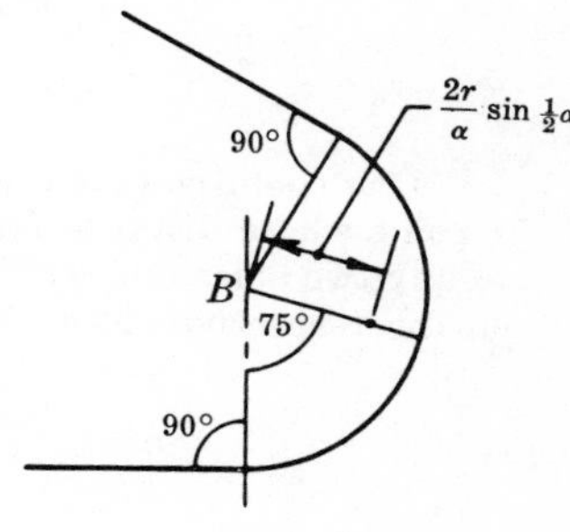

Fig. 9-9

Solution:

The enlarged Fig. 9-9 indicates the trigonometry needed to locate the centroid of the arc. It is on the axis of symmetry which makes an angle of 75° with the vertical and at a distance from the center of the arc equal to $2r/\alpha \sin \frac{1}{2}\alpha$, where $\alpha = 150\pi/180$ radians. (Refer to Problem 5.)

Hence the distance along the radius to the centroid is $\frac{2(1)}{2.62} \sin 75° = 0.738$ in. The $\bar{y}$ distance for the arc is therefore $-0.738 \sin 15° = -0.191$ in.

The length of one arc is $r\alpha = 1(2.62) = 2.62$ in.

Fig. 9-10 below indicates the method of determining the length of the sloping side.

$$DF = AB = 7 \text{ in.}$$
$$BF = BC\cos 30^\circ = 1(0.866) = 0.866 \text{ in.}$$
$$FC = BC\sin 30^\circ = 1(0.500) = 0.500 \text{ in.}$$
$$DC = DF + FC = 7.5 \text{ in.}$$
$$EC = \frac{DC}{\cos 30^\circ} = \frac{7.5}{0.866} = 8.66 \text{ in.}$$

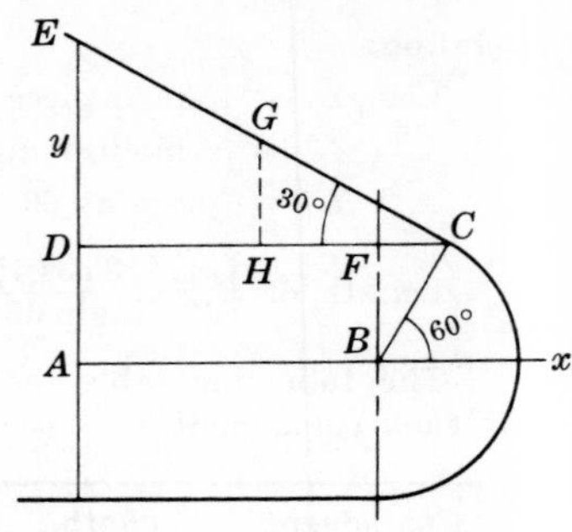

Fig. 9-10

The centroid of EC is at G, at a distance above the x-axis equal to $\bar{y}_{\text{slope}}$, where

$$\bar{y}_{\text{slope}} = GH + FB = \tfrac{1}{2}(8.66\sin 30^\circ) + 0.866 = 3.03 \text{ in.}$$

The centroid of the horizontal bar is below the x-axis. Hence its $\bar{y}$ is -1 inch.

By symmetry, $\bar{x}$ for the composite figure is 0.

To determine $\bar{y}$ for the composite figure, apply the following equation.

$$\bar{y} = \frac{2L_{\text{arc}}\,\bar{y}_{\text{arc}} + L_{\text{hor}}\,\bar{y}_{\text{hor}} + 2L_{\text{slope}}\,\bar{y}_{\text{slope}}}{2L_{\text{arc}} + L_{\text{hor}} + 2L_{\text{slope}}}$$

$$\bar{y} = \frac{2(2.62)(-0.191) + 14(-1) + 2(8.66)(3.03)}{2(2.62) + 14 + 2(8.66)} = 1.02 \text{ inches}$$

8. Locate the centroid of a triangle.

Solution:

Choose the differential element of area as shown in Fig. 9-11, where $dA = s\,dy$.

Note that $s/b = (h-y)/h$.

$$\bar{y} = \frac{Q_x}{A} = \frac{\int y\,dA}{A} = \frac{\int_0^h y\,s\,dy}{\frac{1}{2}bh}$$

$$= \frac{\int_0^h y\left[\frac{b}{h}(h-y)\right]dy}{\frac{1}{2}bh} = \frac{h}{3}$$

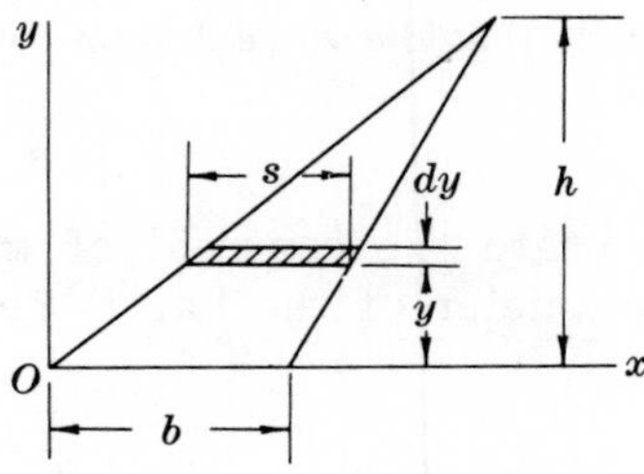

Fig. 9-11

One could now determine $\bar{x}$, but ordinarily it is sufficient to know that the centroid is located at a point whose distance from the base is one-third of the altitude. Draw two lines which are parallel to any two sides and which are at distances from those sides equal to one-third of the altitudes erected on the respective sides. The intersection of these two lines is the centroid.

9. Determine the centroid of the sector of a circle.

Solution:

Choose the x-axis as the axis of symmetry. The angle subtended is 2α. In this case, choose the differential element as shown in Fig. 9-12.

$$\bar{x} = \frac{Q_y}{A} = \frac{\int x\,dA}{\int dA} = \frac{\int_{-\alpha}^{\alpha}\int_0^r \rho\cos\theta\,\rho\,d\rho\,d\theta}{\int_{-\alpha}^{\alpha}\int_0^r \rho\,d\rho\,d\theta}$$

$$= \frac{\int_{-\alpha}^{\alpha}[\rho^3/3]_0^r\cos\theta\,d\theta}{\int_{-\alpha}^{\alpha}[\rho^2/2]_0^r\,d\theta} = \frac{(r^3/3)[\sin\alpha - \sin(-\alpha)]}{(r^2/2)[\alpha-(-\alpha)]}$$

$$= \frac{2r\sin\alpha}{3\alpha}$$

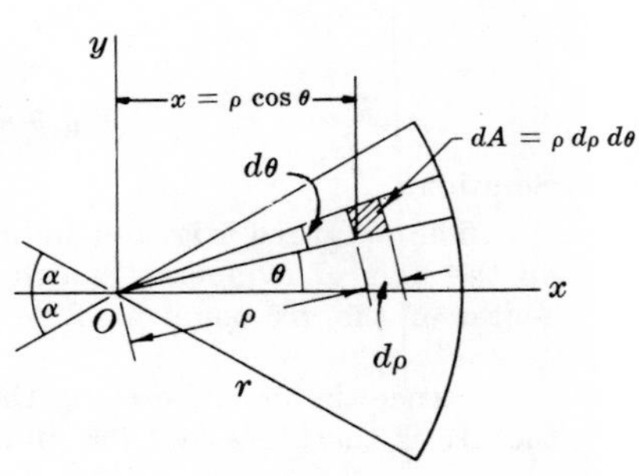

Fig. 9-12

For a semicircular sector, $2\alpha = \pi$ radians and $\bar{x} = \dfrac{2r \sin \pi/2}{3\pi/2} = \dfrac{4r}{3\pi}$.

Of course, $\bar{y} = 0$ by the symmetry of the figure.

10. Rework Problem 9 using as the elementary area the triangle shown in the adjacent Fig. 9-13.

Solution:

The centroid of the triangle is $\frac{2}{3}$ of the altitude from the vertex at the origin.

Note that $\bar{x}$ for the triangle is $\frac{2}{3}r\cos\theta$.

$$\bar{x} = \frac{Q_y}{A} = \frac{\int_{-\alpha}^{\alpha}(\frac{1}{2}r\,d\theta\,r)(\frac{2}{3}r\cos\theta)}{\int_{-\alpha}^{\alpha}\frac{1}{2}r\,d\theta\,r}$$

$$= \frac{\frac{1}{3}r^3\int_{-\alpha}^{\alpha}\cos\theta\,d\theta}{\frac{1}{2}r^2\int_{-\alpha}^{\alpha}d\theta} = \frac{2r\sin\alpha}{3\alpha}$$

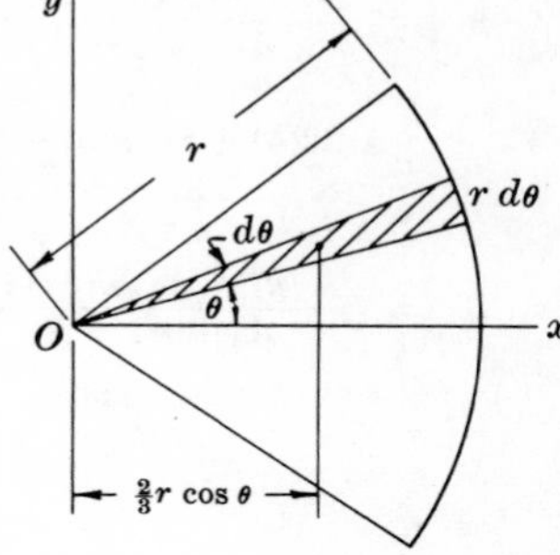

Fig. 9-13

11. Determine the centroid for the area bounded by the parabola $y^2 = 4ax$ and the lines $x = 0$, $y = b$.

Solution:

Choose the differential strip parallel to the x-axis, as shown in Fig. 9-14 below.

$$\bar{x} = \frac{Q_y}{A} = \frac{\int_0^b(\frac{1}{2}x)\,x\,dy}{\int_0^b x\,dy} = \frac{\frac{1}{2}\int_0^b x^2\,dy}{\int_0^b x\,dy} = \frac{\frac{1}{2}\int_0^b(y^4/16a^2)\,dy}{\int_0^b(y^2/4a)\,dy} = \frac{3b^2}{40a}$$

Similarly, $$\bar{y} = \frac{Q_x}{A} = \frac{\int_0^b y\,x\,dy}{b^3/12a} = \frac{\int_0^b(y^3/4a)\,dy}{b^3/12a} = \frac{b^4/16a}{b^3/12a} = \frac{3}{4}b$$

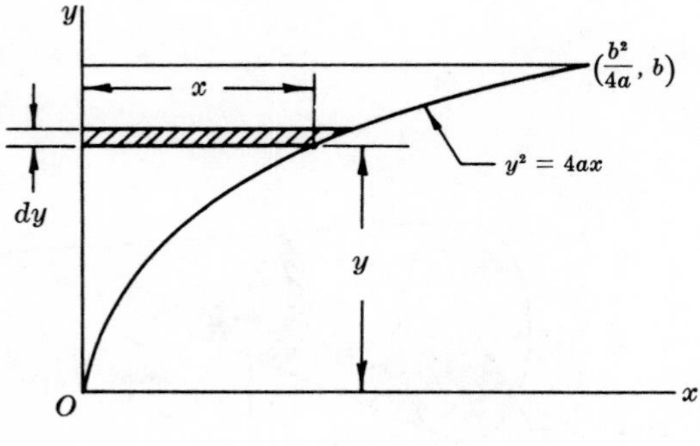

Fig. 9-14

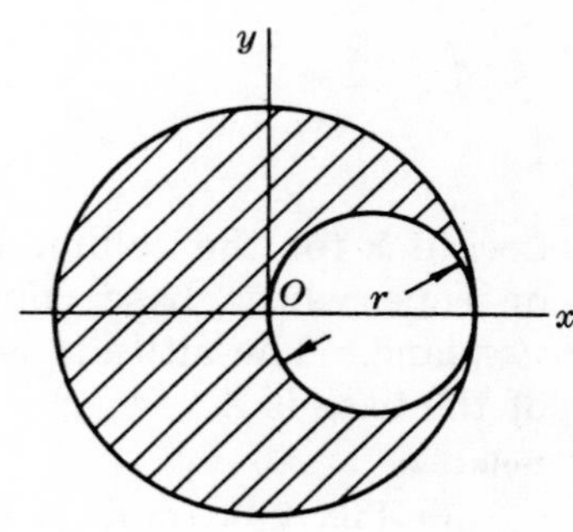

Fig. 9-15

12. Determine the centroid of the area remaining after a circle of diameter r is removed from a circle of radius r, as shown in Fig. 9-15 above.

Solution:

By symmetry $\bar{y} = 0$, i.e., the centroid is on the x-axis.

Using the formula for composite areas where A_L is the area of the large circle and A_S is the area of the small circle, we have $\bar{x} = \dfrac{A_L\bar{x}_L - A_S\bar{x}_S}{A_L - A_S} = \dfrac{\pi r^2(0) - (\pi r^2/4)(r/2)}{\pi r^2 - \pi r^2/4} = -\dfrac{1}{6}r$.

Hence the centroid is on the x-axis and at a distance of $\frac{1}{6}r$ to the left of the y-axis.

13. A semicircular area is removed from the trapezoid as shown in the adjacent figure. Determine the centroid of the remaining area.

Solution:

The shaded area consists of (1) a rectangle plus (2) a triangle minus (3) a semicircular area.

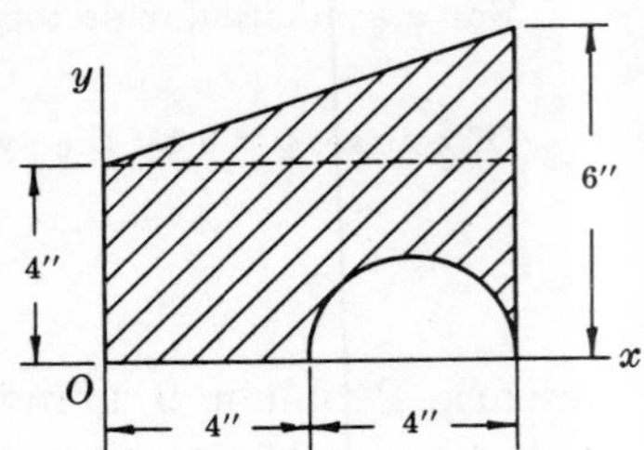

Fig. 9-16

$$\bar{x} = \frac{A_1\bar{x}_1 + A_2\bar{x}_2 - A_3\bar{x}_3}{A_1 + A_2 - A_3}$$

$$= \frac{32(4) + 8(16/3) - (\pi 4/2)(6)}{32 + 8 - 2\pi} = 3.94 \text{ in.}$$

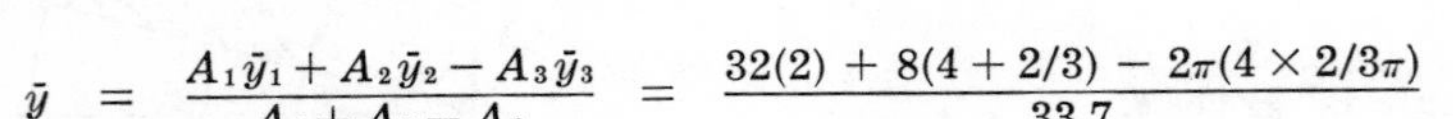

$$\bar{y} = \frac{A_1\bar{y}_1 + A_2\bar{y}_2 - A_3\bar{y}_3}{A_1 + A_2 - A_3} = \frac{32(2) + 8(4 + 2/3) - 2\pi(4 \times 2/3\pi)}{33.7} = 2.85 \text{ in.}$$

14. The area in Fig. 9-17(a) is revolved about the y-axis. Determine the centroid of the resulting volume shown in Fig. 9-17(b).

Solution:

By symmetry in Fig. 9-17(b), $\bar{x} = 0$, $\bar{z} = 0$. Choose a differential volume parallel to the xz-plane. Its thickness or height is dy and it is at a distance y above the xz-plane. Its radius is x.

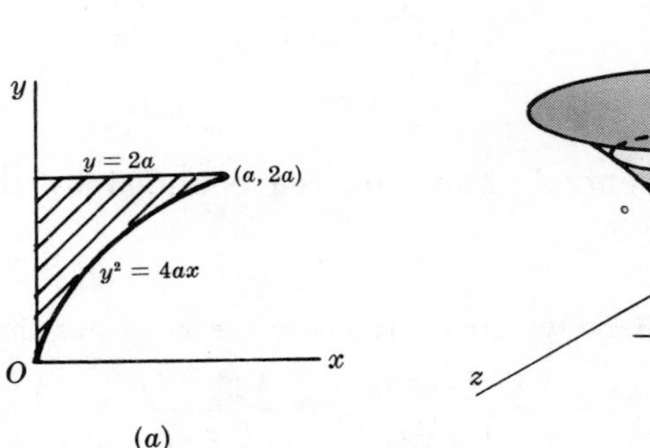

Fig. 9-17

$$\bar{y} = \frac{Q_{xz}}{V} = \frac{\int_0^{2a} y(\pi x^2\,dy)}{\int_0^{2a} \pi x^2\,dy} = \frac{\pi\int_0^{2a} y(y^4/16a^2)\,dy}{\pi\int_0^{2a} (y^4/16a^2)\,dy} = \frac{y^6/6]_0^{2a}}{y^5/5]_0^{2a}} = \frac{5}{3}a$$

15. Locate $\bar{x}$ for the volume of any pyramid or cone whose base coincides with the yz-plane. The altitude is h and the area of the base is A.

Solution:

In Fig. 9-18 choose a differential volume at a distance x from the yz-plane. Its area varies with x. Call the area A_x and the thickness dx.

From geometrical considerations,

$$\frac{A_x}{A} = \frac{(h-x)^2}{h^2}$$

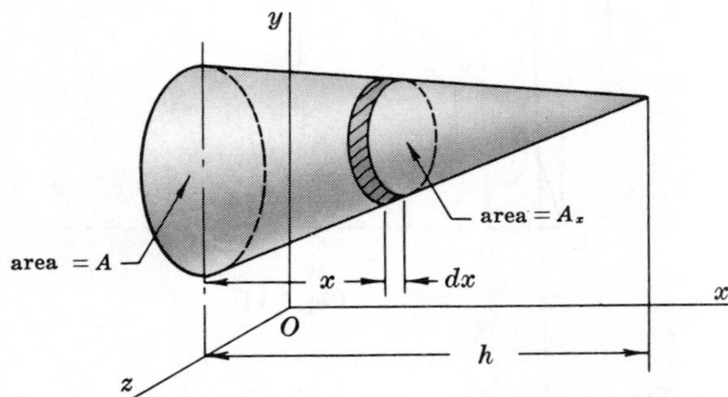

Fig. 9-18

Then, $$\bar{x} = \frac{Q_{yz}}{V} = \frac{\int x\,dV}{\int dV} = \frac{\int_0^h x\,A_x\,dx}{\int_0^h A_x\,dx} = \frac{\int_0^h x(A/h^2)(h-x)^2\,dx}{\int_0^h (A/h^2)(h-x)^2\,dx} = \frac{h}{4}$$

Thus the centroid of any pyramid or cone is at a distance from the base equal to one-fourth the altitude.

16. Locate the centroid of the volume of the fourth of the right circular cylinder shown in Fig. 9-19.

Solution:

It is only necessary to derive the value of $\bar{x}$, since $\bar{z} = \bar{x}$ and of course $\bar{y}$ is one-half the height h.

Choose the differential volume dV parallel to the yz-plane.

$$dV = z\,h\,dx$$

However, since the section cut from the solid by any plane parallel to the xz-plane yields a quarter circle, the relationship between x and z is $x^2 + z^2 = r^2$.

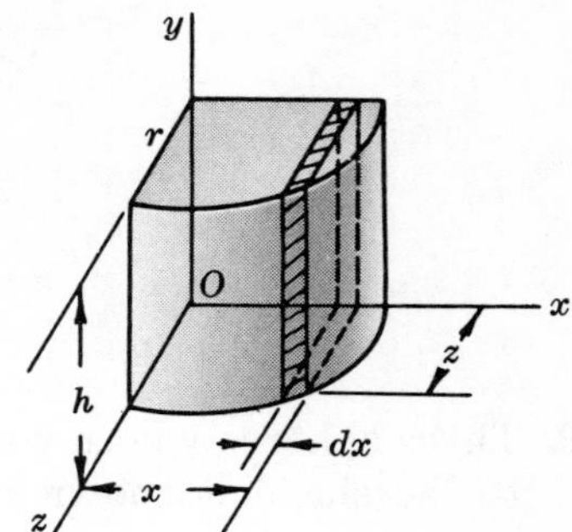

Fig. 9-19

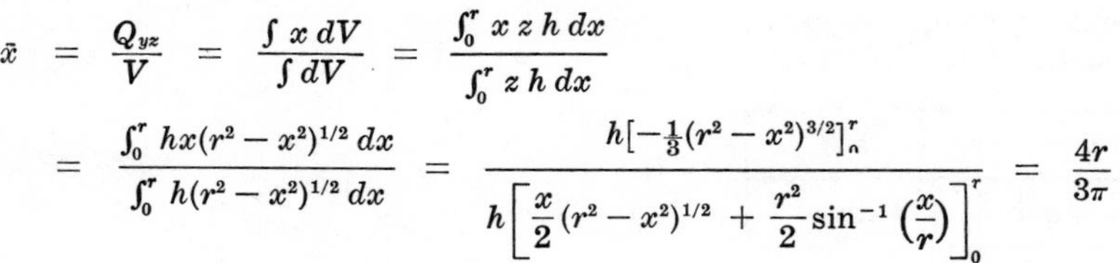

$$\bar{x} = \frac{Q_{yz}}{V} = \frac{\int x\,dV}{\int dV} = \frac{\int_0^r x\,z\,h\,dx}{\int_0^r z\,h\,dx}$$

$$= \frac{\int_0^r hx(r^2 - x^2)^{1/2}\,dx}{\int_0^r h(r^2 - x^2)^{1/2}\,dx} = \frac{h[-\frac{1}{3}(r^2 - x^2)^{3/2}]_0^r}{h\left[\frac{x}{2}(r^2 - x^2)^{1/2} + \frac{r^2}{2}\sin^{-1}\left(\frac{x}{r}\right)\right]_0^r} = \frac{4r}{3\pi}$$

Note that the result is exactly the same as for the centroid of the quadrant of a circle. This should be expected since the only factor which might influence the centroidal position in the solid as compared with the area is the altitude h which was shown to be a common term in both numerator and denominator.

17. A sphere of radius r is cut from a larger sphere of radius R. The distance between their centers is a. Locate the centroid of the remaining volume.

Solution:

This is an example to illustrate the technique employed for composite volumes.

$$V_R = \frac{4}{3}\pi R^3, \qquad V_r = \frac{4}{3}\pi r^3$$

Assume the origin of the x, y, z axes is at the center of the larger sphere and that the positive x-axis is the line of centers of the two spheres. Then $\bar{x}_R = 0$, $\bar{x}_r = a$.

Applying the formula,

$$\bar{x} = \frac{V_R\bar{x}_R - V_r\bar{x}_r}{V_R - V_r} = \frac{\frac{4}{3}\pi R^3(0) - \frac{4}{3}\pi r^3(a)}{\frac{4}{3}\pi R^3 - \frac{4}{3}\pi r^3} = \frac{-ar^3}{R^3 - r^3}$$

This means that the centroid is on the line of centers and to the left of the yz-plane a distance of $ar^3/(R^3 - r^3)$.

18. Locate the centroid of the adjacent Fig. 9-20 shown with a 2 inch diameter hole drilled in the center of the top face and normal to it.

Solution:

From symmetry of figure, $\bar{z} = 4$ in.

Let the parallelepiped be denoted by 1, the triangular piece by 2, and the cylinder by 3.

	V	$\bar{x}$	$\bar{y}$
1	64	2	1
2	16	4.67	0.67
3	6.28	2	1

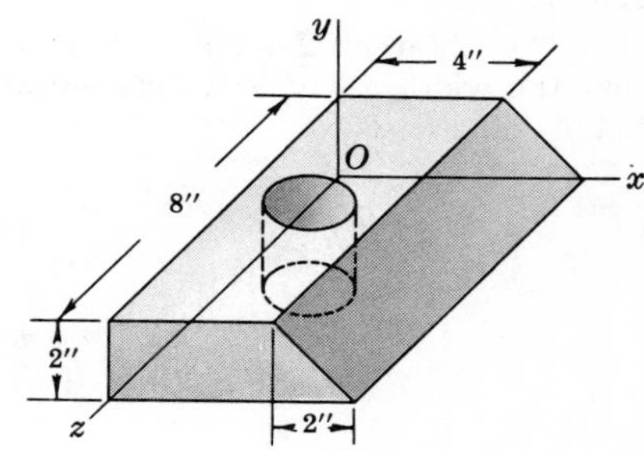

Fig. 9-20

The above table indicates the values needed in the formulas below.

$$\bar{x} = \frac{V_1\bar{x}_1 + V_2\bar{x}_2 - V_3\bar{x}_3}{V_1 + V_2 - V_3} = \frac{64(2) + 16(4.67) - 6.28(2)}{64 + 16 - 6.28} = 2.58 \text{ in.}$$

$$\bar{y} = \frac{V_1\bar{y}_1 + V_2\bar{y}_2 - V_3\bar{y}_3}{V_1 + V_2 - V_3} = \frac{64(1) + 16(0.67) - 6.28(1)}{73.72} = 0.93 \text{ in.}$$

19. Three spheres whose volumes are 10, 15, and 25 cubic inches are located with reference to the shaft as shown in Fig. 9-21. Locate the centroid of the three volumes.

Solution:

Assume that the x-axis is along the shaft. Use a tabular form to list the data.

V	$\bar{x}$	$\bar{y}$	$\bar{z}$
10	+4	$+ 4 \cos 30°$	$- 4 \sin 30°$
15	+12	$- 6 \cos 45°$	$+ 6 \sin 45°$
25	+24	$- 4 \cos 60°$	$- 4 \sin 60°$

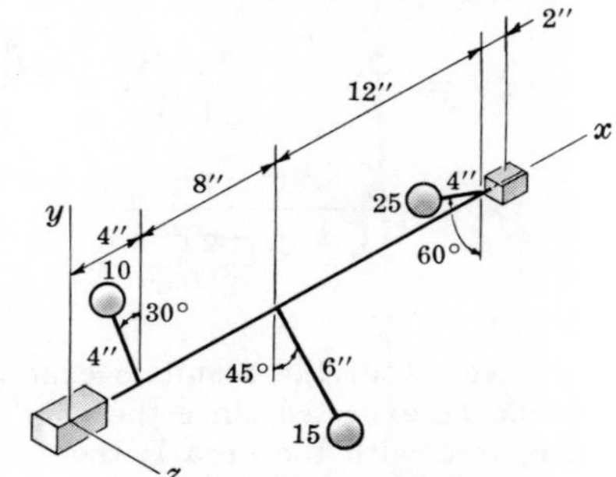

Fig. 9-21

$$\bar{x} = \frac{10 \times 4 + 15 \times 12 + 25 \times 24}{10 + 15 + 25} = 16.4 \text{ inches}$$

$$\bar{y} = \frac{10 \times 4 \times 0.866 - 15 \times 6 \times 0.707 - 25 \times 4 \times 0.500}{50} = -1.58 \text{ inches}$$

$$\bar{z} = \frac{-10 \times 4 \times 0.500 + 15 \times 6 \times 0.707 - 25 \times 4 \times 0.866}{50} = -0.86 \text{ inches}$$

Incidentally, this same procedure should be followed if the numbers 10, 15, and 25 represented weights or masses concentrated at the centers of the corresponding spheres.

20. Determine the centroid of the surface of a hemisphere with respect to its base.

Solution:

Refer to Fig. 9-22. The xz-plane is chosen in the base of the hemisphere.

The differential strip of area dS is chosen as in the figure. Note that the width dL of this differential surface is *not* vertical but sloping. Let $\theta = \tan^{-1} y/x$.

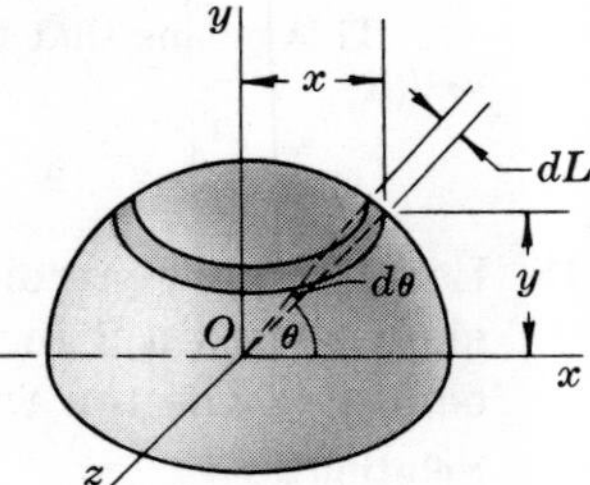

Fig. 9-22

$$\overline{dL}^2 = \overline{dx}^2 + \overline{dy}^2 = \left(\frac{\overline{dx}^2}{\overline{dx}^2} + \frac{\overline{dy}^2}{\overline{dx}^2}\right)\overline{dx}^2 \quad \text{or} \quad dL = \sqrt{1 + (dy/dx)^2}\, dx$$

$$\bar{y} = \frac{Q_{xz}}{S} = \frac{\int y\, dS}{\int dS} = \frac{\int y(2\pi\, x\, dL)}{\int 2\pi\, x\, dL} = \frac{\int_0^r 2\pi\, y\, x\sqrt{1 + (dy/dx)^2}\, dx}{\int_0^r 2\pi\, x\sqrt{1 + (dy/dx)^2}\, dx}$$

In the xy-plane the equation of the circle is $x^2 + y^2 = r^2$.

Then $y = \sqrt{r^2 - x^2}$. Differentiating, $dy/dx = -x(r^2 - x^2)^{-1/2}$. Making these substitutions,

$$\bar{y} = \frac{2\pi r \int_0^r x\, dx}{2\pi r \int_0^r x(r^2 - x^2)^{-1/2}\, dx} = \frac{r}{2}$$

21. Determine the centroid of the surface of a right circular cone with respect to its base. The altitude is h. Refer to Fig. 9-23.

Solution:

In Fig. 9-23, the x-axis is chosen along the altitude of the cone. The differential element of surface dS is $2\pi\, y\, dL$.

$$\bar{x} = \frac{Q_{yz}}{S} = \frac{\int x\, dS}{\int dS} = \frac{\int x\, 2\pi\, y\, dL}{\int 2\pi\, y\, dL}$$

If r is the radius of the base, then by similar triangles in the xy-plane, $y/r = (h-x)/h$.

Hence $dy/dx = -r/h$ and $dL = dx\sqrt{1+(dy/dx)^2} = dx\sqrt{1+r^2/h^2}$.

Substituting and simplifying, $\bar{x} = \dfrac{\int_0^h (hx - x^2)\, dx}{\int_0^h (h-x)\, dx} = \dfrac{h}{3}$.

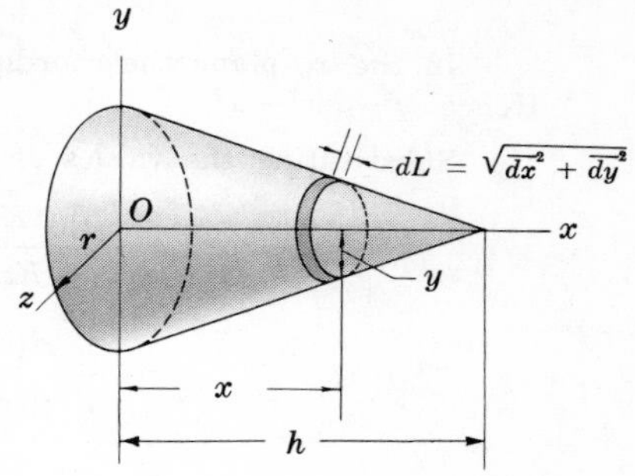

Fig. 9-23

22. Find the center of mass of the first octant of the homogeneous ellipsoid $\dfrac{x^2}{a^2} + \dfrac{y^2}{b^2} + \dfrac{z^2}{c^2} = 1$.

Solution:

Choose the differential mass parallel to the yz-plane as shown in Fig. 9-24.

Let the density be δ.

Note that $dV = A\, dx$, where A is the area of a quarter of an ellipse with axes of length $2y$ and $2z$. Hence $A = \frac{1}{4}\pi y z$.

$$dm = \delta\, dV = \delta\, \tfrac{1}{4}\pi\, y\, z\, dx$$

Then
$$\bar{x} = \frac{Q_{yz}}{m} = \frac{\int x\, dm}{\int dm}$$

$$= \frac{\delta\, \frac{1}{4}\pi \int_0^a x\, y\, z\, dx}{\delta\, \frac{1}{4}\pi \int_0^a y\, z\, dx}$$

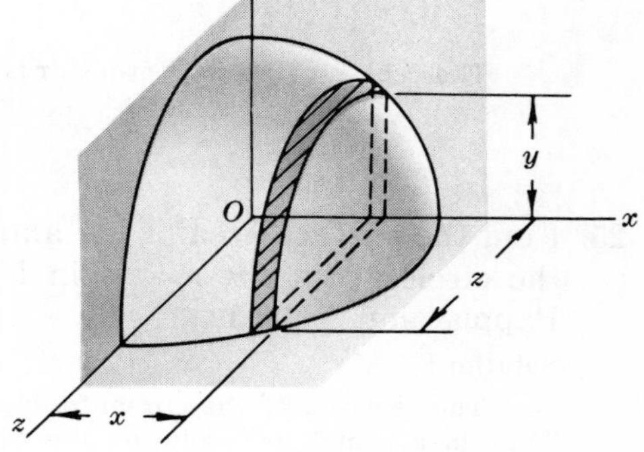

Fig. 9-24

To determine y in terms of x, let $z = 0$ in the equation of the ellipsoid, yielding

$$\frac{x^2}{a^2} + \frac{y^2}{b^2} = 1 \quad \text{or} \quad y = \frac{b}{a}\sqrt{a^2 - x^2}$$

Similarly, with $y = 0$, $\dfrac{x^2}{a^2} + \dfrac{z^2}{c^2} = 1$ or $z = \dfrac{c}{a}\sqrt{a^2 - x^2}$.

Making these substitutions and simplifying, $\bar{x} = \dfrac{\int_0^a x(a^2 - x^2)\, dx}{\int_0^a (a^2 - x^2)\, dx} = \dfrac{3}{8}a$.

Similar reasoning yields $\bar{y} = \frac{3}{8}b$ *and* $\bar{z} = \frac{3}{8}c$.

23. Find the center of mass of a hemisphere whose density varies as the square of the distance from the base.

Solution:

Let the base of the hemisphere be the yz-plane as shown in Fig. 9-25 below. Then the density varies with x^2, or $\delta = Kx^2$.

Choose dm parallel to the yz-plane and at a distance x from the yz-plane.

$$\bar{x} \;=\; \frac{Q_{yz}}{m} \;=\; \frac{\int x\,dm}{\int dm} \;=\; \frac{\int_0^r x\,\delta\,\pi y^2\,dx}{\int_0^r \delta\,\pi y^2\,dx}$$

In the xy-plane the coordinates (x, y) lie on a circle of radius r. Hence $y^2 = r^2 - x^2$.

Substituting the values of y^2 and δ, the equation becomes

$$\bar{x} \;=\; \frac{\int_0^r x\,Kx^2\,\pi(r^2 - x^2)\,dx}{\int_0^r Kx^2\,\pi(r^2 - x^2)\,dx}$$

$$=\; \frac{\int_0^r x^3(r^2 - x^2)\,dx}{\int_0^r x^2(r^2 - x^2)\,dx} \;=\; \frac{\frac{1}{12}r^6}{\frac{2}{15}r^5} \;=\; \frac{5}{8}r$$

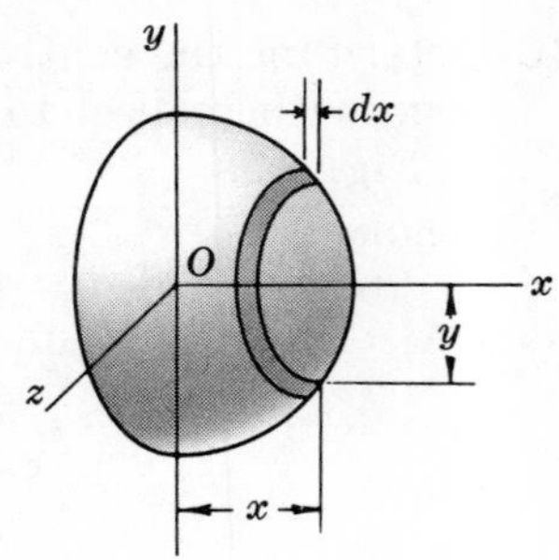

Fig. 9-25

24. The density at any point of a slender rod varies with the first power of the distance of the point from one end of the rod. Where is the mass-center?

Solution:

The density is proportional to the distance x along the rod from the one end chosen as origin, i.e., $\delta = Kx$ where δ equals mass per unit length.

To find dm, multiply the differential length dx at the point x by the density at that point. This yields the equation $dm = \delta\,dx = Kx\,dx$. Then

$$\bar{x} \;=\; \frac{\int x\,dm}{\int dm} \;=\; \frac{\int_0^l x\,Kx\,dx}{\int_0^l Kx\,dx} \;=\; \frac{2}{3}l$$

The center of mass is two-thirds of the length from the end chosen as a reference.

25. Find the surface area of the annular torus formed by revolving the circle about the x-axis in Fig. 9-26. Use the theorems of Pappus and Guldinus.

Solution:

The centroid of the circumference is at a distance d from the x-axis. Thus in a complete revolution the centroid moves on a circular path of radius d. The distance it travels is $2\pi d$. The length $2\pi r$ of the generating curve is the circumference of the circle. Hence the surface of the torus is $2\pi d \times 2\pi r = 4\pi^2 rd$.

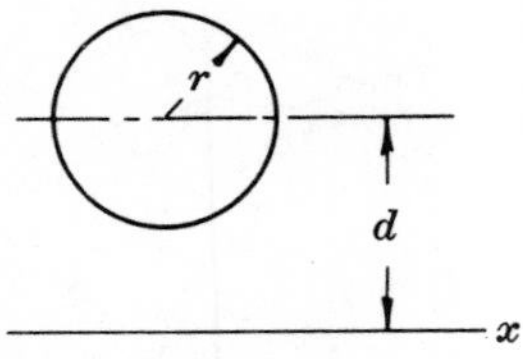

Fig. 9-26

26. Determine the centroid of a quadrant of a circle using the theorems of Pappus and Guldinus.

Solution:

This is really an application of the theorems in reverse. The area in Fig. 9-27 when revolved generates a hemisphere whose volume is known to be $2\pi r^3/3$.

The length of the path of the centroid is the volume divided by the area of the quadrant: $l = \dfrac{2\pi r^3/3}{\pi r^2/4} = \dfrac{8r}{3}$.

But the length of the path of the centroid is $2\pi\bar{y}$. Hence $2\pi\bar{y} = 8r/3$, or $\bar{y} = 4r/3\pi$.

This value was derived previously by the methods of the calculus. See Problem 9 above.

Of course, $\bar{x} = \bar{y}$ for the quadrant of the circle.

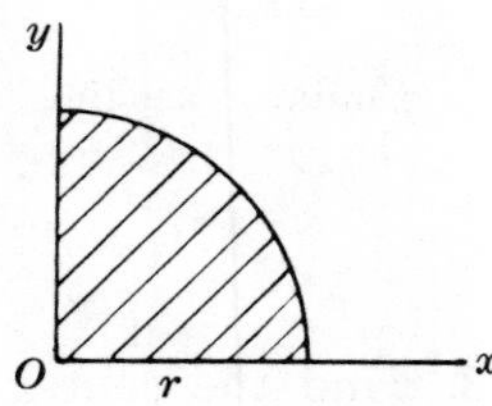

Fig. 9-27

27. In Fig. 9-28, a box with dimensions l, b, and h ft is shown half full of gravel weighing w lb/ft³. Assuming that the height of the gravel varies linearly from zero at the left end to h at the right, determine how far the center of pressure is from the left end.

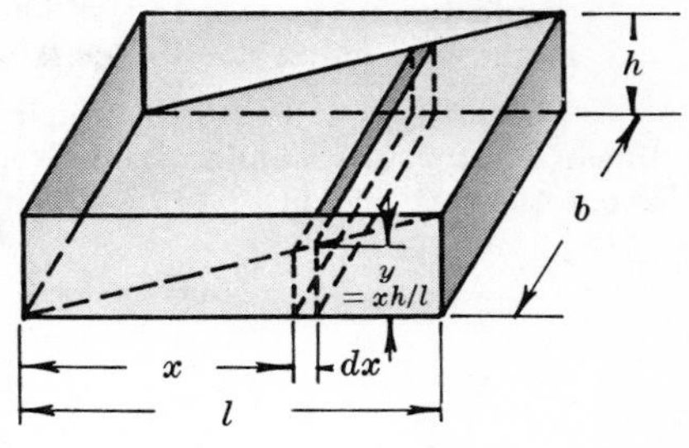

Fig. 9-28

Solution:

At a distance x from the left end select a differential volume as shown. The height $y = xh/l$ and the differential weight $dW = w(xh/l)b\,dx$. To locate the center of pressure use

$$\bar{x} \;=\; \frac{\int x\,dW}{\int dW} \;=\; \frac{(whb/l)\int_0^l x^2\,dx}{(whb/l)\int_0^l x\,dx} \;=\; \frac{2}{3}l$$

The center of pressure of the gravel is on a vertical line located $\frac{2}{3}l$ from the left end and $\frac{1}{2}b$ back from the front wall. The entire weight of the gravel if placed along this vertical line would induce in the box supports the same forces as the distributed load does.

28. (*a*) Refer to Fig. 9-29 below. A beam carries material which weighs w lb/ft³ and has a height y that varies in a known fashion with the distance x from the left end. Determine each right reaction R_R. Assume that b is constant and that the height y for a selected strip does not vary along the b distance.

(*b*) In part (*a*) suppose $w = 150$ lb/ft³ and that the height of the load varies linearly from zero at the left end to 2 ft at the right end as shown in Fig. 9-30 below. The span is 8 ft and the distance b is 2 ft. Determine each of the two right supports, neglecting the weight of the beam.

Solution:

(*a*) The weight of a differential volume at a distance x from the left end is $dW = wby\,dx$. The moment of this dW with respect to the left end (the vertical plane perpendicular to l) is $x\,dW = wby\,x\,dx$.

The sum of the moments of the weights of all such differential volumes about the left end must be balanced by the moments of the two right supports R_R. Thus

$$\int_0^l wby\,x\,dx \;=\; 2\,R_R\,l$$

The integration may be performed directly if y is a convenient function of x. If not, other means must be employed.

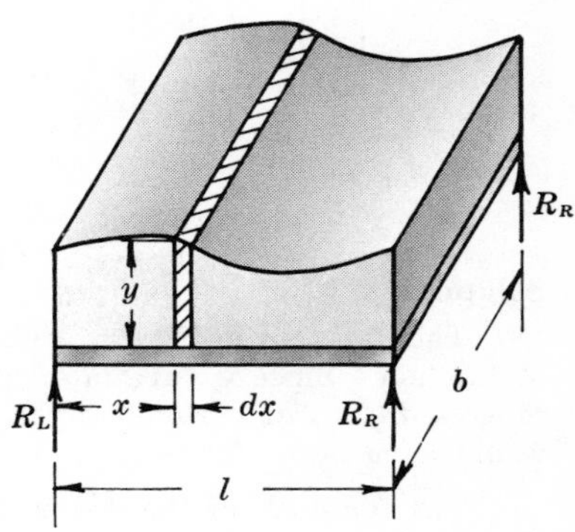

Fig. 9-29

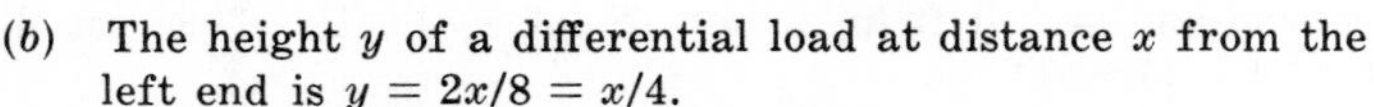

(*b*) The height y of a differential load at distance x from the left end is $y = 2x/8 = x/4$.

The moment of the entire load relative to the vertical plane containing the two left reactions is

$$M \;=\; \int x\,dW \;=\; \int x(150\,dV) \;=\; \int_0^8 x[150(x/4)(2)\,dx]$$
$$= \;12{,}800 \text{ ft-lb}$$

The moment about the left end of the two right reactions must equal the moment of the load. Hence

$$2R_R(8) \;=\; 12{,}800 \qquad\text{or}\qquad R_R \;=\; 800 \text{ lb}$$

The moment of the load may also be found by using the load in lb per linear ft along the beam. If the load at the right end is visualized as 2 ft high, 2 ft back

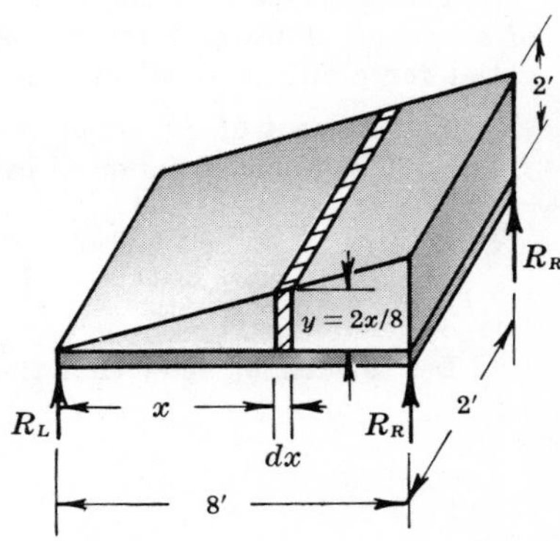

Fig. 9-30

along the beam and 1 ft in the x direction centered at the right edge, the magnitude of the load would be $2 \times 2 \times 1 \times 150 = 600$ lb per linear ft. Thus in this problem the load varies from zero at the left end to 600 lb per ft at the right end.

The load on a length dx which is at a distance x from the left end is $p_x\,dx$, where p_x = load per linear ft at x. By similar triangles, $p_x/x = 600/8$ or $p_x = 75x$. Then the moment of the entire load about the left end is

$$M = \int_0^8 x\,p_x\,dx = \int_0^8 x(75x)\,dx = 12{,}800 \text{ ft-lb}$$

29. The tank shown in Fig. 9-31 is full of water weighing 62.4 lb/ft³. The plate covers a rectangular hole 1 ft high by 2 ft wide. Determine the force induced in each of the two top bolts B by the water acting against the plate.

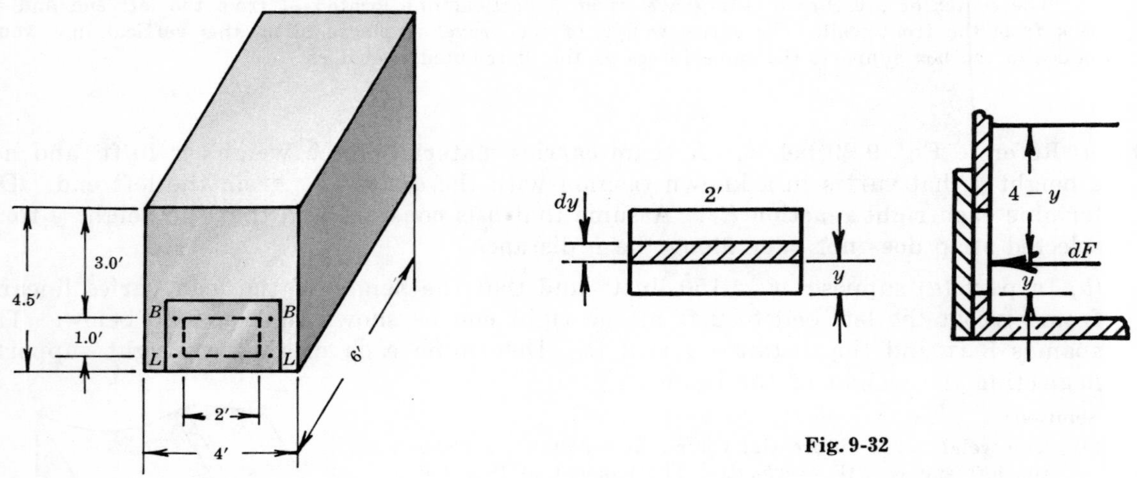

Fig. 9-31

Fig. 9-32

Solution:

The diagram in Fig. 9-32 shows a differential element of the plate a distance y above the bottom of the hole. Since we are interested only in the area of the plate that the water touches, the element of area dA is $2\,dy$. [Note: If the width of the plate had been a function of y then the differential area would have been $f(y)\,dy$.]

The force dF of the water on the selected differential element is the product of the area dA and the pressure at that distance y above the bottom of the hole.

The pressure p at the point y ft above the bottom of the hole is numerically equal to the weight of a column of water 1 ft² in cross section and $(4-y)$ ft high: $p = 62.4(4-y)\text{lb/ft}^2$. Then the differential force on the small area is $dF = 62.4(4-y)(2)dy$.

The moment of dF about the bottom of the hole is $y\,dF$. The moment of the total water force is $\int y\,dF$. This must be balanced by the moment of the bolt forces B about the bottom of the hole. Hence

$$2B(1) = \int_0^1 y\,62.4(4-y)(2)dy \qquad \text{from which} \qquad B = 104 \text{ lb}$$

Let the reader show that the force induced in each bottom bolt is 114 lb.

Supplementary Problems

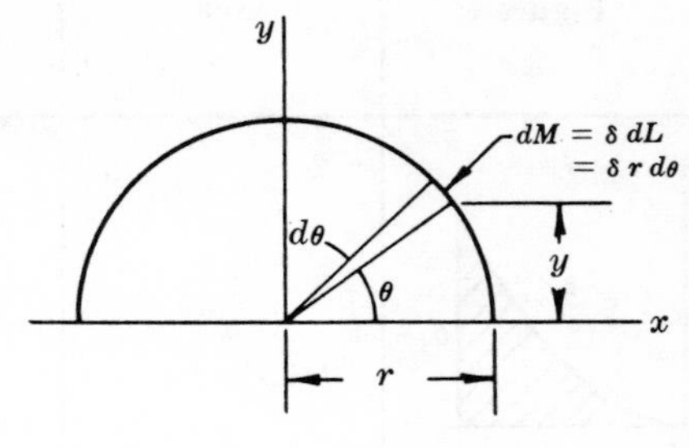

Fig. 9-33

30. Find Q_x for the mass m of half the rim of a wheel shown in Fig. 9-33. Assume the thickness of the wheel is small compared with the radius. Use polar coordinates as indicated. Density is δ units of mass per unit length. *Ans.* $Q_x = 2\delta r^2 = 2rm/\pi$

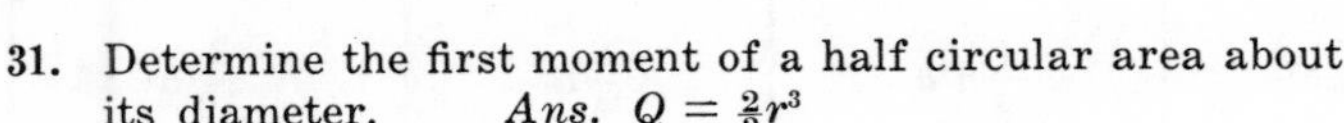

31. Determine the first moment of a half circular area about its diameter. *Ans.* $Q = \frac{2}{3}r^3$

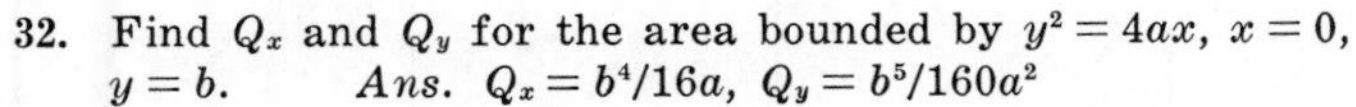

32. Find Q_x and Q_y for the area bounded by $y^2 = 4ax$, $x = 0$, $y = b$. *Ans.* $Q_x = b^4/16a$, $Q_y = b^5/160a^2$

33. Determine the first moment of the volume of a right circular cone about its base. Radius of base is r and height is h. *Ans.* $Q = \pi r^2 h^2/12$

34. Apply the results of Problem 4 of this chapter to determine Q about the base of a hemisphere whose radius is 6 inches. *Ans.* $Q = 1020 \text{ in}^4$

35. Determine Q_x and Q_y for an ellipse whose major and minor axes are 4 inches and 3 inches respectively. *Ans.* $Q_x = 1.50 \text{ in}^3$, $Q_y = 2.00 \text{ in}^3$

36. Apply the results of Problem 33 to find the first moment of a right circular cone about its base whose radius is 3 inches. The height is 4 inches. *Ans.* $Q = 37.7 \text{ in}^4$

37. A uniform bar 20 inches long is bent at its midpoint at an angle of 90°. Determine its centroid using the sides as axes. *Ans.* $\bar{x} = \bar{y} = 2.5$ in.

38. Suppose the bar in Problem 37 was bent so that the angle at its midpoint is 70°, how far is its centroid from the line joining its ends? *Ans.* $d = 4.1$ in.

39. A wire is bent as shown in Fig. 9-34 below. Find its center of gravity. *Ans.* $\bar{x} = 1.36$ in., $\bar{y} = 2.08$ in.

40. Find the centroid of the figure built up of the lines as shown in Fig. 9-35 below. *Ans.* $\bar{x} = -3.1$ in., $\bar{y} = 2.7$ in.

41. Determine the centroid of a uniform rod bent into triangular shape. See Fig. 9-36 below. *Ans.* $\bar{x} = 2.52$ in., $\bar{y} = 3.11$ in.

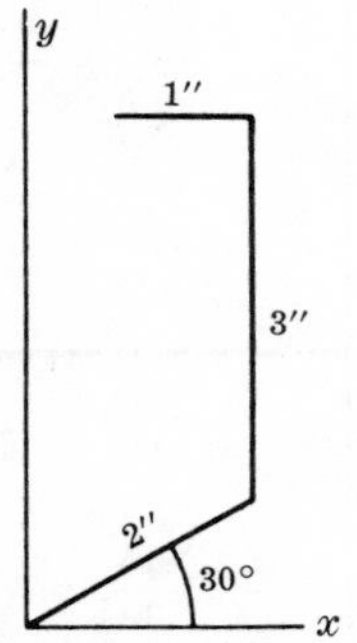

Fig. 9-34

Fig. 9-35

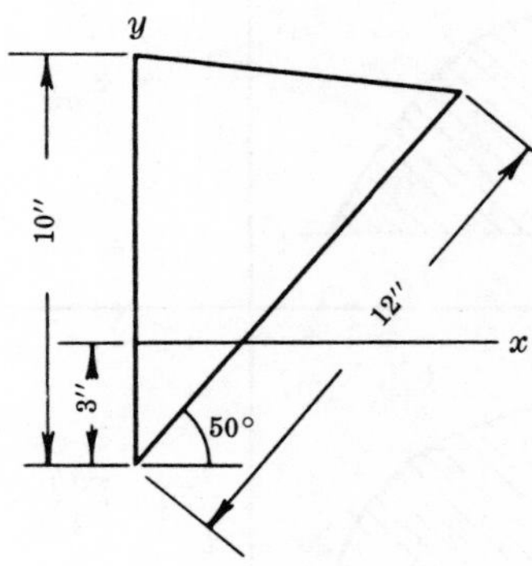

Fig. 9-36

42. Show that the following values are true for the indicated areas.

Figure	Area	Q_X	Q_Y	Centroid	
				$\bar{x}$	$\bar{y}$
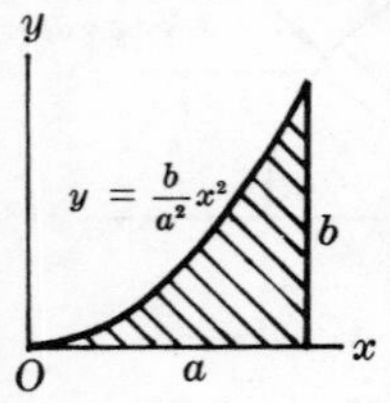 $y = \frac{b}{a^2}x^2$; O, a, b, x, y	$\frac{1}{3}ab$	$\frac{1}{10}ab^2$	$\frac{1}{4}a^2b$	$\frac{3}{4}a$	$\frac{3}{10}b$
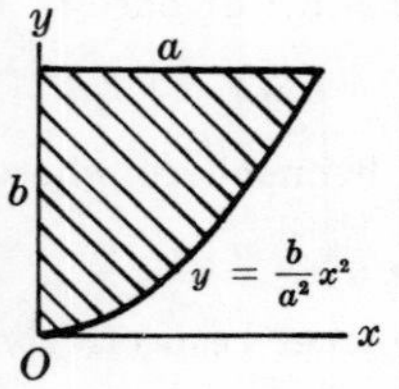 $y = \frac{b}{a^2}x^2$; O, a, b, x, y	$\frac{2}{3}ab$	$\frac{2}{5}ab^2$	$\frac{1}{4}ba^2$	$\frac{3}{8}a$	$\frac{3}{5}b$
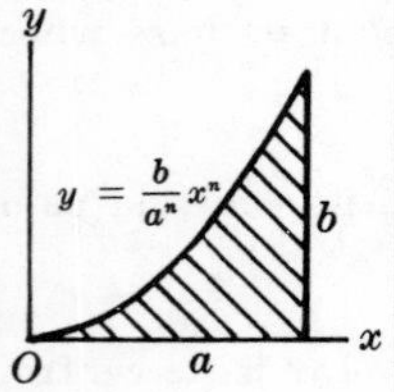 $y = \frac{b}{a^n}x^n$; O, a, b, x, y	$\frac{ab}{n+1}$	$\frac{ab^2}{2(2n+1)}$	$\frac{a^2b}{n+2}$	$\frac{n+1}{n+2}a$	$\frac{n+1}{2(2n+1)}b$
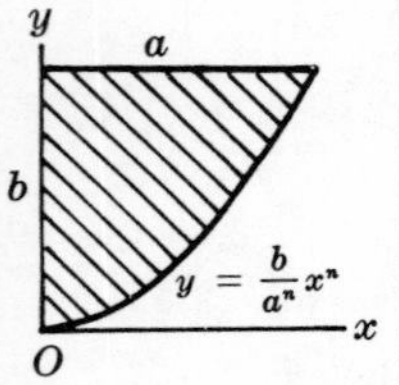 $y = \frac{b}{a^n}x^n$; O, a, b, x, y	$\frac{n}{n+1}ab$	$\frac{n}{2n+1}ab^2$	$\frac{n}{2(n+2)}a^2b$	$\frac{n+1}{2(n+2)}a$	$\frac{n+1}{2n+1}b$
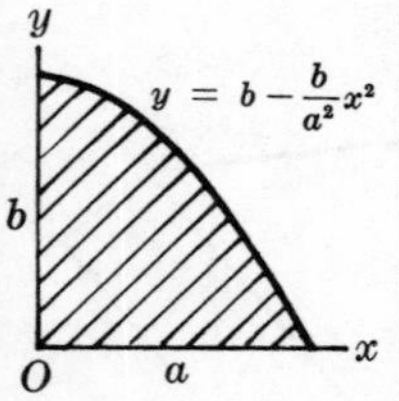 $y = b - \frac{b}{a^2}x^2$; O, a, b, x, y	$\frac{2}{3}ab$	$\frac{4}{15}ab^2$	$\frac{1}{4}a^2b$	$\frac{3}{8}a$	$\frac{2}{5}b$
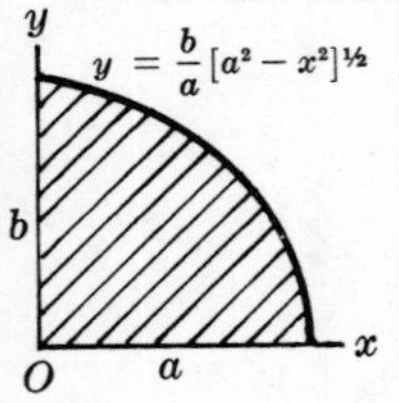 $y = \frac{b}{a}[a^2 - x^2]^{1/2}$; O, a, b, x, y	$\frac{\pi ab}{4}$	$\frac{ab^2}{3}$	$\frac{a^2b}{3}$	$\frac{4a}{3\pi}$	$\frac{4b}{3\pi}$

43. Determine the centroid for the area bounded by the parabola $y^2 = 4ax$ and the lines $x = 0$, $y = 0$.
Ans. $\bar{x} = 3b^2/40a$, $\bar{y} = 3b/4$

44. Compute $\bar{x}$ for the area between the semicubical parabola $ay^2 = x^3$ and the line $x = a$. *Ans.* $\bar{x} = 5a/7$

45. Locate the centroid of the area bounded by $y^2 = 2x$, $x = 3$, $y = 0$. *Ans.* $\bar{x} = 1.8$, $\bar{y} = \frac{3}{8}\sqrt{6}$

46. Locate the centroid of the area between the parabola $y^2 = 4ax$ and the line $y = bx$.
Ans. $\bar{x} = 8a/5b^2$, $\bar{y} = 2a/b$

47. Determine $\bar{x}$ and $\bar{y}$ for the area bounded by a quadrant of a circle whose radius is r and the lines $x = 0$, $y = 0$. *Ans.* $\bar{x} = \bar{y} = 4r/3\pi$

48. Locate the centroid of the area bounded by the parabolas $y^2 = bx$ and $x^2 = by$.
Ans. $\bar{x} = \bar{y} = 9b/20$

49. Determine the y coordinate of the centroid of the area between the x-axis and the curve $y = \sin x$. Use the interval $0 \leqq x \leqq \pi$. *Ans.* $\bar{y} = \pi/8$

50. Determine $\bar{x}$ for the area bounded by the hyperbola $xy = c^2$ and the lines $x = a$, $x = b$, $y = 0$.
Ans. $\bar{x} = \dfrac{b - a}{\log_e b - \log_e a}$

51. Locate the centroid for the area between the ellipse $x^2/a^2 + y^2/b^2 = 1$ and the lines $x = a$, $y = b$.
Ans. $\bar{x} = 0.776a$, $\bar{y} = 0.776b$

52. Find the centroid of the composite area in Fig. 9-37 below. *Ans.* $\bar{x} = 2.61$ in., $\bar{y} = 3.15$ in.

53. Compute the distance below the top line to the centroid of the T section in Fig. 9-38 below.
Ans. $d = 2.97$ in.

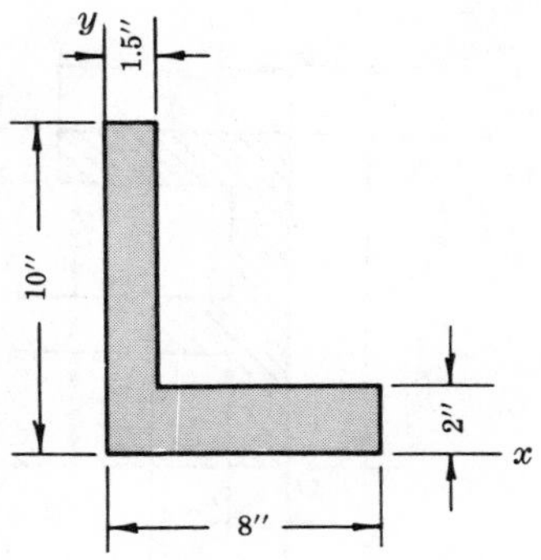

Fig. 9-37

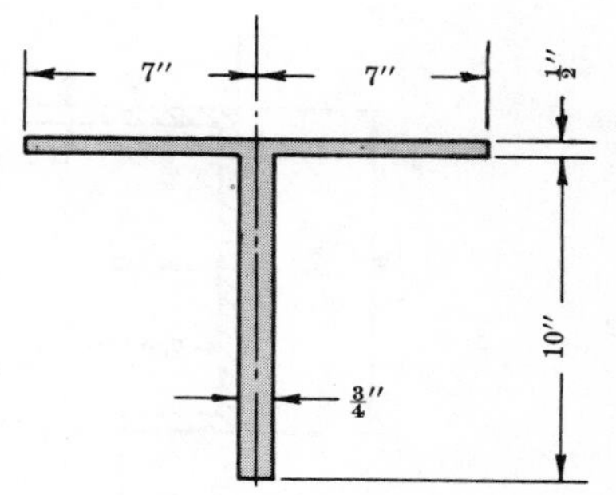

Fig. 9-38

54. Locate the centroid of the shaded area formed by removing the triangle from the semicircular area in Fig. 9-39 below. *Ans.* $\bar{x} = 0$, $\bar{y} = 2.34$ in.

55. Determine the coordinates of the centroid of the shaded area in Fig. 9-40 below. The area removed is semicircular. *Ans.* $\bar{x} = 3.77$ in., $\bar{y} = 2.69$ in.

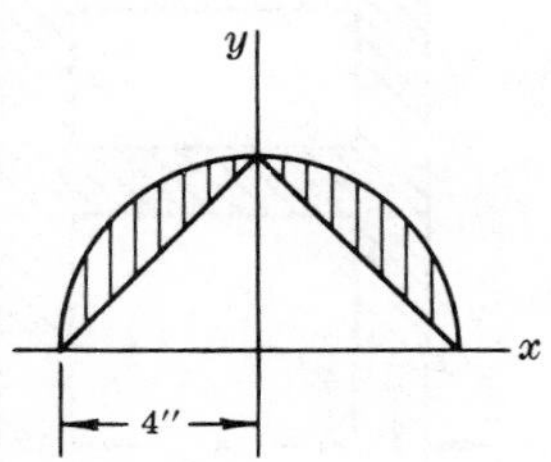

Fig. 9-39

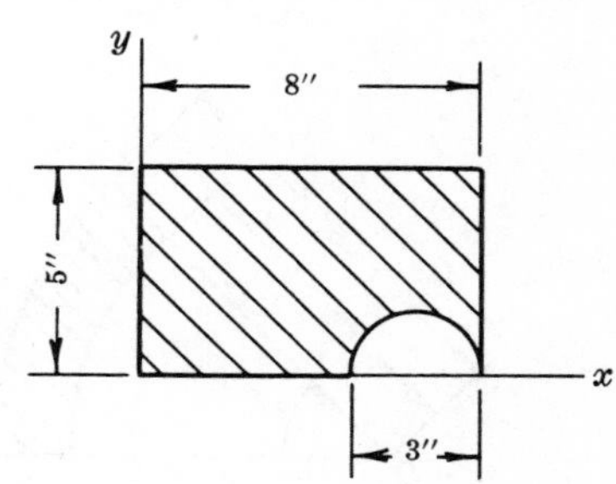

Fig. 9-40

Locate the centroids of the following figures with respect to the axes shown.

56. Fig. 9-41. *Ans.* $\bar{x} = 0.50$ in., $\bar{y} = 5.50$ in.
57. Fig. 9-42. *Ans.* $\bar{x} = 0$, $\bar{y} = 4.37$ in.
58. Fig. 9-43. *Ans.* $\bar{x} = 0.449$ in., $\bar{y} = 0$
59. Fig. 9-44. *Ans.* $\bar{x} = 1.32$ in., $\bar{y} = 3.59$ in.
60. Fig. 9-45. *Ans.* $\bar{x} = 4.07$ in., $\bar{y} = 1.69$ in.
61. Fig. 9-46. *Ans.* $\bar{x} = 3$ in., $\bar{y} = 3.40$ in.

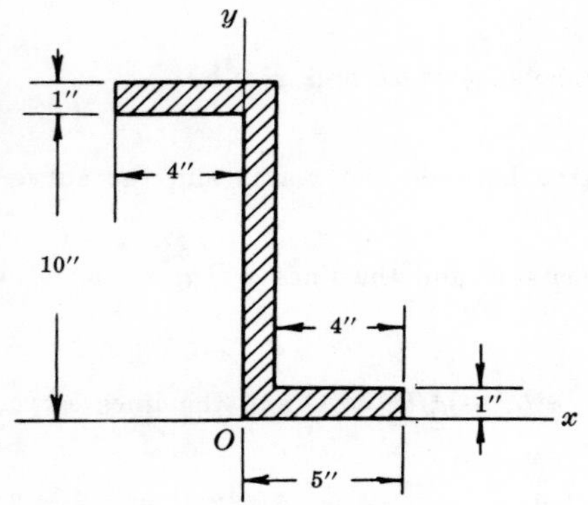

Fig. 9-41

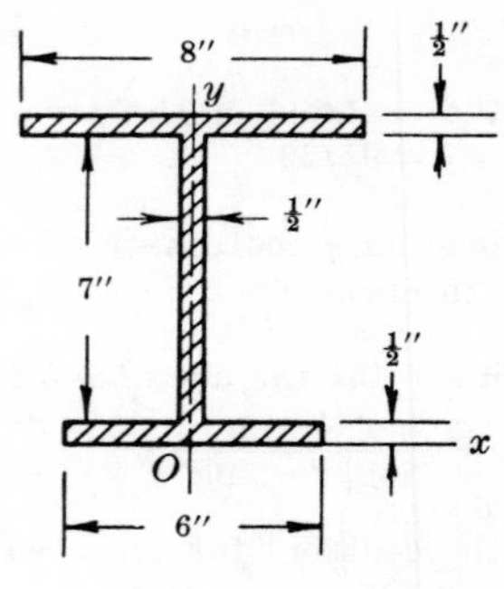

Fig. 9-42

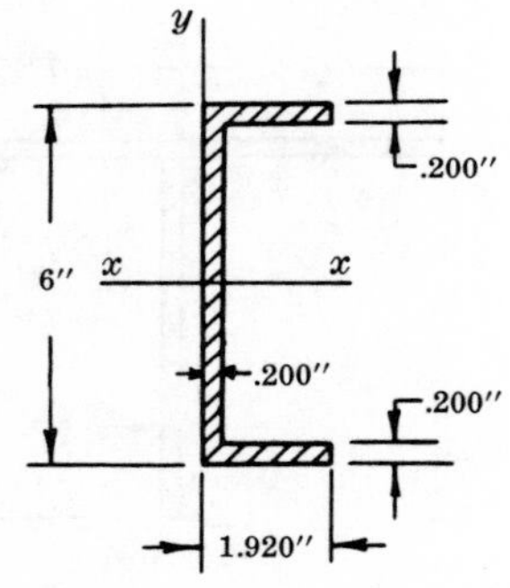

Fig. 9-43

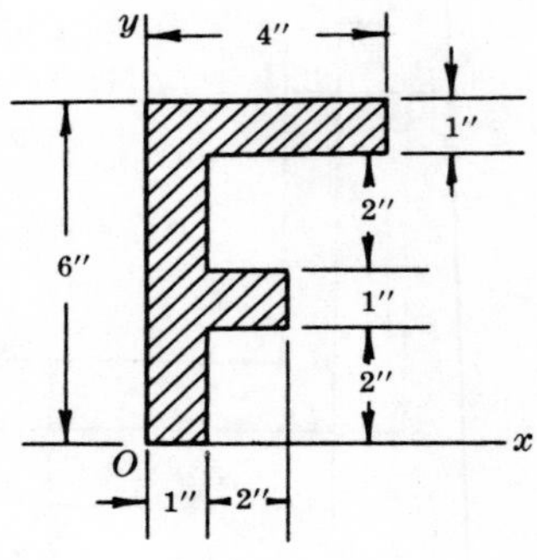

Fig. 9-44

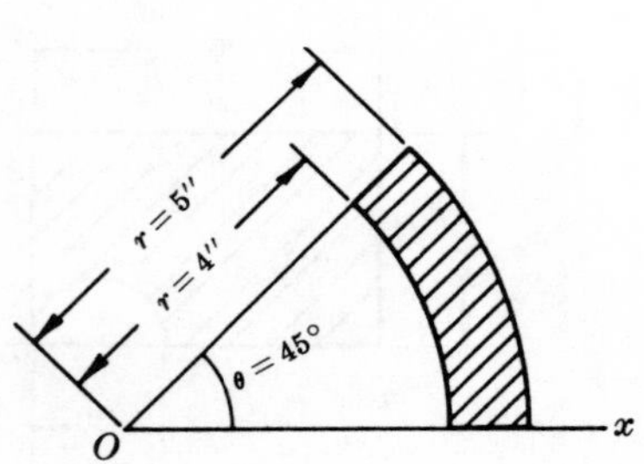

Fig. 9-45

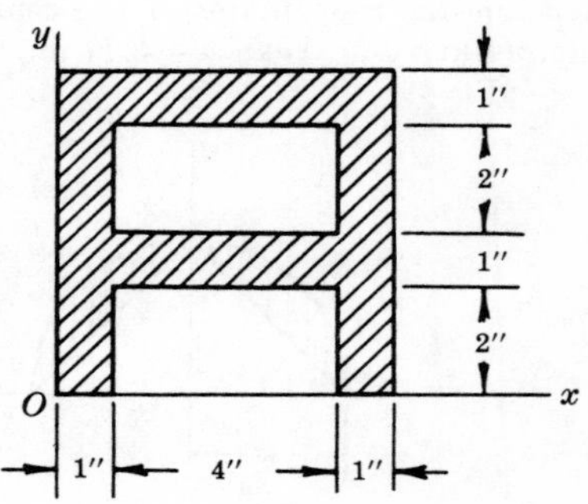

Fig. 9-46

62. By integration show that the centroid of the volume of a right circular cone is at a distance from the base equal to one-fourth the altitude.

63. A cone is formed by revolving the area bounded by the lines $by = ax$, $y = 0$, and $x = b$ around the x-axis. Determine $\bar{x}$ for the volume of the cone. *Ans.* $\bar{x} = 3b/4$

64. By integration show that the centroid of the volume of a hemisphere is at a distance from the base equal to three-eighths the radius.

65. The area formed by the parabola $y^2 = 4ax$ and the lines $x = b$, $y = 0$ is revolved about the x-axis. Show that $\bar{x}$ for the volume of the paraboloid so formed equals $2b/3$.

66. The first quadrant of the ellipse $x^2/a^2 + y^2/b^2 = 1$ is revolved about the x-axis. Show that $\bar{x}$ for the volume generated equals $3a/8$.

67. The area bounded by the hyperbola $x^2/a^2 - y^2/b^2 = 1$ and the lines $y = 0$, $x = 2a$ is revolved about the x-axis. Compute $\bar{x}$ for the solid so generated. *Ans.* $\bar{x} = 27a/16$

68. The area bounded by the curve $y = \sin x$ and lines $y = 0$, $x = 0$, $x = \pi/2$ is revolved about the x-axis. Compute $\bar{x}$ for the volume generated. *Ans.* $\bar{x} = \pi/4 + 1/2\pi$

69. The area formed by the curve $y = e^x$ and lines $x = 0$, $x = b$, $y = 0$ is revolved about the x-axis. Locate the centroid of the volume formed. *Ans.* $\bar{x} = (be^{2b} - \frac{1}{2}e^{2b} + \frac{1}{2})/(e^{2b} - 1)$

70. The area formed by the curve $x^2/a^2 + y^2/b^2 = 1$ and the lines $x = a$, $y = b$ is revolved about the x-axis. Determine $\bar{x}$ for the generated volume. *Ans.* $\bar{x} = 3a/4$

71. Determine the centroid of volume for a right circular cone with a base diameter of 8 inches and an altitude of 16 inches. *Ans.* $d = 4$ in.

72. A right circular cone with altitude 8 inches and radius of base 6 inches is welded to a hemisphere 12 inches in diameter so that their bases coincide. Locate the centroid of the total volume.
Ans. Distance from vertex of cone = 8.55 in.

73. A right circular cone of altitude 10 inches and base of diameter 7 inches rests on top of a right circular cylinder of the same base and 12 inches high. Locate the centroid of the composite volume.
Ans. Distance from vertex of cone = 14.2 in.

74. A hemisphere of radius a rests on a right circular cylinder whose base has radius a also. If the height of the cylinder is a, locate the centroid of the composite volume.
Ans. Distance from bottom of cylinder = $0.85a$

75. The frustum of a right circular cone has an altitude of 2 ft. The radii of the bases are 1 ft and 2 ft respectively. Locate the centroid of its volume. *Ans.* Distance from larger base = 0.786 ft

76. From a hemisphere of radius a is cut a cone of the same base and altitude. Show that the centroid of the remaining volume is at a distance from the base equal to $a/2$.

77. A block is 2 ft × 2 ft × 4 ft. A hole 6 inches in diameter is cut normal to the top 2 by 2 face at its center. If the depth of the hole is $1\frac{1}{2}$ ft, locate the centroid of the remaining volume.
Ans. Distance from bottom = 1.98 ft

78. Fig. 9-47 illustrates a piece turned down in a lathe. Compute its centroid with reference to the left end. *Ans.* $d = 5.48$ in.

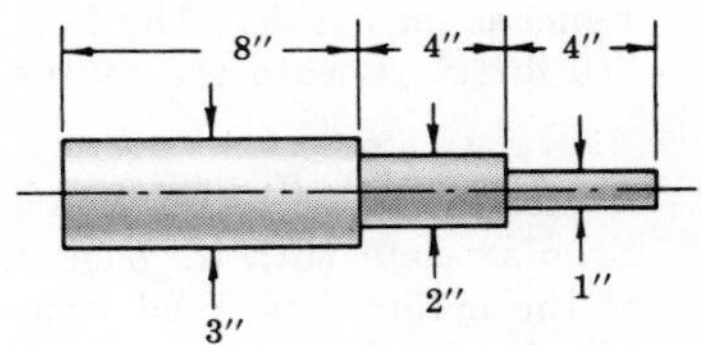

Fig. 9-47

79. Two spheres of volumes 5 in^3 and 15 in^3 are connected by a small rod whose volume is 10 in^3. If the distance between centers is 10 inches, how far from the center of the sphere of 5 in^3 is the centroid? *Ans.* $d = 6.67$ in.

80. A sphere of radius 10 in. has a spherical cavity of radius 2 in. If the distance between the centers of the spheres is 4 in., where is the centroid? Assume that the positive axis passes from the center of the 10 inch sphere through the center of the cavity. *Ans.* $d = -.032$ in.

81. Rework Problem 21 placing the vertex at the origin of the axes. *Ans.* $d = -2h/3$

82. A tank consists of a cylinder 20 ft in diameter and 20 ft high. It has a hemispherical bottom and a cone-shaped cover 5 ft high. Locate the center of gravity of the empty tank by assuming that it is the same as the center of gravity of the outer surface. *Ans.* $\bar{y} = 17.6$ ft from bottom

83. A hemisphere whose radius is 3 in. rests on a cone with the same size base and an altitude of 4 in. Where is the centroid of the surface, assuming full opening between the two?
Ans. Distance from vertex = 4.22 in.

84. An open tank is made of a right circular cylinder 36 in. high with a diameter of 20 in. The bottom extends below the cylinder for 4 in. as a cone with base coinciding with the cylinder. Determine the centroid of the surface. *Ans.* Distance from vertex = 19.5 in.

85. The radius of the base of a right circular cone is 8 inches. Its altitude is 10 inches. Locate the centroid (*a*) of its curved surface and (*b*) of its volume with respect to its base.
Ans. (*a*) $d = 3.33$ in., (*b*) $d = 2.50$ in.

86. In Problem 22 above, show that $\bar{y} = 3b/8$.

87. Locate the mass-center of a hemisphere in which the density at any point varies (*a*) with the first power of the distance of the point from the base, and (*b*) with the square of the radial distance from the center. *Ans.* (*a*) $d = 8r/15$, (*b*) $d = 0.417r$

88. Locate the mass-center of a right circular cone in which the density at any point varies directly as the distance of the point from the base. *Ans.* $d = 2h/5$

89. A mallet has a cylindrical wooden head and a cylindrical wooden handle. The head has a diameter of 4 in. and a length of 6 in. The handle is $1\frac{1}{4}$ in. in diameter and 12 in. long. Where is the center of mass with respect to the free end of the handle? The wood weighs 50 lb/ft^3. *Ans.* $d = 12.7$ in.

90. A cylinder 2 inches in diameter and 2 inches high is cut out of a right circular cone with 8 inch diameter base and 10 inch height. The base of the cylinder lies in the plane of the base of the cone. If the cone is made of steel, density 490 lb/ft^3, how far is the center of mass of the remaining mass above the base? *Ans.* $d = 2.51$ in.

91. A tank 8 ft in diameter has a center of mass 4 ft above the bottom when empty. It is filled to a height of 6 ft with oil (density 55 lb/ft^3). Locate the center of mass of the tank and oil if the tank weighs 8000 lb empty. *Ans.* d above base = 3.33 ft

92. A hole 1 inch in diameter and 3 inches deep is drilled into the center of the top face of a brass cube 4 inches on a side. The hole normal to the face is filled with lead. Brass weighs 525 lb/ft^3 and lead 710 lb/ft^3. Locate the center of mass of the two metals with respect to the bottom. *Ans.* 2.00 in.

93. A steel cylinder has a 4 inch diameter base and a height of 6 inches. A hole in the shape of a right circular cone with its base coincident with the base of the cylinder and axis coincident with the axis of the cylinder is filled with lead. The diameter of the base of the cone is 2 inches and its height is 1 inch. Steel weighs 490 lb/ft^3 and lead 710 lb/ft^3. Locate the centroid with respect to the base.
Ans. $d = 2.97$ in.

94. In Fig. 9-48 below, compute the distance from the left side of the composite mass to its center of mass. The large cylinder weighs 490 lb/ft^3 and the small cylinder 530 lb/ft^3. *Ans.* $d = 11.8$ in.

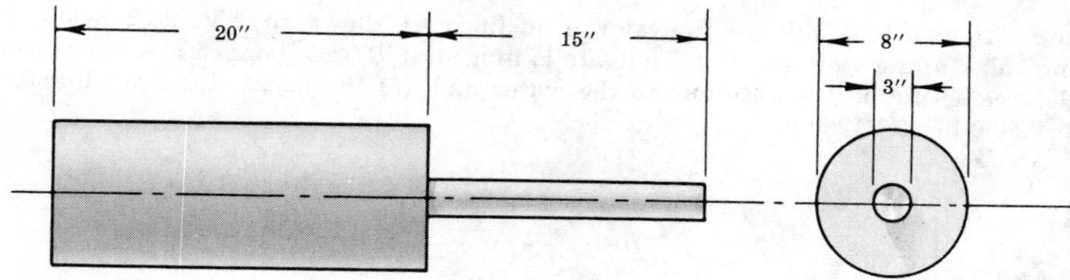

Fig. 9-48

95. In Fig. 9-49 below, determine the surface generated by revolving the arc of the semicircle of radius r about the y-axis. Note that the centroid of the arc of the semicircle is at distance $\bar{x} = d + 2r/\pi$. *Ans.* $S = 2\pi^2 rd + 4\pi r^2$

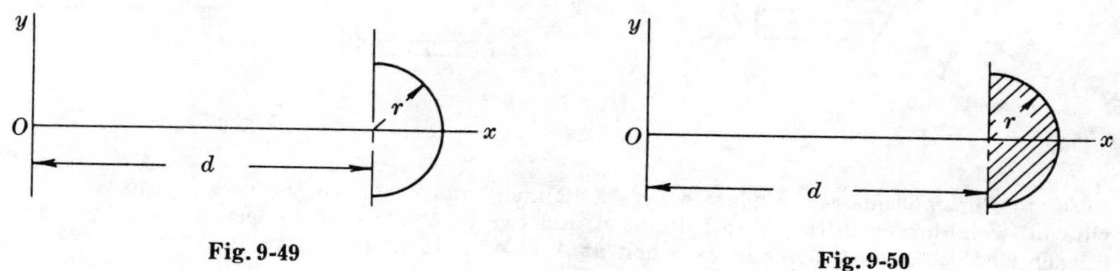

Fig. 9-49 **Fig. 9-50**

96. Determine the volume generated by revolving a semicircular area of radius r about the y-axis. Refer to Fig. 9-50 above. *Ans.* $V = \pi^2 r^2 d + 4\pi r^3/3$

97. Find the volume generated by revolving the ellipse $x^2/a^2 + y^2/b^2 = 1$ about the line $x = 2a$. *Ans.* $V = 4\pi^2 a^2 b$

98. In the adjacent Fig. 9-51 is shown the flywheel on a small compressor. A cross section of the rim is assumed to be rectangular with a semicircular groove of $\frac{1}{2}$ inch radius cut therein. Compute the distance from the center to the centroid of the cross sectional area shown at *A-A*. Using this centroidal distance, determine the volume of the rim only. *Ans.* $V = 95$ in^3

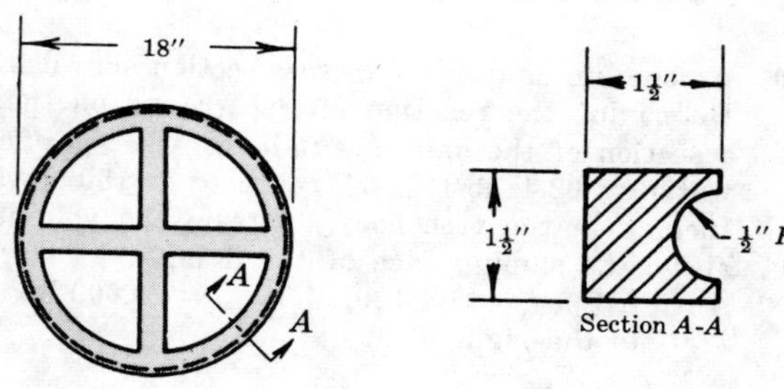

Fig. 9-51

99. A right triangle of base h and altitude r is revolved about its base through 360°. What volume does the triangular area generate? *Ans.* $V = \pi r^2 h/3$

100. A wall 10 ft high is subjected to water pressure on a vertical face. Where is the center of pressure of the water on the wall? *Ans.* 6.67 ft below the surface

101. In Problem 100, what is the turning moment about the base of the water on a section of the wall 20 ft long and 2 ft thick? Water weighs 62.4 lb/ft^3. *Ans.* 208,000 lb-ft

102. A beam 12 ft long carries a uniform unit force of 100 lb per linear ft for a distance of 4 ft from the left end. From that unit force of 100 lb per linear ft the force increases uniformly to a maximum of 300 lb per linear ft at the right end. What are the reactions at the ends? *Ans.* $R_L = 780$ lb, $R_R = 1220$ lb

103. A beam is subjected to a uniform unit force of w lb per ft for one-third the length. The unit force then decreases uniformly to zero at the right end. What is the total force acting, and how is it distributed to the supports? *Ans.* $P = \frac{2}{3}wl$ lb, $R_L = \frac{23}{54}wl$ lb, $R_R = \frac{13}{54}wl$ lb

104. The gate AB in a tank filled with water is inclined as shown in Fig. 9-52 below and is 2 ft wide (perpendicular to the view shown). The gate is hinged at B and clamped at A. Determine (*a*) the total normal force acting on the gate due to the water and (*b*) the horizontal clamping force needed at A. *Ans.* (*a*) 2400 lb, (*b*) 1230 lb

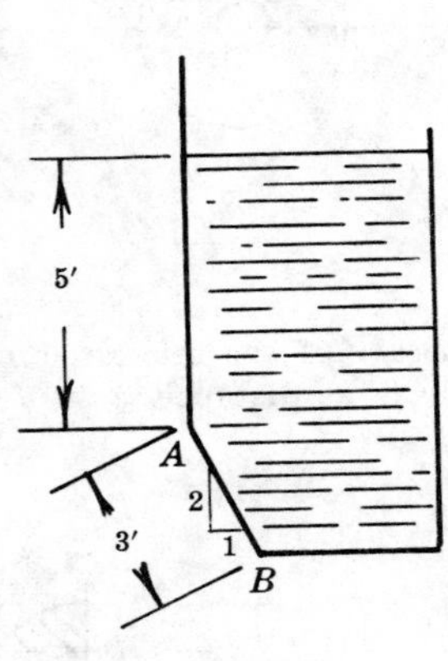

Fig. 9-52

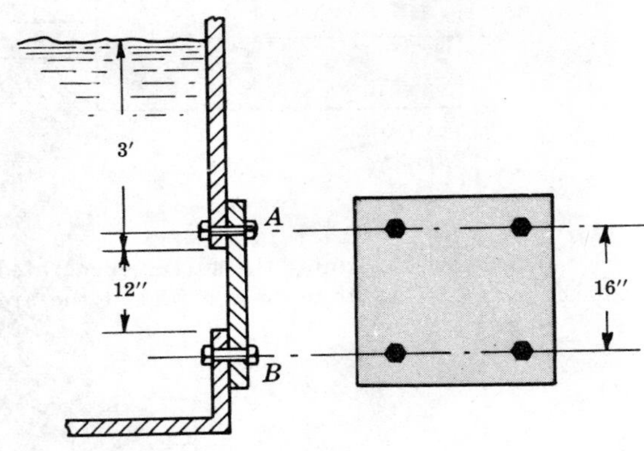

Fig. 9-53

105. Refer to Fig. 9-53 above. A plate covers a 12 inch square hole in the side of the tank which is filled with oil weighing 50 lb/ft³. Two bolts at A and two at B are used to secure the plate. Find the force in each bolt. *Ans.* Force in each bolt at A = 42.1 lb, at B = 45.3 lb

106. Suppose the plate in the preceding problem covers a circular hole 12 inches in diameter. Find the force in each bolt.
Ans. Force in each bolt at A = 33.0 lb, at B = 35.8 lb

107. A concrete dam has the cross section shown in Fig. 9-54. Determine the reaction of the ground on the bottom of a section of the dam one ft long. Use specific weight of water = 62.4 lb/ft³, of concrete = 150 lb/ft³. [Hint: Include in the free body diagram the volume of water above the sloping face of the dam.]
Ans. $R_{horizontal}$ = 7020 lb, $R_{vertical}$ = 29,000 lb at 5.64 ft to the right of A

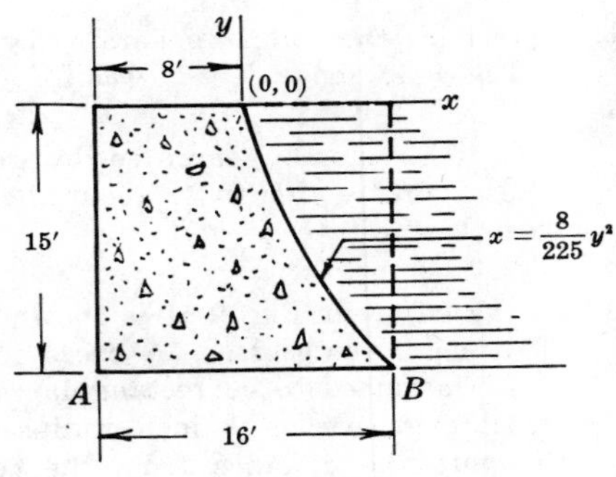

Fig. 9-54

Chapter 10

Kinematics of a Particle

KINEMATICS

Kinematics is the study of motion without regard to the forces or other factors which influence the motion.

RECTILINEAR MOTION

Rectilinear motion is motion of a point P along a straight line which for convenience here will be chosen as the x-axis. Vector symbols are omitted in this part.

(*a*) **The Position** of point P at any time t is expressed in terms of its distance x from a fixed origin O on the x-axis. This distance x is positive or negative according to the usual sign convention.

(*b*) **The Average Velocity** v_{av} of point P during the time interval between t and $t+\Delta t$ during which its position changes from x to $x+\Delta x$ is the quotient $\Delta x/\Delta t$. Mathematically this is written

$$v_{av} = \frac{\Delta x}{\Delta t} \tag{1}$$

(*c*) **The Instantaneous Velocity** v of point P at time t is the limit of the average velocity (defined above) as the increment of time approaches zero as a limit. Mathematically this is written

$$v = \lim_{\Delta t \to 0} \frac{\Delta x}{\Delta t} = \frac{dx}{dt} \tag{2}$$

(*d*) **The Average Acceleration** a_{av} of point P during the time interval between t and $t+\Delta t$ during which its velocity changes from v to $v+\Delta v$ is the quotient $\Delta v/\Delta t$. Mathematically this is written

$$a_{av} = \frac{\Delta v}{\Delta t} \tag{3}$$

(*e*) **The Instantaneous Acceleration** a of point P at time t is the limit of the average acceleration (defined above) as the increment of time approaches zero as a limit. Mathematically this is written

$$a = \lim_{\Delta t \to 0} \frac{\Delta v}{\Delta t} = \frac{dv}{dt} = \frac{d^2x}{dt^2} \tag{4}$$

(*f*) **For Constant Acceleration** $a = k$ the following formulas are valid.

$$v = v_o + kt \tag{5}$$

$$v^2 = v_o^2 + 2ks \tag{6}$$

$$s = v_o t + \tfrac{1}{2}kt^2 \tag{7}$$

$$s = \tfrac{1}{2}(v + v_o)t \tag{8}$$

where v_o = initial velocity, v = final velocity,
k = constant acceleration, t = time, s = displacement.

(*g*) **Simple Harmonic Motion** is rectilinear motion in which the acceleration is negatively proportional to the displacement. Mathematically this is written

$$a = -K^2 x \tag{9}$$

As an example, equation (*9*) is satisfied by a point vibrating so that its displacement x is given by the equation

$$x = b \sin \omega t \tag{10}$$

where b = amplitude in linear measure, ω = constant *circular* frequency in radians per second, and t = time in seconds.

Thus, since $x = b \sin \omega t$, then $v = dx/dt = \omega b \cos \omega t$ and $a = d^2x/dt^2 = -\omega^2 b \sin \omega t = -\omega^2 x$. That is, $a = -K^2 x$, where $K = \omega$, a constant, and the motion is simple harmonic.

CURVILINEAR MOTION

Curvilinear motion in a plane is motion along a plane curve (path). The velocity and acceleration of a point on such a curve will be expressed in (*a*) rectangular components, (*b*) tangential and normal components, and (*c*) radial and transverse components.

RECTANGULAR COMPONENTS

The position vector $\mathbf{r}$ of a point P on such a curve in terms of the unit vectors, $\mathbf{i}$ and $\mathbf{j}$ along the x and y axes respectively is written

$$\mathbf{r} = x\mathbf{i} + y\mathbf{j}$$

As P moves, $\mathbf{r}$ changes and the velocity $\mathbf{v}$ can be expressed as

$$\mathbf{v} = \frac{d\mathbf{r}}{dt} = \frac{dx}{dt}\mathbf{i} + \frac{dy}{dt}\mathbf{j}$$

Using $dx/dt = \dot{x}$ and $dy/dt = \dot{y}$ and $d\mathbf{r}/dt = \dot{\mathbf{r}}$ as convenient symbols, we have

$$\mathbf{v} = \dot{\mathbf{r}} = \dot{x}\mathbf{i} + \dot{y}\mathbf{j} \tag{11}$$

The speed of the point is the magnitude of the velocity $\mathbf{v}$, that is,

$$|\mathbf{v}| = \sqrt{\dot{x}^2 + \dot{y}^2}$$

If θ is the angle which the vector $\mathbf{v}$ makes with the x-axis, we can write

$$\theta = \tan^{-1}\frac{\dot{y}}{\dot{x}} = \tan^{-1}\frac{dy/dt}{dx/dt} = \tan^{-1}\frac{dy}{dx}$$

Thus the velocity vector $\mathbf{v}$ is tangent to the path at point P (see Fig. 10-1).

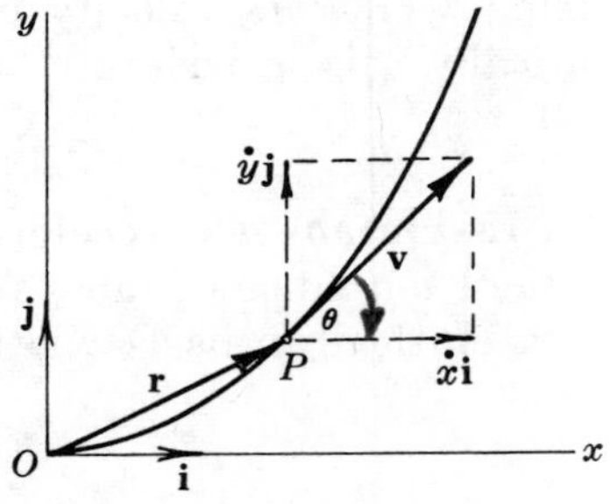

Fig. 10-1

The acceleration vector $\mathbf{a}$ is the time rate of change of $\mathbf{v}$, that is,

$$\mathbf{a} = \frac{d\mathbf{v}}{dt} = \frac{d^2\mathbf{r}}{dt^2} = \frac{d^2x}{dt^2}\mathbf{i} + \frac{d^2y}{dt^2}\mathbf{j}$$

Using the symbolic notation $\mathbf{a} = \dot{\mathbf{v}} = \ddot{\mathbf{r}}$, $\ddot{x} = d^2x/dt^2$ and $\ddot{y} = d^2y/dt^2$, we can write

$$\mathbf{a} = \dot{\mathbf{v}} = \ddot{\mathbf{r}} = \ddot{x}\mathbf{i} + \ddot{y}\mathbf{j} \tag{12}$$

The magnitude of the acceleration vector **a** is

$$|\mathbf{a}| = \sqrt{\ddot{x}^2 + \ddot{y}^2}$$

Don't make the mistake of assuming that **a** is tangent to the path at point P.

TANGENTIAL and NORMAL COMPONENTS

In the preceding discussion the velocity vector **v** and acceleration vector **a** were expressed in terms of the orthogonal unit vectors **i** and **j** along the x and y axes respectively. The following discussion shows how to express the same vector **v** and the same vector **a** in terms of the unit vector **T** tangent to the path at point P and the unit vector **N** at right angles to T.

In Fig. 10-2, point P is shown on the curve at a distance s along the curve from a reference point P_0. The position vector **r** of point P is a function of the scalar quantity s. To study this relationship let Q be a point on the curve near P.

The position vectors $\mathbf{r}(s)$ and $\mathbf{r}(s) + \Delta\mathbf{r}(s)$ for points P and Q respectively are shown as well as the change $\Delta\mathbf{r}(s)$ which is the directed straight line PQ. The distance along the curve from P to Q is Δs. The derivative of $\mathbf{r}(s)$ with respect to s is written

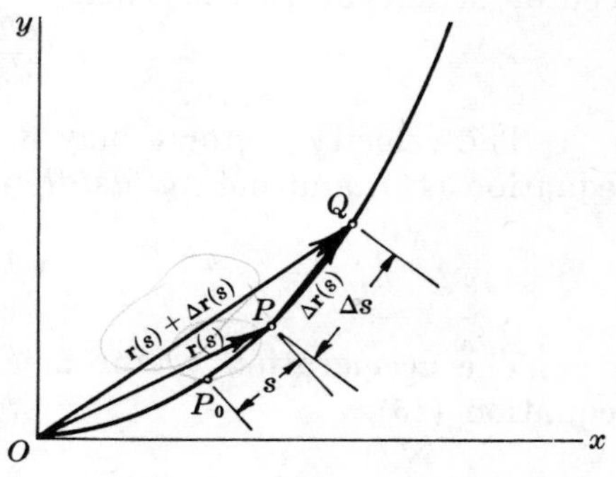

Fig. 10-2

$$\frac{d\mathbf{r}(s)}{ds} = \lim_{\Delta s \to 0} \frac{\mathbf{r}(s) + \Delta\mathbf{r}(s) - \mathbf{r}(s)}{\Delta s} = \lim_{\Delta s \to 0} \frac{\Delta\mathbf{r}(s)}{\Delta s}$$

As Q approaches P, the ratio of the magnitude of the straight line $\Delta\mathbf{r}(s)$ to the arc length Δs approaches unity. Also, in direction the straight line $\Delta\mathbf{r}(s)$ approaches the tangent to the path at P. Thus, in the limit, a unit vector **T** is defined as follows

$$\boxed{\frac{d\mathbf{r}(s)}{ds} = \mathbf{T}} \qquad (13)$$

Next consider how **T** changes with s. As shown in Fig. 10-3(*a*) below, the center of curvature C is a distance ρ from P.

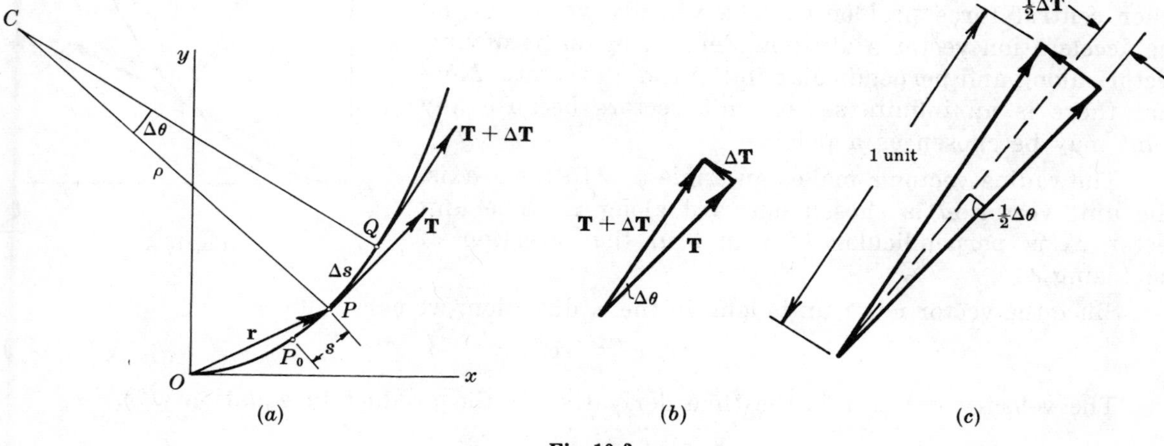

Fig. 10-3

If we can assume point Q close to P, the unit tangent vectors at P and Q are **T** and $\mathbf{T} + \Delta\mathbf{T}$ respectively. Since the tangents at P and Q are perpendicular to the radii drawn to C, the angle between $\mathbf{T}$ and $\mathbf{T}+\Delta\mathbf{T}$ as shown in Fig. 10-3(*b*) above is also $\Delta\theta$. Because $\mathbf{T}$ and $\mathbf{T}+\Delta\mathbf{T}$ are unit vectors, $\Delta\mathbf{T}$ represents only a change in direction (not magnitude). Thus the triangle in Fig. 10-3(*b*) is isosceles and is shown drawn to a larger scale in Fig. 10-3(*c*) above. From Fig. 10-3(*c*) it should be evident that

$$\frac{|\tfrac{1}{2}\Delta\mathbf{T}|}{1} = \sin\tfrac{1}{2}\Delta\theta \approx \tfrac{1}{2}\Delta\theta \quad \text{from which} \quad |\Delta\mathbf{T}| \approx \Delta\theta$$

But from Fig. 10-3(*a*), $\Delta s = \rho\,\Delta\theta$; hence we can write $\Delta s \approx \rho|\Delta\mathbf{T}|$. Thus

$$\lim_{\Delta s\to 0}\frac{|\Delta\mathbf{T}|}{\Delta s} = \frac{1}{\rho}$$

Also, in the limit $\Delta\mathbf{T}$ is perpendicular to $\mathbf{T}$ and is directed towards the center of curvature C. Let $\mathbf{N}$ be the unit vector which is perpendicular to $\mathbf{T}$ and directed towards the center of curvature C. Then

$$\frac{d\mathbf{T}}{ds} = \lim_{\Delta s\to 0}\frac{|\Delta\mathbf{T}|}{\Delta s}\mathbf{N} = \frac{1}{\rho}\mathbf{N} \tag{14}$$

The velocity vector $\mathbf{v}$ may now be given in terms of the unit vectors $\mathbf{T}$ and $\mathbf{N}$. Using equation (*13*), and noting $ds/dt = \dot{s}$ is the speed of P along the path, we can write

$$\mathbf{v} = \frac{d\mathbf{r}}{dt} = \frac{d\mathbf{r}}{ds}\frac{ds}{dt} = \dot{s}\mathbf{T} \tag{15}$$

The acceleration vector $\mathbf{a}$ is the time derivative of the velocity vector $\mathbf{v}$ defined in equation (*15*).

$$\mathbf{a} = \frac{d\mathbf{v}}{dt} = \ddot{s}\mathbf{T} + \dot{s}\frac{d\mathbf{T}}{dt}$$

But $\dfrac{d\mathbf{T}}{dt} = \dfrac{d\mathbf{T}}{ds}\dfrac{ds}{dt}$ and from (*14*) this may be written $\dfrac{d\mathbf{T}}{dt} = \dfrac{\dot{s}}{\rho}\mathbf{N}$. Then

$$\mathbf{a} = \ddot{s}\mathbf{T} + \frac{\dot{s}^2}{\rho}\mathbf{N} \tag{16}$$

RADIAL and TRANSVERSE COMPONENTS

The point P on the curve may be located with polar coordinates in terms of any point chosen as a pole. Fig. 10-4 shows the origin O as the pole. These polar coordinates are useful in studying the motion of planets and other central force problems. The velocity vector $\mathbf{v}$ and the acceleration vector $\mathbf{a}$ are now derived in terms of unit vectors along and perpendicular to the radius vector. Note that there is an infinite set of unit vectors because any point may be chosen as a pole.

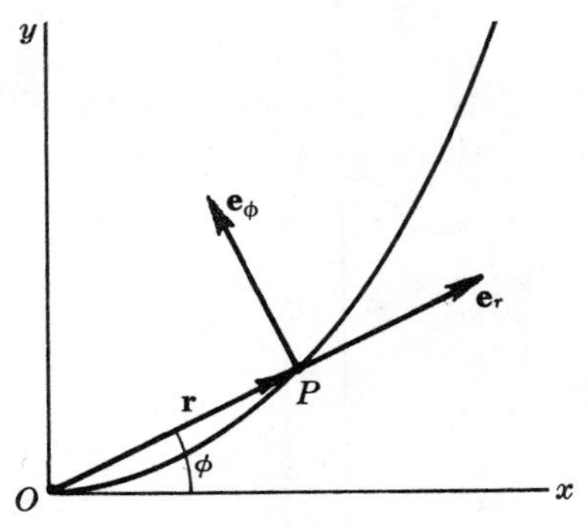

Fig. 10-4

The radius vector $\mathbf{r}$ makes an angle ϕ with the x-axis. The unit vector $\mathbf{e}_r$ is chosen outward along $\mathbf{r}$. The unit vector $\mathbf{e}_\phi$ is perpendicular to $\mathbf{r}$ and in the direction of increasing ϕ.

Since the vector $\mathbf{r}$ is r units long in the $\mathbf{e}_r$ direction, we can write

$$\mathbf{r} = r\mathbf{e}_r \tag{17}$$

The velocity vector $\mathbf{v}$ is the time derivative of the product in equation (*17*).

$$\mathbf{v} = \dot{\mathbf{r}} = \dot{r}\mathbf{e}_r + r\dot{\mathbf{e}}_r$$

where $\dot{\mathbf{e}}_r = d\mathbf{e}_r/dt$.

To evaluate $\dot{\mathbf{e}}_r$ and $\dot{\mathbf{e}}_\phi$, allow P to move to a nearby point Q with a corresponding set of unit vectors $\mathbf{e}_r+\Delta\mathbf{e}_r$ and $\mathbf{e}_\phi+\Delta\mathbf{e}_\phi$ as shown in Fig. 10-5(*a*) below.

Fig. 10-5(*b*) and (*c*) illustrate these unit vectors. Since the triangles are isosceles, we can deduce the following conclusions by reasoning similar to that used in the explanation of **T** and **N** vectors: $d\mathbf{e}_r$ in the limit has a magnitude $d\phi$ in the $\mathbf{e}_\phi$ direction, and $d\mathbf{e}_\phi$ in the limit has a magnitude $d\phi$ in the negative $\mathbf{e}_r$ direction. Hence

$$\dot{\mathbf{e}}_r = \frac{d\mathbf{e}_r}{d\phi}\frac{d\phi}{dt} = \dot{\phi}\mathbf{e}_\phi,$$

$$\dot{\mathbf{e}}_\phi = \frac{d\mathbf{e}_\phi}{d\phi}\frac{d\phi}{dt} = -\dot{\phi}\mathbf{e}_r$$

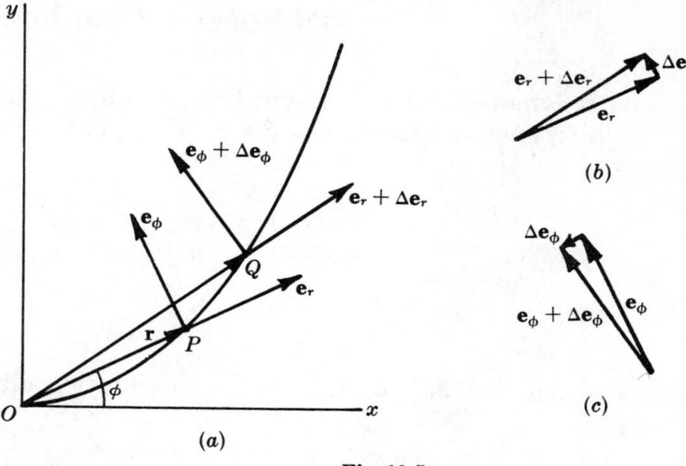

Fig. 10-5

where $\dot{\phi}$ is the time rate of change of the angle ϕ which the radius vector **r** makes with the x-axis.

The velocity vector **v** may now be written

$$\mathbf{v} = \dot{r}\mathbf{e}_r + r\dot{\phi}\mathbf{e}_\phi \tag{18}$$

The acceleration vector **a** is the time derivative of the terms in equation (*18*).

$$\begin{aligned}\mathbf{a} &= \ddot{r}\,\mathbf{e}_r + \dot{r}\dot{\mathbf{e}}_r + \dot{r}\dot{\phi}\mathbf{e}_\phi + r\ddot{\phi}\mathbf{e}_\phi + r\dot{\phi}\dot{\mathbf{e}}_\phi \\ &= \ddot{r}\,\mathbf{e}_r + \dot{r}\dot{\phi}\mathbf{e}_\phi + \dot{r}\dot{\phi}\mathbf{e}_\phi + r\ddot{\phi}\mathbf{e}_\phi - r\dot{\phi}^2\mathbf{e}_r\end{aligned}$$

where $\ddot{\phi}$ is the angular acceleration (time derivative of the angular velocity $\dot{\phi}$). Collecting terms this becomes

$$\mathbf{a} = (\ddot{r} - r\dot{\phi}^2)\mathbf{e}_r + (2\dot{r}\dot{\phi} + r\ddot{\phi})\mathbf{e}_\phi \tag{19}$$

As a special case of curvilinear motion consider a point moving in a circular path of radius R. Substituting R for r in equations (*18*) and (*19*), noting $\dot{R} = \ddot{R} = 0$, we obtain

$$\mathbf{v} = R\dot{\phi}\mathbf{e}_\phi \quad \text{(tangent to the path)} \tag{20}$$

$$\mathbf{a} = -R\dot{\phi}^2\mathbf{e}_r + R\ddot{\phi}\mathbf{e}_\phi \tag{21}$$

Thus the acceleration has a tangential component of magnitude $R\ddot{\phi}$ and a normal component directed towards the center of magnitude $R\dot{\phi}^2$.

UNITS

Units have been purposely omitted in the foregoing discussion. The following table lists three common systems: the engineering system, the centimeter-gram-second (cgs) system, and the meter-kilogram-second (mks) system.

Symbol	Engineering Units	CGS Units	MKS Units
s, ρ, R, x, y	ft	cm	m
$v, \dot{x}, \dot{y}, \dot{s}$	ft/sec or fps	cm/sec	m/sec
$a, \ddot{x}, \ddot{y}, \ddot{s}$	ft/sec²	cm/sec²	m/sec²
θ, ϕ	radians (rad)	radians (rad)	radians (rad)
$\omega, \dot{\theta}, \dot{\phi}$	rad/sec	rad/sec	rad/sec
$\alpha, \ddot{\theta}, \ddot{\phi}$	rad/sec²	rad/sec²	rad/sec²

Solved Problems

1. A point P moves along a straight line according to the equation $x = 4t^3 + 2t + 5$, where x is in ft, t in sec. Determine the displacement, velocity and acceleration when $t = 3$ sec.

Solution:

$$x = 4t^3 + 2t + 5 = 4(3^3) + 2(3) + 5 = 119 \text{ ft}$$
$$v = dx/dt = 12t^2 + 2 = 12(3^2) + 2 = 110 \text{ ft/sec}$$
$$a = dv/dt = 24t = 24(3) = 72 \text{ ft/sec}^2$$

2. In Problem 1, what is the average acceleration during the fourth second?

Solution:

Velocity at end of fourth second is $v = 12(4^2) + 2 = 194$ ft/sec. Hence the change in velocity during the fourth second is $\Delta v = 194 \text{ ft/sec} - 110 \text{ ft/sec} = 84$ ft/sec.

The average acceleration is $a_{av} = \dfrac{\Delta v}{\Delta t} = \dfrac{84 \text{ ft/sec}}{1 \text{ sec}} = 84 \text{ ft/sec}^2$.

3. A point moves along a straight line such that its displacement is $s = 8t^2 + 2t$, where s is in ft, t in sec. Plot the displacement, velocity and acceleration against time. These are called s-t, v-t, a-t diagrams.

Solution:

Differentiating $s = 8t^2 + 2t$ yields $v = ds/dt = 16t + 2$ and $a = dv/dt = d^2s/dt^2 = 16$. This shows that the acceleration is constant, 16 ft/sec².

To determine values for plotting, use the tabular form listed below, where t is in sec, s in ft, and v in ft/sec.

t	t^2	$8t^2$	$2t$	$s = 8t^2 + 2t$	$16t$	$v = 16t + 2$
0	0	0	0	0	0	2
1	1	8	2	10	16	18
2	4	32	4	36	32	34
3	9	72	6	78	48	50
4	16	128	8	136	64	66
5	25	200	10	210	80	82
10	100	800	20	820	160	162

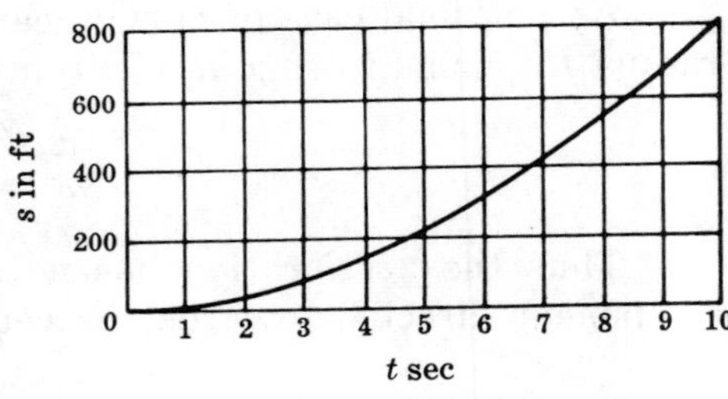

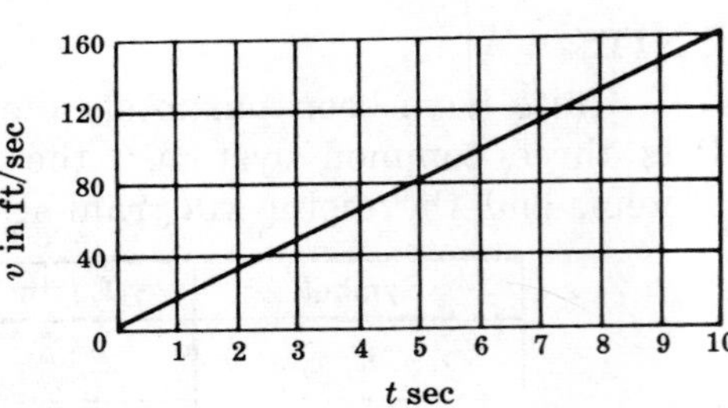

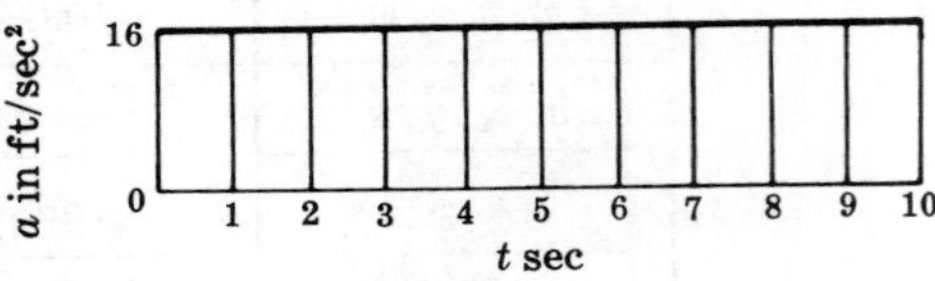

Some valuable relationships may be deduced from these diagrams. The slope of the s-t curve at any time t is the height or ordinate of the v-t curve at time t. This follows since $v = ds/dt$.

Again, the slope of the v-t curve (in this particular case the slope is the same at any point of the straight line, i.e., 16 ft/sec²) at any time t is the ordinate of the a-t curve at any time t. This follows since $a = dv/dt$.

The two equations just given may also be written as

$$a\,dt = dv \quad \text{and} \quad v\,dt = ds$$

Integration between proper limits yields

$$\int_{t_o}^{t} a\,dt = \int_{v_o}^{v} dv = v - v_o \qquad \text{and} \qquad \int_{t_o}^{t} v\,dt = \int_{s_o}^{s} ds = s - s_o$$

where $\int_{t_o}^{t} a\,dt$ = area under the a-t diagram for the time interval from t_o to t,

$\int_{t_o}^{t} v\,dt$ = area under the v-t diagram for the time interval from t_o to t,

$v - v_o$ = change in velocity in the same time interval t_o to t,

$s - s_o$ = change in displacement in the same time interval t_o to t.

The first above equation states that the change in the ordinate of the v-t diagram for any time interval is equal to the area under the a-t diagram within that time interval. A similar statement may be made for the change in the ordinate of the s-t diagram.

4. An automobile accelerates uniformly from rest to 60 mi/hr in 28 sec. Find its acceleration and displacement during this time.

Solution:

The following data are given: $v_0 = 0$, $v = 60$ mi/hr $= 88$ ft/sec, $t = 28$ sec.

To determine the acceleration, which is a constant k, apply the formula $v = v_0 + kt$.

$$k = \frac{v - v_o}{t} = \frac{(88 - 0)\text{ ft/sec}}{28\text{ sec}} = 3.14\text{ ft/sec}^2$$

To determine the displacement using only the original data,

$$s = \frac{v + v_o}{2}t = \frac{(88 + 0)\text{ ft/sec}}{2} \times 28\text{ sec} = 1230\text{ ft}$$

5. A train changes its speed uniformly from 60 mph to 30 mph in a distance of 1500 ft. What is its acceleration?

Solution:

Given: $v_o = 60$ mph $= 88$ ft/sec, $v = 30$ mph $= 44$ ft/sec, $s = 1500$ ft.

The acceleration k is constant because the speed changes uniformly in a straight line.

$$v^2 = v_o^2 + 2ks \quad \text{or} \quad k = \frac{v^2 - v_0^2}{2s} = \frac{(44\text{ ft/sec})^2 - (88\text{ ft/sec})^2}{2(1500\text{ ft})} = -1.94\text{ ft/sec}^2$$

This is sometimes written as a deceleration of 1.94 ft/sec².

6. A balloon is rising with a velocity of 5 ft/sec when a bag of sand is released. If its height at the time of release is 420 ft, how long does it take for the sand to reach the ground?

Solution:

The data given for the sand are: $v_o = +5$ ft/sec, $s = +420$ ft, $g = a = -32.2$ ft/sec².

Assume that displacements, velocities and accelerations in an upward direction are positive.

$$s = v_o t + \tfrac{1}{2}gt^2, \quad 420\text{ ft} = (5\text{ ft/sec})t - \tfrac{1}{2}(32.2\text{ ft/sec}^2)t^2, \quad t = 5.28\text{ sec}$$

7. A ball is projected vertically upward with a velocity of 80 ft/sec. Two seconds later a second ball is projected vertically upward with a velocity of 60 ft/sec. At what point above the surface of the earth will they meet?

Solution:

Let t be the time after the first ball is projected that the two meet. The second ball will then have been traveling for $(t - 2)$ sec. The displacements for both balls will be the same at time t.

Let s_1 and s_2 be the displacements of the first and second balls, respectively. Then

$$s_1 = (v_o)_1 t - \tfrac{1}{2}gt^2 \qquad \text{and} \qquad s_2 = (v_o)_2(t-2) - \tfrac{1}{2}g(t-2)^2$$

Equating s_1 and s_2 and substituting the given values of $(v_o)_1$ and $(v_o)_2$, we obtain

$$80t - 16.1t^2 = 60(t-2) - 16.1(t-2)^2 \qquad \text{or} \qquad t = 4.14 \text{ sec}$$

Substituting this value of t in the equation for s_1 (or s_2), the displacement is

$$s_1 = 80 \text{ ft/sec} \times 4.14 \text{ sec} - \tfrac{1}{2}(32 \text{ ft/sec}^2)(4.14 \text{ sec})^2 = 55.3 \text{ ft}$$

8. The acceleration of a point in rectilinear motion is given by the equation $a = -32.2$. It is known that the velocity v is zero and the displacement x is $+25$ when t equals zero. Determine the equation of the displacement.

Solution:

As is often the case in physical problems, the acceleration may be observed and substituted in the second order differential equation. The solution is simple since the variables x and t can be separated. Write the given equation as follows.

$$a = dv/dt = -32.2$$

Then $dv = -32.2\,dt$. Integrating, $v = -32.2t + C_1$.

The constant C_1 may be evaluated because $v=0$ when $t=0$. Hence $C_1 = 0$.

The equation for velocity is therefore $v = dx/dt = -32.2t$. Then $dx = -32.2t\,dt$.

Integrating, $x = -32.2t^2/2 + C_2$. Substituting $x = +25$ when $t=0$ will determine $C_2 = 25$.

The equation for displacement is therefore $x = -16.1t^2 + 25$.

9. A particle moves along a horizontal straight line with an acceleration $a = 6\sqrt[3]{s}$. When $t = 2$ sec, its displacement $s = +27$ ft and its velocity $v = +27$ ft/sec. Calculate the velocity and acceleration of the point when $t = 4$ sec.

Solution:

Since the acceleration is given as a function of the displacement, use the differential equation $a\,ds = v\,dv$. Then

$$\int 6s^{1/3}\,ds = \int v\,dv \qquad \text{or} \qquad \tfrac{9}{2}s^{4/3} = \tfrac{1}{2}v^2 + C_1$$

Since $v = +27$ when $s = +27$, $C_1 = 0$ and $v = 3s^{2/3}$.

Next use $v = ds/dt$ to obtain $ds/s^{2/3} = 3dt$; from this $3s^{1/3} = 3t + C_2$. Substitute the condition $s = +27$ when $t=2$ to obtain $C_2 = 3$ and $s = (t+1)^3$.

The equations are therefore $s = (t+1)^3$, $v = 3(t+1)^2$, $a = 6(t+1)$. When $t = 4$ sec, $s = 125$ ft, $v = 75$ ft/sec, and $a = 30$ ft/sec^2.

A plot of these quantities against time is shown. Note that the ordinate of the v-t curve at any time t is the slope of the s-t curve at the same time. Also, the ordinate of the a-t curve at any time t is the slope of the v-t curve at that time.

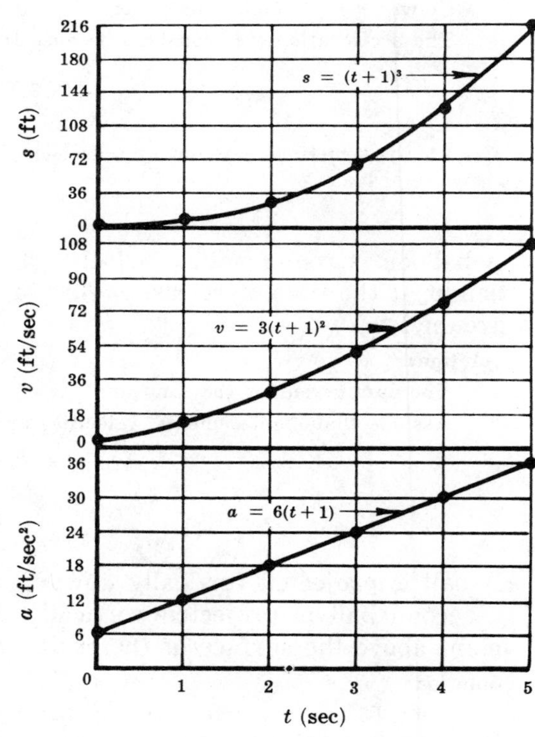

10. A particle moves on a vertical line with an acceleration $a = 2\sqrt{v}$. When $t = 2$ sec, its displacement $s = 64/3$ ft and its velocity $v = 16$ ft/sec. Determine the displacement, velocity and acceleration of the particle when $t = 3$ sec.

Solution:

Since $a = dv/dt$, then $2\sqrt{v} = dv/dt$. Separating the variables, $2dt = dv/v^{1/2}$. Integrating, $2t + C_1 = 2v^{1/2}$. But $v = 16$ ft/sec when $t = 2$ sec; hence $C_1 = 4$.

The equation becomes $t + 2 = v^{1/2}$ or $v = (t+2)^2 = ds/dt$. Then $ds = (t+2)^2 dt$. Integrating, $s = \frac{1}{3}(t+2)^3 + C_2$. But $s = 64/3$ ft when $t = 2$ sec; hence $C_2 = 0$.

The equations are therefore $s = \frac{1}{3}(t+2)^3$, $v = (t+2)^2$, and $a = 2(t+2)$.

When $t = 3$ sec, $s = 41.7$ ft, $v = 25$ ft/sec, and $a = 10$ ft/sec^2.

11. The acceleration of a point moving on a vertical line is given by the equation $a = 12t - 20$. It is known that its displacement $s = -10$ ft at time $t = 0$ and that its displacement $s = +10$ ft at time $t = 5$ sec. Derive the equation of its motion.

Solution:

Integrate $a = dv/dt = 12t - 20$ to obtain $v = 6t^2 - 20t + C_1$.

Integrate this once more to obtain $s = 2t^3 - 10t^2 + C_1 t + C_2$.

The constants of integration may now be evaluated. Substitute the known values of s and t.

$$-10 = 2(0)^3 - 10(0)^2 + C_1(0) + C_2 \quad \text{or} \quad C_2 = -10$$
$$+10 = 2(5)^3 - 10(5)^2 + C_1(5) - 10 \quad \text{or} \quad C_1 = +4$$

The equation of motion is $s = 2t^3 - 10t^2 + 4t - 10$.

12. In the system shown in Fig. 10-6(*a*) determine the velocity and acceleration of block 2 at the instant.

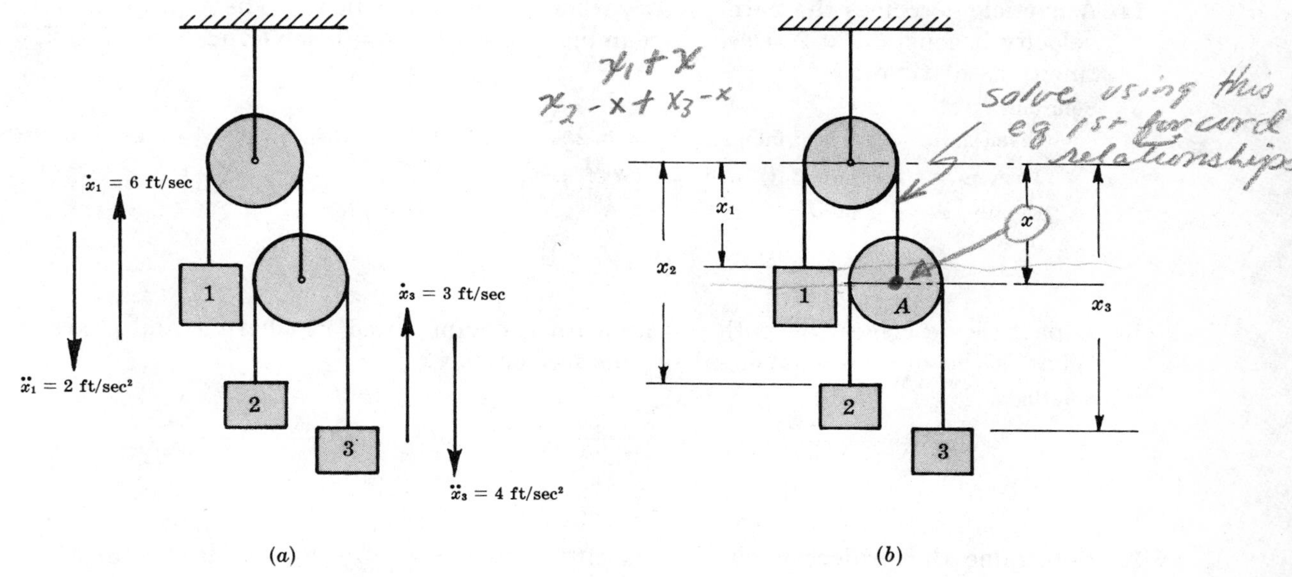

Fig. 10-6

Solution:

Fig. 10-6(*b*) is drawn to show the position of each weight relative to the fixed support. The length of the cord between the weight 1 and point A is a constant and equals one-half the circumference of the top pulley plus $x_1 + x$. The length of the cord between weights 2 and 3 is a constant and equals one-half the circumference of pulley A plus $x_2 - x + x_3 - x$.

Thus $x_1 + x$ = constant; $x_2 + x_3 - 2x$ = constant. Time derivatives then show

$$(1)\ \dot{x}_1 + \dot{x} = 0 \qquad (3)\ \dot{x}_2 + \dot{x}_3 - 2\dot{x} = 0$$
$$(2)\ \ddot{x}_1 + \ddot{x} = 0 \qquad (4)\ \ddot{x}_2 + \ddot{x}_3 - 2\ddot{x} = 0$$

Calling the upward direction positive and substituting $\dot{x}_1 = 6$ ft/sec into equation (1), we find $\dot{x} = -6$. Substituting this value together with $\dot{x}_3 = 3$ ft/sec into equation (3), we find $\dot{x}_2 = 2\dot{x} - \dot{x}_3 = 2(-6) - (3) = -15$ ft/sec (down).

Similar reasoning for the accelerations shows

$$\ddot{x} = +2 \quad \text{and} \quad \ddot{x}_2 = 2\ddot{x} - \ddot{x}_3 = 2(+2) - (-4) = 8 \text{ ft/sec}^2$$

13. Show that the curvature of a plane curve at point P may be expressed as

$$\frac{1}{\rho} = \frac{\dot{x}\ddot{y} - \ddot{x}\dot{y}}{(\dot{x}^2 + \dot{y}^2)^{3/2}}$$

where ρ is the radius of curvature, $\dot{x}$ and $\dot{y}$ are the x and y components of the speed of P, and $\ddot{x}$ and $\ddot{y}$ are the x and y components of the magnitude of the acceleration of P.

Solution:

From the calculus, the curvature of any curve $y = f(x)$ at point P is

$$\frac{1}{\rho} = \frac{d^2y/dx^2}{[1 + (dy/dx)^2]^{3/2}} \qquad (1)$$

But $\dfrac{dy}{dx} = \dfrac{dy}{dt}\dfrac{dt}{dx} = \dfrac{\dot{y}}{\dot{x}}$ and $\dfrac{d^2y}{dx^2} = \dfrac{d}{dx}\left(\dfrac{dy}{dx}\right) = \dfrac{d}{dt}\left(\dfrac{dy}{dx}\right)\dfrac{dt}{dx} = \dfrac{d}{dt}\left(\dfrac{\dot{y}}{\dot{x}}\right)\dfrac{1}{\dot{x}} = \dfrac{\dot{x}\ddot{y} - \dot{y}\ddot{x}}{\dot{x}^2\dot{x}}$.

Substituting in (1), we obtain the required equation.

14. A particle describes the path $y = 4x^2$, where x and y are in ft. The x projection of the velocity is constant at 2 ft/sec. Assuming $x = y = 0$ at $t = 0$, solve for $x, y, \dot{x}, \dot{y}, \ddot{x}, \ddot{y}$ as functions of time t.

Solution:

Given $dx/dt = 2$ ft/sec; integrating, $x = 2t + C_1$. Since $x = 0$ when $t = 0$, $C_1 = 0$ and $x = 2t$ ft.

Then $y = 4(2t)^2 = 16t^2$ ft, and $\dot{y} = 32t$ ft/sec.

Finally, $\ddot{x} = 0$ and $\ddot{y} = 32$ ft/sec^2.

15. A particle describes the path $y = 4x^2$ with constant speed v, where x and y are in ft. What is the normal component of the acceleration?

Solution:

$$\frac{1}{\rho} = \frac{d^2y/dx^2}{[1 + (dy/dx)^2]^{3/2}} = \frac{8}{[1 + (8x)^2]^{3/2}} \quad \text{and} \quad a_n = \frac{v^2}{\rho} = \frac{8v^2}{[1 + 64x^2]^{3/2}}$$

16. Determine the projection of the velocity vector $\mathbf{v} = 4\mathbf{i} - 6\mathbf{j} + \mathbf{k}$ in the direction of the vector $\mathbf{n} = 3\mathbf{i} + 2\mathbf{j} - 5\mathbf{k}$. Distances are in ft and time in sec.

Solution:

The unit vector $\mathbf{e}_L$ in the desired direction is $\mathbf{e}_L = \dfrac{3\mathbf{i} + 2\mathbf{j} - 5\mathbf{k}}{\sqrt{3^2 + 2^2 + (-5)^2}}$.

Hence the projection of $\mathbf{v}$ on $\mathbf{n}$ is $\mathbf{v} \cdot \mathbf{e}_L = \dfrac{12 - 12 - 5}{\sqrt{38}} = -0.81$ ft/sec.

17. A particle moves along the path whose equation is $r = 2\theta$ ft. If the angle $\theta = t^2$ radians, determine the velocity of the particle when θ is 60°. Use two methods.

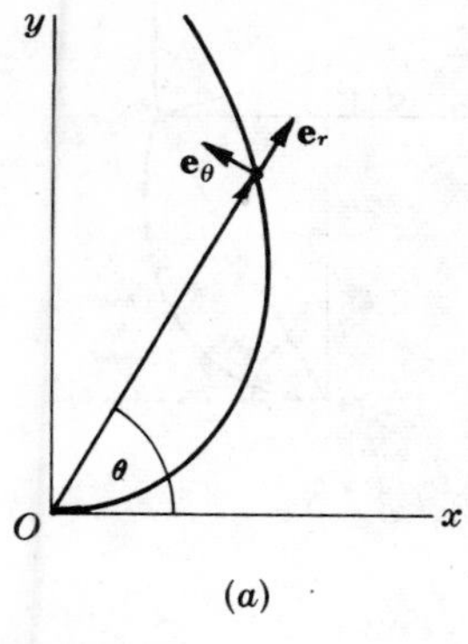

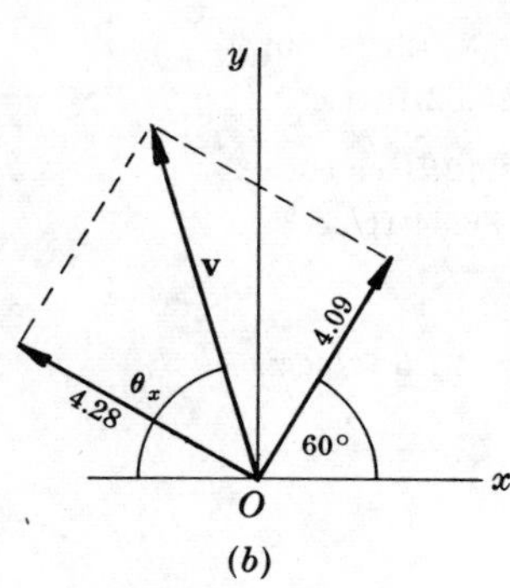

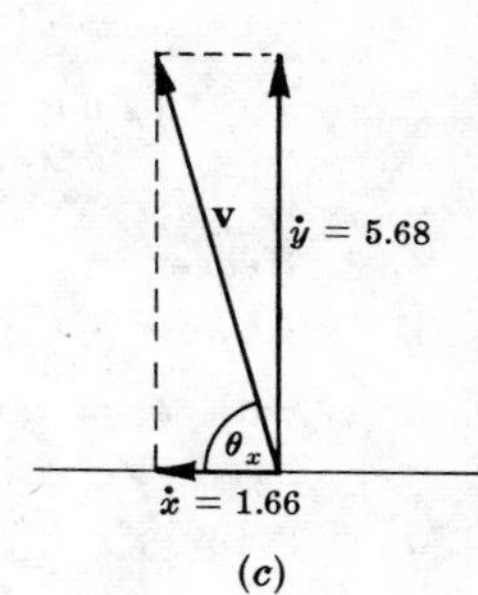

Fig. 10-7

Solution:

A plot of the path is shown in Fig. 10-7(*a*) with unit vectors $\mathbf{e}_r$ along $\mathbf{r}$ and $\mathbf{e}_\theta$ perpendicular to $\mathbf{r}$ and in the direction of increasing θ.

(*a*) **Polar Coordinates.** See Fig. 10-7(*b*).

Since $\theta = t^2$, $\dot{\theta} = 2t$; since $r = 2\theta = 2t^2$, $\dot{r} = 4t$.

The velocity vector $\mathbf{v}$ at $\theta = \pi/3$ radians is found as follows:

$$\theta = \pi/3 = t^2 \qquad \text{or} \qquad t = 1.023 \text{ sec}$$

$$\begin{aligned} \mathbf{v} &= \dot{r}\mathbf{e}_r + r\dot{\theta}\mathbf{e}_\theta = 4(1.023)\mathbf{e}_r + [2(1.047)][2(1.023)]\mathbf{e}_\theta \\ &= 4.09\mathbf{e}_r + 4.28\mathbf{e}_\theta \end{aligned}$$

and $v = \sqrt{(4.09)^2 + (4.28)^2} = 5.92$ ft/sec with $\theta_x = 30° + \tan^{-1} 4.09/4.28 = 73.7°$.

(*b*) **Cartesian Coordinates.** See Fig. 10-7(*c*).

$$x = r\cos\theta = 2\theta\cos\theta = 2t^2\cos t^2 \qquad y = r\sin\theta = 2\theta\sin\theta = 2t^2\sin t^2$$

Then

$$\dot{x} = 4t\cos t^2 + 2t^2(-\sin t^2)(2t) = -1.66 \text{ ft/sec}$$
$$\dot{y} = 4t\sin t^2 + 2t^2(\cos t^2)(2t) = +5.68 \text{ ft/sec}$$

at $t = 1.023$ sec ($\cos t^2 = \cos \pi/3$, $\sin t^2 = \sin \pi/3$). Hence

$$v = \sqrt{(-1.66)^2 + (5.68)^2} = 5.92 \text{ ft/sec with } \theta_x = \tan^{-1}\frac{5.68}{1.66} = 73.7°$$

18. In the preceding problem determine the acceleration of the particle using the same two methods.

Solution:

(*a*) **Polar Coordinates.** See Fig. 10-8.

$$\begin{aligned} \mathbf{a} &= (\ddot{r} - r\dot{\theta}^2)\mathbf{e}_r + (2\dot{r}\dot{\theta} + r\ddot{\theta})\mathbf{e}_\theta \\ &= -4.77\,\mathbf{e}_r + 20.94\,\mathbf{e}_\theta \end{aligned}$$

since $\theta = t^2$, $\dot{\theta} = 2t$, $\ddot{\theta} = 2$; $r = 2\theta = 2t^2$, $\dot{r} = 4t$, $\ddot{r} = 4$; $t = 1.023$ sec at $\theta = \pi/3$. Thus

$$\begin{aligned} a &= \sqrt{(-4.77)^2 + (20.94)^2} \\ &= 21.5 \text{ ft/sec}^2 \end{aligned}$$

with $\theta_x = 30° - 12.8° = 17.2°$

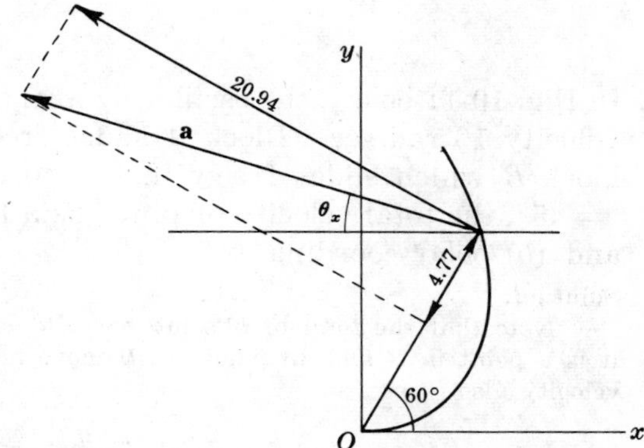

Fig. 10-8

(b) **Cartesian Coordinates.** See Fig. 10-9.

Continuing the time derivatives and evaluating at $t = 1.023$ ($t^2 = \pi/3$),

$$\ddot{x} = 4\cos t^2 + 4t(-\sin t^2)(2t) + 12t^2(-\sin t^2) - 4t^3(\cos t^2)(2t) = -20.52 \text{ ft/sec}^2$$

$$\ddot{y} = 4\sin t^2 + 4t(\cos t^2)(2t) + 12t^2(\cos t^2) + 4t^3(-\sin t^2)(2t) = +6.34 \text{ ft/sec}^2$$

Hence

$$a = \sqrt{(-20.52)^2 + (6.34)^2} = 21.5 \text{ ft/sec}^2$$

with $\theta_x = \tan^{-1} 6.34/20.52 = 17.2°$

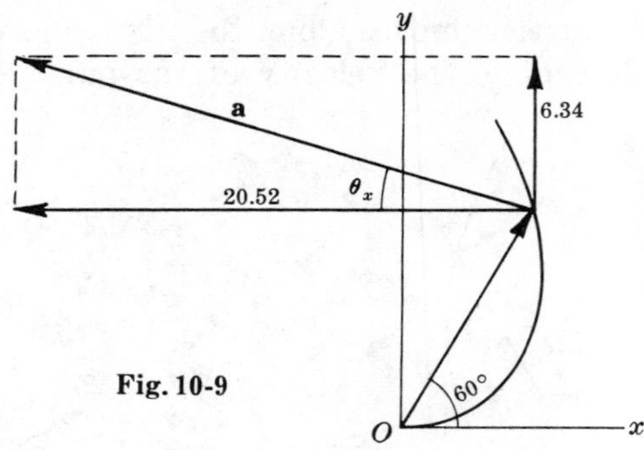

Fig. 10-9

19. In the Scotch yoke shown in Fig. 10-10, crank OA is turning with a constant angular velocity ω rad/sec. Derive the expressions for the displacement, velocity, and acceleration of the sliding member.

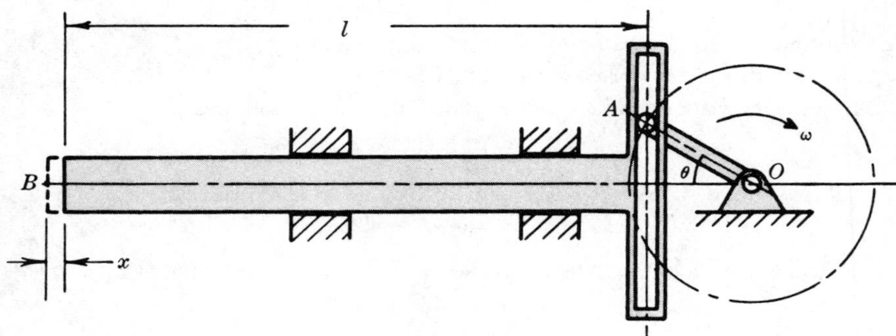

Fig. 10-10

Solution:

Let B represent the position of the left end of the slider when $\theta = 0°$. The displacement x is written $x = OB - l - OA\cos\theta$. When the crank is horizontal, $OB = l + OA$ and hence

$$x = l + OA - l - OA\cos\theta = OA(1 - \cos\theta)$$

Let $OA = R$. Also, since the crank is turning with constant angular velocity ω, the expression ωt may be substituted for θ. Differentiating $x = R(1 - \cos\omega t)$ yields

$$v = dx/dt = R\omega\sin\omega t \quad \text{and} \quad a = dv/dt = R\omega^2\cos\omega t$$

20. In Fig. 10-11 below, the oscillating arm OD is rotating clockwise with a constant angular velocity 10 rad/sec. Block A slides freely in the slot in the arm OD and is pinned to block B which slides freely in the horizontal slot in the framework. Determine, for $\theta = 45°$, the total velocity of pin P as a point in block B using (a) Cartesian coordinates and (b) polar coordinates.

Solution:

Note that the total or absolute velocity of P as a point in block B can only be horizontal since it is a point in B and all points in B move horizontally. But as a point in block A it has this same velocity also.

(a) Let x = distance of P from C. Then $x = 3\tan\theta$ and $v_P = dx/dt = 3(\sec^2\theta)d\theta/dt$. When $\theta = 45°$, $v_P = 3(\sec^2 45°)10 = 60$ ft/sec.

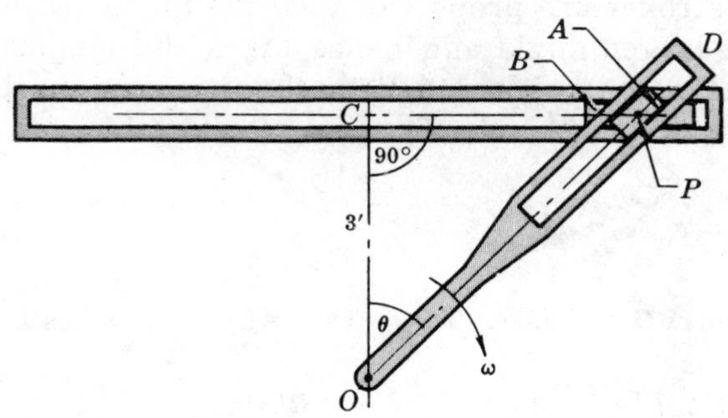

Fig. 10-11

(b) Let ρ = distance of P from O which will be used as a pole in studying the motion. Then $\rho = 3 \sec \theta$.

The radial component of the velocity along OP is $d\rho/dt = 3 \sec \theta \tan \theta \, d\theta/dt$. For $\theta = 45°$, this becomes $3 \sec 45° \tan 45° \times 10 = 42.4$ ft/sec. This component is directed outward along OP.

The transverse component of the velocity is $\rho \, d\theta/dt = 3 \sec \theta \, d\theta/dt$. For $\theta = 45°$, this becomes $3 \sec 45° \times 10 = 42.4$ ft/sec. This component is perpendicular to the arm OP and acts down to the right because ω is clockwise.

The two components are shown to the right of the figure. Hence $v_P = \sqrt{(d\rho/dt)^2 + (\rho \, d\theta/dt)^2} = 60$ ft/sec and is horizontal.

21. The bar AB shown in Fig. 10-12 moves so that its lowest point A travels horizontally to the right with constant velocity $v_A = 5$ ft/sec. What is the velocity of point B when $\theta = 70°$? The length of the bar is 6.24 ft.

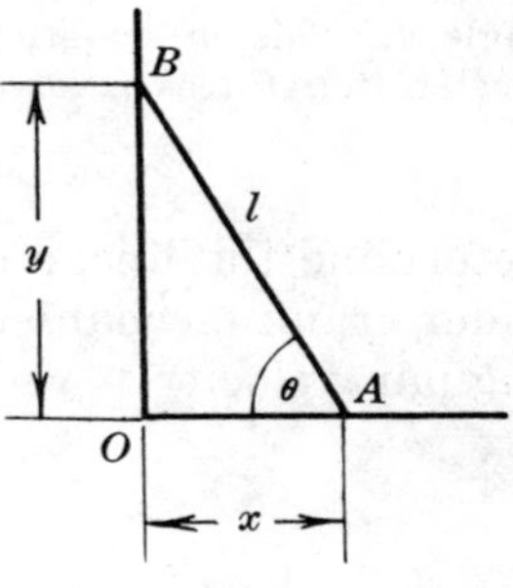

Fig. 10-12

Solution:

Let x and y be the distances of A and B from point O at any time during the motion. Since $x^2 + y^2 = l^2$, then

$$2x\frac{dx}{dt} + 2y\frac{dy}{dt} = 0$$

and $v_B = \dfrac{dy}{dt} = -\dfrac{x \, dx}{y \, dt} = -(\cot \theta)(v_A) = -1.82$ ft/sec.

The minus sign indicates that B is traveling down. Note that v_B is independent of l.

22. In the quick-return mechanism shown in Fig. 10-13, the crank AB is driven at a constant angular velocity ω rad/sec. The slider at B slides along the rod OP, also causing it to oscillate about the hinge O. In turn, OP slides in and out of the slider at P and also moves the second slider pinned at P along the horizontal slot. A cutting tool attached to this second slider would be subjected to a reciprocating motion. The cutting tool reaches the extremes of its horizontal travel when OP is tangent to the crank circle at B_1 and B_2. Cutting occurs while the crank pin moves from B_2 counterclockwise to B_1. The return stroke occurs in the remaining arc from B_1 to B_2. Since the speed is constant, the

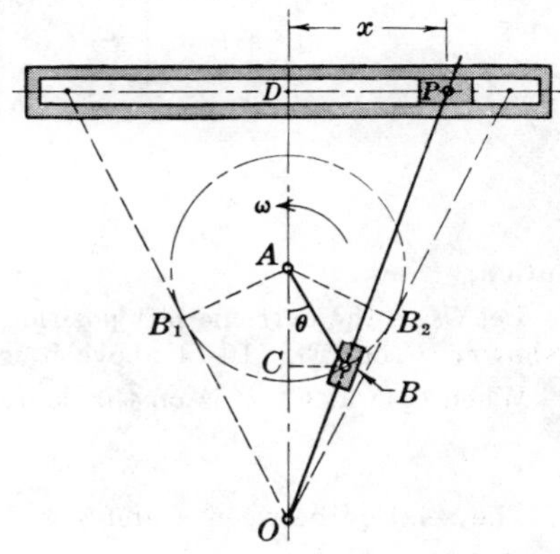

Fig. 10-13

times of the working and return strokes are proportional to the angles traversed. The cutting stroke occurs during the larger angle and hence takes the longer time. The return stroke is faster; hence the name quick-return mechanism is appropriate. Determine the expressions for displacement, velocity, and acceleration of the cutting tool at P.

Solution:

From the figure, $\dfrac{x}{BC} = \dfrac{OD}{OC}$.

Let $OD = l$, $AB = R$, $OA = d$. Then $BC = R\sin\theta$, $OC = OA - AC = d - R\cos\theta$. Substituting in the original expression in x,

$$\frac{x}{R\sin\theta} = \frac{l}{d - R\cos\theta} \quad \text{or} \quad x = \frac{Rl\sin\theta}{d - R\cos\theta} \tag{1}$$

and

$$v = \frac{dx}{dt} = \frac{Rl[(d - R\cos\theta)\cos\theta\, d\theta/dt - \sin\theta\,(R\sin\theta\, d\theta/dt)]}{(d - R\cos\theta)^2}$$

Since $d\theta/dt$ is the angular velocity ω, this equation becomes

$$v = \frac{Rl\omega(d\cos\theta - R\cos^2\theta - R\sin^2\theta)}{(d - R\cos\theta)^2} \quad \text{or} \quad v = Rl\omega\frac{(d\cos\theta - R)}{(d - R\cos\theta)^2} \tag{2}$$

and

$$a = \frac{dv}{dt} = Rl\omega\frac{(d - R\cos\theta)^2(-d\sin\theta\, d\theta/dt) - (d\cos\theta - R)[2(d - R\cos\theta)R\sin\theta\, d\theta/dt]}{(d - R\cos\theta)^4}$$

$$= \frac{-Rl\omega^2\sin\theta\,(d^2 - 2R^2 + Rd\cos\theta)}{(d - R\cos\theta)^3} \tag{3}$$

Equations (*1*), (*2*) and (*3*) give the displacement, velocity, and acceleration for any value of the angle θ. This information is necessary to design the members of the mechanism to withstand the accelerating forces involved.

23. Determine the linear displacement, velocity and acceleration of the crosshead C in the slider crank mechanism for any position of the crank R which is rotating at a constant angular velocity ω rad/sec.

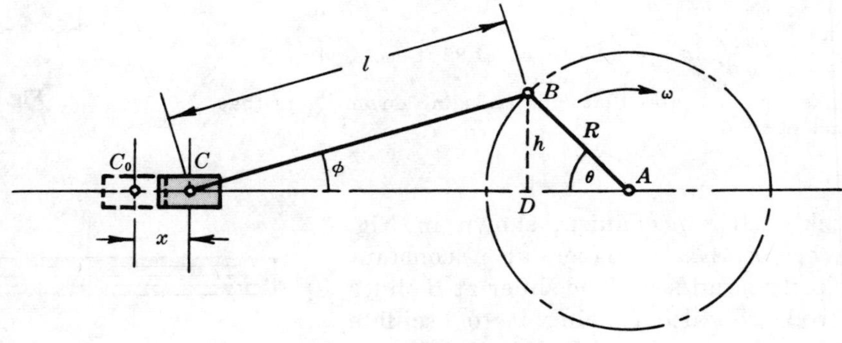

Fig. 10-14

Solution:

Let C_0 be the extreme left position of the crosshead which travels horizontally along the centerline as shown. From Fig. 10-14 above it is evident that $x = C_0A - CA$ and $CA = CD + DA$.

When C is at C_0, B is on the centerline; hence $C_0A = l + R$. Also, $CA = l\cos\phi + R\cos\theta$. Then

$$x = l + R - l\cos\phi - R\cos\theta$$

The relation between ϕ and θ is derived from the right triangles ADB and DCB:

$$h = l\sin\phi = R\sin\theta$$

Then $\quad \sin\phi = (R/l)\sin\theta \quad$ and $\quad \cos\phi = \sqrt{1 - \sin^2\phi} = \sqrt{1 - (R^2/l^2)\sin^2\theta}$

and the displacement is

$$x = l + R - l\sqrt{1 - (R^2/l^2)\sin^2\theta} - R\cos\theta$$

Differentiation of this expression with respect to time is somewhat involved because of the radical. However, an approximation of the radical which is sufficiently accurate when $R/l < \frac{1}{4}$ is obtained by using the first two terms of the power series expansion of the square root term.

$$\sqrt{1 - (R^2/l^2)\sin^2\theta} \approx 1 - \tfrac{1}{2}(R^2/l^2)\sin^2\theta$$

Making this substitution, the displacement becomes

$$x = l + R - l + (R^2/2l)\sin^2\theta - R\cos\theta = R(1-\cos\theta) + (R^2/2l)\sin^2\theta$$

Differentiation yields

$$v = \frac{dx}{dt} = R\sin\theta\frac{d\theta}{dt} + \frac{R^2}{2l}2\sin\theta\cos\theta\frac{d\theta}{dt} = R\omega\left(\sin\theta + \frac{R}{2l}\sin 2\theta\right)$$

and $$a = \frac{dv}{dt} = R\omega\left(\cos\theta\frac{d\theta}{dt} + \frac{R}{2l}2\cos 2\theta\frac{d\theta}{dt}\right) = R\omega^2\left(\cos\theta + \frac{R}{l}\cos 2\theta\right)$$

24. A point P moves on a circular path with a constant speed (magnitude of its linear velocity) of 12 ft/sec. If the radius of the path is 2 ft, study the motion of the projection of the point on a horizontal diameter.

Solution:

Point A is the projection on the horizontal diameter of point P. Assume the origin at the center of the circle. The displacement x of the point A is the projection of the radius vector OP on the x-axis (along the horizontal diameter).

Since the line OP sweeps out equal angles in equal times (the angular velocity is constant), the expression for θ may be written $\theta = \omega t$. The x coordinate of A is therefore

$$x = OP\cos\theta = 2\cos\omega t$$

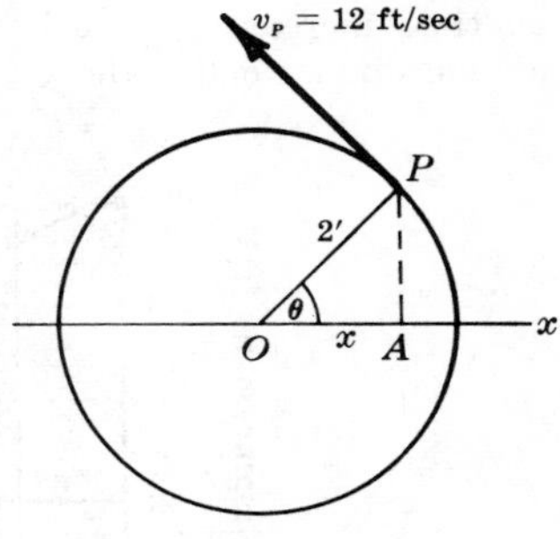

Fig. 10-15

The angular velocity of the radius is $\omega = v/r = 12/2 = 6$ rad/sec. Then $x = 2\cos 6t$ and

$$v = dx/dt = -12\sin 6t \quad \text{and} \quad a = dv/dt = -72\cos 6t$$

The equation for a may be rewritten $a = -(36)(2\cos 6t) = -36x$. But this means that point A moves so that its acceleration a is negatively proportional to its displacement x. Since this is the requirement of Simple Harmonic Motion, it is evident that if point P moves on a circular path with constant speed, the projection of point P on a diameter moves with Simple Harmonic Motion.

25. Plot the displacement, velocity, and acceleration of point A in Problem 24 against time.

Solution:

The amplitude of the displacement x is the maximum value which occurs when the $\cos 6t$ assumes its maximum value of unity (either plus or minus). The amplitude is therefore 2.

The amplitude of velocity v is 12, and the amplitude of acceleration a is 72.

The period is the time T to complete one cycle. It is evident that the motion will repeat itself after the radius vector OP has gone through a complete revolution $\theta = 2\pi$ radians. Equating the function $6t$ to the value of $\theta = 2\pi$ for a cycle yields

$$6\text{ rad/sec} \times T = 2\pi\text{ rad} \quad \text{or} \quad T = 2\pi/6\text{ sec} = 1.05\text{ sec}$$

It is now possible to correlate the angle θ with time t. For example, when $\theta = \pi/2$ or 90°, t is $\frac{1}{4}$ of a revolution, i.e., $\frac{1}{4} \times 1.05$ sec = 0.263 sec. Proceeding in this fashion, draw up a table as follows.

θ	t	$\cos 6t$ or $\cos\theta$	$\sin 6t$ or $\sin\theta$	$x = 2\cos 6t$	$v = -12\sin 6t$	$a = -72\cos 6t$
0°	0	+1.000	0	+2.00	0	−72.0
30°	0.088	+0.866	+0.500	+1.73	−6.00	−62.3
60°	0.175	+0.500	+0.866	+1.00	−10.4	−36.0
90°	0.263	0	+1.000	0	−12.0	0
120°	0.350	−0.500	+0.866	−1.00	−10.4	+36.0
150°	0.438	−0.866	+0.500	−1.73	−6.00	+62.3
180°	0.525	−1.000	0	−2.00	0	+72.0
210°	0.612	−0.866	−0.500	−1.73	+6.00	+62.3
240°	0.700	−0.500	−0.866	−1.00	+10.4	+36.0
270°	0.788	0	−1.000	0	+12.0	0
300°	0.875	+0.500	−0.866	+1.00	+10.4	−36.0
330°	0.962	+0.866	−0.500	+1.73	+6.00	−62.3
360°	1.05	+1.000	0	+2.00	0	−72.0

The plotting of these points provides a visual picture of the motion during one cycle. Naturally, as time increases, these curves duplicate in succeeding cycles. The following graphs indicate the motion during one cycle.

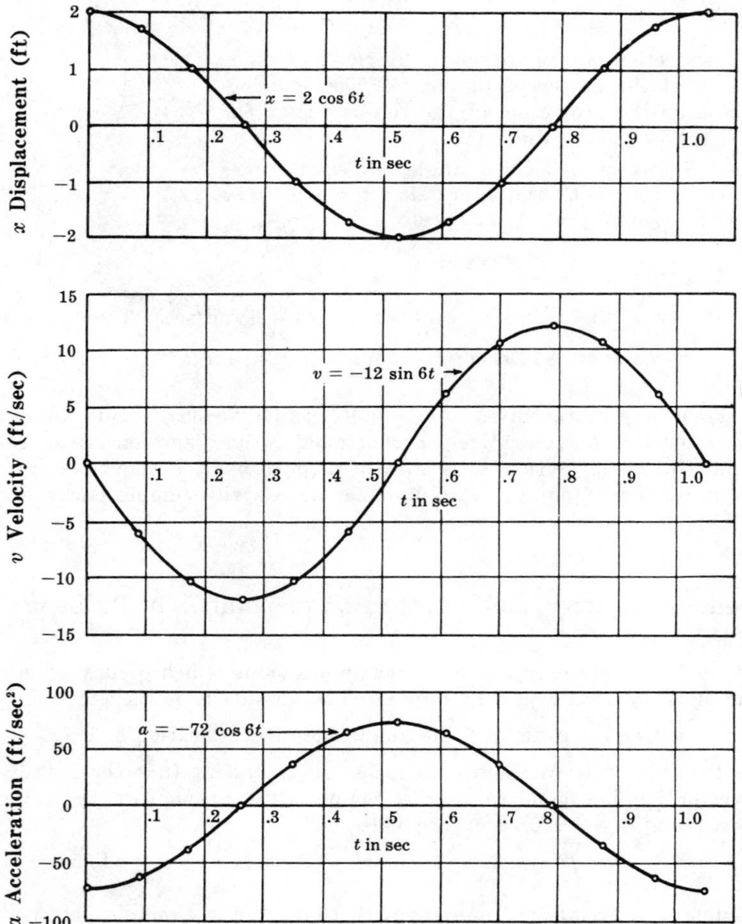

26. A flywheel 4 ft in diameter accelerates uniformly from rest to 2000 rpm in 20 sec. What is its angular acceleration?

Solution:

In the analysis of this problem, first note that a constant angular acceleration is involved. This means that the formulas for constant acceleration may be used. These are similar in angular motion to those in rectilinear motion, i.e., ω replaces v, α replaces k, and θ replaces s.

The wheel starts from rest; hence $\omega_0 = 0$. The three known quantities are ω_0, ω, t. The quantity sought is angular acceleration α. The formula involving these four symbols is

$$\omega = \omega_0 + \alpha t$$

A word of warning is in order regarding units. It is safer (not always necessary of course) to use ω in rad/sec and α in rad/sec² when t is in sec.

$$\omega_0 = 0, \qquad \omega = 2000 \text{ rpm} = \frac{2000}{60}\frac{\text{rev}}{\text{sec}} \times 2\pi \frac{\text{rad}}{\text{rev}} = 209 \frac{\text{rad}}{\text{sec}}$$

Hence $$\alpha = \frac{\omega - \omega_0}{t} = \frac{209 \text{ rad/sec} - 0 \text{ rad/sec}}{20 \text{ sec}} = 10.5 \frac{\text{rad}}{\text{sec}^2}$$

27. In Problem 26, how many revolutions does the flywheel make in attaining its speed of 2000 rpm?

Solution:

To determine the number of revolutions θ, select the equation expressing the relation between θ and the three given quantities ω_0, ω, t. Of course, a formula may be used involving the angular acceleration α just determined but it is advisable to proceed with data given in the problem to derive the value θ independently of the α which could by chance have been found incorrectly.

$$\theta = \tfrac{1}{2}(\omega + \omega_0)t = (209 \text{ rad/sec} + 0 \text{ rad/sec})(20 \text{ sec}) = 2090 \text{ rad}$$

To express θ in revolutions: $\theta = \dfrac{2090 \text{ rad}}{2\pi \text{ rad/rev}} = 333$ revolutions.

The same result is obtained using ω in rev/sec as follows:

$$\theta = \frac{\dfrac{2000}{60}\dfrac{\text{rev}}{\text{sec}} + 0\dfrac{\text{rev}}{\text{sec}}}{2} \times 20 \text{ sec} = 333 \text{ revolutions}$$

28. Determine the linear velocity and linear acceleration of a point on the rim of the flywheel in Problem 26, 0.6 sec after it has started from rest.

Solution:

The velocity of a point on the rim is found by multiplying the radius by the angular velocity.

The angular velocity is $\omega = \omega_0 + \alpha t = 0 + (10.5 \text{ rad/sec}^2)(0.6 \text{ sec}) = 6.30$ rad/sec.

The magnitude of the linear velocity of a point on the rim when $t = 0.6$ sec is

$$v = r\omega = (2 \text{ ft})(6.30 \text{ rad/sec}) = 12.6 \text{ ft/sec, tangent to the rim.}$$

To determine the acceleration completely, use the normal and tangential components. The tangential component a_t is $a_t = r\alpha = (2 \text{ ft})(10.5 \text{ rad/sec}^2) = 21.0 \text{ ft/sec}^2$. The normal component a_n is $a_n = r\omega^2 = (2 \text{ ft})(6.3 \text{ rad/sec})^2 = 79.4 \text{ ft/sec}^2$. Fig. 10-16 illustrates these components for any point P on the rim.

The total acceleration a is the vector sum of the two components a_t and a_n. Let ϕ be the angle between the total acceleration and the radius. The normal acceleration a_n is directed toward the center of the circle.

$$a = \sqrt{a_t^2 + a_n^2} = \sqrt{(21.0)^2 + (79.4)^2} = 82.1 \text{ ft/sec}^2$$

$$\phi = \tan^{-1} a_t/a_n = \tan^{-1} 21.0/79.4 = 14.8°$$

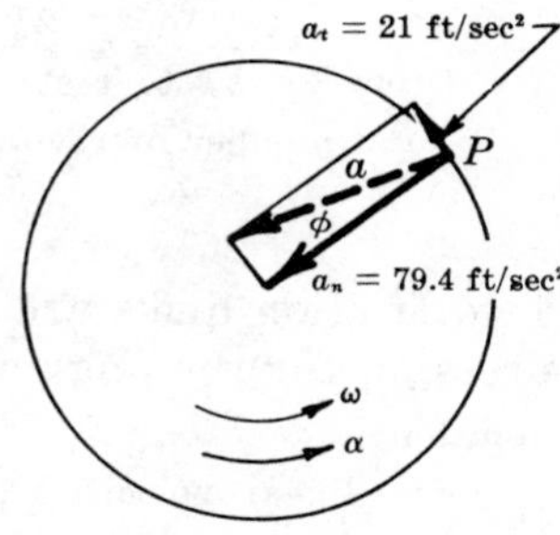

Fig. 10-16

29. A bar 6 ft long and of small cross section rotates in a horizontal plane about a vertical axis through one end. It accelerates uniformly from 20 rev/sec to 30 rev/sec in a 5 sec interval. What is the linear velocity of its midpoint at the beginning and end of that interval of time?

Solution:

The midpoint describes a circular arc. Its velocity is given by $v = r\omega$.

The velocities at the beginning and end of the time interval are respectively

$$v_B = r\omega_B = (3\text{ ft})(20 \times 2\pi\text{ rad/sec}) = 377\text{ ft/sec}$$
$$v_E = r\omega_E = (3\text{ ft})(30 \times 2\pi\text{ rad/sec}) = 565\text{ ft/sec}$$

30. In Problem 29, find the normal and tangential components of the acceleration of the midpoint of the bar 3 sec after acceleration begins.

Solution:

The distance r from the center of rotation to the midpoint is 3 ft.

The uniform angular acceleration α at any time during the 5 sec interval is

$$\alpha = \frac{\omega - \omega_0}{t} = \frac{60\pi\text{ rad/sec} - 40\pi\text{ rad/sec}}{5\text{ sec}} = 4\pi\text{ rad/sec}^2$$

The angular velocity ω after 3 sec is

$$\omega = \omega_0 + \alpha t = 40\pi\text{ rad/sec} + (4\pi\text{ rad/sec}^2)(3\text{ sec}) = 52\pi\text{ rad/sec}$$

The components of acceleration are: $a_t = r\alpha = (3\text{ ft})(4\pi\text{ rad/sec}^2) = 37.7\text{ ft/sec}^2$

$a_n = r\omega^2 = (3\text{ ft})(52\pi\text{ rad/sec})^2 = 80{,}100\text{ ft/sec}^2$.

31. A wheel 8 inches in diameter coasts to rest from a speed of 800 rpm in 10 minutes. Determine the angular acceleration.

Solution:

Given: $\omega_0 = 800\text{ rpm} = 1600\pi\text{ rad/min}$, $\omega = 0$, $t = 10$ min. Then

$$\alpha = \frac{\omega - \omega_0}{t} = \frac{-1600\pi\text{ rad/min}}{10\text{ min}} = -502\text{ rad/min/min (deceleration)}$$

The acceleration is negative. This means that the angular velocity is in one direction, while the angular acceleration is oppositely directed, thereby indicating a slowing down of the wheel.

32. A wheel accelerates uniformly from rest to a speed of 200 rpm in $\frac{1}{2}$ sec. It then rotates at that speed for 2 sec before decelerating to rest in $\frac{1}{3}$ sec. How many revolutions does it make during the entire time interval?

Solution:

From $t = 0$ to $t = \frac{1}{2}$: $\theta_1 = \frac{1}{2}(\omega_0 + \omega)t = \frac{1}{2}(0 + 200/60\text{ rev/sec})(\frac{1}{2}\text{ sec}) = 0.83$ rev.

From $t = \frac{1}{2}$ to $t = 2\frac{1}{2}$: $\theta_2 = \omega t = (200/60\text{ rev/sec})(2\text{ sec}) = 6.67$ rev.

From $t = 2\frac{1}{2}$ to rest: $\theta_3 = \frac{1}{2}(\omega_0 + \omega)t = \frac{1}{2}(200/60\text{ rev/sec} + 0)(\frac{1}{3}\text{ sec}) = 0.56$ rev.

Total number of revolutions $\theta = \theta_1 + \theta_2 + \theta_3 = 8.06$ rev.

33. Two friction disks are shown in Fig. 10-17. Derive the expression for the angular velocity ratio in terms of the radii.

Solution:

The linear velocities of the mating points A and B on the two wheels are equal. If this were not true, the wheels would slip relative to one another.

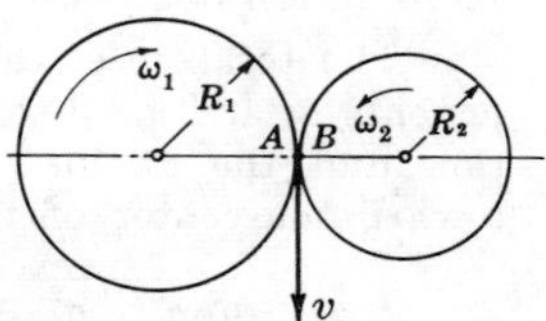

Fig. 10-17

The linear velocities of points A and B are respectively

$$v_A = R_1\omega_1, \quad v_B = R_2\omega_2$$

But $v_A = v_B$ if the drive is positive, i.e., without slip. Then,

$$R_1\omega_1 = R_2\omega_2 \quad \text{or} \quad \omega_1/\omega_2 = R_2/R_1$$

34. A point P moves on a circular path in a counterclockwise direction so that the length of arc it sweeps out is $s = t^3 + 3$. The radius of the path is 12 ft. The units of s and t are ft and sec respectively. Determine the axial components of velocity (v_x, v_y) when $t = 2$ sec.

Solution:

The distance AP is traversed in 2 sec, or $AP = s = 2^3 + 3 = 11$ ft. By inspection $x = 12\cos\theta$ and $y = 12\sin\theta$. Differentiating, $v_x = (-12\sin\theta)\,d\theta/dt$, and $v_y = (12\cos\theta)\,d\theta/dt$.

These may be evaluated provided that θ is found as a function of time. The relation $s = r\theta$ yields $\theta = s/r = (t^3+3)/12$, where θ must be in radians. Differentiate to obtain $d\theta/dt = \frac{1}{4}t^2$. When $t = 2$ sec, $\theta = \frac{11}{12}$ rad and $d\theta/dt = 1$ rad/sec.

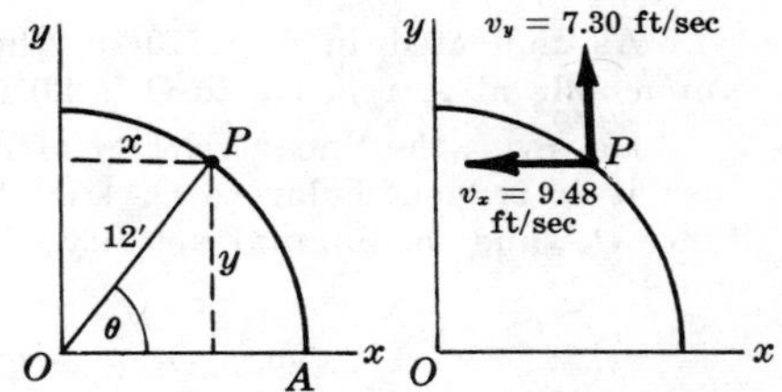

Fig. 10-18

Substitution yields $v_x = -9.48$ ft/sec and $v_y = 7.30$ ft/sec. The negative sign indicates that the x component of the velocity is directed to the left. The y component of the velocity is directed up.

The total velocity $v = \sqrt{(v_x)^2 + (v_y)^2} = 12.0$ ft/sec. This could be obtained directly from $v = r\,d\theta/dt = 12(\frac{1}{4}t^2)$ when $t = 2$ sec, or from $s = t^3 + 3$ and hence $v = ds/dt = 3t^2$.

35. In Problem 34, determine the axial components of the acceleration a_x and a_y when $t = 2$ sec.

Solution:

Differentiate the expression for v_x to obtain $a_x = -12\cos\theta\,(d\theta/dt)^2 - 12\sin\theta\,d^2\theta/dt^2$. Since $d\theta/dt = \frac{1}{4}t^2$, then $d^2\theta/dt^2 = \frac{1}{2}t$.

At $t = 2$ sec, therefore,

$$a_x = -12(\cos\tfrac{11}{12})(1)^2 - 12(\sin\tfrac{11}{12})(1) = -16.8 \text{ ft/sec}^2, \text{ i.e., to the left.}$$

Similarly, $a_y = -12\sin\theta\,(d\theta/dt)^2 + 12\cos\theta\,d^2\theta/dt^2 = -2.19$ ft/sec², i.e., down.

The total acceleration $a = \sqrt{(a_x)^2 + (a_y)^2} = 17.0$ ft/sec².

This could be obtained also by combining the tangential component a_t and the normal component a_n of the acceleration. These are $a_t = r\alpha = r\,d^2\theta/dt^2 = 12(\frac{1}{2}t)$ or 12 ft/sec², and $a_n = r\omega^2 = r(d\theta/dt)^2 = 12(1)^2$ or 12 ft/sec². Hence $a = \sqrt{(a_t)^2 + (a_n)^2} = 17.0$ ft/sec².

Note that $a_t = d^2s/dt^2 = 6t$ and $a_n = v^2/r = 9t^4/12$ give the same results with $t = 2$.

36. The x and y components of the displacement in ft of a point are given by the equations

$$x = 10t^2 + 2t, \quad y = t^3 + 5$$

Determine the velocity and acceleration of the point when $t = 3$ seconds.

Solution:

The velocity components obtained by differentiation are

$$v_x = dx/dt = 20t + 2, \qquad v_y = dy/dt = 3t^2$$

At $t = 3$ sec, $v_x = 62$ ft/sec and $v_y = 27$ ft/sec. Hence

$$v = \sqrt{(v_x)^2 + (v_y)^2} = 67.6 \text{ ft/sec} \quad \text{and} \quad \theta_x = \tan^{-1} v_y/v_x = \tan^{-1} 27/62 = 23.5°$$

where θ_x is the angle between the total velocity and the x-axis.

A second differentiation yields the acceleration components: $a_x = dv_x/dt = 20$, $a_y = 6t$. At $t = 3$ sec, $a_x = 20$ ft/sec² and $a_y = 6(3) = 18$ ft/sec². Hence

$$a = \sqrt{(a_x)^2 + (a_y)^2} = 26.9 \text{ ft/sec}^2 \quad \text{and} \quad \phi_x = \tan^{-1} a_y/a_x = \tan^{-1} 18/20 = 42.0°$$

37. An automobile is moving South with an absolute velocity of 20 mph. An observer O is stationed 50 ft to the East of the line of travel. When the automobile is directly West of the observer, what is the angular velocity relative to the observer? After moving 50 ft South, what is its angular velocity relative to the observer at O?

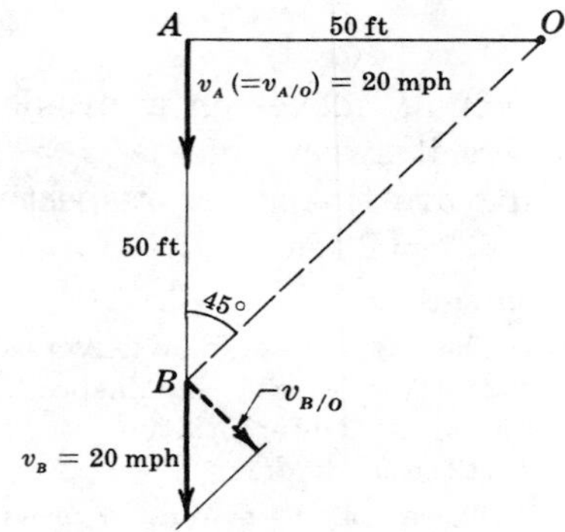

Fig. 10-19

Solution:

As indicated in Fig. 10-19, the velocity $v_{A/O}$ of the automobile at A relative to O is 20 mph or 29.3 ft/sec.

However, the linear velocity of A relative to O (in this case it is perpendicular to OA) is the product of the distance OA and the angular velocity of A relative to O. Then

$$v_{A/O} = OA \times \omega_{A/O}$$

$$29.3 \text{ ft/sec} = 50 \text{ ft} \times \omega_{A/O} \quad \text{and} \quad \omega_{A/O} = 0.588 \text{ rad/sec}$$

For the next part of the problem note that the absolute velocity v_B of the vehicle at B is still South 20 mph (29.3 ft/sec). The component $v_{B/O}$ (velocity of B relative to O) is perpendicular to the arm BO; hence $v_{B/O} = 29.3 \cos 45° = 20.8$ ft/sec. Then

$$v_{B/O} = OB \times \omega_{B/O}, \quad 20.8 \text{ ft/sec} = (50\sqrt{2} \text{ ft}) \times \omega_{B/O} \quad \text{and} \quad \omega_{B/O} = 0.294 \text{ rad/sec}$$

38. A boat is travelling 12 mi/hr due East. An observer is stationed 100 ft South of the line of travel. Determine the angular velocity of the boat relative to the observer when in the position shown in Fig. 10-20.

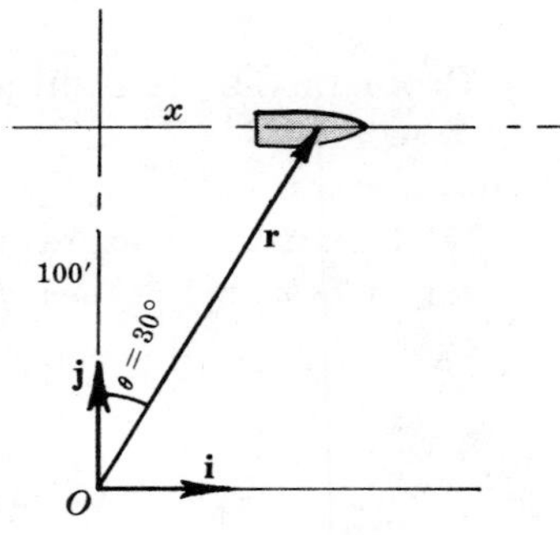

Fig. 10-20

Solution:

Select unit vectors **i** and **j** in the East and North directions respectively. Let **r** be the position vector of the boat relative to the observer O. Then

$$\mathbf{r} = x\mathbf{i} + 100\mathbf{j} = 100 \tan\theta\,\mathbf{i} + 100\mathbf{j}$$

The velocity **v** of the boat is

$$\mathbf{v} = \dot{\mathbf{r}} = 100(\sec^2\theta)(\dot{\theta})\mathbf{i} + 0\mathbf{j}$$

Since the speed $v = 12$ mi/hr $= 17.6$ ft/sec and $\theta = 30°$,

$$17.6 = 100(\sec^2 30°)\dot{\theta} \quad \text{or} \quad \omega = \dot{\theta} = 0.132 \text{ rad/sec clockwise}$$

39. The motion of a point is described by the following equations:

$$v_x = 20t + 5, \quad v_y = t^2 - 20$$

In addition it is known that $x = 5$ ft and $y = -15$ ft when $t = 0$. Determine the displacement, velocity and acceleration when $t = 2$ sec.

Solution:

Rewriting the given equations as $v_x = dx/dt = 20t + 5$, $v_y = dy/dt = t^2 - 20$ and integrating, the expressions for x and y are $x = 10t^2 + 5t + C_1$, $y = \frac{1}{3}t^3 - 20t + C_2$.

To evaluate C_1, substitute $x = 5$ and $t = 0$ in the x equation. Then $C_1 = 5$.

To evaluate C_2, substitute $y = -15$ and $t = 0$ in the y equation. Then $C_2 = -15$.

Substituting the values of C_1 and C_2, the equations for displacement become

$$x = 10t^2 + 5t + 5, \quad y = \tfrac{1}{3}t^3 - 20t - 15$$

Differentiate v_x and v_y to obtain the equations for acceleration.

$$a_x = dv_x/dt = 20, \qquad a_y = dv_y/dt = 2t$$

Substituting $t = 2$ sec in the expressions for displacement, velocity and acceleration, the following values are obtained: $x = 55$ ft, $y = -52.3$ ft; $v_x = 45$ ft/sec, $v_y = -16$ ft/sec; $a_x = 20$ ft/sec², $a_y = 4$ ft/sec².

The magnitudes and directions of the total displacement, velocity and acceleration can be found by combining their components as before.

40. The 4 inch diameter pulley on a generator is being turned by a belt moving 60 ft/sec and accelerating 20 ft/sec². A fan with an outside diameter of 6 inches is attached to the pulley shaft. What are the linear velocity and acceleration of the tip of the fan?

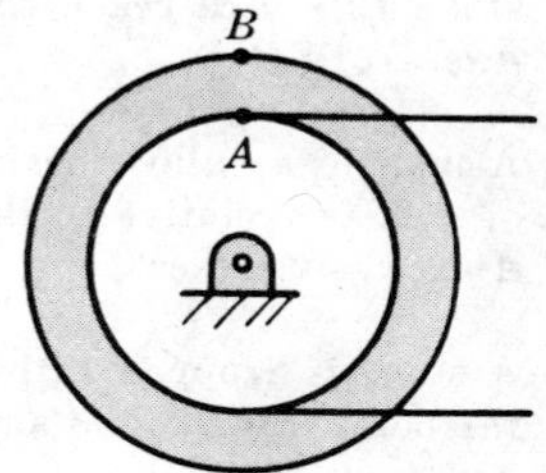

Fig. 10-21

Solution:

In the diagram (Fig. 10-21) point A on the pulley has the same velocity as the belt with which it coincides at the instant. Hence the angular velocity ω of the pulley (and also of the fan keyed to the same shaft) is equal to $v/r = 60 \div 2/12 = 360$ rad/sec. The linear velocity of the fan tip $v_B = (3/12)(360) = 90$ ft/sec.

The tangential component of the linear acceleration of point A is equal to the acceleration of the belt, i.e., $a_t = r\alpha$ or $20 = (2/12)\alpha$. From this, the angular acceleration α of the system is 120 rad/sec². Then the tangential component of the acceleration of point B is $(3/12)(120) = 30$ ft/sec².

Of course it has a normal component which equals $r\omega^2 = (3/12)(360)^2 = 32{,}400$ ft/sec².

Hence the magnitude of the linear acceleration is $a = \sqrt{(32{,}400)^2 + (30)^2} = 32{,}400$ ft/sec².

Supplementary Problems

41. An automobile travels a distance of 230 miles in a period of 7 hours. Determine the average velocity of the vehicle in mi/hr and in ft/sec. *Ans.* 32.9 mi/hr, 48.2 ft/sec

42. If a body moves at the rate of 30 mi/hr for 6 minutes, then 60 mi/hr for 10 minutes and finally 5 mi/hr for 3 minutes, what is the average velocity in the total interval? *Ans.* 61.3 ft/sec

43. A particle has straight line motion according to the equation $x = t^3 - 3t^2 - 5$ with x in ft and t in sec. What is the change in displacement while the velocity changes from 8 ft/sec to 40 ft/sec? *Ans.* $\Delta x = 41.6$ ft

44. A body moves along a straight line so that its displacement from a fixed point on the line is given by $s = 3t^2 + 2t$. Find the displacement, velocity, and acceleration at the end of 3 sec. *Ans.* 33 ft, 20 ft/sec, 6 ft/sec²

45. The motion of a particle is defined by the relation $s = t^4 - 3t^3 + 2t^2 - 8$ where s is in ft and t in sec. Determine the velocity $\dot{s}$ and the acceleration $\ddot{s}$ when $t = 2$ sec. *Ans.* $\dot{s} = +4$ ft/sec, $\ddot{s} = +16$ ft/sec²

46. A car travels along a straight track between two points at a mean speed of 60 ft per sec. It returns at a mean speed of 40 ft per sec. What is the mean speed for the round trip? *Ans.* 48 ft/sec

47. An automobile accelerates uniformly from rest to 45 mph and then the brakes are applied so that it decelerates uniformly to a stop. If the total time is 15 sec, what distance was travelled?
Ans. $d = 495$ ft

48. A bullet is fired wtih a muzzle velocity of 2000 ft/sec. If the length of barrel is 30 inches, what is the average acceleration? *Ans.* 800,000 ft/sec^2

49. An automobile is decelerating from a speed of 40 mph at the rate of 5 ft per sec per sec. How long will it take to come to rest and how far will it have gone? *Ans.* 11.7 sec, 344 ft

50. A stone is dropped from a balloon that is ascending at a uniform rate of 30 ft/sec. If it takes the stone 10 sec to reach the ground, how high was the balloon at the instant the stone was dropped?
Ans. 1310 ft

51. A man in a balloon rising with a constant velocity of 12 ft/sec propels a ball upward with a velocity of 4 ft/sec relative to the balloon. After what time interval will the ball return to the balloon?
Ans. $t = 0.25$ sec

52. A stone is dropped with zero velocity into a well. The sound of the splash is heard 3.63 sec later. How far below the ground surface is the surface of the water? Assume the velocity of sound is 1090 ft/sec.
Ans. $s = 193$ ft

53. Boy A throws a ball vertically up with a speed of 30 ft/sec from the top of a shed 8 ft high. Boy B on the ground at the same instant throws a ball vertically up with a speed of 40 ft/sec. Determine the time at which the two balls will be at the same height above the ground. What is the height?
Ans. $t = 0.8$ sec, $h = 21.7$ ft

54. Body A starts from rest at point O and travels along a straight line with an acceleration of 2 ft/sec/sec. Body B starts from rest at O, 4 sec later and travels along the same path with an acceleration of 3 ft/sec/sec. How far from O will they be when B overtakes A? *Ans.* 475 ft

55. A radar equipped police car notes a car travelling 70 mph. The police car starts pursuit 30 sec after the observation and accelerates to 100 mph in 20 sec. Assuming the speeds are maintained on a straight road, how far from the observation post will the chase end? *Ans.* $s = 13{,}700$ ft

56. An automobile accelerates uniformly from rest on a straight level road. A second automobile starting from the same point 6 sec later with initial velocity zero accelerates at 6 ft/sec^2 to overtake the first automobile 400 ft from the starting point. What is the acceleration of the first automobile?
Ans. $a = 2.62$ ft/sec^2

57. Plane A leaves an airport and flies north at 120 mph. Plane B leaves the same airport 20 minutes later and flies north at 150 mph. How long will it take B to overtake A?
Ans. $t = 1.33$ hr

58. A particle moves on a straight line with the acceleration shown in the graph in Fig. 10-22. Determine the velocity and displacement at time $t = 1, 2, 3$ and 4 sec. Assume the initial velocity is +3 ft/sec and the initial displacement is zero.
Ans. $v_1 = +1$ ft/sec, $s_1 = +2$ ft; $v_2 = +3$ ft/sec, $s_2 = +4$ ft;
$v_3 = -1$ ft/sec, $s_3 = +5$ ft; $v_4 = -3$ ft/sec, $s_4 = +3$ ft.

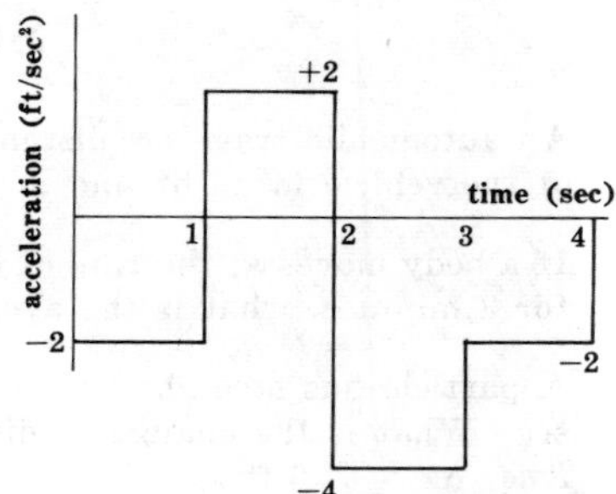

Fig. 10-22

59. A particle of dirt falls from an elevator that is moving up with a velocity of 10 ft/sec. If the particle reaches the bottom in 1.85 sec, how high above the bottom was the elevator when the particle started falling? *Ans.* $s = 36.6$ ft

60. A bullet is fired vertically upward with a velocity of 2200 ft/sec. Theoretically, to what height will the bullet ascend? *Ans.* 75,000 ft

61. A ball is thrown vertically upward with a speed of 30 ft/sec from the edge of a cliff 50 ft above sea level. What is the highest point above sea level reached? How long does it take to hit the water? With what velocity does it hit the water? *Ans.* $h = 64.0$ ft, $t = 2.93$ sec, $v = 64.3$ ft/sec

62. A particle moves with an acceleration $a = -6v$, where a is in ft/sec² and v is in ft/sec. When $t = 0$ sec, the displacement $s = 0$ and the velocity $v = 9$ ft/sec. Find the displacement, velocity, and acceleration when $t = 0.5$ sec. *Ans.* $s = 1.43$ ft, $\dot{s} = v = 0.448$ ft/sec, $\ddot{s} = a = 2.69$ ft/sec²

63. The speed of a particle is given by $v = 2t^3 + 5t^2$. What distance does it travel while its speed increases from 7 ft/sec to 99 ft/sec? *Ans.* $s = 83.3$ ft

64. A particle moves to the right from rest with an acceleration of 6 ft/sec² until its velocity is 12 ft/sec to the right. It is then subjected to an acceleration of 12 ft/sec² to the left until its total distance travelled is 36 ft. Determine the total elapsed time. *Ans.* $t = 5.24$ sec

65. Water drips from a faucet at the rate of 6 drops per sec. The faucet is 8 inches above the sink. When one drop strikes the sink, how far is the next drop above the sink? *Ans.* $h = 7.75$ in.

66. A particle moving with a velocity of 6 ft/sec upward is subjected to an acceleration of 3 ft/sec² downward until its displacement is 2 ft below its position when the acceleration began. The acceleration then ceases for 3 sec. The particle is then subjected to an acceleration of 4 ft/sec² upward for 5 sec. Determine the displacement and the distance travelled.
Ans. $s = -7.2$ ft relative to start, $d = 50.2$ ft

67. The velocity-time curve for a point moving on a straight line is shown in the graph in Fig. 10-23. How far does the point move in the 2 sec? *Ans.* $x = 5.09$ ft

68. An object moves on a straight line with constant acceleration 2 m/sec². How long will it take to change its speed from 5 to 8 m/sec? What change in displacement takes place during this time interval? *Ans.* $t = 1.5$ sec, $s = 9.75$ m

69. The motion of a particle is given by the acceleration $a = t^3 - 2t^2 + 7$, where a is in ft/sec² and t in sec. The velocity is 3.58 ft/sec when $t = 1$ sec and the displacement is +9.39 ft when $t = 1$ sec. Calculate the displacement, velocity, and acceleration when $t = 2$ sec. *Ans.* $s = 15.9$ ft, $v = 9.67$ ft/sec, $a = 7$ ft/sec²

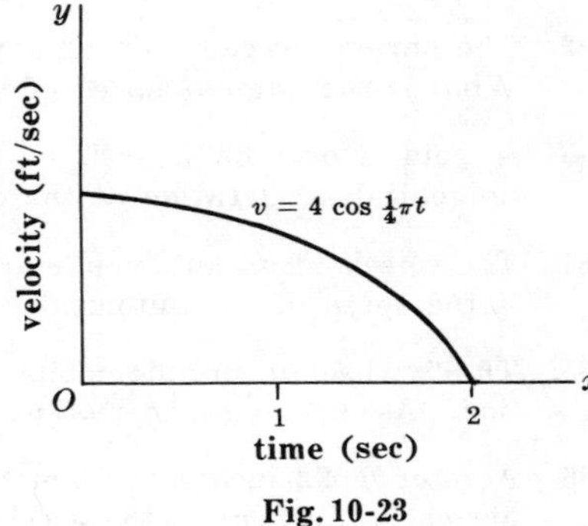

Fig. 10-23

70. The motion of a particle is defined by the relation $v = 4t^2 + 3t - 5$. Knowing the displacement $x = -2$ ft when $t = 0$ sec, determine the displacement and acceleration when $t = 3$ sec.
Ans. $x = +32.5$ ft, $a = \ddot{x} = 27$ ft/sec²

71. In the system shown in Fig. 10-24, determine the velocity and acceleration of block 3 at the instant considered.
Ans. $v_3 = 10.5$ ft/sec up, $a_3 = 5.0$ ft/sec² up

72. A point moves along the path $y = \frac{1}{3}x^2$ with a constant speed of 8 ft/sec. What are the x and y components of the velocity when $x = 3$? What is the acceleration of the point when $x = 3$?
Ans. $\dot{x} = 3.58$ ft/sec, $\dot{y} = 7.16$ ft/sec,
$a = 3.82$ ft/sec²

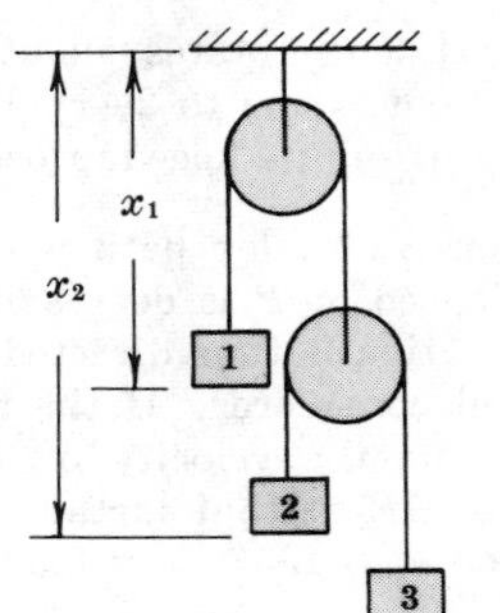

$v_1 = \dot{x}_1 = 4.0$ ft/sec↑
$a_1 = \ddot{x}_1 = 1.5$ ft/sec²↑
$v_2 = \dot{x}_2 = 2.5$ ft/sec↓
$a_2 = \ddot{x}_2 = 2.0$ ft/sec²↓

Fig. 10-24

73. Determine the projection of the velocity vector $\mathbf{v} = 1.62\mathbf{i} - 3.87\mathbf{j} + 2.83\mathbf{k}$ on the line originating at point (2, 3, 5) and passing through point (4, −2, 6).
Ans. +4.64

74. A point P moves at a constant speed v in a counterclockwise direction along a circle of radius a as shown in Fig. 10-25. Selecting a pole O at the left end of the horizontal diameter, derive expressions for the radial and transverse components of the acceleration. Hint: $r = 2a\cos\theta$.
Ans. $a_r = -(v^2/a)\cos\theta$ $\quad a_\theta = -(v^2/a)\sin\theta$

75. In the preceding problem show that total acceleration is v^2/a, which is the normal component for circular motion of a point with constant speed (there is no tangential component).

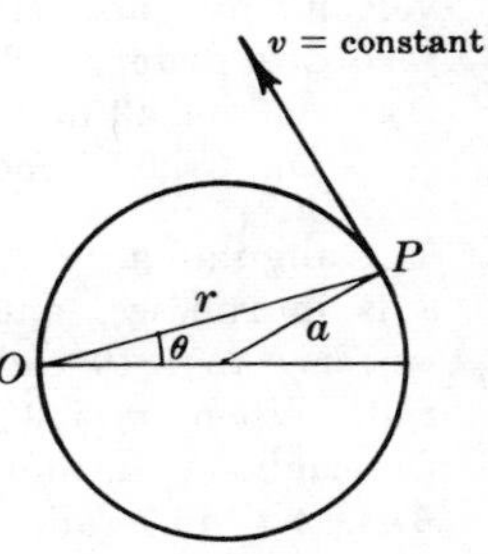

Fig. 10-25

76. In the slider crank mechanism shown in Fig. 10-26 the crank is rotating 200 rpm. What is the velocity of the crosshead when $\theta = 30°$?
Ans. 25.5 ft/sec

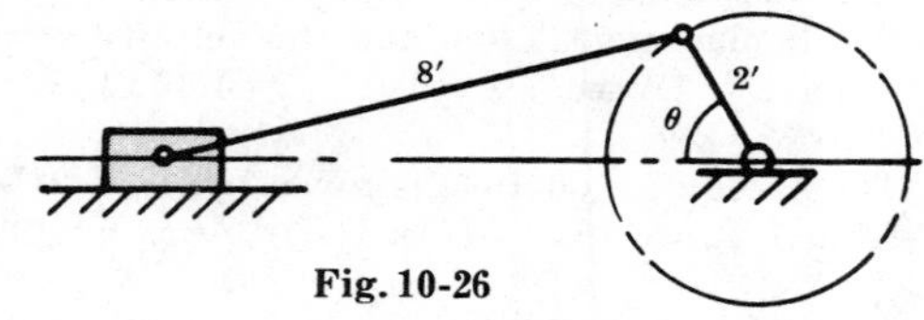

Fig. 10-26

77. In Problem 76, determine the acceleration of the crosshead in the 30° phase.
Ans. 870 ft/sec²

78. A particle oscillates with an acceleration $a = -kx$. Determine k if the velocity $v = 2$ ft/sec when the displacement $x = 0$, and $v = 0$ when $x = +2$ ft. *Ans.* $k = +1$

79. A particle moves with simple harmonic motion with a frequency of 30 cycles per min and an amplitude of 0.25 in. Determine the maximum velocity and acceleration.
Ans. $v_{max} = 0.785$ in/sec, $a_{max} = 2.46$ in/sec²

80. A body having simple harmonic motion has a period of 6 sec and an amplitude of 4 ft. Determine the maximum velocity and acceleration of the body. *Ans.* $4\pi/3$ ft/sec, $4\pi^2/9$ ft/sec²

81. A particle having a constant speed of 15 ft/sec moves around a circle 10 ft in diameter. What is the normal acceleration? *Ans.* 45 ft/sec²

82. The normal acceleration of a point on the rim of a 10 ft diameter flywheel is a constant 45 ft/sec². What is the angular speed of the flywheel? *Ans.* 3 rad/sec

83. A point moves on a path 10 ft in diameter so that the distance traversed is $s = 3t^2$. What is the tangential acceleration at the end of 2 sec? *Ans.* 6 ft/sec²

84. The wheels of an automobile are 30 inches in diameter and have an angular speed of 17 rad/sec. What is the speed of the automobile in miles per hour? *Ans.* 14.5 mph

85. The flywheel of an automobile acquires a speed of 2000 rpm in 0.75 minute. Find its angular acceleration. Assume uniform motion. *Ans.* 4.65 rad/sec²

86. A rotor 0.592 inches in diameter is spinning 2,000,000 rpm in a high vacuum chamber. What is the normal component of the acceleration of a point on the rim? *Ans.* $a_n = 1.08 \times 10^9$ ft/sec²

87. A point moves on a circular path with its position from rest defined by $s = t^3 + 5t$, where s and t are measured in ft and sec respectively. The magnitude of the acceleration is 8.39 ft/sec² when $t = 0.66$ sec. What is the diameter of the path? *Ans.* $d = 10.8$ ft

88. A horizontal bar 3 ft long rotates about a vertical axis through its midpoint. Its angular velocity changes uniformly from 5 rpm to 20 rpm in 20 sec. What is the linear acceleration of a point on the end of the bar 5 sec after the speedup occurs? *Ans.* $a_t = 0.118$ ft/sec², $a_n = 1.26$ ft/sec²

89. Particle P travels on a circular path with radius 8 ft as shown in Fig. 10-27. The speed of P is decreasing (the tangential component of the acceleration is thus directed opposite to the velocity vector) at the instant considered. If the total acceleration vector is as shown, determine the velocity of P and the angular acceleration of the line OP at that instant.
Ans. $v = 19.7$ ft/sec, $\theta_x = 135°$; $\alpha = 3.5$ rad/sec²

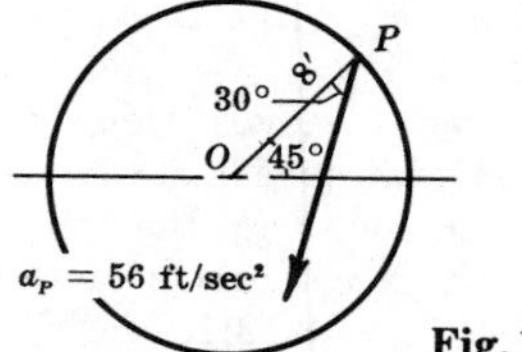

Fig. 10-27

90. Disk A drives disk B without slip occurring. Determine the velocity and acceleration of the weight D which is connected by a cord to drum C which is keyed to disk B as shown in Fig. 10-28.
Ans. $v_D = 2.40$ in/sec up
$a_D = 3.60$ in/sec² up

91. The angular acceleration of a rotor is given by $\alpha = Kt^{-1/2}$, where α is in rad/sec² and t is in sec. When $t = 1$ sec, the angular velocity $\omega = 10$ rad/sec and the angular displacement $\theta = 3.33$ rad. When $t = 0$ sec, the angular displacement $\theta = -4$ rad. Determine θ, ω, and α when $t = 4$ sec.
Ans. $\theta = 46.7$ rad,
$\omega = 18$ rad/sec,
$\alpha = 2$ rad/sec²

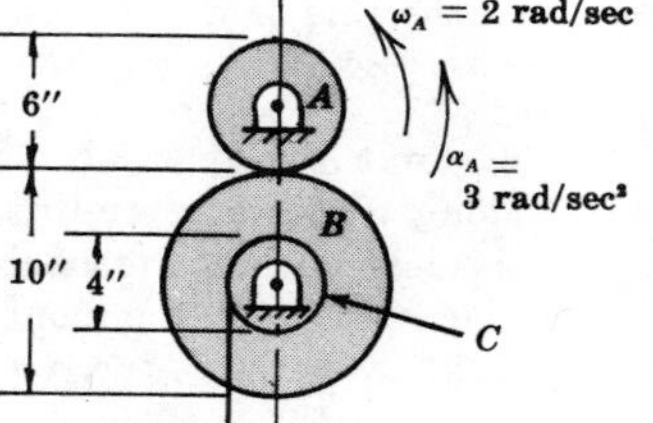

Fig. 10-28

92. The displacement of a point is described in terms of x and y components as follows:

$$x = 2t^2 + 5t, \quad y = 16.1t^2$$

Determine the velocity and acceleration at the end of 4 sec.
Ans. $v_x = 21$ ft/sec, $v_y = 129$ ft/sec; $a_x = 4$ ft/sec^2, $a_y = 32.2$ ft/sec^2

93. A bicycle rider travels 250 ft North and then 160 ft Northwest. What is his displacement? What distance does he cover? *Ans.* $s = 381$ ft, 72.7° North of West; $d = 410$ ft

94. Car A is moving Northwest with a speed of 60 mph. Car B is moving East with a speed of 40 mph. Determine the velocity of A relative to B. Determine the velocity of B relative to A.
Ans. $v_{A/B} = -82.4\mathbf{i} + 42.4\mathbf{j}$ mph, $v_{B/A} = 82.4\mathbf{i} - 42.4\mathbf{j}$ mph

95. Body A has a velocity of 10 mph from West to East relative to body B which in turn has a velocity of 30 mph from Northeast to Southwest relative to body C. Determine the velocity of A relative to C.
Ans. $v_{A/C} = 24.0$ mph, at 62.2° South of West

96. Refer to Fig. 10-29. A rotating spotlight is at a perpendicular distance l from a horizontal floor. The light revolves a constant N revolutions per minute about a horizontal axis perpendicular to the paper. Derive expressions for the velocity and acceleration of the light spot travelling along the floor. Let θ be the angle between the vertical line l and the light beam at time t.
Ans. $\dot{x} = 0.105\, l N \sec^2\theta$, $\ddot{x} = 0.022\, l N^2 \sec^2\theta \tan\theta$

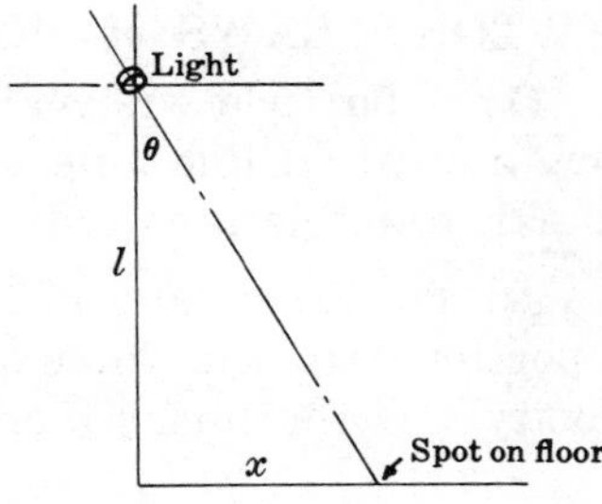

Fig. 10-29

Chapter 11

Dynamics of a Particle

NEWTON'S LAWS of MOTION

(1) A particle will maintain its state of rest or of uniform motion (at constant speed) along a straight line unless compelled by some force to change that state. In other words, a particle accelerates only if an unbalanced force acts on it.

(2) The time rate of change of the product of the mass and velocity of a particle is proportional to the force acting on the particle. The product of the mass m and the velocity $\mathbf{v}$ is the linear momentum $\mathbf{G}$. Thus the second law states

$$\mathbf{F} \quad = \quad K\frac{d(m\mathbf{v})}{dt} \quad = \quad K\frac{d\mathbf{G}}{dt}$$

If m is constant the above equation becomes

$$\mathbf{F} \quad = \quad Km\frac{d\mathbf{v}}{dt} \quad = \quad Km\mathbf{a}$$

If suitable units are chosen so that the constant of proportionality $K=1$, these equations are

$$\mathbf{F} = \frac{d\mathbf{G}}{dt} \qquad \text{or} \qquad \mathbf{F} = m\mathbf{a}$$

(3) To every action, or force, there is an equal and opposite reaction, or force. In other words, if a particle exerts a force on a second particle, then the second particle exerts a numerically equal and oppositely directed force on the first particle.

UNITS

Units depend on the system. In most engineering work the value of K in the above formulas is made equal to unity by proper selection of units. The two fundamental units assigned are *lb* for force and *ft per sec per sec* or *ft/sec*^{2} for acceleration. The unit of mass is then the derived unit in terms of these two. In the case of a freely falling particle near the earth's surface, the only force acting is its weight W. Its acceleration is the acceleration of gravity g (assumed to be 32.2 ft/sec^2 for most localities in the United States). The equation of the second law is then written (vector notation deleted – straight line motion)

$$W = Kma \text{ or } W = (1)mg. \quad \text{Then } m = \frac{W \text{ lb}}{g \text{ ft/sec}^2} = \frac{W}{g}\frac{\text{lb-sec}^2}{\text{ft}}.$$

This derived unit of mass is sometimes called a *slug*.

ACCELERATION

Acceleration of a particle may now be determined by the vector equation

$$\Sigma\mathbf{F} = m\mathbf{a} = m\ddot{\mathbf{r}}$$

where

$$\Sigma\mathbf{F} = \text{vector sum of all the forces acting on the particle}$$
$$m = \text{mass of the particle}$$
$$\mathbf{a} = \ddot{\mathbf{r}} = \text{acceleration}$$

PROBLEMS in DYNAMICS

The solutions of problems in Dynamics vary with the type of force system. Many problems involve forces which are constant; Problems 1-16 are examples of this type. In other problems the forces vary with distance (rectilinear or angular); Problems 17-22 are examples of this type. The subject of Vibrations (Chapter 20) is built on force systems which vary not only with distance but velocity. Problems 23 and 24 deal with forces which vary with the first and second powers of the velocity.

The subject of Ballistics is introduced in an elementary manner in Problem 25 which deals with the motion of a projectile under the action of the constant force of gravity. To this solution may be added retarding forces which vary with the velocity of the projectile.

Some aspects of Central Force motions (satellites and planets are common illustrations of such force systems) are discussed in Problems 26-29.

Solved Problems

In solving problems, the vector equation $\mathbf{F} = m\mathbf{a}$ is replaced by scalar equations using components. In the diagrams vectors are designated by their magnitudes when the directions are apparent.

1. A particle weighing 2 lb is pulled up a smooth plane by a force F as shown in Fig. 11-1(a) below. Determine the force of the plane on the particle and the acceleration along the plane.

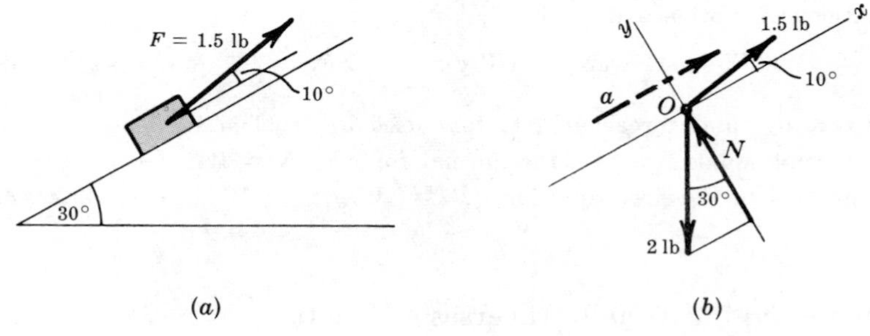

Fig. 11-1

Solution:

The free body diagram is shown in Fig. 11-1(b). The acceleration a is shown as a dashed vector acting parallel to the plane and upwards. If the value obtained is negative, this indicates that the acceleration acts parallel to the plane but downwards.

It is important to keep in mind that the force system shown acting on the particle is not in equilibrium. If it were in equilibrium the particle would not accelerate.

Applying Newton's laws, there result two equations along the x and y axes chosen respectively parallel and perpendicular to the plane.

$$\Sigma F_x = (W/g)a_x \quad \text{or} \quad 1.5 \cos 10° - 2 \sin 30° = (2/32.2)a_x$$
$$\Sigma F_y = (W/g)a_y \quad \text{or} \quad 1.5 \sin 10° - 2 \cos 30° + N = 0$$

Assuming that the particle does not leave the plane, its velocity in the y direction is zero. Therefore a_y must also be zero.

The second equation yields the force of the plane on the particle, $N = 1.47$ lb. From the first equation, $a_x = 7.68$ ft/sec².

2. A particle weighing 10 lb starts from rest and attains a speed of 12 ft/sec in a horizontal distance of 36 ft. Assuming a coefficient of friction of 0.25 and uniformly accelerated motion, what is the smallest value a constant horizontal force P may have to accomplish this?

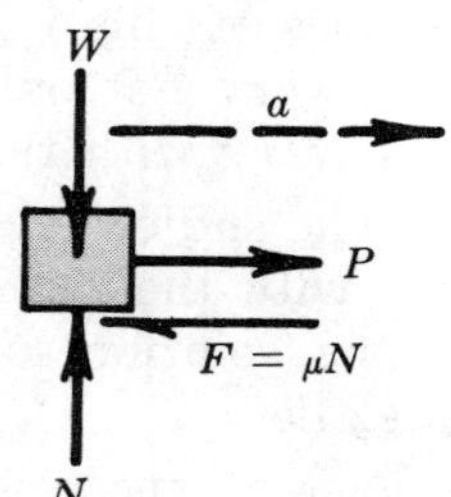

Fig. 11-2

Solution:

Equation of motion in the horizontal direction: $\Sigma F = P - 0.25N = (W/g)a$.

By inspection, $N = W = 10$ lb.

To determine the acceleration a, apply the Kinematic equation

$$v^2 = v_0^2 + 2as; \quad \text{hence} \quad a = \frac{(12 \text{ ft/sec})^2}{2(36 \text{ ft})} = 2 \text{ ft/sec}^2.$$

Substituting into the original equation, $P = (10/32.2)(2) + 0.25 \times 10 = 3.12$ lb.

3. A particle indicated by the block rests on a rough plane. The coefficient of kinetic friction between the plane and the block is μ. What force P parallel to the plane is necessary to give the particle an acceleration a_x parallel to the plane?

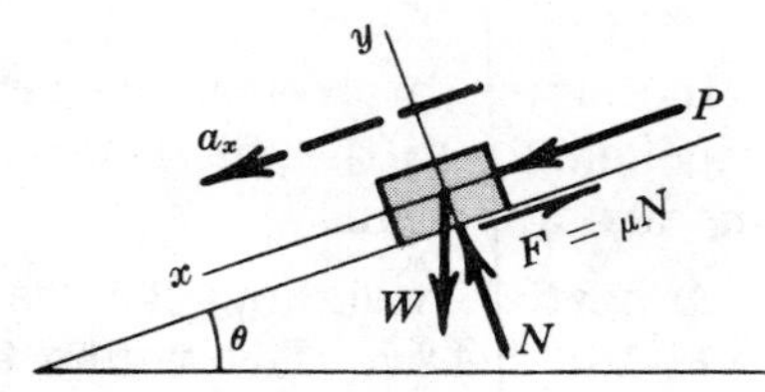

Fig. 11-3

Solution:

Choosing the x-axis positive to the lower left (see Fig. 11-3), it is noted that W has a component acting in the positive x direction and that the force of friction $F = \mu N$ opposes motion. The equations of motion are

$$\Sigma F_x = P + W \sin\theta - \mu N = (W/g)a_x \qquad \Sigma F_y = N - W\cos\theta = (W/g)a_y = 0$$

The a_y is zero by similar reasoning to that used in Problem 1.

From the second equation above, the normal force is $N = W\cos\theta$.

Substituting this in the first equation, $P = (W/g)a_x - W\sin\theta + \mu W\cos\theta$.

4. An automobile weighing 3200 lb traverses a 1500 ft curve at a constant speed of 30 mph. Assuming no banking of the curve, what must be the force of the tires on the road to maintain motion along the curve?

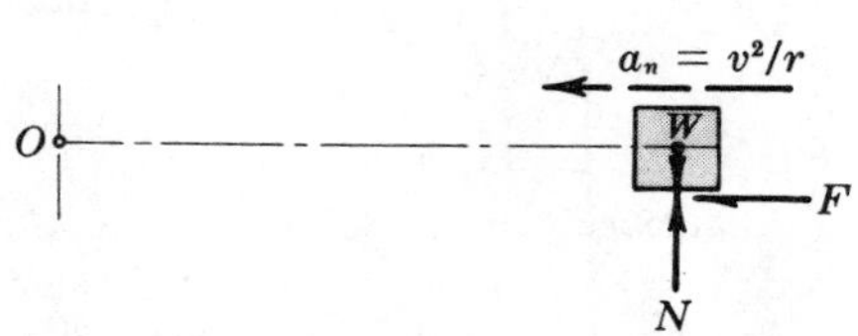

Fig. 11-4

Solution:

Let O be the center of the curve, 1500 ft away from the car. The forces acting are the weight, the normal force which by inspection is equal to the weight, and the lateral push F of the road on the tires and directed toward the center O. See Fig. 11-4. Sum forces along the radius (assume forces and acceleration to the left as positive).

$$\Sigma F = \frac{W}{g}a_n = \frac{W}{g} \times \frac{v^2}{r} \qquad \text{or} \qquad F = \frac{3200 \text{ lb}}{32.2 \text{ ft/sec}^2} \times \frac{(44 \text{ ft/sec})^2}{1500 \text{ ft}} = 128 \text{ lb}$$

5. A small mass of weight W rests on a rotating turntable at a distance r ft from the center as shown in Fig. 11-5. Assuming a coefficient of friction μ between the mass and the turntable, what is the maximum linear velocity the mass may have without slipping?

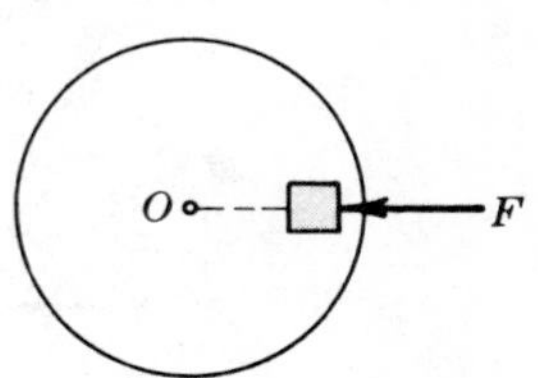

Fig. 11-5

Solution:

The only force acting horizontally is the friction F which equals μN.
Sum forces along the radius: $\Sigma F = (W/g)a_n$ or $F = (W/g)a_n$.
Since the normal force N is equal to W and $a_n = v^2/r$,

$$\mu W = (W/g)(v^2/r) \qquad \text{or} \qquad v = \sqrt{\mu g r}$$

6. Fig. 11-6 indicates a particle of weight W which can move in a circular path about the y-axis. The plane of the circular path is horizontal and perpendicular to the y-axis. As its angular velocity ω increases, however, the particle rises, which means that the radius r of its circular path also increases. The weight and cord are called a spherical (sometimes conical) pendulum. Derive the relationship between θ and ω for constant angular velocity, and find the frequency in terms of θ.

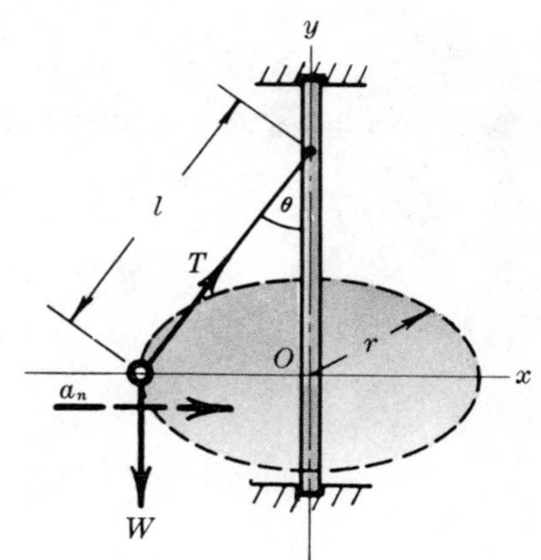

Fig. 11-6

Solution:

Assume that the constant angular velocity of the particle (or of its cord l) is ω rad/sec. Angle θ is the angle between the cord and the y-axis. The forces acting on the particle are its weight W and the tension T in the cord.

Since the particle is moving with constant angular velocity, its only linear acceleration is the normal component a_n directed toward the center of the path (i.e., toward the intersection of the horizontal plane of travel and the y-axis).

Summing forces along this normal, $\Sigma F_n = T \sin\theta = \dfrac{W}{g}a_n$.

Since $a_n = r\omega^2 = (l \sin\theta)\omega^2$, this equation becomes $T \sin\theta = \dfrac{W}{g}(l \sin\theta)\omega^2$. (1)

Summing forces in the y direction, $\Sigma F_y = T\cos\theta - W = \dfrac{W}{g}a_y = 0$ or $T = \dfrac{W}{\cos\theta}$.

Substituting this value T in (1), $\dfrac{W}{\cos\theta}\sin\theta = \dfrac{W}{g}(l \sin\theta)\omega^2$ or $\omega = \sqrt{\dfrac{g}{l\cos\theta}}$.

If θ is known, this equation can be solved for the necessary angular velocity ω to maintain θ constant. Or if ω is known, θ may be found.

Since ω is constant for a given angle θ, frequency $f = \dfrac{\omega \text{ rad/sec}}{2\pi \text{ rad/rev}} = \dfrac{1}{2\pi}\sqrt{\dfrac{g}{l\cos\theta}}$ rev/sec. This is the frequency about the y-axis.

7. A block, assumed to be a particle and weighing 10 lb, rests on a smooth plane which can turn about the y-axis. See Fig. 11-7(a) below. The length of the cord l is 2 ft. What is the tension in the cord when the angular velocity of the plane and block is 10 revolutions per minute?

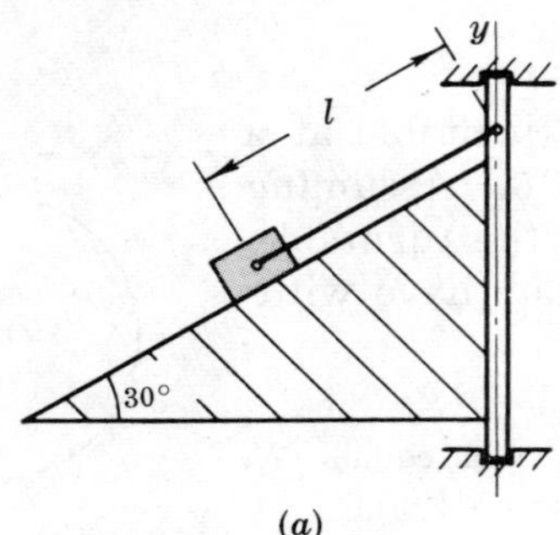

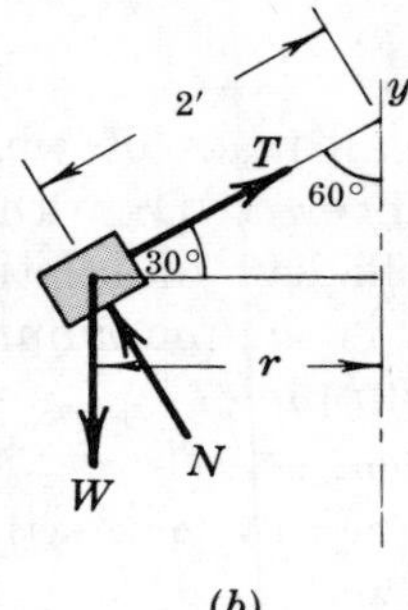

Fig. 11-7

Solution:

From the free body diagram of the block shown in Fig. 11-7(b), $r = 2 \cos 30° = 1.732$ ft. The only acceleration present is the normal component a_n directed horizontally toward the y-axis.

$$a_n = r\omega^2 = (1.732 \text{ ft})\left(\frac{10 \text{ rev/min} \times 2\pi \text{ rad/rev}}{60 \text{ sec/min}}\right)^2 = 1.91 \text{ ft/sec}^2$$

Sum forces horizontally along the radius r and along the y-axis to obtain the following equations.

$$\Sigma F_n = T \cos 30° - N \sin 30° = \frac{W}{g} a_n = \frac{10}{32.2} \times 1.91 \qquad (1)$$

$$\Sigma F_y = N \cos 30° + T \sin 30° - 10 = \frac{W}{g} a_y = 0 \qquad (2)$$

Solve equation (2) for $N = \dfrac{10}{\cos 30°} - T\dfrac{\sin 30°}{\cos 30°}$. Substituting into (1),

$$T \cos 30° - \left(\frac{10}{\cos 30°} - T\frac{\sin 30°}{\cos 30°}\right) \sin 30° = \frac{10}{32.2} \times 1.91 \quad \text{or} \quad T = 5.52 \text{ lb}$$

8. In Problem 7, what is the angular velocity necessary to cause the weight to float just above the plane? What is the tension in the cord under these conditions?

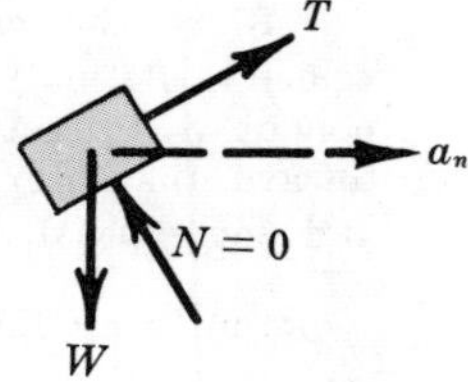

Fig. 11-8

Solution:

The normal force of the plane on the weight for this condition is zero. Summation of forces yields the following equations.

$$\Sigma F_n = T \cos 30° = \frac{W}{g} a_n = \frac{10}{32.2} \times 1.732\omega^2$$

$$\Sigma F_y = T \sin 30° - 10 = 0$$

Hence $T = \dfrac{10}{\sin 30°} = 20$ lb, and $\omega^2 = T\dfrac{\cos 30° \times 32.2}{17.32} = 32.2$ or $\omega = 5.67$ rad/sec.

9. In a device known as Atwood's machine, two equal weights W are connected by a very light (negligible weight) tape passing over a frictionless pulley as shown in Fig. 11-9(a) below. A weight w whose magnitude is much less than W is added to one side, causing that weight to fall and the other of course to rise. The time is recorded by an inked stylus resting on the tape and vibrating. Study the motion.

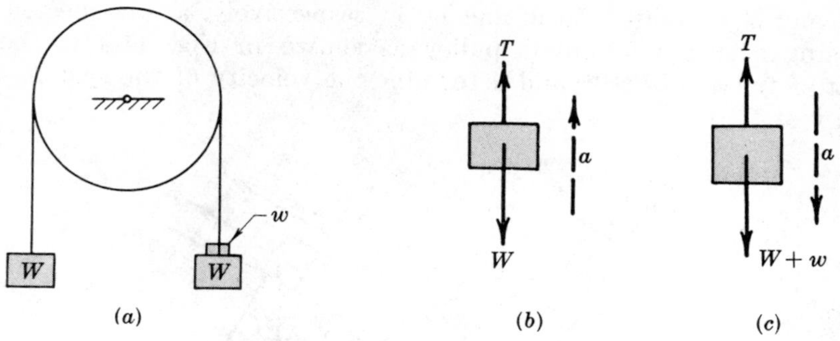

Fig. 11-9

Solution:

The free body diagrams of the two mass systems are shown in Fig. 11-9(*b*) and (*c*). The same tension T is acting on each system through the tape because the friction of the pulley is assumed negligible.

The equations of motion, using the same acceleration (otherwise the tape would have broken or become slack) for the two free body diagrams are

$$\Sigma F = T - W = \frac{W}{g}a \tag{1}$$

$$\Sigma F = W + w - T = \frac{W+w}{g}a \tag{2}$$

Add equations (*1*) and (*2*) to eliminate the tension T and obtain

$$w = \frac{W}{g}a + \frac{W+w}{g}a = \frac{2W+w}{g}a \qquad \text{or} \qquad a = \frac{w}{2W+w}g$$

This expresses the relation between the acceleration of gravity g at the locality where the experiment is performed and the acceleration a of the weights as determined by measurement of distance and time on the tape.

10. Fig. 11-10 below indicates a 10 lb weight resting on a smooth plane inclined 30° with the horizontal. A cord passes from this weight over a frictionless weightless pulley to a weight of 20 lb which when released will drop vertically down. What will be the velocity of the 20 lb weight 3 sec after it is released from rest?

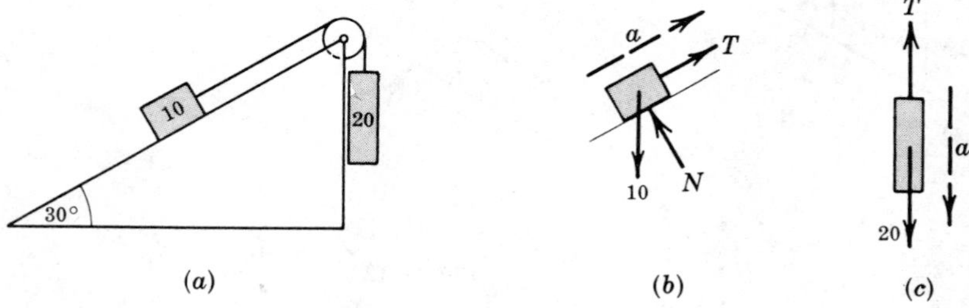

Fig. 11-10

Solution:

The free body diagrams of the two weights are shown in Fig. 11-10(*b*) and (*c*) above. Note that the acceleration a is common to both diagrams, as in the tension T in the cord.

The equations of motion for the two weights are as follows:

$$\Sigma F = T - 10\sin 30^\circ = \frac{10}{32.2}a$$

$$\Sigma F = 20 - T = \frac{20}{32.2}a$$

Adding the two equations, we find $a = 16.1$ ft/sec^2. Then $v = v_0 + at = 0 + 16.1(3) = 48.3$ ft/sec.

11. Blocks A and B, weighing 20 lb and 60 lb respectively, are connected by a weightless rope passing over a frictionless pulley as shown in Fig. 11-11(*a*) below. Assume a coefficient of friction of 0.30 and determine the velocity of the system 4 sec after starting from rest.

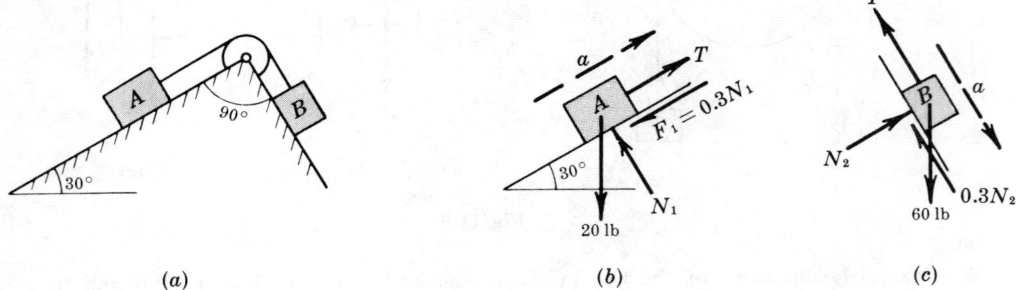

Fig. 11-11

Solution:

Free body diagrams are drawn for bodies A and B [see Fig. 10-11(*b*) and (*c*)]. Summing forces perpendicular and parallel to the planes, the equations of motion are:

$$(1)\quad \Sigma F_\perp = N_1 - 20\cos 30° = (20/32.2)(0) = 0$$

$$(2)\quad \Sigma F_{\parallel} = T - 20\sin 30° - 0.30N_1 = (20/32.2)a$$

$$(3)\quad \Sigma F_\perp = N_2 - 60\cos 60° = (20/32.2)(0) = 0$$

$$(4)\quad \Sigma F_{\parallel} = 60\sin 60° - T - 0.30N_2 = (60/32.2)a$$

Solve equations (*1*) and (*3*) for N_1 and N_2. Substitute these values into equations (*2*) and (*4*) and add the two equations to eliminate T. This yields an acceleration $a = 11.1$ ft/sec^2.

Applying the Kinematics equation $v = v_0 + at$, $v = 0 + 11.1(4) = 44.4$ ft/sec.

12. Refer to Fig. 11-12(*a*). Determine the least coefficient of friction between A and B so that slip will not occur. A weighs 100 lb, B weighs 40 lb, and F is 120 lb parallel to the plane which is smooth.

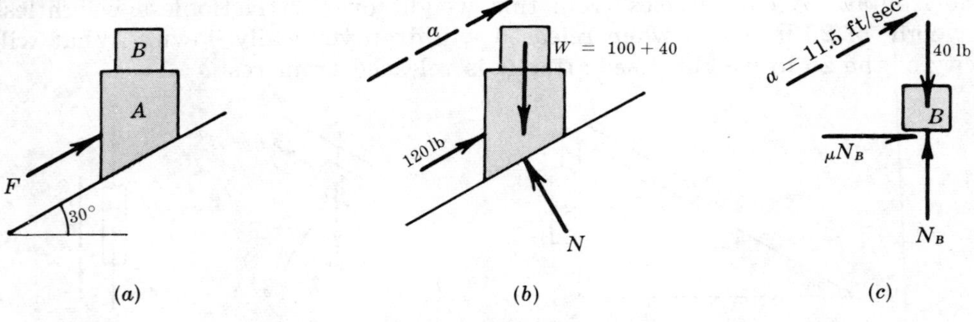

Fig. 11-12

Solution:

To determine the acceleration a of the system, draw a free body diagram of the two bodies taken as a unit as shown in Fig. 11-12(*b*). Summing forces along the plane, $120 - 140\sin 30° = (140/32.2)a$ or $a = 11.5$ ft/sec^2.

Draw a free body diagram of B as shown in Fig. 11-12(*c*). Sum forces along the acceleration vector and perpendicular to it to obtain:

$$(1)\quad \Sigma F_\perp = -40\cos 30° + N_B\cos 30° - \mu N_B \sin 30° = 0$$

$$(2)\quad \Sigma F_{\parallel} = \mu N_B \cos 30° + N_B \sin 30° - 40\sin 30° = (40/32.2)(11.5)$$

Multiply the first equation by cos 30° and the second by sin 30°. Then add to obtain $N_B = 47.2$ lb. Substituting into either equation (*1*) or (*2*), $\mu = 0.263$.

13. A horizontal force $P = 18$ lb is exerted on weight $A = 36$ lb as shown in Fig. 11-13(*a*). The coefficient of friction between A and the horizontal plane is 0.25. B weighs 8 lb and the coefficient of friction between it and the plane is 0.50. The cord between the two weights makes an angle of 10° with the horizontal. What is the tension in the cord?

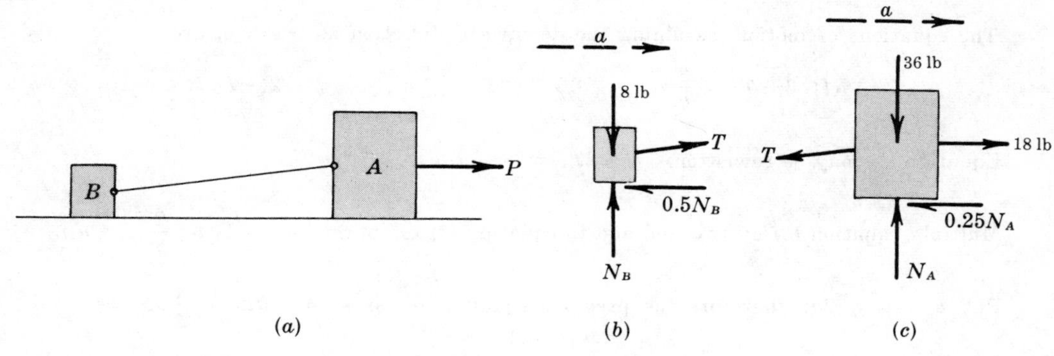

Fig. 11-13

Solution:

The equations of motion from the two free body diagrams are [see Fig. 11-13(*b*) and (*c*)]:

(*1*) $\Sigma F_h = 18 - T\cos 10° - 0.25N_A = \dfrac{36}{32.2}a$ (*3*) $\Sigma F_h = T\cos 10° - 0.50N_B = \dfrac{8}{32.2}a$

(*2*) $\Sigma F_v = N_A - 36 - T\sin 10° = 0$ (*4*) $\Sigma F_v = N_B - 8 + T\sin 10° = 0$

Substitute N_A in terms of T from equation (*2*) into equation (*1*). Substitute N_B in terms of T from equation (*4*) into equation (*3*). Eliminate a between these two new equations to obtain $T = 4.62$ lb.

14. A box is dropped on a conveyor belt moving at 10 ft/sec. If the box is initially at rest and the coefficient of friction between the box and belt is $\frac{1}{3}$, how long will it take before slipping stops?

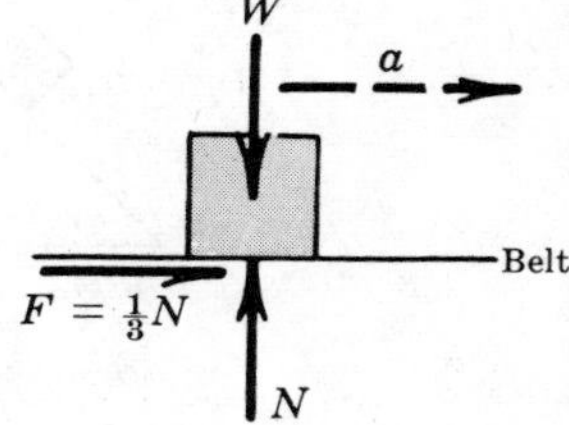

Fig. 11-14

Solution:

The equation of motion for the box is $\Sigma F_h = \dfrac{W}{g}a$, or $\dfrac{1}{3}N = \dfrac{W}{g}a$. See Fig. 11-14. But by inspection $N = W$; hence $a = g/3$ ft/sec².

To find the time required to bring the box from rest up to the belt speed, use $v = v_0 + at$; thus $10 = 0 + \frac{1}{3}gt$ or $t = 0.93$ sec.

15. In the system of pulleys and weights shown in Fig. 11-15, let x_1, x_2, x_3 be the positions of the 1, 2, and 3 lb weights respectively during any phase of the motion after the system is released. Neglect the masses of the pulleys and the cords, and assume no friction. Determine the tensions T_1 and T_2.

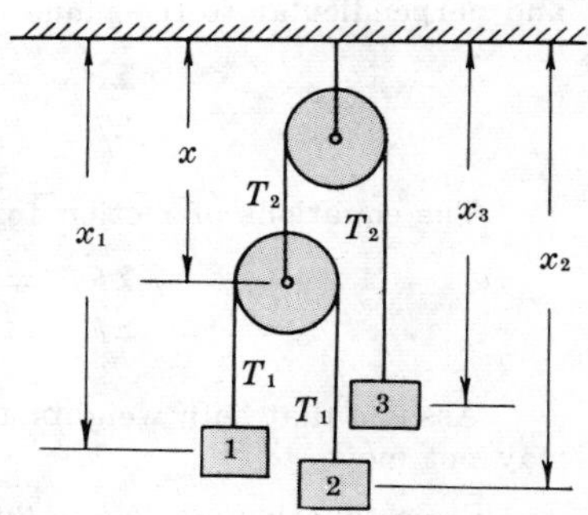

Fig. 11-15

Solution:

By continuity and the assumption of weightless pulleys and cords, $2T_1 = T_2$.

Also, the cord over the lower pulley remains constant in length; therefore,

$$x_1 - x + x_2 - x = \text{a constant}$$

The second derivative with respect to time yields the following relation among the accelerations a_1, a_2 and a:

$$a_1 + a_2 - 2a = 0$$

Similarly, since $x + x_3 =$ a constant,

$$a + a_3 = 0$$

The equations of motion, assuming the downward direction as positive, are:

$$(1)\ \ 3 - T_2 = \frac{3}{g}a_3, \qquad (2)\ \ 1 - T_1 = \frac{1}{g}a_1, \qquad (3)\ \ 2 - T_1 = \frac{2}{g}a_2$$

Equation (1) may be rewritten: $3 - 2T_1 = -\dfrac{3}{g}a.$

Multiply equation (2) by two and add to equation (3) to obtain $4 - 3T_1 = \dfrac{2}{g}(a_1 + a_2).$

But $a_1 + a_2 = 2a$; therefore the previous equation becomes $4 - 3T_1 = \dfrac{4}{g}a.$

Eliminate a between the two equations to obtain $T_1 = 1.41$ lb and $T_2 = 2.82$ lb.

16. Two weights of 15 lb and 30 lb connected by a flexible inextensible cord rest on a smooth plane inclined 45° with the horizontal as shown in Fig. 11-16(*a*) below. When the weights are released, what will be the tension T in the cord? Assume the coefficient of friction between the plane and the 30 lb weight is $\frac{1}{4}$, and between the plane and the 15 lb weight $\frac{3}{8}$.

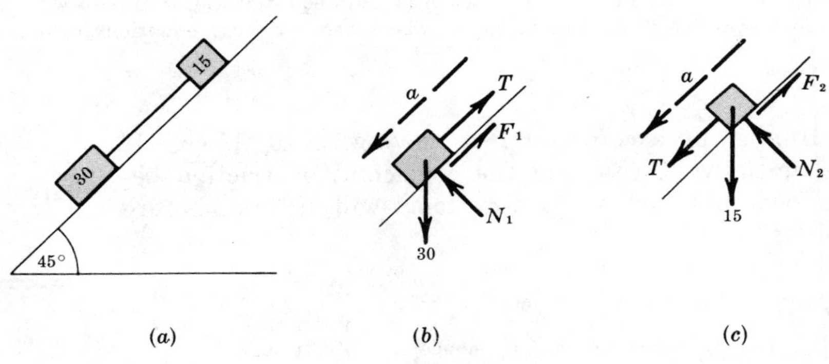

Fig. 11-16

Solution:

Free body diagrams of the two weights are shown in Fig. 11-16(*b*) and (*c*) above.

The equations of motion for the 30 lb weight are as follows, where the summations are parallel and perpendicular to the plane.

$$\Sigma F_{\parallel} = 30 \sin 45° - F_1 - T = (30/32.2)a \qquad (1)$$

$$\Sigma F_{\perp} = N_1 - 30 \cos 45° = 0 \qquad (2)$$

The equations of motion for the 15 lb weight are as follows.

$$\Sigma F_{\parallel} = T + 15 \sin 45° - F_2 = (15/32.2)a \qquad (3)$$

$$\Sigma F_{\perp} = N_2 - 15 \cos 45° = 0 \qquad (4)$$

Assume that both weights are moving. Of course, if the friction is great enough the upper weight may not move.

From equation (2), $N_1 = 30 \times 0.707.$ Then $F_1 = \frac{1}{4}N_1 = 5.30.$

From equation (4), $N_2 = 15 \times 0.707.$ Then $F_2 = \frac{3}{8}N_2 = 3.98.$

Substituting these values into equations (1) and (3), there result the following equations.

$$30 \times 0.707 - 5.30 - T = (30/32.2)a \qquad (5)$$
$$T + 15 \times 0.707 - 3.98 = (15/32.2)a \qquad (6)$$

Multiply equation (6) by two and subtract from equation (5) to obtain $T = 0.887$ lb.

17. A particle of weight W is suspended on a cord of length l as shown in Fig. 11-17(a). Determine the period and frequency of this simple pendulum.

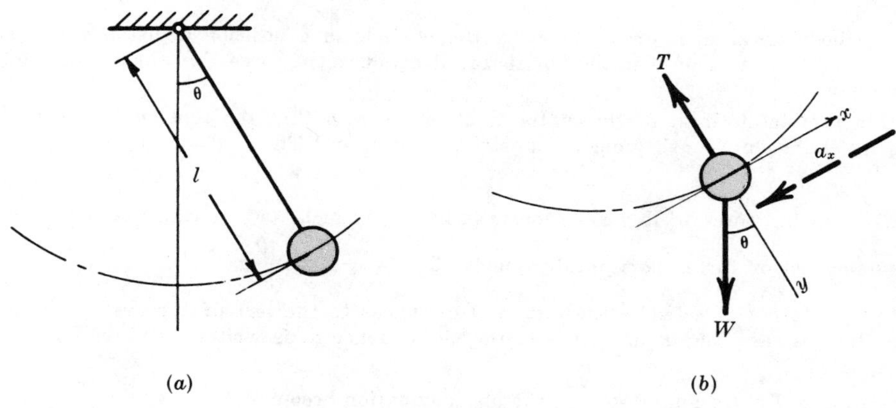

Fig. 11-17

Solution:

The only forces acting on the particle are its weight vertically down and the tension T in the cord. The position of the particle at any time t may be specified in terms of the angle θ.

Choosing as the x-axis the tangent to the path of the particle in the position shown in Fig. 11-17(b), the equation of motion becomes

$$\Sigma F_x = -W \sin\theta = \frac{W}{g} a_x$$

Thus the acceleration is zero when θ is zero, i.e., at the lowest position of the particle.

The above differential equation may be solved by noting that a_x is tangent to the path and may therefore be written

$$a_x = l\alpha$$

where α = angular acceleration of the cord and particle. Then

$$-W \sin\theta = \frac{W}{g} l\alpha = \frac{W}{g} l \frac{d^2\theta}{dt^2} \qquad \text{or} \qquad -\frac{g}{l}\sin\theta = \frac{d^2\theta}{dt^2}$$

The solution of this differential equation is simplified by using a series expansion for $\sin\theta$.

$$\sin\theta = \theta - \frac{\theta^3}{3!} + \frac{\theta^5}{5!} - \frac{\theta^7}{7!} + \cdots$$

For small angular displacements, $\sin\theta$ is approximately equal to θ expressed in radians.

The equation of motion becomes for small displacements, $-\frac{g}{l}\theta = \frac{d^2\theta}{dt^2}$.

The solution as derived in the theory of differential equations is in the form of sines and cosines:

$$\theta = A \sin\sqrt{g/l}\, t + B \cos\sqrt{g/l}\, t$$

The constants A and B can be evaluated in a given problem by using the boundary conditions. The frequency ω in radians/sec is $\sqrt{g/l}$. The frequency f in cycles/sec is $\frac{1}{2\pi}\sqrt{g/l}$.

Incidentally, a cycle is the motion of the particle from a starting point through all possible positions back to the same point. A cycle is complete when the particle moves, let us say, from its top left position to its top right position and back to its top left position.

The time to complete one cycle is called the period T. This is then the reciprocal of the frequency f.

$$T = \frac{1}{f \text{ cycles/sec}} = \frac{1}{f} \text{ sec/cycle}$$

18. Consider the motion of a particle of weight W resting on a smooth horizontal plane as shown in Fig. 11-18(*a*). It is attached to a spring which has a constant of K lb per inch. Displace the weight a distance x_0 from its equilibrium position (spring tension or compression is zero at equilibrium position) and then release it with zero velocity.

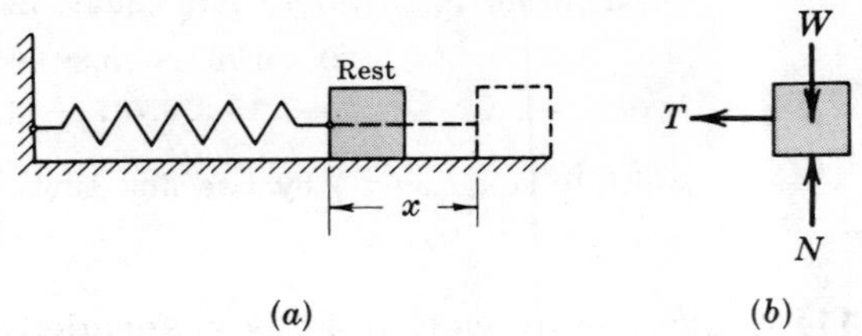

Fig. 11-18

Solution:

A free body diagram is drawn showing the particle in a position a distance x from equilibrium. See Fig. 11-18(*b*). Acting on it in the horizontal direction is the force T in the spring which is stretched the distance x.

Within the elastic limit of the material, it is assumed that the tension in the spring is proportional to its change in length from the unstressed position. Then

$$T = Kx$$

where T = spring force in lb, K = constant in lb per inch, x = change of length in inches.

A summation of forces horizontally yields $\Sigma F_x = -T = \frac{W}{g} a_x$.

Note that forces directed to the left, and distances to the left of the rest position, are assumed negative. In this case the distance x is to the right; hence a_x is written positive. Tension T is to the left or negative.

Substituting $T = Kx$ and $a_x = \frac{d^2x}{dt^2}$, the above equation becomes $-Kx = \frac{W}{g}\frac{d^2x}{dt^2}$ (a simple harmonic motion). This differential equation is similar to the one found in Problem 17. Its solution is in the form of sines and cosines:

$$x = A \sin\sqrt{Kg/W}\,t + B\cos\sqrt{Kg/W}\,t$$

The values of A and B can now be calculated. The value of x is x_0 when t is zero; hence

$$x_0 = A\sin\sqrt{Kg/W}\,0 + B\cos\sqrt{Kg/W}\,0 = 0 + B \qquad \text{and} \qquad B = x_0$$

and

$$x = A\sin\sqrt{Kg/W}\,t + x_0\cos\sqrt{Kg/W}\,t$$

To evaluate A it is necessary to differentiate x with respect to time, since the other known condition is that the velocity v is zero when the time t is zero.

$$v = dx/dt = A\sqrt{Kg/W}\cos\sqrt{Kg/W}\,t - x_0\sqrt{Kg/W}\sin\sqrt{Kg/W}\,t$$

When $t = 0$, $v = 0$ and $\sin\sqrt{Kg/W}\,0 = 0$. Then $0 = A\sqrt{Kg/W}\cos 0 - 0$, and $A = 0$.

Of course if the initial velocity had some value other than zero, say v_0, then $A = \frac{v_0}{\sqrt{Kg/W}}$.

The equation of motion for this problem is $x = x_0\cos\sqrt{Kg/W}\,t$.

19. In Problem 18, find the frequency f and period T of the system if the weight W is 12 oz and the spring constant is 2 oz per inch.

Solution:

$$f = \frac{1}{2\pi}\sqrt{\frac{Kg}{W}} = \frac{1}{2\pi}\sqrt{\frac{2/16\text{ lb/in} \times 12(32.2)\text{ in/sec}^2}{12/16\text{ lb}}} = 1.28\text{ cycles/sec}$$

$$T = \frac{1}{f} = \frac{1}{1.28\text{ cycles/sec}} = 0.782\text{ sec/cycle}$$

Of course, the units of K could be oz/in provided that W is in oz units.

20. A particle of mass m rests on the top of a smooth sphere of radius r as shown in Fig. 11-19(a) below. Assuming that the particle starts to move from rest, at what point will it leave the sphere?

Solution:

Let θ be the angular displacement at any time t during its travel. The free body diagram indicates the only two forces acting on the particle, i.e., the plane reaction along the radius and the weight mg [see Fig. 11-19(b)].

The equations of motion found by summing forces along the radius and the tangent are:

$$\Sigma F_r = -N + mg\cos\theta = mr\omega^2 \qquad (1)$$

$$\Sigma F_t = mg\sin\theta = mr\alpha \qquad (2)$$

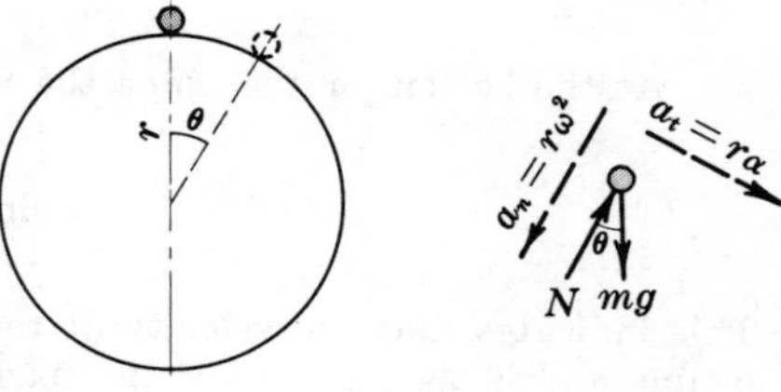

Fig. 11-19

A third equation is necessary for a solution. This is the one expressing a relationship among θ, ω, and α; and for review it is here derived.

Eliminating dt from $\omega = d\theta/dt$ and $\alpha = d\omega/dt$ yields

$$\alpha\, d\theta = \omega\, d\omega \qquad (3)$$

From equation (2), $\alpha = \frac{g}{r}\sin\theta$. Substituting into (3), $\frac{g}{r}\sin\theta\, d\theta = \omega\, d\omega$. Then

$$\int_0^\theta \frac{g}{r}\sin\theta\, d\theta = \int_0^\omega \omega\, d\omega \quad \text{or} \quad \frac{g}{r}(1-\cos\theta) = \frac{\omega^2}{2} \qquad (4)$$

At the position of departure from the sphere, the normal reaction N will be zero. Equation (1) now becomes

$$mg\cos\theta = mr\omega^2 \quad \text{or} \quad \frac{g}{r}\cos\theta = \omega^2 \qquad (5)$$

Replacing ω^2 in equation (4) by its value in equation (5), we obtain

$$\frac{g}{r}(1-\cos\theta) = \frac{g}{2r}\cos\theta, \qquad \cos\theta = \tfrac{2}{3}, \qquad \theta = 48.2°$$

21. A particle of mass m slides down a frictionless chute and enters a "loop-the-loop" of diameter d. What should be the height h at the start in order that the particle may make a complete circuit in the loop?

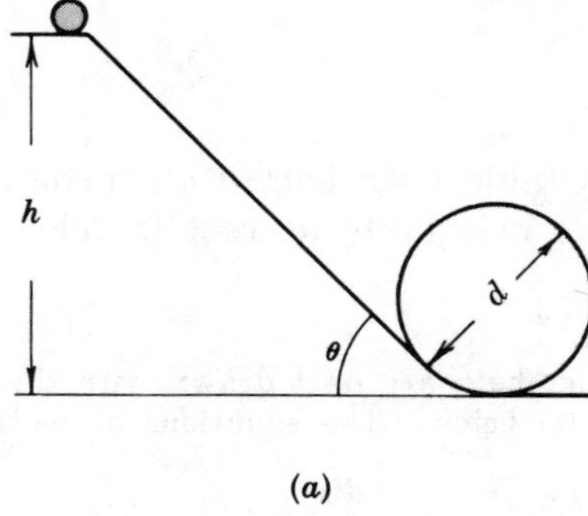

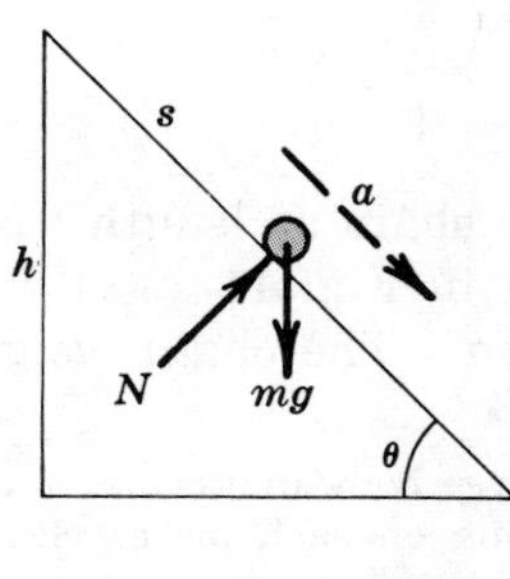

Fig. 11-20

Solution:

A free body diagram of the particle at any time in its travel down is shown in Fig. 11-20(b). The equations of motion parallel and perpendicular to the plane are:

$$\Sigma F_{\parallel} = mg\sin\theta = ma \quad (1) \qquad \text{and} \qquad \Sigma F_{\perp} = N - mg\cos\theta = 0 \quad (2)$$

From equation (*1*), the acceleration is $a = g \sin\theta$. This value of a is now substituted into the Kinematics equation $a\,ds = v\,dv$, where s refers to displacement along the plane. Then $g \sin\theta\, ds = v\,dv$ and

$$\int_0^s g \sin\theta\, ds = \int_0^v v\,dv \qquad \text{or} \qquad g \sin\theta\, s = \frac{v^2}{2} \tag{3}$$

At the bottom of the plane the velocity is found by substituting $\dfrac{h}{\sin\theta}$ for s in (*3*).

$$g \sin\theta \frac{h}{\sin\theta} = \frac{v^2}{2} \qquad \text{or} \qquad v^2 = 2gh \tag{4}$$

This indicates that the velocity at the bottom of the smooth incline is independent of the slope of the incline and is the same as if the particle fell vertically downward.

Next draw a free body diagram of the particle at the top of the loop (see Fig. 11-21).

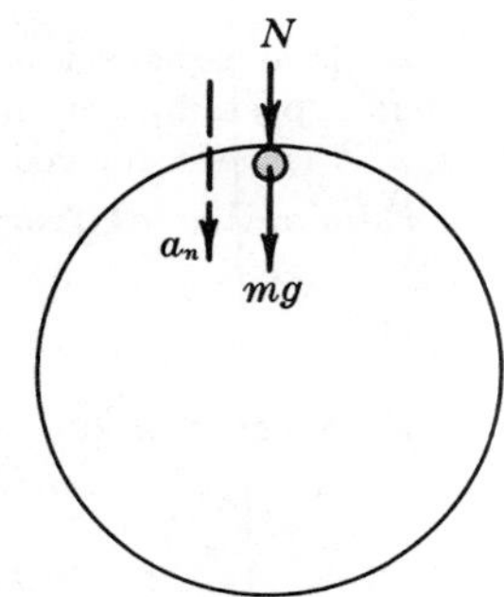

Fig. 11-21

The forces acting are the weight mg and the force of the loop along the vertical radius. Since the minimum velocity is the one with which we are concerned, the value of N in this case is zero. Expressed somewhat differently, as velocity v increases, the normal acceleration a_n increases. To provide this increasing value of a_n the reaction N must increase also, because $\Sigma F_n = ma_n$.

In determining the velocity at the top of the loop, use the fact just determined that the motion on a smooth path (such as the plane or the side of the loop) is equivalent to vertical motion under the acceleration of gravity only. Hence the particle loses velocity in moving up the side of the loop in amount equal to the loss in straight line vertical motion.

$$v_{\text{top}}^2 = v_{\text{bot}}^2 - 2gd$$

But the velocity at bottom v_{bot} has been expressed in (*4*) in terms of height h. Then

$$v_{\text{top}}^2 = 2gh - 2gd = 2g(h-d)$$

A summation of forces along the vertical radius when the particle is at the top results in the following equation since $N = 0$.

$$\Sigma F_n = mg = \frac{mv_{\text{top}}^2}{d/2} \qquad \text{or} \qquad mg = \frac{2m}{d} \times 2g(h-d)$$

Hence $h = 5d/4$. This means that the particle must leave at this height (or above) in order that a minimum velocity be obtained. Otherwise the particle will not follow the circular path at the top but rather "jump across".

22. A flexible chain of length l rests on a smooth table with length c overhanging the edge as shown in Fig. 11-22(*a*) below. The system originally at rest is released. Describe the motion. The chain weighs w lb per ft.

Solution:

The free body diagrams of the two pieces of the chain are next drawn with the same tension T shown acting on each piece. See Fig. 11-22(*b*) and (*c*) below. The equations of motion are:

$$T = \frac{w(l-x)}{g} a = \frac{w(l-x)}{g} \frac{d^2x}{dt^2} \tag{1}$$

$$wx - T = \frac{wx}{g} a = \frac{wx}{g} \frac{d^2x}{dt^2} \tag{2}$$

Add (*1*) and (*2*) to obtain: $wx = \dfrac{wl}{g}\dfrac{d^2x}{dt^2}$, or $\dfrac{d^2x}{dt^2} = \dfrac{g}{l}x$. (This is not Simple Harmonic Motion. Why?)

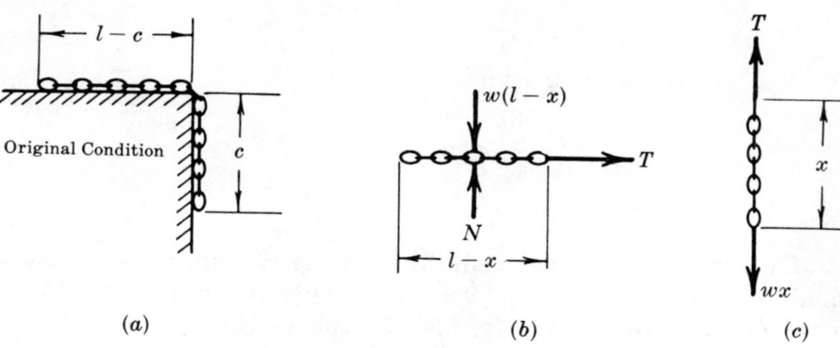

Fig. 11-22

The solution of this second order differential equation is $x = Ae^{\sqrt{g/l}\,t} + Be^{-\sqrt{g/l}\,t}$.

To evaluate the constants A and B note the initial conditions, namely $x = c$ and $v = 0$. Substituting $x = c$ when $t = 0$, we get $c = A + B$.

Differentiate x with respect to t: $v = dx/dt = A\sqrt{g/l}\,e^{\sqrt{g/l}\,t} - B\sqrt{g/l}\,e^{-\sqrt{g/l}\,t}$.

Substituting the condition $v = 0$ when $t = 0$, we get $A - B = 0$.

Solving simultaneously $A - B = 0$ and $c = A + B$, we get $A = B = \frac{1}{2}c$.

The solution of the problem is $x = \frac{1}{2}ce^{\sqrt{g/l}\,t} + \frac{1}{2}ce^{-\sqrt{g/l}\,t}$.

The exponential functions should be evaluated for any given time to determine the length x of the overhang.

23. An object of weight W falls in a medium with the resistance R proportional to the velocity. This is approximately true for slowly moving objects. Study the motion.

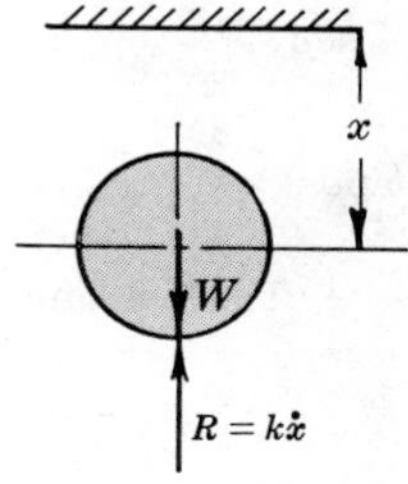

Fig. 11-23

Solution:

The free body diagram of the object shows the weight W acting down and the retarding force R acting up (see Fig. 11-23). Assuming x, $\dot{x}$ and $\ddot{x}$ are positive in the downward direction, the differential equation of the motion is

$$+W - k\dot{x} = \frac{W}{g}\ddot{x} = \frac{W}{g}\frac{d\dot{x}}{dt}$$

Rewriting this equation and then integrating,

$$\frac{d\dot{x}}{W/k - \dot{x}} = \frac{kg}{W}dt \qquad \text{and} \qquad -\ln(W/k - \dot{x}) = \frac{kg}{W}t + C_1$$

Assuming $\dot{x} = \dot{x}_0$ when $t = 0$, $-\ln(W/k - \dot{x}_0) = C_1$ and

$$-\ln(W/k - \dot{x}) = \frac{kg}{W}t - \ln(W/k - \dot{x}_0) \qquad \text{or} \qquad \frac{W/k - \dot{x}}{W/k - \dot{x}_0} = e^{-(kg/W)t}$$

from which

$$\dot{x} = \frac{W}{k}(1 - e^{-(kg/W)t}) + \dot{x}_0 e^{-(kg/W)t} \tag{1}$$

Since the limiting velocity $\dot{x}_{max}$ occurs as $t \to \infty$,

$$\dot{x}_{max} = W/k \tag{2}$$

To determine distance x as a function of time, write (*1*) as

$$dx = \frac{W}{k}(1 - e^{-(kg/W)t})dt + \dot{x}_0 e^{-(kg/W)t}\,dt$$

Integrating,

$$x = \frac{W}{k}t - \frac{W}{k}(-W/kg)e^{-(kg/W)t} + \dot{x}_0(-W/kg)e^{-(kg/W)t} + C_2$$

Assuming $x = 0$ when $t = 0$, $0 = 0 + W^2/k^2g - W\dot{x}_0/kg + C_2$ and

$$x = \frac{W}{k}t + \frac{W^2}{k^2g}e^{-(kg/W)t} - \frac{W\dot{x}_0}{kg}e^{-(kg/W)t} - \frac{W^2}{k^2g} + \frac{W\dot{x}_0}{kg}$$

Using $W/k = \dot{x}_{max}$, this expression is regrouped as

$$x = \frac{W}{k}t + \frac{1}{g}(\dot{x}_{max}\dot{x}_0 - \dot{x}^2_{max})(1 - e^{-(kg/W)t}) \qquad (3)$$

24. An object of weight W falls in a medium with the resistance R proportional to the square of the velocity. This is approximately true for high speed objects. Study the motion.

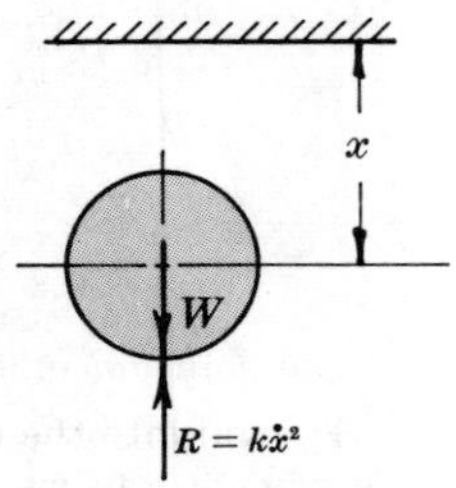

Fig. 11-24

Solution:

The free body diagram of the object shows the weight W acting down and the retarding force R acting up. Assuming x, $\dot{x}$ and $\ddot{x}$ are positive in the downward direction, the differential equation of the motion is

$$+W - k\dot{x}^2 = \frac{W}{g}\ddot{x} = \frac{W}{g}\frac{d\dot{x}}{dx}\frac{dx}{dt} = \frac{W}{g}\frac{d\dot{x}}{dx}\dot{x}$$

Rewriting this equation and integrating,

$$\frac{\dot{x}\,d\dot{x}}{W/k - \dot{x}^2} = \frac{kg}{W}dx \qquad \text{and} \qquad -\tfrac{1}{2}\ln(W/k - \dot{x}^2) = \frac{kg}{W}x + C_1$$

Assuming $\dot{x} = \dot{x}_0$ when $x = 0$, $-\frac{1}{2}\ln(W/k - \dot{x}_0^2) = C_1$ and

$$-\tfrac{1}{2}\ln(W/k - \dot{x}^2) = (kg/W)x - \tfrac{1}{2}\ln(W/k - \dot{x}_0^2)$$

Then

$$-\frac{2kg}{W}x = \ln\frac{W/k - \dot{x}^2}{W/k - \dot{x}_0^2} \qquad \text{or} \qquad W/k - \dot{x}^2 = (W/k - \dot{x}_0^2)e^{-(2kg/W)x}$$

and

$$\dot{x}^2 = \frac{W}{k}(1 - e^{-(2kg/W)x}) + \dot{x}_0^2 e^{-(2kg/W)x} \qquad (1)$$

The limiting velocity occurs as $x \to \infty$; thus

$$\dot{x}_{max} = \sqrt{W/k} \qquad (2)$$

25. A projectile of mass m is given an initial velocity v_0 at an angle θ with the horizontal. Determine the range, the maximum height, and the time of flight, assuming the projectile hits on the same horizontal plane from which it is fired. Neglect air resistance in this solution.

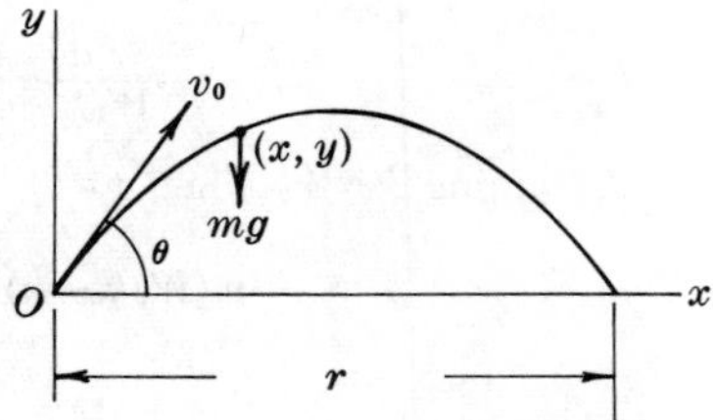

Fig. 11-25

Solution:

The horizontal line is chosen as the x-axis. The range is r. The projectile is shown in Fig. 11-25 at some point (x, y) along the path. The only force acting is its weight mg.

The equations of motion in the x and y directions are

$$\Sigma F_x = 0 = m\ddot{x} \qquad \text{and} \qquad \Sigma F_y = -mg = m\ddot{y}$$

from which

$$\ddot{x} = 0 \qquad \text{and} \qquad \ddot{y} = -g$$

Integrating,

$$\dot{x} = C_1 \qquad \text{and} \qquad \dot{y} = -gt + C_2$$

But $\dot{x}$ has constant value $v_0 \cos\theta$; and at $t = 0$, $\dot{y} = v_0 \sin\theta$. Hence

$$\dot{x} = v_0 \cos\theta \qquad \text{and} \qquad \dot{y} = -gt + v_0 \sin\theta$$

Another integration yields

$$x = (v_0 \cos\theta)t + C_3 \qquad \text{and} \qquad y = (v_0 \sin\theta)t - \tfrac{1}{2}gt^2 + C_4$$

Since $x = 0$ and $y = 0$ when $t = 0$, $C_3 = C_4 = 0$ and the equations of motion are

$$x = (v_0 \cos\theta)t \qquad \text{and} \qquad y = (v_0 \sin\theta)t - \tfrac{1}{2}gt^2$$

To obtain the equation of the path in Cartesian coordinates, eliminate t from the equations. This is most easily done by solving the x equation for t and substituting its value in the y equation.

Substituting $t = \dfrac{x}{v_0 \cos\theta}$ in the y equation yields

$$y = v_0 \sin\theta \frac{x}{v_0 \cos\theta} - \tfrac{1}{2}g\frac{x^2}{v_0^2 \cos^2\theta} = x \tan\theta - \frac{g}{2v_0^2 \cos^2\theta}x^2$$

which is the equation of a parabola.

To determine the time of flight, equate y to zero. The height is zero at the beginning of the motion and also when the projectile hits the ground.

$$y = 0 = (v_0 \sin\theta)t - \tfrac{1}{2}gt^2 = t(v_0 \sin\theta - \tfrac{1}{2}gt)$$

The values of t which satisfy this equation are $t = 0$ (beginning of flight) and $t = \dfrac{2v_0 \sin\theta}{g}$ (time of flight) obtained from $v_0 \sin\theta = \tfrac{1}{2}gt$.

The range may be calculated by substituting this value of t in the x equation of motion.

$$x = r = v_0 \cos\theta \left(\frac{2v_0 \sin\theta}{g}\right) = \frac{v_0^2}{g}\sin 2\theta \qquad \text{(range)}$$

Note that the maximum range for a given initial velocity v_0 is achieved when $\sin 2\theta = 1$, or when $2\theta = 90°$. Thus θ should be 45° for maximum range for a given v_0.

The maximum height can be determined either by assuming that it occurs at half the time of flight (theoretical considerations only) or by solving for the time t when the y component of velocity is equated to zero. Assume that the time is half the time of flight. Substituting this value in the y equation there results

$$h = v_0 \sin\theta \left(\frac{v_0 \sin\theta}{g}\right) - \tfrac{1}{2}g\left(\frac{v_0 \sin\theta}{g}\right)^2 = \frac{v_0^2 \sin^2\theta}{2g} \qquad \text{(maximum height)}$$

26. A particle of mass m moves under the attraction of another mass M which we will assume to be at rest. Study the motion of the mass m, noting that the distance between the two particles is not necessarily constant.

Solution:

The particle m has radius vector $\mathbf{r}$ with respect to mass M. The unit vectors $\mathbf{e}_r$ and $\mathbf{e}_\phi$ for the polar coordinate system are shown. A free body diagram of mass m contains only the attractive force $\mathbf{F}$ which is in the negative $\mathbf{e}_r$ direction [see Fig. 11-26(*b*)]. This central force according to Newton's law of gravitation is

$$\mathbf{F} = -G\frac{Mm}{r^2}\mathbf{e}_r \qquad (1)$$

where G = universal gravitational constant. $G = 6.66 \times 10^{-8}$ cm³/g-sec² in the cgs system and $G = 3.43 \times 10^{-8}$ lb-ft²/(slug)² in the engineering system.

In polar coordinates the acceleration $\mathbf{a}$ is (see Chapter 10)

$$\mathbf{a} = (\ddot{r} - r\dot{\phi}^2)\mathbf{e}_r + (r\ddot{\phi} + 2\dot{r}\dot{\phi})\mathbf{e}_\phi \qquad (2)$$

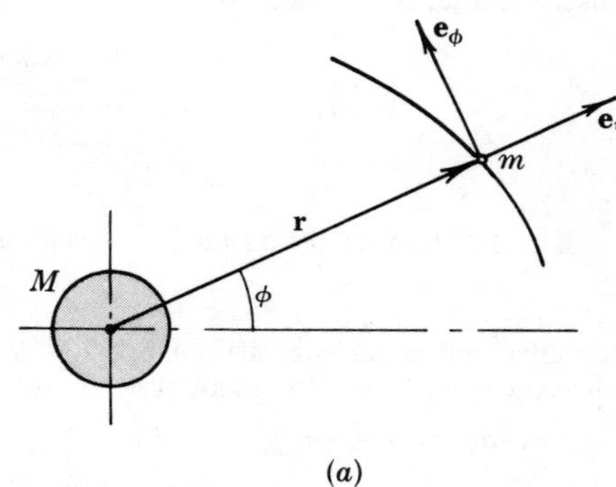

(*a*)

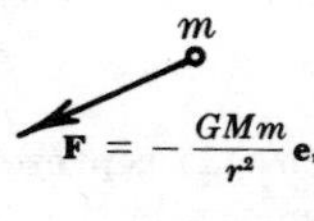

(*b*)

Fig. 11-26

Substituting from equations (*1*) and (*2*) into

$$\mathbf{F} = m\mathbf{a} \tag{3}$$

gives

$$\frac{-GMm}{r^2}\mathbf{e}_r = +m(\ddot{r} - r\dot{\phi}^2)\mathbf{e}_r + m(r\ddot{\phi} + 2\dot{r}\dot{\phi})\mathbf{e}_\phi \tag{4}$$

Equating coefficients of $\mathbf{e}_\phi$ and then of $\mathbf{e}_r$,

$$r\ddot{\phi} + 2\dot{r}\dot{\phi} = 0 \tag{5}$$

and

$$\frac{-GM}{r^2} = (\ddot{r} - r\dot{\phi}^2) \tag{6}$$

Equation (*5*) is identical with

$$\frac{1}{r}\frac{d}{dt}(r^2\dot{\phi}) = 0 \tag{7}$$

Integrating (*7*),

$$r^2\dot{\phi} = C, \quad \text{a constant} \tag{8}$$

In Fig. 11-27 note that the radius vector $\mathbf{r}$ sweeps out area dA as it rotates through the angle $d\phi$. The differential area dA is approximately equal to $\frac{1}{2}r(r\,d\phi) = \frac{1}{2}r^2\,d\phi$. Dividing by dt,

$$\frac{dA}{dt} = \frac{1}{2}r^2\frac{d\phi}{dt} = \frac{1}{2}r^2\dot{\phi} = \frac{1}{2}C$$

Thus the radius vector $\mathbf{r}$ sweeps out equal areas in equal times.

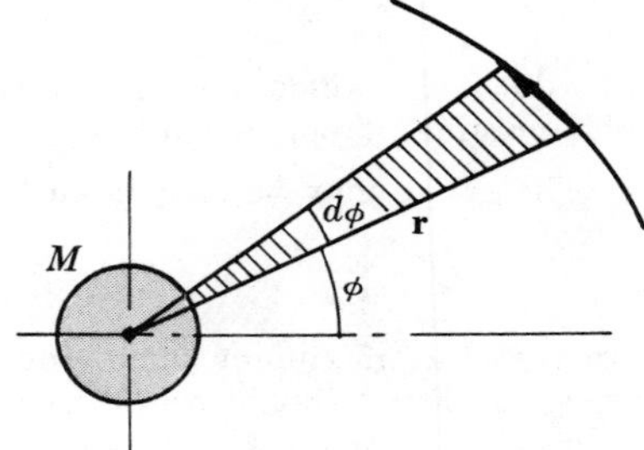

Fig. 11-27

Next solve equation (*6*) to determine the path of mass m. For a convenient solution let $r = 1/u$; then $dr/du = -1/u^2$ and

$$\dot{r} = \frac{dr}{du}\frac{du}{d\phi}\frac{d\phi}{dt} = \frac{-1}{u^2}\frac{du}{d\phi}\dot{\phi}$$

But from equation (*8*), $\dot{\phi} = C/r^2 = Cu^2$; then

$$\dot{r} = \frac{-1}{u^2}\frac{du}{d\phi}Cu^2 = -C\frac{du}{d\phi}$$

and

$$\ddot{r} = \frac{d\dot{r}}{dt} = \frac{d\dot{r}}{d\phi}\frac{d\phi}{dt} = \left(-C\frac{d^2u}{d\phi^2}\right)(Cu^2) = -C^2u^2\frac{d^2u}{d\phi^2}$$

Substituting into (*6*), we obtain

$$-GMu^2 = -C^2u^2\frac{d^2u}{d\phi^2} - C^2u^3$$

or

$$\frac{d^2u}{d\phi^2} + u = \frac{GM}{C^2} \tag{9}$$

The solution of (*9*) as an intelligent guess might be of the form

$$u = A\cos\phi + B \tag{10}$$

This form must be substituted into (*9*) to determine the necessary conditions to make it a solution; thus

$$du/d\phi = -A\sin\phi, \qquad d^2u/d\phi^2 = -A\cos\phi$$

and $\quad -A\cos\phi + A\cos\phi + B = GM/C^2$

Hence $B = GM/C^2$ and the solution of (*9*) is

$$u = A\cos\phi + GM/C^2, \quad \text{where } u = 1/r \tag{11}$$

Equation (*11*) represents a conic section, which from analytic geometry is the path of a point which moves so that the ratio of its distance from a fixed point (a focus) to its perpendicular distance from a fixed line (the directrix) is a constant e (the eccentricity). Fig. 11-28 illustrates the distances involved in the definition

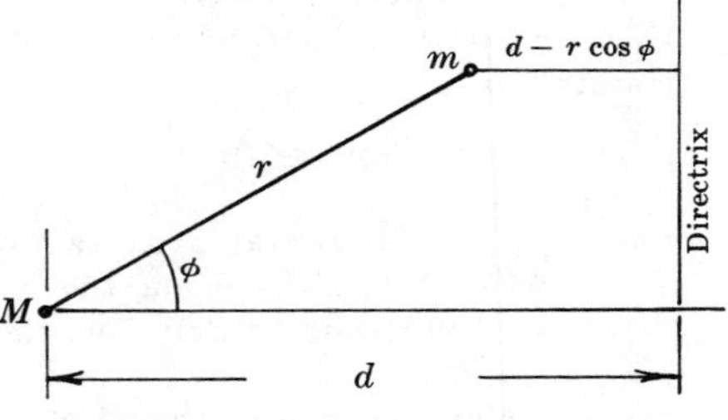

Fig. 11-28

$$e = \frac{r}{d - r\cos\phi} \tag{12}$$

Equation (*12*) can be solved for $1/r$ yielding

$$\frac{1}{r} = \frac{1}{d}\cos\phi + \frac{1}{ed} \tag{13}$$

The table lists the types of conics (curves) with their e values.

e	0	<1	1	>1
Curve	circle	ellipse	parabola	hyperbola

Comparison of equations (*11*) and (*13*) indicates that

$$1/ed = GM/C^2 \qquad \text{or} \qquad ed = C^2/GM \tag{14}$$

The equation of motion is

$$u = \frac{1}{r} = \frac{1}{d}\cos\phi + \frac{GM}{C^2} \tag{15}$$

27. In the preceding problem the path of a planet m around the sun M is elliptical (eccentricity $e < 1$). The sun is at one of the foci of the ellipse. Show that the period T (time for a complete revolution of the planet around the sun) is given by

$$T = \frac{2\pi a^{3/2}}{\sqrt{GM}}$$

where a = semi-major axis of ellipse
G = gravitational constant
M = mass of the sun.

Solution:

The path is an ellipse with semi-major axis a and semi-minor axis b. Fig. 11-29 shows that

$$f + g = 2a \tag{1}$$

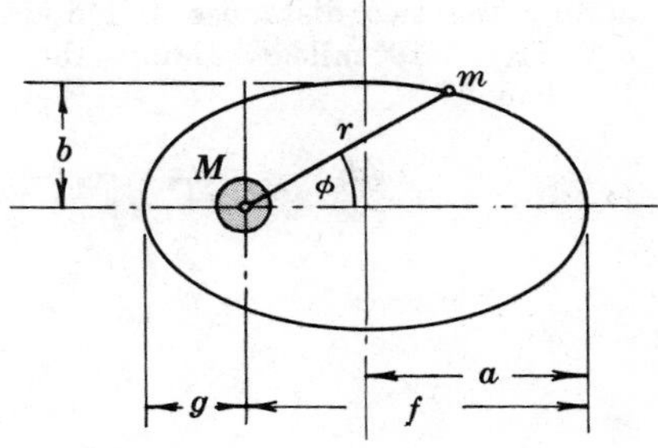

Fig. 11-29

Also, from equation (*13*) of Problem 26,

$$\frac{1}{r} = \frac{1}{d}\cos\phi + \frac{1}{ed} \tag{2}$$

When $\phi = 0°$, $r = f$; and when $\phi = 180°$, $r = g$. Thus (*2*) becomes

$$\frac{1}{f} = \frac{1}{d}\cos 0° + \frac{1}{ed} = \frac{1+e}{ed} \qquad \text{and} \qquad \frac{1}{g} = \frac{1}{d}\cos 180° + \frac{1}{ed} = \frac{1-e}{ed}$$

Solving these two equations for f and g and substituting into (*1*), we find

$$\frac{ed}{1+e} + \frac{ed}{1-e} = 2a \qquad \text{or} \qquad 1 - e^2 = \frac{ed}{a} \tag{3}$$

But by equation (*14*) of Problem 26, $ed = C^2/GM$. Hence (*3*) becomes

$$1 - e^2 = \frac{C^2}{GMa}$$

The area A of the ellipse is $A = \pi ab = \pi a^2\sqrt{1-e^2}$. Thus

$$A = \pi a^2 \frac{C}{\sqrt{GMa}} = \frac{2\pi a^{3/2}}{\sqrt{GM}}\frac{C}{2} \tag{4}$$

Equation (*4*) gives the area swept out in one complete revolution or in time T. This area is also the product of the constant rate dA/dt and time T: $A = (dA/dt)T$. Finally, substituting $dA/dt = C/2$ as determined in Problem 26, we have

$$A = \frac{C}{2}T = \frac{2\pi a^{3/2}}{\sqrt{GM}}\frac{C}{2} \qquad \text{or} \qquad T = \frac{2\pi a^{3/2}}{\sqrt{GM}}$$

28. Convert the gravitational constant $G = 6.658 \times 10^{-8}\ \text{cm}^3/\text{g-sec}^2$ into the equivalent in engineering units.

Solution:

$$G = 6.658 \times 10^{-8}\frac{\text{cm}^3}{\text{g sec}^2} = 6.658 \times 10^{-8}\frac{(3.281 \times 10^{-2}\ \text{ft})^3}{(2.205/32.2)10^{-3}\ \text{slug sec}^2} = 3.43 \times 10^{-8}\frac{\text{ft}^3}{\text{slug sec}^2}$$

Since 1 slug $= 1\dfrac{\text{lb sec}^2}{\text{ft}}$, $\quad 1\dfrac{\text{ft}^3}{\text{slug sec}^2} = 1\dfrac{\text{ft}^3}{(\text{lb sec}^2/\text{ft})\ \text{sec}^2} = 1\dfrac{\text{ft}^4}{\text{lb sec}^2}$.

Thus $G = 3.43 \times 10^{-8}\dfrac{\text{ft}^4}{\text{lb sec}^4}$.

29. Given that the period T for the passage of the earth around the sun is one year and that the earth at its perigee is 91,340,000 miles from the sun and at its apogee is 94,450,000 miles from the sun, determine the mass M of the sun. Assume 1 year = 365 days, and ignore the effects of the other planets.

Solution:

Fig. 11-30 shows the earth in both its near and far positions (perigee and apogee respectively). The semi-major axis a of the elliptical orbit is obtained by adding the two distances and dividing by 2; this gives $a = 92.9 \times 10^6$ miles. Using the formula derived in Problem 27,

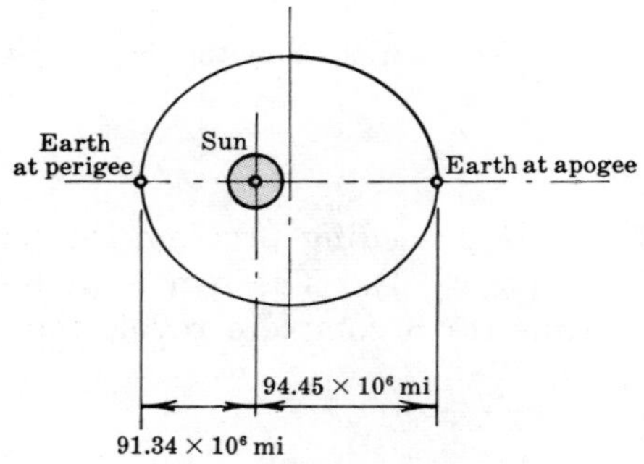

Fig. 11-30

$$M = \frac{4\pi^2a^3}{GT^2} = \frac{4\pi^2(92.9 \times 10^6 \times 5280)^3}{(3.43 \times 10^{-8})(365 \times 24 \times 3600)^2} = 13.7 \times 10^{28}\ \text{slugs}$$

Supplementary Problems

30. An automobile weighing 4000 lb is accelerated at the rate of 3 ft/sec² along a horizontal roadway. What constant force F (parallel to the ground) is required to produce this acceleration?
Ans. 373 lb

31. A body is projected up a 25° plane with an initial velocity of 40 ft/sec. If the coefficient of friction between the body and the plane is 0.25, determine how far the body will move up the plane and the time required to reach the highest point. *Ans.* 38.4 ft, 1.92 sec

32. A meteorite weighing 1000 lb is found buried 60 ft in the earth. Assuming a striking velocity of 1000 ft/sec, what was the average retarding force of the earth on the meteorite?
Ans. $F = 265{,}000$ lb

33. Two particles of same weight are released from rest on a 25° incline when they are 30 ft apart. The coefficient of friction between the upper one and the plane is 0.15, and between the lower one and the plane it is 0.25. Find the time required for the upper one to overtake the lower one.
Ans. $t = 4.53$ sec

34. Refer to Fig. 11-31. An automobile weighing 2800 lb and traveling 30 mi/hr hits a depression in the road which has a radius of curvature of 50 ft. What is the total force to which the springs are subjected? *Ans.* $N = 6170$ lb

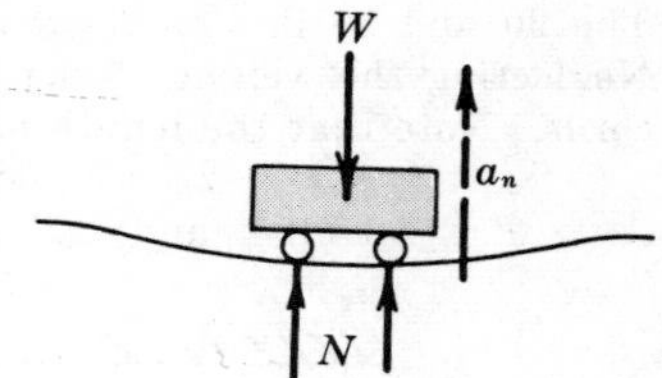

Fig. 11-31

35. A small bob weighing 1 oz is whirled in an assumed horizontal circular path on a string 16 inches long as shown in Fig. 11-32. What is the tension in the string when the constant angular speed is 30 radians/sec? *Ans.* $T = 2.33$ lb

36. A 90 lb weight has a velocity of 30 ft/sec horizontally on a smooth surface. Determine the value of the horizontal force that will bring the weight to rest in 4 sec. *Ans.* 21.0 lb

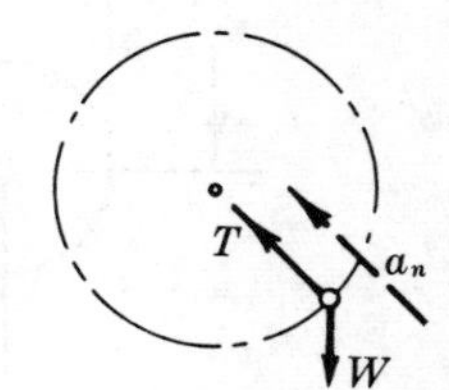

Fig. 11-32

37. The ball of a conical pendulum weighing 10 lb hangs at the end of an 8 ft string and describes a circular path in a horizontal plane. If the weight is swung so that the string makes an angle of 30° with the vertical, what is the linear speed of the ball? *Ans.* 8.63 ft/sec

38. Two weights, of 90 lb and 80 lb respectively, are attached by a cord that passes over a frictionless pulley. If the weights start from rest, find the distance covered by the 90 lb weight in 6 sec. *Ans.* 34.1 ft

39. A 100 lb weight is dragged along the surface of a table by means of a cord attached to a 30 lb weight which passes over a frictionless pulley at the edge of the table. If the coefficient of friction between the 100 lb weight and the table is 0.15, determine the acceleration of the system and the tension in the cord. *Ans.* 3.72 ft/sec², 26.6 lb

40. Block A weighs 20 lb and is at rest on the frictionless horizontal surface. A 10 lb weight B is attached to the rope as shown in Fig. 11-33 below. Determine the acceleration of the weight B and the tension in the cord. The pulley is frictionless. *Ans.* $a = 10.7$ ft/sec² down $T = 6.67$ lb

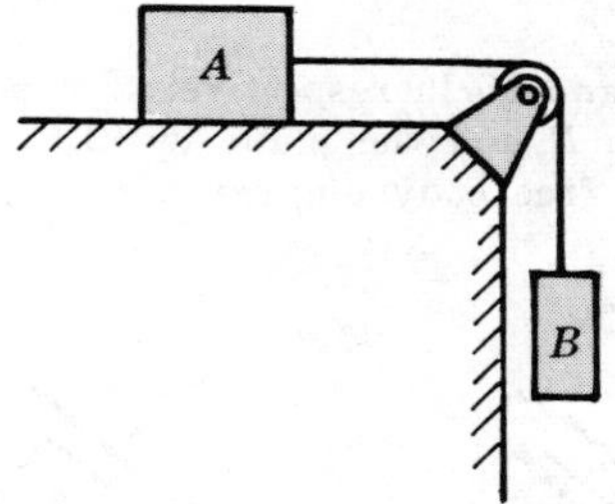

Fig. 11-33

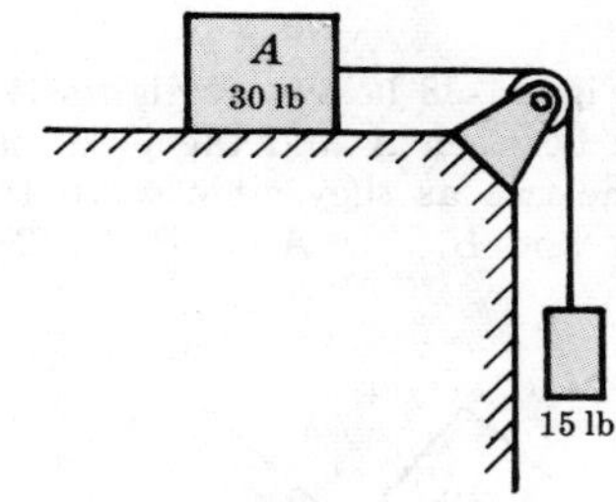

Fig. 11-34

41. Two blocks are connected as shown in Fig. 11-34 above. The coefficient of friction between block A and the plane is 0.30. The pulley is frictionless. Determine the tension in the rope. *Ans.* $T = 13.0$ lb

42. B rests on A which is being pulled by force P to the right as shown in Fig. 11-35. If the coefficient of friction between A and B is μ and the floor is smooth, determine the expression for the acceleration a that will cause slip between A and B. *Ans.* $a = \mu g$

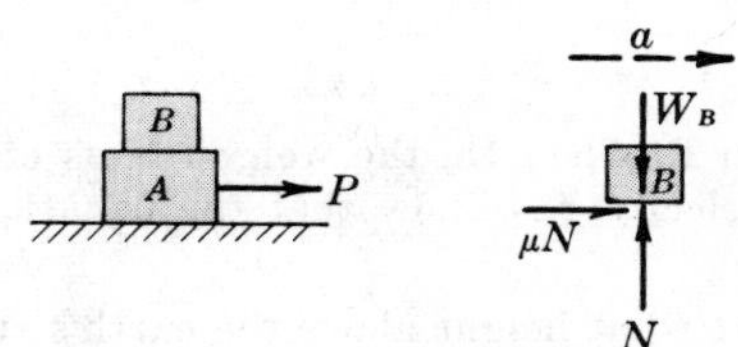

Fig. 11-35

43. A 20 lb box is dropped onto the body of a truck moving 30 mph horizontally. If the coefficient of friction is 0.5, calculate how far the truck will move before the box stops slipping. *Ans.* $s = 120$ ft

44. The 90 and 80 lb weights are attached to ropes passing over pulleys as shown in Fig. 11-36 below. Neglecting the weight of the pulleys and cords and assuming no friction, compute the tensions in the cords. Note that the length of the cord passing over the pulleys is constant.
$x_2 + (x_2 - c) + x_1 =$ constant $-$ 2 half circumferences.
Ans. $T_1 = 52.8$ lb and $T_2 = 106$ lb

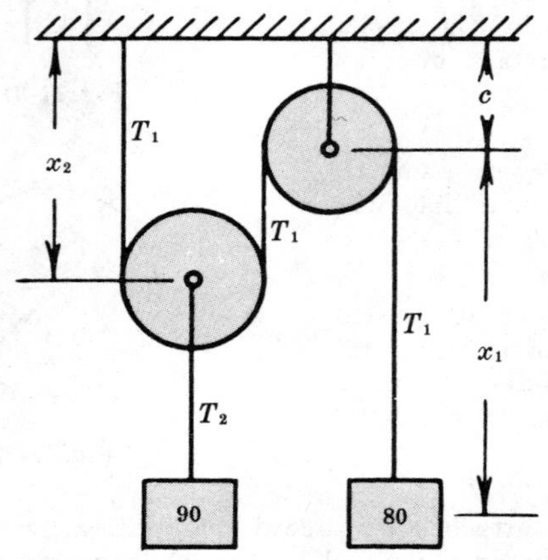

Fig. 11-36

Fig. 11-37

45. In the system shown in Fig. 11-37 above, the pulleys are to be considered weightless and frictionless. The weights in lb are the numbers 1, 2, 3 and 4. Determine the acceleration of each weight and the tension in either moving cord.
Ans. $\ddot{x}_1 = 29.6$ ft/sec² up
$\ddot{x}_2 = 1.29$ ft/sec² down
$\ddot{x}_3 = 11.6$ ft/sec² down
$\ddot{x}_4 = 16.7$ ft/sec² down
$T = 1.92$ lb

46. Refer to Fig. 11-38 below. Weights A and B are 15 lb and 55 lb respectively. Assume the coefficient of friction between A and the plane is 0.25 and between B and the plane is 0.10. What is the force between the two as they slide down the incline? In the free body diagrams, P is the unknown force between A and B. *Ans.* $P = 1.35$ lb

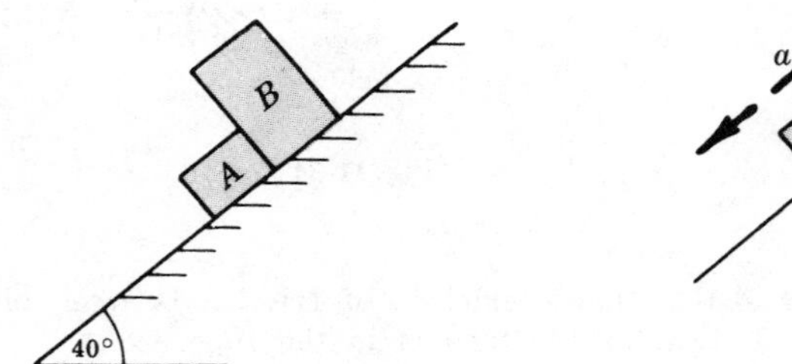

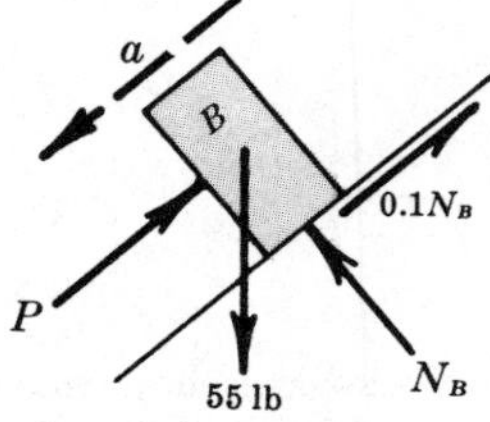

Fig. 11-38

47. In Problem 18, the weight W is displaced at a distance x_0 from equilibrium and released with initial velocity v_0. Show that the equation of motion is $x = v_0\sqrt{W/kg}\,\sin\sqrt{Kg/W}\,t + x_0 \cos\sqrt{Kg/W}\,t$.

48. At what height above the earth's surface must a 6 ft pendulum be in order to have a period of 2.8 sec? Assume $g = 32.2$ ft/sec/sec at the earth's surface and the gravitational force to vary inversely as the square of the distance from the earth's center. Use the radius of the earth as 4000 miles.
Ans. 130 miles

49. In Problem 21, assume the weight is 30 lb and the diameter of the loop is 40 ft. Determine the vertical reaction of the track on the weight when the weight is at the top of the loop for (a) $h = 50$ ft and (b) $h = 60$ ft. *Ans.* (a) $N = 0$, (b) $N = 30$ lb down

50. A chain 32.2 ft long is stretched with one half its length on a smooth horizontal table and the other half hanging freely. If the chain starts from rest, find the time for the chain to leave the table. See figure of Problem 22. *Ans.* $t = 1.32$ sec

51. A body weighing 3 lb falls in a medium where the resistance k is 0.05 lb-sec/ft. What is the terminal velocity? *Ans.* $\dot{x}_{max} = 60$ ft/sec

52. A particle of mass m is projected vertically upward with a velocity v_0 in a medium whose resistance is kv. Determine the time for the particle to come to rest. *Ans.* $t = (m/k) \ln (1 + kv_0/mg)$

53. A particle of mass m is projected vertically upward with a velocity v_0 in a medium whose resistance is kv^2. Determine the time for the particle to come to rest. *Ans.* $t = \sqrt{m/kg} \tan^{-1} v_0\sqrt{k/mg}$

54. Refer to Fig. 11-39. An airplane flying 250 mi/hr horizontally, accidentally loses a rivet when it is 6000 ft above the ground. Determine the location of point B where the rivet will land if air resistance is neglected. *Ans.* $AB = 7080$ ft

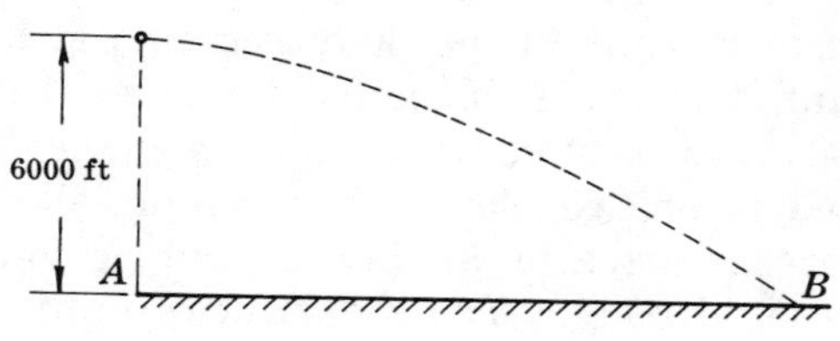

Fig. 11-39

55. A toy launcher projects a missile with an initial speed of 60 ft/sec. If the missile lands 40 ft away at the same level, what must have been the angle of elevation of the launcher?
Ans. $\theta = 10.5°$ or $55.5°$

56. A ball thrown horizontally from the top of a 150 ft high building hits the horizontal ground 60 ft from the base of the building. What was the initial velocity of the ball?
Ans. $v = 19.7$ ft/sec horizontally

57. A ball is projected from A with a speed of 10 ft/sec at an angle of 25° as shown in Fig. 11-40. Determine the coordinates of point B at which the ball will hit the plane which is 25° below the horizontal.
Ans. $x = +4.74$ ft, $y = -2.21$ ft

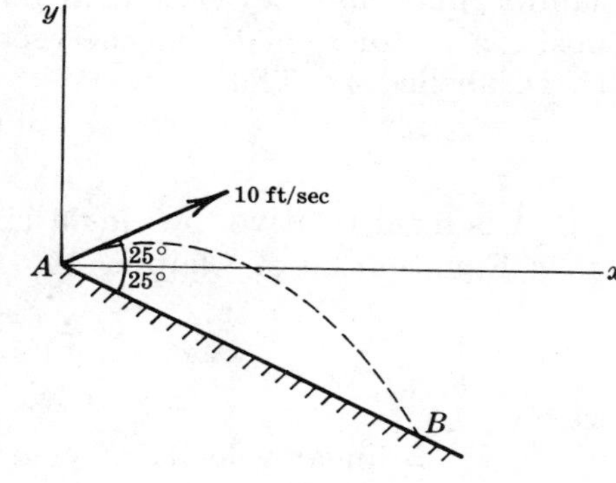

Fig. 11-40

58. A ball thrown with a speed of 40 ft/sec at an angle of 60° with the building strikes the ground 37.2 ft horizontally from the foot of the building. Determine the height of the building. See Fig. 11-41. *Ans.* $h = 40.0$ ft

59. The planet Mars at its apogee in its orbit is 154.8×10^6 miles from the Sun. At its perigee it is 128.8×10^6 miles away. Show that the time for one complete revolution by the methods of this chapter is close to the actual value of 687 days.

60. The Russian satellite Sputnik I weighed 184 lb and orbited on a path around the earth that varied between 140 and 560 miles above the surface of the earth. Using the radius of the earth as 3960 miles, show that the time for a revolution is about 1.5 hours.

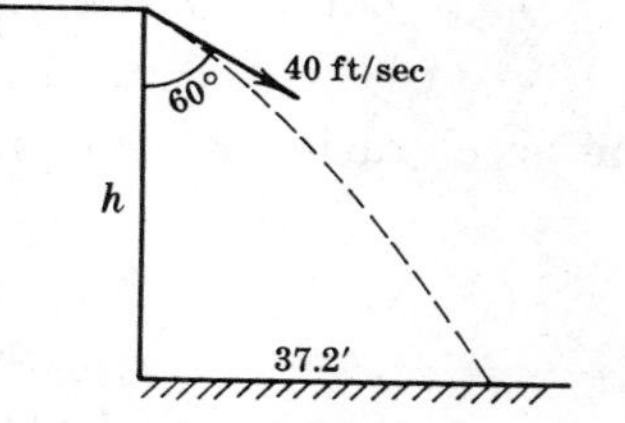

Fig. 11-41

61. The spherical plastic balloon Satellite Echo I was orbited on a path that varied from 945 to 1049 miles from the earth's surface. Show the time for a revolution was initially about 2 hours.

62. At what point on a journey from the Earth to the moon will the attractive forces of the two masses on the space ship be equal? Use the mass of the moon as 0.012 that of the mass of the Earth and the distance from the Earth to the moon as 239,000 miles. *Ans.* $d = 216{,}000$ miles from Earth's center

Chapter 12

Kinematics of a Rigid Body in Plane Motion

PLANE MOTION of a RIGID BODY

Plane motion of a rigid body takes place if every point in the body remains at a constant distance from a fixed plane. In Fig. 12-1 it is assumed that the xy plane is the fixed reference plane. The lamina shown is representative of all lamina which compose the rigid body. The z distance to any point in the lamina remains constant as the lamina moves.

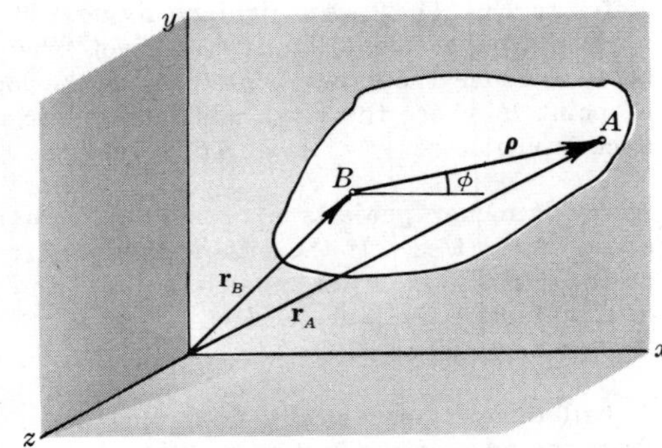

Fig. 12-1

It is customary to select an arbitrary point B in the lamina as a base point or pole.

The position vector $\mathbf{r}_A$ of any point A in the lamina may now be written in terms of the position vector $\mathbf{r}_B$ of B and the vector **BA** which is labelled $\boldsymbol{\rho}$. Thus

$$\mathbf{r}_A = \mathbf{r}_B + \boldsymbol{\rho}$$

If A is fixed relative to B (as a rigid body implies) the time rate of change of $\boldsymbol{\rho}$ is that given in Equation 20 of Chapter 10. Thus we can write

$$\mathbf{v}_A = \dot{\mathbf{r}}_A = \dot{\mathbf{r}}_B + \dot{\boldsymbol{\rho}} = \dot{\mathbf{r}}_B + \rho\omega\mathbf{e}_\phi \tag{1}$$

where

$\dot{\mathbf{r}}_B$ = linear velocity of pole B relative to the fixed axes x, y, and z

ω = magnitude of the angular velocity of $\boldsymbol{\rho}$ about any line parallel to the z-axis

$\mathbf{e}_\phi$ = unit vector perpendicular to $\boldsymbol{\rho}$ and in the direction of increasing ϕ (as indicated by the right hand rule this is counterclockwise about the z-axis.

The acceleration $\mathbf{a}_A$ may next be found by applying equation (*21*) in Chapter 10,

$$\mathbf{a}_A = \dot{\mathbf{v}}_A = \ddot{\mathbf{r}}_A = \ddot{\mathbf{r}}_B - \rho\omega^2\mathbf{e}_r + \rho\alpha\mathbf{e}_\phi \tag{2}$$

where

$\ddot{\mathbf{r}}_B$ = linear acceleration of pole B relative to the fixed axes x, y, and z

$\mathbf{e}_r$ = unit vector along $\boldsymbol{\rho}$ directed from B towards A

$\mathbf{e}_\phi$ = unit vector as specified in equation (*1*) above

α = magnitude of the angular acceleration of $\boldsymbol{\rho}$ about any line parallel to the z-axis.

An alternate method of writing equations (*1*) and (*2*) may be evolved as follows.

$$\boldsymbol{\omega} = \dot{\phi}\mathbf{k} = \omega\mathbf{k} \qquad \text{and} \qquad \boldsymbol{\alpha} = \ddot{\phi}\mathbf{k} = \dot{\omega}\mathbf{k} = \alpha\mathbf{k}$$

where ω and α will be positive if they are in a counterclockwise direction about the z-axis as viewed from the positive end of the axis (right hand rule).

Then in equation (*1*) the term $\rho\omega\mathbf{e}_\phi$ may be replaced by the cross product $\boldsymbol{\omega}\times\boldsymbol{\rho}$ which is identical with it. (See Problem 1.) Likewise in equation (*2*) the component of the acceleration $\rho\alpha\mathbf{e}_\phi$ is equivalent to $\boldsymbol{\alpha}\times\boldsymbol{\rho}$. The component $-\rho\omega^2\mathbf{e}_r$ is identical with $\boldsymbol{\omega}\times(\boldsymbol{\omega}\times\boldsymbol{\rho})$. These substitutions give

$$\mathbf{v}_A = \mathbf{v}_B + \boldsymbol{\omega}\times\boldsymbol{\rho} \tag{3}$$

$$\mathbf{a}_A = \mathbf{a}_B + \boldsymbol{\omega}\times(\boldsymbol{\omega}\times\boldsymbol{\rho}) + \boldsymbol{\alpha}\times\boldsymbol{\rho} \tag{4}$$

Equations (*3*) and (*4*) may also be written

$$\mathbf{v}_A = \mathbf{v}_B + \mathbf{v}_{A/B} \tag{5}$$

$$\mathbf{a}_A = \mathbf{a}_B + \mathbf{a}_{A/B} \tag{6}$$

where $\mathbf{v}_{A/B}$ and $\mathbf{a}_{A/B}$ are the relative velocity and relative acceleration that A possesses as it rotates around B, which as we have seen is moving with respect to the x, y, and z reference frame.

TRANSLATION

Translation is motion in which the line $\boldsymbol{\rho}$ from B to A does not rotate. Thus as the lamina moves, every straight line in the lamina is always parallel to its original direction.

ROTATION

Rotation is motion in which the base point B is fixed. Extending this to the rigid body, there is a line through B parallel to the z-axis which is fixed in the x, y, and z system. To describe this motion the velocity $\mathbf{v}_B$ and acceleration $\mathbf{a}_B$ in equations (*1*) and (*2*) or in equations (*3*) and (*4*) are equated to zero.

INSTANTANEOUS AXIS of ROTATION

The instantaneous axis of rotation is that line in a body in plane motion (or the body extended) that is the locus of points of zero velocity. It is perpendicular to the plane of motion (parallel to the z-axis in our system). All other points in the rigid body rotate about that line at that *instant*. It is important to realize that the position of this line of zero velocity in general changes continuously. To locate the instant center I for a lamina which has an angular velocity $\boldsymbol{\omega}$ write the velocity expressions for any two

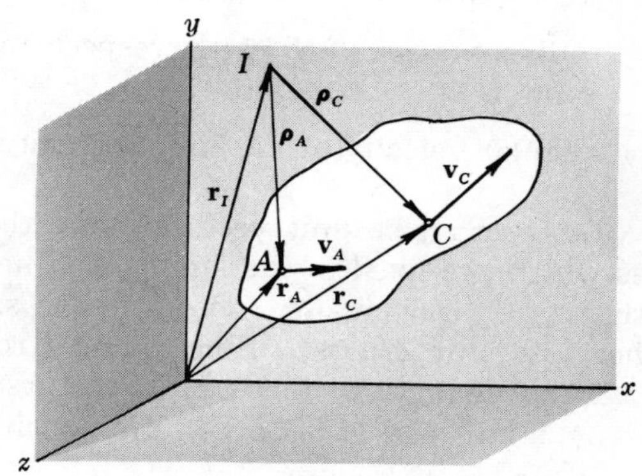

Fig. 12-2

points A and C in the body in terms of I as the base point. (See Fig. 12-2.) Thus

$$\mathbf{v}_A = \mathbf{v}_I + \boldsymbol{\omega} \times \boldsymbol{\rho}_A$$
$$\mathbf{v}_C = \mathbf{v}_I + \boldsymbol{\omega} \times \boldsymbol{\rho}_C$$

But $\mathbf{v}_I$ is zero because I is the instant center. Hence $\mathbf{v}_A = \boldsymbol{\omega} \times \boldsymbol{\rho}_A$ and $\mathbf{v}_C = \boldsymbol{\omega} \times \boldsymbol{\rho}_C$. These equations mean $\boldsymbol{\rho}_A$ is perpendicular to $\mathbf{v}_A$ (I is on $\boldsymbol{\rho}_A$) and that $\boldsymbol{\rho}_C$ is perpendicular to $\mathbf{v}_C$ (I is on $\boldsymbol{\rho}_C$). Therefore the instant center I is the intersection of the perpendiculars to $\mathbf{v}_A$ and $\mathbf{v}_C$.

CORIOLIS' LAW

Coriolis' law applies to the motion of a particle P along a path in the lamina as the lamina (and hence also the path) moves. The total acceleration

$$\mathbf{a}_P = \mathbf{a}_{P/\text{path}} + \mathbf{a}_M + 2\boldsymbol{\omega} \times \mathbf{v}_{P/\text{path}} \tag{7}$$

where $\mathbf{a}_{P/\text{path}}$ = acceleration of P relative to the path considered as fixed – use components tangent and normal to the path.

$\mathbf{a}_M$ = acceleration of that point M which is on the path and with which P coincides at the instant.

$\mathbf{v}_{P/\text{path}}$ = velocity of P relative to the point M which is on the path and with which P coincides at the instant. This velocity can only be tangent to the path along which the particle is moving in the body.

$\boldsymbol{\omega}$ = angular velocity of the path (or lamina).

$2\boldsymbol{\omega} \times \mathbf{v}_{P/\text{path}}$ = Coriolis' component, the direction of which is obtained by visualizing the rotation in the plane of the lamina of $\mathbf{v}_{P/\text{path}}$ through a right angle in the same sense as $\boldsymbol{\omega}$.

The proof of Coriolis' law follows.

PROOF of CORIOLIS' LAW

In the lamina shown parallel to the fixed XY plane in Fig. 12-3, let

$\mathbf{R}$ = position vector of base point B

$\mathbf{r}$ = position vector of point P which is moving on some path on the lamina

$\boldsymbol{\rho}$ = radius vector of P with respect to the base point B

$\boldsymbol{\omega}$ = angular velocity of the lamina about the Z-axis.

Let $\mathbf{i}$ and $\mathbf{j}$ be unit vectors along the x and y axes which are fixed in the lamina and hence rotate with angular velocity $\omega\mathbf{k}$ (note $\boldsymbol{\omega}$ is the same about either z axis or Z axis). Since $\mathbf{i}$ and $\mathbf{j}$ rotate there is a time change in these unit vectors; thus $\dot{\mathbf{i}} = \omega\mathbf{j} = \boldsymbol{\omega} \times \mathbf{i}$ and $\dot{\mathbf{j}} = -\omega\mathbf{i} = \boldsymbol{\omega} \times \mathbf{j}$. Now write

$$\mathbf{r} = \mathbf{R} + \boldsymbol{\rho} \tag{8}$$

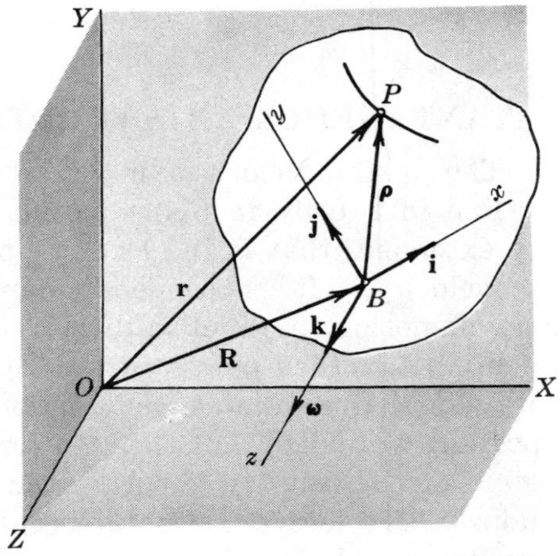

Fig. 12-3

Next take the time derivative of equation (8) obtaining

$$\dot{\mathbf{r}} = \dot{\mathbf{R}} + \dot{\boldsymbol{\rho}} \tag{9}$$

But in terms of the moving axes

$$\boldsymbol{\rho} = x\mathbf{i} + y\mathbf{j} \tag{10}$$

and

$$\dot{\boldsymbol{\rho}} = \dot{x}\mathbf{i} + \dot{y}\mathbf{j} + x\dot{\mathbf{i}} + y\dot{\mathbf{j}} = \mathbf{v}_{P/\text{path}} + \boldsymbol{\omega}\times\boldsymbol{\rho}$$

where $\dot{x}\mathbf{i} + \dot{y}\mathbf{j} = \mathbf{v}_{P/\text{path}}$ and $x\dot{\mathbf{i}} + y\dot{\mathbf{j}} = x(\boldsymbol{\omega}\times\mathbf{i}) + y(\boldsymbol{\omega}\times\mathbf{j}) = \boldsymbol{\omega}\times(x\mathbf{i}+y\mathbf{j}) = \boldsymbol{\omega}\times\boldsymbol{\rho}$.

Thus we can rewrite equation (9) as

$$\dot{\mathbf{r}} = \dot{\mathbf{R}} + \boldsymbol{\omega}\times\boldsymbol{\rho} + \mathbf{v}_{P/\text{path}} \tag{11}$$

Note that the first two terms on the right side of equation (11) give the absolute velocity of the point M which is fixed on the path but coincides with P at the instant (refer to equation (3)). Equation (11) may be expressed as

$$\mathbf{v}_P = \mathbf{v}_M + \mathbf{v}_{P/\text{path}} \tag{12}$$

To derive the expression for the acceleration of P, take the time derivative of equation (9) but express $\dot{\boldsymbol{\rho}}$ as $\dot{x}\mathbf{i} + \dot{y}\mathbf{j} + \boldsymbol{\omega}\times\boldsymbol{\rho}$.

$$\frac{d(\dot{\mathbf{r}})}{dt} = \frac{d(\dot{\mathbf{R}})}{dt} + \frac{d}{dt}(\dot{x}\mathbf{i} + \dot{y}\mathbf{j} + \boldsymbol{\omega}\times\boldsymbol{\rho})$$

Thus

$$\begin{aligned}\ddot{\mathbf{r}} &= \ddot{\mathbf{R}} + \ddot{x}\mathbf{i} + \ddot{y}\mathbf{j} + \dot{x}\dot{\mathbf{i}} + \dot{y}\dot{\mathbf{j}} + \dot{\boldsymbol{\omega}}\times\boldsymbol{\rho} + \boldsymbol{\omega}\times\dot{\boldsymbol{\rho}}\\ &= \ddot{\mathbf{R}} + \mathbf{a}_{P/\text{path}} + \boldsymbol{\omega}\times(\dot{x}\mathbf{i}+\dot{y}\mathbf{j}) + \boldsymbol{\alpha}\times\boldsymbol{\rho} + \boldsymbol{\omega}\times(\dot{x}\mathbf{i} + \dot{y}\mathbf{j} + \boldsymbol{\omega}\times\boldsymbol{\rho})\\ \ddot{\mathbf{r}} &= \ddot{\mathbf{R}} + \mathbf{a}_{P/\text{path}} + \boldsymbol{\omega}\times\mathbf{v}_{P/\text{path}} + \boldsymbol{\alpha}\times\boldsymbol{\rho} + \boldsymbol{\omega}\times\mathbf{v}_{P/\text{path}} + \boldsymbol{\omega}\times(\boldsymbol{\omega}\times\boldsymbol{\rho}) \qquad (13)\end{aligned}$$

The 1st, 4th, and 6th terms on the right side of equation (13) give the absolute acceleration of the point M which is fixed on the path and which coincides with P at the instant (refer to equation (4)). Equation (13) can be written

$$\ddot{\mathbf{r}} = \mathbf{a}_M + \mathbf{a}_{P/\text{path}} + 2\boldsymbol{\omega}\times\mathbf{v}_{P/\text{path}}$$

which is equivalent to equation (7).

Solved Problems

As in previous chapters, vectors in the diagrams are identified only by their magnitudes when the directions are evident by inspection.

1. Determine the linear velocity $\mathbf{v}$ of any point P in the rigid body rotating with angular velocity $\boldsymbol{\omega}$ about an axis as shown in Fig. 12-4.

Solution:

Select *any* reference point O on the axis of rotation. The radius vector $\boldsymbol{\rho}$ for point P relative to O is shown. The velocity vector $\mathbf{v}$ for point P is tangent to a circle which is in the plane perpendicular to the axis of rotation. The magnitude of the vector $\mathbf{v}$ is the product of the radius of the circle and the angular speed ω. From the figure it is apparent that the radius of the circle is $\rho \sin\theta$. Hence the magnitude of $\mathbf{v}$ is $\rho\omega \sin\theta$.

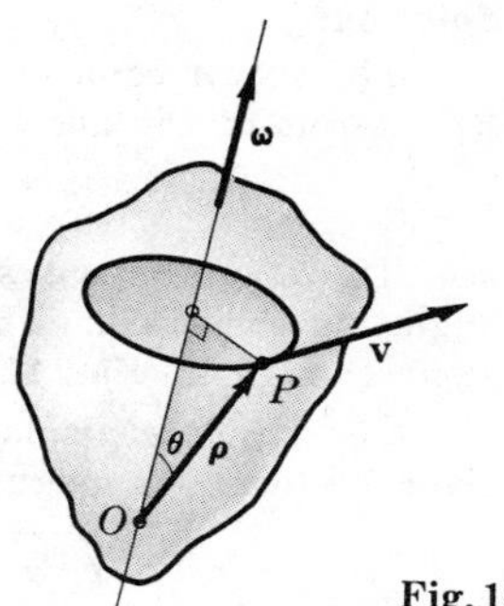

Fig. 12-4

By the definition of the cross product, $\boldsymbol{\omega} \times \boldsymbol{\rho}$ has this same magnitude $\rho\omega \sin\theta$ and is perpendicular to the plane containing $\boldsymbol{\rho}$ and $\boldsymbol{\omega}$. Hence we conclude

$$\mathbf{v} = \boldsymbol{\omega} \times \boldsymbol{\rho}$$

In the plane motion of the lamina shown at the beginning of this chapter, $\boldsymbol{\omega}$ is perpendicular to the plane of the lamina and $\boldsymbol{\rho}$ is in the plane of the lamina; hence $\mathbf{v} = \boldsymbol{\omega} \times \boldsymbol{\rho}$ is in the plane of the lamina and is perpendicular to $\boldsymbol{\rho}$.

2. A flywheel 18 inches in diameter is brought uniformly from rest up to a speed of 280 rpm in 15 sec. Find the velocity and acceleration of a point on the rim one second after starting from rest.

Solution:

This problem illustrates the application of the equations of rotation.

First determine the magnitude of the angular acceleration $\boldsymbol{\alpha}$ which obtains until operating speed is attained.

$$\alpha = \frac{\omega - \omega_0}{t} = \frac{2\pi(280/60)\text{ rad/sec} - 0}{15\text{ sec}} = 1.96\text{ rad/sec}^2$$

Next determine the angular speed of the wheel one sec after starting.

$$\omega_1 = \omega_0 + \alpha t = 0 + (1.96\text{ rad/sec}^2)(1\text{ sec}) = 1.96\text{ rad/sec}$$

The speed of a point on the rim is

$$v = r\omega = \tfrac{3}{4}\text{ ft} \times 1.96\text{ rad/sec} = 1.47\text{ ft/sec}$$

The magnitude of the normal acceleration of a point on the rim is

$$a_n = r\omega^2 = (\tfrac{3}{4}\text{ ft})(1.96\text{ rad/sec})^2 = 2.88\text{ ft/sec}^2$$

The magnitude of the tangential acceleration of a point on the rim is

$$a_t = r\alpha = (\tfrac{3}{4}\text{ ft})(1.96\text{ rad/sec}^2) = 1.47\text{ ft/sec}^2$$

The magnitude of the total acceleration of a point on the rim is

$$a = \sqrt{a_n^2 + a_t^2} = \sqrt{2.88^2 + 1.47^2} = 3.23\text{ ft/sec}^2$$

The angle between the total acceleration vector and the radius to the point is

$$\theta = \cos^{-1} a_n/a = \cos^{-1} 2.88/3.23 = 26.9°$$

Fig. 12-5 indicates the acceleration result.

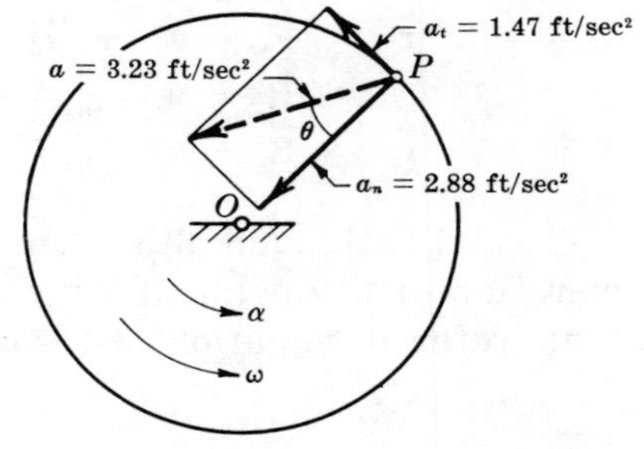

Fig. 12-5

3. A ball rolls 6 ft across a flat car in a direction perpendicular to the path of the car. In the same time interval during which the ball is rolling, the car moves at a constant speed on a horizontal straight track for a distance of 8 ft. What is the absolute displacement of the ball?

Solution:

The vector equation for the absolute displacement $\mathbf{s}_B$ of the ball B in terms of the absolute displacement $\mathbf{s}_C$ of the car C is

$$\mathbf{s}_B = \mathbf{s}_{B/C} + \mathbf{s}_C$$

The displacement $\mathbf{s}_{B/C}$ of the ball to the car is 6 ft at right angles to the track. The absolute displacement of the car is 8 ft along the track. Fig. 12-6 indicates these relations.

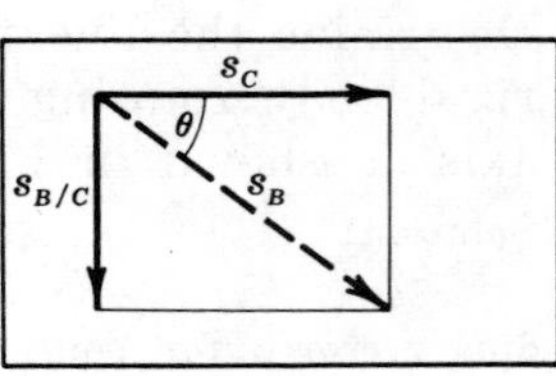

Fig. 12-6

The absolute displacement $\mathbf{s}_B$ of the ball is the sum of the two given vectors. Its magnitude is

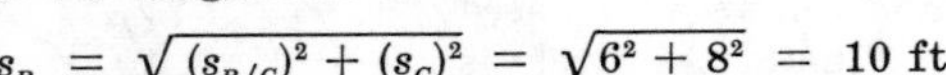

$$s_B = \sqrt{(s_{B/C})^2 + (s_C)^2} = \sqrt{6^2 + 8^2} = 10\text{ ft}$$

The angle which the vector $\mathbf{s}_B$ makes with the track is $\theta = \tan^{-1} s_{B/C}/s_C = \tan^{-1} 6/8 = 36.9°$.

4. Automobile A is traveling 20 mph along a straight road headed northwest. Automobile B is traveling 70 mph along a straight road headed 60° south of west. See Fig. 12-7(a) below. What is the relative velocity of A to B? Of B to A?

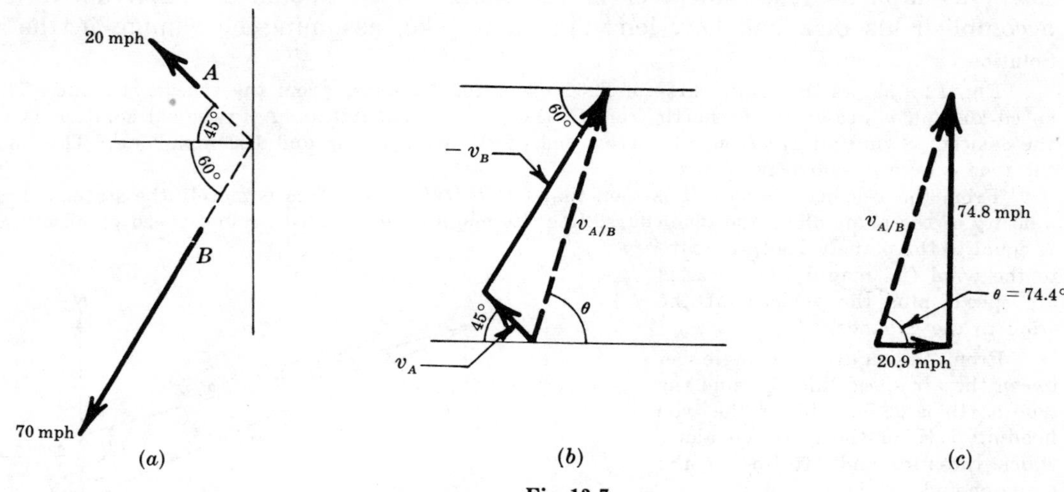

(a) (b) (c)

Fig. 12-7

Solution:

The vector equation relating the velocities is

$$\mathbf{v}_A = \mathbf{v}_{A/B} + \mathbf{v}_B$$

The equation may also be written as follows by subtracting $\mathbf{v}_B$ from both sides.

$$\mathbf{v}_{A/B} = \mathbf{v}_A - \mathbf{v}_B$$

The vector subtraction is performed by adding the negative of $\mathbf{v}_B$ to $\mathbf{v}_A$ as indicated in Fig. 12-7(b) above.

Use the following tabular setup to find the magnitudes of the x and y components of $\mathbf{v}_{A/B}$.

Vector	x component	y component
$\mathbf{v}_A$	-20×0.707	$+20 \times 0.707$
$-\mathbf{v}_B$	$+70 \times 0.500$	$+70 \times 0.866$

The magnitude of the x component of $\mathbf{v}_{A/B}$ is +20.9 mph and the magnitude of the y component is +74.8 mph. See Fig. 12-7(c) above. Then

$$v_{A/B} = \sqrt{(20.9)^2 + (74.8)^2} = 77.7 \text{ mph}$$

The angle which the $\mathbf{v}_{A/B}$ makes with the easterly direction is $\theta = \tan^{-1} 74.8/20.9 = 74.4°$.

To determine the velocity of B relative to A, it is necessary to subtract the $\mathbf{v}_A$ from $\mathbf{v}_B$ as indicated in the vector equation.

$$\mathbf{v}_{B/A} = \mathbf{v}_B - \mathbf{v}_A$$

Fig. 12-8 indicates the subtraction.

Again use a tabular setup.

Vector	x component	y component
$\mathbf{v}_B$	-70×0.500	-70×0.866
$-\mathbf{v}_A$	$+20 \times 0.707$	-20×0.707

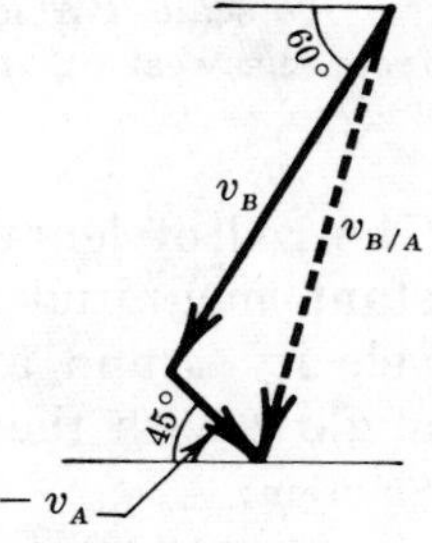

Fig. 12-8

The magnitude of the x component of $\mathbf{v}_{B/A}$ is −20.9 mph and the magnitude of the y component is −74.8 mph. It is apparent that $\mathbf{v}_{B/A}$ is the negative of $\mathbf{v}_{A/B}$.

5. An airplane pilot determines from a sectional map that his true course over the ground should be 295° to a destination 95 miles away. There is a south wind of 20 mph (blows from the south). His air speed (speed relative to the air which is moving north 20 mph as just indicated) is 120 mph. What should be his true heading to accomplish his task and how long should it take, assuming no change in the wind?

Solution:

Fig. 12-9 shows the true course of 295° measured clockwise from the true north line. The wind speed 20 mph is drawn to the north from O, the point of departure. A graphical solution is perhaps the easiest. Swing an arc from the arrow end of the wind vector and 120 mph long. This intersects the true course line in point C.

From the velocity triangle it is seen that $\mathbf{OC} = \mathbf{OW} + \mathbf{WC}$. This is merely the statement that the velocity of the plane along the true course line (its magnitude is called ground speed or absolute speed) is equal to the plane velocity relative to the wind (its magnitude is called air speed) plus the velocity of the wind to the ground.

From the figure the angle between the air speed line WC and the true north is 286°. This is the true heading T.H. or the direction along which the fore and aft line of the plane should be placed.

The ground speed OC scales 127 mph. Hence the estimated time of flight is 95 miles divided by 127 mph, or 45 minutes.

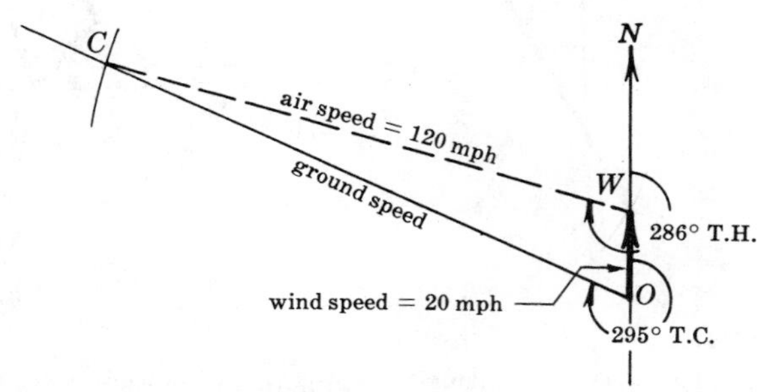

Fig. 12-9

6. Correcting for magnetic deviation and variation, a pilot calculates his true heading to be 58°. His airspeed is 250 mph. Several check points indicate that the plane is making good a true course of 63° at a ground speed of 295 mph. Determine the wind direction and magnitude.

Solution:

Draw the true heading line at an angle of 58° with the true north line as shown in Fig. 12-10. Along it draw $\mathbf{OP}$ to scale as 250 mph. Draw the true course line at an angle of 63° with the true north line through O. Along it draw $\mathbf{OW}$ to scale as the ground speed 295 mph. But $\mathbf{OW} = \mathbf{OP} + \mathbf{PW}$. This is interpreted as follows: the absolute velocity of the plane equals its velocity relative to the air plus the velocity of the air (wind vector).

To scale, $\mathbf{PW}$ is approximately a 50 mph wind from the west or from 270°.

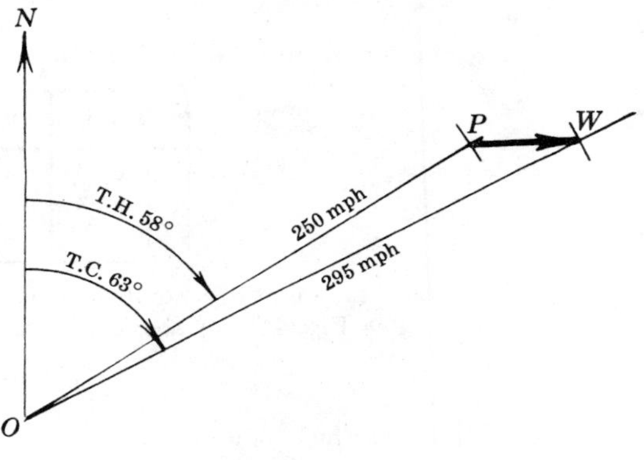

Fig. 12-10

7. The rod of length l moves so that the velocity of point A is of constant magnitude and directed to the left. Determine the angular velocity $\boldsymbol{\omega}$ and angular acceleration $\boldsymbol{\alpha}$ of the rod when it makes an angle θ with the vertical.

Solution:

The absolute velocity $\mathbf{v}_A$ of point A should be expressed in terms of the relative velocity of A to B ($\mathbf{v}_{A/B}$) because this term will introduce the desired quantity $\boldsymbol{\omega}$.

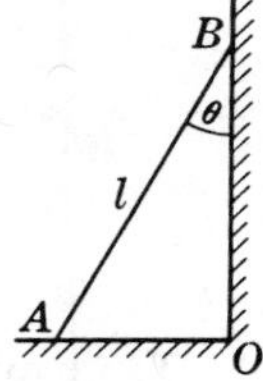

Fig. 12-11

$$\mathbf{v}_A = \mathbf{v}_{A/B} + \mathbf{v}_B$$

The table indicates the known parts of the vectors.

Vector	Direction	Magnitude
$\mathbf{v}_A$	horizontal	v_A
$\mathbf{v}_{A/B}$	$\perp$ rod	$l\omega$ (ω unknown)
$\mathbf{v}_B$	vertical	unknown

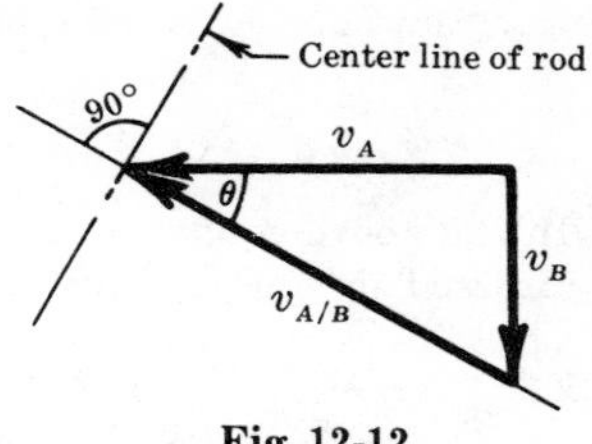

Fig. 12-12

The unknowns are the two magnitudes. Draw a vector triangle starting with $\mathbf{v}_A$ which is known completely. Through one end of $\mathbf{v}_A$ draw a line perpendicular to the rod, and through the other end draw a vertical line to close the triangle as shown in Fig. 12-12 above.

Label the vertical leg v_B and the leg perpendicular to the rod $v_{A/B}$ as indicated. In the right triangle, $v_{A/B} = v_A/\cos\theta$. But $v_{A/B} = l\omega$. Hence $\omega = v_{A/B}/l = v_A/l\cos\theta$.

Since $\mathbf{v}_{A/B}$ is directed up to the left, A must turn clockwise about B, i.e. $\boldsymbol{\omega}$ is clockwise.

Now determine the angular acceleration $\boldsymbol{\alpha}$. The magnitude of the tangential component of A relative to B is $l\alpha$. The equation is $\mathbf{a}_A = (\mathbf{a}_{A/B})_t + (\mathbf{a}_{A/B})_n + \mathbf{a}_B$.

Vector	Direction	Magnitude
$\mathbf{a}_A$	none	0, since v_A = constant
$(\mathbf{a}_{A/B})_t$	$\perp$ rod	$l\alpha$ (α unknown)
$(\mathbf{a}_{A/B})_n$	along the rod	$l\omega^2$ ($\omega = v_A/l\cos\theta$)
$\mathbf{a}_B$	vertical	unknown

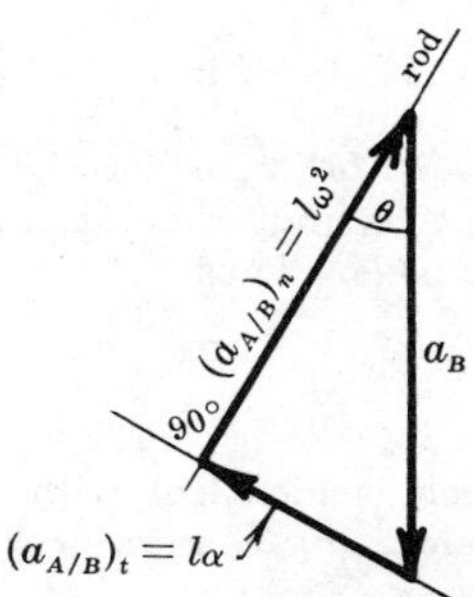

Fig. 12-13

The two unknowns are listed. The sum of the three vector quantities on the right side of the vector equation must equal zero. Start with the known $(\mathbf{a}_{A/B})_n$ and through its end points draw two lines, one perpendicular to the rod and the other vertical, i.e., making an angle θ with $(\mathbf{a}_{A/B})_n$. Note that the acceleration $(\mathbf{a}_{A/B})_n$ must be directed from A to B.

In this right triangle, $(a_{A/B})_t = l\alpha = l\omega^2\tan\theta$. Hence $\alpha = \omega^2\tan\theta = (v_A^2\tan\theta)/(l^2\cos^2\theta)$. This is clockwise because the tangential component $(\mathbf{a}_{A/B})_t$ indicates that A accelerates around B in a clockwise direction.

8. The ladder of length l makes an angle θ with the vertical wall, as shown in Fig. 12-14(*a*). The foot of the ladder moves to the right with constant speed v_A. Determine $\dot\theta$ and $\ddot\theta$ in terms of v_A, l and θ.

Solution:

Let $\mathbf{i}$ and $\mathbf{j}$ be unit vectors along and perpendicular to the ladder; they move with the ladder. In vector notation,

$$\mathbf{r}_A = \mathbf{r}_B + l\mathbf{i}$$

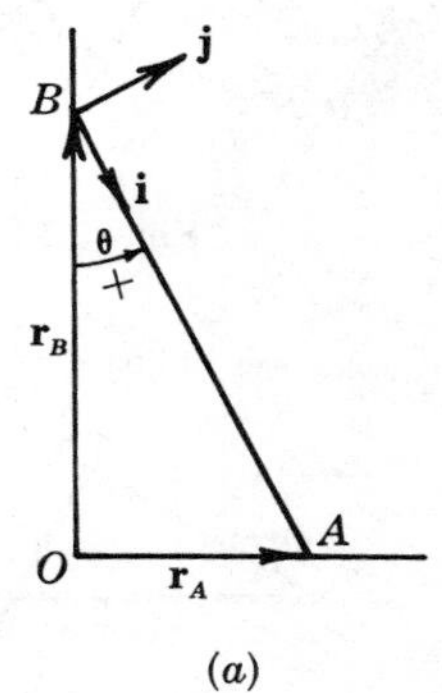

(*a*)

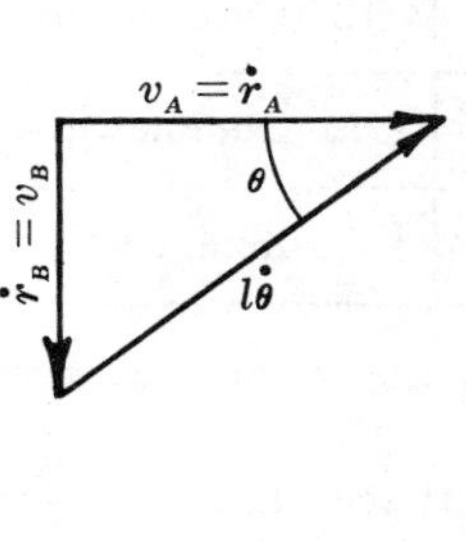

(*b*)

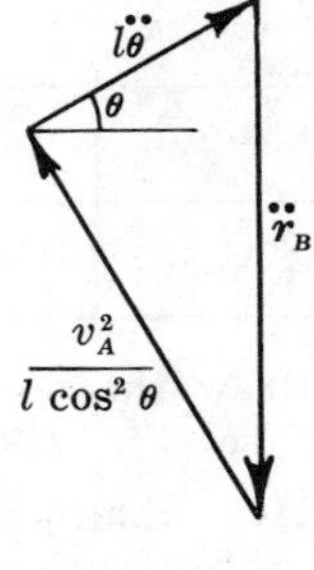

(*c*)

Fig. 12-14

Taking a time derivative, we have

$$\dot{\mathbf{r}}_A = \dot{\mathbf{r}}_B + l\dot{\mathbf{i}}$$

In the above equation $\dot{\mathbf{r}}_A$ is $\mathbf{v}_A$, known completely. Also, $\dot{\mathbf{i}}$ is $\dot{\theta}\mathbf{j}$ (the time derivative of a unit vector was stressed in Chapter 10), and $\dot{\mathbf{r}}_B$ can only be vertical. Fig. 12-14(*b*) shows this relationship. Thus

$$\cos\theta = \frac{v_A}{l\dot{\theta}} \quad \text{or} \quad \dot{\theta} = \frac{v_A}{l\cos\theta} \tag{1}$$

Since $l\dot{\theta}\mathbf{j}$ is in the positive $\mathbf{j}$ direction, $\dot{\theta}$ is positive (ladder is moving counterclockwise).

The derivative of equation (*1*) with respect to time is

$$\ddot{\theta} = \frac{v_A^2 \tan\theta}{l^2\cos^2\theta} \tag{2}$$

$\ddot{\theta}$ could also be obtained by differentiating $\dot{\mathbf{r}}_A = \dot{\mathbf{r}}_B + l\dot{\theta}\mathbf{j}$ which yields

$$\ddot{\mathbf{r}}_A = \ddot{\mathbf{r}}_B + l\ddot{\theta}\mathbf{j} - l\dot{\theta}^2\mathbf{i} \tag{3}$$

The last term in (*3*) is $l\dot{\theta}\dot{\mathbf{j}} = l\dot{\theta}(-\dot{\theta}\mathbf{i}) = -l\dot{\theta}^2\mathbf{i}$.

In (*3*), $\ddot{\mathbf{r}}_A$ is zero because $\dot{\mathbf{r}}_A$ is constant, $\ddot{\mathbf{r}}_B$ can only be vertical, the $\mathbf{j}$ component (magnitude $l\ddot{\theta}$) is positive, and the $\mathbf{i}$ component [magnitude $l\dot{\theta}^2 = v_A^2/(l\cos^2\theta)$] is negative. Fig. 12-14(*c*) shows these relations. Then

$$\tan\theta = \frac{l\ddot{\theta}}{v_A^2/(l\cos^2\theta)} \quad \text{or} \quad \ddot{\theta} = \frac{v_A^2\tan\theta}{l^2\cos^2\theta}$$

which is identical with equation (*2*). Since $l\ddot{\theta}\mathbf{j}$ is in the positive $\mathbf{j}$ direction, $\ddot{\theta}$ is positive (ladder is accelerating counterclockwise).

9. The 8 ft rod slides down the plane shown with $\mathbf{v}_A = 12$ ft/sec to the left and $\mathbf{a}_A = 15$ ft/sec² to the right. Determine the angular velocity $\boldsymbol{\omega}$ and the angular acceleration $\boldsymbol{\alpha}$ of the rod when $\theta = 30°$.

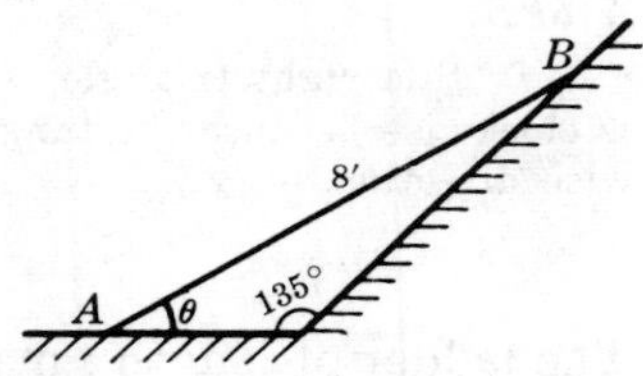

Fig. 12-15

Solution:

As in Problem 7,

$$\mathbf{v}_A = \mathbf{v}_{A/B} + \mathbf{v}_B$$

The table indicates what is known.

Vector	Direction	Magnitude
$\mathbf{v}_A$	horizontal	12 ft/sec to left
$\mathbf{v}_{A/B}$	⊥ rod	$l\omega$ (ω unknown)
$\mathbf{v}_B$	along the 45° line	unknown

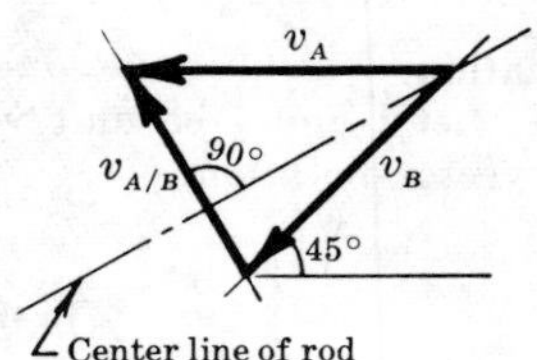

Fig. 12-16

Draw the vector triangle to fit the vector equation given above (see Fig. 12-16).

Measurement yields $v_{A/B} = 8.78$ ft/sec. Hence $\omega = v_{A/B}/l = 1.10$ rad/sec. $\boldsymbol{\omega}$ is clockwise.

To determine $\boldsymbol{\alpha}$, use the vector equation $\mathbf{a}_A = (\mathbf{a}_{A/B})_t + (\mathbf{a}_{A/B})_n + \mathbf{a}_B$ with the following table.

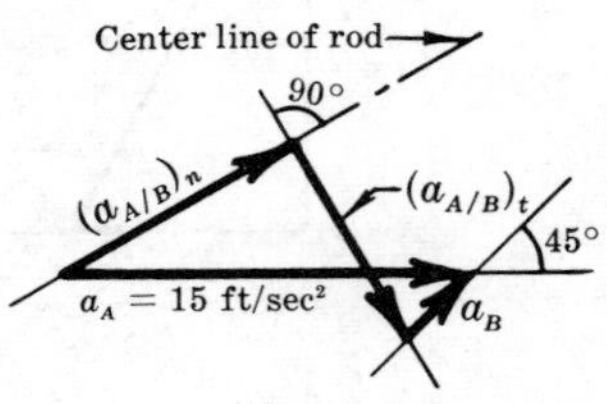

Fig. 12-17

Vector	Direction	Magnitude
$\mathbf{a}_A$	horizontal	15 ft/sec² to right
$(\mathbf{a}_{A/B})_t$	⊥ rod	$l\alpha$ (α unknown)
$(\mathbf{a}_{A/B})_n$	along the rod from A to B	$l\omega^2$ $8(1.10)^2 = 9.68$ ft/sec²
$\mathbf{a}_B$	along the 45° line	unknown

Draw the vector polygon to fit the above table. First draw $\mathbf{a}_A$. (See Fig. 12-17 above.) Then through the tail of this vector draw $(\mathbf{a}_{A/B})_n$. Through the head of the vector $\mathbf{a}_A$ draw a line along a 45° line, and through the head of the vector $(\mathbf{a}_{A/B})_n$ draw a perpendicular to the rod in the 30° phase.

The value of $(a_{A/B})_t$ is 8.4 ft/sec². Hence $\alpha = 8.4/8 = 1.05$ rad/sec². $\boldsymbol{\alpha}$ is counterclockwise.

10. In the slider crank mechanism shown in Fig. 12-18 below, the crank is rotating at a constant speed of 120 rpm. The connecting rod is 24 inches long and the crank is 4 inches long. For an angle of 30°, determine the absolute velocity of the crosshead P by the method of resolution.

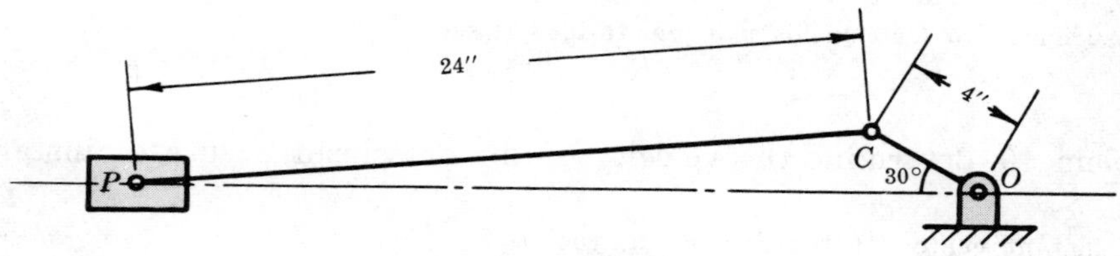

Fig. 12-18

Solution:

The angular speed of the crankpin C is 120 rpm. Then $\omega = 2\pi(120/60) = 4\pi$ rad/sec.

The linear speed of the crankpin C is therefore $v = r\omega = (4/12)(4\pi) = 4.19$ ft/sec.

Fig. 12-19 below indicates this velocity in a direction perpendicular to the crank.

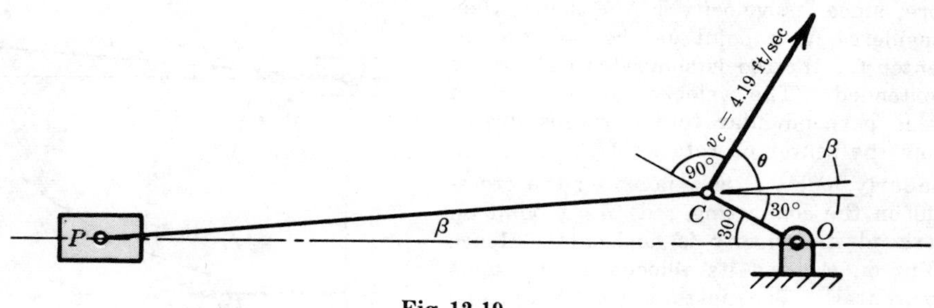

Fig. 12-19

The component of this velocity along the connecting rod PC will next be determined.

Angle θ must first be found. The figure indicates that $\theta = 90° - 30° - \beta$. But β may be found by using the sine law in the triangle PCO.

$$\frac{4}{\sin\beta} = \frac{24}{\sin 30°} \qquad \text{from which} \qquad \beta = 4.8° \qquad \text{and} \qquad \theta = 60° - 4.8° = 55.2°$$

Hence the component of the velocity of C along the connecting rod is $4.19 \cos 55.2° = 2.39$ ft/sec.

But all points on the connecting rod must have the same velocity along the rod, otherwise the rod would be either crushed or pulled apart. Hence point P, which is a point on the rod, has a velocity component of 2.39 ft/sec along the rod. However, its total velocity is along the line of travel of the crosshead. Then

$$v_P = \frac{2.39 \text{ ft/sec}}{\cos\beta} = \frac{2.39 \text{ ft/sec}}{0.9965} = 2.40 \text{ ft/sec}$$

11. In Problem 10, determine the velocity of the crosshead by graphical means.

Solution:

Use the vector equation $\mathbf{v}_P = \mathbf{v}_{P/C} + \mathbf{v}_C$.

In Fig. 12-19 above, $\mathbf{v}_C$ is shown perpendicular to the crank at C. Its length is 4.19 ft/sec.

The absolute velocity of point P is along the line of travel of the crosshead (horizontal in this figure). The velocity $\mathbf{v}_{P/C}$ of P to C is perpendicular to the line joining P and C (the connecting rod). The vector equation for this problem contains three vectors, each of course with magnitude and direction. If four of the six quantities (counting direction and magnitude as two quantities per vector) are known, the other two may be found.

Vector	Direction	Magnitude
$\mathbf{v}_P$	horizontal	?
$\mathbf{v}_{P/C}$	$\perp$ rod	?
$\mathbf{v}_C$	$\perp$ crank	4.19 ft/sec

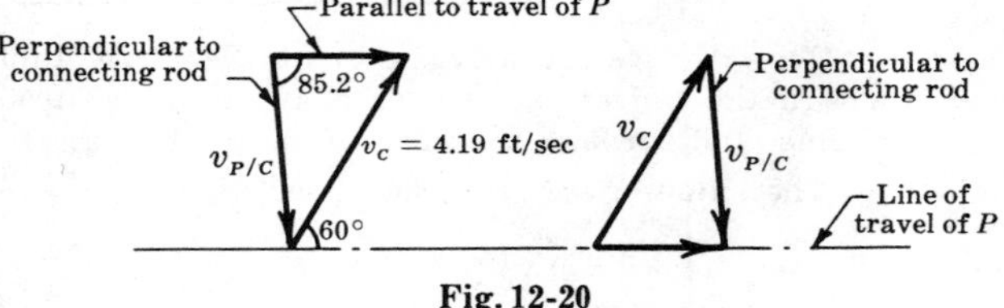

Fig. 12-20

The table indicates that the magnitude of $\mathbf{v}_P$ may be found, since four of the six quantities are known.

First draw the one vector $\mathbf{v}_C$ which is known both in magnitude and direction. Through one end of $\mathbf{v}_C$ draw a horizontal line and through the other end draw a line perpendicular to the connecting rod. The choice of ends is immaterial. Both are shown in Fig. 12-20 above.

Measurement to scale yields v_P equal to 2.40 ft/sec.

12. In Problem 10, determine the velocity of the crosshead by use of instant centers.

Solution:

The instant center of the connecting rod relative to the frame is the point about which all points in the rod appear to rotate at that instant. Point C is a point on the crank and on the rod. Its absolute velocity (relative to the frame) is the same whether it is a point on the crank or the rod. However, as a point on the crank its velocity is perpendicular to the crank. Therefore, since its velocity is the same when it is considered as a point on the rod, the instant center for the rod is somewhere along the crank extended. (The velocity of a point in rotation is perpendicular to the radius drawn to it from the center of rotation.)

Similarly, point P is a point on the crosshead and on the connecting rod. As a point on the crosshead its velocity is horizontal. Hence as a point on the rod its velocity is the same (i.e., horizontal). The instant (instantaneous) center for the rod is therefore on a perpendicular to the line of travel of the crosshead, i.e., on a vertical line through P.

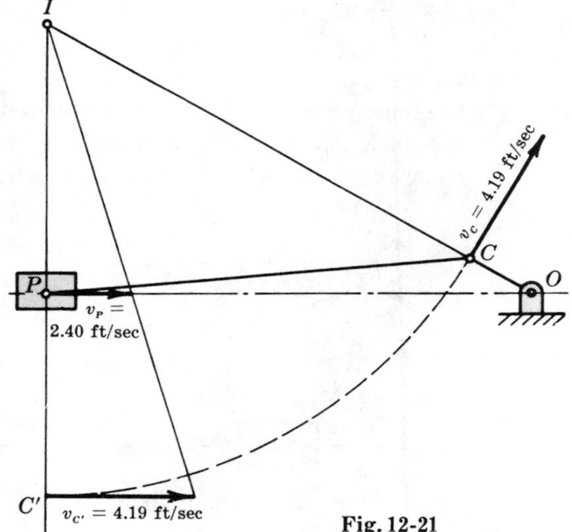

Fig. 12-21

The instant center I is at the intersection of the vertical line through P and the crank extended, as shown in Fig. 12-21 above.

Since I is the center of rotation for all points on the rod, it follows that the linear velocity of a particular point is perpendicular to the line joining I and the point, and the magnitude of the velocity is proportional to the distance of the point from I, i.e., center of rotation.

These facts are utilized in a graphical solution as follows. Swing an arc with radius IC until it intersects the line connecting I with the point whose velocity is desired, i.e., P. IC' equals IC. Draw the vector 4.19 ft/sec perpendicular to IC' at C'. Draw a gauge line from I to the tip of this vector. The velocity of point P is drawn from P perpendicular to IC' out to the gauge line. Its value is 2.40 ft/sec.

13. Determine the angular velocity of the connecting rod, referring to Problem 11.

Solution:

Direct measurement of the figure in Problem 11 yields $v_{P/C} = 3.68$ ft/sec.

Then the angular speed of the rod is found by dividing $v_{P/C}$ by the length of the rod.

$$\omega \quad = \quad \frac{v_{P/C}}{l} \quad = \quad \frac{3.68\text{ ft/sec}}{2\text{ ft}} \quad = \quad 1.84\text{ rad/sec.} \qquad \boldsymbol{\omega}\text{ is counterclockwise}$$

14. Referring to Problem 10, find the acceleration of the crosshead in the slider crank mechanism.

Solution:

Since the angular velocity of the crank is constant, the linear acceleration of point C consists of only the normal component directed toward the center O. Its magnitude is $r\omega^2$.

$$(a_C)_n \quad = \quad r\omega^2 \quad = \quad (4/12)(4\pi)^2 \quad = \quad 52.6\text{ ft/sec}^2$$

The acceleration of P which is horizontal is determined by the following vector equation.

$$(\mathbf{a}_P) \quad = \quad (\mathbf{a}_{P/C}) + (\mathbf{a}_C)$$

The acceleration $\mathbf{a}_{P/C}$ of point P relative to point C is one of rotation. It is well to write it in terms of its tangential and normal components, which are respectively perpendicular and parallel to the connecting rod. The equation now becomes

$$(\mathbf{a}_P) \quad = \quad (\mathbf{a}_{P/C})_n + (\mathbf{a}_{P/C})_t + (\mathbf{a}_C)$$

where n and t denote normal and tangential components. Eight elements are involved, and the equation may be solved if no more than two of these elements are unknown. Tabulate the elements as follows.

Vector	Direction	Magnitude
$\mathbf{a}_C$	along crank	52.6 ft/sec²
$(\mathbf{a}_{P/C})_t$	⊥ connecting rod	?
$(\mathbf{a}_{P/C})_n$	along connecting rod	(length of rod) × $(\omega_{\text{rod}})^2$
$\mathbf{a}_P$	horizontal	?

Note that the magnitude of the normal component $(a_{P/C})_n$ is actually

$$(a_{P/C})_n \quad = \quad (2\text{ ft})(1.84\text{ rad/sec})^2 \quad = \quad 6.77\text{ ft/sec}^2$$

The vector diagram is now drawn starting with the known vectors (both magnitude and direction) $\mathbf{a}_C$ and $(\mathbf{a}_{P/C})_n$ (see Fig. 12-22).

Through the tail end of $\mathbf{a}_C$ draw a horizontal line, and through the arrow end of $(\mathbf{a}_{P/C})_n$ draw a line perpendicular to the connecting rod. These lines meet in a point M which determines the length of $\mathbf{a}_P$ and $(\mathbf{a}_{P/C})_t$. By measurement, $a_P = 50.0$ ft/sec².

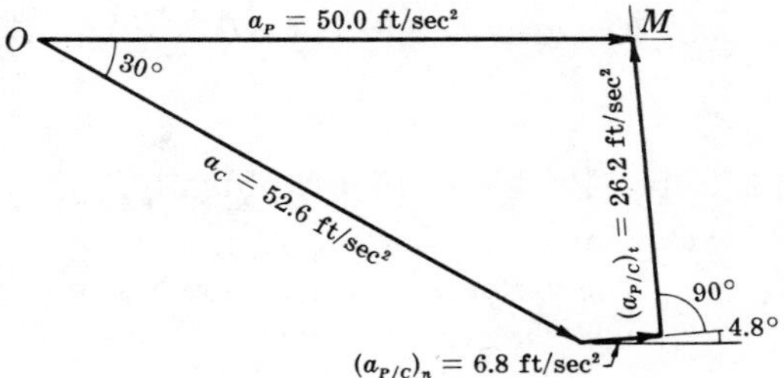

Fig. 12-22

15. If the angular velocity of BC is as shown in the quadric crank mechanism in Fig. 12-23 below, determine the absolute angular and linear velocities of point D for the phase indicated.

Solution:

The velocity of point C as a point on BC is perpendicular to BC and its magnitude is found to be

$$v_C \quad = \quad BC \times \omega_{BC} \quad = \quad (\tfrac{1}{2}\text{ ft})(10\text{ rad/sec}) \quad = \quad 5\text{ ft/sec}$$

To determine the velocity of D by resolution of velocities, first analyze Fig. 12-23 to find angles.

By the cosine law,

$$AC = \sqrt{(AB)^2 + (BC)^2 - 2AB \times BC \cos 45^\circ}$$
$$= \sqrt{10^2 + 6^2 - 2 \times 10 \times 6 \times \cos 45^\circ}$$
$$= 7.15 \text{ inches}$$

By the sine law,

$$\frac{BC}{\sin\beta} = \frac{AC}{\sin 45^\circ} = \frac{AB}{\sin\gamma}$$

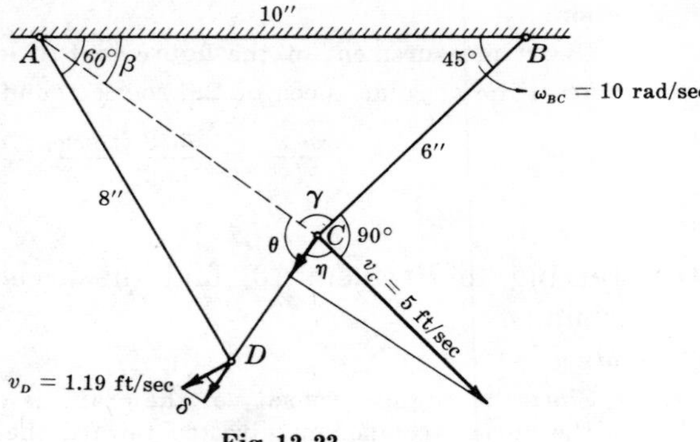

Fig. 12-23

Hence $\sin\beta = \dfrac{BC \sin 45^\circ}{AC} = \dfrac{6 \times 0.7071}{7.15} = 0.5934, \quad \beta = 36.4^\circ$

$\sin\gamma = \dfrac{AB \sin 45^\circ}{AC} = \dfrac{10 \times 0.7071}{7.15} = 0.9888, \quad \gamma = 98.6^\circ$ (γ being obtuse)

In triangle ADC, angle $DAC = 60^\circ - 36.4^\circ = 23.6^\circ$. Applying the cosine law in triangle ADC.

$$CD = \sqrt{(AD)^2 + (AC)^2 - 2AD \times AC \cos 23.6^\circ}$$
$$= \sqrt{8^2 + (7.15)^2 - 2 \times 8 \times 7.15 \times 0.9164} = 3.21 \text{ inches}$$

By the sine law, $\dfrac{CD}{\sin 23.6^\circ} = \dfrac{AD}{\sin\theta} = \dfrac{AC}{\sin D}$ or $\dfrac{3.21}{\sin 23.6^\circ} = \dfrac{8}{\sin\theta} = \dfrac{7.15}{\sin D}$.

Solving, $\sin\theta = 0.9980$ and $\sin D = 0.8922$, from which $\theta = 93.6^\circ$ and $D = 63.1^\circ$.

It is now evident that the angle η between CD and the velocity $\mathbf{v}_C$ is

$$\eta = 360^\circ - (90^\circ + \theta + \gamma) = 360^\circ - (90^\circ + 93.6^\circ + 98.6^\circ) = 77.8^\circ$$

The component of this velocity along the bar CD is $5 \cos 77.8^\circ = 1.06$ ft/sec. Note that this component is directed from C toward D. This is also the component of the velocity of D along CD.

Angle δ between velocity vector of D and bar CD is $\delta = 180^\circ - (90^\circ + 63.1^\circ) = 26.9^\circ$.

The magnitude of the velocity of D is $v_D = 1.06/(\cos 26.9^\circ) = 1.19$ ft/sec. Note that this velocity is directed such that the arm AD turns clockwise whereas the arm BC turns counterclockwise.

The angular speed of AD is $\omega_{AD} = \dfrac{v_D}{AD} = \dfrac{1.19 \text{ ft/sec}}{8/12 \text{ ft}} = 1.78$ rad/sec clockwise.

16. Solve Problem 15 graphically.

Solution:

The vector equation involved is $\mathbf{v}_D = \mathbf{v}_{D/C} + \mathbf{v}_C$. List the six components.

Vector	Direction	Magnitude
$\mathbf{v}_D$	$\perp AD$	?
$\mathbf{v}_{D/C}$	$\perp DC$	?
$\mathbf{v}_C$	$\perp BC$	5 ft/sec

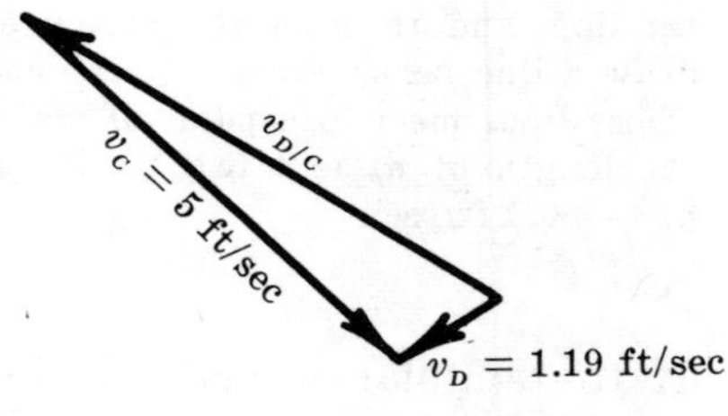

Fig. 12-24

A solution is possible, since only two components are unknown. Draw the vector $\mathbf{v}_C$, since it is known both in direction and magnitude. Through the ends of this vector draw lines parallel to the directions of $\mathbf{v}_D$ and $\mathbf{v}_{D/C}$ until they meet. This determines the velocity $\mathbf{v}_D$ when measured to the scale of the drawing. Hence $v_D = 1.19$ ft/sec.

17. Solve Problem 15 by use of instant centers.

Solution:

Points C and D are points on the arm CD. The instant center of CD relative to the frame is the intersection of the lines drawn perpendicular to the absolute velocities of C and D. The latter velocities, however, are perpendicular to BC and AD. Hence the instant center I is at the intersection of BC and AD as shown in Fig. 12-25 below.

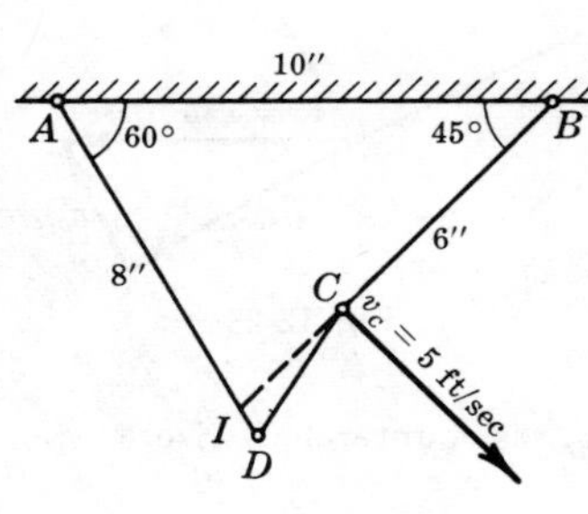

Fig. 12-25

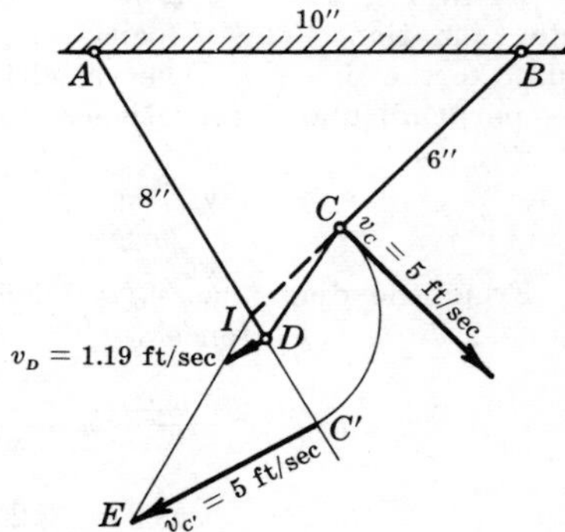

Fig. 12-26

The velocity of C is shown perpendicular to BC and of magnitude 5 ft/sec. In Fig. 12-26 above draw an arc with IC as radius and I as center cutting the line AD in C'.

Draw a vector at C' perpendicular to AD and of magnitude 5 ft/sec. Draw the gauge line IE. The velocity of D is found by erecting a perpendicular to AD at D. Its length to the gauge line is 1.19 ft/sec.

18. In the linkage shown in Fig. 12-27, determine the linear acceleration of point D with the following information given. At the instant, the angular velocity of BC is 6 rad/sec counterclockwise and its angular acceleration is 18 rad/sec² counterclockwise.

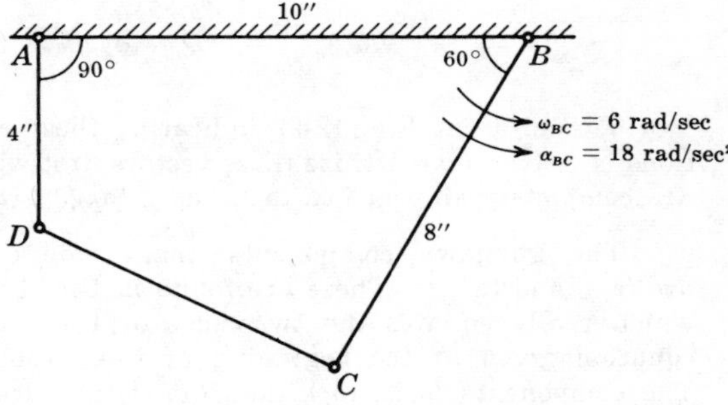

Fig. 12-27

Solution:

The vector equation which applies to this problem is

$$(\mathbf{a}_D)_t + (\mathbf{a}_D)_n = (\mathbf{a}_{D/C})_n + (\mathbf{a}_{D/C})_t + (\mathbf{a}_C)_n + (\mathbf{a}_C)_t$$

List the six vectors and determine what is unknown.

Vector	Direction	Magnitude
$(\mathbf{a}_C)_n$	from C to B	$BC \times (\omega_{BC})^2$
$(\mathbf{a}_C)_t$	$\perp BC$	$BC \times \alpha_{BC}$
$(\mathbf{a}_{D/C})_n$	from D to C	* $CD \times (\omega_{DC})^2$
$(\mathbf{a}_{D/C})_t$	$\perp CD$	?
$(\mathbf{a}_D)_n$	from D to A	* $AD \times (\omega_{AD})^2$
$(\mathbf{a}_D)_t$	$\perp AD$	?

*At the moment ω_{DC} and ω_{AD} are unknown, but may be found as shown in Fig. 12-28 below.

The velocities will be determined by the technique illustrated in Problem 16.

The speed of C is

$$v_C = r\omega = (8/12 \text{ ft})(6 \text{ rad/sec}) = 4 \text{ ft/sec}$$

Draw Fig. 12-28 to illustrate the following vector equation. The velocity $\mathbf{v}_{D/C}$ is perpendicular to the link DC. The velocities $\mathbf{v}_D$ and $\mathbf{v}_C$ are perpendicular respectively to links AD and BC.

$$\mathbf{v}_D = \mathbf{v}_{D/C} + \mathbf{v}_C$$

From the diagram, $v_D = 4.45$ ft/sec and $v_{D/C} = 2.25$ ft/sec. Hence

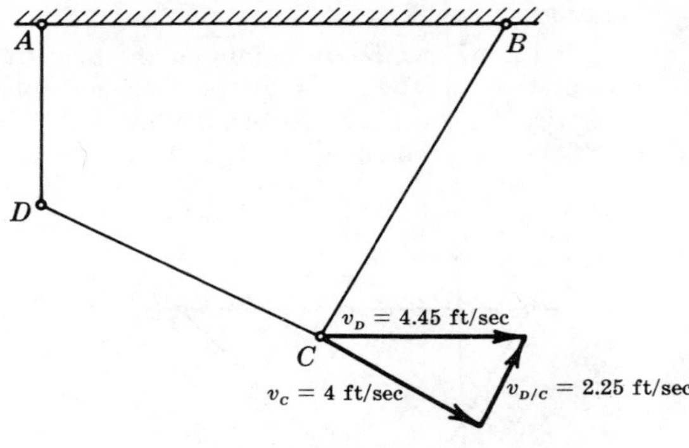

Fig. 12-28

$$\omega_{AD} = \frac{v_D}{r} = \frac{4.45 \text{ ft/sec}}{4/12 \text{ ft}} = 13.4 \text{ rad/sec.} \quad \boldsymbol{\omega}_{AD} \text{ is counterclockwise.}$$

$$\omega_{DC} = \frac{v_{D/C}}{DC} = \frac{2.25 \text{ ft/sec}}{6.65/12 \text{ ft}} = 4.07 \text{ rad/sec.} \quad \boldsymbol{\omega}_{DC} \text{ is clockwise.}$$

Note that the velocity $\mathbf{v}_{D/C}$ of D relative to C is upward and to the right; hence the angular velocity of link DC is clockwise.

Referring back to the tabular setup of the acceleration components, compute the four magnitudes as shown next.

$$(a_C)_n = BC \times (\omega_{BC})^2 = (8/12 \text{ ft})(6 \text{ rad/sec})^2 = 24.0 \text{ ft/sec}^2$$

$$(a_C)_t = BC \times \alpha_{BC} = (8/12 \text{ ft})(18 \text{ rad/sec}^2) = 12.0 \text{ ft/sec}^2$$

$$(a_{D/C})_n = CD \times (\omega_{DC})^2 = (6.65/12 \text{ ft})(4.07 \text{ rad/sec})^2 = 9.17 \text{ ft/sec}^2$$

$$(a_D)_n = AD \times (\omega_{AD})^2 = (4/12 \text{ ft})(13.4 \text{ rad/sec})^2 = 59.8 \text{ ft/sec}^2$$

The adjacent Fig. 12-29 indicating these relations is drawn next. Utilize those vectors first which are completely known, i.e., $(\mathbf{a}_C)_t$, $(\mathbf{a}_C)_n$, $(\mathbf{a}_{D/C})_n$, $(\mathbf{a}_D)_n$.

The unknown components (magnitude only) are $(\mathbf{a}_D)_t$ and $(\mathbf{a}_{D/C})_t$. These are found in Fig. 12-29, which really equates the two sides of the vector equation given at the beginning of this problem. The components $(\mathbf{a}_C)_n$, $(\mathbf{a}_C)_t$, and $(\mathbf{a}_{D/C})_n$ are drawn from O to M. The fourth vector on the right side of the equation is $(\mathbf{a}_{D/C})_t$. Since its direction is known to be perpendicular to the arm DC, draw a line in that direction through M.

Next draw through O the only completely known vector $(\mathbf{a}_D)_n$ on the left side of the equation. This vector terminates in N. However, the other vector $(\mathbf{a}_D)_t$ is perpendicular to the arm AD and must pass through N. Draw a horizontal line (perpendicular to AD) through N. The vector equation states that this line must terminate in the same point that the line through M does. This of course is their intersection P.

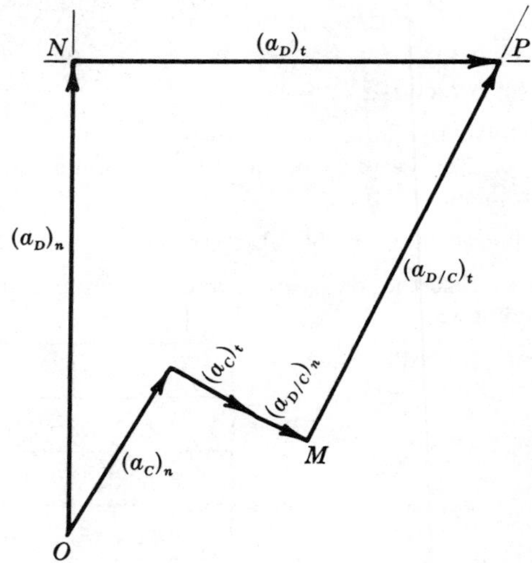

Fig. 12-29

Hence by measurement to scale of NP, it is found that $(a_D)_t = 55$ ft/sec^2. $(\mathbf{a}_D)_t$ is to the right.

The total acceleration of point D may be found by adding vectorially the normal and tangential components both of which are now known, i.e., by measuring to scale OP.

19. In Problem 18, what is the angular acceleration of the arm AD?

Solution:

By direct measurement the tangential component of the acceleration of D is 55 ft/sec². Then

$$\alpha_{AD} = \frac{(a_D)_t}{AD} = \frac{55\text{ ft/sec}^2}{4/12\text{ ft}} = 165\text{ rad/sec}^2$$

$\boldsymbol{\alpha}_{AD}$ is counterclockwise because the tangential component is directed to the right.

20. A wheel 10 ft in diameter rolls to the right on a horizontal plane with an angular velocity of 8 rad/sec (clockwise of course) and an angular acceleration counterclockwise of 4 rad/sec² as shown in Fig. 12-30 below. The latter merely indicates that the angular velocity of the wheel is decreasing. Determine the linear velocity and acceleration of the top point B on the wheel.

Solution:

Draw Fig. 12-31 illustrating the problem. For convenience the center O is chosen as the base point in this relative motion.

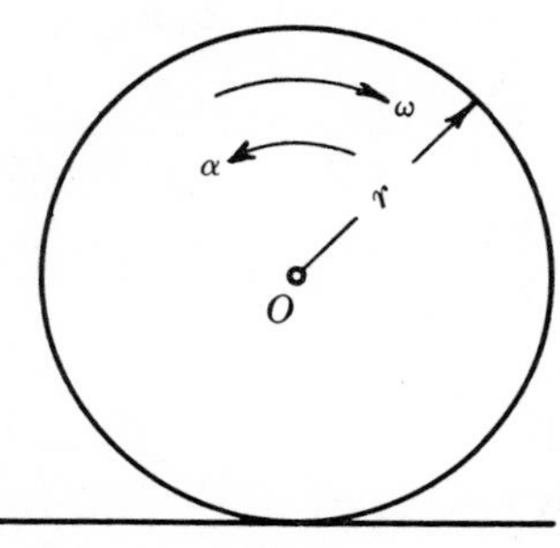

Fig. 12-30

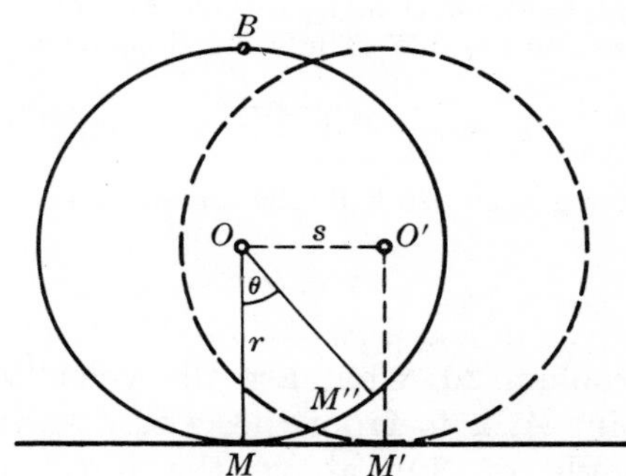

Fig. 12-31

It is necessary to show first that the velocity and acceleration of the center O of a rolling wheel may be expressed in terms of the given $\boldsymbol{\omega}$ and $\boldsymbol{\alpha}$ respectively and the distance from O to the surface on which the wheel rolls (5 ft in this case).

In Fig. 12-31, O is displaced to O' through a distance s. Since rolling takes place, MM'' on the wheel contacts MM' on the horizontal surface. Hence the arc MM'', which is $r\theta$, equals MM' which is equal to OO' or s.

Therefore $s = r\theta$, where s = magnitude of the linear displacement of center O

r = radius of wheel

θ = magnitude of the angular displacement of wheel.

Differentiation yields $v_O = r\dfrac{d\theta}{dt} = r\omega$ and $a_O = r\dfrac{d\omega}{dt} = r\alpha$.

These facts may now be applied to the present problem.

$$v_O = (5\text{ ft})(8\text{ rad/sec}) = 40\text{ ft/sec}$$
$$a_O = (5\text{ ft})(4\text{ rad/sec}^2) = 20\text{ ft/sec}^2$$

$\mathbf{v}_o$ is directed to the right and $\mathbf{a}_o$ is directed to the left.

The velocity vector equation which will be used is

$$\mathbf{v}_B = \mathbf{v}_{B/O} + \mathbf{v}_O$$

The velocity of B relative to O is perpendicular to the radius OB and to the right (since OB is moving clockwise). Its magnitude is

$$v_{B/O} = OB \times \omega = (5\text{ ft})(8\text{ rad/sec}) = 40\text{ ft/sec}$$

The absolute velocity $\mathbf{v}_B$ is therefore made up of two components ($\mathbf{v}_{B/O}$ and $\mathbf{v}_O$) each horizontal to the right and each 40 ft/sec. Hence $\mathbf{v}_B$ has magnitude 80 ft/sec and is directed horizontally to the right.

To determine the absolute acceleration $\mathbf{a}_B$ apply the vector equation

$$\mathbf{a}_B = (\mathbf{a}_{B/O})_t + (\mathbf{a}_{B/O})_n + \mathbf{a}_O$$

The relative acceleration $(\mathbf{a}_{B/O})_t$ is directed horizontally to the left (since the angular acceleration of OB is counterclockwise). Its magnitude is equal to OB times the magnitude of the angular acceleration $\boldsymbol{\alpha}$, or

$$(a_{B/O})_t = (5\text{ ft})(4\text{ rad/sec}^2) = 20\text{ ft/sec}^2\text{ (directed left)}$$

The $(a_{B/O})_n$ is directed toward O at the instant considered and is equal in magnitude to OB times the square of the angular speed ω.

$$(a_{B/O})_n = (5\text{ ft})(8\text{ rad/sec})^2 = 320\text{ ft/sec}^2\text{ (directed down)}$$

Fig. 12-32 will present these facts more clearly.

The acceleration of B may be found graphically or analytically. For an analytical solution, note that the horizontal component is $20 + 20 = 40$ ft/sec² to the left. The vertical component is 320 ft/sec² down. Therefore,

$$a_B = \sqrt{(a_h)^2 + (a_v)^2} = \sqrt{(40)^2 + (320)^2} = 323\text{ ft/sec}^2$$

and $\tan\phi = 40/320 = 0.1250$ or $\phi = 7.1°$.

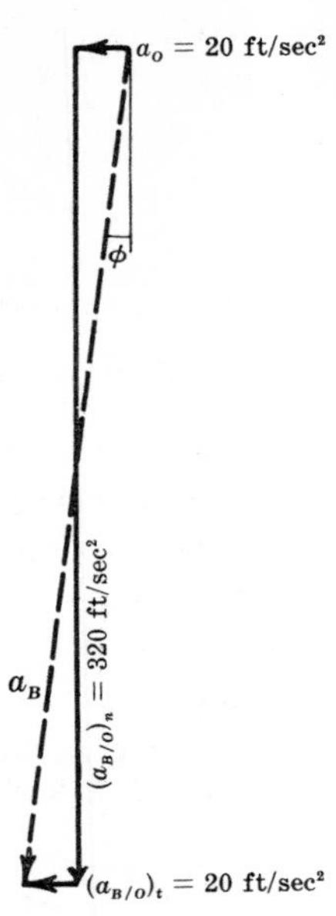

Fig. 12-32

21. In Problem 20, what are the velocity and acceleration of point A, 2 ft from the center and on a line making an angle of 30° above the horizontal radius? See Fig. 12-33.

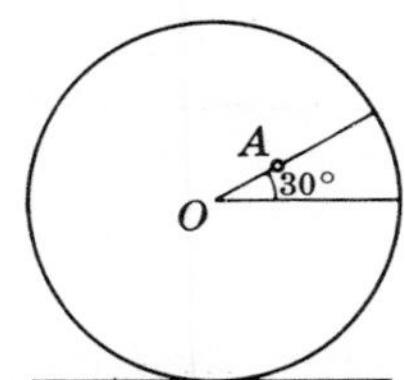

Fig. 12-33

Solution:

The usual vector equations apply.

$$(1)\quad \mathbf{v}_A = \mathbf{v}_{A/O} + \mathbf{v}_O$$

$$(2)\quad \mathbf{a}_A = (\mathbf{a}_{A/O})_t + (\mathbf{a}_{A/O})_n + \mathbf{a}_O$$

Tabulate the six elements in the velocity equation (1).

Vector	Direction	Magnitude
$\mathbf{v}_A$	?	?
$\mathbf{v}_{A/O}$	$\perp OA$	$OA \times \omega = 2\text{ ft} \times 8\text{ rad/sec} = 16\text{ ft/sec}$
$\mathbf{v}_O$	horizontal	40 ft/sec to the right

Fig. 12-34 for the velocity equation (1) follows.

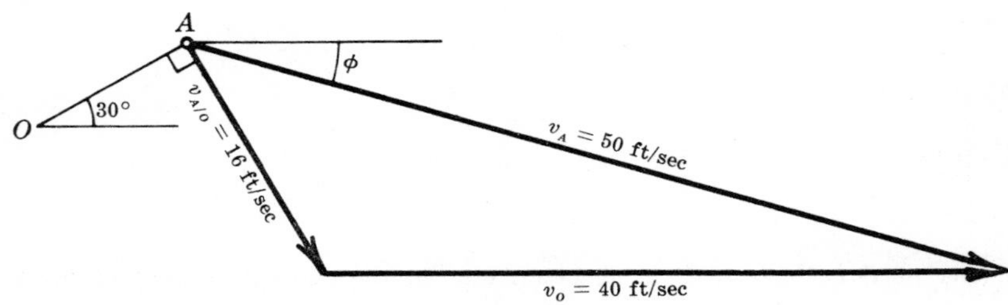

Fig. 12-34

The magnitude of the horizontal component of $\mathbf{v}_A = 16 \sin 30° + 40 = 48.0$ ft/sec.
The magnitude of the vertical component of $\mathbf{v}_A = 16 \cos 30° = 13.9$ ft/sec.
Hence $v_A = \sqrt{(48.0)^2 + (13.9)^2} = 50$ ft/sec and $\phi = \tan^{-1} 13.9/48 = 16.1°$.
Next tabulate the eight elements in the acceleration equation (*2*).

Vector	Direction	Magnitude
$\mathbf{a}_A$	?	?
$(\mathbf{a}_{A/O})_t$	$\perp OA$	$OA \times \alpha = 2 \text{ ft} \times 4 \text{ rad/sec}^2 = 8 \text{ ft/sec}^2$
$(\mathbf{a}_{A/O})_n$	along OA	$OA \times \omega^2 = 2 \text{ ft} \times (8 \text{ rad/sec})^2 = 128 \text{ ft/sec}^2$
$\mathbf{a}_O$	horizontal	20 ft/sec² to the left

Fig. 12-35 for equation (*2*) follows.

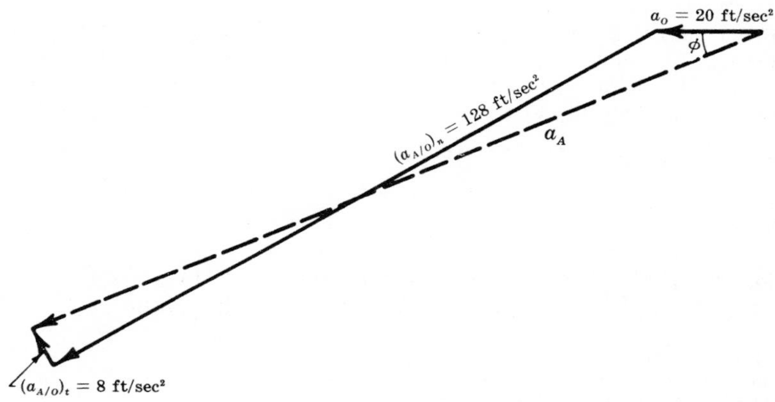

Fig. 12-35

To obtain the value of $\mathbf{a}_A$, note that the magnitudes of its components horizontally and vertically are

$$(a_A)_h = -20 - 128 \cos 30° - 8 \sin 30° = -135 \text{ ft/sec}^2$$
$$(a_A)_v = -128 \sin 30° + 8 \cos 30° = -57.1 \text{ ft/sec}^2$$

Hence $a_A = \sqrt{(-135)^2 + (-57.1)^2} = 147 \text{ ft/sec}^2$ with $\phi = \tan^{-1} 57.1/135 = 22.9°$.

Note that ϕ is below the negative x-axis.

22. The cylinder and axle roll under the influence of the weight W as shown in Fig. 12-36. What is the displacement s_O of the center of the cylinder when the weight is displaced 10 ft down? The pulley is assumed to work in frictionless bearings.

Solution:

Point I is the instant center between the cylinder and the surface on which it is rolling. Hence the magnitude of the absolute displacement of A, which is equal to that of the displacement of W, may be expressed as 5θ where the radius $AI = 5$ ft and θ = magnitude of the angular displacement. Since $5\theta = 10$, then $\theta = 2$ rad and $s_O = 3\theta = 6$ ft (directed to the right).

It should be evident upon further thought that the relative displacement of A to O is $10 - 6 = 4$ ft to the right. This means that 4 ft of cord is unwrapped from the axle during this displacement.

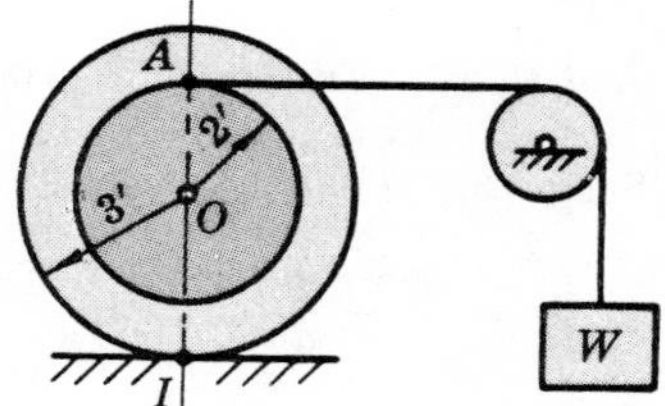

Fig. 12-36

23. Weight W is suspended from the pulley which turns in frictionless bearings. As the pulley turns, the cord AB from the axle of the cylinder is wrapped upon the pulley. The weight descends from rest with a constant acceleration of 16 ft/sec². Determine the displacement, velocity and acceleration of the center O of the cylinder after 3 sec. Cord AB is parallel to the inclined plane.

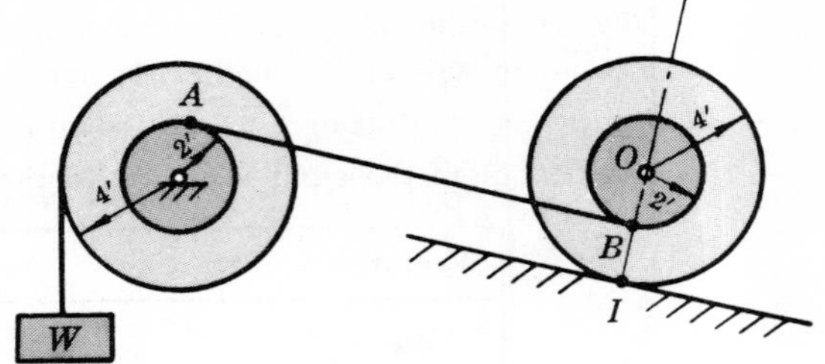

Fig. 12-37

Solution:

From Fig. 12-37 it is apparent that the magnitudes of the linear displacement, velocity and acceleration of a point on the cord AB will be one half of the corresponding values for weight W. For weight W, after 3 sec, $a = 16$ ft/sec²; $v = at = 16 \times 3 = 48$ ft/sec; and $s = \frac{1}{2}at^2 = \frac{1}{2} \times 16 \times 9 = 72$ ft. For point B which is on the cord AB and also on the cylinder, $s_B = 36$ ft; $v_B = 24$ ft/sec; and $a_B = 8$ ft/sec² after 3 sec.

Consider the motion of the cylinder about I, the instant center between the cylinder and the surface on which it is rolling. The magnitude of the angular displacement θ for any point on the cylinder is $\theta = s_B/IB = 36/2 = 18$ rad in the 3 sec interval. Thus $s_O = 18 \times 4 = 72$ ft. This means that 36 ft of cord will be wrapped on the axle of the cylinder as it rolls up the plane.

Similarly, $\omega = v_B/IB = 24/2 = 12$ rad/sec and $v_O = 4 \times 12 = 48$ ft/sec.

Also, $\alpha = (a_B)_t/IB = 8/2 = 4$ rad/sec² and $a_O = r\alpha = 4 \times 4 = 16$ ft/sec².

24. Solve Problem 23 if the cord AB is unwrapping from the top of the axle instead of the bottom. Move the pulley so that AB is parallel to the plane.

Solution:

The distance IB changes from 2 ft to 6 ft. The values of s_B, v_B, and a_B are the same. Hence, after 3 sec, $\theta = s_B/IB = 36/6 = 6$ rad, and $s_O = 4 \times 6 = 24$ ft. This means that 12 ft of cord will be unwrapped from the axle.

Similarly, $v_O = 16$ ft/sec and $a_O = 5.33$ ft/sec².

25. In Fig. 12-38, the cylinder of radius r rolls on the surface of radius R. Study the motion.

Solution:

Let OGP be the original position and $OG'B$ the position after some time has elapsed. Point P has then moved to point P' and since pure rolling is assumed, the arc BCP' of the cylinder must equal the arc PB on the surface.

The angle θ is the angular displacement of the line $G'P'$ with respect to its original position GP (or $G'C$ which is parallel to GP). The angle ϕ is the angular displacement of the line OGP in the same time interval. Hence $\theta + \phi$ is the total angular displacement of the line $G'P'$.

Arc BCP' = arc PB, or $r(\theta + \phi) = R\phi$.

Solve this for $\theta = \dfrac{R-r}{r}\phi$. Hence

$$\frac{d\theta}{dt} = \frac{R-r}{r}\frac{d\phi}{dt} \quad \text{and} \quad \frac{d^2\theta}{dt^2} = \frac{R-r}{r}\frac{d^2\phi}{dt^2}$$

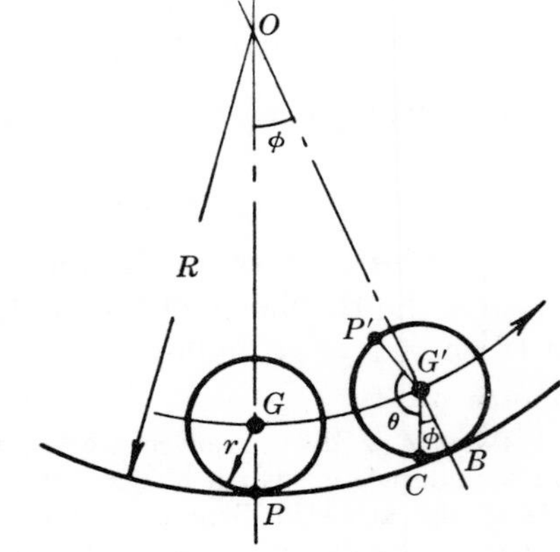

Fig. 12-38

If the linear speed of point G is v_G and the magnitude of its linear acceleration tangent to its circular path (of radius $R - r$) is $(a_G)_t$, the angular speed $d\phi/dt$ and angular acceleration magnitude $d^2\phi/dt^2$ of G are found as follows. Note again that G moves on a circular path of radius $R - r$.

$$\frac{d\phi}{dt} = \frac{v_G}{R-r} \quad \text{and} \quad \frac{d^2\phi}{dt^2} = \frac{(a_G)_t}{R-r}$$

The angular speed $d\theta/dt$ and angular acceleration magnitude $d^2\theta/dt^2$ of any point on the cylinder relative to its center are then

$$\frac{d\theta}{dt} = \frac{R-r}{r}\frac{d\phi}{dt} = \frac{v_G}{r} \quad \text{and} \quad \frac{d^2\theta}{dt^2} = \frac{(a_G)_t}{r}$$

The absolute velocity and the absolute acceleration of any point on the cylinder may now be found by referring the motion of the point first to the center and then adding the motion of the center.

For example, the velocity $\mathbf{v}_B$ of the contact point B equals the sum of the velocity of B relative to the center (magnitude is $r\,d\theta/dt$) and the velocity of the center $\mathbf{v}_G$. If the wheel is rolling up the plane, $d\theta/dt$ is clockwise and $r\,d\theta/dt$ is thus tangent to the cylinder and directed down to the left. Its magnitude is $r\,d\theta/dt$ or $r(v_G/r) = v_G$. Add this to the velocity of G which is of the same magnitude, parallel to it but directed up to the right. The absolute velocity of B is found to be zero, i.e., B is the instant center.

26. In the epicyclic gear train, the arm is moving 6 rad/sec clockwise and has an acceleration of 10 rad/sec² counterclockwise. Determine the linear velocity and acceleration of point B in the phase shown in Fig. 12-39.

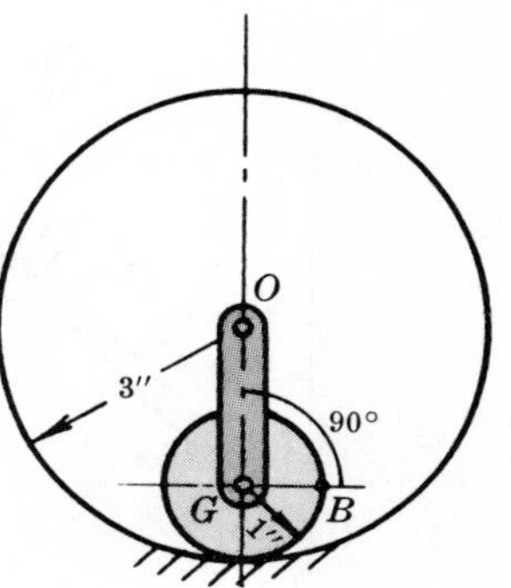

Fig. 12-39

Solution:

In the previous problem, it was shown that the magnitudes of the angular velocity $\boldsymbol{\omega}$ and angular acceleration $\boldsymbol{\alpha}$ of the small wheel about its own center are given by

$$\omega = \frac{d\theta}{dt} = \frac{R-r}{r}\frac{d\phi}{dt} \quad \text{and} \quad \frac{d^2\theta}{dt^2} = \frac{R-r}{r}\frac{d^2\phi}{dt^2}$$

where ϕ is the angular change of the arm OG. Hence $\omega = (3-1)6 = 12$ rad/sec (counterclockwise) and $\alpha = (3-1)10 = 20$ rad/sec² (clockwise). By inspection, clockwise motion of the arm means counterclockwise motion of the small gear.

The velocity $\mathbf{v}_B$ of point B is the vector sum of its relative velocity to G and the velocity of G. $\mathbf{v}_B = \mathbf{v}_{B/G} + \mathbf{v}_G$. The table follows.

Vector	Direction	Magnitude
$\mathbf{v}_B$	?	?
$\mathbf{v}_{B/G}$	vertically up	$r\omega = 12$ in/sec
$\mathbf{v}_G$	horizontally to left	$(R-r)6 = 12$ in/sec

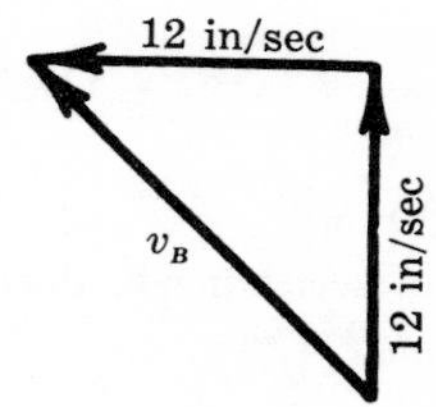

Fig. 12-40

The sum shown graphically in Fig. 12-40 is 17 in/sec.

The acceleration $\mathbf{a}_B$ of point B is the vector sum of its relative acceleration to G (two components: normal and tangent) and the acceleration of G (two components also).

$$\mathbf{a}_B = (\mathbf{a}_{B/G})_t + (\mathbf{a}_{B/G})_n + (\mathbf{a}_G)_t + (\mathbf{a}_G)_n$$

The table follows.

Vector	Direction	Magnitude
$(\mathbf{a}_{B/G})_t$	vertically down	$r\alpha = 20$ in/sec²
$(\mathbf{a}_{B/G})_n$	horizontally to left	$r\omega^2 = 144$ in/sec²
$(\mathbf{a}_G)_t$	horizontally to right	$(R-r)10 = 20$ in/sec²
$(\mathbf{a}_G)_n$	vertically up	$(R-r)6^2 = 72$ in/sec²

A vector diagram need not be drawn. The vertical summation is a net amount of $72 - 20 = 52$ in/sec² up. The horizontal summation is a net amount of 124 in/sec² to the left. The resultant absolute acceleration of B is thus to the left and up and of magnitude 134 in/sec².

Note: The absolute velocity $\mathbf{v}_B$ may be determined also by using the instant center, the common point of tangency of the two circles (see preceding problem). Although the absolute velocity of the instant center is zero, its absolute acceleration is not.

27. In Fig. 12-41 below, the washer is sliding outward on the rod with a velocity of 4 ft/sec when its distance from point O is 2 ft. Its velocity along the rod is increasing at the rate of 3 ft/sec². The angular velocity of the rod is 5 rad/sec counterclockwise and its angular acceleration is 10 rad/sec² clockwise. Determine the absolute acceleration of point P on the washer.

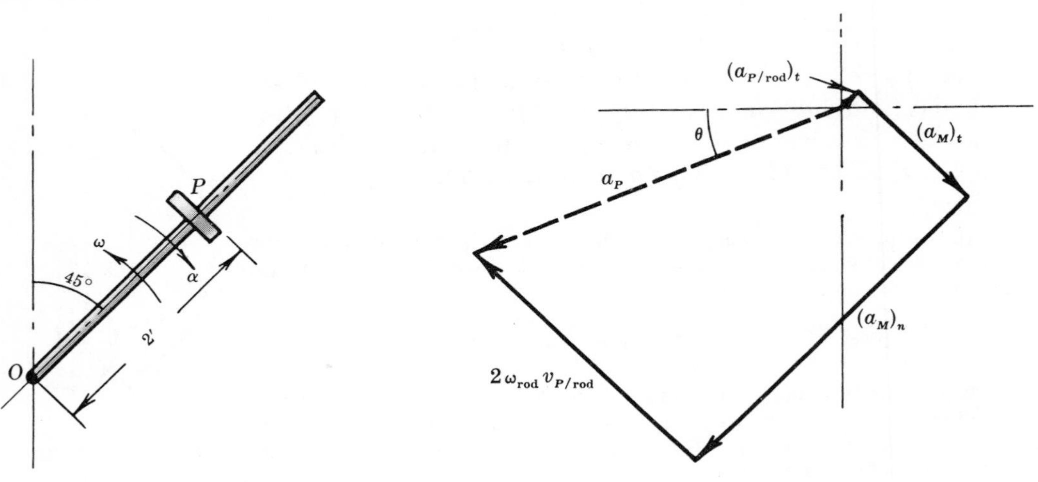

Fig. 12-41 **Fig. 12-42**

Solution:

According to Coriolis' law the absolute acceleration of P is expressed as

$$\mathbf{a}_P = (\mathbf{a}_{P/\text{rod}})_t + (\mathbf{a}_{P/\text{rod}})_n + (\mathbf{a}_M)_t + (\mathbf{a}_M)_n + 2\boldsymbol{\omega} \times \mathbf{v}_{P/\text{rod}}$$

where

$(\mathbf{a}_{P/\text{rod}})_t$ = acceleration of P along its path relative to the rod, i.e., 3 ft/sec² outward along the rod,

$(\mathbf{a}_{P/\text{rod}})_n$ = acceleration of P normal to its path along the rod, i.e., zero here since it is moving on a straight line path,

$(\mathbf{a}_M)_t$ = tangential component of the acceleration of the point M on the rod which coincides with P at the instant involved. Magnitude $r\alpha = 2(10) = 20$ ft/sec²; down to the right,

$(\mathbf{a}_M)_n$ = normal component of the acceleration of the point M on the rod which coincides with P at the instant involved. Magnitude $r\omega^2 = 2(5^2) = 50$ ft/sec²; along the rod towards O,

$2\boldsymbol{\omega} \times \mathbf{v}_{P/\text{rod}}$ = supplementary or Coriolis' acceleration with the magnitude indicated and in a direction obtained by rotating the vector $\mathbf{v}_{P/\text{rod}}$ through a right angle in the same sense as $\boldsymbol{\omega}$ (counterclockwise in this problem). Magnitude $2(5)(4) = 40$ ft/sec²; up to the left.

The vector diagram indicates each of these accelerations and their vector sum $\mathbf{a}_P$ (see Fig. 12-42 above).

Thus $a_P = 51$ ft/sec² with $\theta = 22°$.

28. One vane of an impeller wheel has its center of curvature C located as shown in Fig. 12-43 at a given instant. A particle P, 8 inches from the center O, has a velocity 10 in/sec and an acceleration 20 in/sec² directed outward and tangent to the vane. Determine the acceleration of P if the angular velocity of the wheel is 2 rad/sec counterclockwise and the angular acceleration is 3 rad/sec² clockwise.

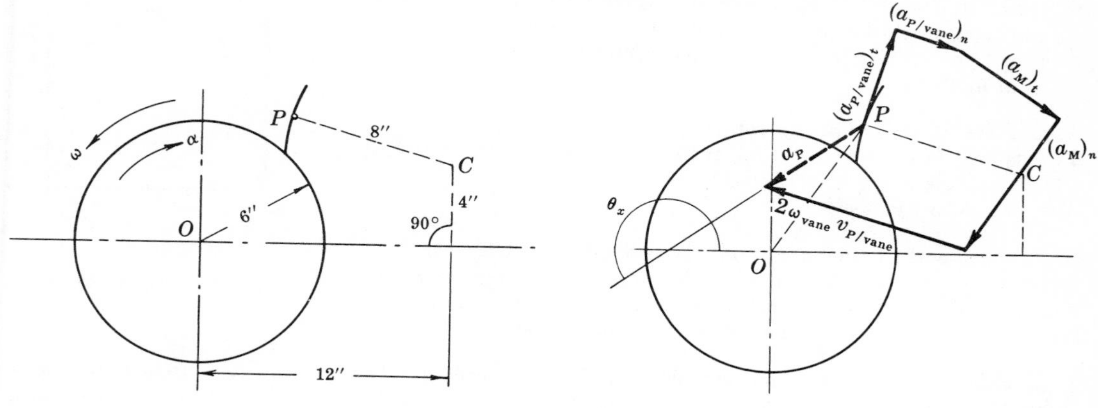

Fig. 12-43

Fig. 12-44

Solution:

According to Coriolis' law the absolute acceleration of P is

$$\mathbf{a}_P = (\mathbf{a}_{P/\text{vane}})_t + (\mathbf{a}_{P/\text{vane}})_n + (\mathbf{a}_M)_t + (\mathbf{a}_M)_n + 2\boldsymbol{\omega}_{\text{vane}} \times \mathbf{v}_{P/\text{vane}}$$

where

$(\mathbf{a}_{P/\text{vane}})_t$ = 20 in/sec² outward and $\perp$ to PC,

$(\mathbf{a}_{P/\text{vane}})_n$ = $(v_{P/\text{vane}})^2 \div PC = (10)^2 \div 8 = 12.5$ in/sec² directed from P to C,

$(\mathbf{a}_M)_t$ = tangential component of the acceleration of the point M on the vane that coincides with P at the instant. Magnitude $= OP \times \alpha = 8 \times 3 = 24$ in/sec²; directed down to the right and $\perp$ to OP,

$(\mathbf{a}_M)_n$ = normal component of the acceleration of the point M on the vane that coincides with P at the instant. Magnitude $= OP \times \omega^2 = 8(2)^2 = 32$ in/sec²; directed from P to O,

$(\mathbf{v}_{P/\text{vane}})$ = 10 in/sec; outward $\perp$ to PC,

$(\boldsymbol{\omega}_{\text{vane}})$ = 2 rad/sec; counterclockwise,

$2\boldsymbol{\omega}_{\text{vane}} \times \mathbf{v}_{P/\text{vane}}$ = 2(2)(10) = 40 in/sec² directed from C towards P. The direction is obtained by rotating $\mathbf{v}_{P/\text{vane}}$, which is $\perp$ to CP and outwards, through 90° in the sense of $\boldsymbol{\omega}_{\text{vane}}$, i.e., counterclockwise in plane of the paper.

The vector sum of these components indicated above yields $a_P = 21$ in/sec² and $\theta_x = 215°$. See Fig. 12-44 above.

Supplementary Problems

29. A rigid body is rotating 12 rad/sec about an axis through the origin and with direction cosines 0.421, 0.365, and 0.831 with respect to the x, y, and z axes respectively. What is the velocity of a point in the body defined by the position vector (with respect to the origin) $\mathbf{r} = -2\mathbf{i} + 3\mathbf{j} - 4\mathbf{k}$?
Ans. $\mathbf{v} = -47.4\mathbf{i} + 0.26\mathbf{j} + 23.9\mathbf{k}$ ft/sec

30. A rigid body is rotating 200 rpm about the line $\mathbf{i} - 3\mathbf{j} + 4\mathbf{k}$. The origin is on the line. What is the linear velocity of the point $P(3, 3, -1)$? *Ans.* $\mathbf{v} = -37.0\mathbf{i} + 53.6\mathbf{j} + 49.3\mathbf{k}$ ft/sec

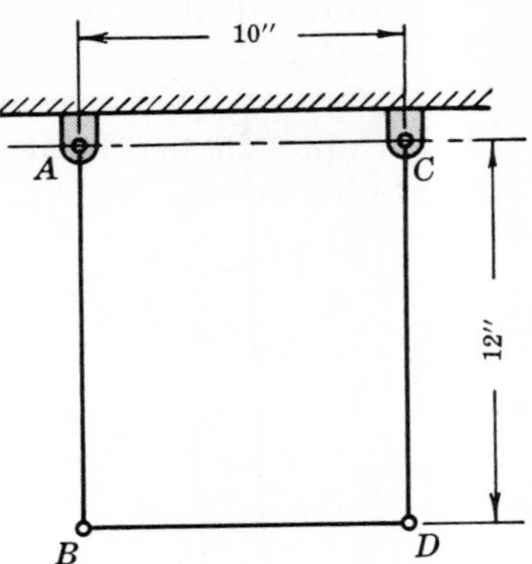

Fig. 12-45

31. Refer to Fig. 12-45. The equal bars AB and CD are free to rotate about pins in the frame. The bar BD is equal in length to the distance AC. The constant angular velocity of AB is 10 rpm counterclockwise. Determine the motion of BD.
Ans. All points on BD have $v = 753$ in/min to right, $a = 47{,}300$ in/min^2 up

32. At a given time a shaft is rotating 50 rpm about a fixed axis. Twenty sec later it is rotating 1050 rpm. What is the average angular acceleration α in rad/sec^2?
Ans. $\alpha = 5.23$ rad/sec^2

33. A flywheel with diameter 50 cm starts from rest with constant angular acceleration 2 rad/sec^2. Determine the tangential and normal components of a point on the rim 3 sec after the motion began. *Ans.* $a_t = 50$ cm/sec^2, $a_n = 900$ cm/sec^2

34. A particle is at rest on a phonograph turntable at a radius r from the center. Determine the magnitudes of the tangential and normal components of the acceleration of the particle at time t assuming the turntable starts from rest and accelerates with a uniform angular acceleration C.
Ans. $a_n = C^2 rt^2$, $a_t = Cr$

35. A rotor decreases uniformly from 1800 rpm to rest in 320 sec. Determine the angular deceleration and the number of radians before coming to rest. *Ans.* $\alpha = -0.589$ rad/sec^2, $\theta = 30{,}100$ rad

36. A bar pivoted at one end and moving 5 rad/sec clockwise is then subjected to a constant angular deceleration. After a certain time interval the bar has an angular displacement of 8 radians counterclockwise and has moved through a total angle of 20.5 rad. What is the angular velocity at the end of the time interval? *Ans.* $\omega = 7.56$ rad/sec

37. The drum shown in Fig. 12-46 below is used to hoist the weight W a distance 6 ft. The drum accelerates uniformly from rest to 15 rpm in 1.5 sec and then moves at a constant speed of 15 rpm. What is the total elapsed time? *Ans.* $t = 6.48$ sec

38. Refer to Fig. 12-47 below. The hoisting mechanism consists of a drum 48 inches in diameter around which is wrapped the cable. Integral with the drum is a gear with pitch diameter 36 inches. This gear is driven by a pinion with 12 inch pitch diameter. The acceleration of the weight W is 18 ft/sec^2 up. What is the angular acceleration of the pinion? *Ans.* $\alpha = 27$ rad/sec^2 clockwise

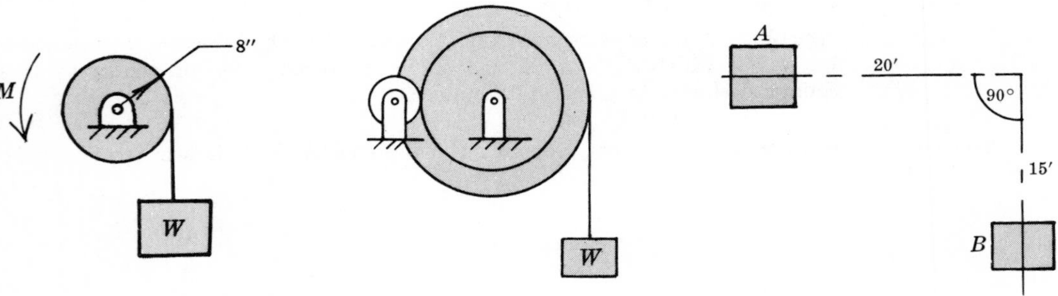

Fig. 12-46 Fig. 12-47 Fig. 12-48

39. Rain is falling vertically down at 5 ft/sec. A woman is walking on level ground with a speed of 4 ft/sec. What is the velocity of the rain relative to the woman and at what angle forward of the vertical should she hold her umbrella? *Ans.* 6.4 ft/sec, 38.6°

40. Mass A moves to the right at 15 ft/sec from the position shown in Fig. 12-48 above. Mass B starts from B at the same instant moving vertically upward at 20 ft/sec. Determine the relative velocity of A to B one-half sec after motion begins. *Ans.* $v_{A/B} = 25$ ft/sec, $\theta_x = 307°$

41. Refer to Fig. 12-49 below. Points O and P are on a thin lamina which has motion in the xy plane. Point O is known to have the velocity shown. Determine the absolute velocity of point P.
Ans. $v_P = 0$

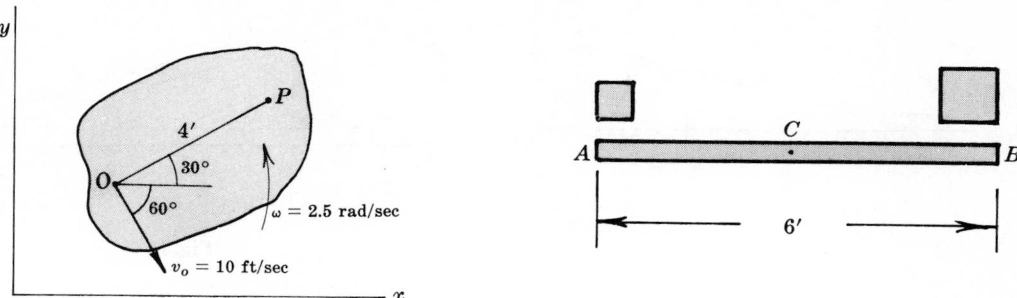

Fig. 12-49

Fig. 12-50

42. The iron bar is attracted by two large but unequal magnets in such a way that end A has a vertical acceleration of 4 ft/sec² and end B has a vertical acceleration of 6 ft/sec². See Fig. 12-50 above. Determine the acceleration of the center and the angular acceleration of the bar.
Ans. a_C = 5 ft/sec² up, α = 0.333 rad/sec² counterclockwise

43. A ladder of length l leans against a vertical wall. The bottom moves away from the wall along a horizontal floor with a constant velocity v_0. Determine the velocity and acceleration of the top of the ladder. (Hint: $x = v_0 t$.)

Ans. $\dot{y} = \dfrac{-v_0 t}{\sqrt{l^2 - v_0^2 t^2}}, \quad \ddot{y} = \dfrac{-v_0^2 l^2}{(l^2 - v_0^2 t^2)^{3/2}}$

44. A ladder of length l makes an angle θ with the horizontal floor and leans against a vertical wall. Show that the center of the ladder moves on a circular path of radius one-half l.

45. The bar AB slides so that its bottom point A has a velocity of 16 in/sec to the left along the horizontal plane. See Fig. 12-51. What is the angular velocity of the bar in the phase shown?
Ans. ω = 0.293 rad/sec clockwise

46. The top of a ladder 10 ft long slides down a smooth wall while its bottom slides along a smooth plane perpendicular to the wall. Show that the velocity of its midpoint is directed along the ladder when the ladder makes an angle of 45° with the horizontal plane.

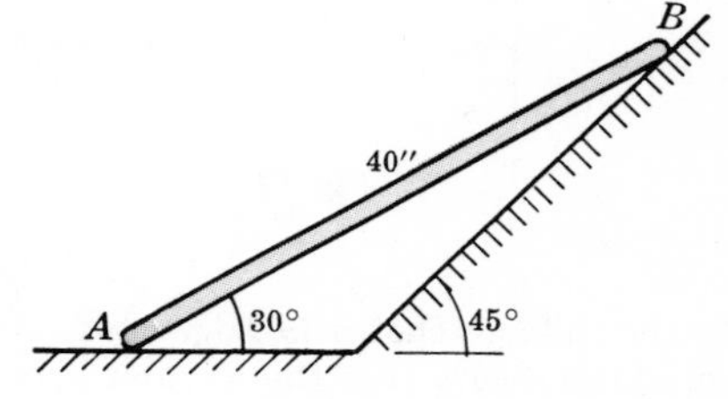

Fig. 12-51

47. In Fig. 12-52 below, the velocity of point A is 5 ft/sec to the right and its acceleration is 8 ft/sec² to the right. What are the velocity and acceleration of point B?
Ans. v_B = 8.66 ft/sec down, a_B = 33.9 ft/sec² down

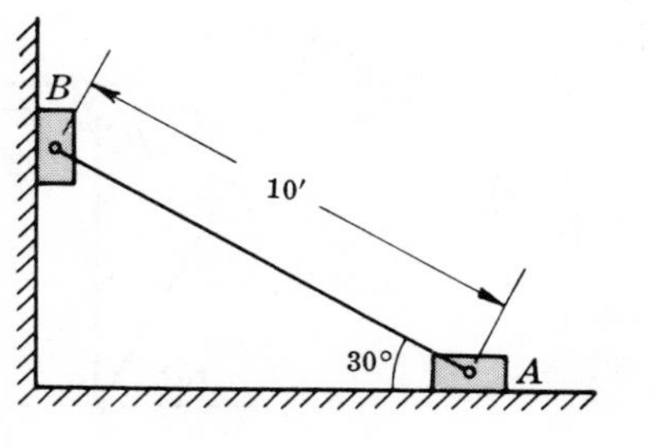

Fig. 12-52

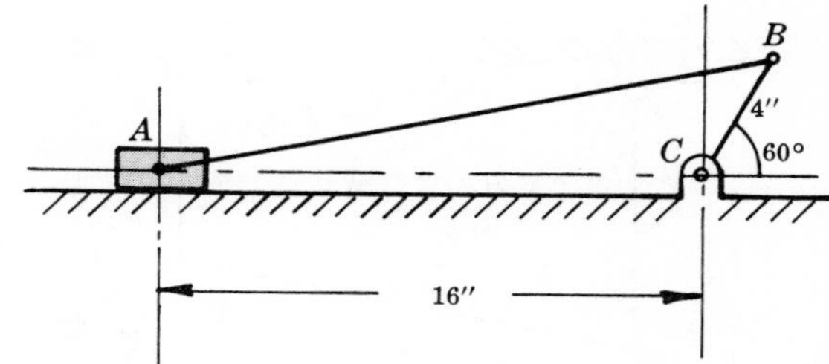

Fig. 12-53

48. The crank CB of the slider crank mechanism is rotating at a constant 30 rpm clockwise. Determine the velocity of the crosshead A in the phase shown in Fig. 12-53 above.
Ans. v_A = 9.68 in/sec to the right

49. In the slider crank mechanism shown in Fig. 12-54 below, the crank is turning clockwise at 120 rpm. What is the velocity of the crosshead when the crank is in the 60° phase?
Ans. 10.3 ft/sec toward the right

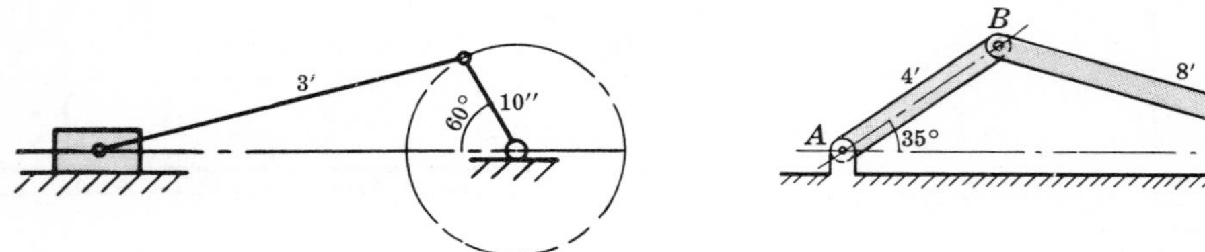

Fig. 12-54 **Fig. 12-55**

50. The linear velocity of the crosshead (slider) is 8 ft/sec to the left along the horizontal plane. Using the instant center method, determine the angular velocity of the crank AB in the phase shown in Fig. 12-55 above. *Ans.* ω = 23.3 rpm counterclockwise

51. The bar slides on the vertical post and is attached to the block A which is moving to the right with a constant velocity C. See Fig. 12-56 below. Determine the angular velocity $\dot{\theta}$ of the bar.
Ans. $\dot{\theta} = -(C/a)\sin^2\theta$

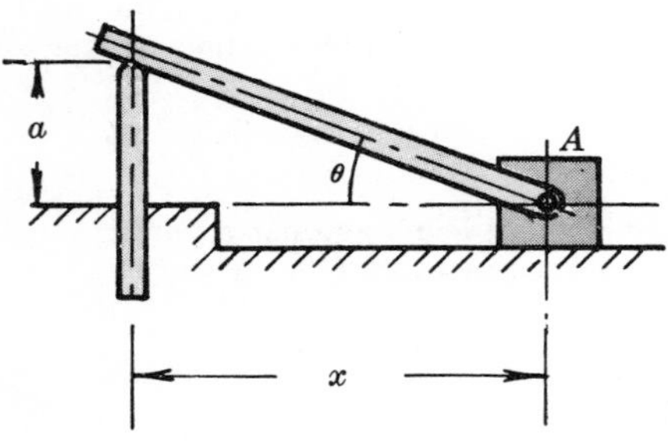

Fig. 12-56

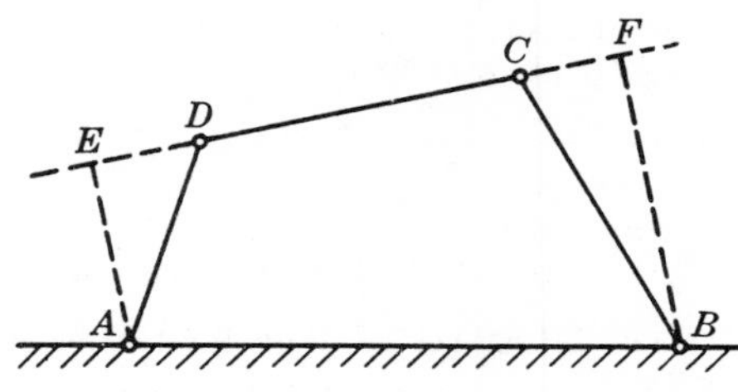

Fig. 12-57

52. In the linkage shown in Fig. 12-57 above, the pins A and B are fixed. Link AD rotates with angular speed ω_A. Prove that the angular speed ω_B of link BC is given by the expression $\omega_B = \omega_A(AE/BF)$, where AE and BF are perpendicular to DC.

53. In the four bar linkage (quadric crank mechanism) shown in Fig. 12-58 below, the angular velocity of AB is 8 rad/sec clockwise. Determine the angular velocity of CD and the angular velocity of BC.
Ans. ω_{CD} = 12.0 rad/sec clockwise, $\omega_{BC} = 0$

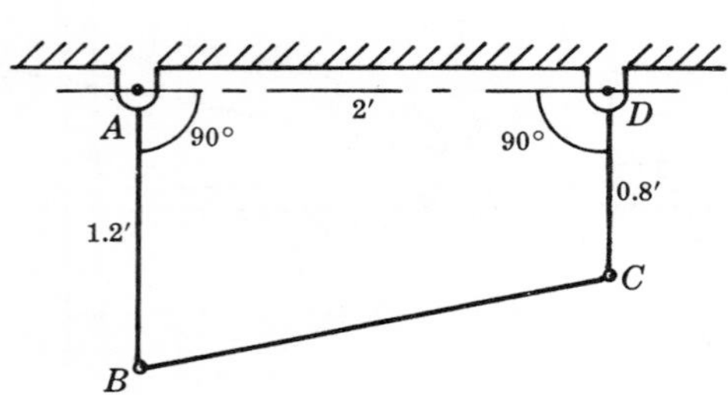

Fig. 12-58

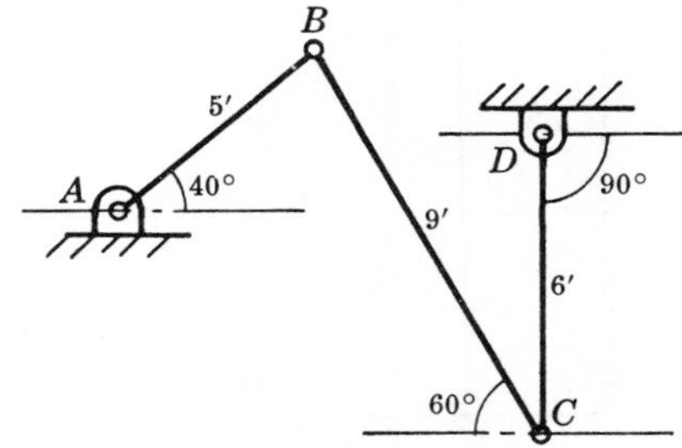

Fig. 12-59

54. Crank AB is moving 2 rpm clockwise in the phase shown in Fig. 12-59 above. What is the angular velocity of the arm CD? *Ans.* ω_{CD} = 3.28 rpm counterclockwise

55. A wheel 10 ft in diameter rolls to the right so that its center has a velocity of 20 ft/sec. What is the velocity of the top point? *Ans.* 40 ft/sec toward the right

56. If the center of the wheel in Problem 55 is accelerating to the right with 15 ft/sec², what is the acceleration of the top point? *Ans.* $a = 85.4$ ft/sec², $\theta_x = 291°$

57. A wheel 30 cm in diameter rolls to the right without slipping on a horizontal plane. Its angular speed is 30 rpm. What is the velocity of (*a*) the top point on the wheel, (*b*) the point on the front end of the horizontal diameter? *Ans.* (*a*) $v = 0.944$ m/sec, $\theta_x = 0°$; (*b*) $v = 0.667$ m/sec, $\theta_x = 315°$

58. The block moves on two 4 inch diameter rollers as shown in the adjacent Fig. 12-60. If the block has a velocity of 3.0 ft/sec and an acceleration of 2.0 ft/sec² both to the right, determine the velocity and acceleration of the center of one of the rollers.
Ans. $v = 1.5$ ft/sec to the right, $a = 1.0$ ft/sec² to the right

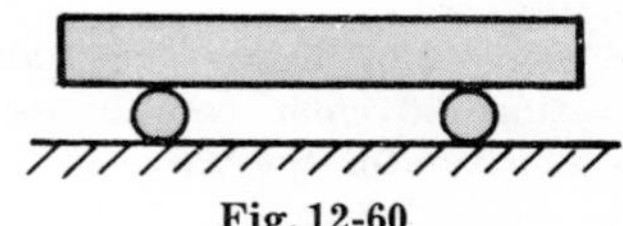

Fig. 12-60

59. The disk rolls without slipping on the horizontal plane. Its center has an acceleration of 4 ft/sec² directed horizontally to the left. At the instant that its center has a velocity of 3 ft/sec to the right, determine the acceleration of point P. Refer to Fig. 12-61 below.
Ans. $a = 7.32$ ft/sec², $\theta_x = 183°$

60. The composite wheel rolls without slipping with angular velocity 30 rpm clockwise. See Fig. 12-62 below. Determine the absolute velocities of points A and B.
Ans. $v_A = 2.22$ ft/sec, $\theta_x = -45°$; $v_B = 4.72$ ft/sec to right

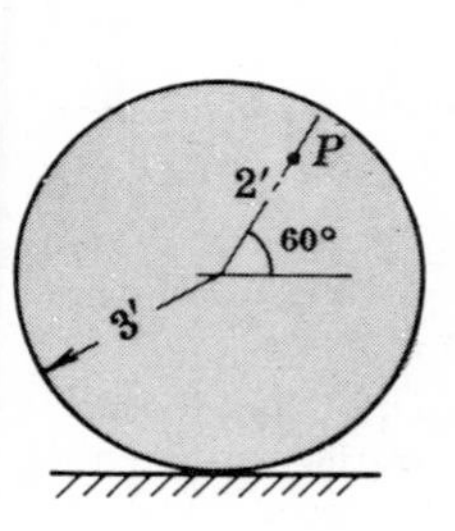

Fig. 12-61

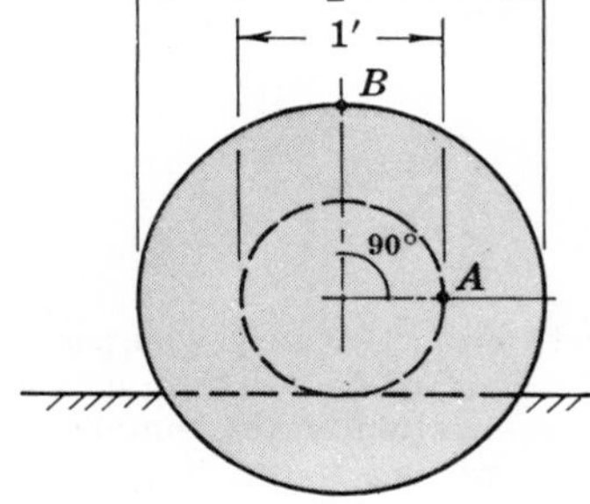

Fig. 12-62

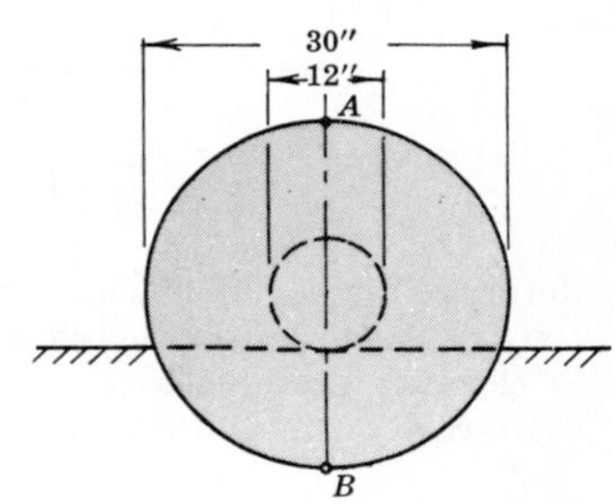

Fig. 12-63

61. The composite wheel rolls without slipping on the horizontal plane shown in Fig. 12-63 above. If the angular velocity of the wheel is 20 rad/sec clockwise, determine the absolute velocities of the top and bottom points of the wheel. *Ans.* $v_A = 35$ ft/sec to right, $v_B = 15$ ft/sec to left

62. The wheel moves such that its center has a velocity of 12 ft/sec horizontally to the right. The angular velocity of the wheel is 4 rad/sec clockwise. Determine the absolute velocity of the points P and Q (see Fig. 12-64 below). *Ans.* $v_P = 16.6$ ft/sec, $\theta_x = 27.0°$; $v_Q = 0$

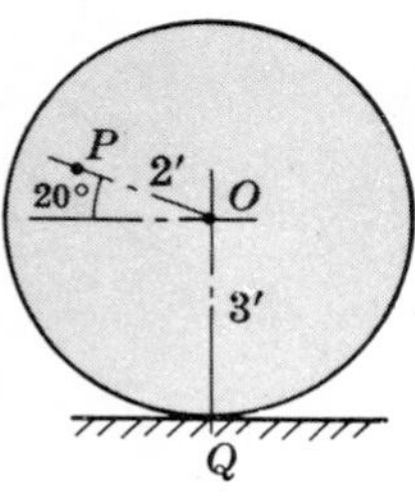

Fig. 12-64

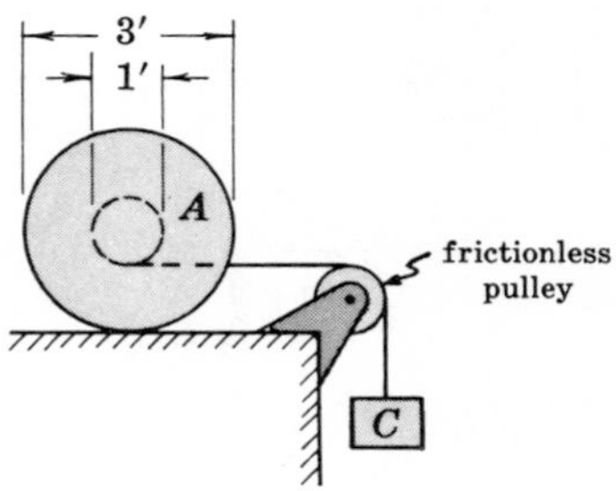

Fig. 12-65

63. The composite wheel A rolls without slipping on the horizontal plane. The cord wrapped around the axle is attached to the weight C as shown in Fig. 12-65 above. The velocity of C changes uniformly from 2 ft/sec downward to 6 ft/sec downward in 2 sec. Determine the angular displacement of A during the interval. *Ans.* $\theta = 8$ rad clockwise

64. The cylinder C is 20 inches in diameter and rolls without slipping on the horizontal plane shown in Fig. 12-66. The pulley B is frictionless. If the displacement of A is 4 inches down, what is the angular displacement of C? *Ans.* $\theta = 0.4$ rad clockwise

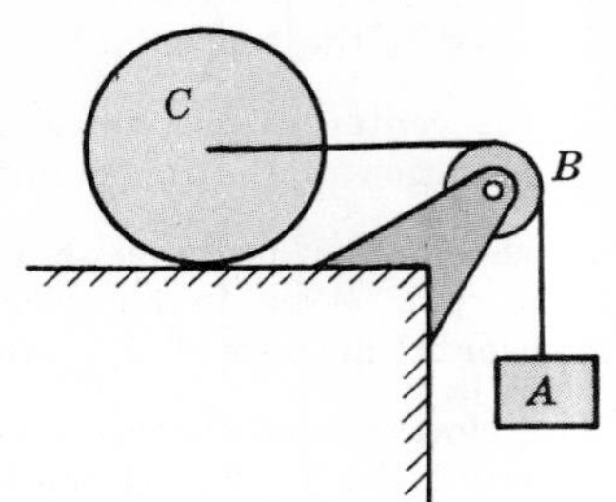

Fig. 12-66

65. In the preceding problem the velocity and acceleration of A are 4 in/sec and 2 in/sec² respectively. What are the angular velocity and angular acceleration of the cylinder C?
Ans. $\omega = 0.4$ rad/sec clockwise, $\alpha = 0.2$ rad/sec² clockwise

66. In Fig. 12-67 below, the valve A is actuated by the eccentric rotating 30 rpm counterclockwise. Express the velocity and acceleration of the valve in terms of the angle ϕ.
Ans. $\dot{x} = 7.85 \sin\phi$ in/sec to left, $\ddot{x} = 24.6 \cos\phi$ in/sec² to left

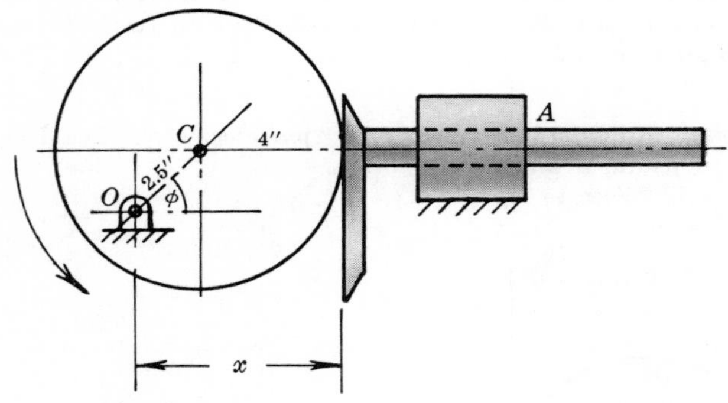

Fig. 12-67

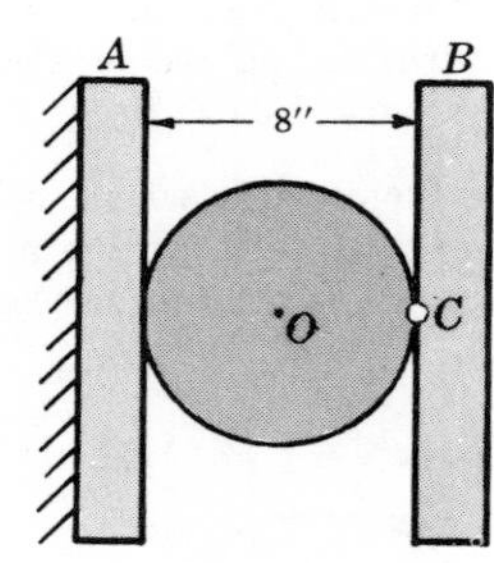

Fig. 12-68

67. Refer to Fig. 12-68 above. The gear rack A is stationary whereas the gear rack B has a velocity of 2 ft/sec down and an acceleration of 1.5 ft/sec² down. Determine the velocity of the gear center, the acceleration of the gear center, and the acceleration of the contact point C on the gear.
Ans. $v_0 = 1$ ft/sec down, $a_0 = 0.75$ ft/sec² down, $a_C = 3.35$ ft/sec² at $\theta_x = 206°$

68. In Fig. 12-69 below, the disk rolls without slipping on the horizontal plane with an angular velocity of 10 rpm clockwise and an angular acceleration of 6 rad/sec² counterclockwise. The bar AB is attached as shown. Line OA is horizontal. Point B moves along the horizontal plane shown. Determine the velocity and acceleration of point B for the phase shown.
Ans. $v_B = 3.66$ ft/sec to right, $a_B = 24.9$ ft/sec² to left

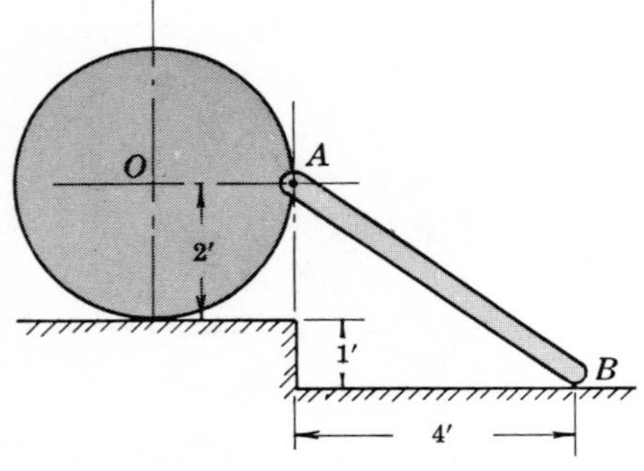

Fig. 12-69

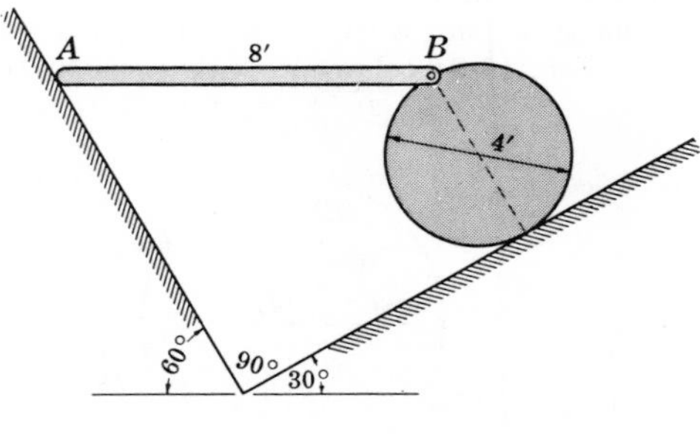

Fig. 12-70

69. The 8 ft bar AB is connected by a frictionless pin at B to the 4 ft diameter cylinder which is rolling with constant center velocity 12 ft/sec down the plane inclined 30° with the horizontal (see Fig. 12-70 above). The end A slides on the frictionless plane which makes an angle of 60° with the horizontal. For the phase shown with the bar horizontal, determine the angular velocity and acceleration of the bar.
Ans. $\omega = 6.0$ rad/sec clockwise, $\alpha = 62.4$ rad/sec² counterclockwise

70. The crank OB is rotating with constant angular velocity 6 rad/sec clockwise. Disk C rolls without slipping inside the fixed circle. Determine (a) the angular velocity of disk C and (b) the absolute acceleration of point P in the phase shown in Fig. 12-71 where crank OB is horizontal and P is the top point on disk C.
Ans. ω_C = 18 rad/sec counterclockwise, a_P = 341 ft/sec^2 at $\theta_x = 252°$

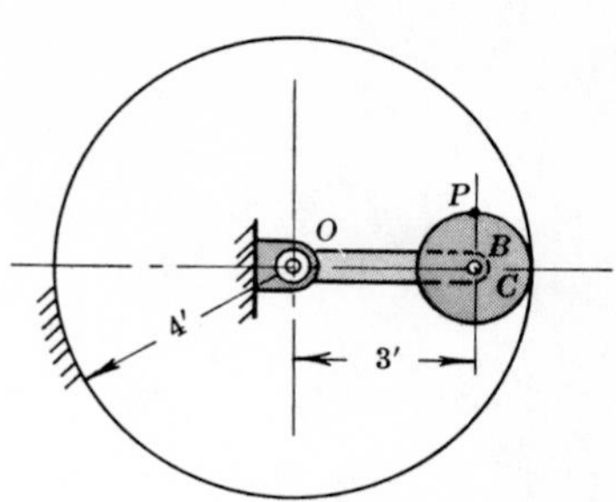

Fig. 12-71

71. A bead moves along a rod such that $r = 0.5t^2$. At the same time the rod moves such that $\theta = t^2 + t$. Determine the radial and transverse components of the velocity and acceleration of the bead at time $t = 2$ sec.
Ans. $v_r = 2$ ft/sec, $v_\theta = 10$ ft/sec, $a_r = -49$ ft/sec^2, $a_\theta = 24$ ft/sec^2

72. A rod is rotating in a horizontal plane at a constant rate of 4 rad/sec clockwise about a vertical axis through one end. A washer is sliding along the rod with a constant speed of 3 ft/sec. At the instant when the washer is 6 inches from the vertical axis, determine the radial and transverse components of the acceleration. *Ans.* $a_r = -8$ ft/sec^2, $a_\theta = 24$ ft/sec^2

73. In Fig. 12-72 below, the ball P moves with a constant velocity 4 ft/sec down along the smooth slot cut as shown in the disk rotating with angular velocity 3 rad/sec clockwise and angular acceleration 8 rad/sec^2 counterclockwise. Determine the acceleration of P in the phase shown.
Ans. $a = 34.0$ ft/sec^2 with $\theta_x = 194°$

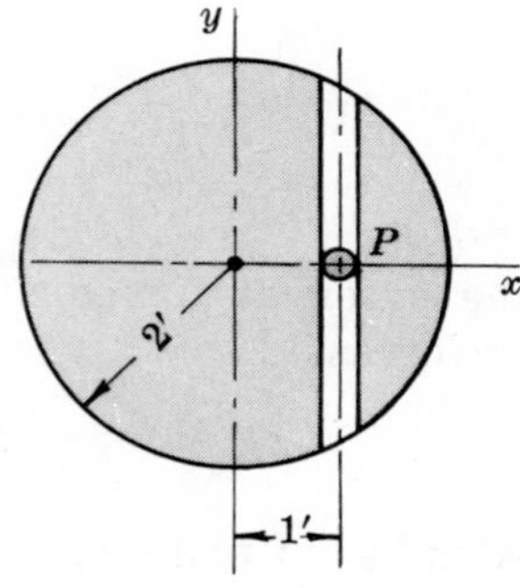

Fig. 12-72

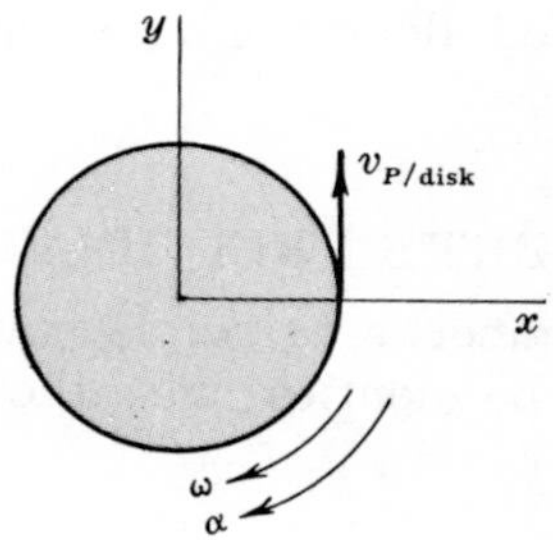

Fig. 12-73

74. A particle moves with speed 2 ft/sec around the circumference of a 4 ft diameter disk which is rotating in the opposite direction with angular velocity 2 rad/sec (clockwise). See Fig. 12-73 above. The disk has angular acceleration 4 rad/sec^2 clockwise. Determine the acceleration of the particle in the phase shown. *Ans.* $a = 8.24$ ft/sec^2 with $\theta_x = 256°$

75. A bug moves with constant speed v ft/sec along the circumference of a disk r ft in radius. The disk rotates in the opposite direction with angular velocity ω rad/sec. What is the absolute acceleration of the bug? *Ans.* $(v - r\omega)^2/r$

76. A bug moves with constant speed v along a radius r of a disk rotating with constant angular velocity ω. What is the absolute acceleration of the bug as it reaches the rim of the disk?
Ans. $a = \omega\sqrt{r^2\omega^2 + 4v^2}$

Chapter 13

Dynamics of a Rigid Body in Translation

ANY PARTICLE of a rigid body in translation has the same acceleration as any other particle.

EFFECTIVE FORCE on a PARTICLE

The effective force on a particle is the product of its mass dm and its acceleration $\mathbf{a}$. It is directed along the acceleration vector $\mathbf{a}$.

D'ALEMBERT'S PRINCIPLE

D'Alembert's principle states that the resultant of the external forces acting on a body must be identical with the resultant of all the effective forces.

FOR TRANSLATION the scalar equations expressing D'Alembert's principle are:

$$\Sigma F_x = \Sigma(dm)a_x = a_x \Sigma dm = ma_x$$

$$\Sigma F_y = \Sigma(dm)a_y = a_y \Sigma dm = ma_y$$

$$\Sigma \bar{M} = 0$$

where ΣF_x, ΣF_y = algebraic sums of the components of the external forces in the x and y directions respectively,

m = mass of the body,

a_x, a_y = components of the acceleration of the body in the x and y directions respectively,

$\Sigma \bar{M}$ = sum of the moments of the external forces about the mass center of the body.

INERTIA-FORCE METHOD

The inertia-force method of solution may be applied provided the effective forces for the body are reversed and added to the external force system. This has the effect of holding the system in equilibrium for the instant for study purposes only. The reversed effective (inertia) forces do *not* actually exist. The equations of static equilibrium may then be used.

Solved Problems

As in previous chapters, vectors in the diagrams are identified only by their magnitudes when the directions are evident by inspection.

1. A 50 lb homogeneous door is supported, as shown in Fig. 13-1 below, on rollers A and B resting on a horizontal track. A constant force P of 10 lb is applied. What will be the velocity of the door 5 sec after starting from rest? What are the reactions of the rollers? Assume that roller friction is negligible.

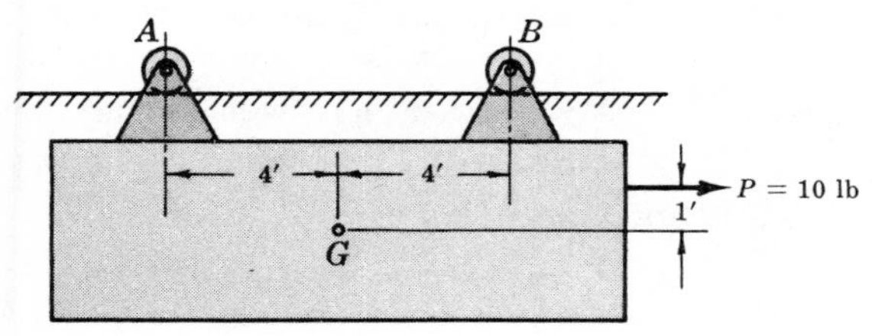

Fig. 13-1

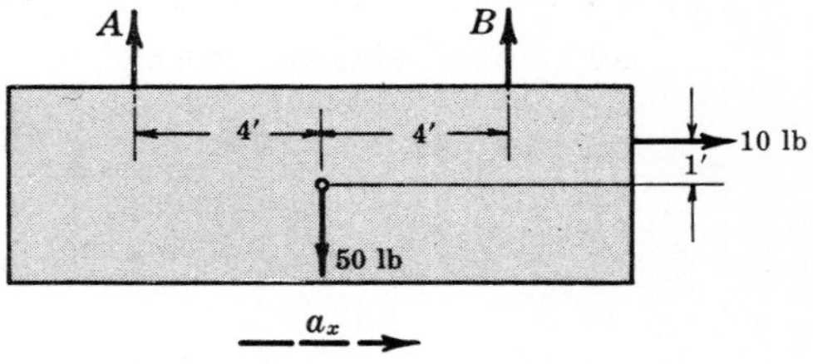

Fig. 13-2

Solution:

The acceleration a_x is assumed in the x direction in the free body diagram shown in Fig. 13-2 above.

Apply the equations $\Sigma F_x = ma_x$, $\Sigma F_y = ma_y$, $\Sigma \bar{M} = 0$, to obtain the following:

$$(1) \quad \Sigma F_x = 10 = (50/32.2)a_x$$
$$(2) \quad \Sigma F_y = A + B - 50 = (50/32.2)0$$
$$(3) \quad \Sigma \bar{M} = B \times 4 - A \times 4 - 10 \times 1 = 0$$

From equation (1), $a_x = 6.44$ ft/sec². Since a_x is constant, the formula involving the known quantities v_0, a, t and the unknown v should be used.

$$v = v_0 + at = 0 + (6.44 \text{ ft/sec}^2)(5 \text{ sec}) = 32.2 \text{ ft/sec}$$

Solve equations (2) and (3) simultaneously to obtain $A = 23.7$ lb and $B = 26.3$ lb.

2. Solve Problem 1 by using the inertia-force method.

Solution:

In this method the reversed effective force ma_x is added to the system at the mass center as shown in Fig. 13-3. The problem is then considered as an exercise in statics.

The equations of equilibrium are:

$$(1) \quad \Sigma F_x = 10 - ma_x = 0$$
$$(2) \quad \Sigma M_{A'} = B \times 8 - 10 \times 1 - 50 \times 4 = 0$$
$$(3) \quad \Sigma M_{B'} = -A \times 8 + 50 \times 4 - 10 \times 1 = 0$$

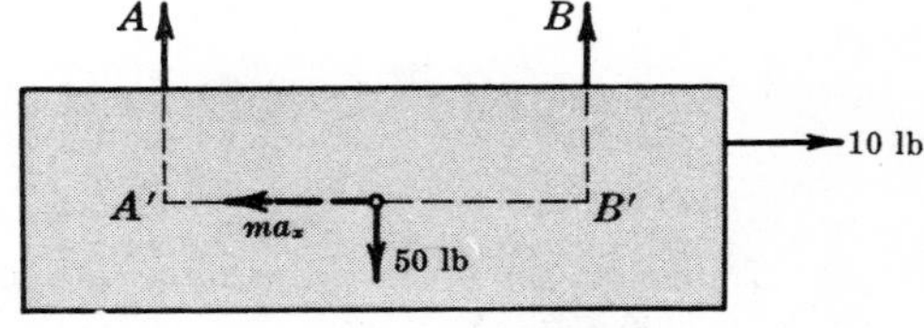

Fig. 13-3

The first equation yields the same value of a_x as in Problem 1.

Note, however, the advantage gained in equations (2) and (3) in which A and B may be found independently. The choice of moment center is no longer limited to only the mass center. Since the system is in equilibrium (for study purposes only) any moment center may be utilized.

From equation (2), $B = (10 + 200)/8 = 26.3$ lb. From equation (3), $A = 190/8 = 23.7$ lb.

The sum of A and B is 50 lb, thereby checking the result.

3. A freight car starts from rest at the top of a 1.5 percent grade one mile long. Assuming a resistance of 8 lb per ton, how far along the level track at the bottom of the grade will it roll before coming to rest?

Solution:

On the incline the forces acting parallel to the track are the component of the weight $0.015W$ and the resistance of $(8/2000)W = 0.004W$ lb. The first accelerates the car, but the latter decelerates it.

The equation of motion is $\Sigma F_{||} = 0.015W - 0.004W = (W/g)a$.

The resulting acceleration is $a = 0.011g$. The velocity at the bottom of the grade is found from

$$v^2 = v_0^2 + 2as = 0 + 2(0.011g)(5280)$$

To determine the distance s along the level, apply the same formulas noting that the only horizontal force acting is a constant decelerating one (because of train resistance), that is, $-0.004W$.

As before, $\Sigma F_{||} = -0.004W = (W/g)a$; hence $a = -0.004g$. Also, v_0^2 now becomes the value of v^2 found above. Thus

$$v^2 = v_0^2 + 2as, \quad \text{or} \quad 0 = 2(0.011g)(5280) - 2(0.004gs)$$

from which $s = 14{,}500$ ft along the level.

4. A block A rests on a cart B as shown in Fig. 13-4. The coefficient of friction between the cart and the block is 0.30. If A weighs 161 lb, investigate the maximum acceleration the cart may have.

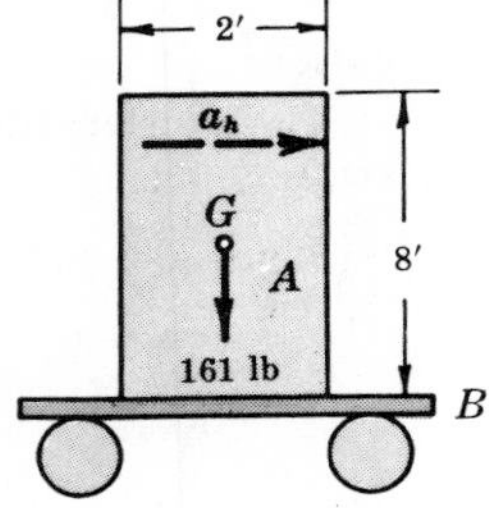

Fig. 13-4

Solution:

The two possibilities to consider are (a) the acceleration which will cause the block to slide off the cart and (b) the acceleration which will cause tipping.

In case (a) the friction is a maximum and is equal to the product of the coefficient of friction and the normal force between the cart and the block.

In case (b) the necessary friction may be (1) less than maximum, (2) maximum, or (3) greater than maximum. If the friction necessary to cause tipping is greater than maximum, naturally it cannot be obtained and therefore the block will slide at the value of the acceleration calculated in (a).

The free body diagram for case (a) is shown in Fig. 13-5. Note that the friction may be applied at any point along the bottom in this representation. The friction is shown acting to the right, since the cart is being accelerated to the right and pulls the block with it because of friction. Only two equations are necessary.

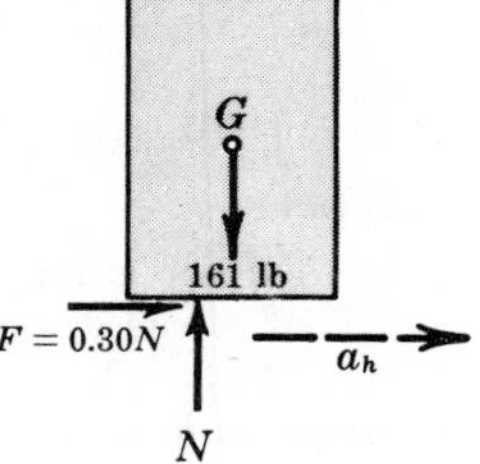

Fig. 13-5

(1) $\Sigma F_h = ma_h$, i.e. $F = ma_h$

(2) $\Sigma F_v = ma_v = 0$, i.e. $N - 161 = 0$

From equation (2), $N = 161$ lb.

Equation (1) becomes $0.30 \times 161 \text{ lb} = \dfrac{161 \text{ lb}}{32.2 \text{ ft/sec}^2} a_h$ or $a_h = 9.66$ ft/sec².

Accelerations above 9.66 ft/sec² will cause the block to slide.

The free body diagram for the case of tipping is shown next in Fig. 13-6. In this case motion impends about the back edge of the block. Therefore the normal and frictional forces are shown there.

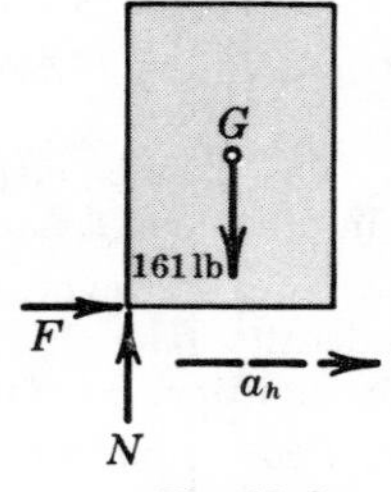

Fig. 13-6

Note that F is not placed equal to $0.30N$. The limiting value of friction is used only if slipping impends, as in case (a). The equations are:

(3) $\Sigma F_h = ma_h$, (4) $\Sigma F_v = ma_v = 0$, (5) $\Sigma \bar{M} = 0$

These become: (*3'*) $F = (161/32.2)a_h$

(*4'*) $N - 161 \text{ lb} = 0$

(*5'*) $-N \times 1 + F \times 4 = 0$

From (*4'*), $N = 161$ lb. Substituting into (*5'*), $F = 40.3$ lb.

Substituting $F = 40.3$ lb into (*3'*), $a_h = 8.06$ ft/sec^2.

Thus at an acceleration of 8.06 ft/sec^2 the block tends to tip. Since at this acceleration the force of friction is 40.3 lb, which is less than the limiting value $(0.30 \times 161 \text{ lb})$, the tipping will occur before sliding can exist.

Or looking at the problem slightly differently, an acceleration of 8.06 ft/sec^2 will cause tipping whereas an acceleration of 9.66 ft/sec^2 will cause sliding. Since the acceleration 8.06 ft/sec^2 is reached first, tipping will occur before sliding.

It can also be deduced directly from equation (*5'*) that tipping will occur first, since the ratio of F to N in this case is 0.25, which is less than the maximum ratio of 0.30.

5. An elevator weighing 1200 lb is ascending with an acceleration of 12 ft/sec^2. The operator weighs 140 lb when at rest. Assuming that he is standing on scales during the ascension, what is the scale reading? What is the tension in the cables?

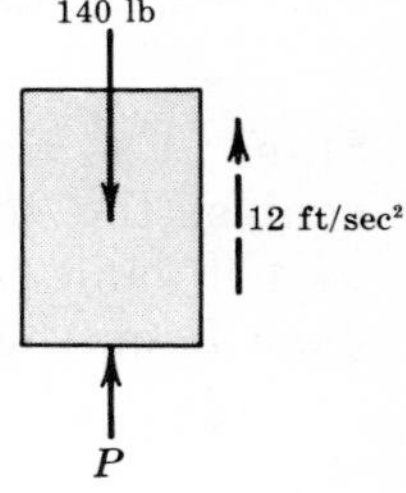

Fig. 13-7

Solution:

A free body diagram of the man shows his weight down, the scale push P upward, and the acceleration dotted at the side of the diagram (see Fig. 13-7).

The push P is of course indicated by the scale reading. Assume upward direction as positive. Any forces acting up are then positive. It is well to assume the positive direction as that direction in which you think the acceleration acts. Of course, here the acceleration is definitely known to be acting up.

$$\Sigma F_v = P - 140 \text{ lb} = \frac{140 \text{ lb}}{32.2 \text{ ft/sec}^2} \times 12 \text{ ft/sec}^2$$

Hence $P = 140 \text{ lb} + 52.2 \text{ lb} = 192.2$ lb. There is an apparent increase in weight of 52.2 lb.

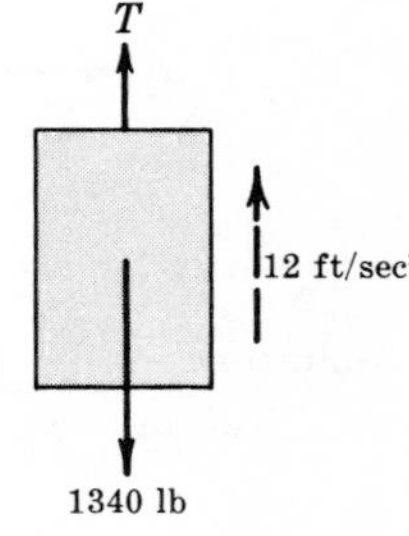

Fig. 13-8

To determine the tension in the cables, draw a free body diagram of both the man and the elevator (see Fig. 13-8). The equation which applies is $\Sigma F_v = ma_v$, or

$$T - 1340 \text{ lb} = \frac{1340 \text{ lb}}{32.2 \text{ ft/sec}^2} \times 12 \text{ ft/sec}^2$$

Hence $T = 1340 \text{ lb} + 500 \text{ lb} = 1840$ lb. The tension in the cables is therefore 500 lb greater than the dead weight.

6. Block B is accelerated along the horizontal plane by means of a weight A attached to it by a flexible, inextensible, weightless rope passing over a smooth pulley as shown in Fig. 13-9. Assume that the coefficient of friction between B which weighs 6 lb and the plane is 0.20. Discuss the motion if A weighs 2 lb.

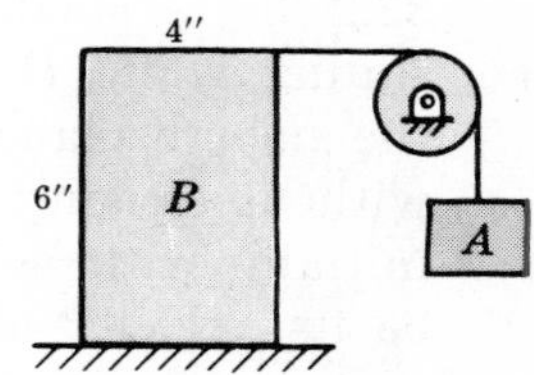

Fig. 13-9

Solution:

Draw the free body diagrams of A and B (see Fig. 13-10 below). Note that the force N is shown acting at a distance x from the vertical line through the mass center of block B. An acceleration a of block B is assumed to the right; therefore the reversed effective force $(6/g)a$ is added to hold the block in "equilibrium for study purposes".

Sum the vertical forces acting on block A to obtain

$$(1) \quad 2 - T = (2/32.2)a$$

By inspection, $N = 6$ lb; therefore $F = \mu N = 1.2$ lb.

Sum the horizontal forces acting on block B to obtain

$$(2) \quad T - 1.2 - (6/32.2)a = 0$$

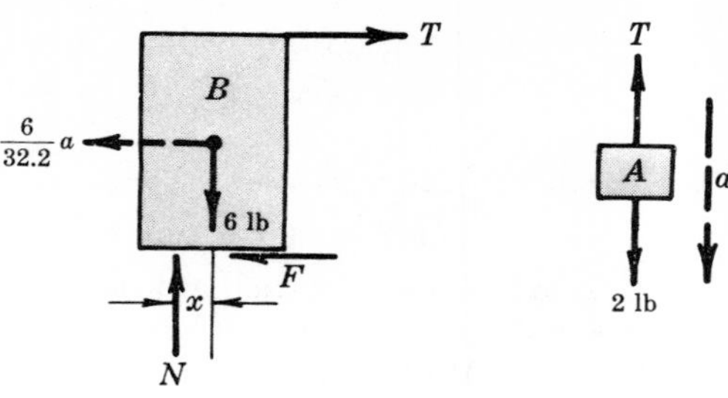

Fig. 13-10

Solve equations (1) and (2) simultaneously to obtain $a = 3.22$ ft/sec². From (2), $T = 1.8$ lb.

To determine the distance x, sum moments about the mass center to obtain

$$(3) \quad T \times 3 + F \times 3 + N \times x = 0$$

Substituting the values of T, F and N, we obtain $x = -1.5$ inches. The minus sign indicates that the force N is acting to the right of the mass center instead of to the left as assumed.

7. Stand A is accelerated to the left 8 ft/sec². Bar B is pinned at P and its top rests against the smooth vertical surface at H as shown in Fig. 13-11 below. A weighs 64.4 lb, and B weighs 16.1 lb. Determine the horizontal force F and the horizontal push H on the top of the bar.

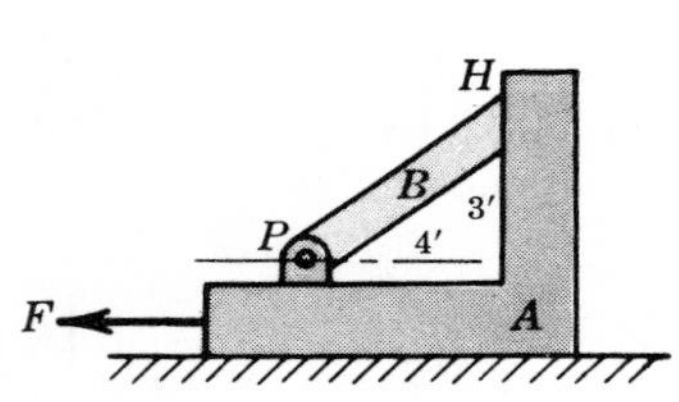

Fig. 13-11

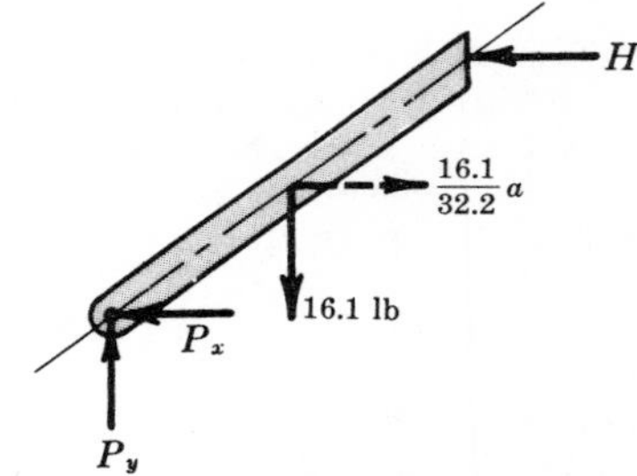

Fig. 13-12

Solution:

Taking the entire figure as a free body, sum the forces horizontally to obtain

$$F = \frac{W}{g}a = \frac{16.1 \text{ lb} + 64.4 \text{ lb}}{32.2 \text{ ft/sec}^2} \times 8 \text{ ft/sec}^2 = 20 \text{ lb}$$

Next draw the free body diagram of bar B showing the reversed effective force $(16.1/32.2)8 = 4$ lb (see Fig. 13-12 above). Sum moments about point P to obtain

$$4 \times 1.5 + 16.1 \times 2 - H \times 3 = 0 \qquad \text{or} \qquad H = 12.7 \text{ lb}$$

8. Each of the bodies A and B in Fig. 13-13 weighs 32.2 lb. A small strip is nailed to B to prevent A from sliding while P accelerates the system to the left on the smooth plane. Determine the maximum value of P without causing A to tip. If the system was moving initially with velocity 6 ft/sec to the right, what will be its velocity after moving 20 ft?

Solution:

Draw a free body diagram of A (see Fig. 13-14). The normal force N is shown acting up at the extreme right since the block

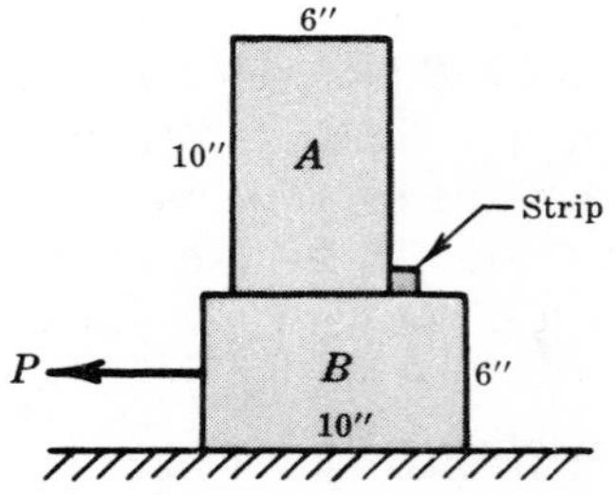

Fig. 13-13

is about to tip. Sum moments about the lower right corner to obtain the maximum value of a.

$$-(32.2/32.2)a \times 5 + 32.2 \times 3 = 0 \quad \text{or} \quad a = 19.3 \text{ ft/sec}^2$$

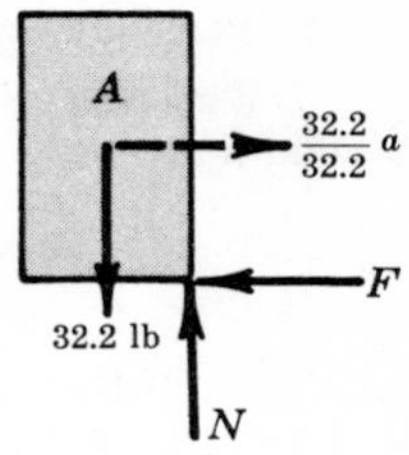

Fig. 13-14

Next consider the entire system as a free body and sum forces horizontally.

$$\Sigma F_h = (W/g)a \quad \text{or} \quad P = (64.4/32.2)19.3 = 38.6 \text{ lb}$$

To determine the velocity after traversing 20 ft, first consider the distance moved to the right before the velocity becomes zero. Applying the kinematics equation

$$v^2 = v_0^2 + 2as, \quad 0 = (+6)^2 - 2 \times 19.3s \quad \text{and} \quad s = 0.932 \text{ ft}$$

The remainder of the 20 ft will be to the left, i.e., 19.068 ft. Using the same equation and assuming the positive direction to the left, $v^2 = 0 + 2 \times 19.3 \times 19.07$. Hence the final velocity is to the left and equals 27.2 ft/sec.

9. Block A weighing 64.4 lb rests on block B weighing 96.6 lb as shown in Fig. 13-15 below. The coefficient of friction between the blocks is $\frac{1}{3}$. The coefficient of friction between block B and the horizontal plane is $\frac{1}{10}$. Determine the maximum value of P that will cause A neither to slide nor tip.

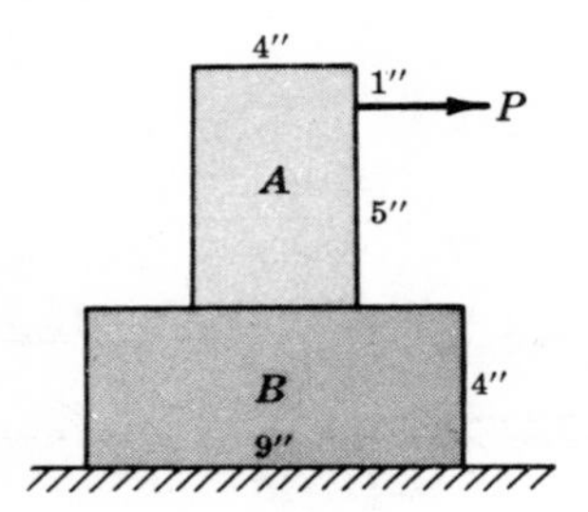

Fig. 13-15

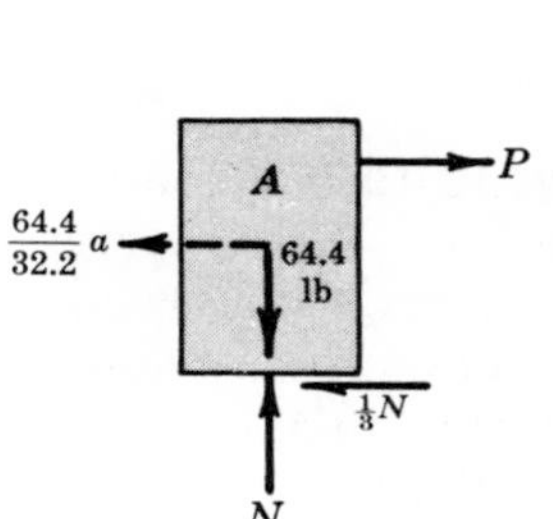

Fig. 13-16

Solution:

First determine the force P to cause A to slip on B. Draw the free body diagrams using friction as $\frac{1}{3}N$ between the blocks (see Fig. 13-16 above).

The location of the reaction N is unknown but is also immaterial. By inspection its value is 64.4 lb. Likewise by inspection $N_1 = 161$ lb and the friction between B and the plane is $\frac{1}{10}N_1 = 16.1$ lb. Sum the forces horizontally in both figures to obtain:

$$(1)\ P - 2a - \tfrac{1}{3}(64.4) = 0 \qquad (2)\ \tfrac{1}{3}(64.4) - 3a - 16.1 = 0$$

From (2), $a = 1.80$ ft/sec². Substitute this into (1) to find $P = 25.1$ lb to cause sliding.

Next draw free body diagrams to determine the force P to cause tipping (see Fig. 13-17). Now the normal component N of the reaction must be drawn at the lower right corner. The frictional component F of the reaction is unknown.

By inspection $N = 64.4$ lb, $N_1 = 161$ lb, and the friction at the bottom is again 16.1 lb.

For block A sum forces horizontally and take moments about the lower right corner to obtain

$$(3)\quad P - 2a - F = 0$$

$$(4)\quad -5P + 2a \times 3 + 64.4 \times 2 = 0$$

A third equation results from a summation of horizontal forces for block B.

$$(5)\quad F - 3a - 16.1 = 0$$

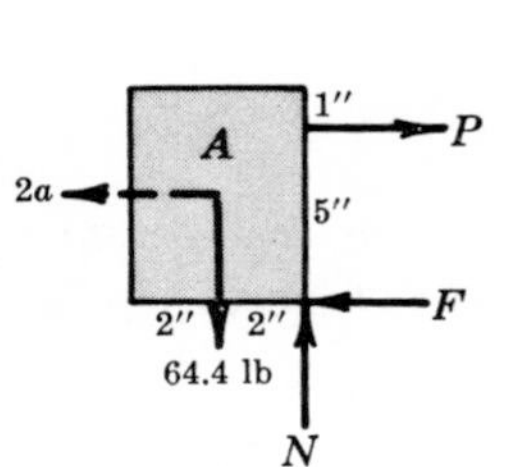

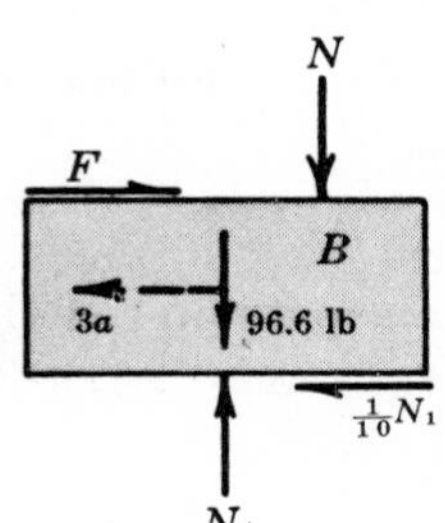

Fig. 13-17

Solve equations (*3*), (*4*) and (*5*) simultaneously to obtain $P = 28.8$ lb.

The smaller of the two values of P, i.e. 25.1 lb, is the maximum value which will cause A neither to slide nor tip.

10. Find the minimum time for the lift truck shown in Fig. 13-18 to attain its rated speed of 5 ft/sec without causing the 6 cartons, 54 in. high, to tip. Each carton is 9 in. high with a base 15 in. on each side. Assume sufficient friction present so that sliding will not occur.

Solution:

The free body diagram of the cartons on the truck indicates motion to the right. Thus the reversed effective force $(W/g)\bar{a}$ is applied to the left through the mass center to hold the system in "equilibrium for study purposes". Tipping will tend to occur about the left lower edge.

Moments about that edge yields

$$W \times 7.5 = (W/g)\bar{a} \times 27$$

The maximum acceleration is thus $\bar{a} = 8.94$ ft/sec². Since $v = v_0 + at$, minimum time $t = 0.56$ sec.

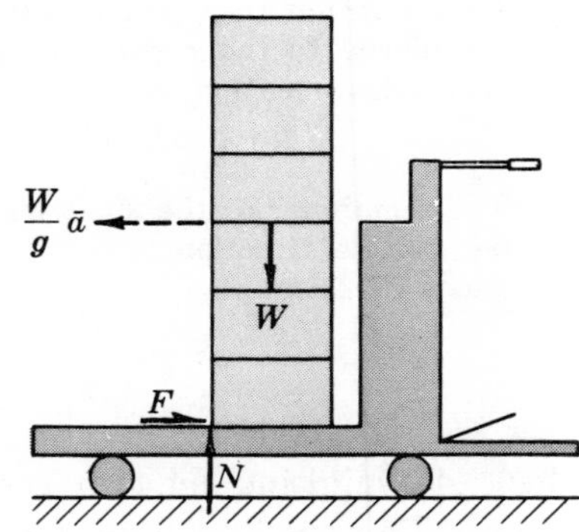

Fig. 13-18

11. The car shown in Fig. 13-19 is a conventional rear wheel drive. What force must the rear wheels exert on the ground to cause an acceleration $\bar{a}$ to the right? Neglecting the rotational inertia effect of the wheels, what is the maximum acceleration possible for a given coefficient of friction μ between the tires and the pavement?

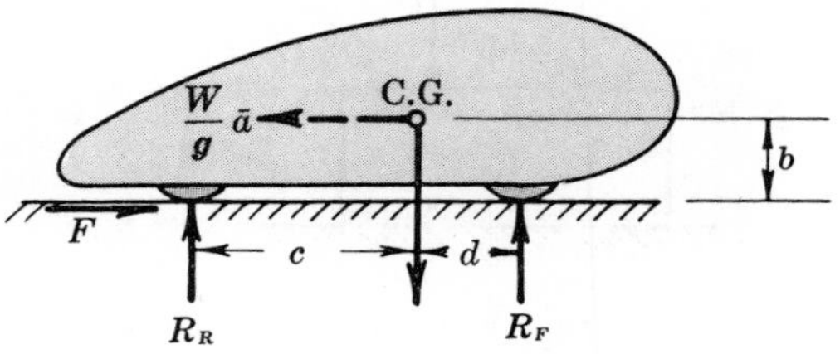

Solution:

Apply the reversed effective force $(W/g)\bar{a}$ to hold the car in "equilibrium for study purposes".

Fig. 13-19

Summing forces horizontally, $\Sigma F_h = F - (W/g)\bar{a} = 0$. From this equation, $F = (W/g)\bar{a}$.

To find the maximum acceleration possible, solve for R_R from which the greatest friction may be determined from $F = \mu R_R$.

Take moments about R_F to obtain $-R_R \times (c+d) + W \times d + (W/g)\bar{a} \times b = 0$.

From this equation, $R_R = \dfrac{Wd}{c+d} + \dfrac{Wb\bar{a}}{g(c+d)}$. Hence, $F = \mu R_R = \mu\dfrac{Wd}{c+d} + \mu\dfrac{Wb\bar{a}}{g(c+d)}$.

Substitute this value of F into $F = (W/g)\bar{a}$ and then solve for $\bar{a} = \mu dg/(c + d - \mu b)$.

12. Assuming sufficient friction to be available, what is the acceleration required to cause the car in Problem 11 to start to tip backwards?

Solution:

In this case the reaction R_F on the front wheels is zero. Sum moments about the line of contact of the rear wheels with the ground to obtain $\Sigma M_{R_R} = 0 = (W/g)\bar{a} \times b - W \times c$. Hence the acceleration to cause tipping is $\bar{a} = cg/b$.

Supplementary Problems

13. A homogeneous door weighing 200 lb is free to roll on the horizontal track on frictionless wheels at A and B as shown in Fig. 13-20 below. What horizontal force P will reduce the vertical reaction at A to zero? What will be the acceleration of the door under the action of this force P? What will be the reaction at B? *Ans.* $P = 450$ lb to left, $a = 72.5$ ft/sec² to left, $B = 200$ lb up

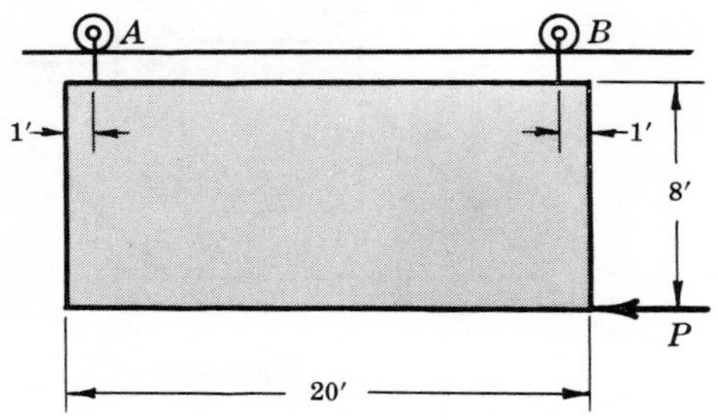

Fig. 13-20

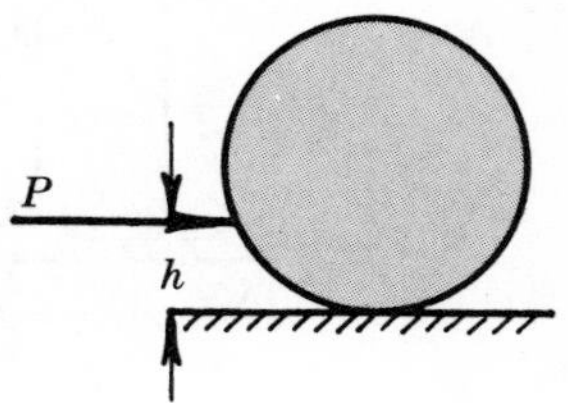

Fig. 13-21

14. A cylinder weighing 20 lb with a radius of 2 ft is pushed to the right without rotation and with acceleration 8 ft/sec². See Fig. 13-21 above. Determine the magnitude and location of the force P if the coefficient of friction is 0.20. *Ans.* $P = 8.97$ lb to right, $h = 1.11$ ft above surface

15. A homogeneous lamina with a face that is an isosceles triangle is pulled to the right with an acceleration of 4 ft/sec². See Fig. 13-22 below. Assuming the supports at A and B are frictionless, determine the force P and the reactions at A and B if the weight is 6 lb.
Ans. $P = 0.746$ lb to right, $A = 3.75$ lb up, $B = 2.25$ lb up

16. A homogeneous sphere of weight W lb and radius R is acted upon by a horizontal force P as shown in Fig. 13-23 below. The coefficient of sliding friction between the plane and sphere is μ. Where must the force P be applied in order that the sphere skids without rotating? What is the acceleration?
Ans. $d = R(1 - \mu W/P)$, $a = g(P/W - \mu)$ to right

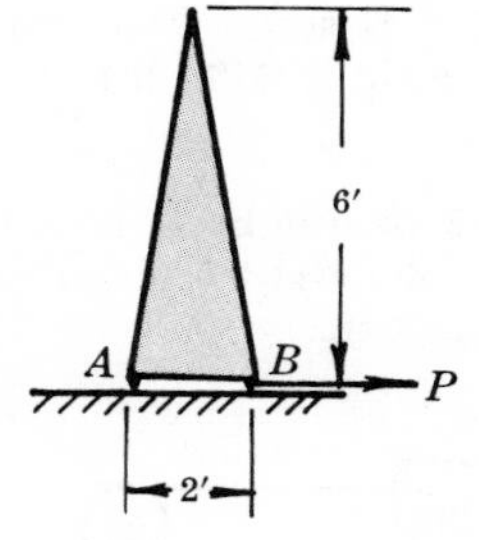

Fig. 13-22

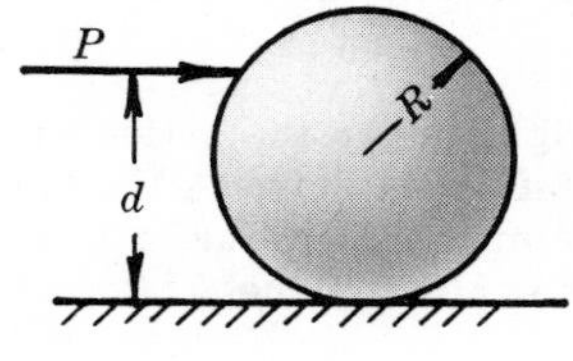

Fig. 13-23

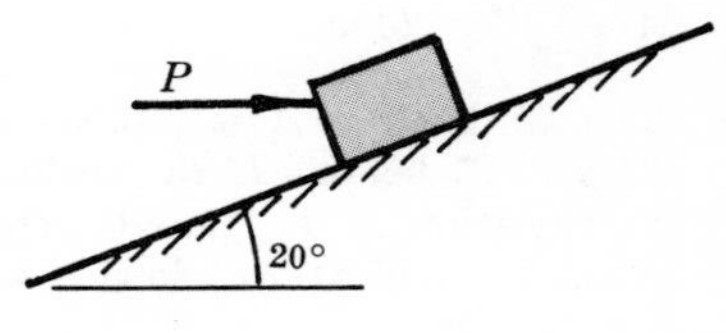

Fig. 13-24

17. Refer to Fig. 13-24 above. What horizontal force is needed to give the 100 lb block an acceleration of 8.6 ft/sec² up the 20° plane? Assume a coefficient of friction between the block and the plane of 0.25.
Ans. $P = 98.7$ lb to right

18. A body is projected up a 30° plane with initial velocity 30 ft/sec. If the coefficient of friction between the body and the plane is 0.20, how far will the body move up the plane and how long will it take to reach this point? *Ans.* 20.7 ft, 1.38 sec

19. An object weighing W lb starts from rest at the top of a plane which has a slope of 50° with the horizontal. After travelling 20 ft down the plane the object slides 30 ft across the horizontal floor and comes to rest. Determine the coefficient of friction between the object and the surfaces.
Ans. $\mu = 0.357$

20. A dish slides 18 inches on a level table before coming to rest. If the coefficient of friction between the dish and the table is 0.12, what was the time of travel? *Ans.* $t = 0.88$ sec

21. The 9.3 lb block is prevented from slipping by means of a strip as shown in Fig. 13-25 below. What is the greatest acceleration the cart may have without tipping the block?
Ans. $a = 12.1$ ft/sec^2 to left

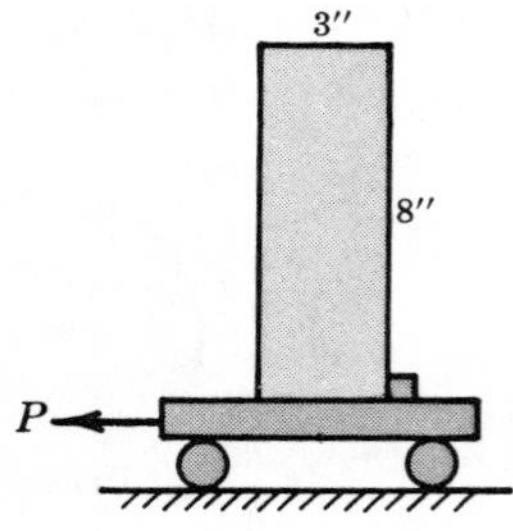

Fig. 13-25

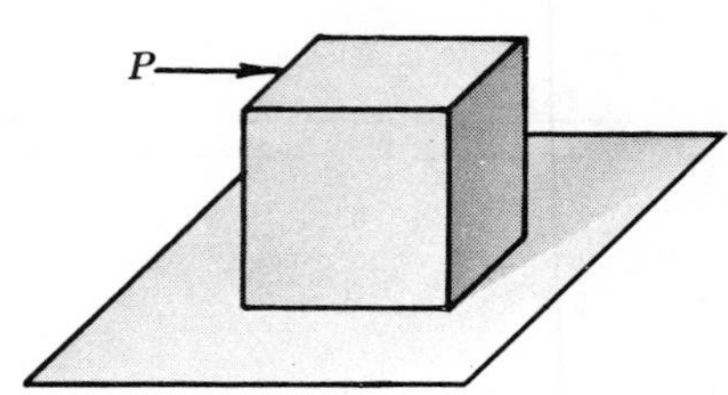

Fig. 13-26

22. The homogeneous cube is moved along the horizontal plane by a horizontal force P as shown in Fig. 13-26 above. If the coefficient of friction is 0.2, what is the maximum acceleration the block can have if it slides but is on the verge of tipping? *Ans.* $a = 19.3$ ft/sec^2 to right

23. A cylindrical shaft 6 ft in diameter and 35 inches high stands on end in a flat car that has a constant velocity of 35 mph. If the car is brought to rest with uniform deceleration in a distance of 40 ft, will the cylinder tip? *Ans.* No

24. A 20 lb weight is being lowered by a rope that can safely support 16 lb. What is the minimum acceleration the weight can have under these conditions? *Ans.* 6.44 ft/sec^2 down

25. An elevator requires 2 sec from rest to acquire a downward velocity of 10 ft/sec. Assuming uniform acceleration, what is the force of the floor during this time on an operator whose normal weight is 150 lb? *Ans.* 127 lb

26. A 10 lb weight hangs from a spring balance in an elevator which is accelerating upward at 8.05 ft/sec^2. Determine the spring balance reading. *Ans.* 12.5 lb

27. An elevator weighs 1500 lb. It is designed for a maximum acceleration of 6 ft/sec^2. What maximum load may be placed in the elevator if the allowable load in the supporting cable is 4150 lb?
Ans. $W = 2000$ lb

28. The 12 lb block A is pulled to the right under the action of the falling 12 lb weight B as shown in Fig. 13-27 below. If the coefficient of friction between the block A and the plane is 0.5, determine the acceleration of B. Also determine the maximum value of h if block A is not to tip.
Ans. $a = 8.05$ ft/sec^2 down, $h = 2.33$ ft above plane

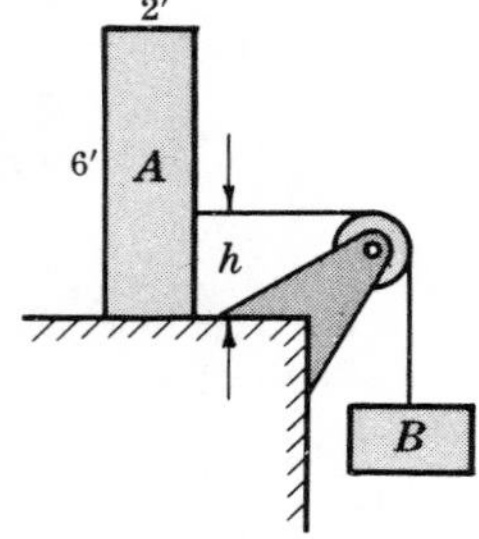

Fig. 13-27

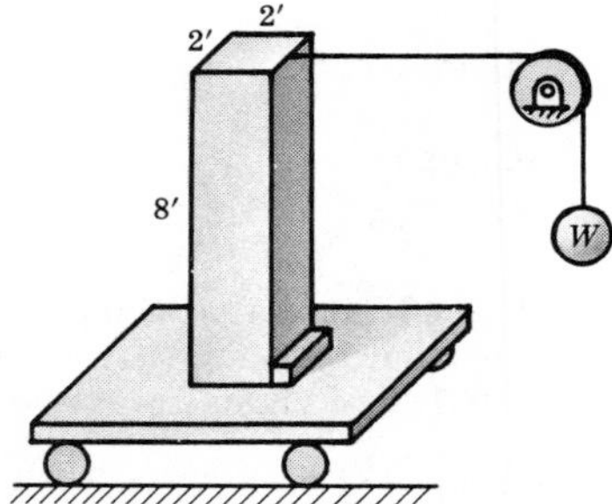

Fig. 13-28

29. The 1000 lb block stands on the truck as shown in Fig. 13-28 above. The weight W falls, accelerating the block and the 300 lb truck. A strip nailed to the platform prevents sliding. What is the maximum acceleration the system may have without tipping the block? Find W. Assume rolling of the truck and neglect the inertia effects of the wheels. *Ans.* $a = 5.03$ ft/sec^2 to right, $W = 241$ lb

30. In Fig. 13-29 below, the homogeneous bar AB weighs 10 lb and is fastened by a frictionless pin at A and rests against the smooth vertical part of the cart. The cart weighs 100 lb and is pulled to the right by horizontal force P. At what value of P will the bar exert no force on the cart at B?
Ans. $P = 131$ lb

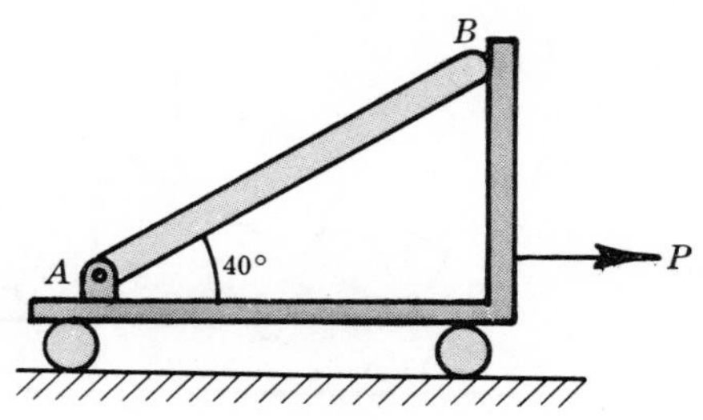

Fig. 13-29

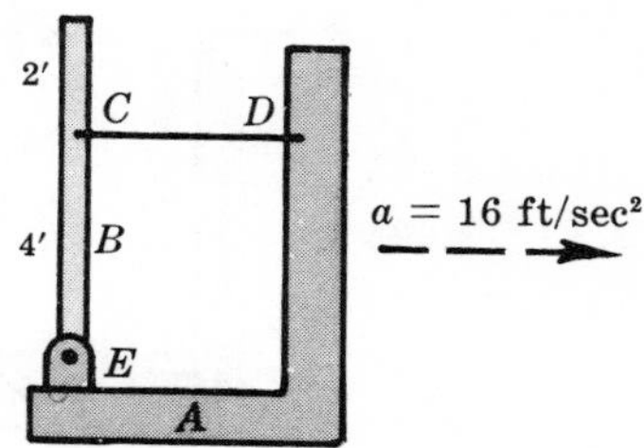

Fig. 13-30

31. The platform A accelerates to the right on a horizontal plane as shown in Fig. 13-30 above. The bar B is held in the vertical position by the horizontal cord CD. Determine the tension in the cord and the pin reactions at E. The bar is uniform and weighs 18 lb.
Ans. $T = 6.71$ lb, $E_y = 18.0$ lb up, $E_x = 2.24$ lb to right

32. In Fig. 13-31, a block is moving to the right with acceleration 5 ft/sec². A homogeneous bar 3 ft long and weighing 2 lb is suspended as shown by means of a frictionless pin at A. What is the angle θ? *Ans.* $\theta = 8.83°$

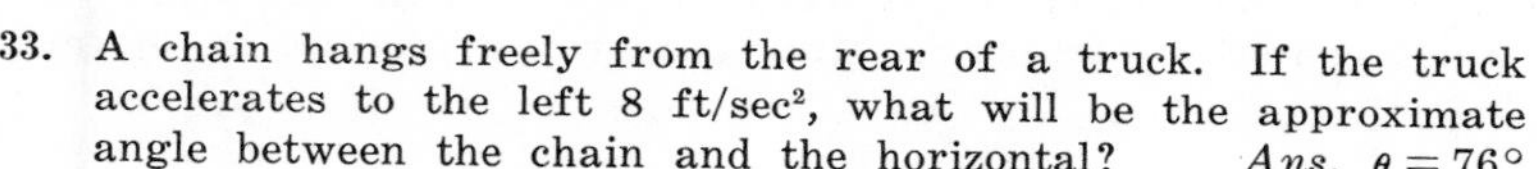

33. A chain hangs freely from the rear of a truck. If the truck accelerates to the left 8 ft/sec², what will be the approximate angle between the chain and the horizontal? *Ans.* $\theta = 76°$

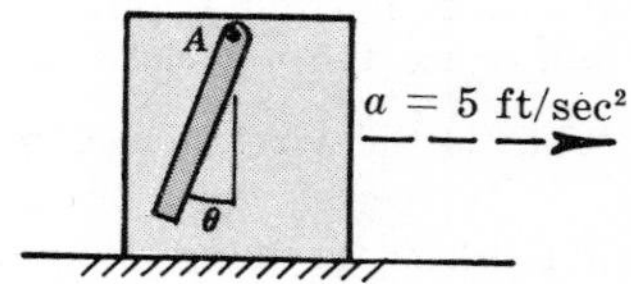

Fig. 13-31

34. A helicopter carrying an 80 ft transmission line tower by means of a sling accelerates 6 ft/sec² horizontally. What angle does the centerline of the tower make with the vertical? *Ans.* $\theta = 10.5°$

35. A small block moves with an acceleration a along the horizontal surface shown in Fig. 13-32 below. The bob at the end of the string maintains a constant angle θ with the vertical. Show that θ is a measure of the acceleration a in this simple accelerometer. *Ans.* $a = g \tan\theta$

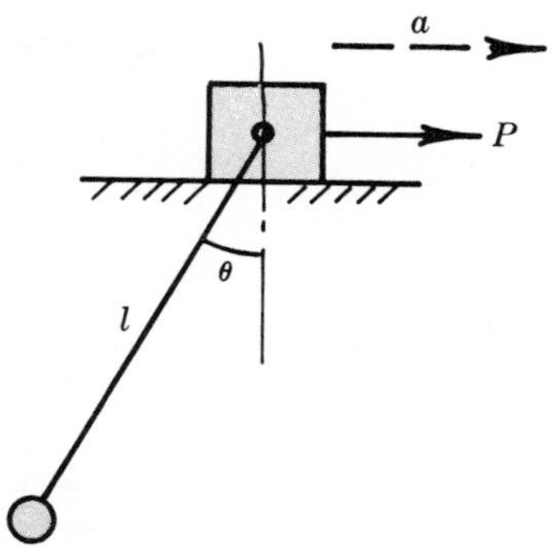

Fig. 13-32

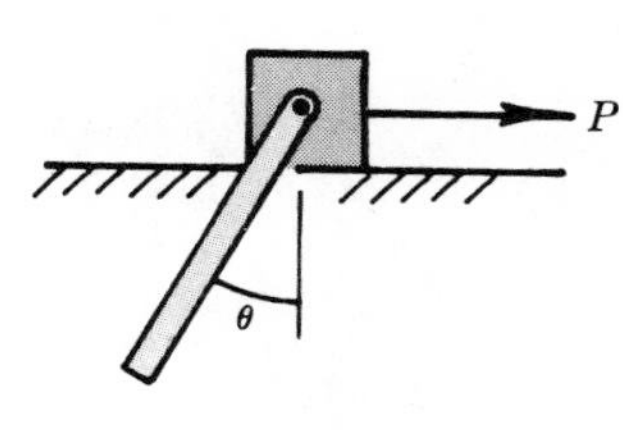

Fig. 13-33

36. In Fig. 13-33 above, the 100 lb block is subjected to an acceleration of 10 ft/sec² horizontally to the right under the action of a horizontal force P. The homogeneous slender bar weighs 20 lb and is 3 ft long. Assuming a frictionless surface, determine P and θ. What are the horizontal and vertical components of the pin reaction on the bar?
Ans. $P = 37.3$ lb, $\theta = 17.3°$, $V = 20.0$ lb, $H = 6.22$ lb

37. In the preceding problem what are the values of P and θ if the coefficient of friction between the block and the plane is 0.25? *Ans.* $P = 67.3$ lb, $\theta = 17.3°$

38. A box weighing 1000 lb is shown on a truck in Fig. 13-34 below. The coefficient of friction between the box and the truck is 0.30. Determine the forward acceleration at which the box tips or slides. Assume homogeneity. *Ans.* Tips at $a = 8.05$ ft/sec^2 to right

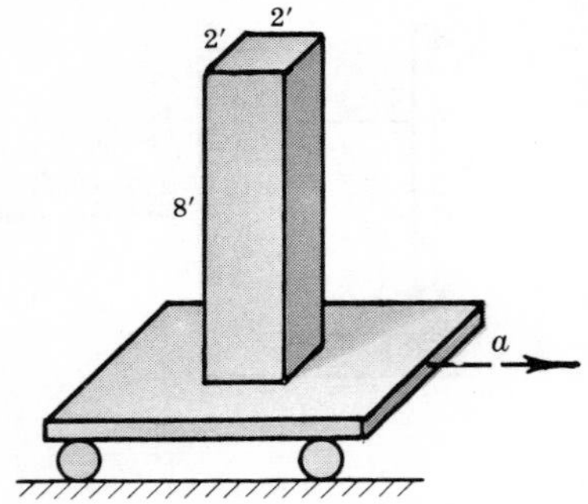

Fig. 13-34

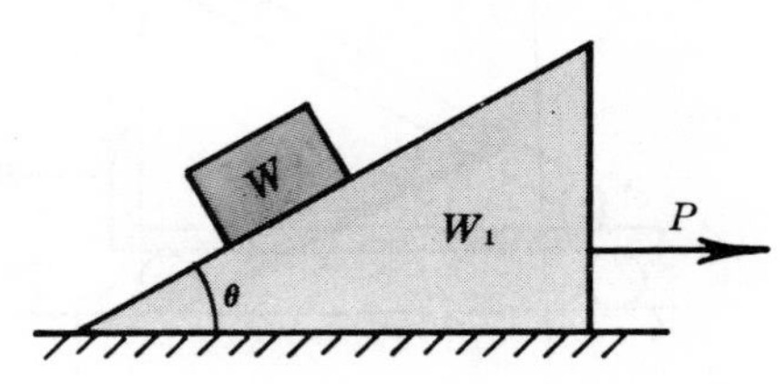

Fig. 13-35

39. The block of weight W is on a surface (the coefficient of friction is μ) which is inclined at angle θ with the horizontal as shown in Fig. 13-35 above. This surface is part of a triangular block of weight W_1 lb. A horizontal force P causes the system to have an acceleration a to the right. What value of P will cause the top block to move relative to the surface? Assume no friction on the bottom surface. What is the acceleration? *Ans.* $a = \dfrac{g(\mu - \tan\theta)}{1 + \mu\tan\theta}$, $P = \dfrac{(W + W_1)(\mu - \tan\theta)}{1 + \mu\tan\theta}$

40. A truck is traveling along a level road at a constant speed. Its body weighs W lb and contains a box weighing w lb. Show that if the box drops off the truck, the truck body will experience an upward acceleration of wg/W.

41. The 2500 lb automobile in the figure at the right is brought to rest from a speed of 60 mph. If the car is equipped with four wheel brakes and the coefficient of friction between the tires and the road is 0.6, what will be the time required to bring the car to rest? What will be the distance covered in that time?
Ans. 4.56 sec, 200 ft

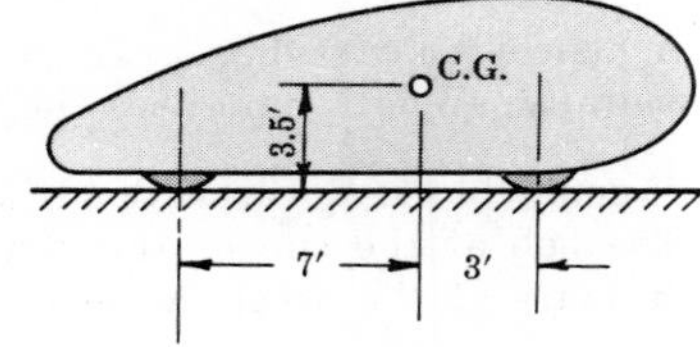

Fig. 13-36

42. In Problem 41, determine the reaction on the two front tires and on the two rear tires.
Ans. $R_F = 2275$ lb, $R_R = 225$ lb

Chapter 14

Area Moments of Inertia

AXIAL MOMENT of INERTIA of an ELEMENT of AREA

The axial moment of inertia I of an element of area about an axis in its plane is the product of the area of the element and the square of its distance from the axis.

POLAR MOMENT of INERTIA of an ELEMENT of AREA

The polar moment of inertia J of an element of area about an axis perpendicular to its plane is the product of the area of the element and the square of its distance from the axis.

In Fig. 14-1, the moments of inertia are

$$I_x = y^2\,dA$$
$$I_y = x^2\,dA$$
$$J = \rho^2\,dA = (x^2+y^2)dA = I_x + I_y$$

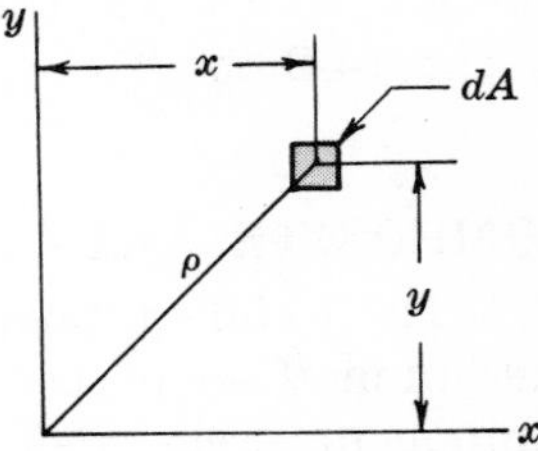

Fig. 14-1

PRODUCT of INERTIA of an ELEMENT of AREA

The product of inertia of an element of area in the figure is defined as follows:

$$I_{xy} = xy\,dA$$

AXIAL MOMENT of INERTIA of an AREA

The axial moment of inertia of an area is the summation of the axial moments of inertia of its elements.

$$I_x = \int y^2\,dA$$
$$I_y = \int x^2\,dA$$

POLAR MOMENT of INERTIA of an AREA

The polar moment of inertia of an area is the summation of the polar moments of inertia of its elements.

$$J = \int \rho^2\,dA$$

PRODUCT of INERTIA of an AREA

The product of inertia of an area is the summation of the products of inertia of its elements.

$$I_{xy} = \int xy\,dA$$

PARALLEL AXIS THEOREM

The parallel axis thoerem states that the axial or polar moment of inertia of an area about any axis equals the axial or polar moment of inertia of the area about a parallel axis through the centroid of the area plus the product of the area and the square of the distance between the two parallel axes.

In Fig. 14-2, x and y are any axes through O, while x' and y' are coplanar parallel axes through the centroid G.

$$I_x = I_{x'} + Am^2$$
$$I_y = I_{y'} + An^2$$
$$J_O = \bar{J} + Ar^2$$

The product of inertia of an area with respect to any two axes equals the product of inertia about two parallel centroidal axes plus the product of the area and the distances from respective axes.

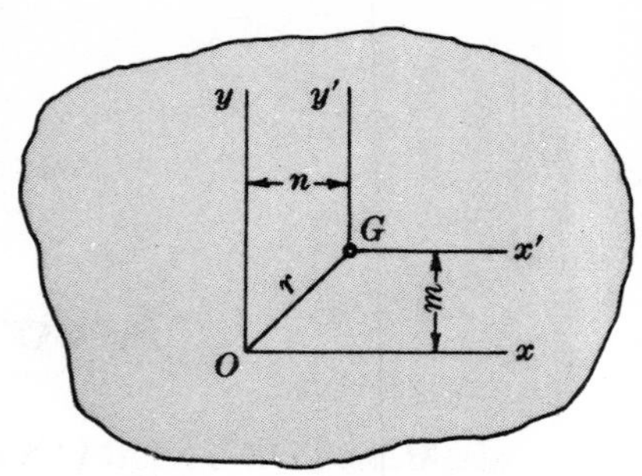

Fig. 14-2

$$I_{xy} = I_{x'y'} + Amn$$

where m and n are the perpendicular distances between the x, y and x', y' axes as shown in Fig. 14-2 above.

COMPOSITE AREA

The axial or polar moment of inertia, or product of inertia, of a composite area is the summation of the axial or polar moments of inertia, or products of inertia, of the component areas making up the whole.

Units of any of the foregoing are the fourth power of a length, usually in^4.

ROTATED SET of AXES

The moments of inertia of any area with respect to a rotated set of axes (x', y') may be expressed in terms of the moments and product of inertia with respect to the (x, y) axes as follows:

$$I_{x'} = \tfrac{1}{2}(I_x + I_y) + \tfrac{1}{2}(I_x - I_y)\cos 2\theta - I_{xy}\sin 2\theta$$
$$I_{y'} = \tfrac{1}{2}(I_x + I_y) - \tfrac{1}{2}(I_x - I_y)\cos 2\theta + I_{xy}\sin 2\theta$$
$$I_{x'y'} = \tfrac{1}{2}(I_x - I_y)\sin 2\theta + I_{xy}\cos 2\theta$$

where I_x, I_y = moments of inertia with respect to the x, y axes

$I_{x'}, I_{y'}$ = moments of inertia with respect to the x', y' axes which have the same origin as the x, y axes but are rotated through an angle θ

I_{xy} = product of inertia with respect to the x, y axes

$I_{x'y'}$ = product of inertia with respect to the x', y' axes.

For a proof of this refer to Problem 21.

Maximum moments of inertia of any area occur with respect to principal axes: a special set of x', y' axes for which $2\theta' = \tan^{-1} -2I_{xy}/(I_x - I_y)$.

$$I_{x'} = \tfrac{1}{2}(I_x + I_y) \pm \sqrt{\tfrac{1}{4}(I_x - I_y)^2 + I_{xy}^2}$$
$$I_{y'} = \tfrac{1}{2}(I_x + I_y) \mp \sqrt{\tfrac{1}{4}(I_x - I_y)^2 + I_{xy}^2}$$

For details refer to Problem 22.

MOHR'S CIRCLE

Mohr's circle is a device to eliminate memorizing of the formulas connected with rotation of axes. Refer to Problems 23 and 24.

Solved Problems

1. Determine the axial moment of inertia of the rectangle with base b and altitude h about a centroidal axis parallel to the base. Refer to Fig. 14-3 below.

Solution:

Choose an element of area dA parallel to the base and at a distance y from the centroidal x-axis.

$$I_x = \int y^2\, dA = \int_{-\frac{1}{2}h}^{\frac{1}{2}h} y^2 b\, dy = \tfrac{1}{12}bh^3$$

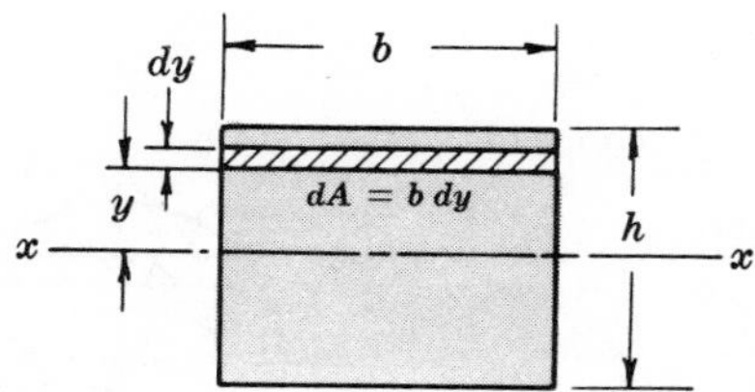

Fig. 14-3

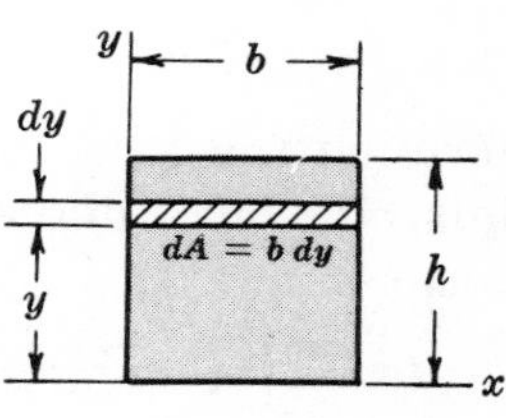

Fig. 14-4

2. Determine the axial moment of inertia of the rectangle in Problem 1 with respect to the base. Refer to Fig. 14-4 above.

Solution:

$$I_x = \int y^2\, dA = \int_0^h y^2 b\, dy = \tfrac{1}{3}bh^3$$

3. Determine the axial moment of inertia of a rectangle with respect to its base by means of the parallel axis theorem. See the adjacent Fig. 14-5. Assume that the results of Problem 1 are known.

Solution:

$$I_x = \bar{I} + A(\tfrac{1}{2}h)^2 = \tfrac{1}{12}bh^3 + bh(\tfrac{1}{4}h^2)$$
$$= \tfrac{1}{3}bh^3$$

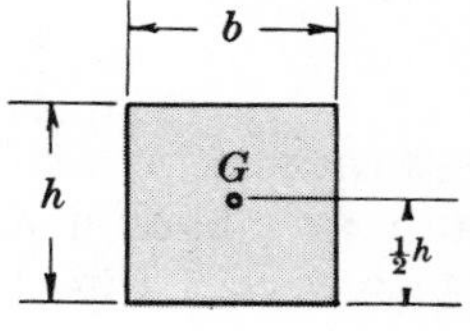

Fig. 14-5

4. Determine the axial moment of inertia for a triangle of base b and altitude h about a centroidal axis parallel to the base. Refer to Fig. 14-6 below.

Solution:

$$I_{mn} = \int y^2\, dA = \int_{-h/3}^{2h/3} y^2 z\, dy = \int_{-h/3}^{2h/3} y^2(b/h)(\tfrac{2}{3}h - y)dy = \tfrac{1}{36}bh^3$$

since by similar triangles, $b/h = z/(\frac{2}{3}h - y)$ or $z = (b/h)(\frac{2}{3}h - y)$.

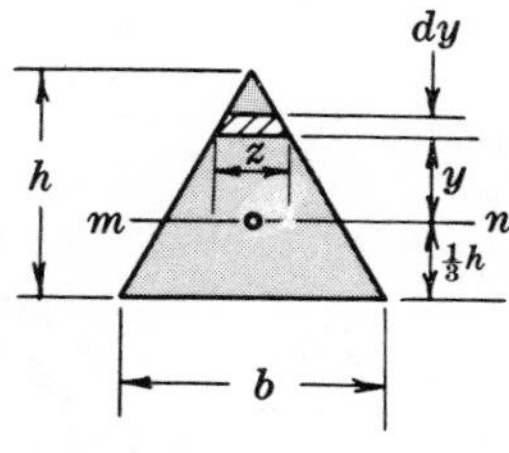

Fig. 14-6

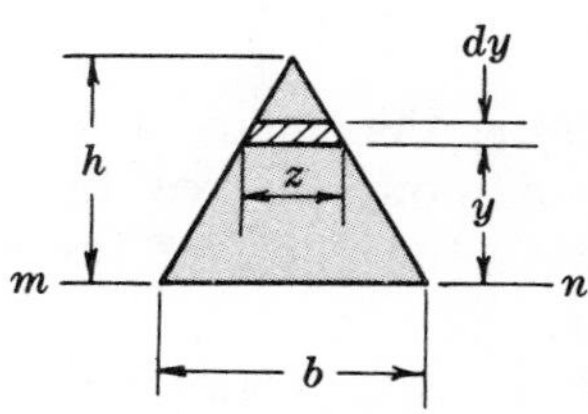

Fig. 14-7

5. Determine the axial moment of inertia for a triangle of base b and altitude h about the base. Refer to Fig. 14-7 above.

Solution:

$$I_{mn} = \int y^2\, dA = \int_0^h y^2 z\, dy = \int_0^h y^2(b/h)(h - y)dy = \tfrac{1}{12}bh^3$$

since by similar triangles, $z/(h - y) = b/h$ or $z = (b/h)(h - y)$.

6. Knowing the results of Problem 5 (a comparatively simple integration problem), find the axial moment of inertia of the triangle about a centroidal axis parallel to the base. See Fig. 14-8.

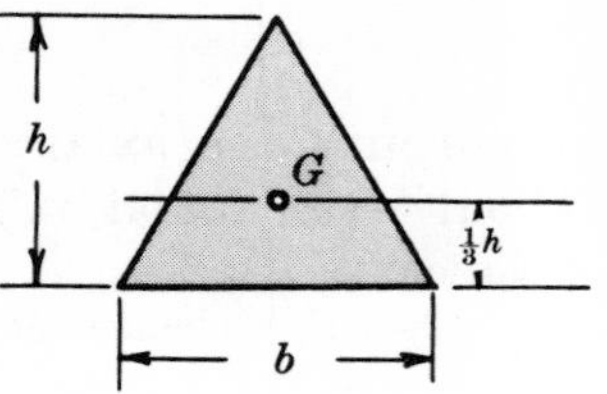

Fig. 14-8

Solution:

This is really an application of the parallel axis theorem in reverse.

$$\bar{I} = I_b - A(\tfrac{1}{3}h)^2 = \tfrac{1}{12}bh^3 - (\tfrac{1}{2}bh)(\tfrac{1}{3}h)^2 = \tfrac{1}{36}bh^3$$

7. Determine the axial moment of inertia for a circle of radius r about a diameter. Refer to Fig. 14-9.

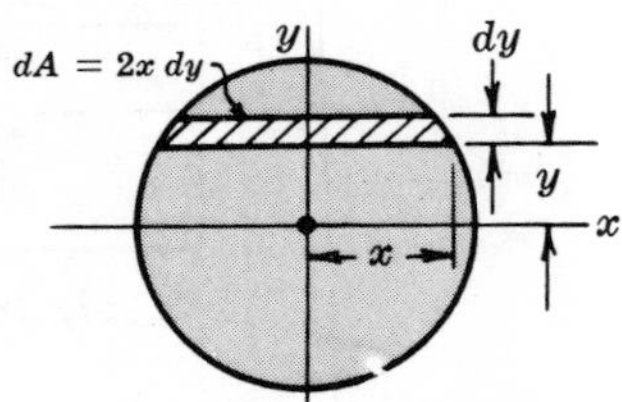

Fig. 14-9

Solution:

$$\begin{aligned} I_x &= \int y^2\,dA = 2\int_0^r y^2(2x\,dy) = 4\int_0^r y^2\sqrt{r^2-y^2}\,dy \\ &= 4[-\tfrac{1}{4}y\sqrt{(r^2-y^2)^3} + \tfrac{1}{8}r^2(y\sqrt{r^2-y^2} + r^2\sin^{-1}y/r)]_0^r \\ &= 4[0 + \tfrac{1}{8}r^2(0 + r^2\sin^{-1}1) + 0 - \tfrac{1}{8}r^2(0+0)] \\ &= 4(\tfrac{1}{8}r^4)(\tfrac{1}{2}\pi) = \tfrac{1}{4}\pi r^4 \end{aligned}$$

Note that this integral could have been evaluated from $-r$ to r instead of twice the integral from 0 to r. This is permissible since the moment of inertia of the two halves of the area is equal to the moment of inertia of the whole.

8. Determine the axial moment of inertia of a circle of radius r about a diameter, using the differential area dA shown in Fig. 14-10.

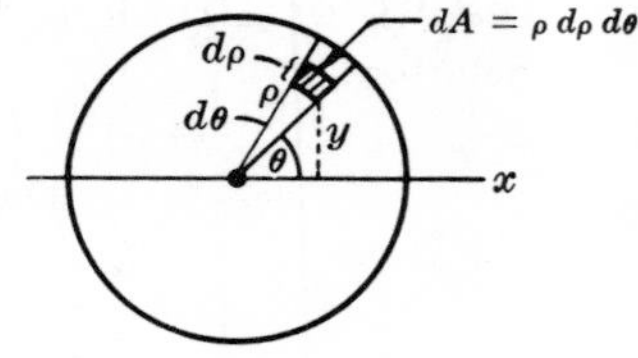

Fig. 14-10

Solution:

$$I_x = \int y^2\,dA$$

where $y = \rho\sin\theta$ and $dA = \rho\,d\rho\,d\theta$. Then,

$$\begin{aligned} I_x &= \int_0^{2\pi}\int_0^r \rho^3\,d\rho\,\sin^2\theta\,d\theta = \int_0^{2\pi}\sin^2\theta\,d\theta\left[\frac{\rho^4}{4}\right]_0^r \\ &= \tfrac{1}{4}r^4[\tfrac{1}{2}\theta - \tfrac{1}{4}\sin 2\theta]_0^{2\pi} = \tfrac{1}{4}r^4(\pi - \tfrac{1}{4}\sin 4\pi - 0 + \tfrac{1}{4}\sin 0) = \tfrac{1}{4}\pi r^4 \end{aligned}$$

This problem illustrates the ease of finding moments of inertia if the proper choice of element of area is made.

9. Determine the polar moment of inertia for a circle of radius r about an axis through its center and perpendicular to the plane of the circle. Refer to Fig. 14-11.

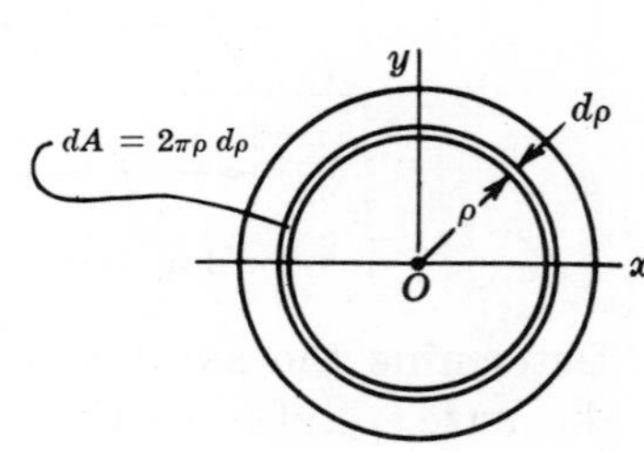

Fig. 14-11

Solution:

Choose as the differential area an annular ring of radius ρ and thickness $d\rho$. Hence dA equals the circumference, $2\pi\rho$, times the thickness $d\rho$.

$$\bar{J} = \int \rho^2\,dA = \int_0^r \rho^2 2\pi\rho\,d\rho = 2\pi[\tfrac{1}{4}\rho^4]_0^r = \tfrac{1}{2}\pi r^4$$

At this point it is possible to derive the value of the axial moment of inertia about a diameter. Since $\bar{J} = I_x + I_y$ and $I_x = I_y$, then $I_x = \tfrac{1}{2}\bar{J} = \tfrac{1}{4}\pi r^4$.

10. Determine the axial and polar moments of inertia for the ellipse shown in the adjacent Fig. 14-12.

Solution:

$$I_x = \int y^2\,dA = \int_{-b}^{b} y^2(2x\,dy)$$

where

$x^2/a^2 + y^2/b^2 = 1$ or $x = (a/b)\sqrt{b^2 - y^2}$. Then

Fig. 14-12

$$I_x = \int_{-b}^{b} y^2(2a/b)\sqrt{b^2-y^2}\,dy$$

$$= (2a/b)\left[-\tfrac{1}{4}y\sqrt{(b^2-y^2)^3} + \tfrac{1}{8}b^2\left(y\sqrt{b^2-y^2} + b^2\sin^{-1}y/b\right]_{-b}^{b} = \tfrac{1}{4}\pi ab^3$$

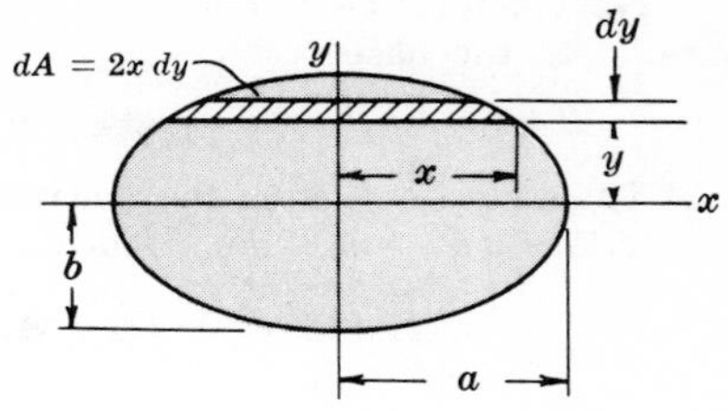

A similar integration would yield $I_y = \frac{1}{4}\pi a^3 b$. Of course, $\bar{J} = I_x + I_y = \frac{1}{4}\pi ab[a^2 + b^2]$.

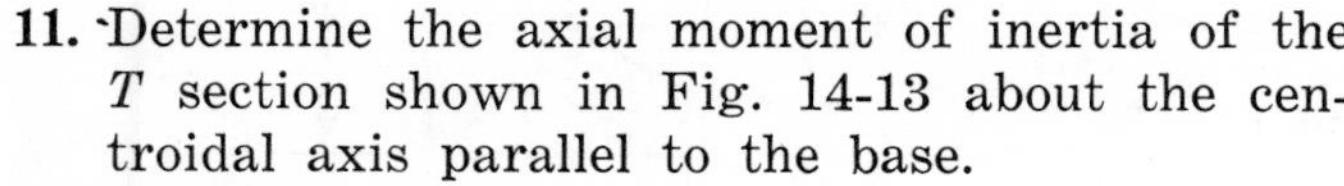

11. Determine the axial moment of inertia of the T section shown in Fig. 14-13 about the centroidal axis parallel to the base.

Solution:

The first step is to locate the centroid G using the two subdivisions of area shown.

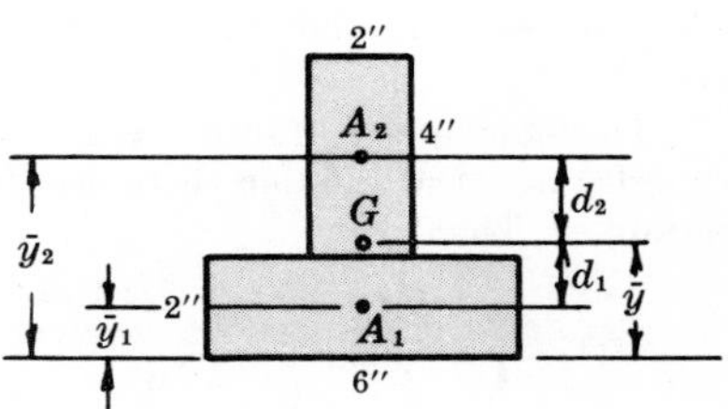

Fig. 14-13

Subdivision	Area	Distance from base to centroid
A_1	12	1
A_2	8	4

Hence $(A_1 + A_2)\bar{y} = A_1\bar{y}_1 + A_2\bar{y}_2$, $(12+8)\bar{y} = 12\times 1 + 8\times 4$, and $\bar{y} = 2.2$ in.

Next determine I for each subdivision about its own centroidal axis parallel to the base, using $I = \frac{1}{12}bh^3$.

$$I_1 = \tfrac{1}{12}(6)(2^3) = 4 \text{ in}^4, \qquad I_2 = \tfrac{1}{12}(2)(4^3) = 10.7 \text{ in}^4$$

The final step is to transfer from each subdivision's centroidal axis to the axis through G to determine $\bar{I}$ for the entire area. By the parallel axis theorem, ($d_1 = 2.2 - 1 = 1.2$; $d_2 = 4 - 2.2 = 1.8$),

$$\bar{I} = (I_1 + A_1 d_1^2) + (I_2 + A_2 d_2^2) = (4 + 12\times 1.44) + (10.7 + 8\times 3.24) = 57.9 \text{ in}^4$$

12. Determine the axial moment of inertia about a centroidal axis parallel to the base of the angle shown in Fig. 14-14.

Solution:

First locate the centroidal axis parallel to the base. Consider the component areas *1* and *2*.

Component	Area	Distance from base to centroid
1	6	3
2	3	$\frac{1}{2}$

The equation of first moments is

$$(A_1 + A_2)\bar{y} = A_1\bar{y}_1 + A_2\bar{y}_2$$

Hence $(6+3)\bar{y} = 6\times 3 + 3\times\frac{1}{2}$ and $\bar{y} = 2.17$ in.

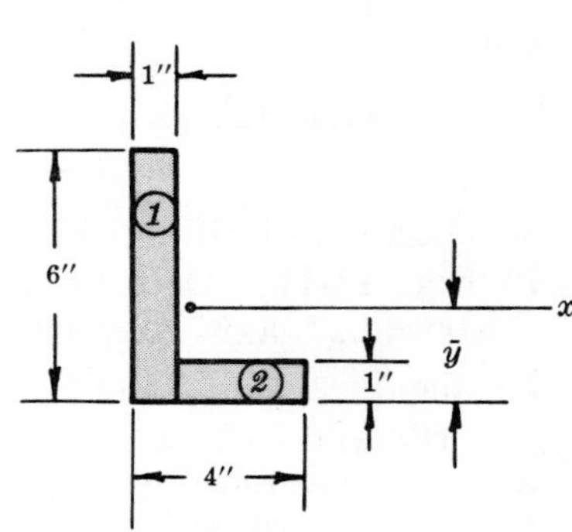

Fig. 14-14

Next determine I_1 and I_2 for the components about axes through their individual centroids and parallel to the base.

$$I_1 = \tfrac{1}{12}b_1h_1^3 = \tfrac{1}{12}\times 1\times 6^3 = 18.0 \text{ in}^4, \qquad I_2 = \tfrac{1}{12}b_2h_2^3 = \tfrac{1}{12}\times 3\times 1^3 = 0.25 \text{ in}^4$$

The distance d_1 from the centroid of 1 to the common centroid is $d_1 = 3 - 2.17 = 0.83$ in. Similarly, $d_2 = 2.17 - 0.50 = 1.67$ in. Hence

$$\bar{I} = (I_1 + A_1d_1^2) + (I_2 + A_2d_2^2) = (18.0 + 6\times\overline{0.83}^2) + (0.25 + 3\times\overline{1.67}^2) = 30.8 \text{ in}^4$$

13. Determine the axial moment of inertia for the channel shown in Fig. 14-15 about a centroidal axis parallel to the base b.

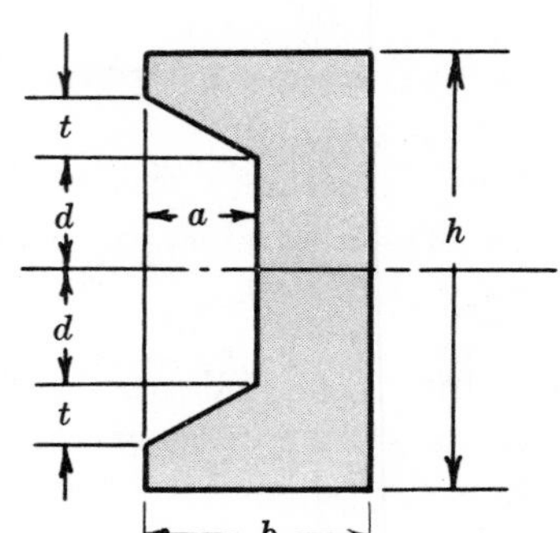

Fig. 14-15

Solution:

In this case the centroidal axis is, by symmetry, at half the height.

Consider the channel as made up of a rectangle of base b and altitude h from which has been deleted two triangles of altitude t and base a together with a rectangle of base a and height $2d$. Refer to Fig. 14-16 below.

To determine I_x for the channel, subtract I_x of the triangles and the smaller rectangle from I_x for the large rectangle. That is,

$$I_x = I_1 - (I_2 + I_3 + I_4)$$

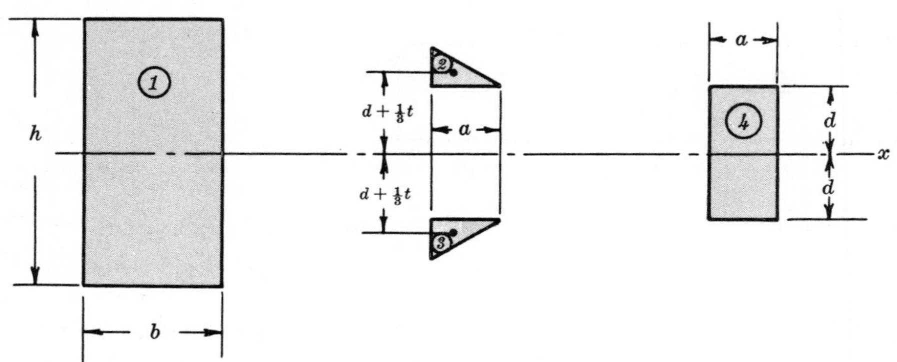

Fig. 14-16

Then

$$I_1 = \tfrac{1}{12}bh^3$$
$$I_2 = I_3 = \tfrac{1}{36}at^3 + \tfrac{1}{2}at(d+\tfrac{1}{3}t)^2 = \tfrac{1}{36}at^3 + \tfrac{1}{2}atd^2 + \tfrac{1}{3}at^2d + \tfrac{1}{18}at^3$$
$$I_4 = \tfrac{1}{12}a(2d)^3 = \tfrac{2}{3}ad^3$$

and

$$I_x = \tfrac{1}{12}bh^3 - atd^2 - \tfrac{2}{3}at^2d - \tfrac{1}{6}at^3 - \tfrac{2}{3}ad^3$$

Of course, this result will be less formidable when numerical dimensions are assigned.

14. A column is built up of 2-inch planks (actual size) as shown in Fig. 14-17. Determine the axial moment of inertia about a centroidal axis parallel to a side.

Solution:

The centroid is located by inspection at the midpoint. To determine $\bar{I}$ it is only necessary to double the summation of the axial moments of areas *1* and *2* about a line through the midpoint.

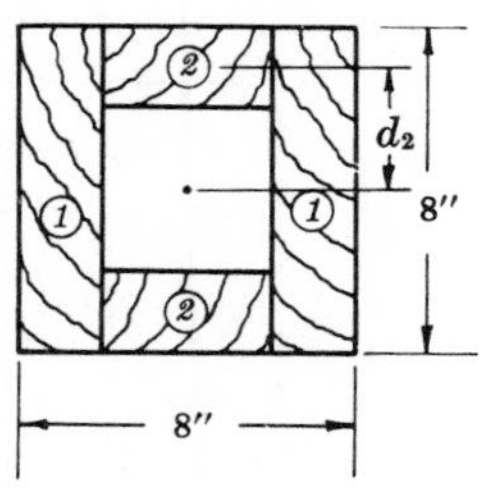

Fig. 14-17

$$I_1 = \tfrac{1}{12}b_1h_1^3 = \tfrac{1}{12}(2)(8^3) = 85.3 \text{ in}^4$$
$$I_2 = \tfrac{1}{12}b_2h_2^3 + A_2d_2^2 = \tfrac{1}{12}(4)(2^3) + (8)(3^2) = 74.7 \text{ in}^4$$
$$\bar{I} = 2(85.3 + 74.7) = 320 \text{ in}^4$$

Another technique is to subtract I for the inner square from I for the outer square, both about an axis through the midpoint and parallel to a side.

$$\bar{I} = I_o - I_i = \tfrac{1}{12}b_o h_o^3 - \tfrac{1}{12}b_i h_i^3 = \tfrac{1}{12}(8)(8^3) - \tfrac{1}{12}(4)(4^3) = 320 \text{ in}^4$$

15. Determine the axial moment of inertia about the centroidal axis of the Z section shown in the adjacent Fig. 14-18.

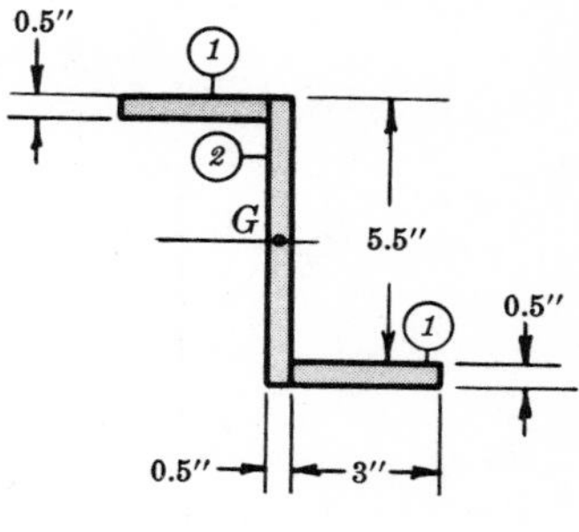

Fig. 14-18

Solution:

The centroidal axis is an axis of symmetry in this case. The axial moment of inertia $\bar{I}$ is then equal to the summation of the axial moments of area *2* and two areas *1*.

$$I_2 = \tfrac{1}{12}b_2 h_2^3 = \tfrac{1}{12}(0.5)(6.0)^3 = 9 \text{ in}^4$$

$$I_1 = \tfrac{1}{12}b_1 h_1^3 + A_1 d_1^2 = \tfrac{1}{12}(3)(0.5)^3 + 1.5(2.75)^2 = 11.4 \text{ in}^4$$

$$\bar{I} = 2I_1 + I_2 = 31.8 \text{ in}^4$$

16. What is the product of inertia about two adjacent sides of a rectangle of base b and altitude h?

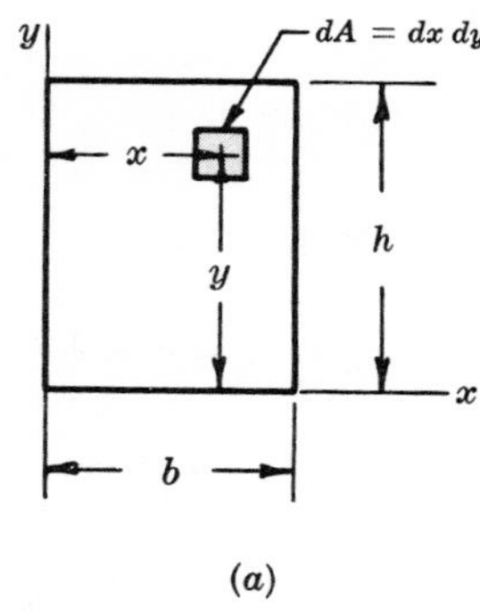

(a)

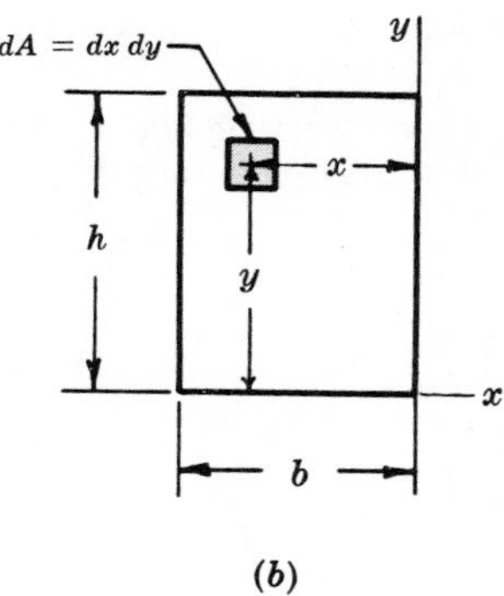

(b)

Fig. 14-19

Solution:

As shown in Fig. 14-19(a) above, the x and y axes are along two adjacent sides. Denote the product of inertia as I_{xy}. Then

$$I_{xy} = \int xy\,dA = \int_0^h \int_0^b xy\,dx\,dy = \tfrac{1}{2}x^2\Big]_0^b \tfrac{1}{2}y^2\Big]_0^h = \tfrac{1}{4}b^2h^2$$

Next choose the y-axis as the right side of the rectangle (see Fig. 14-19(b) above). Then

$$I_{xy} = \int_0^h \int_{-b}^0 xy\,dx\,dy = \tfrac{1}{2}x^2\Big]_{-b}^0 \tfrac{1}{2}y^2\Big]_0^h = (0 - \tfrac{1}{2}b^2)(\tfrac{1}{2}h^2 - 0) = -\tfrac{1}{4}b^2h^2$$

This indicates that the product of inertia may be positive or negative, depending on the location of the area relative to the axes.

17. Determine the product of inertia about two centroidal axes parallel to the sides of a rectangle of base b and altitude h.

Solution:

In Fig. 14-20 below it is seen that the x' limits of integration are from $-b/2$ to $b/2$. The y' limits of integration are from $-h/2$ to $h/2$. Hence

$$I_{x'y'} = \int x'y'\,dA = \int_{-h/2}^{h/2} \int_{-b/2}^{b/2} x'y'\,dx'\,dy' = 0$$

This could also be deduced from the result of Problem 16 using the Parallel Axis Theorem for product of inertia.

$$I_{xy} = I_{x'y'} + A(\tfrac{1}{2}b)(\tfrac{1}{2}h)$$

where $\frac{1}{2}b$ and $\frac{1}{2}h$ are the perpendicular distances between the x,y and x',y' axes.

$$I_{x'y'} = \tfrac{1}{4}b^2h^2 - bh(\tfrac{1}{2}b)(\tfrac{1}{2}h) = 0$$

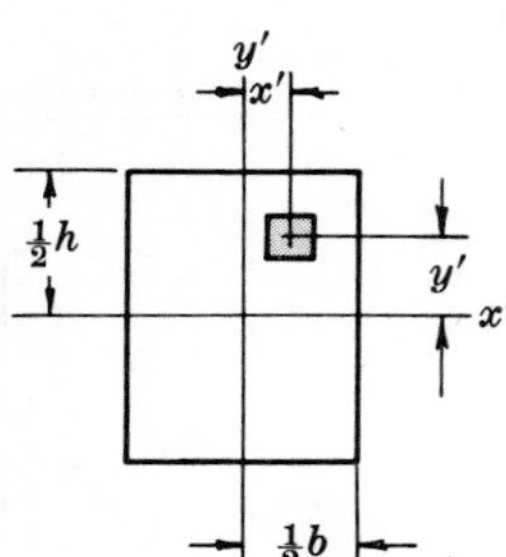

Fig. 14-20

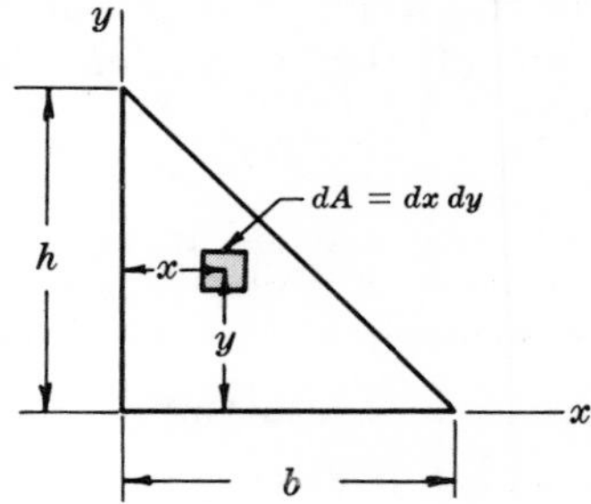

Fig. 14-21

18. Determine the product of inertia with respect to the base and altitude of a right triangle. Refer to Fig. 14-21 above.

Solution:

By definition, $I_{xy} = \int_0^h \int_0^x xy\,dx\,dy.$

The upper limit of the x integration depends on y. Therefore it must be evaluated from the equation of the sloping line which is

$$y = -(h/b)x + h \qquad \text{or} \qquad x = -(b/h)(y-h)$$

$$I_{xy} = \int_0^h \int_0^{-(b/h)(y-h)} xy\,dx\,dy = \int_0^h [\tfrac{1}{2}x^2]_0^{-(b/h)(y-h)}\,y\,dy = \int_0^h \frac{b^2}{2h^2}(y^2 - 2yh + h^2)\,y\,dy = \frac{b^2h^2}{24}$$

19. Determine the product of inertia with respect to the bounding radii of a quadrant of a circle of radius r. See Fig. 14-22.

Solution:

By definition, $I_{xy} = \int_0^r \int_0^x xy\,dx\,dy.$

Here the double integration means a summation first with respect to the variable x which depends on y according to the equation $x^2 + y^2 = r^2$. Substituting,

$$I_{xy} = \int_0^r \int_0^{\sqrt{r^2-y^2}} xy\,dx\,dy = \int_0^r [\tfrac{1}{2}x^2]_0^{\sqrt{r^2-y^2}}\,y\,dy$$

$$= \int_0^r \tfrac{1}{2}(r^2-y^2)y\,dy = \tfrac{1}{8}r^4$$

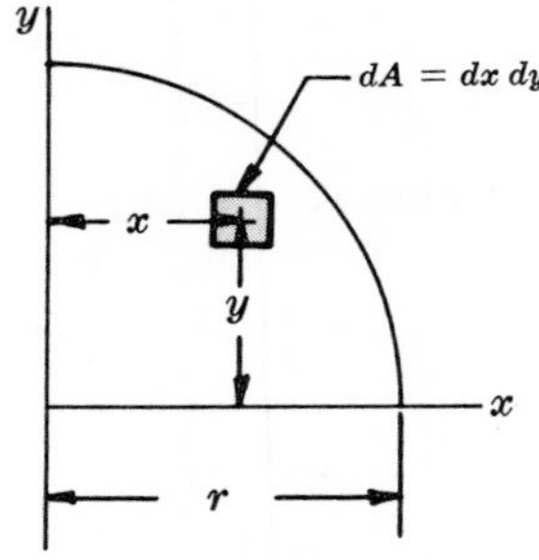

Fig. 14-22

20. Repeat Problem 19 using polar coordinates. See Fig. 14-23.

Solution:

As before, $I_{xy} = \iint xy\,dA = \int_0^r \int_0^{\pi/2} \rho\cos\theta\;\rho\sin\theta\;\rho\,d\rho\,d\theta.$

$$I_{xy} = \int_0^{\pi/2} [\tfrac{1}{4}\rho^4]_0^r \cos\theta\,\sin\theta\,d\theta = \tfrac{1}{4}r^4[\tfrac{1}{2}\sin^2\theta]_0^{\pi/2} = \tfrac{1}{8}r^4$$

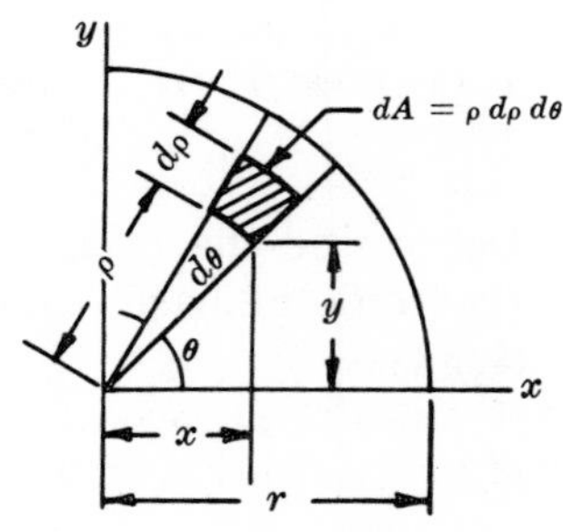

Fig. 14-23

21. Show that the moments of inertia of an area with respect to the rotated set of axes (x', y') may be expressed as follows:

$$I_{x'} = \tfrac{1}{2}(I_x + I_y) + \tfrac{1}{2}(I_x - I_y)\cos 2\theta - I_{xy}\sin 2\theta$$
$$I_{y'} = \tfrac{1}{2}(I_x + I_y) - \tfrac{1}{2}(I_x - I_y)\cos 2\theta + I_{xy}\sin 2\theta$$
$$I_{x'y'} = \tfrac{1}{2}(I_x - I_y)\sin 2\theta + I_{xy}\cos 2\theta$$

where I_x, I_y = moments of inertia with respect to the x, y axes
$I_{x'}, I_{y'}$ = moments of inertia with respect to the x', y' axes
I_{xy} = product of inertia with respect to the x, y axes
$I_{x'y'}$ = product of inertia with respect to the x', y' axes.

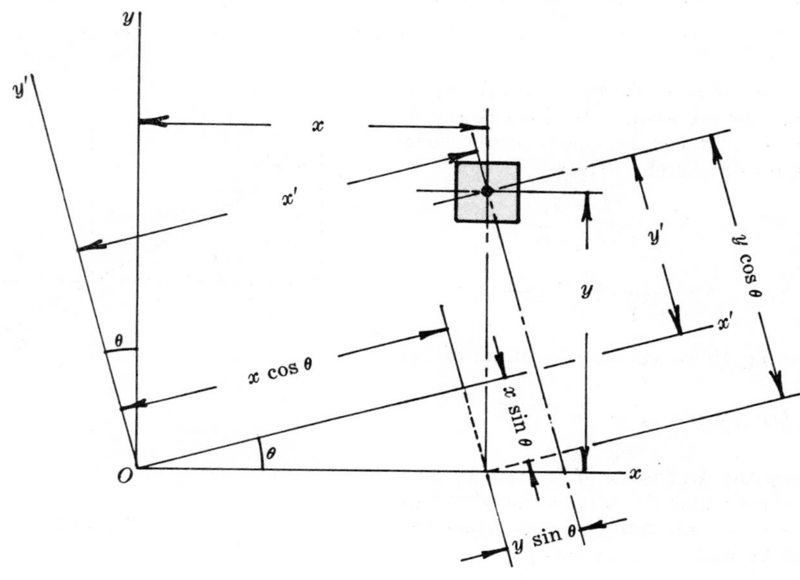

Fig. 14-24

Solution:

Fig. 14-24 above indicates an element dA of the area. By definition,

$$I_{x'} = \int y'^2\,dA, \qquad I_{y'} = \int x'^2\,dA, \qquad I_{x'y'} = \int x'y'\,dA \tag{1}$$

But from the figure, $x' = x\cos\theta + y\sin\theta$ and $y' = -x\sin\theta + y\cos\theta$. Squaring,

$$x'^2 = x^2\cos^2\theta + 2xy\cos\theta\sin\theta + y^2\sin^2\theta$$
$$y'^2 = x^2\sin^2\theta - 2xy\sin\theta\cos\theta + y^2\cos^2\theta$$

Also,
$$x'y' = -x^2\cos\theta\sin\theta - xy\sin^2\theta + xy\cos^2\theta + y^2\cos\theta\sin\theta$$

Substituting into equation (*1*),

$$I_{x'} = \int x^2\sin^2\theta\,dA - \int xy\sin 2\theta\,dA + \int y^2\cos^2\theta\,dA$$

$$I_{y'} = \int x^2\cos^2\theta\,dA + \int xy\sin 2\theta\,dA + \int y^2\sin^2\theta\,dA$$

$$I_{x'y'} = \int -x^2\cos\theta\sin\theta\,dA + \int y^2\sin\theta\cos\theta\,dA + \int xy(\cos^2\theta - \sin^2\theta)\,dA$$

Since the integration over the area is independent of θ, the above equations may be writter (using $I_x = \int y^2\,dA$, $I_y = \int x^2\,dA$, and $I_{xy} = \int xy\,dA$) as

$$I_{x'} = I_y\sin^2\theta - I_{xy}\sin 2\theta + I_x\cos^2\theta$$
$$I_{y'} = I_y\cos^2\theta + I_{xy}\sin 2\theta + I_x\sin^2\theta$$
$$I_{x'y'} = \tfrac{1}{2}(I_x - I_y)\sin 2\theta + I_{xy}\cos 2\theta$$

Using $\cos^2\theta = \frac{1}{2}(1+\cos 2\theta)$ and $\sin^2\theta = \frac{1}{2}(1-\cos 2\theta)$, we finally obtain

$$I_{x'} = \tfrac{1}{2}(I_x+I_y) + \tfrac{1}{2}(I_x-I_y)\cos 2\theta - I_{xy}\sin 2\theta$$
$$I_{y'} = \tfrac{1}{2}(I_x+I_y) - \tfrac{1}{2}(I_x-I_y)\cos 2\theta + I_{xy}\sin 2\theta$$
$$I_{x'y'} = \tfrac{1}{2}(I_x-I_y)\sin 2\theta + I_{xy}\cos 2\theta$$

22. Determine the values of $I_{x'}$ and $I_{y'}$ in Problem 21 with respect to the principal axes (these axes yield maximum or minimum values of I).

Solution:

To determine the value of θ which will make $I_{x'}$ a maximum, take the derivative of $I_{x'}$ with respect to θ and equate the resulting expression to zero. Thus

$$\frac{dI_{x'}}{d\theta} = \tfrac{1}{2}(I_x-I_y)(-2\sin 2\theta) - I_{xy}(2\cos 2\theta) = 0 \qquad \text{or} \qquad \tan 2\theta' = -\frac{2I_{xy}}{I_x-I_y}$$

This is the value θ' of θ which will make $I_{x'}$ a maximum (or minimum). It is necessary to evaluate $\sin 2\theta'$ and $\cos 2\theta'$, most easily done from the adjoining sketch. Thus

$$\cos 2\theta' = \pm\frac{I_x-I_y}{\sqrt{(I_x-I_y)^2+4I_{xy}^2}}$$

$$\sin 2\theta' = \mp\frac{2I_{xy}}{\sqrt{(I_x-I_y)^2+4I_{xy}^2}}$$

Substituting these values and simplifying, we obtain

$$I_{x'} = \tfrac{1}{2}(I_x+I_y) \pm \sqrt{[\tfrac{1}{2}(I_x-I_y)]^2+I_{xy}^2}$$

By taking the derivative of $I_{y'}$ with respect to θ, it can be seen that the same value θ' makes $I_{y'}$ a maximum (or minimum). Substituting the values of $\sin 2\theta'$ and $\cos 2\theta'$ yields

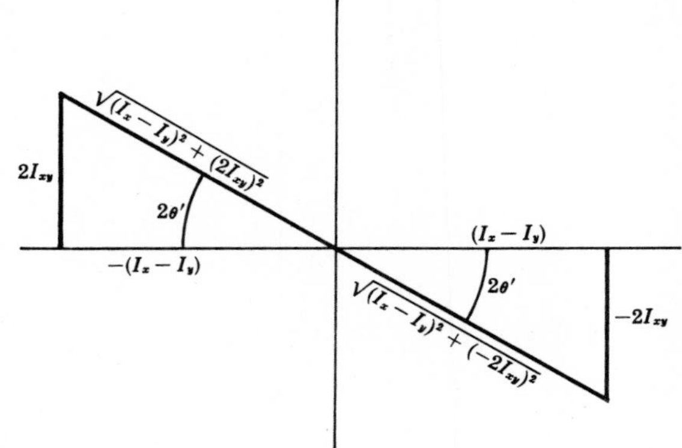

Fig. 14-25

$$I_{y'} = \tfrac{1}{2}(I_x+I_y) \mp \sqrt{[\tfrac{1}{2}(I_x-I_y)]^2+I_{xy}^2}$$

Since $I_{y'}$ has a negative sign before the radical for the value of θ' that gives $I_{x'}$ a positive sign before the radical, we can consolidate the results by stating that with respect to the principal axes x', y' one value say $I_{x'}$ is maximum while the other value $I_{y'}$ is minimum. Note further that for this particular θ' (principal axes),

$$I_{x'y'} = \tfrac{1}{2}(I_x-I_y)\sin 2\theta + I_{xy}\cos 2\theta = 0$$

23. Given I_x, I_y, and I_{xy} for an area with respect to x, y axes. Determine graphically, using Mohr's circle, the values $I_{x'}$, $I_{y'}$, and $I_{x'y'}$ for a set of x', y' axes located counterclockwise at an angle θ with respect to the x, y set of axes.

Solution:

Fig. 14-26(a) shows the orientation of the axes. In Fig. 14-26(b) is drawn an orthogonal set of lines. Any moment of inertia I will be located to the right of the vertical line. Any product of inertia will

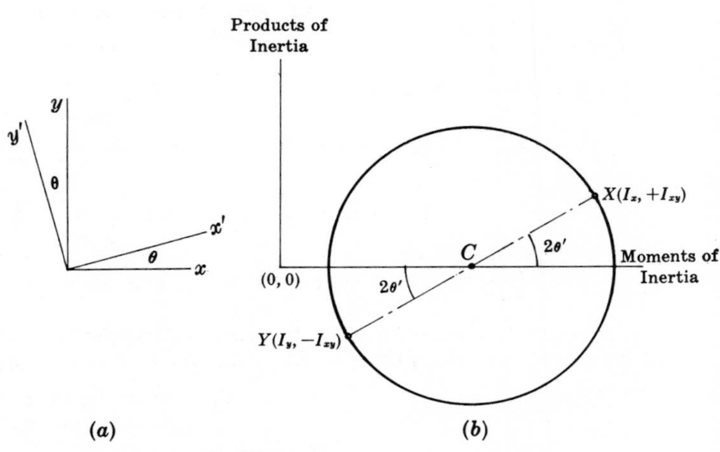

Fig. 14-26

be located above or below the horizontal line.

Assume $I_x > I_y$ and that I_{xy} is positive. Locate the point X which has coordinates $(I_x, +I_{xy})$ and the point Y which has coordinates $(I_y, -I_{xy})$. Draw the line XY which crosses the horizontal line at C as shown. Draw a circle with C as center and containing X and Y (see Fig. 14-27).

Next draw a line at an angle 2θ from the XY line in a counterclockwise direction. The coordinates of X' and Y' are the values of $I_{x'}$, $I_{y'}$ and $I_{x'y'}$.

These values can be seen as follows. Draw vertical lines XA and YB. The distance $BC = CA = \frac{1}{2}(I_x - I_y)$. The distance XA is I_{xy}. The radius of the circle CX is the hypotenuse of the right triangle ACX. As such any radius equals $\sqrt{[\frac{1}{2}(I_x - I_y)]^2 + I_{xy}^2}$.

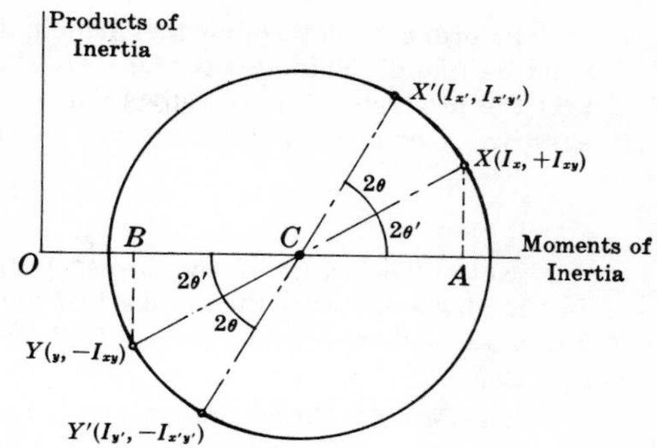

Fig. 14-27

Now the I coordinate of X' equals the distance from O to the center $\frac{1}{2}(I_x + I_y)$ plus the projection of CX' on the horizontal. CX' (a radius) makes an angle $(2\theta + 2\theta')$ with the horizontal I axis. The $2\theta'$ is the angle referred to in Problem 22 which makes $I_{x'}$ a maximum.

$$\text{Projection of } CX' = \sqrt{[\tfrac{1}{2}(I_x - I_y)]^2 + I_{xy}^2}\,[\cos(2\theta + 2\theta')]$$

$$= \sqrt{[\tfrac{1}{2}(I_x - I_y)]^2 + I_{xy}^2}\,[\cos 2\theta \cos 2\theta' - \sin 2\theta \sin 2\theta'] \quad (1)$$

Substituting $\cos 2\theta' = \dfrac{CA}{CX} = \dfrac{\frac{1}{2}(I_x - I_y)}{\sqrt{[\frac{1}{2}(I_x - I_y)]^2 + I_{xy}^2}}$, $\sin 2\theta' = \dfrac{AX}{CX} = \dfrac{I_{xy}}{\sqrt{[\frac{1}{2}(I_x - I_y)]^2 + I_{xy}^2}}$ into equation (1) and simplifying, we find

$$\text{Projection of } CX' = \tfrac{1}{2}(I_x - I_y)\cos 2\theta - I_{xy}\sin 2\theta$$

Thus
$$I_{x'} = \tfrac{1}{2}(I_x + I_y) + \tfrac{1}{2}(I_x - I_y)\cos 2\theta - I_{xy}\sin 2\theta$$

Similarly,
$$I_{y'} = \tfrac{1}{2}(I_x + I_y) - \tfrac{1}{2}(I_x - I_y)\cos 2\theta + I_{xy}\sin 2\theta$$

Note that the maximum I occurs at the right end of the diameter of the circle with a value equal to the distance to the center plus a radius. Hence

$$I_{\max} = \tfrac{1}{2}(I_x + I_y) + \sqrt{[\tfrac{1}{2}(I_x - I_y)]^2 + I_{xy}^2}$$

The minimum value is at the left end of the diameter and is

$$I_{\min} = \tfrac{1}{2}(I_x + I_y) - \sqrt{[\tfrac{1}{2}(I_x - I_y)]^2 + I_{xy}^2}$$

24. Determine the principal centroidal moments of inertia for the unequal angle shown in Fig. 14-28.

Solution:

First locate the centroid of the angle using the $x''y''$ axes as shown. The angle is divided into the parts A and B.

$$\bar{x}'' = \frac{4(1)(3) + 3(1)(0.5)}{4(1) + 3(1)} = 1.93''$$

$$\bar{y}'' = \frac{4(1)(0.5) + 3(1)(1.5)}{7} = 0.93''$$

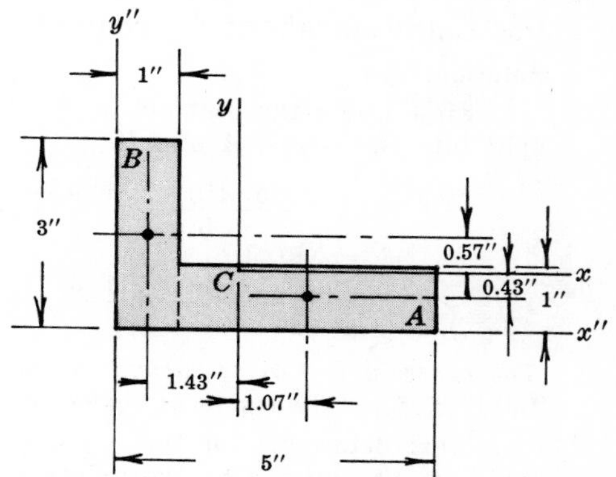

Fig. 14-28

Next draw x, y axes through this centroid C whose coordinates have been found. Then determine I_x and I_y by transferring from the parallel centroidal axes of A and B.

$$I_x = \tfrac{1}{12}(4)(1)^3 + 4(0.43)^2 + \tfrac{1}{12}(1)(3)^3 + 3(0.57)^2 = 4.30 \text{ in}^4$$

$$I_y = \tfrac{1}{12}(1)(4)^3 + 4(1.07)^2 + \tfrac{1}{12}(3)(1)^3 + 3(1.43)^2 = 16.3 \text{ in}^4$$

In order to determine the principal moments of inertia, the value of the product of inertia I_{xy} must be found. This also is transferred from the parallel centroidal axes of A and B, keeping in mind that the products of inertia about the centroidal axes of A and B are both zero because these centroidal axes are axes of symmetry. Hence in terms of the transfer distances the value of I_{xy} becomes

$$I_{xy} = 0 + (4 \times 1)(+1.07)(-0.43) + 0 + (3 \times 1)(-1.43)(+0.57) = -4.28 \text{ in}^4$$

Note: The signs of the transfer distances are important in determining the product of inertia. In the above equation the values 1.07 and -0.43 are the coordinates of the centroid of A relative to the x, y axes. Likewise the values -1.43 and 0.57 are the coordinates of the centroid of B relative to the x, y axes.

The Mohr's Circle analysis will now be used to determine the values of the principal moments and their axes. The points X and Y are located as shown in Fig. 14-29 below. Then using XY as the diameter draw a circle.

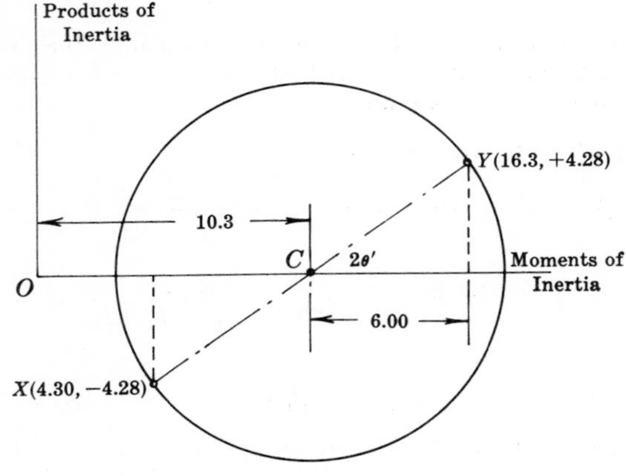

Fig. 14-29

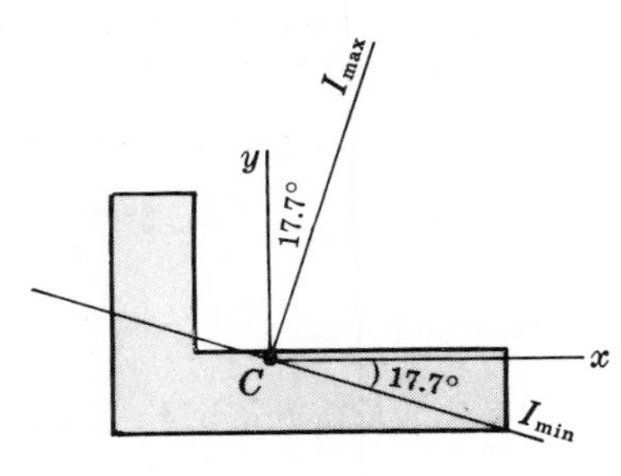

Fig. 14-30

The distance to the center of the circle is $\frac{1}{2}(I_x + I_y) = 10.3$, and the radius of the circle is $\sqrt{(6.00)^2 + (4.28)^2} = 7.36$. Thus the maximum I is $10.3 + 7.36 = 17.7$ in^4 and occurs clockwise from the y-axis at an angle θ' defined by $\theta' = \frac{1}{2}\tan^{-1} 4.28/6.00 = 17.7°$. The minimum value of I is $10.3 - 7.36 = 2.94$ in^4 and is located 17.7° clockwise from the x-axis shown in Fig. 14-30 above.

25. Calculate the principal moments of inertia for the L-section about its centroid.

Solution:

First locate the centroid of the L-section which is split into two areas, A and B.

$$\bar{x}'' = \frac{-1(5)(2.5) - 1(4)(0.5)}{5 + 4} = -1.61''$$

$$\bar{y}'' = \frac{+1(5)(0.5) + 1(4)(3)}{9} = +1.61''$$

The centroid is 1.61 in. above the base and 1.61 in. to the left of its right side as shown in Fig. 14-31.

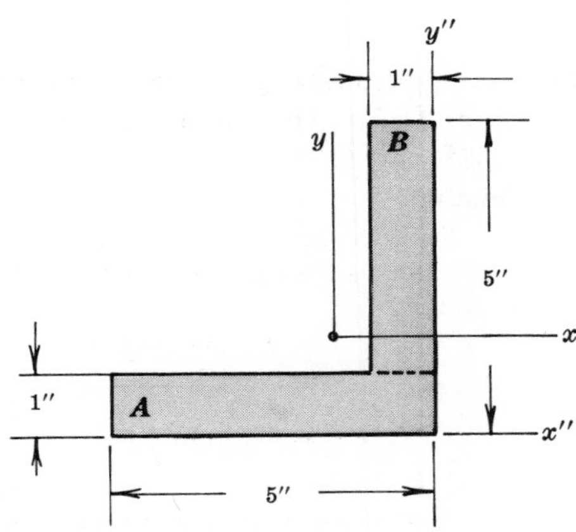

Fig. 14-31

Next determine for the x, y axes through the centroid of the entire area the values I_x, I_y and I_{xy}. This will be done by transferring each value from the centroids of the parts A and B.

$$I_x = \tfrac{1}{12}(5)(1)^3 + 5(1)(1.61 - 0.5)^2 + \tfrac{1}{12}(1)(4)^3 + 4(1)(3.0 - 1.61)^2 = 19.6 \text{ in}^4$$

$$I_y = \tfrac{1}{12}(1)(5)^3 + 5(1)(2.5 - 1.61)^2 + \tfrac{1}{12}(4)(1)^3 + 4(1)(1.61 - 0.5)^2 = 19.6 \text{ in}^4$$

The second computation checks the first; they should be equal for an equal L-section.

$$I_{xy} = 0 + 5(1)(+0.89)(+1.11) + 0 + 4(1)(-1.11)(-1.39) = +11.1 \text{ in}^4$$

Using a Mohr Circle construction with point X as I_x and I_{xy}, i.e. 19.6 and +11.1, and point Y as I_y and $-I_{xy}$, i.e. 19.6 and −11.1, it is evident that the radius is 11.1. See Fig. 14-32. Hence

$$I_{max} = 19.6 + 11.1 = 30.7 \text{ in}^4, \qquad I_{min} = 19.6 - 11.1 = 8.5 \text{ in}^4$$

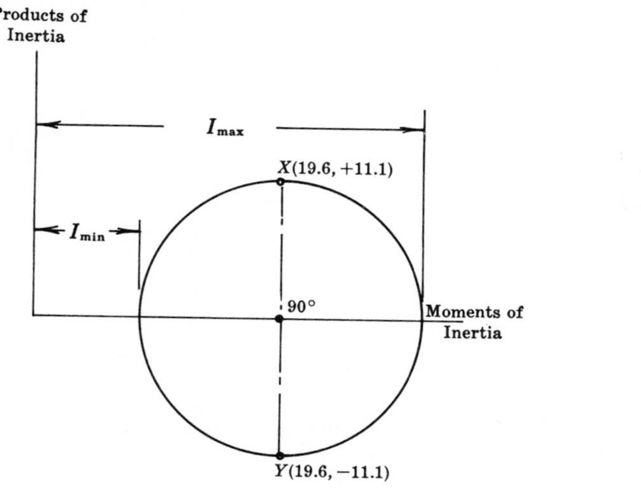

Fig. 14-32

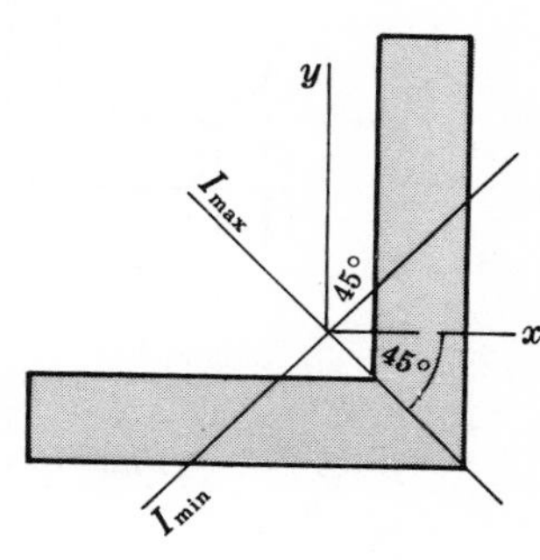

Fig. 14-33

The principal axes are at $2\theta = 90°$ in the Mohr Circle or 45° in the actual figure. The maximum value in the Mohr Circle is clockwise from the X point. Hence in the actual figure it is 45° clockwise from the x-axis. Similarly the minimum value is 90° clockwise from the Y point and in the actual figure it is 45° clockwise from the y-axis as shown in Fig. 14-33.

Supplementary Problems

26. Calculate the moment of inertia of a rectangle having a base 6 inches and a height 24 inches about an axis through its center of gravity and parallel to the base. *Ans.* 6920 in^4

27. Determine the moment of inertia of an isosceles triangle with base of 6 inches and sides of 5 inches about its base. *Ans.* 32 in^4

28. Determine the moment of inertia of a circle of radius 2 ft about a diameter. *Ans.* 12.6 ft^4

29. Find the moment of inertia with respect to the y-axis of the plane area between the parabola $y = 9 - x^2$ and the x-axis. *Ans.* 324/5

30. Find the moment of inertia with respect to each coordinate axis of the area between the curve $y = \sin x$ from $x = 0$ to $x = \pi$ and the x-axis. *Ans.* $I_x = 4/9$, $I_y = \pi^2 - 4$

31. Refer to Fig. 14-34 below. Determine the moment of inertia of the composite figure about an axis through its center of gravity and parallel to the base. *Ans.* 46.3 in^4

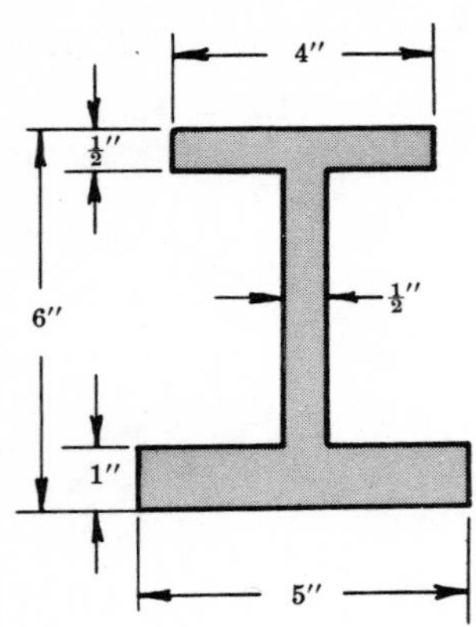

Fig. 14-34

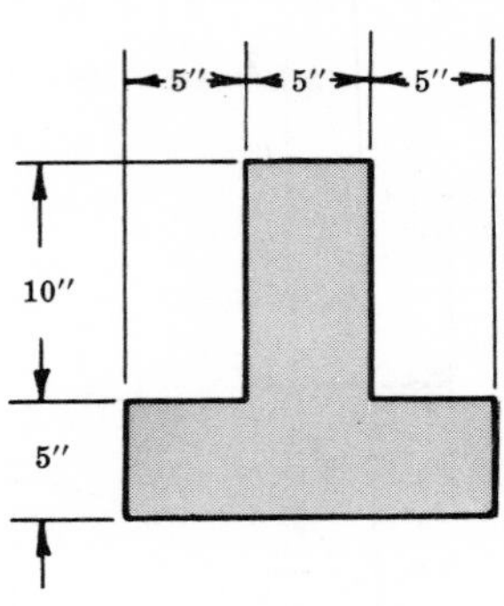

Fig. 14-35

32. Referring to Fig. 14-35 above, determine the moment of inertia of the composite figure about a centroidal axis parallel to the 15 inch edge. *Ans.* 2260 in^4

33. Refer to Fig. 14-36 below. Compute the moment of inertia of the composite figure about a centroidal axis parallel to the 10 inch side. *Ans.* 3.52 in^4

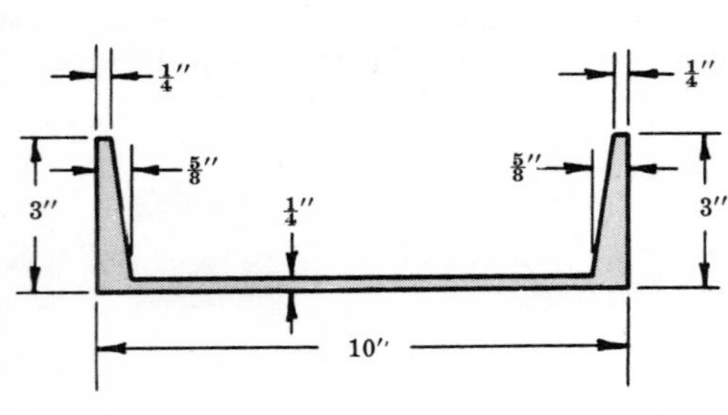

Fig. 14-36

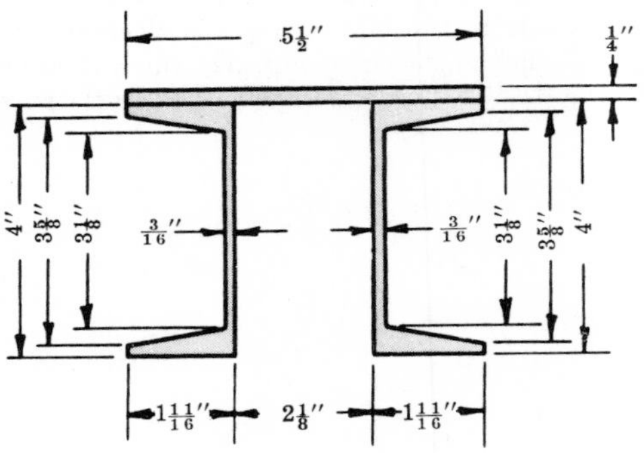

Fig. 14-37

34. Refer to Fig. 14-37 above. Compute the moment of inertia of the composite figure about a centroidal axis parallel to the 5.5 inch side. *Ans.* 12.8 in^4

35. What is the polar moment of inertia of a 4 ft diameter circle about an axis through its center of gravity and perpendicular to its plane? *Ans.* 25.1 ft^4

36. Calculate the product of inertia of a rectangle of base 6 inches and height 4 inches about two adjacent sides. *Ans.* 144 in^4

37. Determine the values of I_x, I_y, and I_{xy} for the area bounded by the x axis, the line $x = a$, and the curve $y = (b/a^n)x^n$. *Ans.* $I_x = \dfrac{ab^3}{3(3n+1)}$, $I_y = \dfrac{a^3b}{n+3}$, $I_{xy} = \dfrac{a^2b^2}{2(2n+2)}$

38. In the preceding problem the area becomes triangular when $n = 1$. Check the values by direct integration. *Ans.* $I_x = \frac{1}{12}ab^3$, $I_y = \frac{1}{4}a^3b$, $I_{xy} = \frac{1}{8}a^2b^2$

39. Determine the values of I_x, I_y, and I_{xy} for the area bounded by the y-axis, the line $y = b$, and the curve $y = (b/a^n)x^n$. *Ans.* $I_x = \dfrac{nab^3}{3n+1}$, $I_y = \dfrac{na^3b}{3(n+3)}$, $I_{xy} = \dfrac{na^2b^2}{4(n+1)}$

40. The figures in Problems 37 and 39 form a rectangle when added together. Add the values of I_x and check with Problem 1 to see if you obtain the moment of inertia of a rectangle about its base. Repeat the process for I_{xy} and check your result with Problem 16.

41. Determine I_y and I_{xy} for the figure shown in Problem 32 where the x, y axes pass through the centroid. *Ans.* $I_y = 1510\text{ in}^4$, $I_{xy} = 0$

42. Determine I_x, I_y, and I_{xy} for the x, y axes that pass through the centroid of the unequal angle shown in Fig. 14-38.
Ans. $I_x = 163\text{ in}^4$, $I_y = 291\text{ in}^4$, $I_{xy} = -120\text{ in}^4$

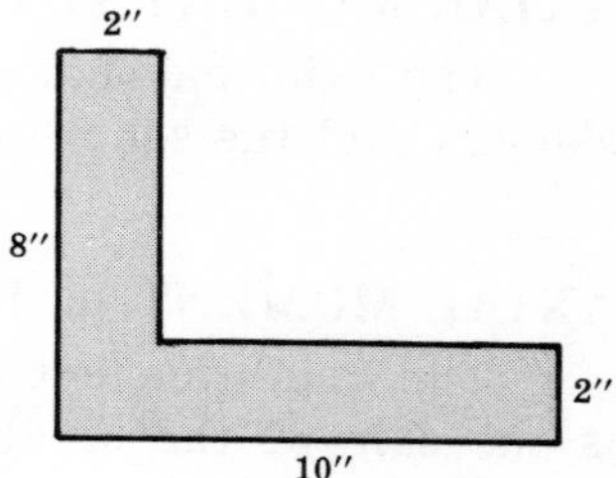

Fig. 14-38

43. In the preceding problem use Mohr's Circle to locate the principal axes and determine the principal moments of inertia (see Fig. 14-39 below).
Ans. $I_{max} = 363\text{ in}^4$ at 30.9° clockwise from y-axis
$I_{min} = 91\text{ in}^4$ at 30.9° clockwise from x-axis

44. Locate the principal axes through the centroid of the area shown in Fig. 14-40 below. Next determine the principal moments of inertia for those axes.
Ans. $I_{max} = 373\text{ in}^4$ at 28.8° clockwise from y-axis,
$I_{min} = 45.0\text{ in}^4$ at 28.8° clockwise from x-axis

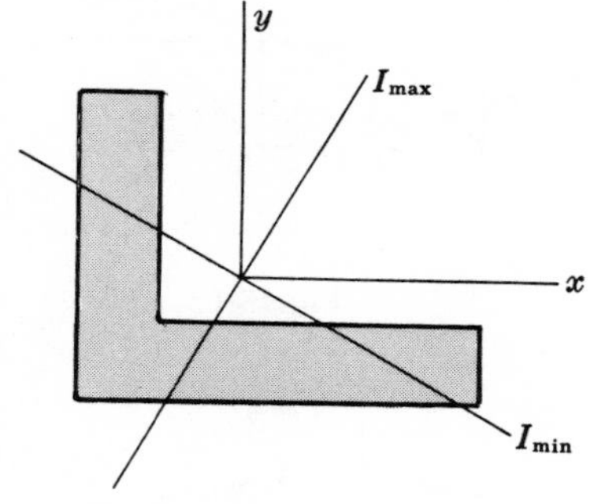

Fig. 14-39

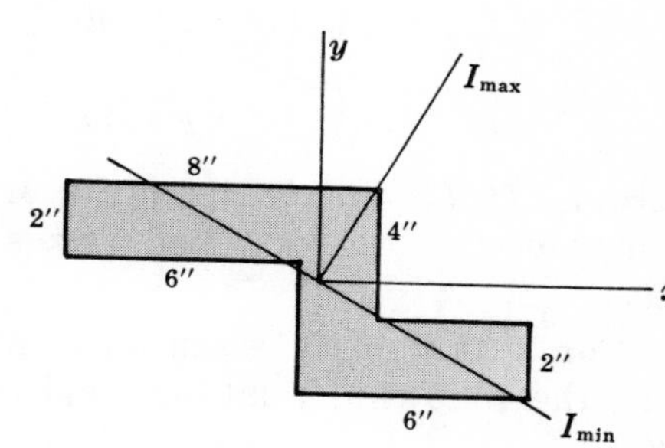

Fig. 14-40

Chapter 15

Mass Moments of Inertia

AXIAL MOMENT of INERTIA of an ELEMENT of MASS

The axial moment of inertia of an element of mass is the product of the mass of the element and the square of the distance of the element from the axis.

AXIAL MOMENT of INERTIA of a MASS

The axial moment of inertia of a mass is the sum of the axial moments of all its elements. Thus for a mass of which dm is one element with coordinates x, y, z the following definitions hold (see Fig. 15-1).

$$I_x = \int (y^2 + z^2)\, dm$$

$$I_y = \int (x^2 + z^2)\, dm$$

$$I_z = \int (x^2 + y^2)\, dm$$

where I_x, I_y, I_z = axial moments of inertia (with respect to the x, y, and z axes respectively).

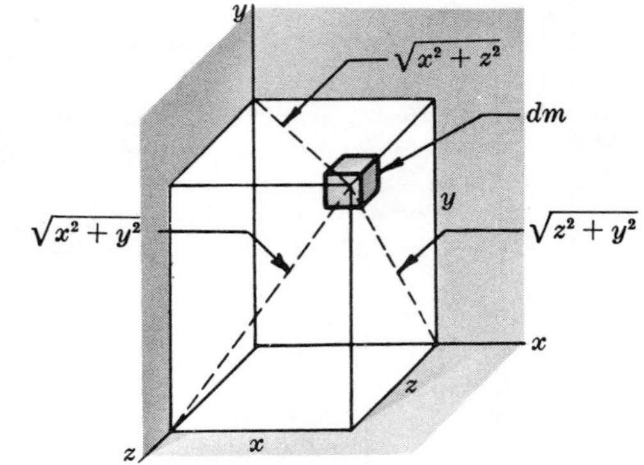

Fig. 15-1

For a thin plate essentially in the xy plane, the following relations hold (see Fig. 15-2).

$$I_x = \int y^2\, dm$$

$$I_y = \int x^2\, dm$$

$$J_O = \int \rho^2\, dm = \int (x^2 + y^2)\, dm$$

$$= I_x + I_y$$

where I_x, I_y = axial moments of inertia about the x and y axes respectively, and J_O = axial moment of inertia about the z-axis, or, as it is also called, the polar moment of inertia.

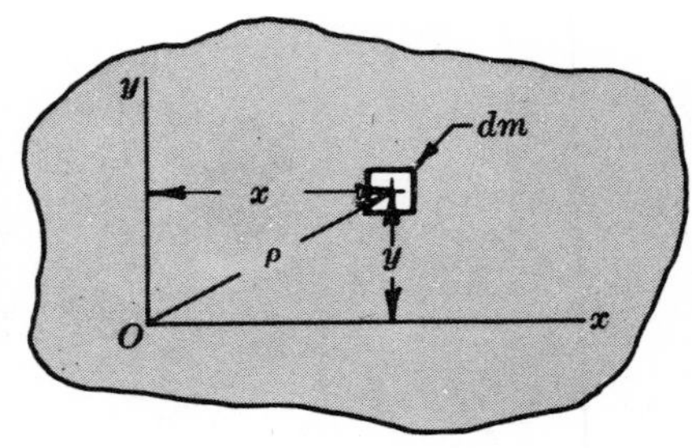

Fig. 15-2

PARALLEL AXIS THEOREM

The parallel axis theorem states that the moment of inertia of a body about an axis is equal to the moment of inertia $\bar{I}$ about a parallel axis through the center of gravity of the body plus the product of the mass of the body and the square of the distance between the two parallel axes.

RADIUS of GYRATION

The radius of gyration of a body with respect to an axis is defined mathematically as the square root of the quotient of its moment of inertia divided by the mass.

UNITS

Units of all the foregoing moments involve those of mass m and the square of a length l, usually $\dfrac{W\text{ lb}}{g\text{ ft/sec}^2}(l\text{ ft})^2 = \left(\dfrac{W}{g}l^2\right)$ lb-sec²-ft in engineering mechanics.

CENTROIDAL MASS MOMENTS of INERTIA

Centroidal mass moments of inertia for some common shapes are listed below with reference to problems in which these results are derived.

Problem	Figure	Name	I_x	I_y	I_z
1		Slender Bar	–	$\frac{1}{12}ml^2$	$\frac{1}{12}ml^2$
3		Rectangular Parallelepiped	$\frac{1}{12}m(a^2+b^2)$	$\frac{1}{12}m(a^2+c^2)$	$\frac{1}{12}m(b^2+c^2)$
4 or 5		Thin Circular Disk	$\frac{1}{4}mR^2$	$\frac{1}{4}mR^2$	$\frac{1}{2}mR^2$
7, 8		Right Circular Cylinder	$\frac{1}{12}m(3R^2+h^2)$	$\frac{1}{12}m(3R^2+h^2)$	$\frac{1}{2}mR^2$
9		Sphere	$\frac{2}{5}mR^2$	$\frac{2}{5}mR^2$	$\frac{2}{5}mR^2$

Solved Problems

1. Derive the expression for the moment of inertia about a centroidal axis perpendicular to a bar of length l, mass m, and small cross section as shown in Fig. 15-3. Find its radius of gyration.

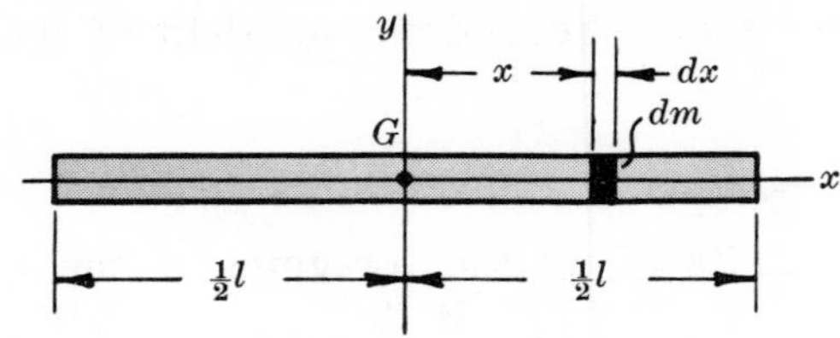

Fig. 15-3

Solution:

By definition, $I_y = \int x^2\,dm$.

But dm is the mass of a portion of the bar of length dx. Its mass dm is dx/l of the entire mass m, i.e., $(dx/l)m = dm$.

Hence $I_y = \int_{-l/2}^{l/2} x^2(m/l)\,dx = \frac{1}{12}ml^2$. The radius of gyration $k = \sqrt{I_y/m} = l/\sqrt{12}$.

2. Derive the expression for the moment of inertia of a bar about an axis through one end and perpendicular to the bar whose length is l. Assume the mass is m and the cross section is small in comparison with the length. Refer to Fig. 15-4.

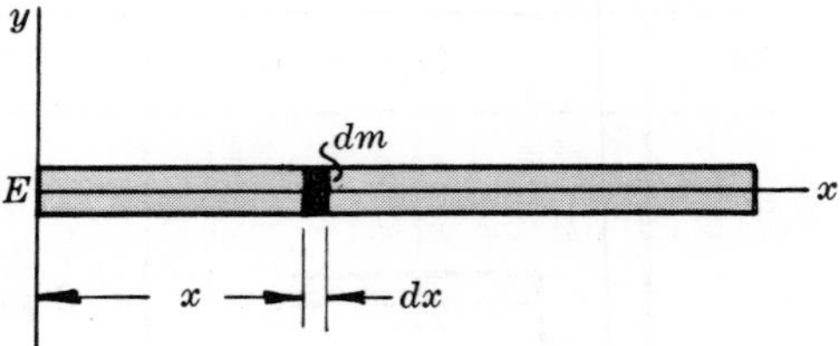

Fig. 15-4

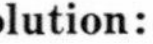

Solution:

As in Problem 1, write

$$I_y = \int x^2\,dm = \int_0^l x^2(m/l)\,dx = \tfrac{1}{3}ml^2$$

The same result could be obtained by use of the parallel axis theorem.

$$I_E = \bar{I} + m(\tfrac{1}{2}l)^2 = \tfrac{1}{12}ml^2 + \tfrac{1}{4}ml^2 = \tfrac{1}{3}ml^2$$

3. What is the moment of inertia about a centroidal axis perpendicular to a face of a rectangular parallelepiped (block)?

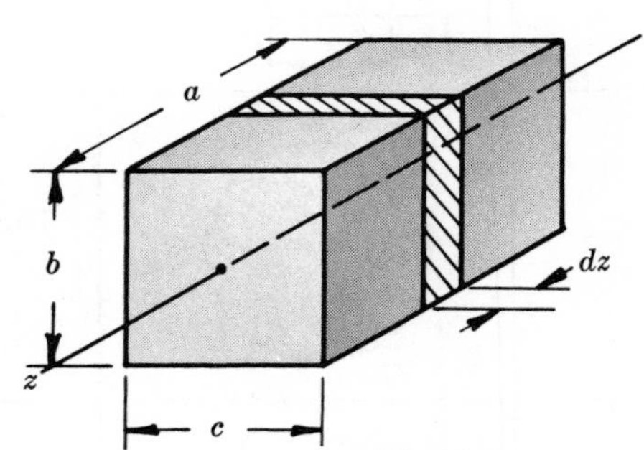

Fig. 15-5

Solution:

As can be seen in Fig. 15-5, the moment of inertia of the block about the z-axis is equal to the summation of a series of thin plates each of thickness dz, cross section b by c, and mass dm.

First determine I_z for a thin plate with cross section b by c, thickness dz, and mass dm (see Fig. 15-6).

Since $I_z = I'_x + I'_y$, find I'_x and I'_y to obtain the result. However, I'_x is really the summation of the centroidal moments of a series of bars of mass dm' and of negligible cross section (dx by dz) and height b. According to Problem 1, this may then be written

$$I'_x = \int \tfrac{1}{12}\,dm'\,b^2 = \tfrac{1}{12}b^2\,dm$$

Similar reasoning yields

$$I'_y = \int \tfrac{1}{12}\,dm'\,c^2 = \tfrac{1}{12}c^2\,dm$$

From this it follows that I_z for a thin plate of mass dm is equal to $I'_x + I'_y$, or $I_z = \frac{1}{12}dm(b^2 + c^2)$.

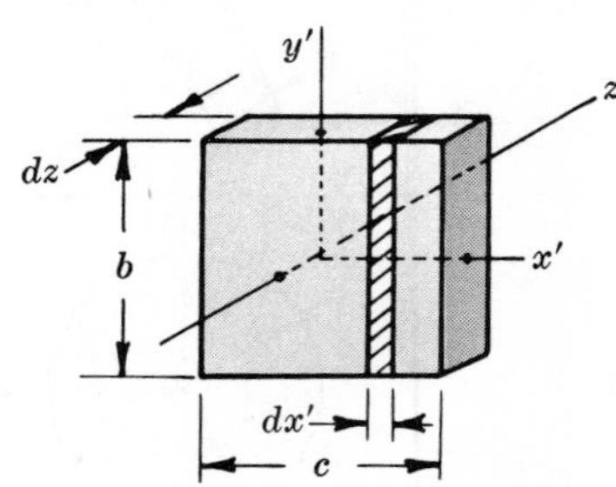

Fig. 15-6

For the entire block it is seen that $I_z = \int \frac{1}{12} dm(b^2+c^2) = \frac{1}{12}m(b^2+c^2)$,

or more rigorously, since $dm = (dz/a)m$, $I_z = \int_0^a \frac{1}{12}(m/a)(b^2+c^2)\,dz = \frac{1}{12}m(b^2+c^2)$.

4. What is the moment of inertia about the x-axis chosen along a diameter of a homogeneous thin circular disk of mass m and thickness t as shown in Fig. 15-7?

Solution:

Consider the disk to be made of a series of thin bars dx by t in cross section and of varying heights $2y$. The mass of such a strip is $(2y\,dx/\pi r^2)m$ from the geometry of the figure.

But the moment of inertia about the x-axis for this strip is indicated in Problem 1.

The total I is the sum of the moments of all such strips. The integration is shown below. Note the substitution of $\sqrt{r^2 - x^2}$ for y as derived from the equation of a circle.

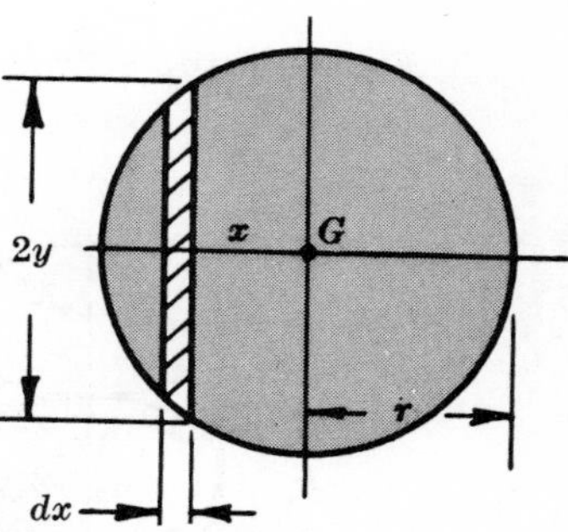

Fig. 15-7

$$I_x = \int_{-r}^{r} \frac{1}{12}\frac{2y\,dx\,m}{\pi r^2}(2y)^2 = \frac{2m}{3\pi r^2}\int_{-r}^{r} y^3\,dx = \frac{2m}{3\pi r^2}\int_{-r}^{r}(\sqrt{r^2-x^2})^3\,dx$$

$$= \frac{2m}{3\pi r^2}\left[\frac{x}{4}(\sqrt{r^2-x^2})^3 + \frac{3r^2x}{8}\sqrt{r^2-x^2} + \frac{3r^4}{8}\sin^{-1}\frac{x}{r}\right]_{-r}^{r} = \frac{mr^2}{4}$$

5. Repeat Problem 4 using a differential mass as shown in Fig. 15-8 below. Assume homogeneity, and density δ in mass per unit volume.

Solution:

$$I_x = \int \rho^2 \sin^2\theta\,dm = \int_0^{2\pi}\int_0^r \rho^2\sin^2\theta\,\delta t\,\rho\,d\rho\,d\theta = \delta t\int_0^{2\pi}\int_0^r \rho^3\sin^2\theta\,d\rho\,d\theta$$

$$= \delta t\int_0^{2\pi}[\tfrac{1}{4}\rho^4]_0^r\sin^2\theta\,d\theta = \delta t(\tfrac{1}{4}r^4)[\tfrac{1}{2}\theta - \tfrac{1}{4}\sin\theta]_0^{2\pi}$$

$$= \delta t(\tfrac{1}{4}r^4)(\tfrac{1}{2}\times 2\pi) = \tfrac{1}{4}\,\delta t\,\pi r^4 = \tfrac{1}{4}(\delta t\pi r^2)r^2 = \tfrac{1}{4}mr^2$$

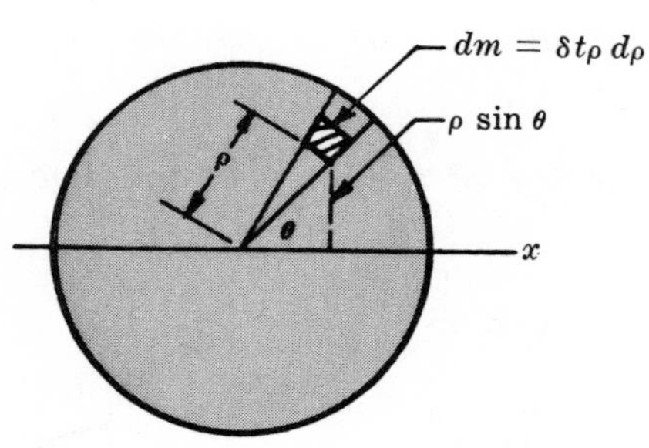

Fig. 15-8

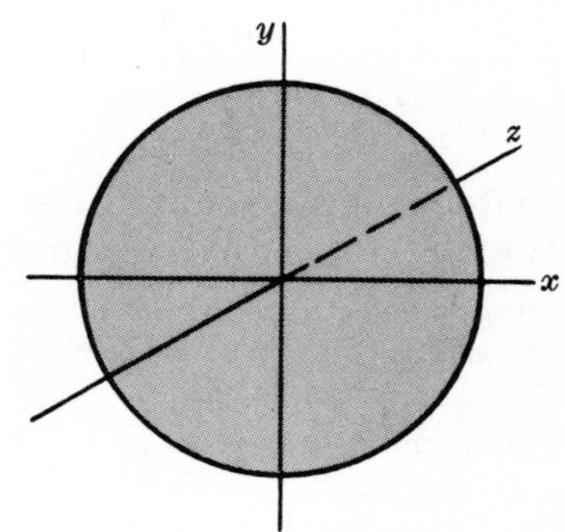

Fig. 15-9

6. Show that the polar moment of inertia for the disk in Problems 4 and 5 is $\frac{1}{2}mr^2$. What is its radius of gyration? Refer to Fig. 15-9 above.

Solution:

Since $I_y = I_x = \frac{1}{4}mr^2$, the polar moment of inertia I_z is

$$I_z = I_x + I_y = \tfrac{1}{2}mr^2$$

The radius of gyration $k = \sqrt{I_z/m} = \sqrt{\frac{1}{2}mr^2/m} = r/\sqrt{2}$.

7. Determine the moment of inertia about a geometrical axis of a right circular cylinder of radius R and mass m. See Fig. 15-10 below.

Solution:

Consider the cylinder to be made of a series of thin disks of height dz as shown.

For a thin disk (Problem 6), $I_z = \frac{1}{2}(\text{mass})R^2 = \frac{1}{2}(m\,dz/h)R^2$.

For the entire cylinder, $I_z = \int_0^h \frac{1}{2}(m\,dz/h)R^2 = \frac{1}{2}mR^2$.

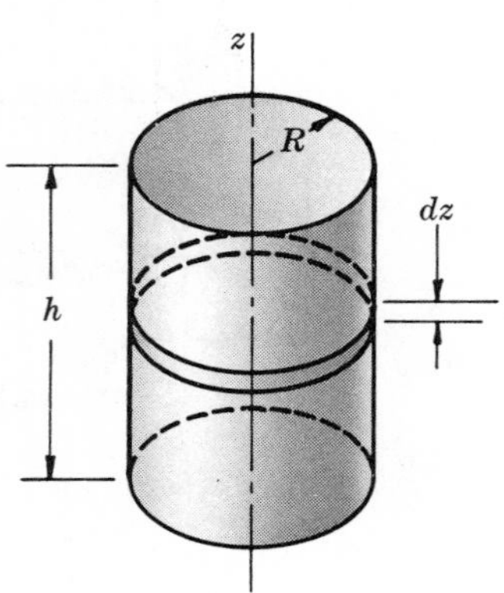

Fig. 15-10

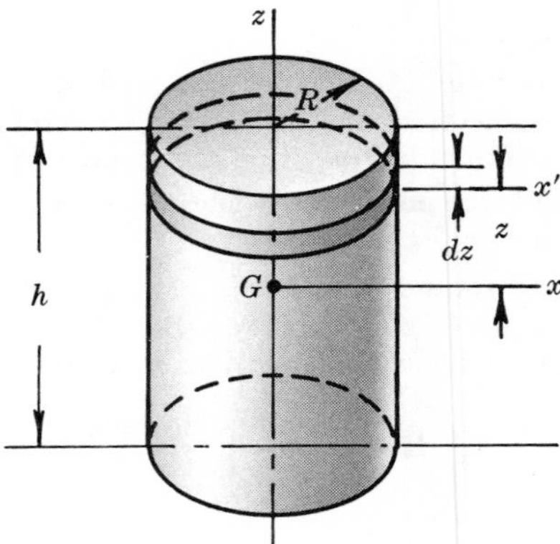

Fig. 15-11

8. Determine the moment of inertia for a right circular cylinder of radius R and mass m about the centroidal x-axis shown in Fig. 15-11 above.

Solution:

As in Prob. 7, consider the cylinder as made up of a series of thin disks of height dz and mass $(m\,dz/h)$. The moment of inertia of the thin disk about its x'-axis parallel to the x-axis is given by Prob. 4 as $I_{x'} = \frac{1}{4}(m\,dz/h)R^2$. Transferring to the x-axis by the parallel axis theorem,

$$I_x = I_{x'} + (m\,dz/h)z^2 = \frac{1}{4}(m\,dz/h)R^2 + (m\,dz/h)z^2$$

To determine I_x for the entire cylinder, sum the I_x for all disks.

$$I_x = \frac{mR^2}{4h}\int_{-\frac{1}{2}h}^{\frac{1}{2}h} dz + \frac{m}{h}\int_{-\frac{1}{2}h}^{\frac{1}{2}h} z^2\,dz = \frac{1}{4}mR^2 + \frac{1}{12}mh^2 = \frac{1}{12}m(3R^2 + h^2)$$

9. Determine the moment of inertia about a diameter of a sphere of mass m and radius R. What is its radius of gyration?

Solution:

Choose a thin disk parallel to the xz-plane as shown in Fig. 15-12. Assume density δ.

The moment of inertia of a thin disk of radius x about the y-axis is $\frac{1}{2}(\text{mass})x^2$. To find I_y for the entire sphere, add the individual moments just indicated, where $dm = \delta\,dV = \delta(\pi x^2\,dy)$.

$$I_y = \int_{-R}^{R} \frac{1}{2}(\delta\pi x^2\,dy)x^2 = \frac{1}{2}\pi\delta\int_{-R}^{R} x^4\,dy$$

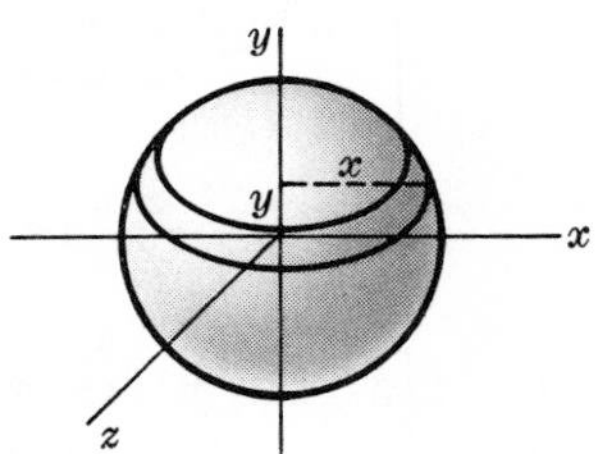

Fig. 15-12

But, from the equation of the cross section of the sphere in the xy-plane (a circle), $x^2 + y^2 = R^2$. Hence

$$I_y = \frac{1}{2}\pi\delta\int_{-R}^{R}(R^2 - y^2)^2\,dy = \frac{8}{15}\delta\pi R^5$$

Since the mass is $m = \frac{4}{3}\pi R^3\delta$, we have $I_y = (\frac{4}{3}\pi R^3\delta)(\frac{2}{5}R^2) = \frac{2}{5}mR^2$.

The radius of gyration $k = \sqrt{I_y/m} = \sqrt{2/5}\,R$.

10. Determine the moment of inertia of a homogeneous hollow right circular cylinder with respect to its geometric axis.

Solution:

For the outer cylinder,

$$(I_z)_o = \tfrac{1}{2}m_o r_o^2 = \tfrac{1}{2}(\pi r_o^2 h\delta)r_o^2$$

For the inner cylinder,

$$(I_z)_i = \tfrac{1}{2}m_i r_i^2 = \tfrac{1}{2}(\pi r_i^2 h\delta)r_i^2$$

For the hollow cylinder,

$$I_z = (I_z)_o - (I_z)_i = \tfrac{1}{2}\pi\delta h(r_o^2 + r_i^2)(r_o^2 - r_i^2)$$

Expanding, $I_z = (\tfrac{1}{2}h\pi\delta r_o^2 - \tfrac{1}{2}h\pi\delta r_i^2)(r_o^2 + r_i^2)$

$$= (\tfrac{1}{2}m_o - \tfrac{1}{2}m_i)(r_o^2 + r_i^2) = \tfrac{1}{2}m(r_o^2 + r_i^2)$$

where m refers to the mass of the hollow cylinder.

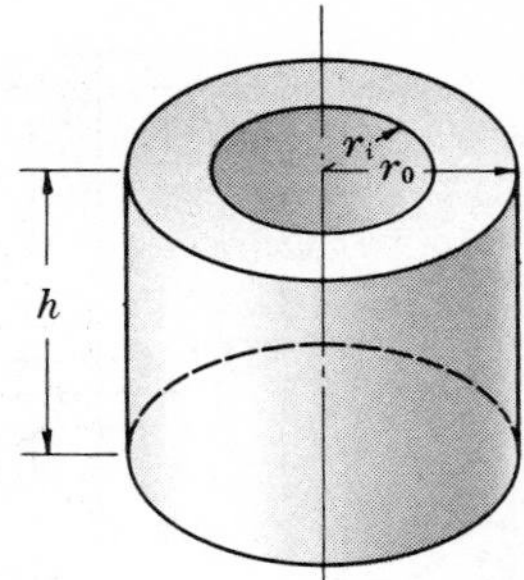

Fig. 15-13

11. Determine the moments of inertia about the x and y axes of the right circular cone of mass m and dimensions shown in Fig. 15-14 below.

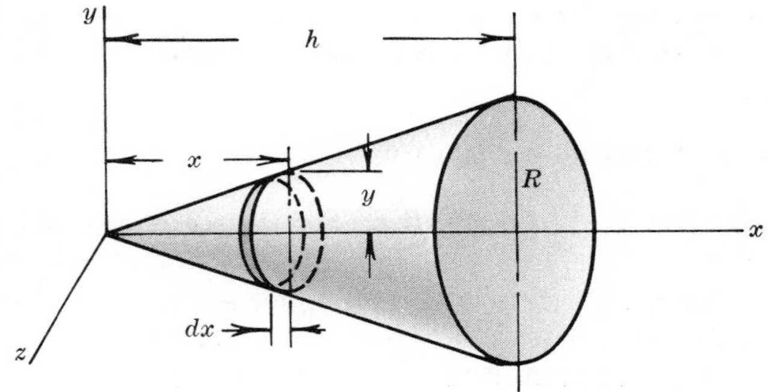

Fig. 15-14

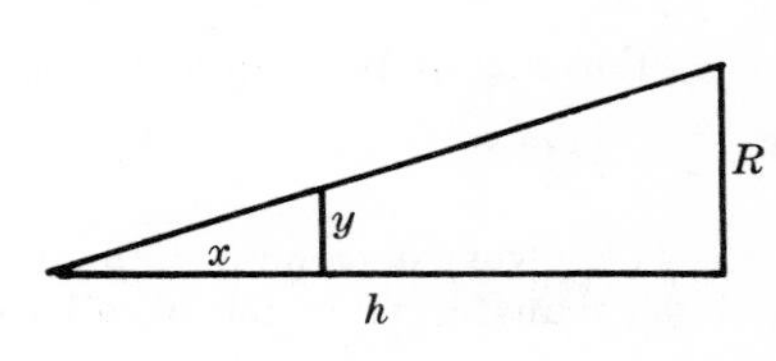

Fig. 15-15

Solution:

To find I_x, choose a thin lamina perpendicular to the x-axis as shown in Fig. 15-14 above. Assume density δ.

The moment of inertia about the x-axis of this lamina of radius y is $\tfrac{1}{2}(\text{mass})y^2$. To find I_x for the entire cone add the individual moments just indicated, noting that the mass of the chosen lamina is $dm = \delta\, dV = \delta(\pi y^2\, dx)$. Since $y = Rx/h$,

$$I_x = \int_0^h \tfrac{1}{2}\delta(\pi y^2\, dx)y^2 = \int_0^h \tfrac{1}{2}\delta\pi(Rx/h)^4\, dx = \tfrac{1}{10}\delta\pi R^4 h$$

But the mass of the entire cone is $\tfrac{1}{3}\pi\delta R^2 h$. Thus we can write

$$I_x = (\tfrac{1}{3}\pi R^2 h\delta)(\tfrac{3}{10}R^2) = \tfrac{3}{10}mR^2$$

To find I_y, which equals I_z, it is necessary to apply the transfer theorem to obtain the moment of inertia for the lamina relative to the y-axis. Since $y = Rx/h$,

$$I_y = \int (\tfrac{1}{4}\,dm\, y^2 + dm\, x^2) = \int_0^h (\delta\pi y^2\, dx)(\tfrac{1}{4}y^2 + x^2)$$

$$= \int_0^h \tfrac{1}{4}\pi\delta(R^4/h^4)x^4\, dx + \int_0^h \pi\delta(R^2/h^2)x^4\, dx = \tfrac{1}{20}\pi\delta R^4 h + \tfrac{1}{5}\pi\delta R^2 h^3$$

Using $m = \tfrac{1}{3}\pi\delta R^2 h$, this expression becomes $I_y = \tfrac{3}{20}mR^2 + \tfrac{3}{5}mh^2$.

12. Compute the moment of inertia of the cast iron flywheel shown in Fig. 15-16 below. Cast iron weighs 450 lb/ft³.

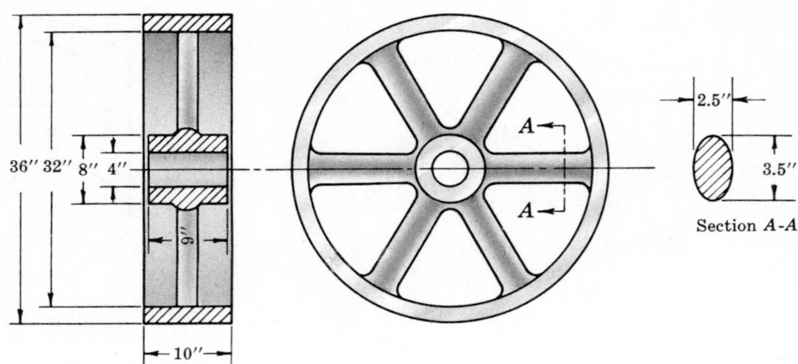

Fig. 15-16

Solution:

In analyzing the problem, consider the hub and the rim as hollow cylinders and the spokes as slender rods. First determine the weights of the components.

$$W_{hub} = (\pi r_o^2 - \pi r_i^2)h\delta = \pi[(4/12)^2 - (2/12)^2](9/12)(450) = 88.2 \text{ lb}$$
$$W_{rim} = (\pi r_o^2 - \pi r_i^2)h\delta = \pi[(18/12)^2 - (16/12)^2](10/12)(450) = 556 \text{ lb}$$

For one spoke, elliptical in cross section,

$$W_{spoke} = \pi abl\delta = \pi(2.5/24)(3.5/24)(12/12)(450) = 21.5 \text{ lb}$$

Next determine I for each component about the axis of rotation. For the spokes this will entail a transfer from the parallel centroidal axis.

$$I_{hub} = \tfrac{1}{2}m(r_o^2 + r_i^2) = \tfrac{1}{2}(88.2/32.2)[(4/12)^2 + (2/12)^2] = 0.19 \text{ lb-sec}^2\text{-ft}$$
$$I_{rim} = \tfrac{1}{2}m(r_o^2 + r_i^2) = \tfrac{1}{2}(556/32.2)[(18/12)^2 + (16/12)^2] = 34.8 \text{ lb-sec}^2\text{-ft}$$
$$I_{spokes} = 6(\tfrac{1}{12}ml^2 + md^2) = 6[\tfrac{1}{12}(21.5/32.2)(12/12)^2 + (21.5/32.2)(10/12)^2]$$
$$= 3.11 \text{ lb-sec}^2\text{-ft}$$
$$I_{wheel} = (0.19 + 34.8 + 3.11) \text{ lb-sec}^2\text{-ft} = 38.1 \text{ lb-sec}^2\text{-ft}$$

Supplementary Problems

13. Determine the moments of inertia of the thin elliptical disk of mass m shown in Fig. 15-17 below. Refer to Problem 4 or 5. *Ans.* $I_x = \tfrac{1}{4}mb^2$, $I_y = \tfrac{1}{4}ma^2$, $I_z = \tfrac{1}{4}m(a^2 + b^2)$

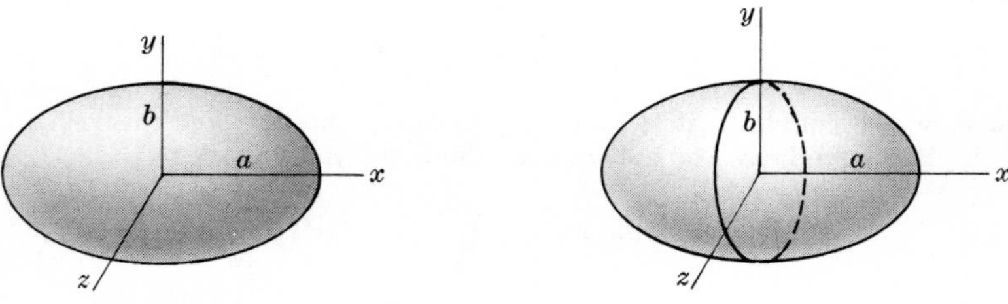

Fig. 15-17 **Fig. 15-18**

14. Determine the moments of inertia of the ellipsoid of revolution with mass m shown in Fig. 15-18 above. *Ans.* $I_x = \tfrac{2}{5}mb^2$, $I_y = I_z = \tfrac{1}{5}m(a^2 + b^2)$

15. Determine the moments of inertia of the paraboloid of revolution with the mass m shown in Fig. 15-19. The equation in the xy plane is $y^2 = -(R^2/h^2)x^2 + R^2$.
Ans. $I_x = \frac{2}{5}mR^2$, $I_y = I_z = \frac{1}{5}m(R^2 + h^2)$

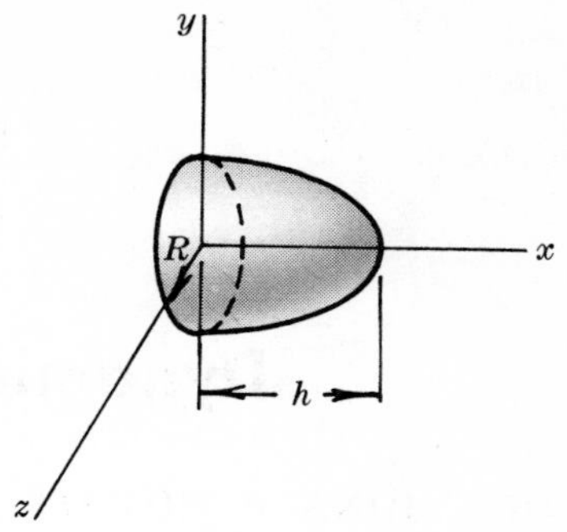

Fig. 15-19

16. Show that the moment of inertia about a diameter for a thin hollow sphere of mass m is $\frac{2}{3}mR^2$.

17. Determine the moment of inertia about a diameter for a hollow sphere of mass m with inner and outer radii R_i and R_0 respectively. *Ans.* $I = \frac{2}{5}m(R_0^5 - R_i^5)/(R_0^3 - R_i^3)$

18. Show that the moment of inertia about a centroidal axis parallel to a side for a cube of mass m is $I = \frac{1}{6}ma^2$, where a is the length of a side.

19. Find the moment of inertia of a steel rod $\frac{1}{2}$ inch in diameter and 4 ft long about an axis through one end and perpendicular to the rod. The steel weighs 490 lb per cubic foot.
Ans. 0.44 slug-ft^2 or lb-sec^2-ft

20. Determine the moment of inertia of a steel pipe with 3.50 in. outside diameter and 2.89 in. inside diameter and 10 ft long with respect to its longitudinal axis. The steel weighs 490 lb/ft^3.
Ans. 0.058 slug-ft^2

21. Find the moment of inertia of a brass cylindrical shaft 3 inches in diameter and 10 ft long with respect to its geometric axis of rotation. Use a weight of 530 lb/ft^3. *Ans.* 0.063 slug-ft^2

22. In Problem 21, what is the moment of inertia of the volume only about the same longitudinal axis? *Ans.* 954 in^5

23. Calculate the moment of inertia of a rectangular prism that is 6 in. high, 4 in. wide, and 10 in. long, with respect to its longitudinal centroidal axis. Use a weight of 40 lb/ft^3. *Ans.* 0.005 slug-ft^2

24. Find the moment of inertia of an aluminum sphere 8 inches in diameter with respect to a centroidal axis. Aluminum weighs 160 lb/ft^3. *Ans.* 0.034 slug-ft^2

25. Determine the moment of inertia of the flywheel shown in Fig. 15-20 (solid web) with respect to its axis of rotation. Cast iron weighs 450 lb/ft^3. *Ans.* 2.12 slug-ft^2

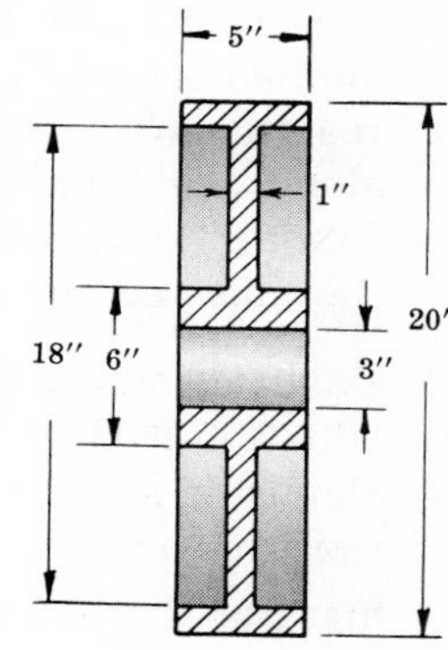

Fig. 15-20

Chapter 16

Dynamics of a Rigid Body in Rotation

EQUATIONS of MOTION

If a body has a plane of symmetry and rotates about a *fixed* axis perpendicular to this plane, the scalar equations of motion of the body under the action of an unbalanced force system are

$$\Sigma F_n = m\bar{r}\omega^2$$
$$\Sigma F_t = m\bar{r}\alpha$$
$$\Sigma M_O = I_O\alpha$$

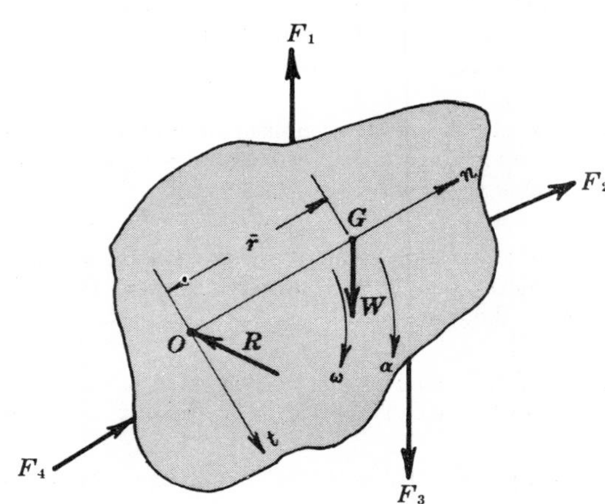

Fig. 16-1

where, with reference to Fig. 16-1,

ΣF_n = algebraic summation of the components of the external forces (which are the weight W of the body, the applied forces F_1, F_2, F_3, etc., and the reaction R of the axis *on* the body) along the n-axis, which is the line joining the center of rotation O with the mass center G.

ΣF_t = algebraic summation of the components of the external forces along the t-axis, which is perpendicular to the n-axis at O.

ΣM_O = algebraic summation of the moments of the external forces about the axis of rotation through O.

m = mass of the body.

$\bar{r}$ = distance from the center of rotation O to the mass center G.

I_O = moment of inertia of the body about the axis of rotation.

ω = angular speed of the body.

α = magnitude of the angular acceleration of the body.

The equations of motion for a body rotating about an axis of symmetry become

$$\Sigma F_x = 0, \quad \Sigma F_y = 0, \quad \Sigma \bar{M} = \bar{I}\alpha$$

where ΣF_x = algebraic summation of the components of the external forces along any line chosen as the x-axis.

ΣF_y = algebraic summation of the components of the external forces along the y-axis.

$\Sigma \bar{M}$ = algebraic summation of the moments of the external forces about the axis of rotation through the mass center G (axis of symmetry).

$\bar{I}$ = moment of inertia of the body about the axis of rotation through the mass center G.

α = magnitude of the angular acceleration of the body.

INERTIA FORCE METHOD

The inertia force method for rotation about *any fixed axis* may be used, provided that the reversed effective forces are applied as shown in Fig. 16-2.

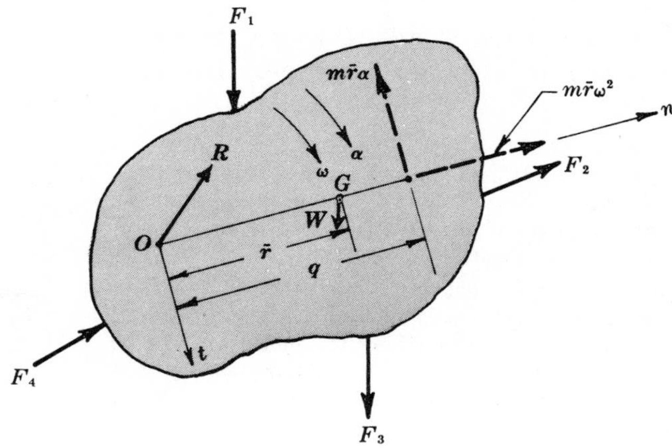

Fig. 16-2

The reversed force $m\bar{r}\omega^2$ must act away from the center of rotation along the n-axis, and the reversed force $m\bar{r}\alpha$ must be introduced at a distance q away from the center of rotation O and parallel to the t-axis but such that the direction of its moment is opposite to the direction of the angular acceleration α.

The distance q is given by

$$q = k_O^2/\bar{r}$$

where k_O^2 = square of the radius of gyration of the body with respect to the axis of rotation through O. Note that $k_O^2 = I_O/m$, I_O being the mass moment of inertia of the body about O and m its total mass.

$\bar{r}$ = distance from the center of rotation O to the mass center G.

Apply the equations of equilibrium to this figure for a solution.

The inertia force method for rotation about an *axis of symmetry* is simple, since the effective forces $m\bar{r}\omega^2$ and $m\bar{r}\alpha$ are each zero. This is true because $\bar{r} = 0$. It is only necessary to apply a reversed couple $\bar{I}\alpha$ to hold the system in equilibrium. The figure illustrates this fact. Note that the couple $\bar{I}\alpha$ is counterclockwise since the angular acceleration α is given clockwise.

This couple, acting with the external forces F_1, F_2, F_3, etc., has the effect of holding the body in equilibrium. Hence the equations of equilibrium may be applied.

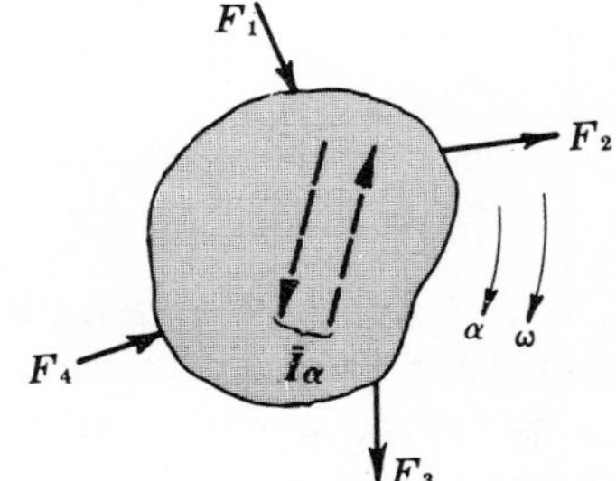

Fig. 16-3

CENTER of PERCUSSION

Center of percussion is that point on the n-axis in the foregoing figures through which the resultant of the effective forces acts. It is at a distance $q = k_O^2/\bar{r}$ from the center of rotation.

Solved Problems

Vectors are designated in the diagrams by their magnitudes when the directions are evident by inspection.

1. A bar pivoted about a horizontal axis through its lower end is allowed to fall from a vertical position. Describe its motion.

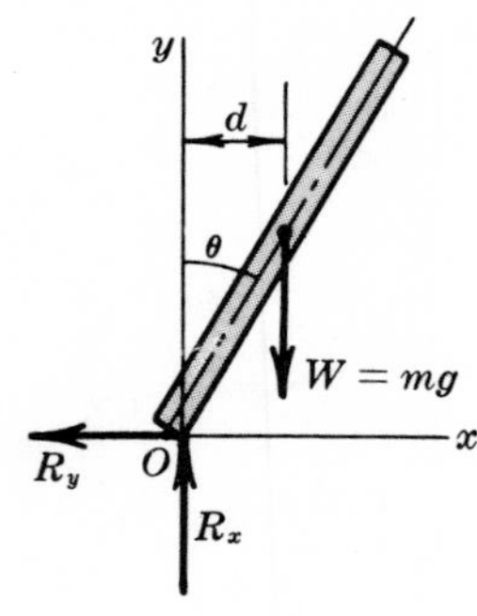

Fig. 16-4

Solution:

Assume that the bar is of length l and mass m. The free body diagram in Fig. 16-4 shows the weight W acting down and the reactions R_x and R_y at the point of support. The bar is assumed to be at a position θ from the rest position (vertical).

The bar is rotating about a fixed axis under the action of an unbalanced force system, the equations of motion being $\Sigma F_n = m\bar{r}\omega^2$, $\Sigma F_t = m\bar{r}\alpha$, $\Sigma M_O = I_O\alpha$. However, in describing the motion it will be sufficient to determine the magnitudes of the angular velocity and acceleration as functions of displacement θ. Hence the only equation needed is $\Sigma M_O = I_O\alpha$.

The only external force with moment about O is weight W. Hence taking clockwise moments as positive, $Wd = I_O\alpha$.

But $d = \frac{1}{2}l\sin\theta$, $I_O = \frac{1}{3}ml^2$, and $\alpha = d^2\theta/dt^2$. Hence

$$mg(\tfrac{1}{2}l\sin\theta) = \tfrac{1}{3}ml^2\frac{d^2\theta}{dt^2} \qquad \text{or} \qquad \frac{d^2\theta}{dt^2} = \frac{3g}{2l}\sin\theta$$

This differential equation may be solved by noting that

$$\frac{d^2\theta}{dt^2} = \frac{d}{dt}\left(\frac{d\theta}{dt}\right) = \frac{d\omega}{dt} = \frac{d\omega}{d\theta}\frac{d\theta}{dt} = \frac{d\omega}{d\theta}\omega$$

The original equation $\dfrac{d^2\theta}{dt^2} = \dfrac{3g}{2l}\sin\theta$ is now written as $\omega\dfrac{d\omega}{d\theta} = \dfrac{3g}{2l}\sin\theta$.

Multiply both sides of the equation by $d\theta$ to obtain $\omega\,d\omega = \dfrac{3g}{2l}\sin\theta\,d\theta$.

Integration yields $\frac{1}{2}\omega^2 = -(3g/2l)\cos\theta + C$, where C is the constant of integration.

To evaluate C note that $\omega = 0$ when $\theta = 0$. Hence $0 = -(3g/2l)(1) + C$ or $C = 3g/2l$. Then

$$\tfrac{1}{2}\omega^2 = -(3g/2l)\cos\theta + 3g/2l \qquad \text{and} \qquad \omega = \sqrt{(3g/l)(1-\cos\theta)}$$

This indicates a method of finding ω in terms of θ and, of course, $\alpha = (3g/2l)\sin\theta$.

2. A uniform bar of length l and mass m is rotating at a constant angular velocity $\boldsymbol{\omega}$ about a vertical axis through a point at a distance a from one end. For the phase shown in Fig. 16-5, when the bar is passing through the plane of the paper, determine the horizontal and vertical components of the reaction of the support on the bar.

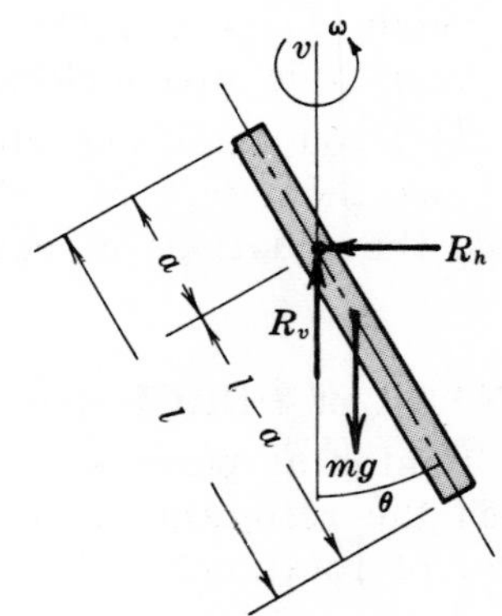

Fig. 16-5

Solution:

Assume a is less than half the length. The weight $W = mg$, and the horizontal and vertical components of the bearing reactions are shown acting on the rod.

Note that, since the angular speed ω is constant, the magnitude α of the angular acceleration (about the vertical axis) is zero. Hence there are no forces acting which have moments about the vertical axis. This conclusion should be evident from the scalar equation $\Sigma M_v = I_v\alpha = I_v(0)$.

Note also that the equations of motion ($\Sigma F_n = m\bar{r}\omega^2$, $\Sigma F_t = m\bar{r}\alpha$, $\Sigma M_O = I_O\alpha$) do *not* apply to the body as a whole since the rod is rotating about an axis which is *not* perpendicular to the plane of symmetry of the rod.

Consider the bar to be composed of a series of thin pieces, each moving with an angular velocity $\boldsymbol{\omega}$ about the vertical axis and at a distance ρ from the axis.

From Fig. 16-6, $\rho = z \sin\theta$ and $dm = \dfrac{dz}{l}m$.

Each differential mass is moving on a circular path (radius ρ). Hence a normal acceleration a_n exists toward the axis along ρ. A normal force dF accompanies this acceleration a_n ($dF = dm \times a_n$). The sum of all such normal forces (all horizontal) must be the horizontal bearing reaction R_h on the bar. Since in the phase shown the longer length of the bar is to the right of the axis, the normal forces directed to the left will be greater than those acting to the right. Hence the reaction R_h which supplies these normal forces acts to the left.

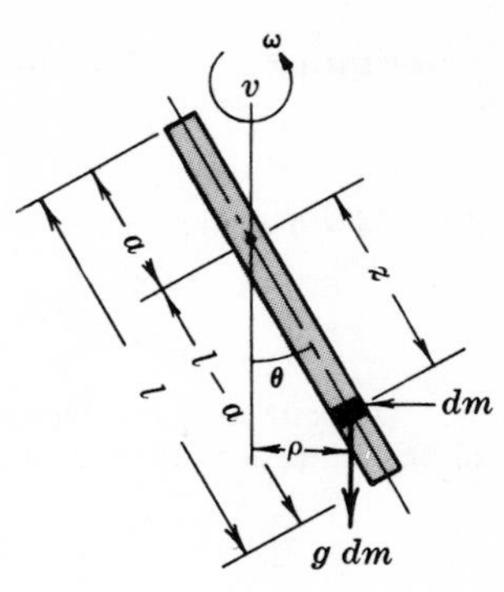

Fig. 16-6

The normal force on the particle dm shown in the figure acts toward the axis, i.e. to the left, which means in a negative direction. Hence

$$dF = -\,dm\,\rho\omega^2 = -\frac{dz}{l}m(z\sin\theta)\omega^2$$

and the total force is

$$F = \int dF = \int_{-a}^{l-a} -\frac{dz}{l}m(z\sin\theta)\omega^2$$

Since θ and ω^2 are constant, $\sin\theta$ and ω^2 may be removed from within the integral sign.

$$F = R_h = -\frac{m\omega^2\sin\theta}{l}\int_{-a}^{l-a} z\,dz = -\,m\omega^2(\tfrac{1}{2}l - a)\sin\theta$$

To determine R_v use the fact that the summation of the vertical forces must equal zero:

$$\Sigma F_v = 0 = R_v - mg \qquad \text{or} \qquad R_v = mg$$

For an explanation of the relation between ω and θ refer to Problem 4.

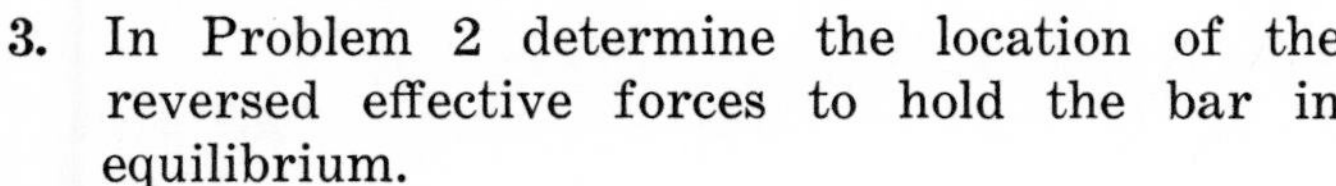

3. In Problem 2 determine the location of the reversed effective forces to hold the bar in equilibrium.

Solution:

Since the rod is rotating about an axis which is not perpendicular to a plane of symmetry, we take elements of mass of the rod which may be considered as rotating in a plane perpendicular to the axis of rotation.

In Fig. 16-7, the reversed effective normal forces $dm\,\rho\,\omega^2 = \dfrac{dz}{l}m\omega^2\rho$ for the masses of rod of length dz are shown acting away from the vertical axis. Also shown is the single force R at distance $\bar{z}$ from the support. This force must equal the sum of the individual forces for the elements and also must have the same moment as they have about the pivot point. Hence

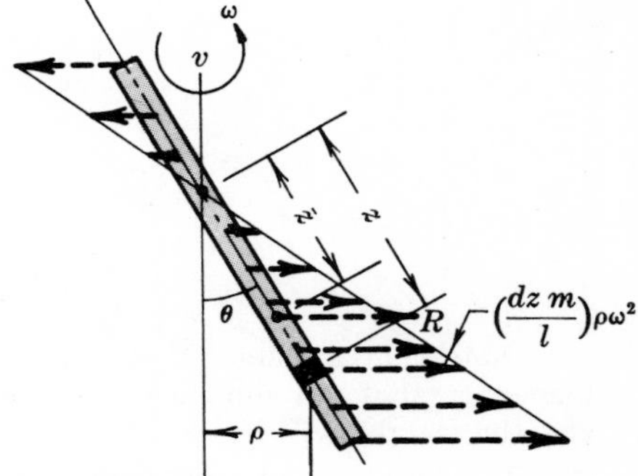

Fig. 16-7

$$R = \int_{-a}^{l-a}\frac{dz}{l}m\omega^2\rho = \int_{-a}^{l-a}\frac{dz}{l}m\omega^2 z\sin\theta = \frac{m\omega^2\sin\theta}{2}(l-2a)$$

This result, of course, is equivalent in magnitude to that derived in Problem 2.

Next take moments of the individual inertia forces about the pivot point and equate them to the moment of R about the pivot point. In all cases the forces are horizontal and the moment arms vertical (therefore equal to the product of the z distance and $\cos\theta$).

$$R\bar{z}\cos\theta = \int_{-a}^{l-a} z\cos\theta\,\frac{dz}{l}m\omega^2\rho$$

Substituting for R its value just derived and for ρ its value $z\sin\theta$, the equation becomes

$$\frac{m\omega^2\sin\theta}{2}(l-2a)\bar{z}\cos\theta = \frac{\cos\theta\, m\omega^2\sin\theta}{l}\int_{-a}^{l-a} z^2\,dz$$

When $\theta \neq 0$ or $90°$, then

$$\frac{\bar{z}}{2}(l-2a) = \frac{1}{3l}[(l-a)^3-(-a)^3] \qquad \text{or} \qquad \bar{z} = \frac{2}{3}\left(\frac{l^2-3la+3a^2}{l-2a}\right)$$

Of course if the bar is turning about an axis through its end, the above equation for $\bar{z}$ reduces to $2l/3$ since a will then be zero. The force R under this condition is

$$R = \tfrac{1}{2}m\omega^2 l\sin\theta$$

4. Rework Problem 2 using the inertia force method.

Solution:

In Problem 3 it was indicated that the reversed effective force ($\frac{1}{2}ml\omega^2\sin\theta$) for a bar of length l rotating about its end should be applied at a distance two-thirds of the length of the bar. Apply reversed effective forces to each of the two parts of the bar as shown in Fig. 16-8 below and solve as a problem in statics.

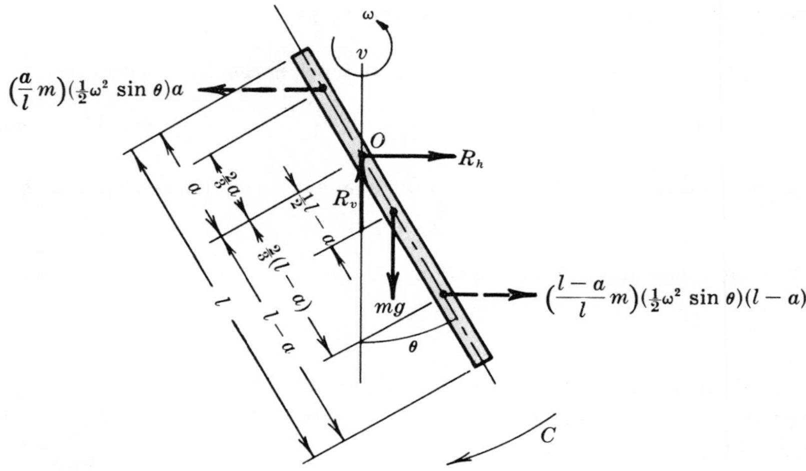

Fig. 16-8

Note that a couple C is applied to the system at the pivot point because it is not evident by inspection that the summation of moments about the pivot point of the two reversed effective forces and the weight will be zero.

The equations of equilibrium are:

$$(1)\quad \Sigma F_h = 0 = +R_h - \left(\frac{a}{l}m\right)(\tfrac{1}{2}\omega^2\sin\theta)a + \left(\frac{l-a}{l}m\right)(\tfrac{1}{2}\omega^2\sin\theta)(l-a)$$

$$(2)\quad \Sigma F_v = 0 = -mg + R_v$$

$$(3)\quad \Sigma M_O = 0 = \left[\left(\frac{l-a}{l}m\right)(\tfrac{1}{2}\omega^2\sin\theta)(l-a)\right][\tfrac{2}{3}(l-a)\cos\theta]$$

$$+ \left[\left(\frac{a}{l}m\right)(\tfrac{1}{2}\omega^2\sin\theta)a\right][\tfrac{2}{3}a\cos\theta] - mg(\tfrac{1}{2}l-a)\sin\theta + C$$

From equation (1), $R_h = -\tfrac{1}{2}\omega^2 m(l-2a)\sin\theta$.

This checks of course with the result in Problem 2. The minus sign indicates that R_h was assumed in the wrong direction and therefore actually acts to the left.

From equation (*2*), $R_v = mg$.

From equation (*3*), $C = m \sin\theta\,[g(\tfrac{1}{2}l - a) - \tfrac{1}{3}\omega^2(l^2 - 3la + 3a^2)\cos\theta]$.

The significance of the couple C may now be appreciated. To hold the bar at a desired angle θ when it is rotating with a given angular speed ω requires a couple C of the magnitude just derived. However if the couple C is not available at the pivot point, then the rod will seek and maintain a definite angle θ for a given angular speed ω. This angle θ may be determined by setting the expression for C equal to zero. Then, since $\sin\theta \neq 0$,

$$\cos\theta = \frac{3g(\tfrac{1}{2}l - a)}{(l^2 - 3la + a^2)\omega^2}$$

Once the value of θ has been determined, the value R_h may be found by substituting the value of $\sin\theta$ in its equation.

5. A cylinder of weight 161 lb (chosen because mass in slugs is then 5.00) rotates from rest in frictionless bearings under the action of a weight of 16.1 lb carried by a rope wrapped around the cylinder as shown in Fig. 16-9(*a*) below. If the diameter is 3 ft, what will be the angular speed of the cylinder two seconds after motion starts?

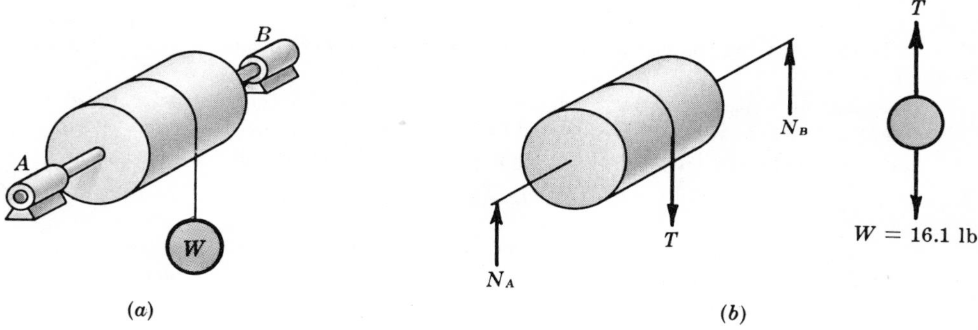

Fig. 16-9

Solution:

Free body diagrams of the cylinder and the weight are shown in Fig. 16-9(*b*). Note that the tension T in the rope is common to both diagrams.

To determine the angular speed ω after 2 seconds, it is necessary to find the magnitude α of the angular acceleration. This is done by the principles of the present chapter. The only equation necessary for the cylinder is the moment equation.

$$\Sigma M_{AB} = \bar{I}_{AB}\,\alpha$$

The subscript AB means *with respect to the axis AB*. Then

$$T \times r = \tfrac{1}{2}mr^2\alpha \qquad \text{or} \qquad (1)\quad T \times 1.5 = \tfrac{1}{2}(161/32.2)(1.5)^2\alpha$$

Since both tension T and angular acceleration α are unknown, another equation is required. Write the equation of motion of the weight W by summing forces in the vertical direction.

$$\Sigma F_v = ma_v \qquad \text{or} \qquad (2)\quad 16.1\text{ lb} - T = (16.1/32.2)a_v$$

Substituting $a_v = r\alpha = 1.5\alpha$ into equation (*2*),

$$(2')\quad 16.1 - T = (16.1/32.2)(1.5\alpha)$$

Solve (*1*) for $T = 3.75\alpha$ and substitute into (*2′*) to obtain $\alpha = 3.58$ rad/sec².

To find ω after 2 sec: $\omega = \omega_0 + \alpha t = 0 + (3.58)(2) = 7.16$ rad/sec.

6. Study the motion of the compound pendulum shown in Fig. 16-10.

Solution:

The compound pendulum differs from the simple pendulum in which only one small particle is considered. Here we are concerned with many particles with various linear velocities and accelerations. The system rotates about an axis perpendicular to the plane of the paper but not a centroidal axis.

Assume the pendulum to be moving counterclockwise, i.e. θ is positive in that direction. Then its weight has a moment about the axis of rotation which tends to retard motion. Hence the equation obtained by taking moments about the axis of rotation is

$$\Sigma M_O = I_O \alpha$$

and since the horizontal moment arm for the weight (force) mg is $\bar{r}\sin\theta$, this is written

$$\Sigma M_O = -(mg)(\bar{r}\sin\theta) = I_O \frac{d^2\theta}{dt^2}$$

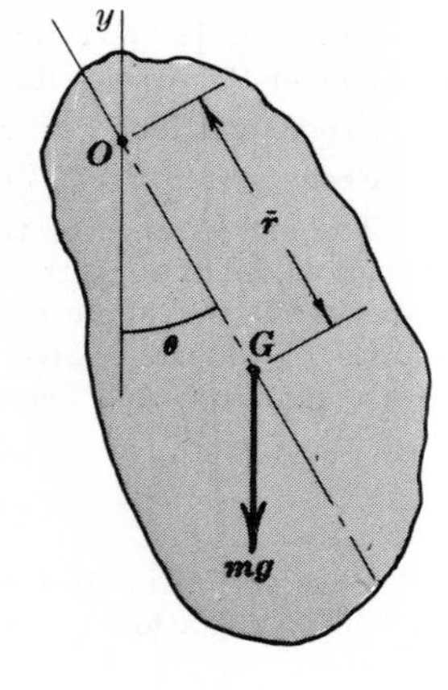

Fig. 16-10

or
$$\frac{d^2\theta}{dt^2} = -\frac{g}{(I_O/m\bar{r})}\sin\theta$$

This is the same type equation as derived for the simple pendulum of length l, i.e.,

$$\frac{d^2\theta}{dt^2} = -\frac{g}{l}\sin\theta$$

Thus it is seen that the compound pendulum has the same period as a simple pendulum whose length l is equal to $I_O/m\bar{r}$.

This may be stated somewhat differently if instead of I_O there is substituted its value in terms of the radius of gyration k_O of the compound pendulum about the axis of rotation.

Since $I_O = mk_O^2$, the length l of the equivalent simple pendulum may be written

$$l = I_O/m\bar{r} = mk_O^2/m\bar{r} = k_O^2/\bar{r}$$

The compound pendulum behaves as a simple pendulum with its mass concentrated at a point $k_O^2/\bar{r}$ from the axis of rotation. This value $k_O^2/\bar{r}$ occurs frequently in problems of rotation.

7. Find the stress in the rim of a flywheel rotating with constant angular speed ω.

Solution:

Draw a free body diagram of one half the rim as shown in Fig. 16-11. Add the reversed normal effective force $m\bar{r}\omega^2$ (where m is the mass of half the rim) to hold the system in equilibrium for study purposes. This is applied along the n-axis drawn through the center of rotation O and the mass center G ($\bar{r} = OG$, and for a *thin* semicircular rim is $2r/\pi$ where r is the radius to the inside of the rim). Some may prefer taking r to the centerline of the rim or the outside, but a thin rim means that the thickness is negligible compared to the radius of the wheel.

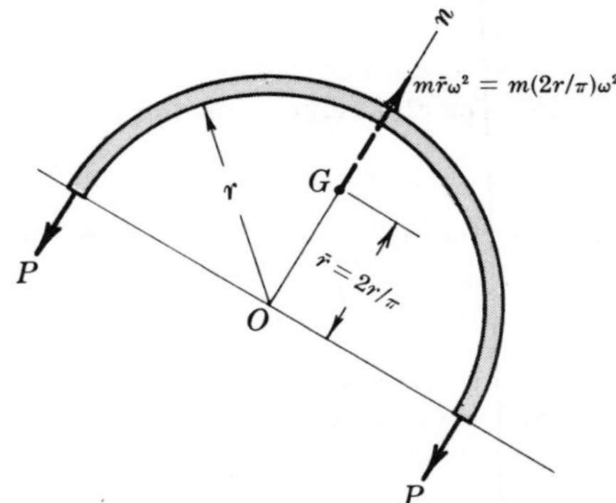

Fig. 16-11

The forces P represent the pull of the other half of the rim exerted as a tensile force (through the rim material) on the free body shown. Since the wheel is rotating with constant angular velocity, there is no angular acceleration and hence no reversed tangential effective force need be shown.

Let δ = density of the material in lb/ft³. Only one equation of motion is necessary:

$$\Sigma F_n = 0 \quad \text{or} \quad -2P + m(2r/\pi)\omega^2 = 0 \quad \text{or} \quad P = mr\omega^2/\pi = (W/g)(r\omega^2/\pi)$$

But the unit tensile stress is the total force P (on one side) divided by the cross sectional area A on which or through which it acts, or $s = P/A = Wr\omega^2/Ag\pi$.

Next express W of the half rim in terms of the rim cross section A, half rim length πr and density δ.

$$W = A\pi r\delta$$

This is, of course, again based on the assumption that the rim thickness is small compared to the rim diameter. Substituting this expression for W in the equation for s, we have

$$s = \delta r^2\omega^2/g$$

Since the rim speed v is equal to $r\omega$, the formula for s may also be written

$$s = \delta v^2/g$$

where s = unit stress in lb/ft², δ = density in lb/ft³, v = rim speed in ft/sec, and g = acceleration of gravity in ft/sec².

8. It is necessary to bank railroad tracks on a curve in such a way that the outer rail is above the inner rail. Since the roadbed is usually at some elevation above sea level, it is customary to call the vertical distance of the outer rail above the inner rail not elevation but the superelevation e. The greater the superelevation e the faster a train may travel around the curve. Determine the value of e in terms of the speed v of the train, the radius r of the curve and the angle of bank. Refer to Fig. 16-12(a) below.

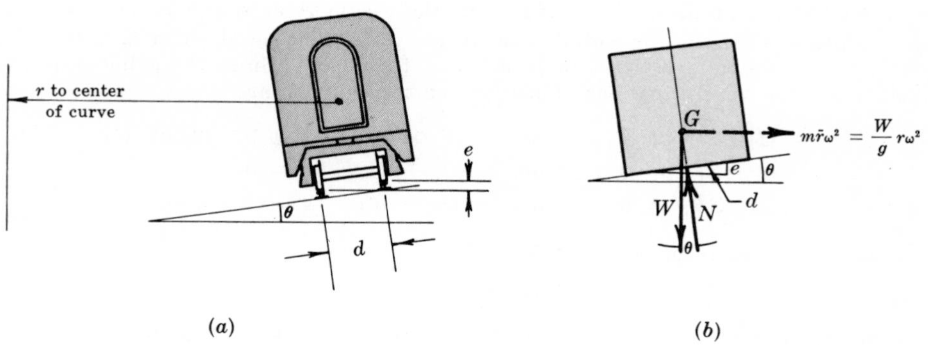

Fig. 16-12

Solution:

This is an example of rotation of a mass which can be considered a particle turning about the center of the curve.

Consider the curve to be properly banked for a definite speed v. This means that the train has no tendency to slide in or out along the radius. Hence the side pressure of either rail against the flange of the wheel is zero. The only reaction on the car is the push, perpendicular to the roadbed, of the tracks on the wheels.

A free body diagram showing the weight W, the reversed effective normal force $m\bar{r}\omega^2$ and a single reaction N of the tracks is now drawn (see Fig. 16-12(b) above). Note that the reversed effective normal force is $(W/g)r\omega^2$, since $\bar{r}$ (distance from center of rotation to center of mass) is also the radius r of the curve. Summing forces along the axis perpendicular to the N force yields

$$\Sigma F = -W \sin\theta + (W/g)r\omega^2 \cos\theta \qquad \text{or} \qquad \tan\theta = r\omega^2/g$$

Since θ is usually a small angle, $\tan\theta$ is approximately equal to $\sin\theta$ (up to about 6°); therefore the above equation may be written

$$\sin\theta \approx \tan\theta = r\omega^2/g = r^2\omega^2/gr$$

But from the figure, $\sin\theta = e/d$. Hence $e/d = r^2\omega^2/gr = v^2/gr$ or $e = dv^2/gr$.

9. In Problem 8 calculate the superelevation necessary for a 2000 ft curve for a speed of 60 mph. Assume standard gauge $d = 4$ ft $8\frac{1}{2}$ in.

Solution:

$$e \;=\; \frac{dv^2}{gr} \;=\; \frac{(4.71\text{ ft})(88\text{ ft/sec})^2}{(32.2\text{ ft/sec}^2)(2000\text{ ft})} \;=\; 0.566\text{ ft} \;=\; 6.80\text{ in.}$$

10. In Fig. 16-13(a) below, weight A is accelerating down 5 ft/sec². It is connected by a weightless, flexible, inextensible rope passing over a smooth drum to a homogeneous cylinder B of weight 161 lb. The cylinder is acted upon by a moment C = 50 lb-ft counterclockwise. Determine the weight of A and the components of the reaction at O on the cylinder B.

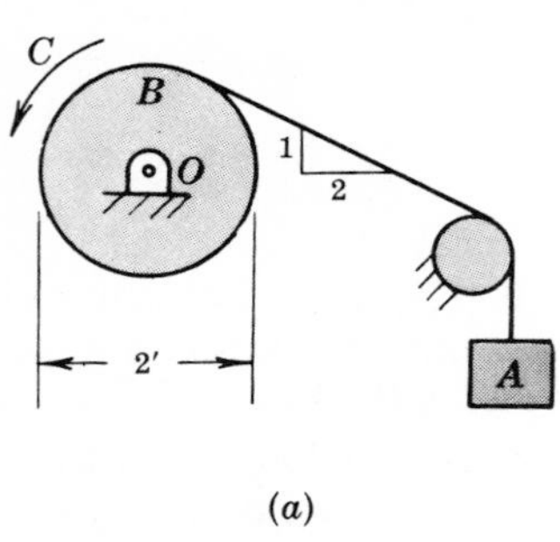

(a)

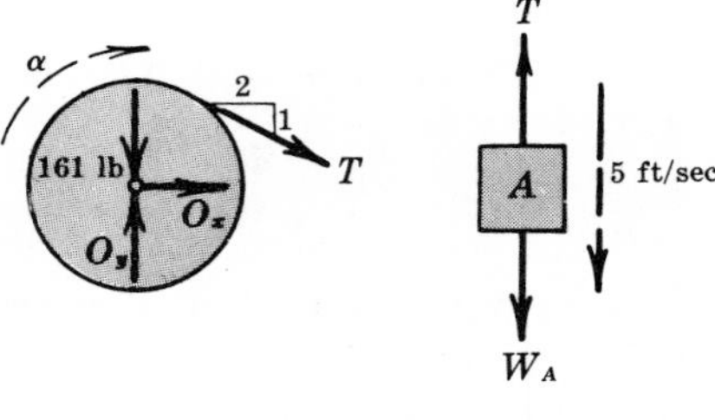

(b)

Fig. 16-13

Solution:

Draw free body diagrams of the cylinder and the weight A as shown in Fig. 16-13(b) above.

The magnitude a_t of the tangential acceleration of a point on the rim is 5 ft/sec². Hence the magnitude α of the angular acceleration is $a_t \div r = 5$ rad/sec². Since the cylinder rotates about an axis of symmetry, the three equations of motion for the cylinder are

$$(1)\quad \Sigma \bar{M} = \bar{I}\alpha \qquad \text{or} \qquad T \times 1 - 50 = \tfrac{1}{2}(161/32.2)(1)^2(5)$$
$$(2)\quad \Sigma F_x = 0 \qquad \text{or} \qquad O_x + T(2/\sqrt{5}) = 0$$
$$(3)\quad \Sigma F_y = 0 \qquad \text{or} \qquad O_y - 161 - T(1/\sqrt{5}) = 0$$

The equation of motion for weight A *is*

$$(4)\quad W_A - T = (W_A/g)(5)$$

Equation (*1*) yields $T = 62.5$ lb. Hence from equation (*4*), $W_A = 74.0$ lb.
Solve equations (*2*) and (*3*) to obtain $O_x = -55.9$ lb and $O_y = +189$ lb.

11. A homogeneous sphere weighing 240 lb is attached to a slender rod weighing 50 lb. In the horizontal position shown in Fig. 16-14 the angular speed of the system is 8 rad/sec. Determine the magnitude of the angular acceleration of the system and the reaction at O on the rod.

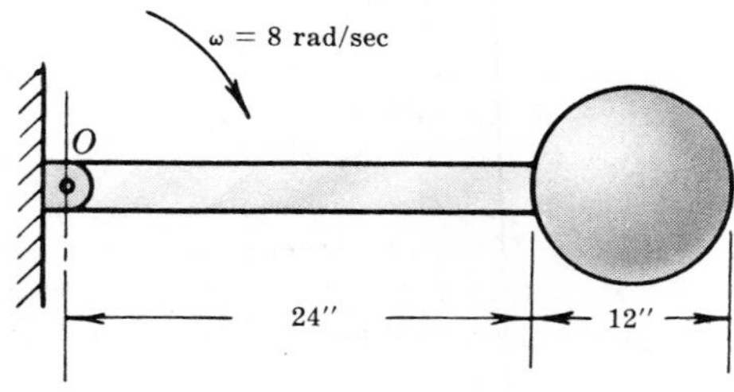

Fig. 16-14

Solution:

Locate the centroid of the system with reference to the pin O.

$$\bar{r} = \frac{50(12) + 240(30)}{290} = 26.9\text{ in.}$$

This problem may be solved as a problem in equilibrium by adding the reversed effective forces through the center of percussion P located $k^2/\bar{r}$ to the right of O, where k^2 equals the moment of inertia of the system about the axis of rotation divided by the total mass.

The total moment of inertia I equals the moment of inertia of the rod about its end ($\frac{1}{3}ml^2$) plus the transferred moment of inertia of the sphere ($\frac{2}{5}mr^2 + md^2$) where $d = 30$ in. $= 2.5$ ft.

$$I = \tfrac{1}{3}(50/32.2)(2)^2 + \tfrac{2}{5}(240/32.2)(\tfrac{1}{2})^2 + (240/32.2)(2.5)^2 = 49.5 \text{ slug-ft}^2$$

Hence $k^2 = I/m = 49.5/(290/32.2) = 5.49$ ft^2.

Point P is $k^2/\bar{r} = 5.49/(26.9/12) = 2.45$ ft to the right of O.

Draw a free body diagram (see Fig. 16-15).

Taking moments about O,

$$m\bar{r}\alpha \times 2.45 - 290 \times 2.24 = 0$$

or $\alpha = 13.1$ rad/sec^2.

Sum forces horizontally to obtain

$$O_x = m\bar{r}\omega^2 = 1290 \text{ lb}$$

Sum forces vertically to obtain

$$O_y = 290 - m\bar{r}\alpha = 25.0 \text{ lb}$$

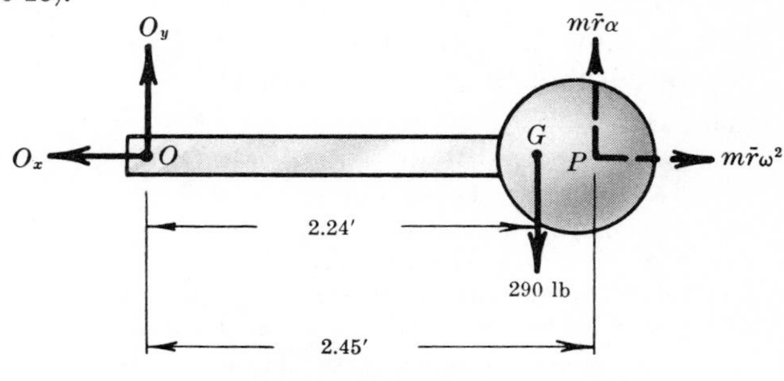

Fig. 16-15

12. An eccentric cylinder used in a vibrator weighs 40 lb and rotates about an axis 2 inches from its geometric center and perpendicular to the top view shown in Fig. 16-16. If the magnitudes ω and α of its angular velocity and angular acceleration are respectively 10 rad/sec and 2 rad/sec^2 in the phase shown, determine the reaction of the vertical shaft on the cylinder.

Solution:

Choose n and t axes as shown. The distance from O to the mass center G is $\bar{r} = 1/6$ ft.

The equations of motion are:

(1) $\Sigma F_n = m\bar{r}\omega^2$ or $O_n = \dfrac{40}{32.2}\,\dfrac{1}{6}\,(10)^2$

(2) $\Sigma F_t = m\bar{r}\alpha$ or $O_t = \dfrac{40}{32.2}\,\dfrac{1}{6}\,(2)$

(3) $\Sigma M_O = I_O\alpha = \left[\dfrac{1}{2}\,\dfrac{40}{32.2}\left(\dfrac{1}{2}\right)^2 + \dfrac{40}{32.2}\left(\dfrac{1}{6}\right)^2\right] \times 2$

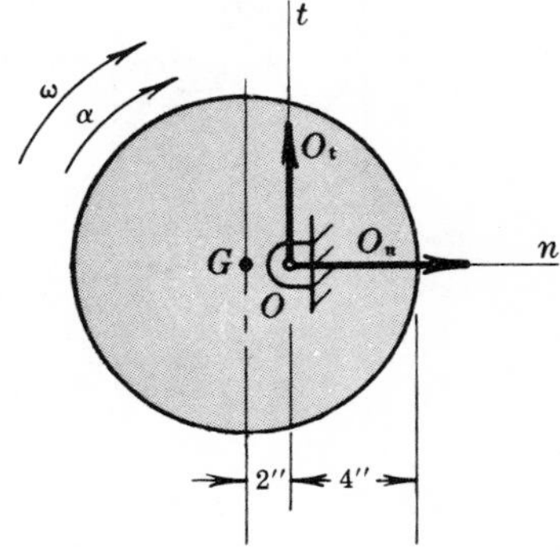

Fig. 16-16

Here the reactions O_n and O_t of the shaft are the external forces acting *on* the cylinder.

Note that I_O is found by the transfer formula. The necessary couple $I_O\alpha$ equals 0.38 lb-ft applied clockwise to the cylinder by the shaft.

The reaction components are $O_n = 20.7$ lb to the right and $O_t = 0.41$ lb up.

13. A homogeneous sphere of diameter 12 in. and weight 260 lb rotates freely about a diameter. What constant moment is necessary to bring it from rest to a speed of 60 rpm in 20 revolutions?

Solution:

Using the kinetic formula $\omega^2 = \omega_0^2 + 2\alpha\theta$, obtain $\alpha = \pi/20$ rad/sec^2.

The moment of inertia $\bar{I} = \frac{2}{5}mr^2 = \frac{2}{5}(260/32.2)(\frac{1}{2})^2 = 0.807$ slug-ft^2.

Hence the necessary moment equals $\bar{I}\alpha = 0.127$ lb-ft.

14. In Fig. 16-17(*a*) the bar AB is held in a vertical position by the weightless cord BC as the system rotates about the vertical y-y axis. The pin at A is smooth and the bar AB weighs 32.2 lb. If the breaking strength of the cord is 120 lb, how fast can the system rotate without breaking the cord?

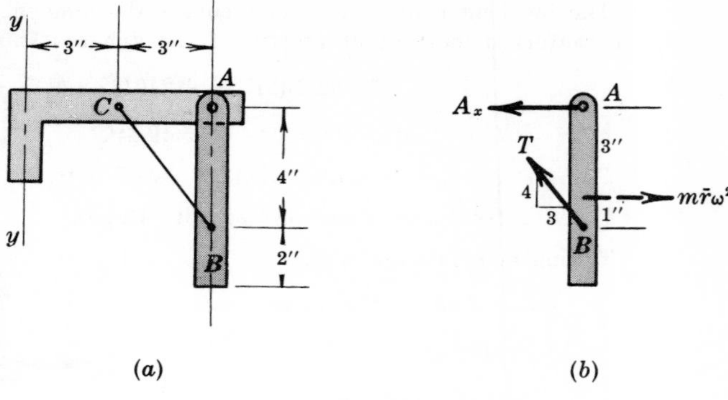

Fig. 16-17

Solution:

Draw a free body diagram of the bar AB, omitting the weight and the vertical component A_y for simplicity. See Fig. 16-17(*b*). The reversed effective force is placed at the center of the bar. This is true only because each horizontal element or slice of the bar is the same distance from y-y as any other element.

Take moments about A to obtain $+m\bar{r}\omega^2 \times 3 - T \times \frac{3}{5} \times 4 = 0$. Note that only the horizontal component of T has a moment about A. Substitute $m = 1$ slug, $T = 120$ lb, and $\bar{r} = 6/12$ ft to obtain $\omega = 13.8$ rad/sec or 132 rpm, the value beyond which the cord will break.

15. The pulley system shown in Fig. 16-18(*a*) weighs 64.4 lb and has radius of gyration 1.50 ft. Determine the tensions in the cords and the angular acceleration of the pulleys when the weights are released.

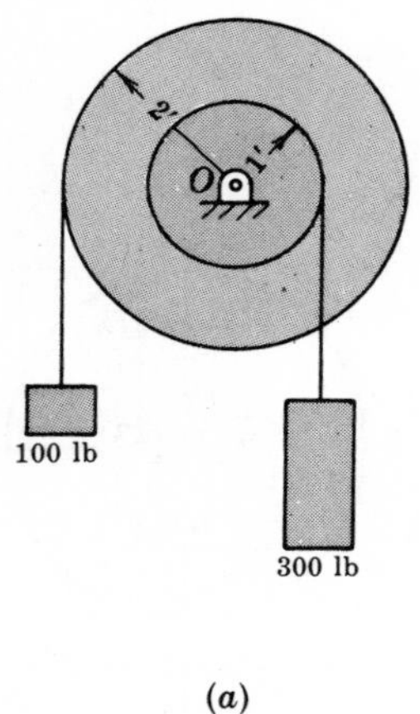

(*a*)

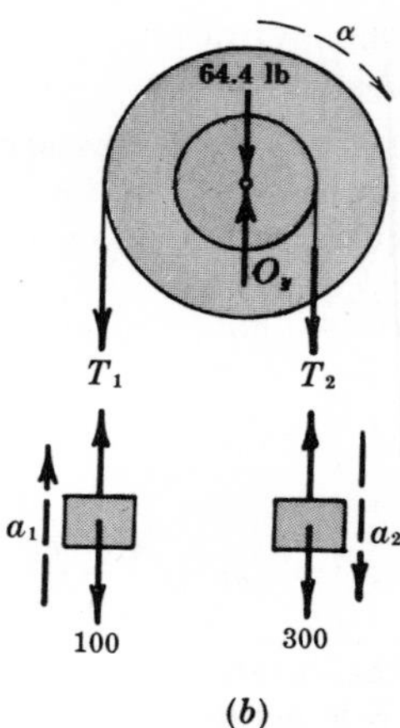

(*b*)

Fig. 16-18

Solution:

Draw free body diagrams of the three components of the system. See Fig. 16-18(*b*) above. Note that $a_1 = 2\alpha$, $a_2 = 1\alpha$.

The equations of motion are:

$$(1)\quad \Sigma F = T_1 - 100 = (100/g)a_1 = (200/g)\alpha$$

$$(2)\quad \Sigma F = 300 - T_2 = (300/g)a_2 = (300/g)\alpha$$

$$(3)\quad \Sigma \bar{M} = \bar{I}\alpha \quad \text{or} \quad T_2 \times 1 - T_1 \times 2 = (64.4/32.2)(1.50)^2\alpha$$

Multiply equation (*1*) by -2 and equation (*2*) by -1 and add to obtain $T_2 - 2T_1 = -(700/g)\alpha + 100$. Equating the right side of this equation to the right side of (*3*),

$$-(700/g)\alpha + 100 = 4.5\alpha \quad \text{or} \quad \alpha = 3.80 \text{ rad/sec}^2$$

From (*1*), $T_1 = 100 + (200/g)\alpha = 124$ lb. From (*2*), $T_2 = 300 - (300/g)\alpha = 265$ lb.

16. In Fig. 16-19(*a*) below, the weight C of 28 lb is moving down with a velocity of 16 ft/sec. The moment of inertia of drum B is 12 ft-lb-sec^2 and it rotates in frictionless bearings. If the coefficient of friction between the brake A and the drum is 0.40, what force P is necessary to stop the system in 2 seconds? What is the reaction at D on the rod?

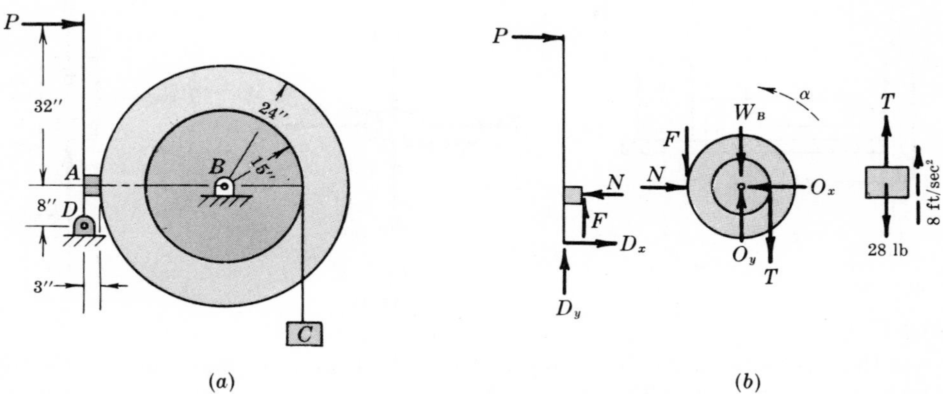

Fig. 16-19

Solution:

Draw the free body diagrams of the bar, the drum and the weight as shown in Fig. 16-19(*b*) above. Next determine the magnitude of the acceleration of the weight knowing the initial speed is 16 ft/sec down, the final speed is zero and the time is 2 sec. The value of acceleration is 8 ft/sec^2 up.

A vertical summation of forces acting on weight C gives $T - 28 = (28/g)(8)$ or $T = 35.0$ lb.

To find the friction F, sum moments of forces about O of the drum. The angular acceleration of the drum is counterclockwise (system is slowing down). Its magnitude may be determined because the linear acceleration of a point 15 inches from O is 8 ft/sec^2. Hence $\alpha = 8/(15/12) = 6.40$ rad/sec^2. The moment equation is $F(2) - T(1.25) = I\alpha$ or $F(2) - 35.0(1.25) = 12(6.40)$, from which $F = 60.3$ lb.

But $F = \mu N$. Hence the necessary normal force $N = 60.3/0.40 = 151$ lb.

Summing moments about D of all forces acting on the bar, $-P(40) + N(8) + F(3) = 0$ or $P = 34.7$ lb.

Sum forces horizontally on the bar to obtain $34.7 - 151 + D_x = 0$ or $D_x = 116$ lb to the right.

Summing forces vertically, $D_y + 60.3 = 0$ or $D_y = 60.3$ lb down.

17. A 20 lb ball A is mounted on a horizontal bar attached to a vertical shaft as shown in Fig. 16-20(*a*). Neglecting the weight of the bar and shaft, what are the reactions at B and C when the system is rotating at a constant speed of 90 rpm?

Solution:

Draw a free body diagram showing the reversed effective force acting horizontally to the right for the phase shown. [See Fig. 16-20(*b*).] Its magnitude is $(20/32.2)(1)(2\pi 90/60)^2 = 55.2$ lb.

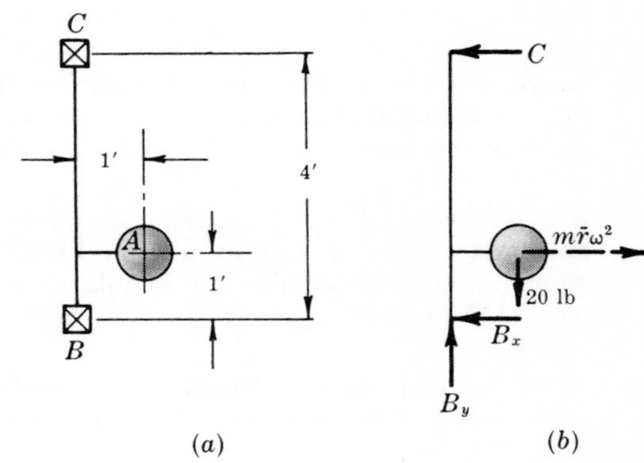

Fig. 16-20

Take moments about B to determine C, i.e., $C(4) - 55.2(1) - 20(1) = 0$ or $C = 18.8$ lb to the left.

By inspection, B_y must equal 20 lb up.

Sum horizontally to obtain $-B_x + 55.2 - 18.8 = 0$ or $B_x = 36.4$ lb to the left.

18. A weighs 32.2 lb and B weighs 96.6 lb. They rest on a rotating surface and are connected by a cord passing around a frictionless pulley as shown in Fig. 16-21(a) below. If the coefficient of friction between the weights and the surface is 0.25, determine the value of ω at which radial sliding will occur.

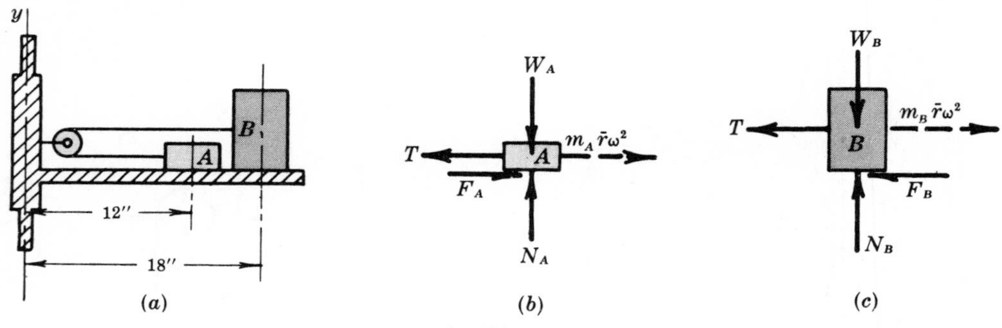

Fig. 16-21

Solution:

Add the reversed effective forces to each free body diagram as shown in Fig. 16-21(b) and (c).
By inspection $N_A = 32.2$ lb and $N_B = 96.6$ lb. Hence $F_A = 8.05$ lb and $F_B = 24.2$ lb.
A horizontal summation of forces for body A yields $T = 8.05 + (32.2/32.2)(12/12)\omega^2$.
Similarly for body B there results $T = (96.6/32.2)(18/12)\omega^2 - 24.2$.
Equate the right members of these equations to find ω: $8.05 + \omega^2 = 4.5\omega^2 - 24.2$.
The desired ω for slipping is thus 3.03 rad/sec or 29 rpm.

Supplementary Problems

19. The 6 ft homogeneous bar of weight W shown in Fig. 16-22 falls from its vertical rest position. Assuming no friction at the pivot, determine the angular velocity of the bar at the time when the tangential component of the acceleration of the unpivoted end is equal to g, the acceleration of gravity. *Ans.* $\theta = 41.8°$, $\dot{\theta} = 2.02$ rad/sec

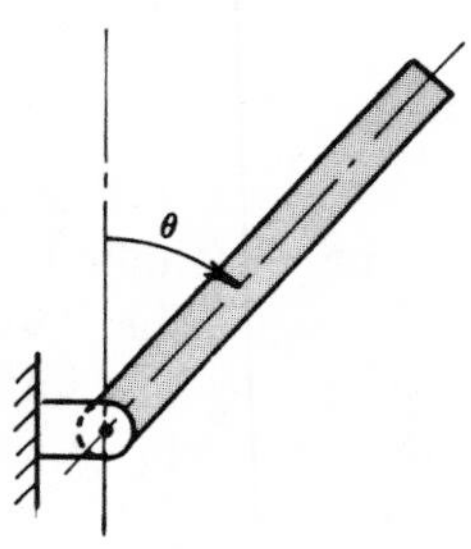

Fig. 16-22

20. A cylinder 2 ft in diameter and weighing 100 lb has an angular acceleration of 2 rad/sec^2 about its geometric axis. What torque is acting to produce this acceleration? *Ans.* 3.11 lb-ft

21. A homogeneous disk 6 ft in diameter weighs 800 lb and is made to revolve about its geometric axis by a force of 80 lb applied tangentially to its circumference. Determine the angular acceleration of the disk. *Ans.* 2.15 rad/sec^2

22. A 400 lb weight hangs vertically downward from the end of a weightless cord which is wrapped around a cylinder 3 ft in diameter. The weight descends 26 ft in 4 sec. What is the weight of the cylinder? *Ans.* 7130 lb

23. A homogeneous bar 10 ft long weighs 30 lb and is suspended vertically from a pivot point located at one end. The bar is struck by a horizontal blow of 90 lb at a point 2 ft below the pivot point. Determine the horizontal reaction at the pivot. What is the angular acceleration of the bar due to the blow? *Ans.* $R_h = 63$ lb, $\alpha = 5.8$ rad/sec^2

24. The slender bar 4 ft long weighs 8 lb. It is hanging vertically at rest when struck by a horizontal force $P = 3$ lb at the lower end as shown in Fig. 16-23. Determine (a) the angular acceleration of the bar and (b) the horizontal and vertical components of the pin reaction O on the bar.
Ans. $\alpha = 9.06$ rad/sec^2 counterclockwise, $O_y = 8$ lb up, $O_x = 1.5$ lb to right

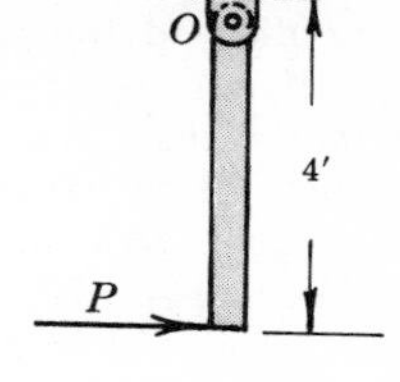

Fig. 16-23

25. In the preceding problem, where should the 3 lb force be applied so that the horizontal component of the pin reaction will be zero?
Ans. 2.67 ft below O.

26. At what point should the horizontal force P be applied to the homogeneous bar, the homogeneous cylinder, and the homogeneous sphere so that the horizontal component of the pin reaction at O is zero? See Fig. 16-24.
Ans. $d_b = \frac{2}{3}d$, $d_c = \frac{3}{4}d$, $d_s = \frac{7}{10}d$

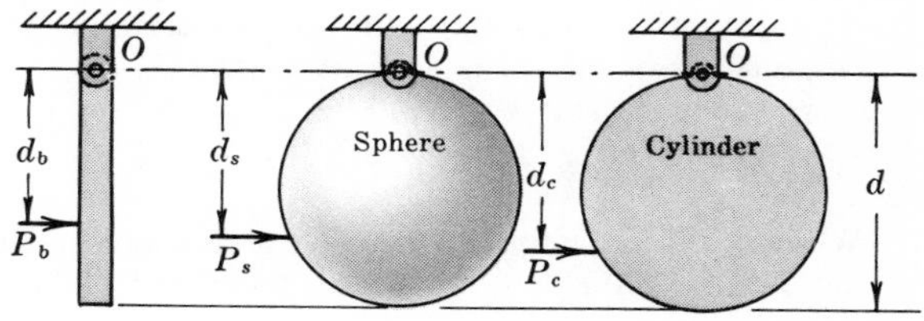

Fig. 16-24

27. The coefficient of friction between the horizontal floor and a trackman's shoes is 0.5. Find the radius of the smallest circular path around which he can travel at constant speed of 16 ft/sec without slipping.
Ans. $r = 15.9$ ft

28. A train weighing 100,000 lb has its center of gravity 5.5 ft above the rails. It rounds an unbanked curve having a radius of 2500 ft with a speed of 30 mph. If the centerlines of the rails are 4 ft $8\frac{1}{2}$ in. apart, find the vertical force on the outer rail. *Ans.* 52,800 lb

29. Determine the angle of banking for a roadway to allow an automobile speed of 40 mph on a curve of 300 ft radius so that there will be no side thrust on the wheels. *Ans.* 19.6°

30. A highway curve with a radius of 2000 ft is to be banked for a 50 mph speed. What should be the angle of bank so that at the design speed there will be no frictional force necessary on the tire in a direction perpendicular to the plane of rolling? *Ans.* $\theta = 4.79°$

31. A 0.06 lb bead slides from rest down a frictionless wire bent as shown in the adjacent Fig. 16-25. Determine the normal force of the wire on the bead at point A. Next find the normal forces at points B and C with the bead on the arc of the circle.
Ans. $N_A = 0.036$ lb, $\theta_x = \tan^{-1} 0.75 = 36.9°$;
$N_B = 0.276$ lb, $\theta_x = 36.9°$; $N_C = 0.348$ lb up

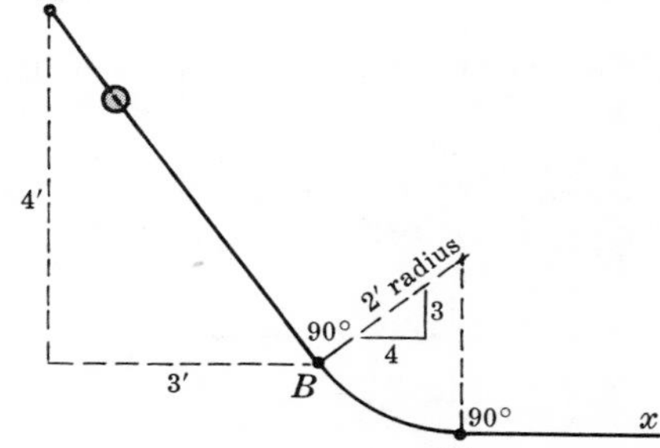

Fig. 16-25

32. A flywheel with radius of gyration k and weight W is acted upon by a constant torque $M = C$. What will be the angular speed after the flywheel has rotated θ radians from rest? *Ans.* $\omega = (1/k)\sqrt{2Cg\theta/W}$

33. A uniform disk, of diameter 6 ft and weighing 34 lb, is released from rest when O and G are on a horizontal line as shown in the adjacent Fig. 16-26. What will be the angular velocity when G is vertically below O? Determine the reactions at O at that time.
Ans. $\omega = 3.78$ rad/sec,
$O_n = 79.3$ lb,
$O_t = 0$

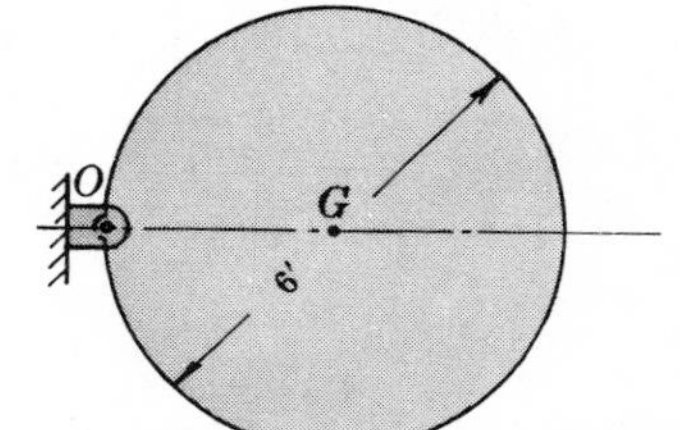

Fig. 16-26

34. If the composite object of Problem 11 is used as a pendulum, determine the frequency of the motion.
Ans. 0.577 cps

35. Refer to Fig. 16-27. A moment M of 45 lb-ft is applied to the uniform disk A which then drives uniform disk B without slip occurring between the two disks. What is the angular acceleration of each disk? Disk A weighs 64.4 lb and disk B weighs 128.8 lb.
Ans. $\alpha_A = 3.75$ rad/sec² counterclockwise, $\alpha_B = 1.88$ rad/sec² clockwise

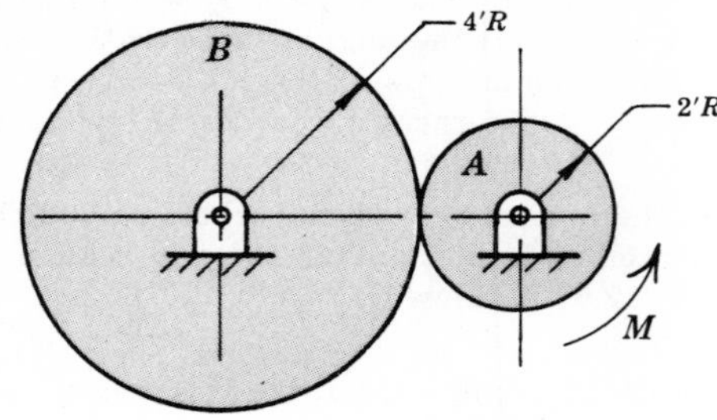

Fig. 16-27

36. A weightless bar 12 ft long rotates in a horizontal plane about its center. From each end of the rod is suspended a 10 lb weight on a 2 ft cord. When the system is revolving 40 rpm, determine the angle between the cords and the vertical.
Ans. $(3 + \sin\theta)/\tan\theta = 0.917$ and $\theta = 77°$ (by trial and error)

37. What is the angular acceleration of the pulley shown in Fig. 16-28 below turning under the action of the two weights? *Ans.* $\alpha = 0.668$ rad/sec² clockwise

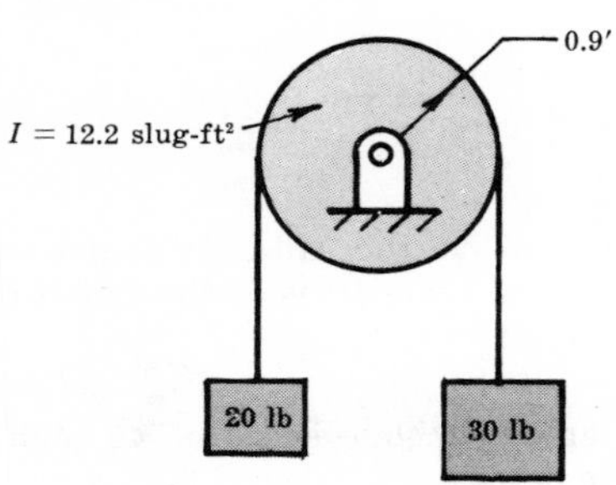

Fig. 16-28

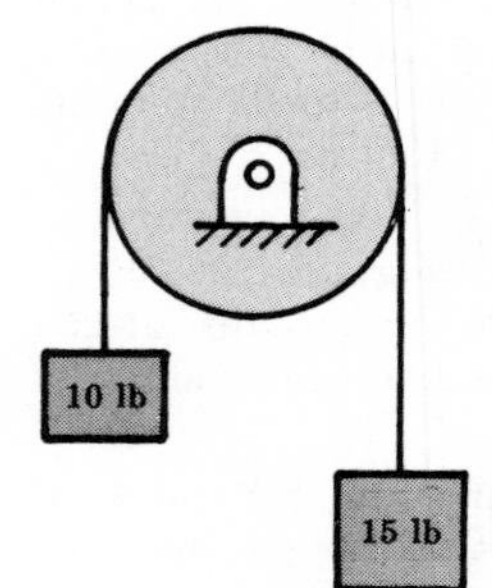

Fig. 16-29

38. In Fig. 16-29 above, the simple Atwood machine has weights of 10 lb and 15 lb measured on Earth with $g = 32.2$ ft/sec². If the machine is moved to the moon where the acceleration of gravity is 0.16 of the Earth's g, what will be the tension in the cord when the masses are released from rest?
Ans. $T = 1.92$ lb

39. Two weights are connected by a light inextensible string which passes over the pulley which rotates in frictionless bearings. See Fig. 16-30 below. At a place where $g = 32.2$ ft/sec² the weights of A, B, and pulley are 30 lb, 20 lb, and 10 lb respectively. The radius of gyration of the pulley is 1.28 ft. Determine the acceleration of the weights if the system is released from rest at another place where $g = 14.7$ ft/sec². *Ans.* $a = 2.56$ ft/sec²

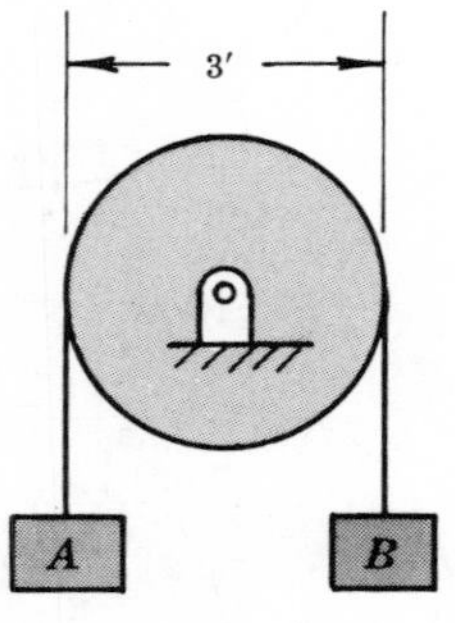

Fig. 16-30

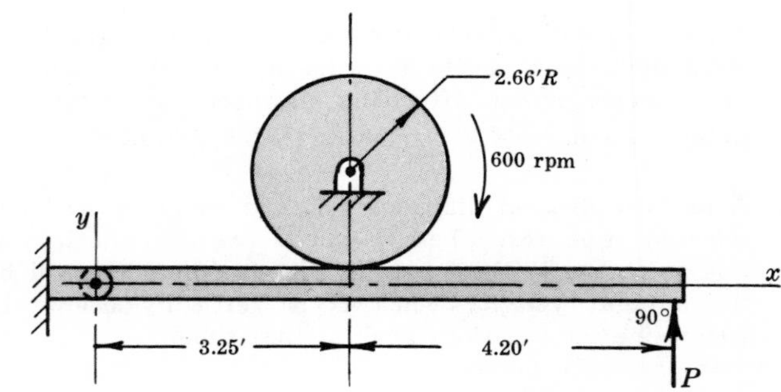

Fig. 16-31

40. In Fig. 16-31 above, the disk of weight 76 lb and radius of gyration 2.34 ft is rotating 600 rpm. What force P must be applied to the braking mechanism to stop the disk in 25 sec? Use a coefficient of friction between the disk and the horizontal member equal to 0.30. *Ans.* $P = 17.8$ lb

41. A thin triangular lamina of weight 20 lb is rotating 30 rpm about a horizontal axis in two frictionless bearings at A and B as shown in Fig. 16-32 below. When the lamina is vertical as shown, determine the bearing reactions at A and B. *Ans.* $A = 8.9$ lb up, $B = 10.7$ lb up

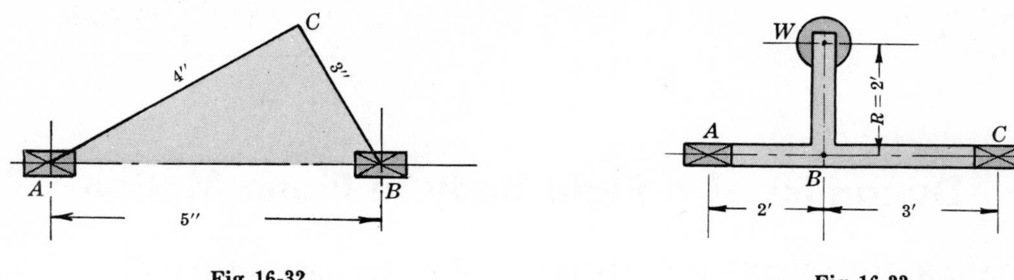

Fig. 16-32

Fig. 16-33

42. In Fig. 16-33 above, $AB = 2$ ft, $BC = 3$ ft, $W = 100$ lb and $R = 2$ ft. If the system is rotating 100 rpm, determine the support reactions due to "centrifugal" force at journals A and C. Neglect the weight of the arm. *Ans.* $A = 406$ lb, $B = 272$ lb

43. Solve Problem 42 if the speed is 1000 rpm.
Ans. $A = 40{,}600$ lb, $B = 27{,}200$ lb

44. A disk rotates uniformly at 15 rpm in a horizontal plane. A 100 lb weight is placed on the disk at a point 6 ft distant from the axis of rotation. If the weight is just about to slip in this position, what is the coefficient of friction between the weight and the disk? *Ans.* $\mu = 0.46$

45. A two pound block is held on a horizontal turntable by a cord which passes over a pulley and down through a hole in the shaft as shown in Fig. 16-34. What tension in the cord is necessary to keep the block 18″ from the center? Assume no friction and that the pin is located as shown. *Ans.* $T = 83.8$ lb

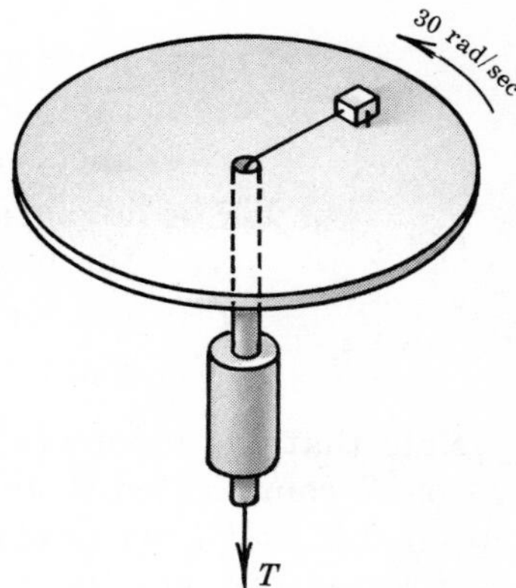

Fig. 16-34

Chap 16 in Book

Chapter 17

Dynamics of a Rigid Body in Plane Motion

THE SCALAR EQUATIONS of MOTION in this general case are

$$\Sigma F_x = m\bar{a}_x, \qquad \Sigma F_y = m\bar{a}_y, \qquad \Sigma\bar{M} = \bar{I}\alpha$$

where $\Sigma F_x, \Sigma F_y$ = algebraic summation of the components of the external forces in the x and y directions respectively

m = mass of the body

$\bar{a}_x, \bar{a}_y$ = components of the linear acceleration of the mass center in the x and y directions respectively

$\Sigma\bar{M}$ = algebraic summation of the moments of the external forces about the mass center

$\bar{I}$ = moment of inertia of the body about the mass center

α = magnitude of the angular acceleration of the body.

Note that the moments are taken about the mass center. As a rule the equation is much more complicated if any other point is chosen. However, it is well to note that the equation $\Sigma M_O = I_O\alpha$ will hold if the point O is a point whose acceleration is either zero or is directed through the mass center.

Note also that the acceleration $\bar{a}_x$ and ΣF_x are in the same direction; $\bar{a}_y$ and ΣF_y are in the same direction; and that the moment of an external force about the mass center is considered positive if it is in the same direction as that assumed for α.

Solved Problems

1. A 600 lb wheel with a diameter of 30 inches rolls without slipping down a plane inclined at an angle of 25° with the horizontal. Determine the friction force F and the acceleration of the mass center. r=15"

Solution:

Fig. 17-1 shows the force system acting on the wheel.

Considerable difficulty is usually experienced in indicating the direction of the friction force F. (F may have any value between $-\mu N$ and μN). In this case the friction must act up the plane; otherwise the wheel would slip down the plane. Also, friction is the only force which has a moment about the mass center and therefore is the force causing the angular acceleration. ($\Sigma\bar{M} = \bar{I}\alpha$).

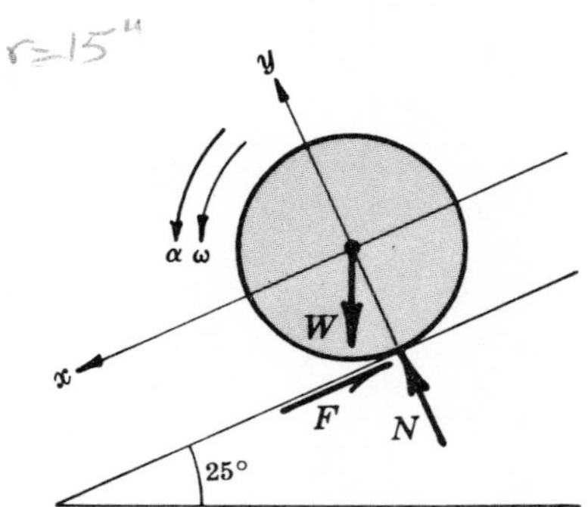

Fig. 17-1

Choose the x-axis parallel to the plane, with the positive direction down. The y-axis is positive up. The mass $m = 600/32.2 = 18.6$ lb-sec²/ft or slugs. The moment of inertia is

$$\bar{I} = \tfrac{1}{2}mr^2 = \tfrac{1}{2}(18.6 \text{ lb-sec}^2/\text{ft})(15/12 \text{ ft})^2 = 14.5 \text{ lb-sec}^2\text{-ft or slug-ft}^2$$

The above units are commonly used in engineering textbooks.

The equations of motion are: (1) $\Sigma F_x = m\bar{a}_x$, (2) $\Sigma F_y = m\bar{a}_y$, (3) $\Sigma \bar{M} = \bar{I}\alpha$. Substituting values, these become

$$(1')\quad 600 \sin 25° - F = 18.6\bar{a}_x$$
$$(2')\quad N - 600 \cos 25° = 18.6\bar{a}_y = 0$$
$$(3')\quad F(15/12) = 14.5\alpha$$

These three equations involve four unknown quantities; hence another equation in the unknowns is needed. Since this is an example of rolling, we will use $\bar{a}_x = r\alpha = (15/12)\alpha$.

Substitute $\alpha = \frac{4}{5}\bar{a}_x$ into equation (3′) to obtain $F = 9.27\bar{a}_x$.

Substitute $F = 9.27\bar{a}_x$ into equation (1′) to obtain $\bar{a}_x = 9.09$ ft/sec². Then $F = 84.3$ lb.

2. The center of a wheel weighing 40 lb and 2 ft in diameter is moving at a certain instant with a speed of 10 ft/sec up a plane inclined 20° with the horizontal (see Fig. 17-2 below). How long will it take to reach the highest point of its travel?

Solution:

The free body diagram shows the friction F acting up the plane. Here, as in Problem 1, friction is the only force with a moment about the mass center. It therefore causes the angular acceleration which must be counterclockwise. Note that the angular velocity $\boldsymbol{\omega}$ is clockwise until the wheel stops at its highest point. Of course, on the way down the $\boldsymbol{\alpha}$ and $\boldsymbol{\omega}$ will be in the same direction — counterclockwise.

The equations of motion are

$$(1)\ \Sigma F_x = m\bar{a}_x,\quad (2)\ \Sigma F_y = m\bar{a}_y,\quad (3)\ \Sigma \bar{M} = \bar{I}\alpha$$

Substitute values to obtain the following equations. Note that the values of m and $\bar{I}$ (assume a cylinder so that $\bar{I} = \frac{1}{2}mr^2$) are not calculated separately as they were in Problem 1.

$$(1')\quad 40 \sin 20° - F = (40/32.2)\bar{a}_x$$
$$(2')\quad N - 40 \cos 20° = (40/32.2)\bar{a}_y = 0$$
$$(3')\quad F \times 1 = \tfrac{1}{2}(40/32.2)(1)^2\alpha$$

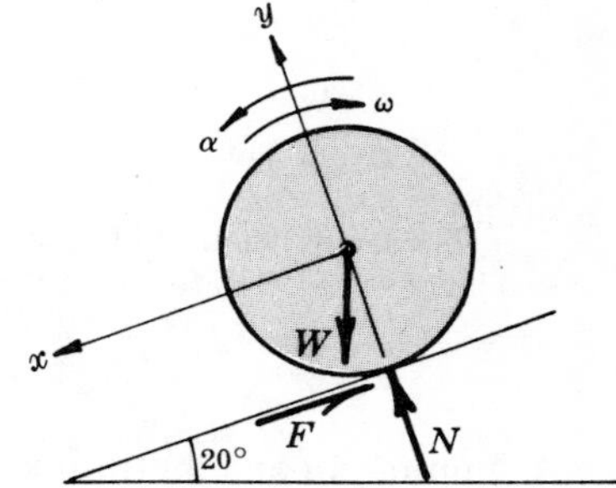

Fig. 17-2

As in Problem 1, $\bar{a}_x = r\alpha = 1\alpha = \alpha$.

From (3′), $F = 0.621\alpha = 0.621\bar{a}_x$. Substitute into (1′) to obtain $\bar{a}_x = 7.33$ ft/sec². The value of $\bar{a}_x$ is positive, i.e., down the plane.

To determine the time to come to rest, i.e., to reach its highest point after the initial speed of 10 ft/sec was observed, apply the kinematics scalar equation $v = v_0 + at$.

Keep in mind that the downward direction is positive. The final speed v is zero, the initial speed v_0 is up the plane and therefore -10 ft/sec. The acceleration $\bar{a}_x$ is down the plane and therefore $+7.33$ ft/sec².

Then $v = v_0 + at$, $0 = -10 + 7.33t$, and $t = 1.36$ sec.

3. Study the motion of a homogeneous cylinder of radius R and weight W which is acted upon by a horizontal force P applied at various positions along a vertical center line as shown in Fig. 17-3 below. Assume movement upon a horizontal plane.

Solution:

The free body diagram shows the force P applied at a distance h above the center.

Assume F acts to the left. The equations of motion are:

$$(1)\quad \Sigma F_x = P - F = (W/g)\bar{a}_x$$
$$(2)\quad \Sigma F_y = N - W = 0$$
$$(3)\quad \Sigma \bar{M} = P\times h + F\times R = \tfrac{1}{2}(W/g)R^2\alpha$$

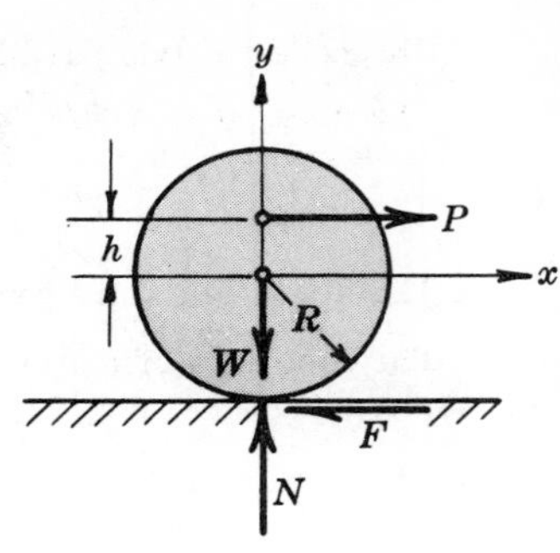

Fig. 17-3

Note that P must be greater than F for motion to ensue to the right. Is it possible for motion to occur with friction F equal to zero?

Substitute $\bar{a}_x = R\alpha$ into (3) to obtain

$$(3')\quad P\times h + F\times R = \tfrac{1}{2}(W/g)R\bar{a}_x$$

Divide equation (3′) by $\frac{1}{2}R$ to obtain

$$(3'')\quad 2Ph/R + 2F = (W/g)\bar{a}_x$$

Equating the left members of (3″) and (1), $2Ph/R + 2F = P - F$ or $3F = P(1-2h/R)$.

It is evident that F will be zero if the term $(1-2h/R)$ is zero, i.e., if $h = \frac{1}{2}R$. Thus if the force P is applied at a point one-half the radius above the center, the frictional force F is zero.

Also note that if h equals R the equation becomes $3F = P(1-2R/R) = -P$. The friction has now reversed itself and acts to the right. With $F = -P/3$, equation (3) becomes for the case where $h = R$,

$$PR - \tfrac{1}{3}PR = \tfrac{1}{2}(W/g)R^2\alpha \qquad \text{or} \qquad \tfrac{2}{3}P = \tfrac{1}{2}(W/g)R\alpha$$

This indicates that α is positive or the cylinder rolls to the right.

Next assume that h becomes smaller until finally P is applied through the mass center where h is zero. Under these conditions $3F = P[1-(2\times 0)/P] = P$.

Naturally at any time in the previous discussion if P becomes too large, F will tend to increase. As soon as the maximum possible value of the friction is exceeded, the cylinder will slip. A new assumption must then be made, i.e., friction F is now equal to the product of the coefficient of friction and the normal force N. The equations of motion are now

$$(4)\quad \Sigma F_x = P - \mu N = (W/g)\bar{a}_x$$
$$(5)\quad \Sigma F_y = N - W = 0$$
$$(6)\quad \Sigma \bar{M} = Ph + \mu NR = \tfrac{1}{2}(W/g)R^2\alpha$$

These equations indicate the existence of both sliding (linear acceleration $\bar{a}_x$) and rolling (angular acceleration α). Refer to Problems 5 and 21.

4. A homogeneous cylinder weighing 32.2 lb has a narrow slot cut in it as shown. A force of 12 lb is exerted on a string wrapped in the slot. If the cylinder rolls without slipping, determine the acceleration of its mass center and the frictional force F. Neglect the effect of the slot.

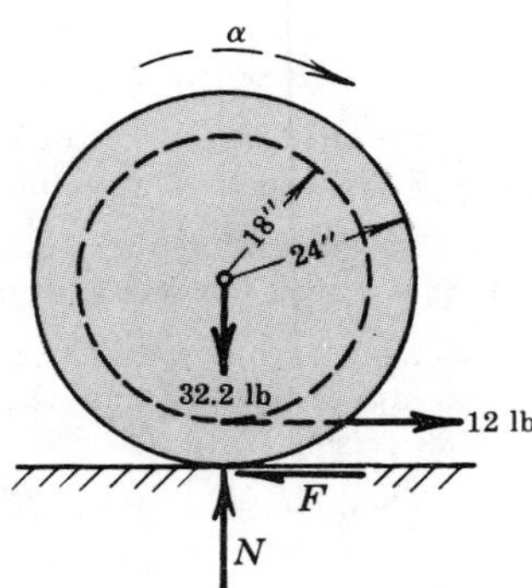

Fig. 17-4

Solution:

Assume rolling to the right. Then the angular acceleration will be clockwise and the mass center acceleration $\bar{a}$ to the right.

The equations of motion assuming the friction acts to the left are:

$$(1)\quad \Sigma F = m\bar{a} \quad \text{or} \quad 12 - F = (32.2/32.2)\bar{a}$$
$$(2)\quad \Sigma \bar{M} = \bar{I}\alpha \quad \text{or} \quad F\times 2 - 12\times 1.5 = \tfrac{1}{2}(1)(2)^2\alpha$$

Since $\bar{a} = r\alpha$, the second equation may be rewritten $2F - 18 = 2\alpha = \bar{a}$. Equate the left members of the two equations to obtain $12 - F = 2F - 18$; hence $F = 10$ lb to the left as assumed. The acceleration $\bar{a}$ is then equal to 2 ft/sec^2 to the right.

5. The coefficient of friction between the homogeneous sphere whose weight is 16.1 lb and the plane is 0.10. Determine the angular acceleration of the sphere and the linear acceleration of its mass center.

Solution:

Draw a free body diagram showing the frictional force F as unknown. See Fig. 17-5. At the beginning it is not known whether the sphere will slip or not; therefore determine F to see whether or not it is greater than μN, i.e., $0.10N$. If it is greater, this means that not enough friction is available and both rolling and sliding occur.

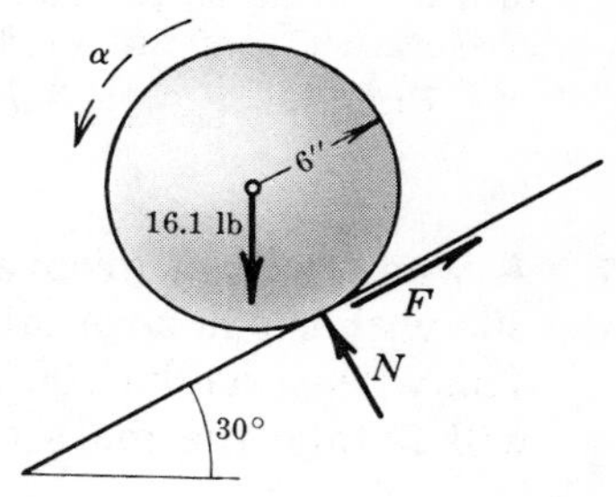

Fig. 17-5

The equations of motion are

(1) $\Sigma F = m\bar{a}$ or $16.1 \sin 30° - F = (16.1/32.2)\bar{a}$

(2) $\Sigma \bar{M} = \bar{I}\alpha$ or $F \times \frac{1}{2} = \frac{2}{5}(16.1/32.2)(\frac{1}{2})^2\alpha$

Note that α is assumed counterclockwise; hence $\bar{a}$ must be positive down the plane. Also, $\bar{a} = r\alpha = \frac{1}{2}\alpha$.

From (1), $8.05 - F = \frac{1}{2}\bar{a}$; from (2), $F = \frac{1}{5}\bar{a}$. Hence $\bar{a} = 11.5$ ft/sec² and $F = 2.30$ lb.

By inspection $N = 16.1 \cos 30° = 13.9$ lb. Maximum friction available $= \mu N = 0.10(13.9) = 1.39$ lb. But the F to prevent sliding (or cause rolling) is 2.30 lb. This means that the problem must be reworked using the maximum friction available, i.e., 1.39 lb instead of F. The relation $\bar{a} = r\alpha$ no longer will hold true. Incidentally, we are also assuming that the coefficients of static and kinetic friction are equal.

Draw a new free body diagram (see Fig. 17-6). The equations of motion are

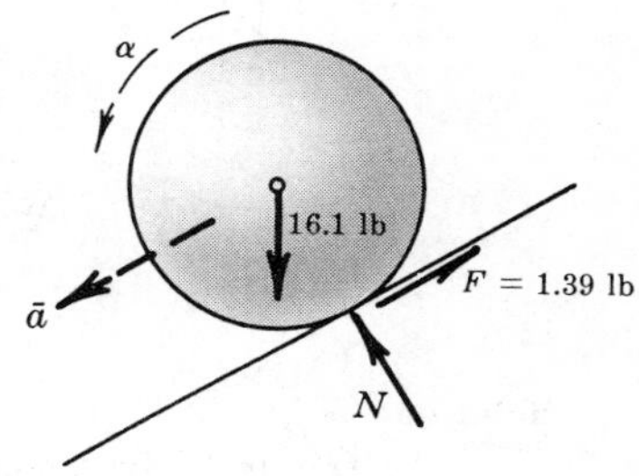

Fig. 17-6

(3) $16.1 \sin 30° - 1.39 = (16.1/32.2)\bar{a}$

(4) $1.39 \times \frac{1}{2} = \frac{2}{5}(16.1/32.2)(\frac{1}{2})^2\alpha$

Hence $\bar{a} = 13.3$ ft/sec² and $\alpha = 13.9$ rad/sec². The sphere will both roll and slip.

6. The cord passes over a frictionless pulley, as shown in Fig. 17-7(a) below, carrying a weight W_1 at one end and wrapped around a cylinder of weight W_2 which rolls on a horizontal plane. What is the acceleration of the weight W_1?

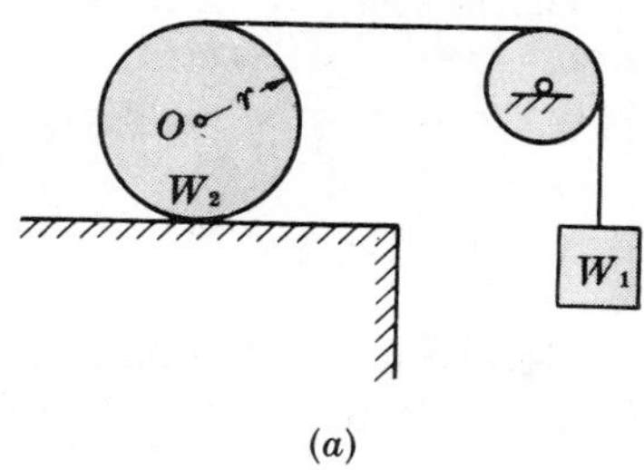

(a)

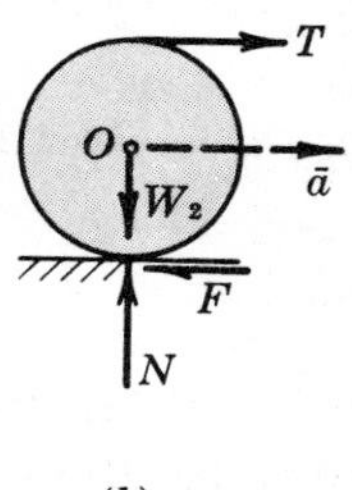

(b)

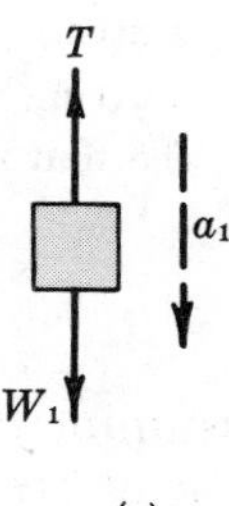

(c)

Fig. 17-7

Solution:

The free body diagrams are shown in Fig. 17-7(b) and (c) above. Note that the magnitude of the acceleration a_1 of W_1 does not equal the magnitude of the acceleration $\bar{a}$ of the center of gravity of the cylinder.

Equations (1) and (2) below apply to the cylinder, and equation (3) to the weight W_1.

(1) $\Sigma F_h = m\bar{a}$ or $T - F = (W_2/g)\bar{a}$

(2) $\Sigma M_O = \bar{I}\alpha$ or $(F + T)r = \frac{1}{2}(W_2/g)r^2\alpha$

(3) $\Sigma F_v = ma_1$ or $W_1 - T = (W_1/g)a_1$

Substitute $\bar{a}/r$ for α in (*2*) and divide through by r to obtain: (*4*) $F + T = \frac{1}{2}(W_2/g)\bar{a}$.

Add equation (*4*) to equation (*1*) to obtain: (*5*) $T = \frac{3}{4}(W_2/g)\bar{a}$.

This may be solved simultaneously with equation (*3*) if the relation between $\bar{a}$ and a_1 is obtained. The horizontal component of the acceleration of the top point of the cylinder is equal to the sum of the acceleration $\bar{a}$ of the center and the product $r\alpha$. In this case, assuming pure rolling, $r\alpha$ is equal to $\bar{a}$ by kinematic considerations. Hence the acceleration of the top point, which is the same as the acceleration a_1 of the weight W_1, is $\bar{a} + r\alpha = 2\bar{a}$.

From (*5*), $T = \frac{3}{4}(W_2/g)(\frac{1}{2}a_1)$. Substituting into (*3*) and solving, $a_1 = W_1 g/(W_1 + \frac{3}{8}W_2)$.

7. A wheel with a groove cut in it as shown in Fig. 17-8(*a*) is pulled up a rail inclined 30° with the horizontal by a rope passing over a pulley and supporting an 80 lb weight. The wheel weighs 100 lb and has a moment of inertia $\bar{I}$ equal to 4 slug-ft^2. How long will it take the mass center to attain a speed of 20 ft/sec starting from rest?

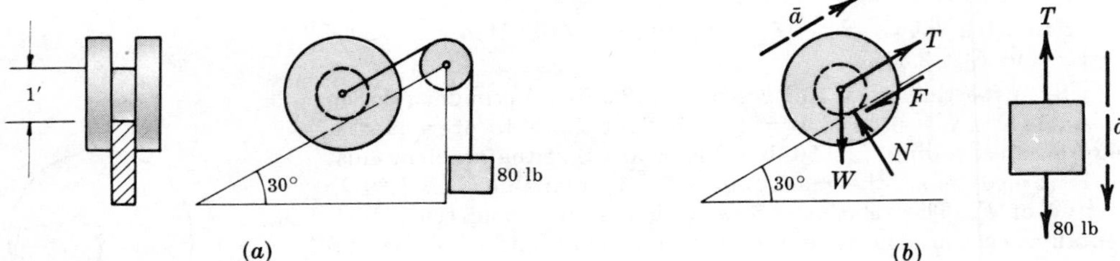

Fig. 17-8

Solution:

The free body diagrams for the wheel and the weight are shown in Fig. 17-8(*b*) above. Note that tension T in the rope is not 80 lb because the weight is accelerating.

The necessary equations for the weight and the wheel are

$$(1)\quad \Sigma F_{\text{weight}} = 80 - T = (80/32.2)\bar{a}$$

$$(2)\quad \Sigma F_{\text{wheel}} = T - F - 100 \sin 30° = (100/32.2)\bar{a}$$

$$(3)\quad \Sigma \bar{M} = F \times \tfrac{1}{2}'' = 4\alpha$$

Since $\bar{a} = r\alpha = \frac{1}{2}\alpha$, equation (*3*) may be written $F \times \frac{1}{2} = 4 \times 2\bar{a} = 8\bar{a}$ or $F = 16\bar{a}$.

Putting $F = 16\bar{a}$ in (*2*) gives $T - 16\bar{a} - 100 \times 0.500 = (100/32.2)\bar{a}$ or $T = 19.1\bar{a} + 50$.

Putting $T = 19.1\bar{a} + 50$ in (*1*) gives $80 - 19.1\bar{a} - 50 = 2.48\bar{a}$ or $\bar{a} = 1.39$ ft/sec^2.

To find the time to attain a speed of 20 ft/sec from rest, apply the kinematics equation

$$v = v_0 + \bar{a}t, \qquad 20 = 0 + 1.39t, \qquad t = 14.4 \text{ sec}$$

8. Assuming the pulley to be weightless and frictionless, determine the smallest coefficient of static friction to cause the cylinder to roll. For the cylinder, $W = 150$ lb and $k_O = 1.25$ ft. Refer to Fig. 17-9(*a*) below.

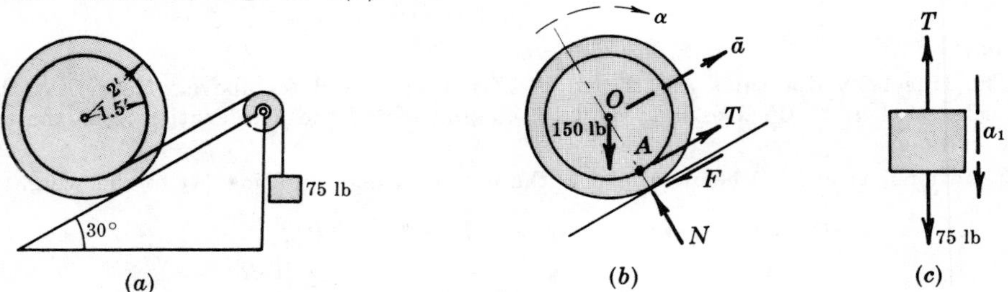

Fig. 17-9

Solution:

Draw free body diagrams of the cylinder and weight as shown in Fig. 17-9(*b*) and (*c*) above. Assume the cylinder is rolling up the plane. The relation $\bar{a} = r\alpha$ holds true since pure rolling is called for. (Acceleration $\mathbf{a}_1$ has the same magnitude as the component of the absolute acceleration of point A parallel to the plane.) By kinematic considerations, $\mathbf{a}_A = \mathbf{a}_{A/O} + \bar{\mathbf{a}}$. Dealing only with components parallel to the plane, the component of $a_{A/O} = OA \times \alpha = 1\frac{1}{2}\alpha$ down the plane since α is clockwise. The value of $\bar{a}$ is 2α up the plane. Hence $a_1 = \frac{1}{2}\alpha$ up the plane.

The equations of motion may now be written. The moment of inertia for the cylinder is $\bar{I} = mk_0^2 = 7.28$ lb-sec²-ft.

(*1*) $\Sigma F = ma_1$ or $75 - T = (75/32.2)a_1$

(*2*) $\Sigma F = m\bar{a}$ or $T - F - 150 \sin 30° = (150/32.2)\bar{a}$

(*3*) $\Sigma \bar{M} = \bar{I}\alpha$ or $F \times 2 - T \times 1.5 = 7.28\alpha$

But $\bar{a} = 2\alpha$ and $a_1 = \frac{1}{2}\alpha$, as has been shown. Eliminate F between (*2*) and (*3*) to obtain $0.5T - 150 = 25.9\alpha$. Solve this simultaneously with (*1*) to find $\alpha = -4.25$ rad/sec² and $T = 80.0$ lb.

Hence $F = 44.6$ lb. By inspection $N = 150 \cos 30° = 130$ lb. Thus the required coefficient of friction $\mu = 44.6/130 = 0.34$.

Note that in this problem the cylinder will roll down the plane and the 75 lb weight will ascend.

9. In Fig. 17-10, the homogeneous cylinder rolls without slipping on the horizontal surface. Determine the acceleration of the mass center and the plane reaction on the cylinder.

Solution:

In the free body diagram assume that the cylinder rolls to the left and that the frictional force F is to the left. By inspection $N = 483$ lb.

The equations of motion are

(*1*) $\Sigma F = m\bar{a}$ or $F + 250 - 200 = (483/32.2)\bar{a}$

(*2*) $\Sigma \bar{M} = \bar{I}\alpha$ or $-F \times 3 - 200 \times 3 = \frac{1}{2}(483/32.2)(3)^2\alpha$

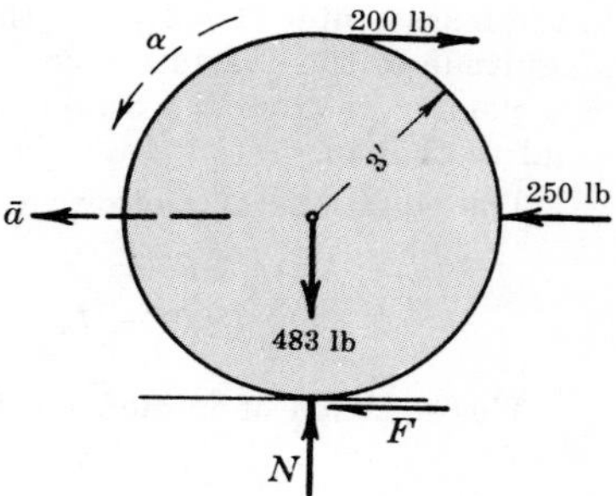

Fig. 17-10

Noting that $\alpha = \bar{a}/3$, solve simultaneously to obtain $\bar{a} = -6.67$ ft/sec² and $F = -150$ lb. The negative signs indicate that the friction is directed to the right instead of the left as assumed and that the wheel rolls to the right instead of to the left as assumed.

10. Assume that the disk A in Fig. 17-11 below rolls without slipping. Determine the tensions in the ropes and the acceleration of the mass center of disk A.

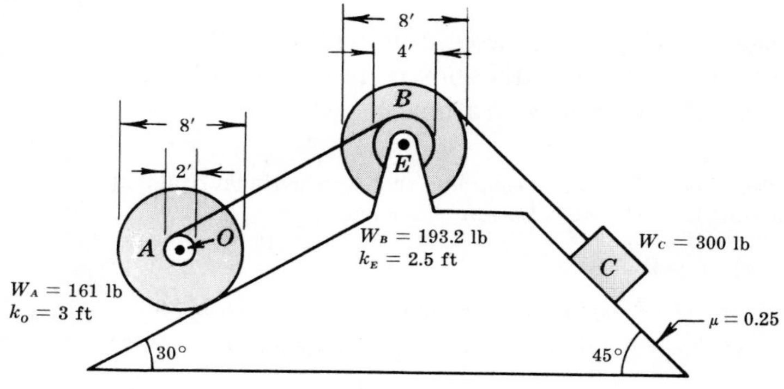

Fig. 17-11

Solution:

$I_A = mk_0^2 = (161/32.2)(3)^2 = 45$ lb-sec²-ft, $I_B = (193.2/32.2)(2.5)^2 = 37.5$ lb-sec²-ft.

Draw free body diagrams of A, B, and C. See Fig. 17-12 below.

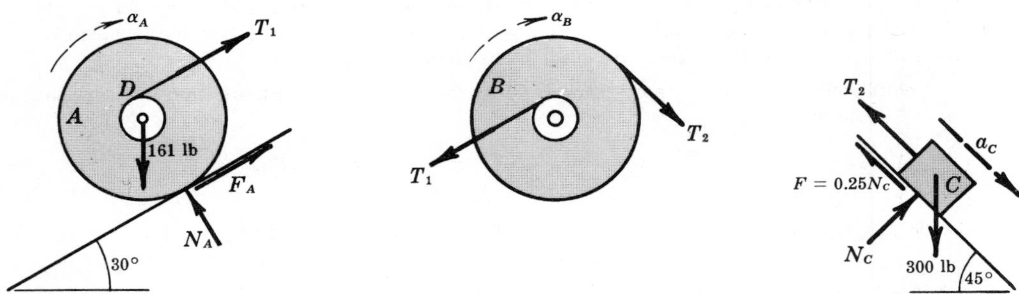

Fig. 17-12

Assume in the diagrams that C moves down the plane and that A rolls up the plane. The following kinematic ideas are needed to solve the problem. First, $\alpha_A = \bar{a}/4$. Second, the acceleration of rope T_1 equals that component of the absolute acceleration of point D (on disk A) which is parallel to the plane. This component has a magnitude which is the sum of the magnitude of the acceleration relative to the mass center $(1 \times \alpha_A)$ and the magnitude of the acceleration of the mass center $(\bar{a})$. Therefore the magnitude of the absolute acceleration of T_1 is $\frac{5}{4}\bar{a}$. Since this is also the magnitude of the acceleration of a point 2 ft from the center of B, $\alpha_B = \frac{5}{4}\bar{a} \div 2 = \frac{5}{8}\bar{a}$. The magnitude of the acceleration of T_2 is equal to that of a_C $(\frac{5}{8}\bar{a} \times 4 = \frac{5}{2}\bar{a})$.

The equations of motion for A are

$$(1) \quad \Sigma F = m\bar{a} \quad \text{or} \quad T_1 + F_A - 161 \sin 30° = (161/32.2)\bar{a}$$

$$(2) \quad \Sigma \bar{M} = \bar{I}\alpha_A \quad \text{or} \quad T_1 \times 1 - F_A \times 4 = 45 \times \tfrac{1}{4}\bar{a}$$

The equation of motion for B is

$$(3) \quad \Sigma \bar{M} = \bar{I}\alpha_B \quad \text{or} \quad T_2 \times 4 - T_1 \times 2 = 37.5 \times \tfrac{5}{8}\bar{a}$$

The equations of motion for C parallel and perpendicular to the plane are

$$(4) \quad \Sigma F_{\parallel} = ma_C \quad \text{or} \quad 300 \times 0.707 - 0.25N - T_2 = (300/32.2)(\tfrac{5}{2}\bar{a})$$

$$(5) \quad \Sigma F_{\perp} = 0 \quad \text{or} \quad N - 300 \times 0.707 = 0$$

Solve equation (5) for N; substitute this into equation (4). Solve equation (4) for T_2 in terms of $\bar{a}$. Eliminate F_A between equations (1) and (2) to find T_1 in terms of $\bar{a}$. Substitute these values of T_1 and T_2 into equation (3) to find $\bar{a}$.

These values are $N = 212$ lb, $T_2 = 159 - 23.3\bar{a}$, $T_1 = 64.4 + 6.25\bar{a}$.

Substitute into equation (3) to obtain: $4(159 - 23.3\bar{a}) - 2(64.4 + 6.25\bar{a}) = 23.4\bar{a}$.

Hence $\bar{a} = 3.93$ ft/sec², $T_1 = 89.0$ lb, $T_2 = 67.4$ lb.

11. A solid homogeneous cylinder weighing 644 lb rolls without slipping on the inclined rails shown in Fig. 17-13. Determine the mass center acceleration.

Solution:

Draw the free body diagram assuming that the cylinder rolls up the plane. Note that here $\bar{a} = r\alpha = 1.5\alpha$ in this case.

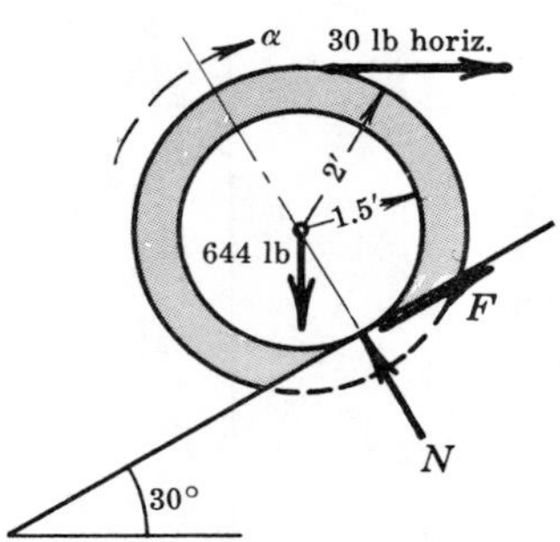

Fig. 17-13

The equations of motion are

$$(1) \quad \Sigma F = m\bar{a} \quad \text{or} \quad F - 644 \sin 30° + 30 \cos 30° = (644/32.2)\bar{a}$$

$$(2) \quad \Sigma \bar{M} = \bar{I}\alpha \quad \text{or} \quad -F \times 1.5 + 30 \times 2 = \tfrac{1}{2}(644/32.2)(2)^2(\bar{a}/1.5)$$

Solve simultaneously to obtain $\bar{a} = -6.78$ ft/sec². This indicates that the cylinder will roll down the plane.

12. A uniform bar of length l and weight W rests on smooth surfaces as shown in Fig. 17-14. Describe its motion. Choose the x-axis along the horizontal surface and the y-axis along the vertical surface.

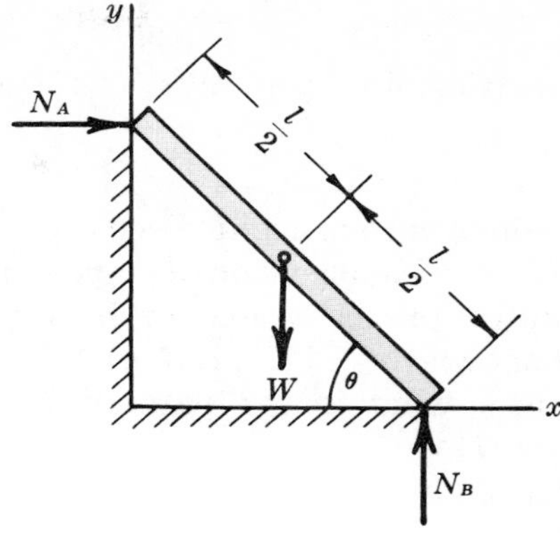

Fig. 17-14

Solution:

The free body diagram shows only normal reactions of the surfaces on the bar because friction is negligible.

Since there is no fixed axis of rotation, we consider plane motion with respect to the center of gravity.

The equations of motion are

(*1*) $\Sigma F_x = N_A = (W/g)\bar{a}_x$

(*2*) $\Sigma F_y = N_B - W = (W/g)\bar{a}_y$

(*3*) $\Sigma \bar{M} = N_A(\tfrac{1}{2}l \sin\theta) - N_B(\tfrac{1}{2}l\cos\theta) = \bar{I}\alpha$

Note that $\bar{I}$ for the rod about its mass center is $\frac{1}{12}(W/g)l^2$, and that α is the second derivative of θ with respect to time.

Since the three equations contain five unknowns, two more equations in the unknowns are required to obtain a solution. The geometry of the figure indicates that

$$\bar{x} = \tfrac{1}{2}l\cos\theta \quad \text{and} \quad \bar{y} = \tfrac{1}{2}l\sin\theta$$

Then $\dfrac{d\bar{x}}{dt} = \dfrac{d\bar{x}}{d\theta}\dfrac{d\theta}{dt} = -\tfrac{1}{2}l\sin\theta\dfrac{d\theta}{dt}$ and $\dfrac{d\bar{y}}{dt} = \dfrac{d\bar{y}}{d\theta}\dfrac{d\theta}{dt} = \tfrac{1}{2}l\cos\theta\dfrac{d\theta}{dt}$.

Differentiate again to obtain: $\bar{a}_x = \dfrac{d^2\bar{x}}{dt^2} = -\tfrac{1}{2}l\left(\cos\theta\dfrac{d\theta}{dt}\right)\dfrac{d\theta}{dt} - \tfrac{1}{2}l\sin\theta\dfrac{d^2\theta}{dt^2}$

and $\bar{a}_y = \dfrac{d^2\bar{y}}{dt^2} = \tfrac{1}{2}l\left(-\sin\theta\dfrac{d\theta}{dt}\right)\dfrac{d\theta}{dt} + \tfrac{1}{2}l\cos\theta\dfrac{d^2\theta}{dt^2}$.

These values will be substituted for $\bar{a}_x$ and $\bar{a}_y$ in the original equations. It is well to substitute first, however, the values $N_A = (W/g)\bar{a}_x$ and $N_B = W + (W/g)\bar{a}_y$ from equations (*1*) and (*2*) respectively into equation (*3*). Thus equation (*3*) becomes, after dividing all terms by $\tfrac{1}{2}Wl$,

$$\frac{\sin\theta}{g}\bar{a}_x - \frac{\cos\theta}{g}\bar{a}_y = \cos\theta + \frac{l}{6g}\frac{d^2\theta}{dt^2}$$

Next substitute the values of $\bar{a}_x$ and $\bar{a}_y$ just determined and simplify to obtain

$$\frac{d^2\theta}{dt^2} = -\frac{3g}{2l}\cos\theta$$

Using the method developed in Problem 1 of Chapter 16, the angular speed ω is found to be

$$\omega = \sqrt{(3g/l)(1-\sin\theta)}$$

If the initial angular speed is zero at $\theta = \theta_0$ instead of $\theta = 90°$, then the constant of integration C becomes $(3g/2l)\sin\theta_0$ and

$$\omega = \sqrt{(3g/l)(\sin\theta_0 - \sin\theta)}$$

13. In the preceding problem determine the value of the angle θ at which the bar no longer touches the vertical wall.

Solution:

This condition means N_A is equal to zero. From equation (*1*), $N_A = 0$ when $\bar{a}_x = 0$; hence set the expression for $\bar{a}_x = 0$ to obtain

$$\bar{a}_x = 0 = -\tfrac{1}{2}l\cos\theta\left(\frac{d\theta}{dt}\right)^2 - \tfrac{1}{2}l\sin\theta\frac{d^2\theta}{dt^2}$$

Substituting $\dfrac{d^2\theta}{dt^2} = -\dfrac{3g}{2l}\cos\theta$ and $\left(\dfrac{d\theta}{dt}\right)^2 = \omega^2 = \dfrac{3g}{l}(1 - \sin\theta)$ into the above equation and simplifying, we obtain $\sin\theta = 2/3$ or $\theta = 41.8°$.

14. A homogeneous right circular cone is precariously balanced on its apex in unstable equilibrium on a smooth horizontal plane as shown in Fig. 17-15(a). If it is disturbed, what is the path of its center of mass G?

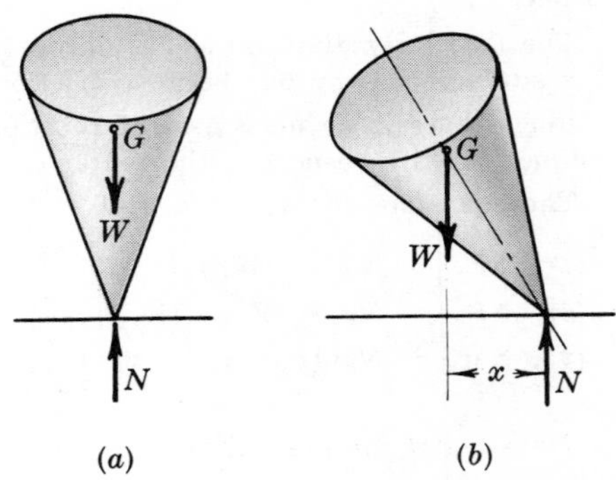

Fig. 17-15

Solution:

Draw a free body diagram to indicate the cone in any position while falling [see Fig. 17-15(b)]. Note that there is no horizontal force present (no friction assumed). Then the sum of the horizontal forces is zero. But the sum of the horizontal forces equals the product of the cone's mass m and the horizontal acceleration $\bar{a}_x$ of the mass center. Since there is a mass, the product $m\bar{a}_x$ can be zero only if $\bar{a}_x$ is zero.

If the initial horizontal speed is zero and the value of $\bar{a}_x$ is zero, the center of mass G can only have motion in a vertical line passing through the apex of the cone in the original undisturbed position.

15. Two homogeneous cylindrical disks are rigidly connected to an axle as shown in Fig. 17-16. Each disk weighs 16 lb and is 3 ft in diameter. The axle is 8 in. in diameter and weighs 20 lb. A string wrapped as shown exerts a force of 10 lb at the middle of the axle. Analyze the motion.

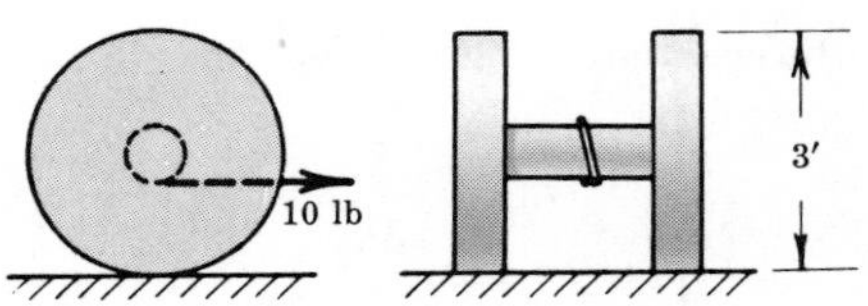

Fig. 17-16

Solution:

A free body diagram is shown indicating only a side view (see Fig. 17-17). Friction F is assumed acting to the left. The equations of motion are

$$(1)\quad \Sigma F_x = 10 - F = m\bar{a}_x$$
$$(2)\quad \Sigma F_y = N - 52 = m\bar{a}_y = 0$$
$$(3)\quad \Sigma \bar{M} = F(3/2) - 10(1/3) = \bar{I}\alpha$$

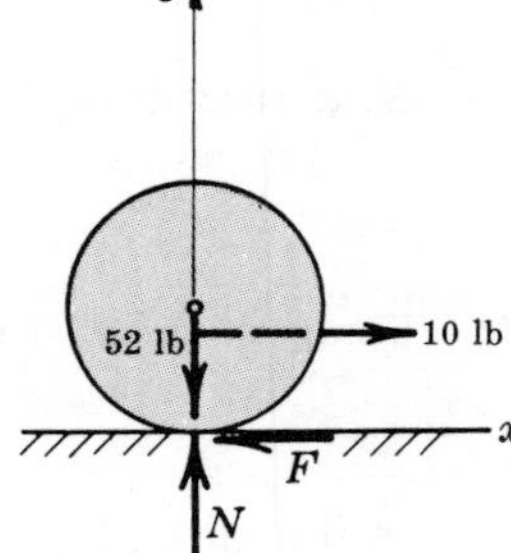

Fig. 17-17

The mass $m = 52/32.2$ slugs. The moment of inertia $\bar{I}$ is the sum of the moments of inertia of the two disks and the axle.

$$\begin{aligned}\bar{I} &= 2(\tfrac{1}{2}\, m_{\text{disk}}\, r^2_{\text{disk}}) + \tfrac{1}{2}\, m_{\text{axle}}\, r^2_{\text{axle}} \\ &= (16/32.2)(3/2)^2 + \tfrac{1}{2}(20/32.2)(4/12)^2 = 1.15 \text{ slug-ft}^2\end{aligned}$$

Substituting these values, the equations (1), (2), (3) become

$$(1')\ 10 - F = 1.62\bar{a}_x \qquad (2')\ N - 52 = 0 \qquad (3')\ \tfrac{3}{2}F - 3.33 = 1.15\alpha$$

Since $\bar{a}_x = 3\alpha/2$ where the distance from the center of rolling to the surface on which the rolling occurs is 3/2 ft, equation (3′) may be written

$$\tfrac{3}{2}F - 3.33 = 1.15(\tfrac{2}{3}\bar{a}_x) \qquad \text{or} \qquad F = 2.22 + 0.512\bar{a}_x$$

Put this value of F into (1′) and obtain $\bar{a}_x = 3.65$ ft/sec². The wheels roll to the right.

16. A homogeneous sphere and a homogeneous cylinder roll, without slipping, from rest at the top of an inclined plane to the bottom. Which reaches the bottom first?

Solution:

From the wording of the problem it is likely that the radii of the solids have no influence. Let the subscripts s and c refer to the sphere and cylinder respectively.

The free body diagram, shown in the adjacent Fig. 17-18, refers to either solid.

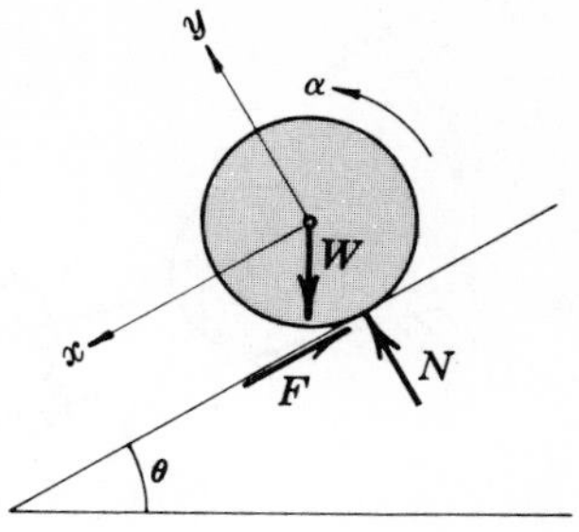

Fig. 17-18

The equations of motion are listed below. The moments of inertia for the sphere and cylinder are respectively $\frac{2}{5}(W_s/g)r_s^2$ and $\frac{1}{2}(W_c/g)r_c^2$

Sphere	Cylinder
(1) $\Sigma F_x = W_s \sin\theta - F_s = (W_s/g)(\bar{a}_x)_s$	(4) $\Sigma F_x = W_c \sin\theta - F_c = (W_c/g)(\bar{a}_x)_c$
(2) $\Sigma F_y = N_s - W_s \cos\theta = 0$	(5) $\Sigma F_y = N_c - W_c \cos\theta = 0$
(3) $\Sigma \bar{M} = F_s r_s = \frac{2}{5}(W_s/g)r_s^2\alpha_s$	(6) $\Sigma \bar{M} = F_c r_c = \frac{1}{2}(W_c/g)r_c^2\alpha_c$
Substitute $r_s\alpha_s = (\bar{a}_x)_s$ into equation (3) and obtain:	Substitute $r_c\alpha_c = (\bar{a}_x)_c$ into (6) and obtain:
(3′) $F_s = \frac{2}{5}(W_s/g)(\bar{a}_x)_s$	(6′) $F_c = \frac{1}{2}(W_c/g)(\bar{a}_x)_c$
Substitute this value into (1) to obtain:	Substitute this value into (4) to obtain:
(1′) $(\bar{a}_x)_s = \frac{5}{7}g \sin\theta$	(4′) $(\bar{a}_x)_c = \frac{2}{3}g \sin\theta$

Thus the sphere, having the larger acceleration, will reach the bottom first.

17. A cylinder 4 inches in diameter has a cord wrapped around it at the midsection. Attach the free end of the cord to a fixed support and allow the cylinder to fall. Analyze the motion.

Solution:

Two equations are useful here: one summing the forces in the vertical direction assuming downward as positive, and the other summing moments about the mass center assuming clockwise moments as positive.

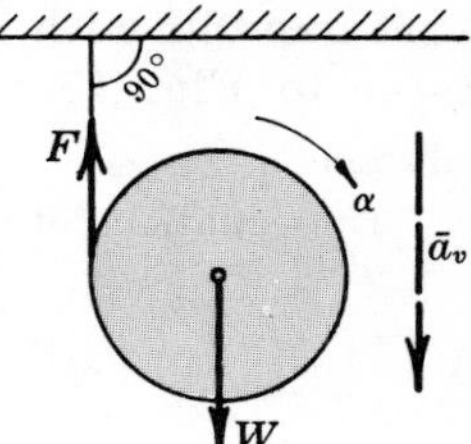

Fig. 17-19

$$(1) \quad \Sigma F_v = W - F = (W/g)\bar{a}_v$$
$$(2) \quad \Sigma \bar{M} = Fr = \bar{I}\alpha$$

F is the tension in the cord. The moment of inertia $\bar{I} = \frac{1}{2}(W/g)r^2$, where $r = 2$ in. $= \frac{1}{6}$ ft. Also, $\alpha = \bar{a}_v/r = 6\bar{a}_v$ since the motion is equivalent to rolling the cylinder on the cord. Equation (2) becomes

$$\tfrac{1}{6}F = \tfrac{1}{2}(W/g)(\tfrac{1}{6})^2(6\bar{a}_v) \qquad \text{or} \qquad F = \tfrac{1}{2}(W/g)\bar{a}_v$$

Substitute this value of F into equation (1) to obtain $\bar{a}_v = 2g/3 = 21.4$ ft/sec².

The yo-yo works on this principle. It is designed so that $\bar{a}_v$ is much smaller than g.

18. A solid sphere and a thin hoop of equal weights W and radii R are harnessed together by a rigging and are free to roll without slipping down the inclined plane shown in Fig. 17-20(*a*) below. Neglecting the mass of the rigging, determine the force in it. Assume frictionless bearings.

Solution:

Draw free body diagrams assuming that C is the compression in the rigging (see Fig. 17-20(*b*) below). If the sign should be negative this merely indicates that C is tension.

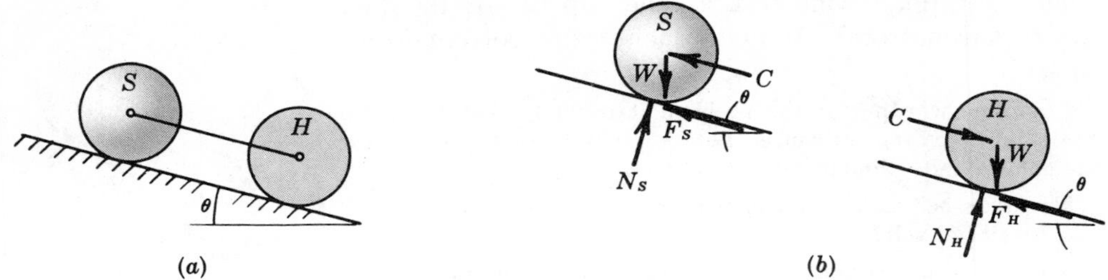

Fig. 17-20

Let the subscripts S and H indicate sphere and hoop respectively. The acceleration $\bar{a}$ is the same for either mass center; and since the radii are equal, the angular acceleration α for each is the same. Sum forces parallel to the plane to obtain the following two equations:

$$(1) \quad W \sin\theta - F_S - C = (W/g)\bar{a}$$
$$(2) \quad W \sin\theta - F_H + C = (W/g)\bar{a}$$

Since there are four unknowns $(F_S, F_H, C, \bar{a})$, two more equations are necessary. These are obtained by taking moments about the mass centers.

$$(3) \quad F_S \times r = I_S\alpha = \tfrac{2}{5}(W/g)r^2\alpha$$
$$(4) \quad F_H \times r = I_H\alpha = (W/g)r^2\alpha$$

The relation $\bar{a} = r\alpha$ holds for either equation. From (3) and (4), $F_S = \frac{2}{5}(W/g)\bar{a}$ and $F_H = (W/g)\bar{a}$. Substitute these values into equations (1) and (2); then add the resulting equations to eliminate C and obtain $\bar{a} = \frac{10}{17}g \sin\theta$. Next solve for $C = \frac{3}{17}W \sin\theta$ (compression).

19. At what height above the table should a billiard ball be hit with a cue held horizontally so that the ball will start moving with no friction between it and the table? Refer to Fig. 17-21.

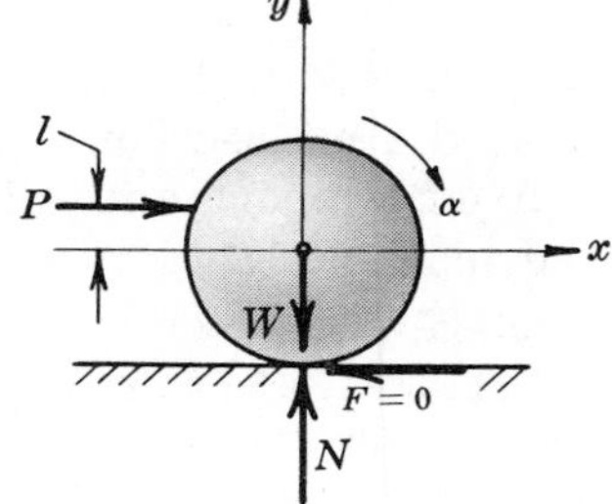

Fig. 17-21

Solution:

The equations of motion are as follows:

$$(1) \quad \Sigma F_x = P = (W/g)\bar{a}_x$$
$$(2) \quad \Sigma F_y = N - W = 0$$
$$(3) \quad \Sigma \bar{M} = Pl = \tfrac{2}{5}(W/g)r^2\alpha$$

Since $\alpha = \bar{a}_x/r$, equation (3) reduces to $P = \frac{2}{5}(Wr/gl)\bar{a}_x$.

Substitute this value of P into equation (1) to obtain $l = \frac{2}{5}r$.

In other words, the proper distance above the table is $r + \frac{2}{5}r = \frac{7}{5}r$.

This same procedure is used to find the proper height of the cushion on a billiard table so that the ball rebounds without causing any friction on the table. In this case the force P of the cushion takes the place of the force of the cue. Of course the answer is the same.

20. The solid homogeneous cylinder weighing 16.1 lb rolls without slipping on the inside of the curved fixed surface. In the phase shown the speed of the mass center of the cylinder is 9 ft/sec down to the right. What is the reaction of the surface on the cylinder?

Solution:

Draw a free body diagram showing the reaction components F and N. See Fig. 17-22(a) below. Fig. 17-22(b) below illustrates any position of the cylinder to study the kinematic relationship necessary to solve the problem. Note that as the cylinder rolls up to the left ϕ and θ increase as indicated. Hence assume the angular acceleration of the cylinder is positive in the counterclockwise direction and that $\bar{a}_t$ is thus positive to the upper left.

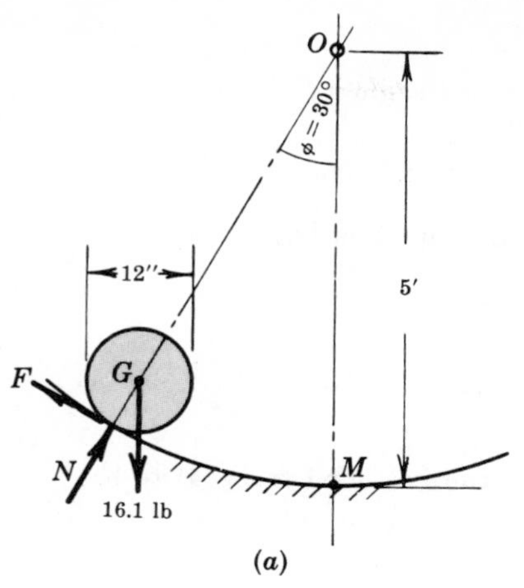

Fig. 17-22

The equations of motion are:

$$(1)\quad \Sigma F_n = m\bar{a}_n \quad \text{or} \quad N - 16.1\cos 30° = (16.1/32.2)\bar{a}_n$$

$$(2)\quad \Sigma F_t = m\bar{a}_t \quad \text{or} \quad -16.1\sin 30° + F = (16.1/32.2)\bar{a}_t$$

$$(3)\quad \Sigma\bar{M} = \bar{I}\alpha \quad \text{or} \quad -F\times\tfrac{1}{2} = \tfrac{1}{2}(16.1/32.2)(\tfrac{1}{2})^2\alpha$$

where $\bar{a}_n$, $\bar{a}_t$ are the magnitudes of the components of the acceleration of the mass center G and α is the magnitude of the angular acceleration of the cylinder about G. The mass center G can be considered as rotating about the center of curvature O. Hence $\bar{a}_n = v^2/OG = 81/4.5 = 18$ ft/sec². The tangential magnitude $\bar{a}_t$ is as yet unknown.

To express α in terms of $\bar{a}_t$ refer to Fig. 17-22(*b*) where D is the point on the cylinder originally in contact with M. Since pure rolling is assumed, the arc BM on the curved surface must equal the arc BD on the cylinder (otherwise slip occurs). Let r = radius of cylinder and R = radius of surface. Then $r(\phi+\theta) = R\phi$ or $\theta = (R/r - 1)\phi$. This relationship holds also for the time derivatives of θ and ϕ. The second derivative of θ with respect to time is the magnitude of the angular acceleration α. The second derivative of ϕ with respect to time may be found in terms of $\bar{a}_t$ since center G is considered to rotate about O, i.e., $\bar{a}_t = (R-r)(d^2\phi/dt^2)$. Hence

$$\alpha = \frac{d^2\theta}{dt^2} = \left(\frac{R-r}{r}\right)\frac{d^2\phi}{dt^2} = \left(\frac{R-r}{r}\right)\left(\frac{\bar{a}_t}{R-r}\right) = \frac{\bar{a}_t}{r}$$

In this case $R = 5$ ft and $r = \frac{1}{2}$ ft; therefore $\alpha = 2\bar{a}_t$ and the equations of motion simplify to

$$(1')\quad N - 13.9 = 9 \qquad (2')\quad -8.05 + F = \tfrac{1}{2}\bar{a}_t \qquad (3')\quad -F = \tfrac{1}{4}\bar{a}_t$$

from which $F = 2.68$ lb, $N = 22.9$ lb.

21. A disk of weight W and radius of gyration k has an angular speed ω_0 clockwise when set on a horizontal floor. See Fig. 17-23(*a*). If the coefficient of friction between the disk and the floor is μ, derive an expression for the time at which skidding stops and rolling occurs.

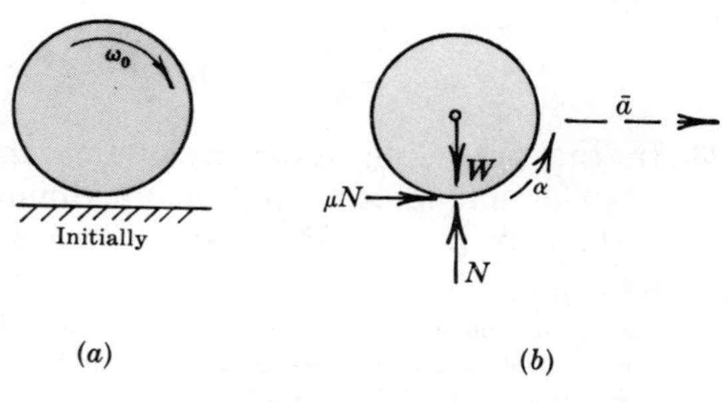

Fig. 17-23

Solution:

A free body diagram shows the friction force μN, the normal force N, and the weight W acting on the disk. These will cause a mass center acceleration $\bar{a}$ to the right and an angular acceleration counterclockwise as shown in Fig. 17-23(*b*).

The equations of motion are

$$(1)\quad \Sigma F_h = m\bar{a} \qquad \text{or} \qquad \mu N = (W/g)\bar{a}$$
$$(2)\quad \Sigma F_v = 0 \qquad \text{or} \qquad N = W$$
$$(3)\quad \Sigma \bar{M} = \bar{I}\alpha \qquad \text{or} \qquad \mu N r = (W/g)k^2\alpha$$

Substituting $N = W$ into equation (1) and then into equation (3), we obtain

$$(4)\quad \bar{a} = \mu g$$
$$(5)\quad \alpha = \mu g r/k^2$$

The speed $\bar{v}$ at any time t can be found from equation (4) to be

$$(6)\quad \bar{v} = \bar{v}_0 + \mu g t = \mu g t$$

since $\bar{v}_0 = 0$. Also, the angular speed ω at any time t can be found from (5) to be

$$(7)\quad \omega = \omega_0 - \mu g r t/k^2$$

where it is assumed that clockwise is positive.

When skidding stops and rolling occurs, $\bar{v} = r\omega$. Hence we multiply (7) by r to obtain $\bar{v}$ which is also given by (6). This yields the required time t':

$$(8)\quad \mu g t' = r\omega_0 - \mu g r^2 t'/k^2 \qquad \text{or} \qquad t' = \frac{r\omega_0}{\mu g(1 + r^2/k^2)}$$

22. A uniform cylinder of radius R is on a platform which is subjected to a constant horizontal acceleration of magnitude a. Assuming no slip determine a_O, the magnitude of the acceleration of the mass center of the cylinder.

Solution:

The free body diagram is shown in Fig. 17-24 with all forces acting on the cylinder.

The angular acceleration of the cylinder is assumed clockwise. Point I is the instant center between the cylinder and the platform and therefore has an acceleration to the right equal to that of the platform.

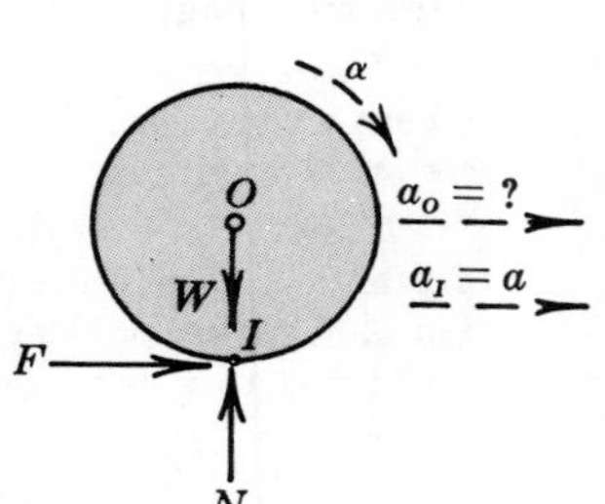

Fig. 17-24

From kinematic concepts we can write

$$\mathbf{a}_O = \mathbf{a}_{O/I} + \mathbf{a}_I$$

Summing in the horizontal direction, this equation yields (to the right is positive)

$$a_O = R\alpha + a \qquad (1)$$

The equations of motion are next applied to the cylinder:

$$\Sigma F_h = m a_O \qquad \text{or} \qquad F = (W/g)a_O \qquad (2)$$
$$\Sigma M_O = I_O\alpha \qquad \text{or} \qquad -FR = \tfrac{1}{2}(W/g)R^2\alpha \qquad (3)$$

The sign on the moment of the frictional force F is negative because the moment is opposite to the assumed α.

Substituting from equation (1) into (2), we find $F = (W/g)(R\alpha + a)$. Now put this value of F into equation (3) to obtain $R\alpha = -\frac{2}{3}a$. Thus by equation (1) the acceleration of the mass center is $a_O = -\frac{2}{3}a + a = \frac{1}{3}a$.

23. In the preceding problem assume the coefficient of friction $\mu = 0.35$ and determine the maximum acceleration the platform may have without slip between cylinder and platform.

Solution:

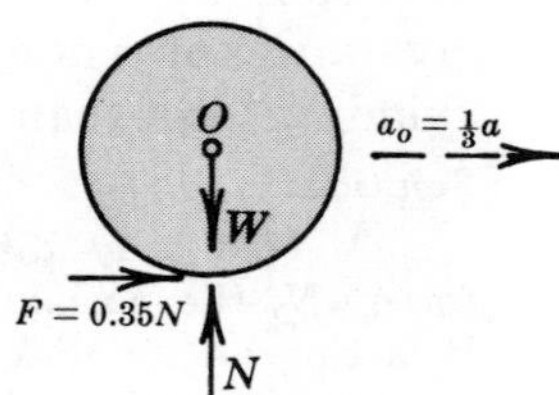

Fig. 17-25

The free body diagram of the cylinder is shown in Fig. 17-25, using the frictional force F as $0.35N$. Summing forces horizontally and then vertically,

$$0.35N = (W/g)(\tfrac{1}{3}a) \qquad \text{and} \qquad N - W = 0$$

from which $a = 0.35(3g) = 33.8 \text{ ft/sec}^2$.

Supplementary Problems

24. A wheel weighing 150 lb is 10 ft in diameter and rolls down a 45 degree plane. If there is no slippage, determine the angular acceleration of the wheel. *Ans.* 3.04 rad/sec^2

25. A cylinder weighing 400 lb and 4 inches in diameter rests on horizontal rails perpendicular to its geometric axis. A force P of 120 lb is applied tangentially to the right at the bottom of the cylinder. Assuming the coefficients of static and kinetic friction are respectively 0.29 and 0.25, what is the motion? Refer to Problem 3.
Ans. cylinder slides to right and rotates counterclockwise; $\bar{a}$ = 1.61 ft/sec^2; α = 19.3 rad/sec^2

26. In Fig. 17-26, a cylinder of weight W and radius of gyration k has a rope wrapped around a groove of radius r. Determine the acceleration of the mass center if pure rolling is assumed. Force P is horizontal, as is the plane. *Ans.* $\bar{a} = PgR(R-r)/W(R^2+k^2)$

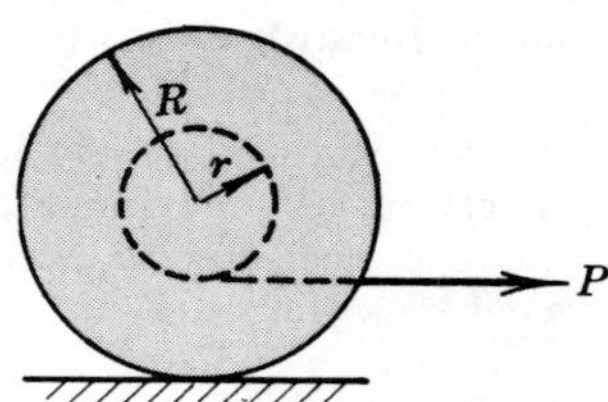

Fig. 17-26

27. A cylindrical wheel 4 ft in diameter weighs 100 lb and is pulled up a 20 degree plane by a cord wrapped around its circumference. If the cord passes over a smooth pulley at the top of the plane and supports a hanging weight of 200 lb, determine the angular acceleration of the wheel. Assume that the cord pulls the top of the wheel and is parallel to the plane. *Ans.* 6.2 rad/sec^2

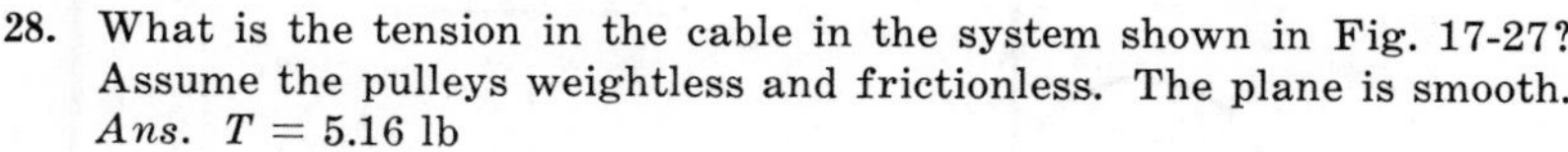

28. What is the tension in the cable in the system shown in Fig. 17-27? Assume the pulleys weightless and frictionless. The plane is smooth. *Ans.* T = 5.16 lb

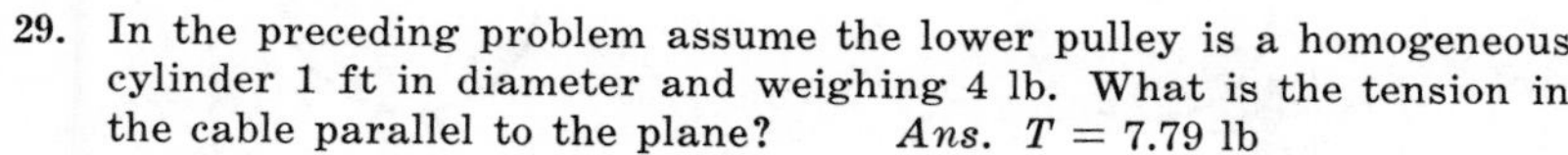

29. In the preceding problem assume the lower pulley is a homogeneous cylinder 1 ft in diameter and weighing 4 lb. What is the tension in the cable parallel to the plane? *Ans.* T = 7.79 lb

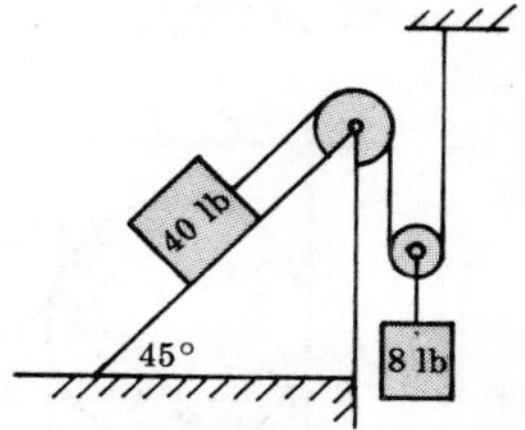

Fig. 17-27

30. A constant force of 50 lb is applied as shown in Fig. 17-28 below. How long will it take the 60 lb block to reach a speed of 4 ft/sec starting from rest? Assume the pulleys are weightless and frictionless. *Ans.* t = 0.186 sec

31. In the system shown in Fig. 17-29 below, the pulleys are to be considered weightless and frictionless. The weights are 1, 2, 3, and 4 pounds. Determine the acceleration of each weight and the tension in either moving cord.
Ans. $\ddot{x}_1$ = 29.6 ft/sec^2 up, $\ddot{x}_2$ = 1.30 ft/sec^2 down, $\ddot{x}_3$ = 11.6 ft/sec^2 down, $\ddot{x}_4$ = 16.8 ft/sec^2 down, T = 1.92 lb

32. In Fig. 17-30 below assume the weight W_2 is moving down. If the pulleys are "weightless" and frictionless, determine the tension in the rope when the weights are released from rest.
Ans. $T = 3W_1W_2/(4W_1+W_2)$

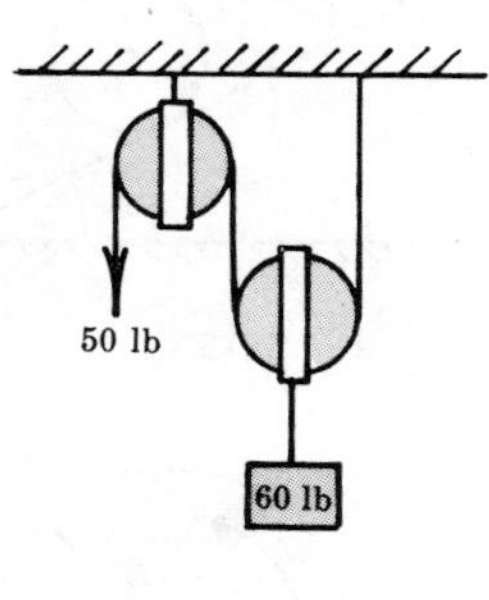

Fig. 17-28

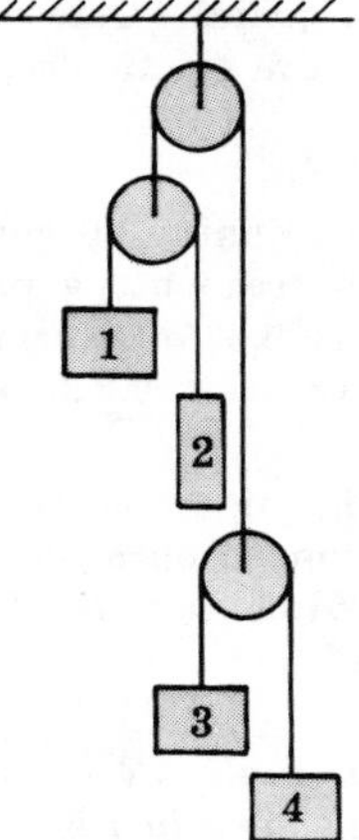

Fig. 17-29

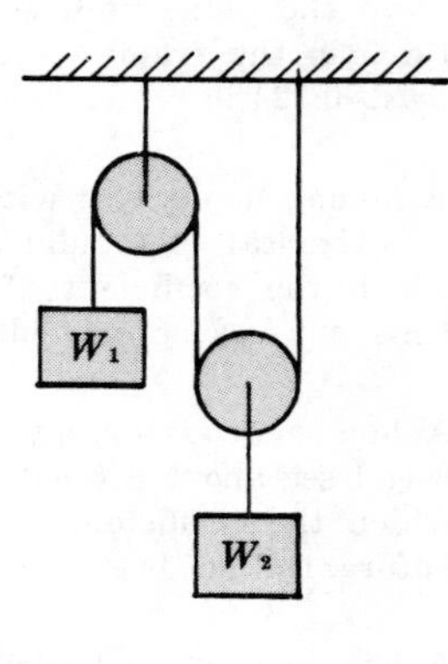

Fig. 17-30

33. A horizontal force of 40 lb is applied tangentially at the top of a 100 lb cylinder having diameter 5 ft. If there is no slippage at the ground, determine the linear acceleration of the center of the cylinder. *Ans.* 17.2 ft/sec^2 horizontally

34. A thin hoop of weight W rolls horizontally under the action of a horizontal force P applied at the top as shown in Fig. 17-31. Express the mass center acceleration $\bar{a}$ in terms of P, W and radius R. Also show that the frictional force of the plane on the hoop is zero. *Ans.* $\bar{a} = Pg/W$

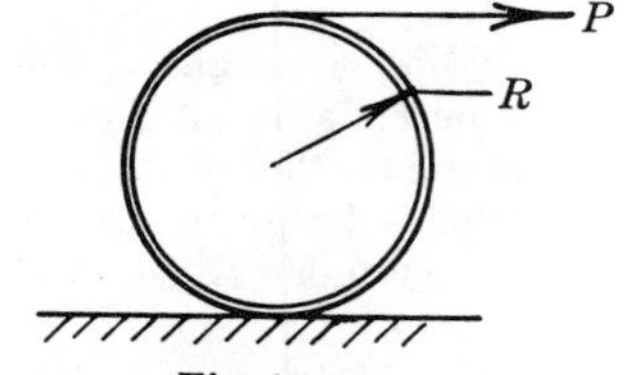

Fig. 17-31

35. In Problem 10, what must be the weight of C so that the acceleration of the disk A is 6 ft/sec^2 up the plane? *Ans.* 1340 lb

36. The 4 lb uniform sphere with radius 3 inches has a rope wrapped around it with one end attached to the ceiling as shown in Fig. 17-32 below. If the sphere is released from rest determine the tension in the rope, the angular acceleration of the sphere, and the speed of the mass center 2 seconds after release. *Ans.* $T = 1.14$ lb, $\alpha = 92.0$ rad/sec^2 clockwise, $\bar{v} = 46.0$ ft/sec down

37. The homogeneous disk 4 ft in diameter and weighing 12 lb has a string wrapped around it as shown in Fig. 17-33 below. If the tension in the string is 8 lb, what is the acceleration of the disk center? *Ans.* $a_C = 10.7$ ft/sec^2 down

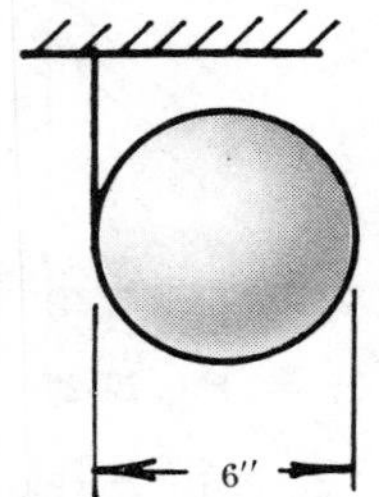

Fig. 17-32

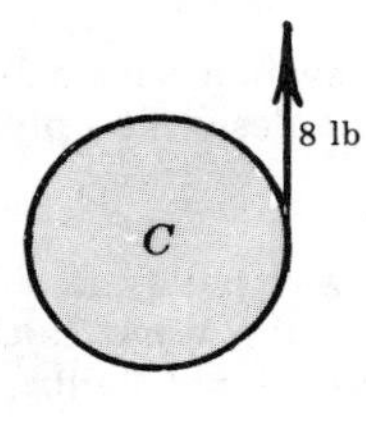

Fig. 17-33

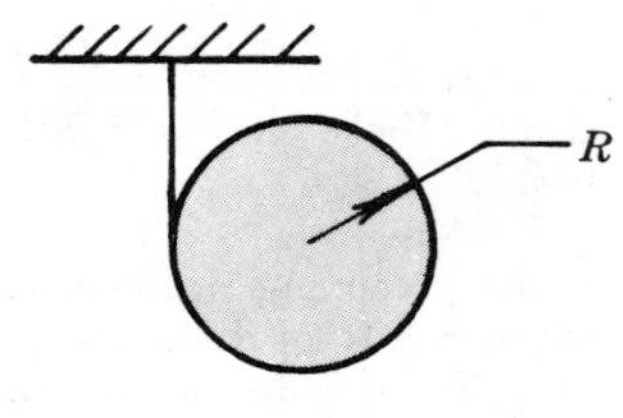

Fig. 17-34

38. A cylinder of weight W and radius of gyration k is released from rest in the position shown in Fig. 17-34 above. If the radius of the cylinder is R, determine the tension in the cord and the acceleration of the mass center. *Ans.* $T = W/(1 + R^2/k^2)$, $\bar{a} = g/(1 + k^2/R^2)$

39. In Fig. 17-35, a homogeneous cylinder and a homogeneous sphere of equal weights $W = 20$ lb and equal radii R are harnessed together by a light frame and are free to roll without slipping down the plane inclined 28° with the horizontal. Determine the force in the frame. Assume the bearings are frictionless. *Ans.* 0.324 lb

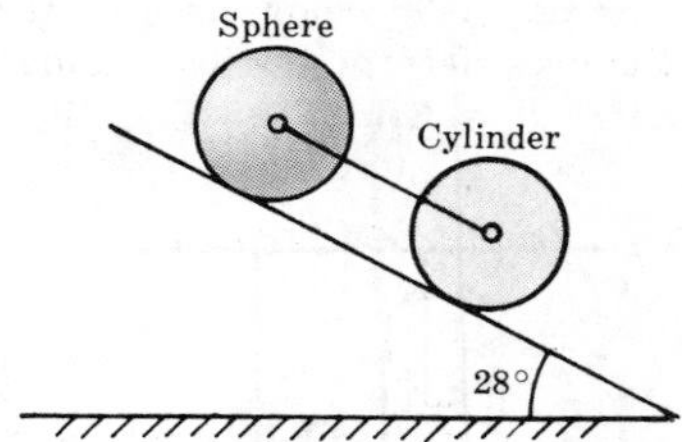

Fig. 17-35

40. A homogeneous ball with diameter 3 ft is spinning 60 rpm about a horizontal centroidal axis when it is lowered onto a plane for which the coefficient of sliding friction is 0.30. Determine the time it takes before rolling begins. *Ans.* $t = 0.279$ sec

41. A homogeneous cylinder weighing 6 lb and of diameter 6 in. is rotating 8 rad/sec about a horizontal axis when dropped on a horizontal plane for which the coefficient of friction is 0.25. How far will the center travel before rolling begins and skidding stops? *Ans.* $d = 0.331$ in.

42. A homogeneous cylinder of weight W and radius R is at rest on a horizontal plane when a couple C is applied as shown in Fig. 17-36. Determine the magnitude of the coefficient of friction between the wheel and the plane so that rolling will occur. *Ans.* $\mu \geqq \frac{2}{3}C/WR$

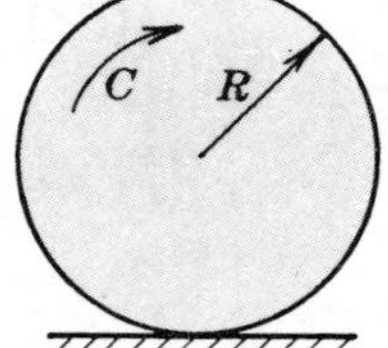

Fig. 17-36

43. In Fig. 17-37 below, the thin-walled cylinder is on a horizontal platform which has an acceleration a. Determine the acceleration of the center assuming rolling without sliding. *Ans.* $\bar{a} = 0.5a$

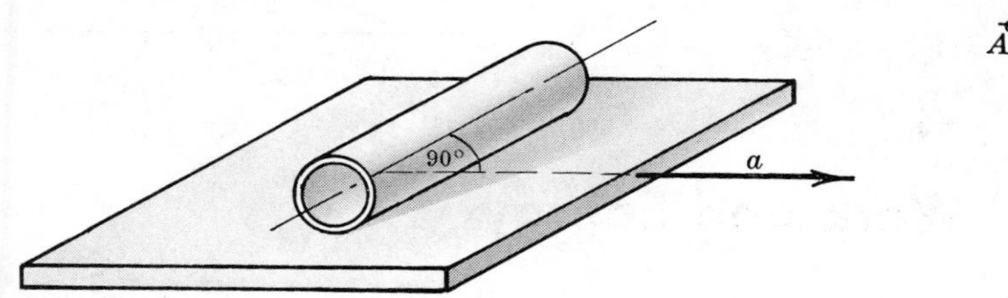

Fig. 17-37

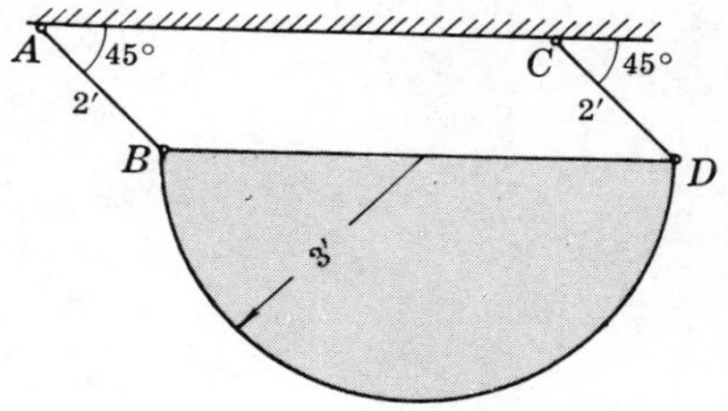

Fig. 17-38

44. The semicircular homogeneous plate of 3 ft radius weighs 60 lb and is released from rest in the position shown in Fig. 17-38 above. What are the forces in the two cords for this position?
Ans. $T_{AB} = 30.2$ lb, $T_{CD} = 12.2$ lb

45. In the preceding problem what will be the tensions when the plate is at its lowest position?
Ans. $T_{AB} = T_{CD} = 47.6$ lb

46. In a 'loop-the-loop' 20 ft in diameter, a car weighing 500 lb leaves the platform 30 ft above the bottom of the loop. What is the normal force of the track on the car at the top of the loop? Assume that the center of gravity of the car is 10 ft from the center of the loop. *Ans.* 500 lb

47. The disk of weight W has its center of mass G at a distance e from the geometric center O. The disk is rolling on the horizontal plane with constant angular velocity ω. In the position shown in Fig. 17-39 determine the normal force and frictional force of the floor on the disk.
Ans. $F = (W/g)e\omega^2 \cos\theta$ to left,
$N = W + (W/g)e\omega^2 \sin\theta$ up

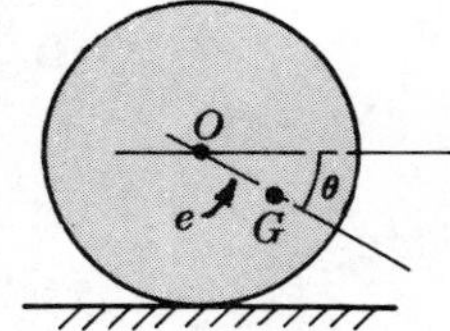

Fig. 17-39

Chapter 18

Work and Energy

WORK U done by a force $\mathbf{F}$ acting on a particle moving along any path is defined as

$$U = \int_{s_1}^{s_2} F_t\,ds$$

where s_1, s_2 = respective distances of the particle from reference point P_0 at the beginning and end of the motion.

F_t = magnitude of the tangential component of the force $\mathbf{F}$ as indicated in Fig. 18-1 below.

ds = infinitesimal change in position of the particle along the path.

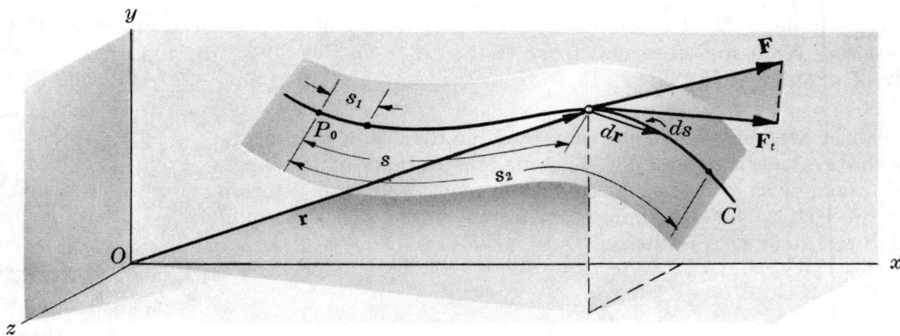

Fig. 18-1

The expression for work may also be written as the line integral from position P_1 at time t_1 to position P_2 at time t_2.

$$U = \int_C \mathbf{F}\cdot d\mathbf{r}$$

where $d\mathbf{r}$ is the infinitesimal change in vector $\mathbf{r}$. Then since

$$\mathbf{F} = F_x\mathbf{i} + F_y\mathbf{j} + F_z\mathbf{k} \qquad \text{and} \qquad d\mathbf{r} = dx\,\mathbf{i} + dy\,\mathbf{j} + dz\,\mathbf{k}$$

we can write

$$U = \int_C (F_x\mathbf{i} + F_y\mathbf{j} + F_z\mathbf{k})\cdot(dx\,\mathbf{i} + dy\,\mathbf{j} + dz\,\mathbf{k}) = \int_C (F_x\,dx + F_y\,dy + F_z\,dz)$$

This can be converted to a time integral as follows:

$$U = \int_{t_1}^{t_2} (F_x\frac{dx}{dt} + F_y\frac{dy}{dt} + F_z\frac{dz}{dt})\,dt$$

Special cases arise as follows:

(1) Force constant in magnitude and with its direction along a straight line.

$$U = Fs$$

where U = work done, F = constant force,
s = displacement during the motion along the straight line.

(2) Force constant in magnitude but at a constant angle with a straight line displacement.

$$U = Fs\cos\theta$$

where U = work done, F = constant force,
s = displacement during the motion along the straight line,
θ = angle between the action line of the force and the displacement.

(3) Forces constituting a couple (rotation occurs).

$$U = \int_{\theta_1}^{\theta_2} M\,d\theta$$

where M = couple, $d\theta$ = differential angular displacement,
θ_1, θ_2 = initial and final angular displacements.

Work is positive if the force acts in the direction of motion. Work is negative if the force acts opposite to the direction of motion.

The work done on a rigid body by two or more concurrent forces during a displacement is equivalent to the work done by the resultant of the force system during the same displacement.

Work is a scalar quantity with units such as ft-lb, newton-meter or joule, etc.

POWER

Power, which is the rate of doing work, equals $dU/dt = \mathbf{F}\cdot d\mathbf{r}/dt = \mathbf{F}\cdot\mathbf{v}$. In the case of a couple, this would be Power $= dU/dt = \mathbf{M}\cdot\boldsymbol{\omega}$.

Units of power are ft-lb/sec, joule/sec or watt, etc.

The *horsepower* (hp) = 550 ft-lb/sec. Two formulas are

$$\text{hp} = \frac{\mathbf{F}\cdot\mathbf{v}}{550} \qquad \text{and} \qquad \text{hp} = \frac{\mathbf{M}\cdot\boldsymbol{\omega}}{550}$$

where $\mathbf{F}$ = force in lb, $\mathbf{v}$ = velocity in ft/sec
$\mathbf{M}$ = moment of couple in lb-ft, $\boldsymbol{\omega}$ = angular velocity in rad/sec.

EFFICIENCY

Efficiency is equal to work output divided by work input for the same period of time. It is also expressed as power output divided by power input. The work output is less than the work input by the work dissipated (usually in overcoming friction).

KINETIC ENERGY

Kinetic energy T of a particle of mass m moving with speed v is defined as $\frac{1}{2}mv^2$.

Units are ft-lb, newton-meter or joule, etc.

WORK DONE on a PARTICLE

Work done on a particle by all forces is equal to the change in the kinetic energy T of the particle.

Proof:

$$\begin{aligned}
U &= \int_{r_1}^{r_2} \mathbf{F}\cdot d\mathbf{r} = \int_{r_1}^{r_2} m\ddot{\mathbf{r}}\cdot d\mathbf{r} = \int_{t_1}^{t_2} m\ddot{\mathbf{r}}\cdot\frac{d\mathbf{r}}{dt}\,dt \\
&= m\int_{t_1}^{t_2} (\ddot{\mathbf{r}}\cdot\dot{\mathbf{r}})\,dt = \tfrac{1}{2}m\int_{t_1}^{t_2} \frac{d}{dt}(\dot{\mathbf{r}})^2\,dt \\
&= \tfrac{1}{2}m[v^2]_{v_1}^{v_2} = \tfrac{1}{2}mv_2^2 - \tfrac{1}{2}mv_1^2 = T_2 - T_1
\end{aligned}$$

where U = work done, m = mass of particle,
v_1, v_2 = initial and final speeds respectively at P_1 and P_2,
T_1, T_2 = initial and final kinetic energy respectively at P_1 and P_2.

KINETIC ENERGY T of a RIGID BODY in TRANSLATION is $T = \frac{1}{2}mv^2$.

Proof: All particles of the rigid body have the same velocity $\mathbf{v}$. Then if m_i = mass of the ith particle,

$$T = \sum_{i=1}^{n} \tfrac{1}{2}m_i v^2 = \tfrac{1}{2}v^2\sum_{i=1}^{n} m_i = \tfrac{1}{2}mv^2$$

where m = mass of the entire body, v = speed of the body.

KINETIC ENERGY T of a RIGID BODY in ROTATION is $T = \frac{1}{2}I_O\omega^2$.

Proof: Let $\mathbf{r}_i$ be the radius vector of the ith particle and $\dot{\mathbf{r}}_i$ its velocity as shown in Fig. 18-2; then

$$\begin{aligned}
T &= \sum_{i=1}^{n} \tfrac{1}{2}m_i(\dot{\mathbf{r}}_i)^2 = \sum_{i=1}^{n} \tfrac{1}{2}m_i(\boldsymbol{\omega}\times\mathbf{r}_i)^2 \\
&= \sum_{i=1}^{n} \tfrac{1}{2}m_i\omega^2 r_i^2 = \tfrac{1}{2}\omega^2\sum_{i=1}^{n} m_i r_i^2 = \tfrac{1}{2}I_O\omega^2
\end{aligned}$$

where I_O = mass moment of inertia of the entire body about the axis of rotation, ω = angular speed.

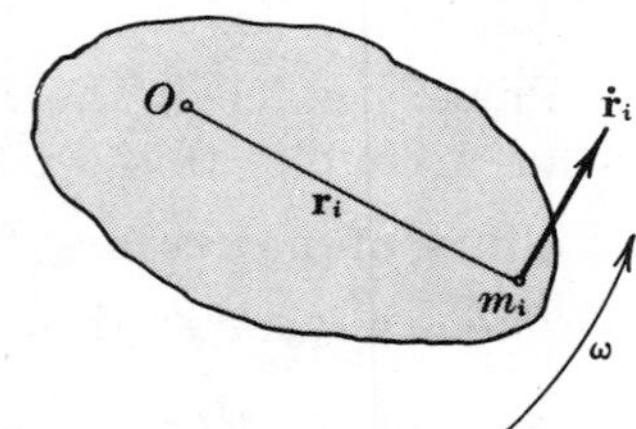

Fig. 18-2

KINETIC ENERGY T of a BODY in PLANE MOTION is $T = \frac{1}{2}m\bar{v}^2 + \frac{1}{2}\bar{I}\omega^2$.

Proof: Study the motion of the ith particle, choosing the mass center as the base point. Refer to Fig. 18-3. Then

$$\mathbf{r}_i = \mathbf{R} + \boldsymbol{\rho}_i \quad\text{and}\quad \dot{\mathbf{r}}_i = \dot{\mathbf{R}} + \dot{\boldsymbol{\rho}}_i$$

$$\dot{\mathbf{r}}_i^2 = (\dot{\mathbf{R}} + \dot{\boldsymbol{\rho}}_i)\cdot(\dot{\mathbf{R}} + \dot{\boldsymbol{\rho}}_i) = \dot{\mathbf{R}}^2 + 2\dot{\mathbf{R}}\cdot\dot{\boldsymbol{\rho}}_i + \dot{\boldsymbol{\rho}}_i^2$$

Then

$$\begin{aligned}
T &= \sum_{i=1}^{n} \tfrac{1}{2}m_i\dot{\mathbf{r}}_i^2 \\
&= \tfrac{1}{2}\sum_{i=1}^{n} m_i\dot{\mathbf{R}}^2 + \sum_{i=1}^{n} m_i\dot{\mathbf{R}}\cdot\dot{\boldsymbol{\rho}}_i + \tfrac{1}{2}\sum_{i=1}^{n} m_i\dot{\boldsymbol{\rho}}_i^2
\end{aligned}$$

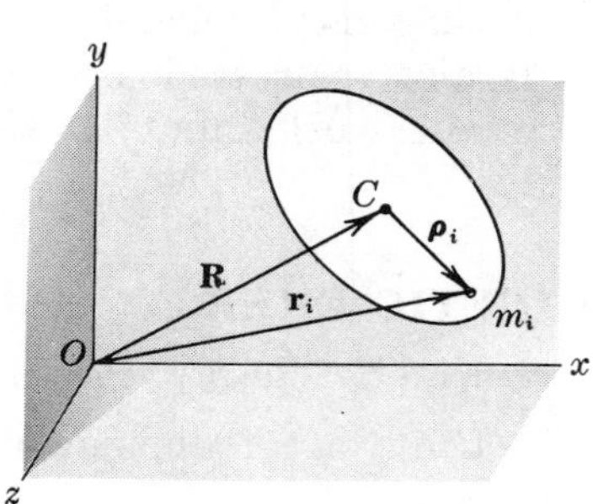

Fig. 18-3

But $\dot{\mathbf{R}}^2 = \bar{v}^2$, $\dot{\boldsymbol{\rho}}_i = \boldsymbol{\omega}\times\boldsymbol{\rho}_i$, and $m_i\dot{\mathbf{R}}\cdot\dot{\boldsymbol{\rho}}_i = m_i\dot{\mathbf{R}}\cdot(\boldsymbol{\omega}\times\boldsymbol{\rho}_i) = \dot{\mathbf{R}}\cdot(\boldsymbol{\omega}\times m_i\boldsymbol{\rho}_i)$. Thus

$$T = \tfrac{1}{2}\bar{v}^2\sum_{i=1}^{n} m_i + \dot{\mathbf{R}}\cdot\left(\boldsymbol{\omega}\times\sum_{i=1}^{n} m_i\boldsymbol{\rho}_i\right) + \tfrac{1}{2}\sum_{i=1}^{n} m_i\omega^2\rho_i^2$$

Since $\sum_{i=1}^{n} m_i \boldsymbol{\rho}_i = 0$ when $\boldsymbol{\rho}_i$ is drawn from an axis through the mass center,

$$T = \tfrac{1}{2}m\bar{v}^2 + 0 + \tfrac{1}{2}\bar{I}\omega^2$$

where $\bar{v}$ = linear speed of the mass center, ω = angular speed,

$\bar{I}$ = moment of inertia about an axis through the mass center parallel to the z axis.

POTENTIAL ENERGY

Potential energy V is the ability to do work which a particle (or body) possesses by virtue of its position or configuration.

Suppose for every position P of the particle there is a force $\mathbf{F}$ acting on the particle, where $\mathbf{F}$ is determined uniquely by the position P. Further suppose the position S is selected as an arbitrary standard or datum position. Then the potential energy V_P at P is defined as the work done by $\mathbf{F}$ as the particle moves from position P to the standard position S over any path (this is true only for conservative force systems, usually meaning in mechanics that no friction is involved). Problems 1 to 3 deal with potential energy $V_P = \int_P^S \mathbf{F}\cdot d\mathbf{r}$.

PRINCIPLE of WORK and ENERGY

Principle of work and energy states that the work done by the external forces acting on a rigid body during a displacement is equal to the change in kinetic energy of the body during the same displacement.

LAW of CONSERVATION of ENERGY

Law of conservation of energy states that if a particle (or body) is acted upon by a conservative force system, the sum of the kinetic energy and potential energy is a constant.

Proof: Let P and Q be respectively the initial and final positions of the particle. The work done by a conservative force $\mathbf{F}$ as the particle moves from P to Q has been defined as $U = \int_P^Q \mathbf{F}\cdot d\mathbf{r}$. But in terms of our standard position S with radius vector $\mathbf{r}_S$ we can write

$$U = \int_P^S \mathbf{F}\cdot d\mathbf{r} + \int_S^Q \mathbf{F}\cdot d\mathbf{r} = \int_P^S \mathbf{F}\cdot d\mathbf{r} - \int_Q^S \mathbf{F}\cdot d\mathbf{r}$$

But potential energy $V_P = \int_P^S \mathbf{F}\cdot d\mathbf{r}$ and $V_Q = \int_Q^S \mathbf{F}\cdot d\mathbf{r}$. Hence

$$U = V_P - V_Q$$

We have also shown in terms of kinetic energy that

$$U = T_Q - T_P$$

Thus

$$V_P - V_Q = T_Q - T_P \quad \text{or} \quad V_P + T_P = V_Q + T_Q$$

Solved Problems

1. Determine the potential energy V of a body of weight W at a height h above a datum plane (standard position).

Solution:

The potential energy V is the work done by the only force acting on the body, its weight $\mathbf{W}$ which is assumed constant throughout the travel from height h to the datum plane. The body can traverse any smooth path as shown in Fig. 18-4(a) below.

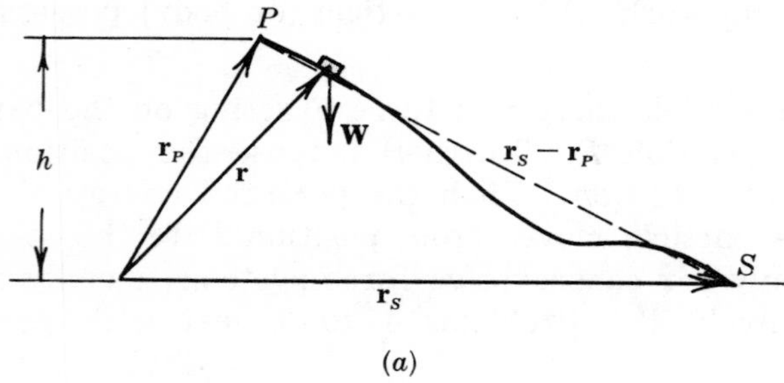

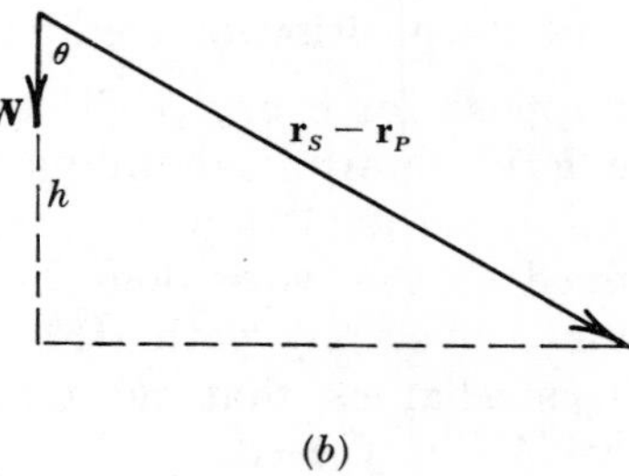

Fig. 18-4

Since $\mathbf{W}$ is constant, we can write

$$V = \int_C \mathbf{W} \cdot d\mathbf{r} = \mathbf{W} \cdot \int_{\mathbf{r}_P}^{\mathbf{r}_S} d\mathbf{r} = \mathbf{W} \cdot (\mathbf{r}_S - \mathbf{r}_P)$$

But Fig. 18-4(b) above indicates that

$$\mathbf{W} \cdot (\mathbf{r}_S - \mathbf{r}_P) = W\,|\mathbf{r}_S - \mathbf{r}_P| \cos\theta = Wh$$

Thus the potential energy V of a body of weight W at a height h above an arbitrary datum plane is Wh.

2. A body on a frictionless horizontal plane is attached to a spring with modulus k lb/in. Refer to Fig. 18-5(a) for top view. When the body is at S, a distance l from the wall, there is no force in the spring ($kl = 0$). This will be considered the standard position. When the body is at P, a distance $l + x$ from the wall, determine the potential energy V_P of the system (actually of the spring because there is no change in the potential energy of the body, which is always on the horizontal plane).

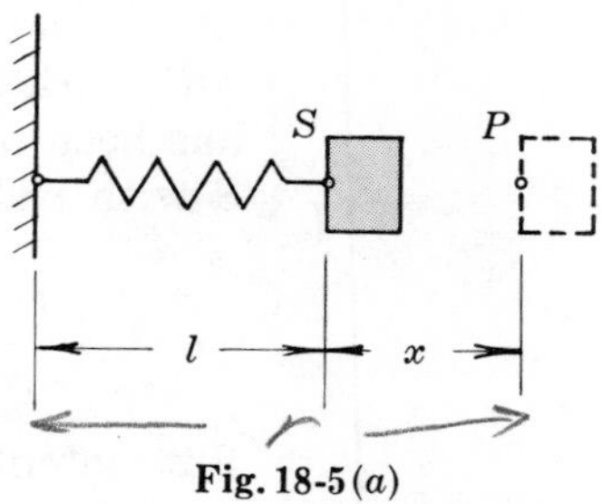

Fig. 18-5 (a)

Solution:

Fig. 18-5(b) shows the body at a distance r from the wall as the origin. The force acting on the body in this position is proportional to the deformation of the spring from the standard unstressed length; thus $\mathbf{F} = -k(r - l)\mathbf{e}_r$. The work done by this force for a differential change $d\mathbf{r}$ (where $d\mathbf{r} = dr\,\mathbf{e}_r$) is $-k(r - l)\mathbf{e}_r \cdot dr\,\mathbf{e}_r = -k(r - l)dr$ since $\mathbf{e}_r \cdot \mathbf{e}_r = 1$. The potential energy is

$$V_P = \int_{l+x}^{l} -k(r - l)\,dr = \tfrac{1}{2}kx^2$$

O
$\mathbf{e}_r$
$\mathbf{F} = -k(r - l)\mathbf{e}_r$

Fig. 18-5 (b)

The body possesses this potential energy because of the pull of the spring which varies with the deformation. In Problem 1 the body possesses potential energy because of the attraction of the earth considered constant for relatively small changes of height h. Problem 3 will introduce an attractive force which varies inversely with the square of the distance.

3. Determine the potential energy of a body attracted by a force which is inversely proportional to the square of the distance r from the source O of the force.

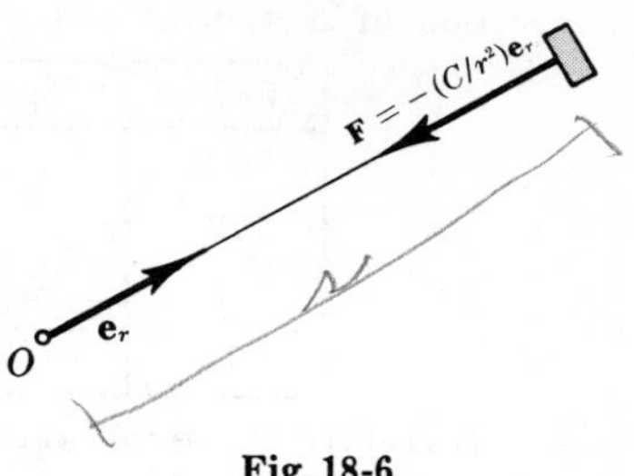

Fig. 18-6

Solution:

The standard position S of the body will be chosen at infinity. (Note: Some students might have tried the source O as the standard position, but V is defined in terms of the work done in moving to the standard position; and since the force tends to become infinite as the body approaches O, the choice of O is seen to be unwise). The body is shown in Fig. 18-6 at a distance r from the origin. The force acting on the body in this position is $\mathbf{F} = -(C/r^2)\mathbf{e}_r$. As in Problem 2, the potential energy is

$$V_P = \int_r^\infty -(C/r^2)\mathbf{e}_r \cdot dr\,\mathbf{e}_r = \int_r^\infty -(C/r^2)\,dr = +C[1/r]_r^\infty = -C/r$$

4. A spring is initially compressed from a free length of 8 in. to a length of 6 in., making a net initial compression of 2 inches. What additional work is done in compressing it 3 inches more (5 inches total) to a 3 inch length? Assume the spring modulus $k = 20$ lb/in.

Solution:

According to definition, the work to compress it from 2 in. to 5 in. is

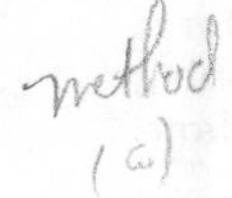

$$U = \int_{s_1}^{s_2} F_t\,ds = \int_2^5 20s\,ds = 210 \text{ in-lb}$$

The same value may be obtained by using Problem 2 to determine the work done in compressing the spring first 2 inches and then 5 inches. Note that both expressions are for work done from the unstressed or zero position. Their difference is the required value.

$$U = \int_0^5 20s\,ds - \int_0^2 20s\,ds = 250 - 40 = 210 \text{ in-lb}$$

5. Solve Problem 4 graphically.

Solution:

To solve graphically, plot the varying force $F = 20s$ against the displacement as shown in Fig. 18-7.

Note that the area included between the 2 inch and 5 inch displacements is in terms of inches horizontally and pounds vertically.

The area is therefore equal to the work done, or equals average height in pounds times change in displacement in inches.

$$U = \tfrac{1}{2}(40 \text{ lb} + 100 \text{ lb})(3 \text{ in.}) = 210 \text{ in-lb}$$

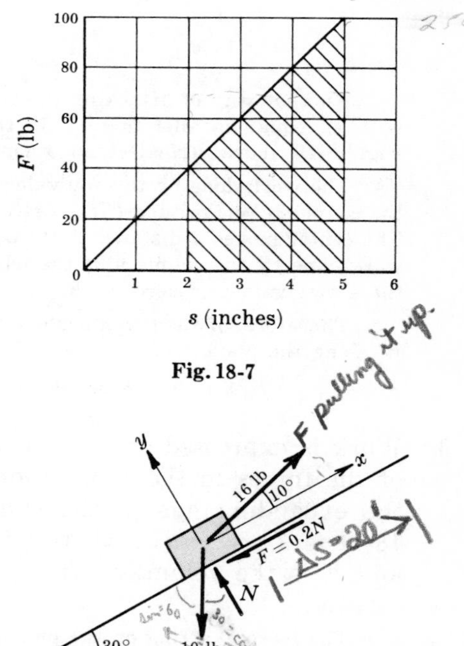

Fig. 18-7

6. Determine the total work done on a body weighing 10 lb that is pulled 20 ft up a rough plane inclined 30° with the horizontal, as shown in Fig. 18-8. Assume coefficient of friction $\mu = 0.20$.

Solution:

From the free body diagram, N may be found by summing forces perpendicular to the plane.

$$\Sigma F_y = 0 = -10\cos 30° + N + 16\sin 10°$$

Solving, $N = 5.88$ lb.

Fig. 18-8

The work done by each force is next determined. The sign is positive if the force acts in the direction of motion of the weight.

Force	Sign of Work Done	Magnitude of Work Done	Result
16 lb	+	$(16 \cos 10°)(20)$	+315 ft-lb
W	−	$(10 \sin 30°)(20)$	−100 ft-lb
N	no work since N acts $\perp$ to direction of motion		0 ft-lb
F	−	$(0.20 \times 5.88)(20)$	−23.5 ft-lb

Therefore the total work done $= +315 - 100 - 23.5 = +191$ ft-lb.

Now find the resultant of all the forces and then determine the work done by the resultant. Since only the x component of the resultant will do work, determine only that.

$$R_x = \Sigma F_x = +16 \cos 10° - 0.20(5.88) - 10 \sin 30° = 9.57 \text{ lb}$$

Since 9.57 lb is a constant force, the total work done $= R_x(20) = 9.57(20) = +191$ ft-lb.

7. Find the work done in rolling a wheel weighing 50 lb a distance 5 ft up a plane inclined 30° with the horizontal as shown in Fig. 18-9 below. Assume a coefficient of friction 0.25. The force P shown is 30 lb.

Solution:

The normal force N does no work because it has no component in the direction of motion.

The force of friction F also does no work, but for a different reason. Its point of contact moves with the wheel and since there is no relative motion (no slipping involved) between the frictional force and the wheel, the work done by the force of friction is zero. This is not the case in the preceding problem where the frictional force did negative work, but of course the body slipped along the plane.

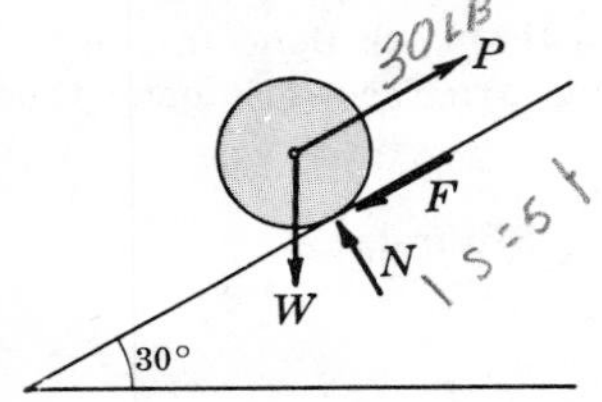

Fig. 18-9

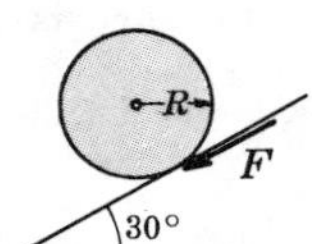

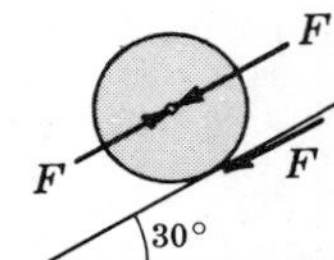

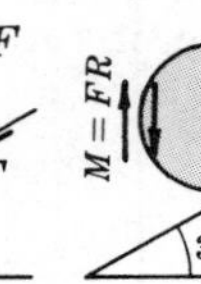

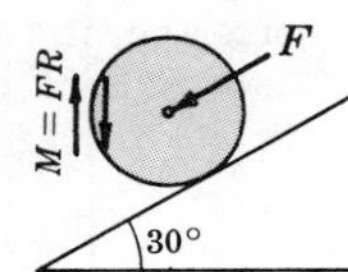

Fig. 18-10

Another way of attacking the friction work in this problem would be to replace the single force F with an equal parallel force F in the same direction through the center and a couple as shown in Fig. 18-10 above. (Neglect all other forces for the time being.)

The work done by the equivalent system is equal to the sum of the work done by the force F and the couple M (magnitude FR). Let us determine the work done for one revolution, i.e., for 2π radians. The center moves a distance $2\pi R$ during one revolution. Hence the work done by F is $-F(2\pi R)$. The work done by the couple aids the wheel in moving up the hill and equals $+M\theta$ or $2\pi FR$. The sum of these two values is zero work.

Therefore, of the given system, the only forces doing work are P and the component of the weight W along the plane (refer to Fig. 18-9 above). The component of W is $50 \sin 30°$.

$$\text{Work done} = +30(5) - (50 \sin 30°)(5) = 25 \text{ ft-lb}$$

8. Work is expressed as $U = \int (\Sigma F_x)\,dx$, where ΣF_x = sum of the forces in the x direction. Show that the work of the expanding gas in the simple engine shown in Fig. 18-11 is $U = \int p\,dV$, where p is the gas pressure in lb/ft² and V is the volume in ft³.

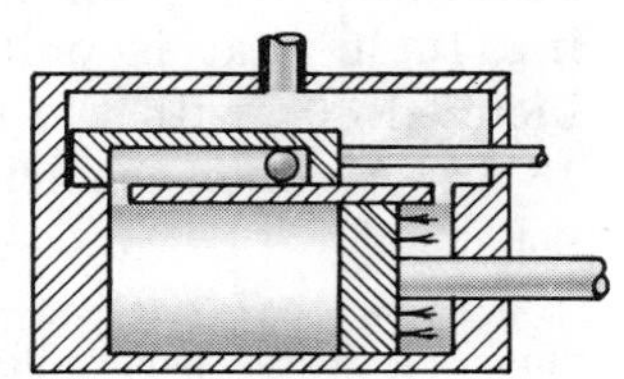

Fig. 18-11

Solution:

The force F acting on the piston pA, where A is the area of the piston in ft². Also the change in volume $dV = A\,dx$. Hence

$$U = \int (\Sigma F_x)\,dx = \int (pA)\frac{dV}{A} = \int p\,dV$$

9. Fig. 18-12 illustrates an indicator card taken on the head end of a steam engine. The horizontal line is proportional to the stroke of the engine and therefore to the volume of the steam in ft^3. The height of the card indicates the internal pressure in either psi or lb/ft^2. Assume that the length of the card is 3.20 in., and the engine is 6 in. by 8 in. The indicator spring scale is 100. This means that a pressure of 100 psi produces a vertical displacement on the card of one inch. The area of the card as measured with a planimeter is 1.82 in^2. Determine the work represented by the card.

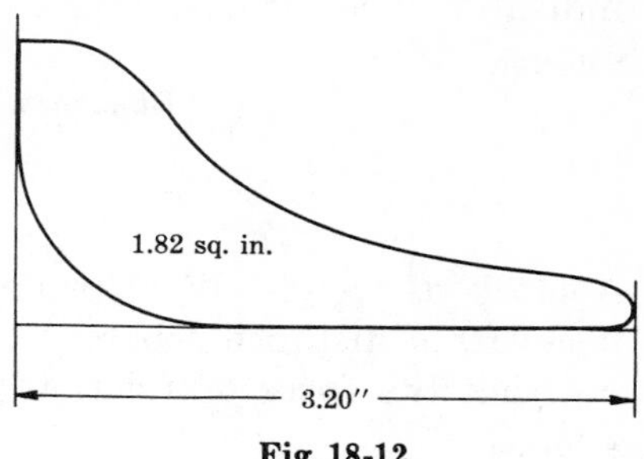

Fig. 18-12

Solution:

The 3.20 inches horizontally represents the volume contained in the engine. Since the bore is 6 in. and the stroke 8 in., this volume is

$$V = Al = \tfrac{1}{4}\pi d^2 l = \tfrac{1}{4}\pi(6/12)^2(8/12)\text{ ft}^3 = 0.131\text{ ft}^3$$

One horizontal inch represents $(0.131\text{ ft}^3)/3.20 = 0.041\text{ ft}^3$.

One vertical inch represents $(100\text{ lb/in}^2)(144\text{ in}^2/\text{ft}^2) = 14{,}400\text{ lb/ft}^2$.

Then 1.82 in^2 of card area represents $1.82(14{,}400)(0.041) = 1075$ ft-lb of work.

10. At a certain instant during acceleration along a level track the draw-bar pull of a locomotive is 20,000 lb. What horsepower is being developed if the speed of the train is 40 mph?

Solution:

$$\text{Hp} = \frac{F(\text{lb}) \times v(\text{ft/sec})}{550\text{ ft-lb/sec}} = \frac{20{,}000\text{ lb} \times 58.7\text{ ft/sec}}{550\text{ ft-lb/sec}} = 2140$$

11. A belt wrapped around a pulley 2 ft in diameter has a tension of 180 lb on the tight side and a tension of 40 lb on the slack side. What power is being transmitted if the pulley is rotating 200 rpm?

Solution:

The torque M is found by taking the algebraic sum of the moments of the tensile forces in the belt about the center of the pulley.

$$\text{Torque } M = 180\text{ lb} \times 1\text{ ft} - 40\text{ lb} \times 1\text{ ft} = 140\text{ lb-ft}$$

$$\text{Hp} = \frac{M(\text{lb-ft}) \times \omega(\text{rad/sec})}{550\text{ ft-lb/sec}} = \frac{140\text{ lb-ft} \times (2\pi \times 200/60)\text{rad/sec}}{550\text{ ft-lb/sec}} = 5.33$$

12. The force P acting on the Prony Brake shown in Fig. 18-13 is 25.6 lb. The moment M transmitted by the shaft of the engine to the brake drum gives it a counterclockwise angular velocity of 600 rpm. Determine the power dissipated by the Prony Brake. Neglect the weight of the Prony Brake.

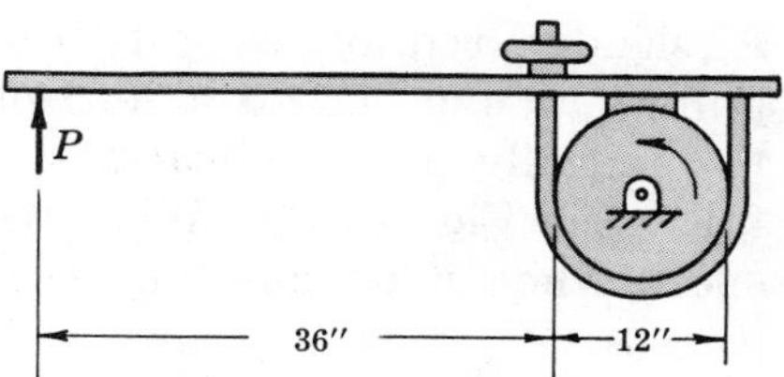

Fig. 18-13

Solution:

The moment M being transmitted by the shaft equals the moment of the force P about the drum center, i.e., $(25.6\text{ lb})(42/12\text{ ft}) = 89.6$ lb-ft.

$$\text{Hp} = \frac{M\omega}{550} = \frac{89.6\text{ lb-ft} \times (2\pi \times 600/60)\text{rad/sec}}{550\text{ ft-lb/sec}} = 10.2$$

Thus the power being transmitted by the shaft is 10.2 hp.

13. The power, as measured by a Prony Brake attached to an engine flywheel, is 6.3 brake horsepower (bhp). The indicated horsepower (ihp) as measured by means of indicator cards is 7.1. What is the efficiency of the engine?

Solution:

$$\text{Efficiency} \;=\; \frac{\text{power output}}{\text{power input}} \;=\; \frac{6.3 \text{ bhp}}{7.1 \text{ ihp}} \;=\; 89\%$$

14. A block of weight W is projected with initial speed v_0 along a horizontal plane. If it covers a distance s before coming to rest, what is the coefficient of friction assuming that the force of friction is proportional to the normal force?

Solution:

The initial kinetic energy is $T_1 = \frac{1}{2}(W/g)v_0^2$. Its final kinetic energy is zero.

The normal force = weight W. Hence the friction = μW and does work = $-\mu Ws$.

$$U = T_2 - T_1 \quad \text{or} \quad -\mu Ws = -\tfrac{1}{2}(W/g)v_0^2, \qquad \text{from which} \quad \mu = v_0^2/2gs$$

15. If the block in Problem 6 starts from rest, what is its speed after traversing the 20 ft?

Solution:

$$\text{Work done on body} \;=\; \text{change in its kinetic energy} \;=\; \tfrac{1}{2}m(v_2^2 - v_1^2)$$

$$191 \text{ ft-lb} \;=\; \tfrac{1}{2}(10/32.2)(v_2^2 - 0) \qquad \text{from which } v_2 = 35.1 \text{ ft/sec}$$

16. A slender rod 6 ft long weighing 8 lb increases its speed about a vertical axis through one end from 20 rpm to 50 rpm in 10 revolutions. Find the constant moment M required to do this.

Solution:

Moment of inertia I_O of the rod about an axis through one end is

$$I_O \;=\; \tfrac{1}{3}ml^2 \;=\; \tfrac{1}{3}(8/32.2)(6^2) \;=\; 2.98 \text{ slug-ft}^2$$

$$\omega_1 = 2\pi(20/60) = 2.09 \text{ rad/sec}, \quad \omega_2 = 2\pi(50/60) = 5.23 \text{ rad/sec}, \quad \theta = 2\pi(10) = 62.8 \text{ rad.}$$

$$\text{Work done} = \text{change in kinetic energy of rotating body}$$

$$M\theta = \tfrac{1}{2}I_O(\omega_2^2 - \omega_1^2)$$

$$M(62.8) = \tfrac{1}{2}(2.98)(\overline{5.23}^2 - \overline{2.09}^2) \qquad \text{or} \qquad M = 0.543 \text{ lb-ft}$$

17. A slender rod of weight W and length l is pinned at one end to a horizontal plane. The rod initially in a vertical position is allowed to fall. See Fig. 18-14. What will be its angular speed when it strikes the floor?

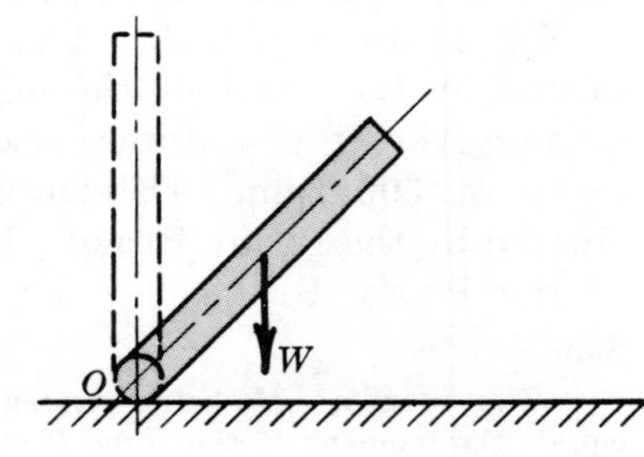

Fig. 18-14

Solution:

The only force doing work is the weight W assumed concentrated at the center of gravity which falls a total vertical distance $l/2$. The work done by gravity is thus $Wl/2$.

The kinetic energy is changed from $T_1 = 0$ to $T_2 = \frac{1}{2}I_O\omega^2 = \frac{1}{2}[\frac{1}{3}(W/g)l^2]\omega^2$. Then

$$U = T_2 - T_1 \quad \text{or} \quad \tfrac{1}{2}Wl = \tfrac{1}{6}Wl^2\omega^2/g, \qquad \text{from which } \omega = \sqrt{3g/l}$$

18. A car and four wheels weighs 1800 lb. Each wheel weighs 200 lb and is 2.50 ft in diameter. The car moving 15 mph coasts to rest on a level track in 2 miles. What is the rolling resistance F?

Solution:

The initial kinetic energy T_1 is the kinetic energy (of translation) of the 1000 lb car and the kinetic energy (both rotation and translation) of the four wheels. The final kinetic energy $T_2 = 0$. The work done by the rolling resistance F equals $F(2)(5280)$ ft-lb. Also, 15 mph = 22 ft/sec. Then

$$U \;=\; T_2 - T_1$$

$$10{,}560\,F \;=\; 0 - \tfrac{1}{2}(1000/32.2)(22)^2 - 4(\tfrac{1}{2})(200/32.2)[(22)^2 + \tfrac{1}{2}r^2\omega^2]$$

Substitute $\omega^2 r^2 = v^2 = (22)^2$ into the expression to obtain $F = -1.57$ lb.

19. A 100 lb slender rod 4 ft long falls from rest in the horizontal line to the 45° position shown in Fig. 18-15. In that phase, what are the bearing reactions at O on the rod?

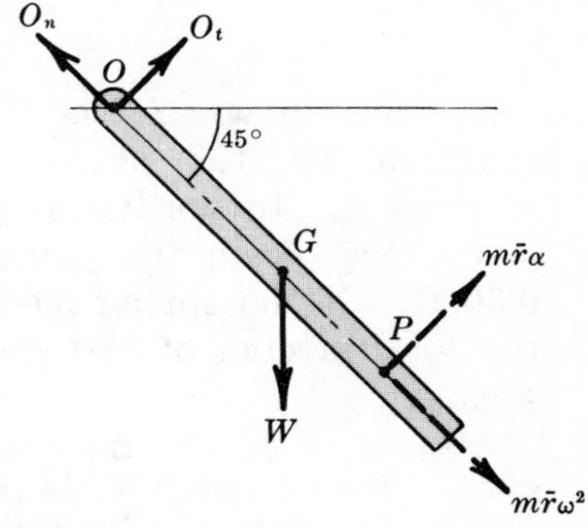

Fig. 18-15

Solution:

Choose n and t axes along and perpendicular to the bar. Apply reversed effective forces through the center of percussion P to hold the bar in "equilibrium for study purposes".

To determine the angular acceleration α in the phase shown, use the equation $\Sigma M_O = 0$ or $-W(2)(0.707) + m\bar{r}\alpha(k_0^2/\bar{r}) = 0$. The distance to P is $k_0^2/\bar{r} = I_O/m\bar{r} = 8/3$ ft. The equation then becomes $-100(2)(0.707) + (100/32.2)(2\alpha)(8/3) = 0$, whence $\alpha = 8.53$ rad/sec².

Summing forces along the t-axis, $O_t + m\bar{r}\alpha - 100\cos 45° = 0$ or $O_t = 17.7$ lb in direction assumed.

To determine O_n it is necessary first to find ω. Use work-energy methods. The only force doing work is the weight whose center G has fallen $2(0.707) = 1.414$ ft. Hence work $= +100(1.414) = 141.4$ ft-lb.

The kinetic energy change is the final value $\frac{1}{2}I_O\omega^2 = \frac{1}{2}[\frac{1}{3}(100/32.2)4^2]\omega^2 = 8.28\omega^2$.

Since U = change in T, we have $141.4 = 8.28\omega^2$ or $\omega^2 = 17.1$.

Summing forces along the bar, $-O_n + 100(0.707) + (100/32.2)(2)(17.1) = 0$ or $O_n = 177$ lb.

20. A solid homogeneous cylinder of radius R and weight W rotates freely from its initial rest position (G vertically above O) about a fixed horizontal axis perpendicular to the plane of the paper. What is the value of its angular speed in the position θ? Refer to Fig. 18-16.

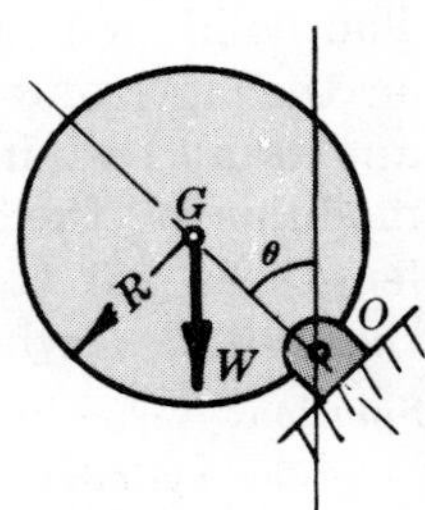

Fig. 18-16

Solution:

By inspection, G falls a vertical distance $R - R\cos\theta$. The work done on the cylinder by gravity is thus equal to $WR(1-\cos\theta)$.

The kinetic energy in its rest position is $T_1 = 0$. The kinetic energy in the θ position is $T_2 = \frac{1}{2}I_O\omega^2$. By the transfer theorem for moments of inertia,

$$I_O \;=\; \bar{I} + mR^2 \;=\; \tfrac{1}{2}(W/g)R^2 + (W/g)R^2 \;=\; \tfrac{3}{2}(W/g)R^2$$

Then $\qquad U = T_2 - T_1, \qquad WR(1-\cos\theta) = \frac{3}{4}(W/g)R^2\omega^2, \qquad \omega = \sqrt{\dfrac{4g(1-\cos\theta)}{3R}}$

21. A sphere, rolling with initial velocity 30 ft/sec, starts up a plane inclined 30° with the horizontal as shown in Fig. 18-17. How far will it roll up the plane?

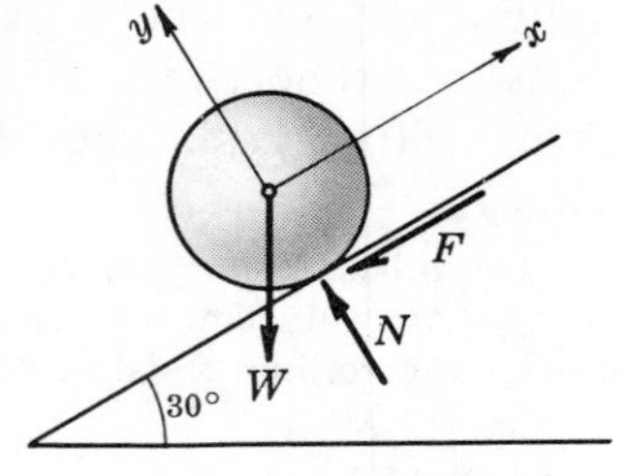

Fig. 18-17

Solution:

The initial kinetic energy T_1 decreases to $T_2 = 0$ at the top of the travel. The only force that does work is the component (negative) of the weight W along the plane.

Work done $= -(W \sin 30°)x$, where x is the required distance.

The initial kinetic energy T_1 for the body in plane motion is

$$T_1 = \tfrac{1}{2}m\bar{v}_1^2 + \tfrac{1}{2}\bar{I}\omega_1^2.$$

Since here $\bar{I} = \tfrac{2}{5}mR^2$ and $v_1 = \omega_1 R$, then $T_1 = \tfrac{1}{2}m\bar{v}_1^2 + \tfrac{1}{5}m\bar{v}_1^2 = \tfrac{7}{10}(W/g)(30)^2$.

Then $\quad U = T_2 - T_1, \quad -(W \sin 30°)x = 0 - \tfrac{7}{10}(W/g)(30)^2, \quad x = 39.2$ ft

22. The wheel shown in Fig. 18-18 weighs 322 lb and has a radius of gyration of 1.20 ft with respect to its center of mass G. In the initial phase shown the velocity of G is 6 ft/sec down the plane and the spring is stretched 0.50 ft. If the spring modulus is 80 lb/ft, what will be the total stretch of the spring?

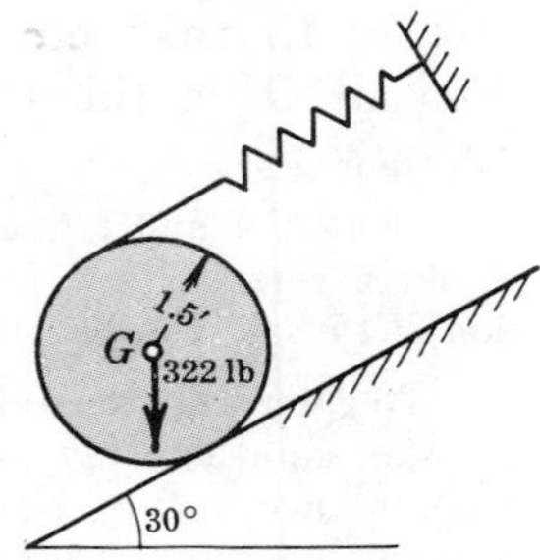

Fig. 18-18

Solution:

The work done on the wheel equals the change in its kinetic energy. The component of the weight along the plane (322 sin 30°) does positive work, and the spring acting on the wheel does negative work of the amount necessary to stretch it from 0.50 ft to its total stretch x.

The initial kinetic energy $T_1 = \tfrac{1}{2}m\bar{v}_1^2 + \tfrac{1}{2}\bar{I}\omega_1^2 = \tfrac{1}{2}m(6)^2 + \tfrac{1}{2}m(1.20)^2\omega_1^2 = 295$ ft-lb, where $m =$ 10 slugs and $\omega_1 = \bar{v}_1/r = 6/1.5 = 4$ rad/sec.

In the final phase the wheel stops and $T_2 = 0$.

Next determine the work done by the component of the weight (161 lb) and the spring. Note that if the length of the spring increases by (x − 0.50 ft), the center of the wheel travels half that distance.

$$\text{Work done} = +161\left(\frac{x - 0.50}{2}\right) - \int_{0.50}^{x} 80x\,dx = 80.5x - 40x^2 - 30.3. \quad \text{Then}$$

$$U = T_2 - T_1, \qquad 80.5x - 40x^2 - 30.3 = 0 - 295, \qquad x = 3.78 \text{ ft}$$

23. The solid homogeneous cylinder rolls without slipping on the horizontal plane. The cylinder weighs 193.2 lb and is at rest in the phase shown in Fig. 18-19. The modulus of the spring is 30 lb/ft and its unstretched length is 2.0 ft. What will be the angular speed of the cylinder when its center has moved 1.5 ft?

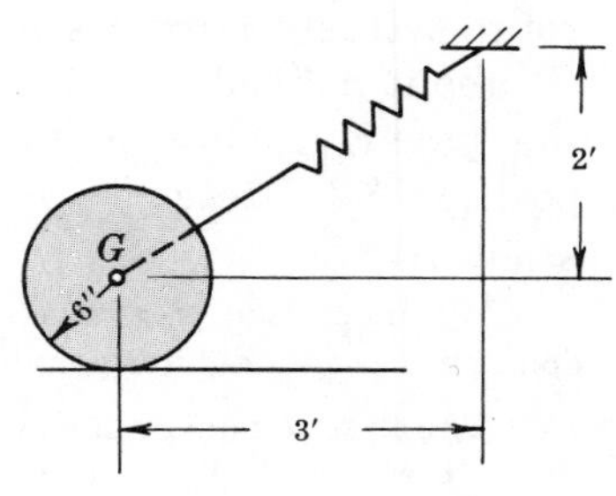

Fig. 18-19

Solution:

The cylinder has initial kinetic energy $T_1 = 0$ and final $T_2 = \tfrac{1}{2}m\bar{v}_2^2 + \tfrac{1}{2}\bar{I}\omega_2^2$. Substituting $\bar{v}_2 = r\omega_2 = \tfrac{1}{2}\omega_2$ and $\bar{I} = \tfrac{1}{2}mr^2 = \tfrac{1}{2}(193.2/32.2)(\tfrac{1}{2})^2 = 0.75$ slug-ft², we obtain $T_2 = 1.13\omega_2^2$.

Since the point of application of the frictional force moves with the cylinder (and hence friction does no work), the forces form a conservative system. The conservation of energy law will be used here. The system possesses potential energy because of the spring configuration (see Problem 2).

The initial and final lengths of the spring are $s_1 = \sqrt{2^2+3^2} = 3.61$ ft and $s_2 = \sqrt{2^2+(1.5)^2} = 2.50$ ft. Then

$$V_1 = \tfrac{1}{2}k(s_1-2)^2 = \tfrac{1}{2}(30)(1.61)^2 = 38.9 \text{ ft-lb}, \qquad V_2 = \tfrac{1}{2}k(s_2-2)^2 = 3.75 \text{ ft-lb}$$

Finally, from the conservation of energy law we can write

$$T_1 + V_1 = T_2 + V_2, \qquad 0 + 38.9 = 1.13\omega_2^2 + 3.75, \qquad \omega_2 = 5.57 \text{ rad/sec}$$

24. In the phase shown in Fig. 18-20, block A is moving down 5 ft/sec. The cylinder B is considered a homogeneous solid and moves in frictionless bearings. The spring is originally compressed 6 inches and has a modulus of 60 lb/ft. What will be the speed v of A after dropping 4 ft?

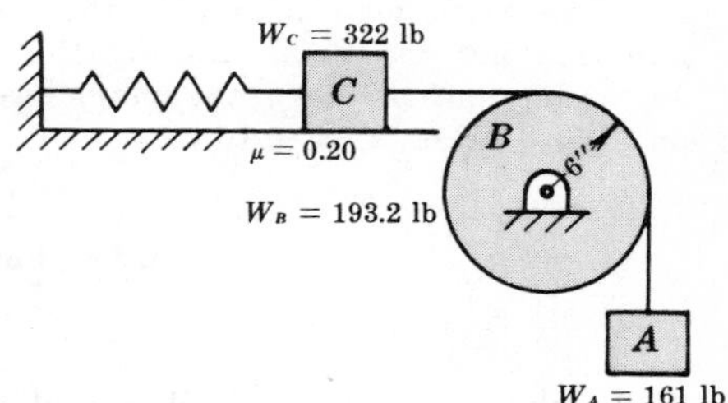

Fig. 18-20

Solution:

Consider the system consisting of bodies A, B, and C as a unit in determining kinetic energy and work done. The initial kinetic energy T_1 is the sum

$$T_1 = \tfrac{1}{2}(W_C/g)v_1^2 + \tfrac{1}{2}\bar{I}_B\omega_1^2 + \tfrac{1}{2}(W_A/g)v_1^2 = 225 \text{ ft-lb}$$

where $\bar{I}_B = \tfrac{1}{2}(W_B/g)(\tfrac{1}{2})^2$, and $\omega_1 = v_1/r = 5/\tfrac{1}{2} = 10$ rad/sec since no slip of rope on cylinder is assumed. The final kinetic energy T_2 in terms of the required speed v is

$$T_2 = \tfrac{1}{2}(W_C/g)v^2 + \tfrac{1}{2}\bar{I}_B(2v)^2 + \tfrac{1}{2}(W_A/g)v^2 = 9.0v^2$$

Work is done on the system as follows: W_A does positive work; friction on C (0.20×322) does negative work; for the first 6 in. or $\tfrac{1}{2}$ ft the spring does positive work, but after that it is being stretched from the neutral position and does negative work. Thus

$$\text{Work } U = +161(4) - 64.4(4) + \int_0^{1/2} ks\,ds - \int_0^{3.5} ks\,ds = +26.4 \text{ ft-lb}$$

Then $U = T_2 - T_1, \qquad 26.4 = 9.0v^2 - 225, \qquad v = \underline{5.29 \text{ ft/sec}}$

25. A cylinder is pulled up a plane by the tension in a rope which passes over a frictionless pulley and is attached to a weight of 150 lb as shown in Fig. 18-21. The cylinder weighs 100 lb and has radius 2 ft. The cylinder moves from rest up a distance of 16 ft. What will be its speed?

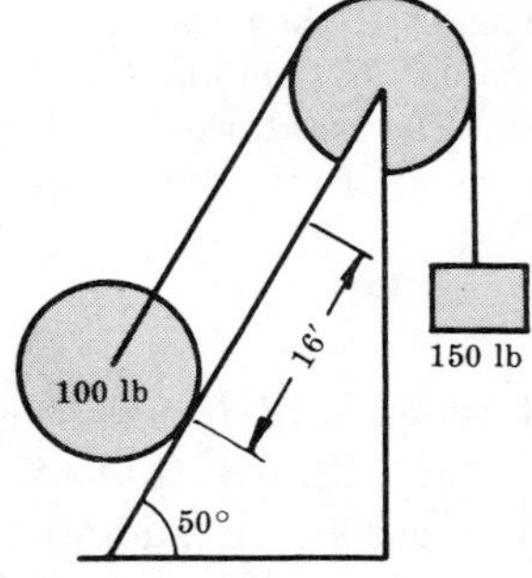

Fig. 18-21

Solution:

The initial kinetic energy of the cylinder and weight is $T_1 = 0$. The final kinetic energy of the system is

$$T_2 = T_c + T_w = (\tfrac{1}{2}m_c\bar{v}^2 + \tfrac{1}{2}\bar{I}_c\omega^2) + (\tfrac{1}{2}m_w\bar{v}^2)$$
$$= 2.33\bar{v}^2 + 2.33\bar{v}^2 = 4.66\bar{v}^2$$

after substituting $m_c = 100/32.2$, $m_w = 150/32.2$, $\bar{I}_c = \tfrac{1}{2}m_cR^2$, and $R^2\omega^2 = \bar{v}^2$.

The initial potential energy for the cylinder will be assumed zero and it will gain potential energy $100(16 \sin 50°)$. The initial potential energy of the weight will be assumed zero and it will lose potential energy $150(16)$. Thus

$$T_1 + V_1 = T_2 + V_2, \qquad 0 + 0 = 4.66\bar{v}^2 + 100(16 \sin 50°) - 150(16), \qquad \bar{v} = 15.9 \text{ ft/sec}$$

26. The 64.4 lb block shown in Fig. 18-22 below rests on the smooth plane. It is attached by a flexible inextensible cord over frictionless, weightless pulleys to the 96.6 lb weight and a support. In what distance will the block on the plane attain a speed of 10 ft/sec?

Solution:

The 96.6 lb block travels half the distance the 64.4 lb block does. Also, its speed will be half that of the 64.4 lb block.

The initial kinetic energy of the system is $T_1 = 0$. Its final kinetic energy is $T_2 = \frac{1}{2}(2)(10)^2 + \frac{1}{2}(3)(10/2)^2 = 137.5$ ft-lb.

The work done is that by the component of the weight along the plane and by the 96.6 lb weight. Assuming motion down the plane, work $U = (64.4 \sin 30°)s - 96.6(\frac{1}{2}s) = -16.1s$.

Then $U = T_2 - T_1$, $-16.1s = 137.5$, and $s = -8.53$ ft (up the plane)

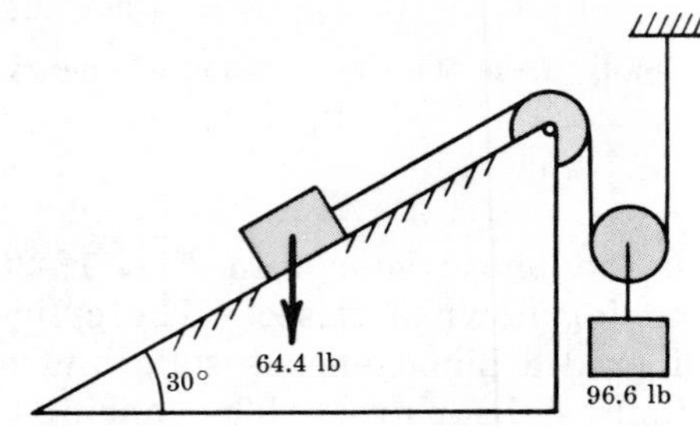

Fig. 18-22

27. A flexible chain of length l and weight w lb/ft is released when it has a free overhang of c ft. What will be its speed when leaving the smooth table?

Solution:

The free overhang of weight wc will fall a distance $l - c$; the work done by gravity on this part will be $wc(l - c)$. The part on the table will fall an average distance $\frac{1}{2}(l - c)$; the work done by gravity on this part is $w(l - c) \times \frac{1}{2}(l - c)$. The total work equals the final kinetic energy $\frac{1}{2}mv^2 = \frac{1}{2}(wl/g)v^2$. Then

$$wc(l - c) + \tfrac{1}{2}w(l - c)^2 = \tfrac{1}{2}(wl/g)v^2 \qquad \text{from which} \qquad v = \sqrt{g(l^2 - c^2)/l}$$

28. Determine the work done in winding up a homogeneous cable which hangs from a horizontal drum if its free length is 20 ft and weighs 100 lb.

Solution:

Fig. 18-23 shows the cable in its original phase. In analyzing the problem, note that any element dx has a weight $100(dx/20) = 5\,dx$.

Assume the element dx is at a distance x from the free end. This element is then raised a distance $(20 - x)$ ft. The work done on it is the product of its weight and the distance raised.

The total work is the integral of the work done on a differential element. Note that x varies from zero to 20 ft.

$$\text{Work} = \int_0^{20} 5(20 - x)dx = 1000 \text{ ft-lb}$$

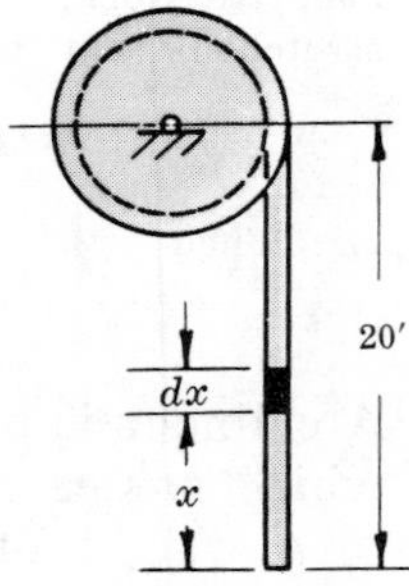

Fig. 18-23

29. If weight W hangs freely it stretches the spring a distance c inches. Show that if the weight (held so that the spring is unstressed) is suddenly released, the spring stretches a distance $2c$ before the weight starts to return upward. Refer to Fig. 18-24.

Solution:

The kinetic energy of the weight at the top (initial position) and the bottom is zero. Hence the total work done on the weight must be zero. But the total work done is the positive work of gravity (Ws) offset by the negative work of the spring ($\frac{1}{2}ks^2$) on the weight. Therefore

$$Ws - \tfrac{1}{2}ks^2 = 0$$

But in the freely hanging position W is balanced by the spring force kc. Putting $W = kc$ in the above equation, we obtain $s = 2c$.

Of course, the tension in the spring for the impact loading is double the weight.

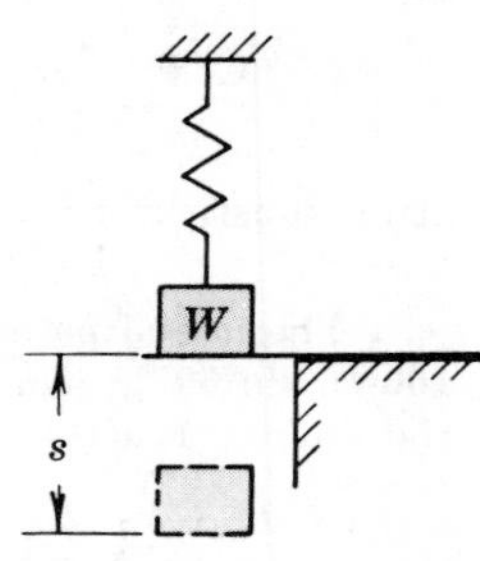

Fig. 18-24

30. In the spring gun shown in Fig. 18-25 below, ball W rests against the compressed spring of constant k. Its initial compression is x_0. What will be the speed of the ball leaving the gun? Assume the spring is unstressed when the bearing plate is at the end of the gun.

Solution:

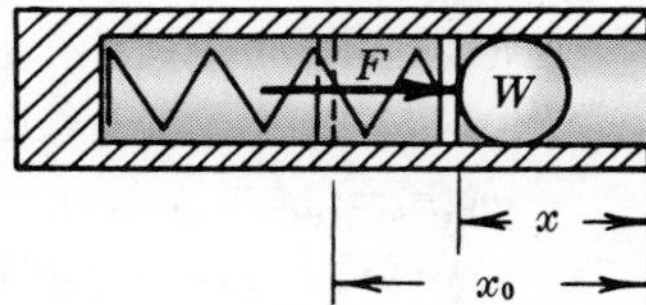

Fig. 18-25

The ball is shown in any position x from the end of the gun. The force doing work is the spring force $F = kx$.

The work done $= \int_0^{x_0} kx\,dx = \frac{1}{2}kx_0^2$. This is equal to the gain in kinetic energy of the ball which is $\frac{1}{2}(W/g)v^2$. Hence $v = x_0\sqrt{kg/W}$.

31. Two balls are connected by a spring whose unstressed length is 18 inches and whose modulus is $\frac{1}{4}$ lb/in. The balls are pushed together (compressing the spring) until they are 6 inches apart. They are then released on a smooth horizontal table. What work is done on the balls in returning them to their original distance apart?

Solution:

The work done on the balls by the varying spring force is equal to the work done in compressing the spring from 18 in. to 6 in., i.e., $U = \int_0^{12} kx\,dx = \frac{1}{2}kx^2\Big]_0^{12} = 18$ in-lb.

32. A rope is wrapped around a 60 lb solid homogeneous cylinder as shown in Fig. 18-26. Find the speed of its center G after it has dropped 6.0 ft from rest.

Solution:

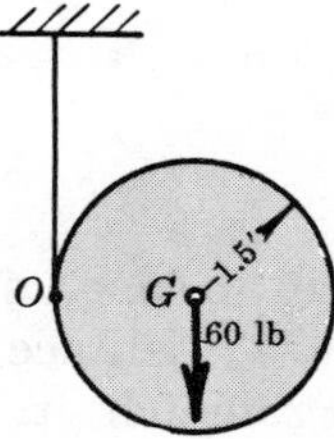

Fig. 18-26

The only force doing work is the weight, 60 lb.

Initial kinetic energy $T_1 = 0$. Final kinetic energy $T_2 = \frac{1}{2}m\bar{v}^2 + \frac{1}{2}\bar{I}\omega^2 = \frac{1}{2}m\bar{v}^2 + \frac{1}{2}(\frac{1}{2}mr^2)(\bar{v}^2/r^2) = \frac{3}{4}m\bar{v}^2 = 1.40\bar{v}^2$.

Then $U = T_2 - T_1$, $60(6.0) = 1.40\bar{v}^2$, and $\bar{v} = 16.0$ ft/sec.

33. Block A initially rests on the spring to which it is connected by a 2 ft inextensible cord which becomes taut after the system is released. What will be the stretch of the spring to bring the system to rest? The cylinder may be considered homogeneous, weighs 161 lb, and rotates in frictionless bearings. Refer to Fig. 18-27.

Solution:

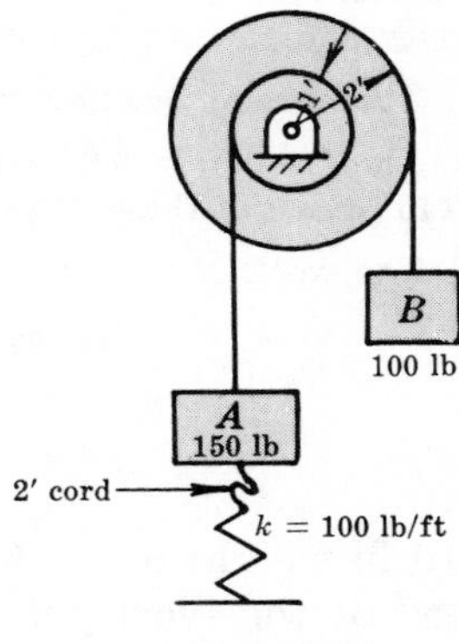

Fig. 18-27

First determine the kinetic energy of the system before the spring comes into play, i.e., while the weight A rises 2.0 ft. The work done $= 100s_B - 150s_A$.

The weight B drops 4.0 ft while A rises 2.0 ft. Hence the kinetic energy when the spring action starts is the work done, or $T_1 = 100(4.0) - 150(2.0) = 100$ ft-lb. The final kinetic energy of the system is $T_2 = 0$.

The work done on the system (A and B) by gravity and by the spring as it stretches a distance x ft is

$$U = +100(2x) - 150(x) - \frac{1}{2}kx^2 = 50x - 50x^2$$

Since $U = T_2 - T_1$, then $50x - 50x^2 = 0 - 100$, $x^2 - x - 2 = 0$, and $x = 2.0$ ft.

34. What is the value of B to cause the cylinder having $\bar{I} = 120$ slug-ft² to attain an angular speed of 2 rad/sec after rotating counterclockwise 10 radians from rest? See Fig. 18-28.

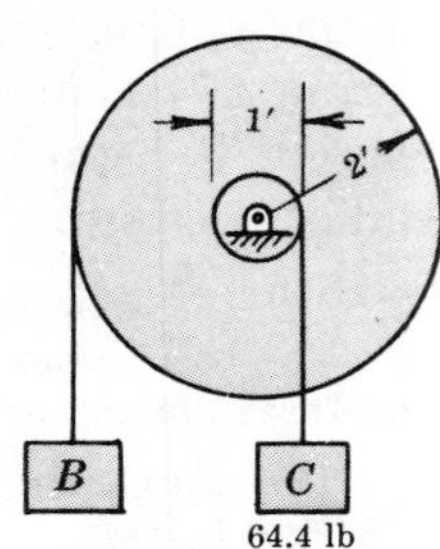

Fig. 18-28

Solution:

The work done in the counterclockwise direction is

$$U \;=\; [W_B(2) - 64.4(\tfrac{1}{2})]\theta \;=\; 20W_B - 322$$

The system has initial kinetic energy $T_1 = 0$ and final kinetic energy $T_2 = \frac{1}{2}m_B v_B^2 + \frac{1}{2}\bar{I}\omega^2 + \frac{1}{2}m_C v_C^2$.

But $v_B = 2\omega$, $v_C = \frac{1}{2}\omega$, and $\omega = 2$ rad/sec; hence

$$T_2 \;=\; \tfrac{1}{2}(W_B/32.2)(4)^2 + \tfrac{1}{2}(120)(2)^2 + \tfrac{1}{2}(2)(1)^2 \;=\; 0.248W_B + 241$$

Then $U = T_2 - T_1$, $20W_B - 322 = 0.248W_B + 241$, and $W_B = 28.5$ lb.

35. Block A weighs 96.6 lb and block B weighs 128.8 lb. The drum has a moment of inertia $\bar{I} = 12$ slug-ft². Through what distance will A fall before it reaches a speed of 6 ft/sec? See Fig. 18-29.

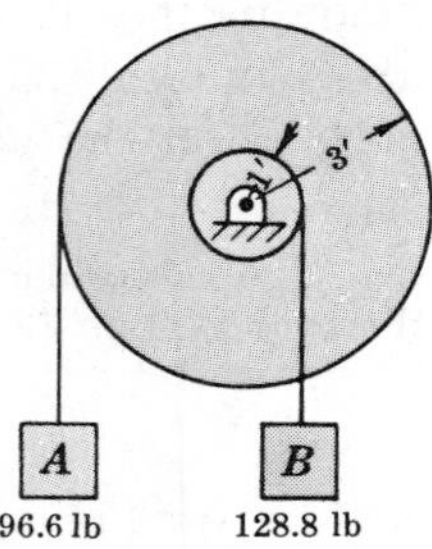

Fig. 18-29

Solution:

Work done $= 96.6s_A - 128.8s_B = 53.7s_A$, since $s_B = \frac{1}{3}s_A$.

The system has initial kinetic energy $T_1 = 0$ and final kinetic energy $T_2 = \frac{1}{2}m_A v_A^2 + \frac{1}{2}\bar{I}\omega^2 + \frac{1}{2}m_B v_B^2 = 86$ ft-lb, since $v_A = 6$ ft/sec, $v_B = 2$ ft/sec, and $\omega = v_A/r = 6/3 = 2$ rad/sec.

Then $53.7s_A = 86$ or $s = 1.60$ ft.

36. The 15 lb block shown in Fig. 18-30 is released from rest and slides a distance s down the inclined plane and strikes the spring which it compresses 3 inches before motion impends up the plane. Assuming the coefficient of friction is 0.25 and the spring constant $k = 16$ lb/in, determine the value of s.

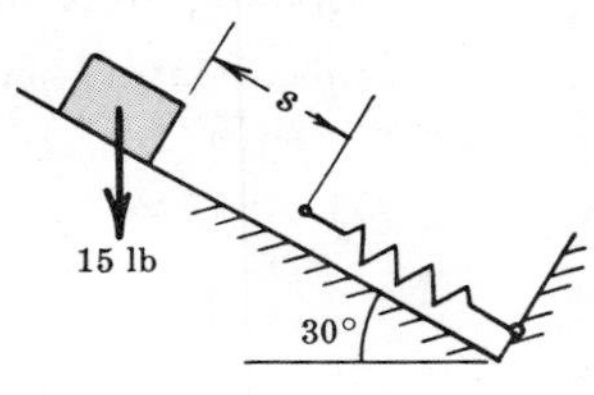

Fig. 18-30

Solution:

The initial kinetic energy and the final kinetic energy (when the block has moved $s + 3$ inches) are zero. Hence the work done by friction, gravity and the spring must be zero.

The normal reaction $= 15 \cos 30° = 13.0$ lb does no work.

The friction $= 0.25 \times 13 = 3.25$ lb. The component of the weight along the plane $= 15 \sin 30° = 7.50$ lb. Each of these forces does work for $s + 3$ inches; frictional work is negative, the other is positive.

The work of the spring is negative and equals $\frac{1}{2}k(3^2)$ or 72 in-lb.

$$U \;=\; (7.50 - 3.25)(s + 3) - 72 \;=\; 0 \qquad \text{and} \qquad s = 13.9 \text{ inches}$$

37. A 10 lb weight drops 6 ft upon a spring whose modulus is 60 lb/in. What will be the speed of the block when the spring is deformed 2 inches?

Solution:

The weight drops $(6 + 2/12)$ ft $= 6.17$ ft. The work done by gravity is 61.7 ft-lb. The work done by the spring on the weight is negative and equals $\frac{1}{2}kx^2 = \frac{1}{2}(60)(2^2)/12 = 10$ ft-lb.

The kinetic energy of the block increases from zero to $\frac{1}{2}mv^2 = \frac{1}{2}(10/32.2)v^2 = 0.155v^2$.

$$U = \text{change of kinetic energy}, \qquad 61.7 - 10 = 0.155v^2, \qquad v = 18.3 \text{ ft/sec}$$

38. A weight dropped from rest through 6 ft on a spring whose modulus is 20 lb/in causes a maximum shortening of 8 in. in the spring. What is the value of the weight?

Solution:

Work done by gravity $= W(72+8) = 80W$ in-lb.

Work done by the spring $= -\frac{1}{2}kx^2 = -\frac{1}{2}(20)(8^2) = -640$ in-lb.

Since the block starts from rest and ends at rest, the change in kinetic energy is zero. Then

$$U = 0, \qquad 80W - 640 = 0, \qquad W = 8 \text{ lb}$$

39. Determine the speed of escape, i.e., the initial speed, which must be given to a particle on the earth's surface to project it to an infinite height.

Solution:

The particle of weight W is shown in Fig. 18-31 at a distance x from the center of the earth of radius R. The earth's attraction F is known to be inversely proportional to the square of the distance x, i.e., $F = -C/x^2$.

To determine C note that the attraction on the earth's surface is the weight W. Thus $-W = -C/R^2$, $C = WR^2$, and hence $F = -WR^2/x^2$.

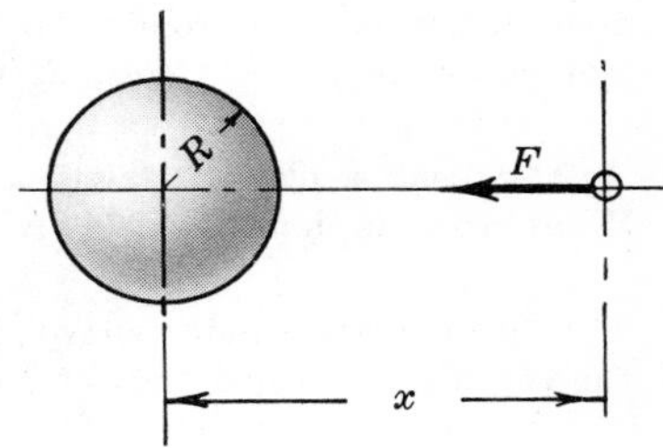

Fig. 18-31

The work done in going from $x = R$ to $x = \infty$ is

$$\int_R^\infty F\,dx = \int_R^\infty -(WR^2/x^2)\,dx = WR^2[1/x]_R^\infty = -WR$$

This work done equals the change in kinetic energy. $T_1 = \frac{1}{2}(W/g)v_0^2$ and $T_2 = 0$ (since $v = 0$ when x becomes infinite). Hence

$$-WR = -\tfrac{1}{2}(W/g)v_0^2 \qquad \text{or} \qquad v_0 = \sqrt{2gR}$$

Assuming the diameter of the earth to be 7900 miles, the required speed of escape is calculated to be 6.93 miles per sec.

Supplementary Problems

40. A cylindrical well is 6 ft in diameter and 40 ft deep. If there is 9 ft of water in the bottom of the well, determine the work done in pumping all of this water to the surface. *Ans.* 565,000 ft-lb

41. Referring to Problem 40, what work must be done by a pump that is 60% efficient?
Ans. 942,000 ft-lb

42. A man, who can push 60 lb, wishes to roll a barrel weighing 200 lb into a truck which is 3 ft above ground. How long a board must he use and how much work will he do in getting the barrel into the truck? *Ans.* 10 ft, 600 ft-lb

43. A 10 lb block slides 4 ft on a horizontal surface. (*a*) If the coefficient of friction is 0.3, what work is done by the block on the surface? (*b*) What work is done by the surface on the block?
Ans. (*a*) $U = 0$, (*b*) $U = 12$ ft-lb

44. The 18 lb block is acted upon by the 26 lb force as shown in the adjacent Fig. 18-32. If the coefficient of sliding friction is 0.30, determine the work done by all forces as the block moves 12 ft to the right. *Ans.* U = 197 ft-lb

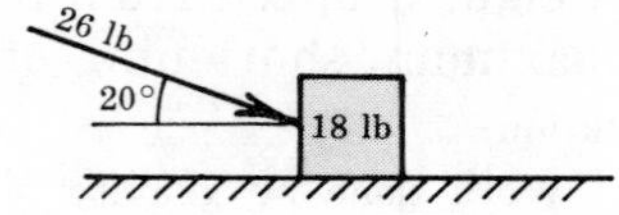

Fig. 18-32

45. A 10 lb block slides 6 ft down a plane inclined 40° with the horizontal. Determine the work done by all forces acting on the block. The coefficient of sliding friction is 0.40.
Ans. U = 20.2 ft-lb

46. A particle moves along the path $x = 2t$, $y = t^3$ where t is in sec and distances in ft. What work is done in the interval from $t = 0$ to $t = 3$ sec by a force whose components are $F_x = 2 + t$ and $F_y = 2t^2$? Forces are in lb. *Ans.* U = 313 ft-lb

47. A 2 oz bead is raised slowly along the frictionless wire from A to B as shown in Fig. 18-33 below. What work is done? *Ans.* U = 0.5 ft-lb

48. The 50 lb solid cylinder shown in Fig. 18-34 below is released from rest. Determine the work done by the earth's pull when the bottom hits the floor. *Ans.* U = 1.06 ft-lb

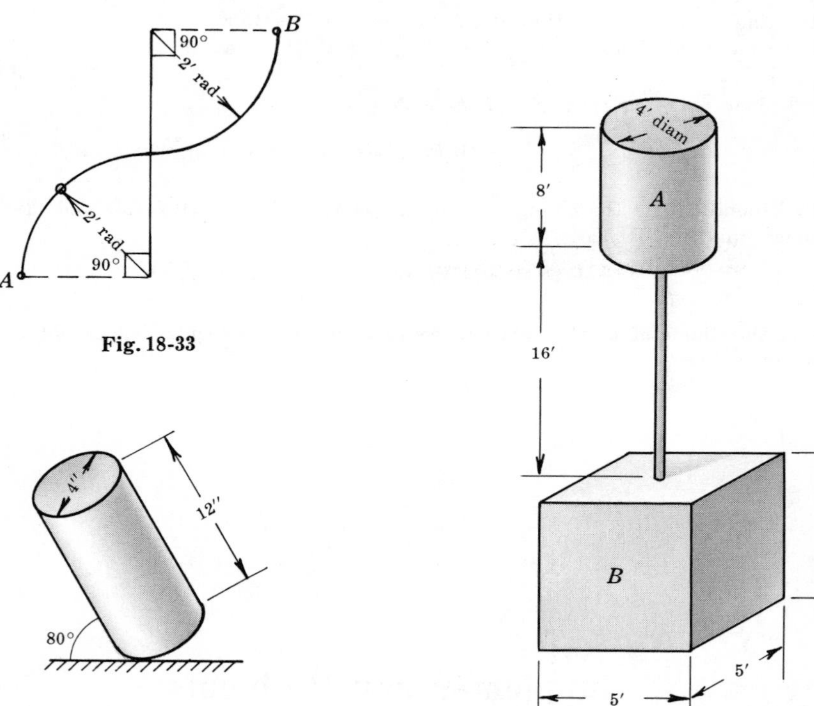

Fig. 18-33

Fig. 18-34

Fig. 18-35

49. The empty cylindrical tank A is filled with water from the cubical tank B. Assume water weighs 62.4 lb/ft³ and that B was filled at the beginning. How much work is done? Refer to Fig. 18-35 above.
Ans. U = 138,000 ft-lb

50. The 40 lb weight drops 8 ft. It is attached by a light rope to a drum which rotates in frictionless bearings. A constant torque M = 80 lb-ft is supplied to the drum. See Fig. 18-36. What work is done on the system?
Ans. $U = 0$

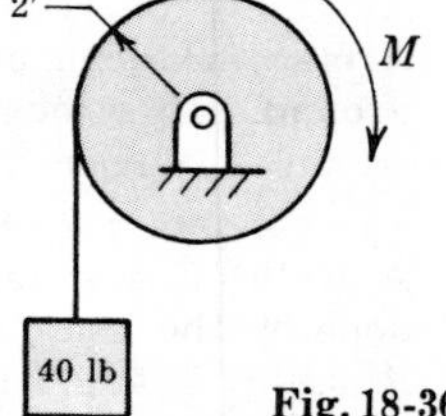

Fig. 18-36

51. The drum shown in Fig. 18-37 below rotates in frictionless bearings. What work is done on the system when the 60 lb weight falls 3 ft?
Ans. U = 60 ft-lb

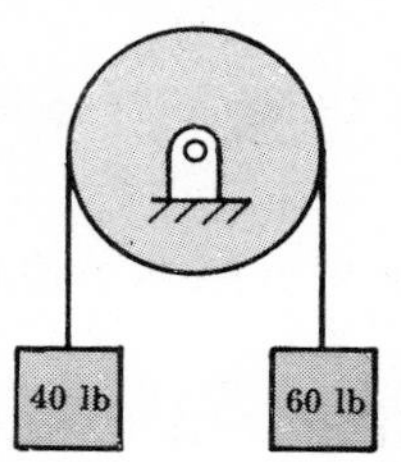

Fig. 18-37

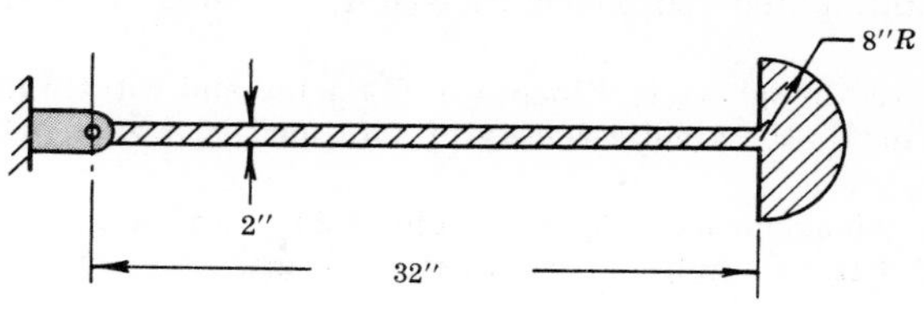

Fig. 18-38

52. The homogeneous object shown in Fig. 18-38 above is one inch thick and weighs 0.283 lb/in.³ It falls from a horizontal to a vertical position. What work is done on the object? *Ans.* $U = 108$ ft-lb

53. A couple $M = 2\theta^3 - \theta$ is applied to a shaft that rotates from $\theta = 0°$ to $\theta = 90°$. Determine the work done if M is in lb-ft. *Ans.* $U = 1.81$ ft-lb

54. A force of 7 lb will stretch an elastic cord 10 inches. If the force required to stretch the cord varies directly as the deformation, what is the work done in stretching the cord 60 inches? *Ans.* 1260 in-lb

55. A force of 200 lb is required to compress a spring through a distance of 4 inches. If the force required to compress the spring varies directly with its deformation, how much work is done in compressing it through 9 inches? *Ans.* 2025 in-lb

56. An automobile weighing 2500 lb climbs a 10% grade at a uniform rate of 15 mi/hr. If the resistance is 20 lb/ton, what hp is the car developing? *Ans.* 11.0 hp

57. A steam engine raises a 4000 lb weight vertically at the rate of 30 ft/sec. What is the hp of the engine assuming an efficiency of 70%? *Ans.* 312 hp

58. A homogeneous cylinder weighing 100 lb and 4 ft in diameter is rotating 100 rpm. A 20 lb-ft torque is needed to keep this speed constant (overcoming friction). What horsepower is required? *Ans.* hp = 0.38

59. A 6 inch diameter pulley is rotating 2000 rpm. The belt driving it has tensions of 1 and 3 lb in the slack and tight sides respectively. What hp is being delivered to the pulley? *Ans.* hp = 0.19

60. A body weighing 4 oz falls 50 ft to the surface of the earth. What is its kinetic energy as it hits the ground? *Ans.* $T = 12.5$ ft-lb

61. A 100 lb body starts from rest and is pulled along the ground by a horizontal force of 75 lb. If the kinetic coefficient of friction is 0.1 and the force acts for a period of 4 sec and then ceases to act, determine the distance required for the body to come to rest. *Ans.* 1090 ft

62. A bullet enters a 2 inch plank with a speed of 2000 ft/sec and leaves with a speed of 800 ft/sec. Determine the greatest thickness of plank that could be penetrated by the bullet. *Ans.* 2.38 in.

63. A 5 lb block slides down a plane inclined 50° with the horizontal. If the coefficient of friction between the block and the plane is 0.25, determine the speed of the block after it has moved 12 ft along the plane after starting with a velocity of 4 ft/sec. *Ans.* $v = 22.0$ ft/sec

64. A block weighing 80 lb is acted upon by a 30 lb horizontal force. If the coefficient of friction is 0.25, determine the speed of the block after it has moved 20 ft from rest. Refer to Fig. 18-39. *Ans.* $v = 12.7$ ft/sec to the right

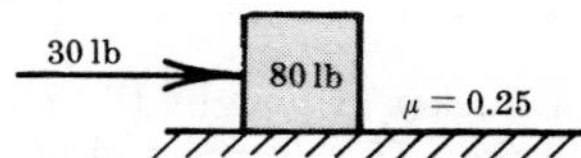

Fig. 18-39

65. Determine the kinetic energy possessed by a 200 lb disk which is 18 in. in diameter, 3 in. thick, and rotating 100 rpm about its center. *Ans.* 95.8 ft-lb

66. A 100 lb sphere is 6 inches in diameter and rotates at 120 rpm about an axis 16 inches from its center. What is the kinetic energy of rotation? *Ans.* 442 ft-lb

67. A homogeneous cylinder weighing 20 lb and with a radius 8 inches is moving with a mass center speed of 6 ft/sec. What is its kinetic energy? *Ans.* $T = 16.8$ ft-lb

68. A sphere weighing 9 lb has radius 4 inches and radius of gyration 2.53 inches. It is rolling on a horizontal plane with angular speed 3 rad/sec. What is the kinetic energy of the sphere? *Ans.* $T = 2.35$ in-lb

69. A solid cylindrical flywheel is 4 ft in diameter and weighs 2000 lb. If the axle is 6 inches in diameter and the coefficient of journal friction is 0.15, find the time required for the flywheel to coast to rest from a speed of 500 rpm. *Ans.* 86.5 sec

70. An electric motor has a rotor weighing 20 lb with radius of gyration $k = 1.83$ inches. A frictional torque of 8 oz-in is present. How many revolutions will the rotor make while coming to rest from a speed of 1800 rpm? *Ans.* $\theta = 979$ rev

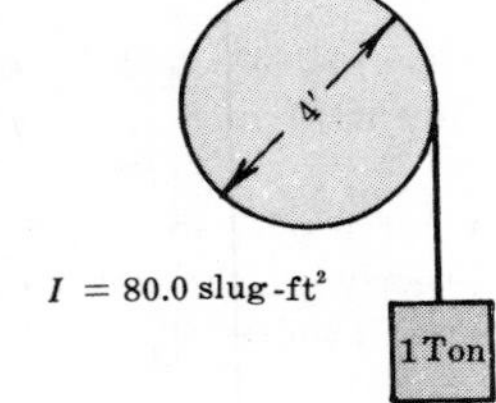

Fig. 18-40

71. The drum rotating 20 rpm is lifting a one ton cage connected to it by a light inextensible cable as shown in Fig. 18-40. If power is cut off, how high will the cage rise before coming to rest? Assume frictionless bearings. *Ans.* $h = 0.36$ ft

72. The car travels down the smooth incline and then moves inside the loop shown in Fig. 18-41. Determine the least value of h so that the car will remain in contact with the track. *Ans.* $h = 75$ ft

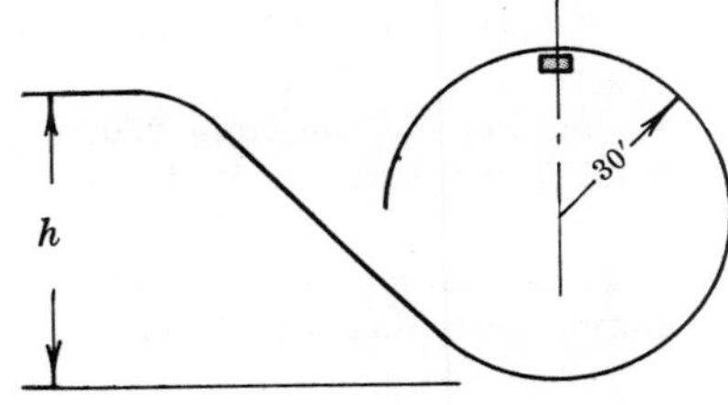

Fig. 18-41

73. A homogeneous slender bar 3 ft long weighs 8 lb. It pivots about one end. When released from rest in the horizontal position, it falls under the action of its weight and a constant retarding torque of 4 lb-ft. What will be its angular speed as it passes through its lowest position? *Ans.* $\omega = 3.92$ rad/sec

74. The drum has a weight of 60 lb and a radius of gyration $k = 2.75$ ft. See Fig. 18-42 below. Assume no friction and determine the speed of the drum after it has made one revolution starting from rest. *Ans.* $\omega = 3.29$ rad/sec

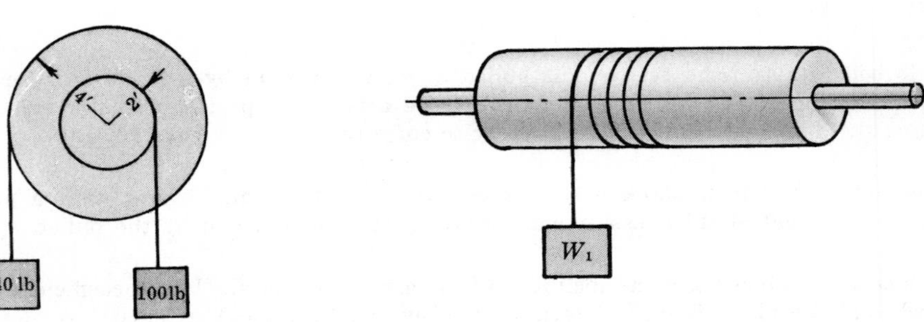

Fig. 18-42 Fig. 18-43

75. A bucket of weight W_1 is attached to a windlass of weight W_2 and radius of gyration k. See Fig. 18-43 above. If r is the radius of the windlass, how long will it take the bucket to drop s ft from rest to the water level? Neglect bearing friction and the weight of the rope. *Ans.* $t = \sqrt{2s(1 + W_2k^2/W_1r^2)/g}$

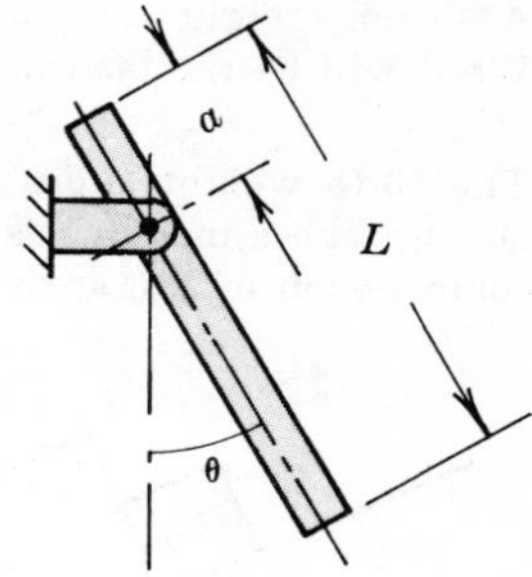

Fig. 18-44

76. A homogeneous bar of length L is pivoted about a point a distance a from one end as shown in Fig. 18-44. If the bar is released from rest in the 30° position, what will be the angular speed when the bar is vertical?

Ans. $\omega^2 = \dfrac{0.402g(L-2a)}{L^2 - 3La + 3a^2}$

77. A 200 lb sphere, 2 ft in diameter, rolls from rest down a 25° plane for a distance of 100 ft. What is its kinetic energy at the end of the 100 ft? *Ans.* 8450 ft-lb

78. In Problem 77, what is the speed of the center of the sphere after it has traveled the 100 ft? *Ans.* 44.2 ft/sec

79. A homogeneous sphere rolls a distance s ft down a plane inclined at angle θ with the horizontal. What is the speed of the sphere if it starts from rest? *Ans.* $v = 6.78\sqrt{s \sin\theta}$

80. A car has a body that weighs 2000 lb, four wheels each weighing 50 lb, and a driver weighing 150 lb. The wheels are 27 inches in diameter and have a radius of gyration equal to 10.2 inches. What will be the speed of the car if it moves from rest 1000 ft down a 5% grade? *Ans.* v = 37.6 mph

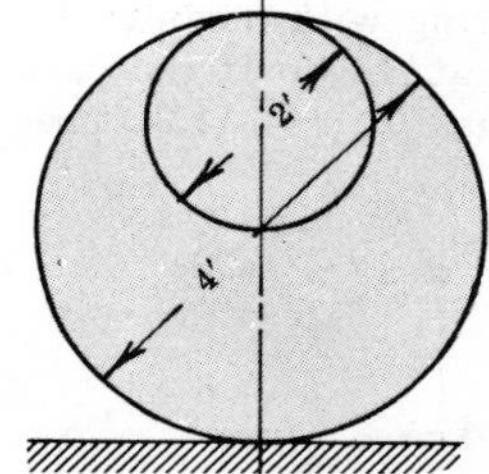

Fig. 18-45

81. The 8 lb homogeneous sphere shown in Fig. 18-45 has a string wrapped around a slot as shown. What will be the speed of the center if it falls 3 ft from the rest position? *Ans.* v = 11.8 ft/sec

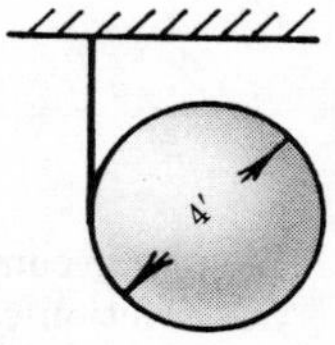

82. A 5 lb homogeneous disk 2 ft in diameter is attached rigidly to a 20 lb homogeneous disk 4 ft in diameter. If the assembly is released from rest as shown in Fig. 18-46, what will be the angular speed when the small disk is at the bottom of its travel?
Ans. ω = 2.25 rad/sec

Fig. 18-46

83. In Fig. 18-47 below, A weighs 14 lb and B 7 lb. If B falls 1.2 ft from rest, determine its speed (a) if no friction exists and (b) if friction between A and the horizontal plane is 0.20.
Ans. (a) v = 5.08 ft/sec, (b) v = 3.93 ft/sec

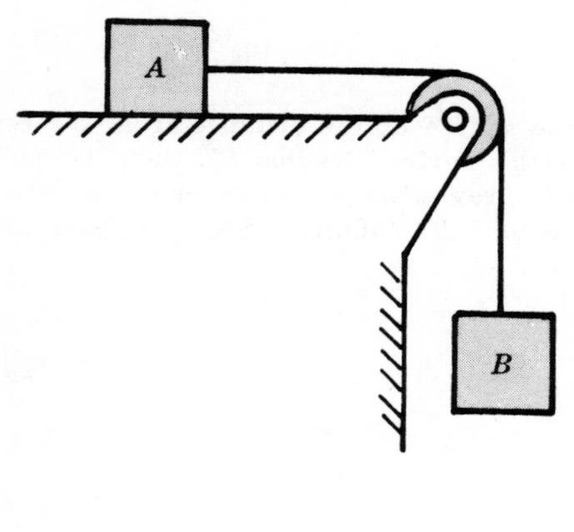

Fig. 18-47

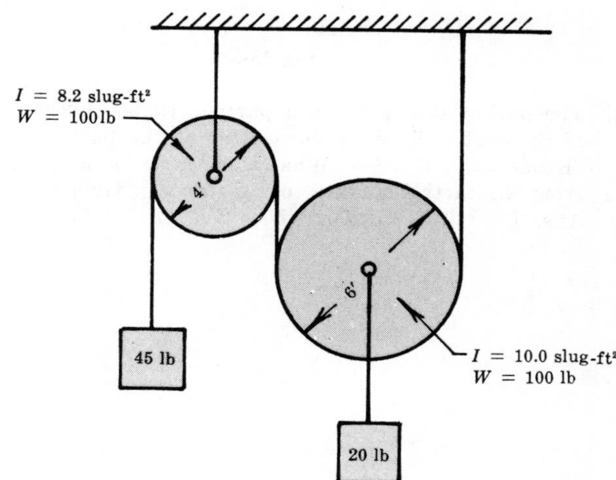

Fig. 18-48

84. In the system shown in Fig. 18-48 above, all ropes are vertical. The 45 lb weight rises 1.6 ft from rest. What will be its speed? *Ans.* v = 3.21 ft/sec

85. A 40 ton freight car moving 5 mph horizontally hits a bumper with a spring constant of 10,000 lb/in. What will be the maximum compression of the spring? *Ans.* $d = 1.05$ ft

86. The 16 lb weight slides 6 inches from rest down the 25° plane shown in Fig. 18-49 below. It hits a spring whose modulus is 10 lb/in. If the coefficient of kinetic friction is 0.20, determine the maximum compression of the spring. *Ans.* $d = 2.57$ in

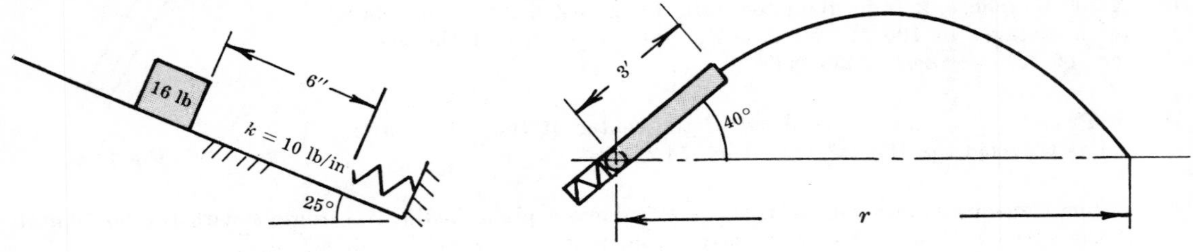

Fig. 18-49

Fig. 18-50

87. A spring compressed 3 inches and with a modulus $k = 30$ lb/in. is used to propel the 2 oz weight in the frictionless tube shown in Fig. 18-50 above. Determine the horizontal distance r at which the weight will be at the same height as initially. Neglect air resistance. *Ans.* $r = 178$ ft

88. The 2 lb weight W slides from rest at A along the frictionless rod bent into a quarter circle. The spring with modulus $k = 1.2$ lb/ft has an unstretched length of 18 inches. (*a*) Determine the speed of W at B. (*b*) If the path is ellipitcal, what is the speed at B? Refer to Fig. 18-51 below.
Ans. (*a*) $v = 11.3$ ft/sec, (*b*) $v = 9.49$ ft/sec

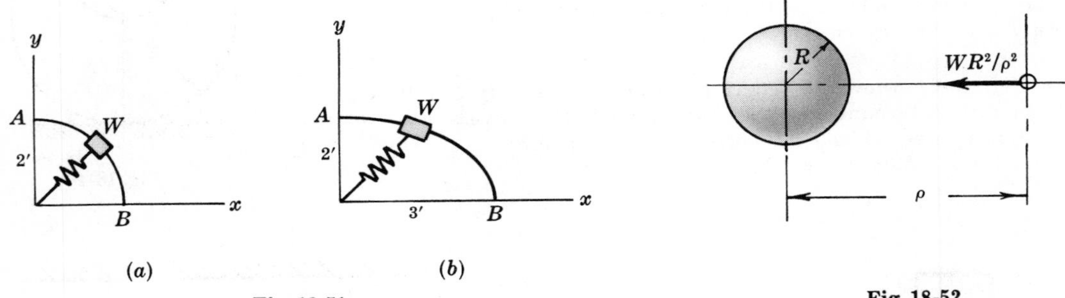

Fig. 18-51

Fig. 18-52

89. The pull of the earth on a particle varies inversely with the square of the distance from the center of the earth. If W is the weight of the particle on the earth's surface (radius R), then the pull at distance ρ is WR^2/ρ^2. What work must be done against this gravitational pull to move the particle from the earth's surface to a distance x from the earth's center? To infinity? See Fig. 18-52 above.
Ans. $U = WR - WR^2/x$, $U = WR$

Chapter 19

Impulse and Momentum

LINEAR MOMENTUM G

The linear momentum **G** *of a particle* was defined in Chapter 11 as

$$\mathbf{G} = m\mathbf{v} \tag{1}$$

where m = mass of particle, $\mathbf{v}$ = velocity of particle.

The vector sum of the external forces acting on a particle was equated in Chapter 11 to the time rate of change of the linear momentum **G**. Thus

$$\Sigma\mathbf{F} = \frac{d(m\mathbf{v})}{dt} = \frac{d\mathbf{G}}{dt} = \dot{\mathbf{G}} \tag{2}$$

which may be integrated as

$$\int_{t_1}^{t_2} \Sigma\mathbf{F}\,dt = \int_{\mathbf{G}_1}^{\mathbf{G}_2} d\mathbf{G} = \mathbf{G}_2 - \mathbf{G}_1 = m\mathbf{v}_2 - m\mathbf{v}_1 \tag{3}$$

The integral on the left side of (*3*) is the *linear impulse* acting in the time interval from t_1 to t_2. Thus this linear impulse is equal to the change in the linear momentum during this time interval.

The vector sum of the external forces acting on an assemblage of n particles equals the time rate of change of the linear momentum of a mass m equal to the sum of the masses of the n particles and with a velocity equal to that of the mass center of the n particles. Thus

$$\Sigma\mathbf{F} = \frac{d(m\bar{\mathbf{v}})}{dt} \tag{4}$$

where $\Sigma\mathbf{F}$ = sum of the external forces acting on the group of particles

$m = \sum_{i=1}^{n} m_i$ = mass of all n particles

$\bar{\mathbf{v}}$ = velocity of the mass center of the group of n particles.

In Chapter 9 the centroid of a group of particles (in this case the mass center) is located by

$$\bar{\mathbf{r}} = \frac{\sum_{i=1}^{n} m_i\mathbf{r}_i}{\sum_{i=1}^{n} m_i}$$

Using $m = \sum_{i=1}^{n} m_i$ and then taking a time derivative, we have

$$m\bar{\mathbf{r}} = \sum_{i=1}^{n} m_i\mathbf{r}_i \qquad \text{and} \qquad m\frac{d\bar{\mathbf{r}}}{dt} = \sum_{i=1}^{n} \frac{d}{dt}(m_i\mathbf{r}_i) \qquad \text{or} \qquad m\bar{\mathbf{v}} = \sum_{i=1}^{n} m_i\mathbf{v}_i$$

Then
$$\frac{d(m\bar{\mathbf{v}})}{dt} = \sum_{i=1}^{n} \frac{d}{dt}(m_i\mathbf{v}_i) = \Sigma\mathbf{F}$$

MOMENT of MOMENTUM $\mathbf{H}_O$

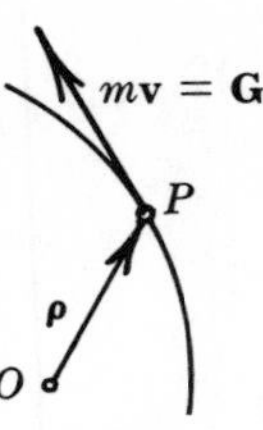

Fig. 19-1

The moment of momentum $\mathbf{H}_O$ (also called angular momentum) is the moment about any point O of the linear momentum vector $\mathbf{G}$. In Fig. 19-1, O can be any point, fixed or moving. Thus

$$\mathbf{H}_O = \boldsymbol{\rho} \times \mathbf{G} = \boldsymbol{\rho} \times (m\mathbf{v}) \tag{5}$$

where $\boldsymbol{\rho}$ = radius vector of particle P relative to O,
$\mathbf{v}$ = absolute velocity of P (tangent to the path).

The sum of the moments about a *fixed* point O of the external forces acting on a *particle* is equal to the time rate of change of the moment of momentum $\mathbf{H}_O$, i.e.,

$$\Sigma\mathbf{M}_O = \frac{d\mathbf{H}_O}{dt} = \dot{\mathbf{H}}_O \tag{6}$$

Proof: From equation (5), $\mathbf{H}_o = \mathbf{r} \times (m\mathbf{v})$, where $\mathbf{r}$ = radius vector in a Newtonian frame of particle m, and $\mathbf{v}$ = absolute velocity of the particle. Taking a time derivative,

$$d\mathbf{H}_O/dt = \dot{\mathbf{r}} \times (m\mathbf{v}) + \mathbf{r} \times (m\dot{\mathbf{v}})$$

Since $\dot{\mathbf{r}} = \mathbf{v}$, $\mathbf{v} \times \mathbf{v} = 0$, $m\dot{\mathbf{v}} = m\mathbf{a} = \Sigma\mathbf{F}$, and $\mathbf{r} \times (m\mathbf{a}) = \mathbf{r} \times (\Sigma\mathbf{F}) = \Sigma\mathbf{M}_O$, we obtain $d\mathbf{H}_O/dt = \Sigma\mathbf{M}_O$.

Equation (6) may be integrated as follows:

$$\int_{t_1}^{t_2} \Sigma\mathbf{M}_O\, dt = \int_{\mathbf{H}_1}^{\mathbf{H}_2} d\mathbf{H}_O = \mathbf{H}_2 - \mathbf{H}_1 = \mathbf{r} \times (m\mathbf{v}_2 - m\mathbf{v}_1) \tag{7}$$

The integral on the left is the *angular impulse* acting throughout the time interval t_1 to t_2, and the right side of (7) is the change that occurs in angular momentum during this time interval.

Equation (6) can be applied as follows to an assemblage of particles. The sum of the moments about point O of the external forces acting on an *assemblage* of n particles is equal to the time rate of change of the moment of momentum $\mathbf{H}_O$ about this point O, only if (*a*) point O is at rest or (*b*) the mass center of the n particles is at rest or (*c*) the velocities of O and the mass center are parallel (certainly true if O is the mass center). See Problem 1 which proves equation (6) holds for an assemblage of particles.

MOMENT of RELATIVE MOMENTUM $\mathbf{H}'_O$

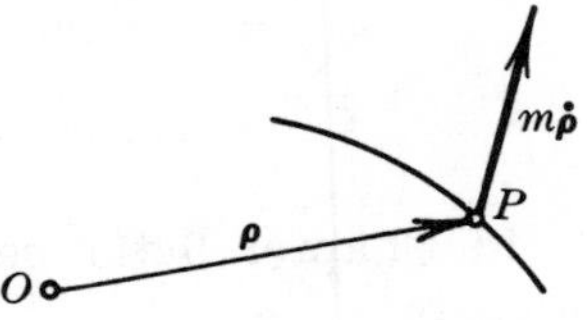

Fig. 19-2

The moment of relative momentum $\mathbf{H}'_O$ is the moment about any moving point O of the product of the mass of the particle and the time rate of change of the radius vector $\boldsymbol{\rho}$ of the particle relative to O (see Fig. 19-2). Thus

$$\mathbf{H}'_O = \boldsymbol{\rho} \times (m\dot{\boldsymbol{\rho}}) \tag{8}$$

where $\boldsymbol{\rho}$ = radius vector of particle P relative to O,
$\dot{\boldsymbol{\rho}}$ = time rate of change of $\boldsymbol{\rho}$.

The sum of the moments about O of the external forces acting on an assemblage of n particles is equal to the time rate of change of the moment of relative momentum $\mathbf{H}'_O$ about this point O, i.e.,

$$\Sigma\mathbf{M}_O = \frac{d\mathbf{H}'_O}{dt} \tag{9}$$

only if (*a*) O is the mass center of the n particles or (*b*) O is a point of constant velocity (or at rest) or (*c*) O is a point with an acceleration vector that passes through the mass center. See Problems 2 and 3.

CORRESPONDING SCALAR EQUATIONS

For a body in translation (all particles have the same velocity), equation (*3*) can be replaced by the scalar equations

$$\Sigma(\text{Imp})_x = \Delta G_x = m(v_x'' - v_x') \tag{10}$$

$$\Sigma(\text{Imp})_y = \Delta G_y = m(v_y'' - v_y') \tag{11}$$

where $\Sigma(\text{Imp})_x$, $\Sigma(\text{Imp})_y$ = linear impulses of external forces in x and y directions

m = mass of body

v_x'', v_y'' = final velocities of body in x and y directions

v_x', v_y' = initial velocities of body in x and y directions.

In these equations, if mass is in slugs and velocity in ft/sec, then linear impulse is in lb-sec.

For a body in rotation about a fixed axis, the foregoing equations become

$$\Sigma(\text{Ang Imp})_O = \Delta H_O = I_O(\omega'' - \omega') \tag{12}$$

where $\Sigma(\text{Ang Imp})_O$ = angular impulse of external forces about the axis of rotation through O

I_O = moment of inertia of body about axis of rotation

ω'' = final angular velocity of body

ω' = initial angular velocity of body.

In these equations, if moment of inertia is in slug-ft^2 and angular velocity in rad/sec, then angular impulse is in lb-sec-ft. (For a proof, see Problem 4.)

For a body in plane motion, the foregoing equations become

$$\Sigma(\text{Imp})_x = \Delta G_x = m(\bar{v}_x'' - \bar{v}_x') \tag{13}$$

$$\Sigma(\text{Imp})_y = \Delta G_y = m(\bar{v}_y'' - \bar{v}_y') \tag{14}$$

$$\Sigma(\text{Ang Imp})_G = \Delta \bar{H} = \bar{I}(\omega'' - \omega') \tag{15}$$

where $\Sigma(\text{Imp})_x$, $\Sigma(\text{Imp})_y$ = linear impulses of external forces in x and y directions

m = mass of body

$\bar{v}_x''$, $\bar{v}_y''$ = final velocities of mass center in x and y directions

$\bar{v}_x'$, $\bar{v}_y'$ = initial velocities of mass center in x and y directions

$\Sigma(\text{Ang Imp})_G$ = angular impulse of external forces about the axis through the mass center G

$\bar{I}$ = moment of inertia of body about mass center G

ω'' = final angular velocity of the body

ω' = initial angular velocity of the body.

In these equations, if mass is in slugs and velocity in ft/sec, then linear impulse is in lb-sec. Likewise, if moment of inertia is in slug-ft^2 and angular velocity in rad/sec, then angular impulse is in lb-sec-ft. (For a proof, see Problem 5.)

CONSERVATION of LINEAR MOMENTUM

Conservation of linear momentum in a given direction occurs if the sum of the external forces in that direction is zero. This follows because there is then no linear impulse in that direction, and hence no change in linear momentum can occur.

CONSERVATION of ANGULAR MOMENTUM

Conservation of angular momentum about an axis occurs if the sum of the moments of the external forces about that axis is zero. This follows because there is then no angular impulse about that axis, and hence no change in angular momentum can occur.

IMPACT

Impact covers the cases where the intervals of time during which the forces act are quite small and usually indeterminate. The time is eliminated from the equations of impulse and momentum.

For *direct central impact* of two bodies the *coefficient of restitution* is the ratio of the relative velocity of separation of the two bodies to their relative velocity of approach.

$$e = \frac{v_2 - v_1}{u_1 - u_2} = -\frac{v_2 - v_1}{u_2 - u_1}$$

where e = coefficient of restitution

u_1, u_2 = velocities of bodies 1 and 2 respectively before impact; $u_1 > u_2$ for collision to occur if both are moving in same direction

v_1, v_2 = velocities of bodies 1 and 2 respectively after impact.

Note: When the velocity is not normal to the striking surface, the normal component of the velocity is used in the above formula.

Since during impact the same force acts on each body (equal and opposite reaction), the sum of the momenta before impact must equal the sum of the momenta after impact. Mathematically this is written

$$m_1 u_1 + m_2 u_2 = m_1 v_1 + m_2 v_2$$

VARIABLE MASS

Suppose at time t a mass m is moving along a straight line with an absolute speed v. Further suppose a mass dm immediately in front of mass m is moving along the same straight line with absolute speed u. If the mass m absorbs the mass dm in a time interval dt, then the combined mass $(m + dm)$ will move with speed $(v + dv)$.

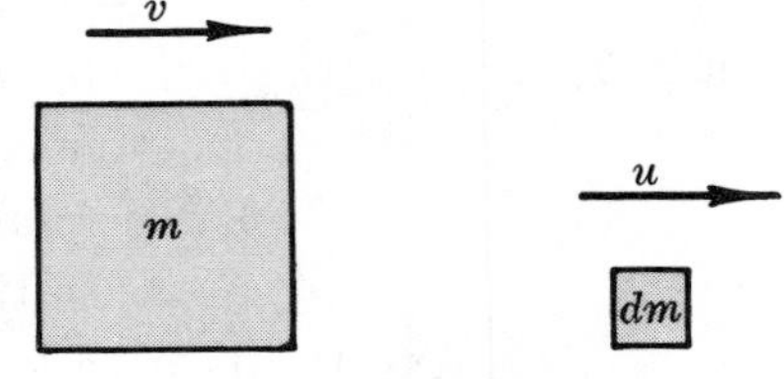

Fig. 19-3

The momentum of the system at time t is $(mv + dm\,u)$. The momentum of the system at time $(t + dt)$ is $(m + dm)(v + dv)$. The change in momentum is then

$$\begin{aligned} dG &= (m + dm)(v + dv) - (mv + dm\,u) \\ &= mv + m\,dv + dm\,v + dm\,dv - mv - dm\,u \end{aligned}$$

Since the magnitude of $(dm\,dv)$ is of second order, we will drop the term. Dividing by dt,

$$\frac{dG}{dt} = m\frac{dv}{dt} + \frac{dm}{dt}(v - u)$$

If mass is released (decrease in mass) then dm/dt will be negative.

The above formula was shown for straight line motion but is of more general nature.

Since the sum of the forces acting equals the time rate of change of momentum,

$$\Sigma F = \frac{dG}{dt} = m\frac{dv}{dt} + \frac{dm}{dt}(v - u) \qquad (16)$$

Solved Problems

1. Given an assemblage of n particles of masses $m_1, m_2, m_3, \ldots, m_n$, show that the sum of the moments about point O of the external forces equals the time rate of change of the moment of momentum of the group of particles about this point O, only if (a) point O is at rest or (b) the mass center of the n particles is at rest or (c) the velocities of O and the mass center are parallel (certainly true if O is the mass center).

Solution:

Fig. 19-4 shows the ith particle of the group with mass m_i. The given point O has a position vector $\mathbf{r}_O$ relative to a Newtonian frame of reference: O' is fixed. The position vector of P, relative to the fixed frame, is $\mathbf{r}_P$. Let $\mathbf{H}_O$ be the angular momentum relative to O of all the particles of which m_i is representative. Thus

$$\mathbf{H}_O = \sum_{i=1}^{n} \boldsymbol{\rho}_i \times (m_i \mathbf{v}_i) \qquad (a)$$

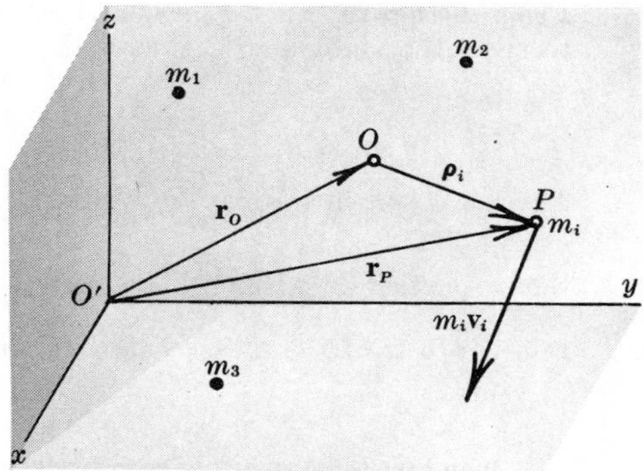

Fig. 19-4

Taking a time derivative of equation (a),

$$\dot{\mathbf{H}}_O = \sum_{i=1}^{n} \dot{\boldsymbol{\rho}}_i \times (m_i \mathbf{v}_i) + \sum_{i=1}^{n} \boldsymbol{\rho}_i \times \frac{d}{dt}(m_i \mathbf{v}_i) \qquad (b)$$

From the figure, $\mathbf{r}_P = \mathbf{r}_O + \boldsymbol{\rho}_i$ and hence $\dot{\mathbf{r}}_P = \dot{\mathbf{r}}_O + \dot{\boldsymbol{\rho}}_i$. Substituting for $\dot{\boldsymbol{\rho}}_i$ into (b),

$$\dot{\mathbf{H}}_O = \sum_{i=1}^{n} (\dot{\mathbf{r}}_P - \dot{\mathbf{r}}_O) \times (m_i \mathbf{v}_i) + \sum_{i=1}^{n} \boldsymbol{\rho}_i \times (m_i \dot{\mathbf{v}}_i) \qquad (c)$$

Expand the first term on the right of (c) into

$$\sum_{i=1}^{n} \dot{\mathbf{r}}_P \times (m_i \mathbf{v}_i) - \sum_{i=1}^{n} \dot{\mathbf{r}}_O \times (m_i \mathbf{v}_i) \qquad (d)$$

Since $\dot{\mathbf{r}}_P$ is the absolute velocity $\mathbf{v}_i$ of P, the first term of (d) is zero. Also, since $\dot{\mathbf{r}}_O$ does not change during the summation and is therefore independent of i, the second term of (d) can be written

$$\dot{\mathbf{r}}_O \times \sum_{i=1}^{n} (m_i \mathbf{v}_i) \qquad \text{or} \qquad \dot{\mathbf{r}}_O \times (m\bar{\mathbf{v}})$$

where $\bar{\mathbf{v}}$ is the velocity of the mass center.

The last term of equation (c) is equal to $\Sigma \mathbf{M}_O$, the sum of the moments of the external forces on all the particles. Thus (c) may now be written as

$$\dot{\mathbf{H}}_O = -\dot{\mathbf{r}}_O \times (m\bar{\mathbf{v}}) + \Sigma \mathbf{M}_O \qquad (e)$$

Equation (e) indicates that the time rate of change of the moment of momentum about O equals $\Sigma \mathbf{M}_O$ only if $-\dot{\mathbf{r}}_O \times (m\bar{\mathbf{v}})$ is zero. This occurs when: (1) O is fixed, i.e. $\dot{\mathbf{r}}_O = 0$; (2) $\bar{\mathbf{v}} = 0$; (3) $\dot{\mathbf{r}}_O$ and $\bar{\mathbf{v}}$ are parallel (cross product of parallel vectors is zero). If O is the mass center, then $\dot{\mathbf{r}}_O = \bar{\mathbf{v}}$ and $\dot{\mathbf{r}}_O \times m\bar{\mathbf{v}} = 0$.

2. Given an assemblage of n particles of masses $m_1, m_2, m_3, \ldots, m_n$, show that the sum of the moments about O of the external forces equals the time rate of change of the moment of *relative* momentum of the group of particles about this point O, only if (a) point O is the mass center of the n particles or (b) point O has constant velocity (or is at rest) or (c) point O has an acceleration vector that passes through the mass center.

Solution:

Fig. 19-5 shows the ith particle of the group with mass m_i. The given point O has a position vector $\mathbf{r}_O$ relative to a

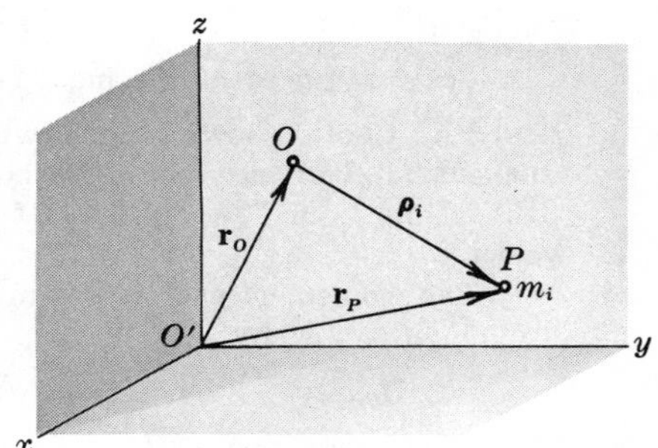

Fig. 19-5

Newtonian (inertial) frame of reference, i.e. O' is fixed. The position vector of P relative to the fixed frame is $\mathbf{r}_P$. Let $\mathbf{H}'_O$ be the moment of relative momentum with respect to O of all the particles of which m_i is representative. Then from equation (*8*),

$$\mathbf{H}'_O \quad = \quad \sum_{i=1}^{n} \boldsymbol{\rho}_i \times (m_i\dot{\boldsymbol{\rho}}_i) \tag{a}$$

Taking a time derivative of equation (*a*),

$$\dot{\mathbf{H}}'_O \quad = \quad \sum_{i=1}^{n} \dot{\boldsymbol{\rho}}_i \times (m_i\dot{\boldsymbol{\rho}}_i) \;+\; \sum_{i=1}^{n} \boldsymbol{\rho}_i \times (m_i\ddot{\boldsymbol{\rho}}_i) \tag{b}$$

From the figure, $\mathbf{r}_P = \mathbf{r}_O + \boldsymbol{\rho}_i$ and hence $\ddot{\mathbf{r}}_P = \ddot{\mathbf{r}}_O + \ddot{\boldsymbol{\rho}}_i$. Substituting for $\ddot{\boldsymbol{\rho}}_i$ in (*b*) and noting that the first term on the right of (*b*) is zero ($\dot{\boldsymbol{\rho}}_i \times \dot{\boldsymbol{\rho}}_i = 0$), we have

$$\dot{\mathbf{H}}'_O \quad = \quad \sum_{i=1}^{n} \boldsymbol{\rho}_i \times (m_i\ddot{\mathbf{r}}_P) \;-\; \sum_{i=1}^{n} \boldsymbol{\rho}_i \times (m_i\ddot{\mathbf{r}}_O) \tag{c}$$

The last term in (*c*) may be written $\left(\sum_{i=1}^{n} m_i\boldsymbol{\rho}_i\right) \times \ddot{\mathbf{r}}_O$, since $\ddot{\mathbf{r}}_O$ does not change as the sum is taken over the n particles. Also, $\sum_{i=1}^{n} \boldsymbol{\rho}_i \times (m_i\ddot{\mathbf{r}}_P) = \Sigma\mathbf{M}_O$ and $\sum_{i=1}^{n} m_i\boldsymbol{\rho}_i = m\bar{\boldsymbol{\rho}}$ where $\bar{\boldsymbol{\rho}}$ is the position vector relative to O of the mass center. Then

$$\dot{\mathbf{H}}'_O \quad = \quad \Sigma\mathbf{M}_O \;-\; m\bar{\boldsymbol{\rho}} \times \ddot{\mathbf{r}}_O \tag{d}$$

The last term in (*d*) is zero if: (1) O is the mass center ($\bar{\boldsymbol{\rho}} = 0$); (2) O has constant velocity ($\ddot{\mathbf{r}}_O = 0$); (3) O has an acceleration $\ddot{\mathbf{r}}_O$ passing through the mass center, i.e. along $\bar{\boldsymbol{\rho}}$ (the cross product of parallel vectors is zero).

3. Four equal masses m are spaced at the quarter points of a thin massless rim of radius R. Show that the moment of momentum relative to the mass center O using absolute velocities is the same as that obtained by using the relative velocities of the masses to the mass center O.

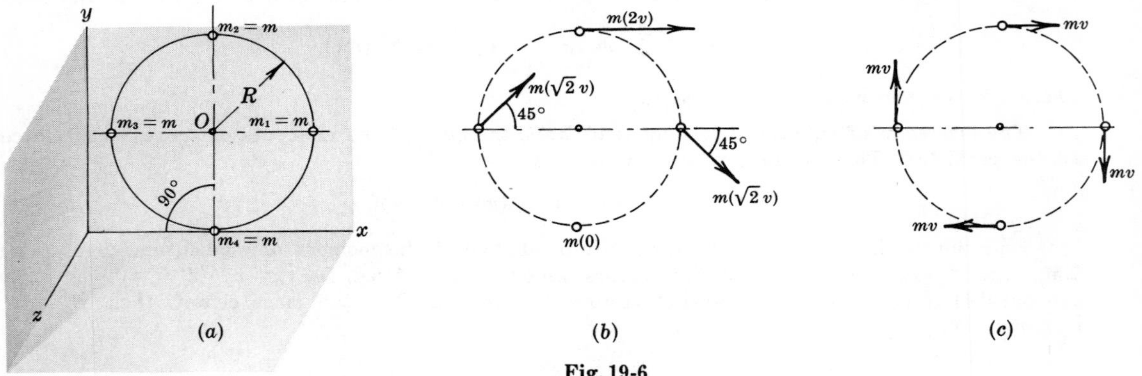

Fig. 19-6

Solution:

Let v = speed of the mass center as the rim rolls to the right.

Fig. 19-6(*a*) shows the rim with the masses m_2 and m_4 in a vertical line. In Fig. 19-6(*b*) the linear momentum $\mathbf{G}$ of each mass is shown using the absolute velocities of each mass. Thus v_1 is $\sqrt{2}\,v$ at 315°, while v_3 is $\sqrt{2}\,v$ at 45°. Of course, v_2 is $2v$ and v_4 is that of the instant center with no absolute velocity.

The moment of each linear momentum is in the negative z direction. Thus summing in 1, 2, 3 and 4 order, we can write

$$\mathbf{H}_O \quad = \quad -\left[m_1(\sqrt{2}\,v)(\tfrac{1}{2}\sqrt{2}\,R) \;+\; m_2(2v)R \;+\; m_3(\sqrt{2}\,v)(\tfrac{1}{2}\sqrt{2}\,R) \;+\; 0\right]\mathbf{k} \quad = \quad -(4mvR)\mathbf{k}$$

Fig. 19-6(*c*) illustrates the velocity of each mass relative to the center multiplied by the mass. The moments about O of these relative momentum vectors are

$$\mathbf{H}_O \quad = \quad -(4mvR)\mathbf{k} \qquad \text{as before}$$

4. For a body rotating about a fixed axis which is through O and perpendicular to the plane of the paper, show that the sum of the angular impulses of the external forces about the fixed axis is equal to the change in $I_O\omega$, where $I_O\omega$ is the angular momentum $\mathbf{H}_O$ of the entire body.

Solution:

In Fig. 19-7, dm represents any differential mass with position vector $\boldsymbol{\rho}$ in the plane of the paper. The angular momentum of dm is $\boldsymbol{\rho} \times (dm\,\mathbf{v})$. The angular momentum $\mathbf{H}_O$ for the entire body is

$$\mathbf{H}_O = \int \boldsymbol{\rho} \times (dm\,\mathbf{v})$$

But $\mathbf{v} = \boldsymbol{\omega} \times \boldsymbol{\rho}$ is in the plane of the paper and has magnitude $\rho\omega$ because the vectors $\boldsymbol{\omega}$ and $\boldsymbol{\rho}$ are at right angles. Also, $\boldsymbol{\rho} \times (\boldsymbol{\omega} \times \boldsymbol{\rho})$ is directed out of the paper and has magnitude $\rho^2\omega$ because $\boldsymbol{\rho}$ and $\mathbf{v}$ are at right angles. Thus

$$\mathbf{H}_O = \int \rho^2 \omega\, dm\, \mathbf{k}$$

where $\mathbf{k}$ is the unit vector perpendicular to the paper and directed toward the reader.

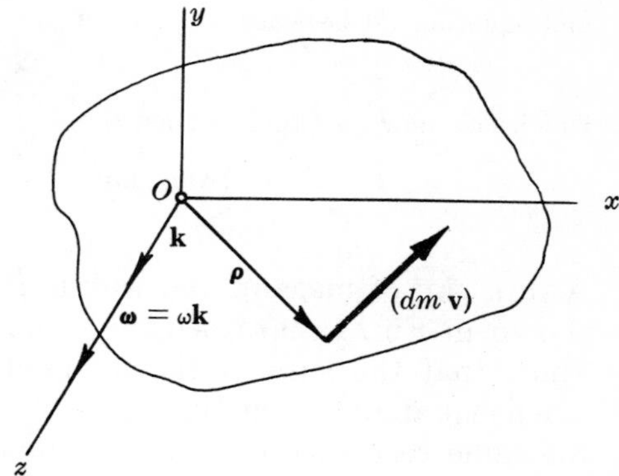

Fig. 19-7

Since ω and $\mathbf{k}$ do not change with dm, they may be taken outside the integral sign. Also, $\int \rho^2\, dm = I_O$. Hence

$$\mathbf{H}_O = I_O \omega \mathbf{k}$$

Then using equation (6), we have

$$\Sigma \mathbf{M}_O = \frac{d\mathbf{H}_O}{dt} = \frac{d(I_O\omega)}{dt}\mathbf{k}$$

Since the moments of forces and $\boldsymbol{\omega}$ are in the $\mathbf{k}$ direction, the equation can be written in scalar form as

$$\int (\Sigma M_O\, dt) = \Delta H_O = I_O(\omega'' - \omega')$$

where ω', ω'' = initial and final angular speeds, and $\int (\Sigma M_O\, dt) = \Sigma(\text{Ang Imp})_O$.

5. For a rigid body of mass m in plane motion (assume the plane of the paper is the plane of motion), show that equations (13), (14) and (15) at the beginning of this chapter are true. Refer to Fig. 19-8.

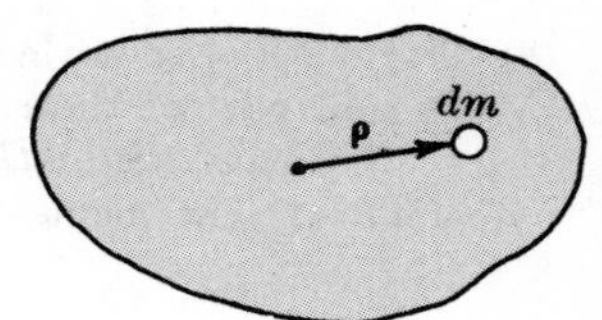

Fig. 19-8

Solution:

Since a rigid body is an assemblage of particles that remain at constant distances from each other, equation (4) of this chapter applies. Thus

$$\Sigma \mathbf{F} = \frac{d(m\bar{\mathbf{v}})}{dt} \quad \text{or} \quad \int_{t_1}^{t_2} \Sigma \mathbf{F}\, dt = \Delta \overline{\mathbf{G}} = m\bar{\mathbf{v}}_2 - m\bar{\mathbf{v}}_1$$

This vector equation is equivalent to the two scalar equations

$$\Sigma(\text{Imp})_x = m(\bar{v}_x'' - \bar{v}_x') \qquad (13)$$

$$\Sigma(\text{Imp})_y = m(\bar{v}_y'' - \bar{v}_y') \qquad (14)$$

To derive equation (15), make use of equation (9). As indicated in Problem 2, the mass center is one of the points which can be selected in order to apply equation (9). Then $\mathbf{H}_O'$ becomes $\overline{\mathbf{H}}'$, the moment of relative momentum of the rigid body about the mass center, and

$$\overline{\mathbf{H}}' = \int \boldsymbol{\rho} \times (dm\,\dot{\boldsymbol{\rho}})$$

Because of the rigid body constraint, the vector $\boldsymbol{\rho}$ (from the mass center to the element of mass dm) can change only in direction but not in magnitude. Hence $\dot{\boldsymbol{\rho}}$ is perpendicular to $\boldsymbol{\rho}$ (it is in the plane of the paper) and of magnitude $\rho\omega$. Then $\boldsymbol{\rho} \times (dm\,\dot{\boldsymbol{\rho}})$ is perpendicular to the plane of the paper and of magnitude $\rho^2\omega$. Thus the vector equation may be replaced by the scalar equation

$$\bar{H}' = \omega \int \rho^2\, dm = \bar{I}\omega$$

and equation (9) becomes

$$\Sigma M_G = \dot{\bar{H}}' = \frac{d(\bar{I}\omega)}{dt}$$

which can now be expressed as

$$(\text{Ang Imp})_G = \int \Sigma M_G\, dt = \Delta\bar{H} = \bar{I}(\omega'' - \omega') \tag{15}$$

6. A thin rim of mass m and radius R is rolling without slipping on a horizontal plane as shown in Fig. 19-9(*a*) below. A horizontal force of magnitude P is applied at the top. Show that the sum of the moments of the external forces about the mass center G, when equated to the time rate of change of the relative momentum about G, yields the same result as obtained by equating the sum of the moments of the external forces about the instant center A to the time rate of change of the relative momentum about A.

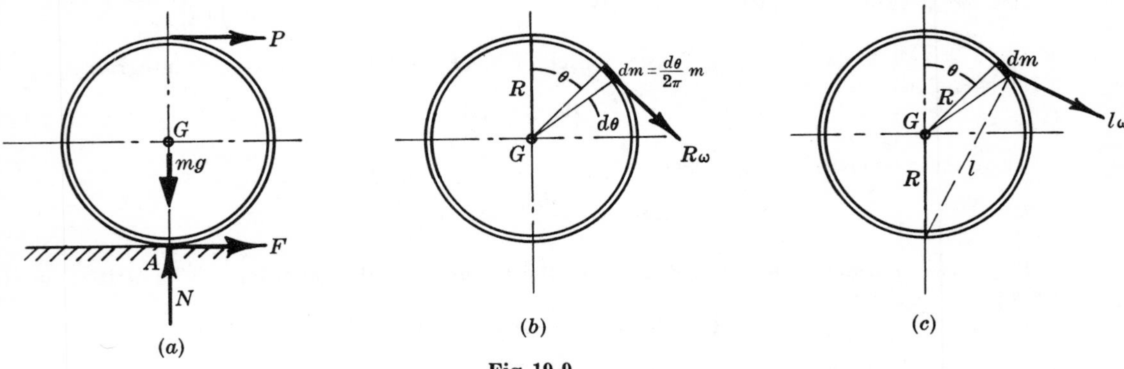

Fig. 19-9

Solution:

The free body diagram in Fig. 19-9(*a*) shows the normal force N, the friction force F, the applied force P and the weight mg concentrated at the mass center G.

Scalar equations will be used in this discussion.

(*a*) Fig. 19-9(*b*) shows a differential mass dm of the rim at an angle θ with the vertical. The mass dm is that part of the rim subtended by the angle $d\theta$; hence $dm = m\,d\theta/2\pi$. The speed of dm relative to the mass center G is $R\omega$ as shown. The moment of relative momentum about G of dm is thus $dm\,R^2\omega$. The moment of relative momentum of the entire rim is

$$H_G' = \int_0^{2\pi} mR^2\omega\, d\theta/2\pi = mR^2\omega$$

The sum of the moments of the external forces about G (considering clockwise positive) is

$$\Sigma M_G = PR - FR$$

But for any group of particles (in this case the rim with mass center speed $v = R\omega$),

$$\Sigma F_x = \dot{G}_x \quad \text{or} \quad P + F = \frac{d}{dt}(mR\omega)$$

From this, $F = -P + \frac{d}{dt}(mR\omega)$. The equation $\Sigma M_G = \dot{H}_G'$ yields

$$PR - \left[-P + \frac{d}{dt}(mR\omega)\right]R = \frac{d}{dt}(mR^2\omega) \quad \text{or} \quad 2PR = \frac{d}{dt}(2mR^2\omega)$$

(*b*) To use the instant center A as the center, let l be the distance from A to the mass dm as shown in Fig. 19-9(*c*). The speed of dm relative to A (which is at rest) is $l\omega$. It is perpendicular to the line l as shown. The line has a length

$$l = \sqrt{R^2 + R^2 + 2RR\cos\theta}$$

The moment of relative momentum of dm about A is $dm(2R^2 + 2R^2 \cos\theta)\omega$. Using $dm = m\,d\theta/2\pi$,

$$H'_A = \int_0^{2\pi} mR^2\omega(2 + 2\cos\theta)\,d\theta/2\pi = 2mR^2\omega$$

The sum of the moments of the external forces about A is

$$\Sigma M_A = 2PR = \dot{H}'_A = \frac{d}{dt}(2mR^2\omega)$$

This is the same result as determined in part (a).

It is interesting to note that since $2mR^2$ is the moment of inertia I_A of the thin rim about the instant center, the above equation can be written

$$\Sigma M_A = \frac{d}{dt}(I_A\omega) = I_A\alpha$$

7. If the mass center of an assemblage of particles is at rest, equation (6) is true and any point O may be used. For two particles of equal mass m mounted on a weightless rim rotating about the center of the rim, show that $\Sigma M_O = \dfrac{d}{dt}(2mR^2\omega)$. Refer to Fig. 19-10.

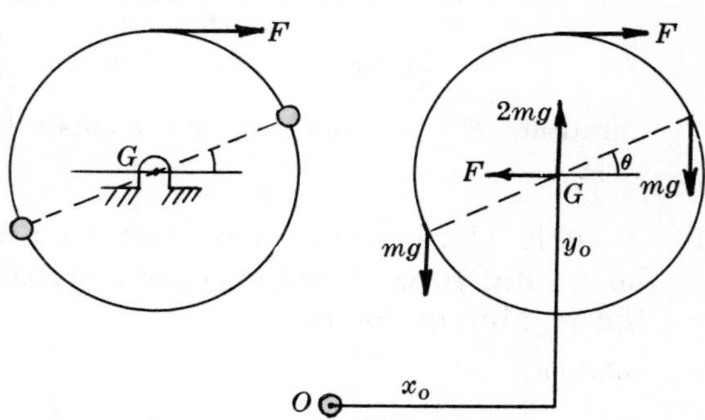

Fig. 19-10

Solution:

The bearing reactions are F to the left and $2mg$ up. Summing moments about any point O, $\Sigma M_O = FR$ and is independent of the moment center O.

$$H_O = m(R\omega\cos\theta)(x_O + R\cos\theta) + m(R\omega\sin\theta)(y_O + R\sin\theta) - m(R\omega\cos\theta)(x_O - R\cos\theta) - m(R\omega\sin\theta)(y_O - R\sin\theta) = 2mR^2\omega$$

Hence $FR = \dfrac{d}{dt}(2mR^2\omega)$, the same equation as when moments are taken relative to mass center G.

8. A block weighing 30 lb slides from rest down a plane inclined 25° with the horizontal. Assuming a coefficient of kinetic friction between the block and the plane of $\frac{1}{4}$, what will be the speed of the block at the end of 3 sec?

Solution:

This problem of course may be solved by previous methods. However, with time as one of the quantities, the impulse-momentum method is the simplest. This is a problem of translation.

Draw a free body diagram indicating all external forces acting on the block (see Fig. 19-11). Since only motion along the plane is considered, one impulse-momentum will suffice.

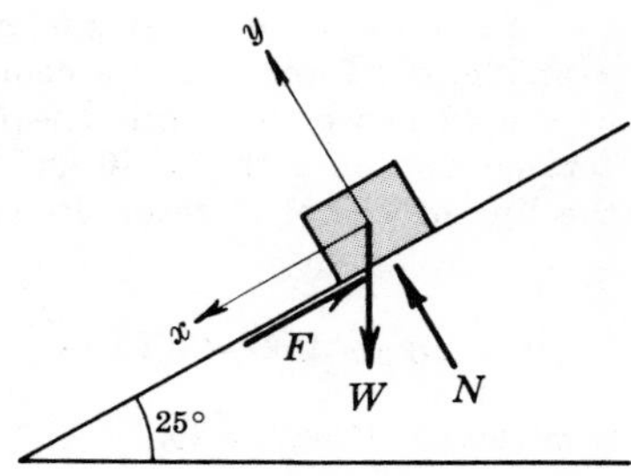

Fig. 19-11

$$\Sigma(\text{Imp})_x = \Delta G_x$$

or

$$(\Sigma F_x)(t) = m(v''_x - v'_x)$$

where t = elapsed time,
m = mass of block,
v''_x = final speed, and v'_x = initial speed.

Assume downward direction as positive, the equation becomes

$$(+W\sin 25° - \mu N)(3) = (30/32.2)(v''_x - 0) \qquad (A)$$

To determine N it is only necessary to sum forces perpendicular to the plane and equate to zero, because no motion in this y direction is assumed.

$$\Sigma F_y = 0 = N - W\cos 25° \quad \text{or} \quad N = 30(0.9063) = 27.2 \text{ lb}$$

Substituting in (A) for N and W, we obtain $v''_x = 19.0$ ft/sec.

9. A block weighing 120 lb rests on a horizontal floor. It is acted upon by a horizontal force that varies from zero according to the law $F = 15t$. If the force acts for 10 sec, what is the speed of the block? Assume that the coefficient of static and kinetic friction is 0.25.

Solution:

As the force F increases, the friction increases to equal it; this will be true until the limiting friction is reached, i.e. $F = \mu N = \mu W = 30$ lb. Up to this time the force F has been balanced by the friction; the net impulse on the system is zero and no motion has ensued. The time for this to occur may be found from $F = 30.0 = 15t$ or $t = 2$ sec.

After $t = 2$ sec, the linear impulse horizontally must equal the change in momentum horizontally. Note that F increases but the friction remains at 30 lb.

$$\Sigma(\text{Imp})_x = \Delta G_x \qquad \text{or} \qquad \int_2^{10} F\,dt - \int_2^{10} 30\,dt = (120/32.2)(v - 0)$$

Substitute $F = 15t$ and integrate to obtain the speed after 10 sec, $v = 129$ ft/sec.

10. A 50 lb block starts from rest on a smooth horizontal plane under the action of a horizontal force F which varies according to the equation $F = 3t - 5t^2$. Determine the maximum speed.

Solution:

$$\Sigma(\text{Imp})_x = \Delta G_x \qquad \text{or} \qquad \int_0^t F\,dt = \int_0^t (3t - 5t^2)dt = (50/32.2)(v - 0)$$

from which $v = 0.966t^2 - 1.072t^3$. To find the maximum value of v, determine the value of t at which $dv/dt = 1.932t - 3.216t^2 = 0$. This gives $t = 0.602$ sec, and hence maximum $v = 0.116$ ft/sec.

11. The 90 lb block shown in Fig. 19-12 is moving up initially with a speed of 8 ft/sec. What constant value of P will give it an upward speed of 16 ft/sec in 12 sec? Assume that the pulleys are frictionless and that the coefficient of friction between the blocks and the plane is 0.10.

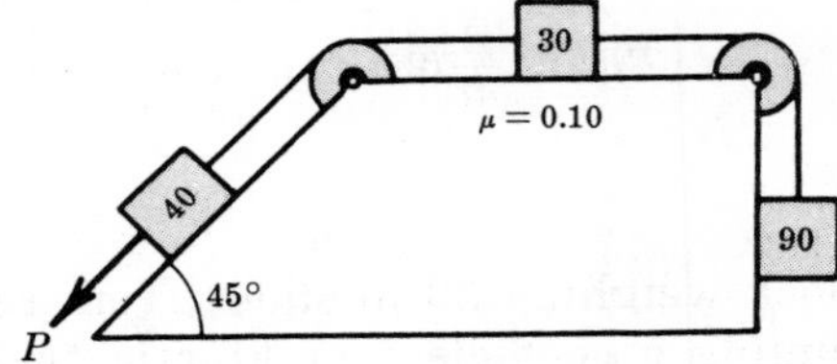

Fig.19-12

Solution:

Solve the problem by summing impulses and momenta along the line of travel of the system. For example, the force P and a component of the 40 lb weight have positive impulses in the line of travel of the 40 lb weight, while the friction acting on the 40 lb weight has a negative impulse. The cord tension acts on both the 40 and 30 lb weights in opposite directions; therefore its linear impulse along the line of travel is zero. Proceeding in this way the impulse-momentum equation becomes

$$\Sigma(\text{Imp}) = \Delta G$$

$$[P + 40 \sin 45° - 0.10(40 \cos 45°) - 0.10(30) - 90](12) \text{ lb-sec} = \frac{40 + 30 + 90}{32.2}(16 - 8) \text{ lb-sec}$$

from which $P = 70.8$ lb.

12. A 4 lb weight rests on a smooth horizontal plane. It is struck with a 2 lb blow that lasts 0.02 sec. Three seconds after the start of the first blow a second blow of -2 lb is delivered. This lasts for 0.01 sec. What will be the speed of the body after 4 sec?

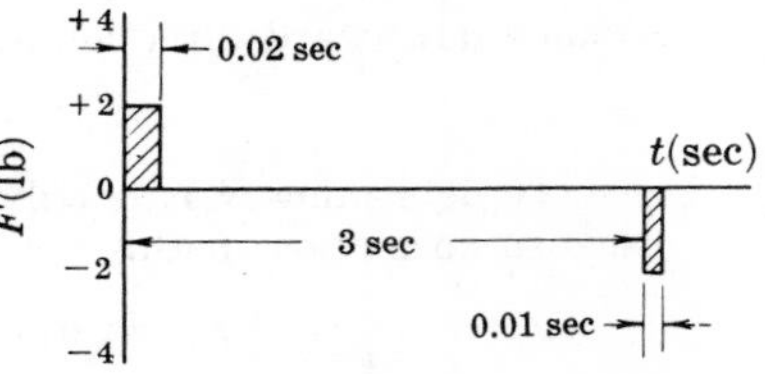

Fig. 19-13

Solution:

The forces in this problem may be represented graphically in a plot against time as shown in Fig. 19-13.

The linear impulse for any length of time greater than 3.01 sec is the algebraic sum of the two areas, i.e., $+2(0.02) - 2(0.01) = 0.02$ lb-sec. Then

$$\Sigma(\text{Imp}) = \Delta G, \qquad 0.02 = (4/32.2)(v - 0), \qquad \text{and} \quad v = 0.16 \text{ ft/sec}$$

13. In Fig. 19-14, the mercury in the left column of the manometer is falling at the rate of 1 inch per second. The left column is 18 inches long and the right column is 22 inches long. What is the vertical momentum of the mercury? The manometer is made of quarter inch glass (inside diameter). Mercury weighs 850 lb/ft^3.

Solution:

Neglecting the bend in the tube, it is evident that the momentum of the left column is down while that of the right column is up. The net upward momentum is that of a 4 inch column, $\frac{1}{4}$ inch in diameter and moving $\frac{1}{12}$ ft/sec.

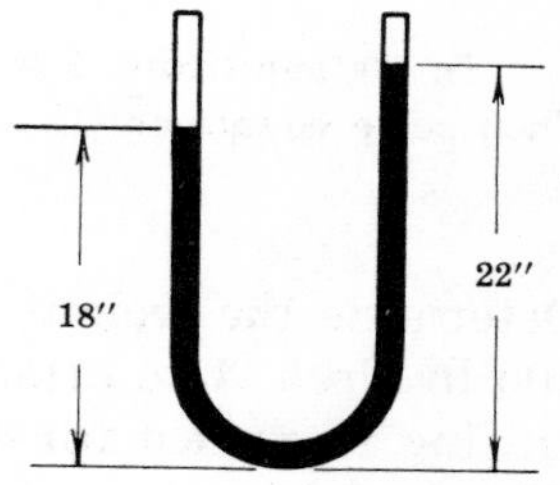

Fig. 19-14

$$\text{Momentum} = \{\tfrac{1}{4}\pi(\tfrac{1}{4})^2(4)/(12)^3 \text{ ft}^3\}\{(850/32.2) \text{ slug/ft}^3\}\{\tfrac{1}{12} \text{ ft/sec}\} = +0.00025 \text{ lb-sec.}$$

14. A flywheel, of mass 120 slugs and radius of gyration 3.6 ft, rotates about a fixed center O from rest to an angular speed of 200 rpm in 150 sec. What moment M is necessary?

Solution:

$$\Sigma(\text{Ang Imp})_O = \Delta H_O = I_O(\omega_2 - \omega_1)$$

$$M(150 \text{ sec}) = [120(3.6)^2 \text{ slug-ft}^2]\,[(400\pi/60 - 0) \text{ rad/sec}], \qquad M = 217 \text{ lb-ft}$$

15. The pendulum consists of a bob of mass m and a slender rod of negligible mass. See Fig. 19-15. Show that the differential equation of the motion is $\ddot{\theta} + (g/L)\sin\theta = 0$.

Solution:

The free body diagram shows the pendulum displaced an angle θ from the vertical position. The angular momentum $\mathbf{H}_O$ for the bob relative to the support is $I\dot{\theta}\mathbf{k}$, where $\mathbf{k}$ is the unit vector perpendicular to the page, arrow pointing toward the reader.

The only force with a moment is the weight of the bob; the moment is clockwise or negative. Then

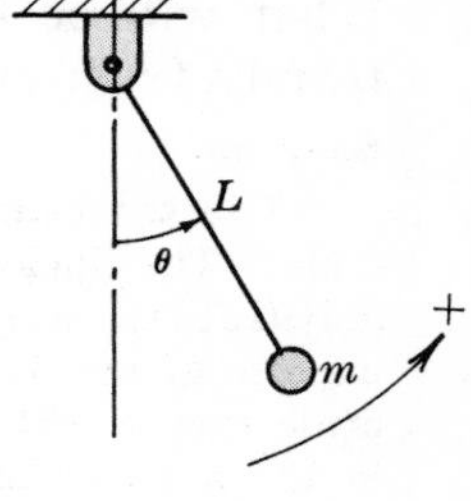

Fig. 19-15

$$\Sigma\mathbf{M}_O = \frac{d\mathbf{H}_O}{dt} \qquad \text{or} \qquad -mg(L\sin\theta)\mathbf{k} = \frac{d}{dt}(I\dot{\theta}\mathbf{k})$$

Since for the bob $I = mL^2$, we obtain the scalar equation $\ddot{\theta} + (g/L)\sin\theta = 0$.

16. In Fig. 19-16, a weightless rope carries two weights of 10 lb and 15 lb when hanging on a pulley of weight 10 lb, radius 2 ft, and radius of gyration 1.5 ft. How long will it take to change the speed of the weights from 10 ft/sec to 20 ft/sec?

Solution:

Draw a free body diagram.

Let T_{10} and T_{15} be the tensions in the rope supporting the 10 lb and 15 lb weights respectively.

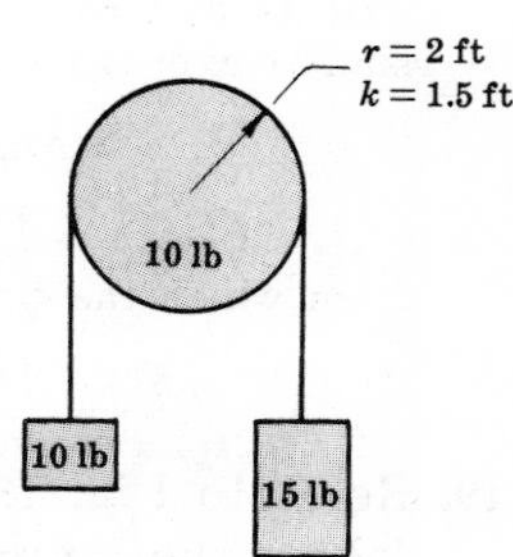

Fig. 19-16

The useful impulse-momentum equations are as follows where equation (*1*) applies to the 10 lb weight, equation (*2*) applies to the 15 lb weight, and equation (*3*) applies to the rotation of the pulley.

$$(1)\quad (T_{10} - 10)t = (10/32.2)(20 - 10)$$
$$(2)\quad (15 - T_{15})t = (15/32.2)(20 - 10)$$
$$(3)\quad [(T_{15} - T_{10})2]t = \bar{I}(\omega_{\text{final}} - \omega_{\text{initial}})$$

In (*3*), substitute $\bar{I} = mk^2 = (10/32.2)(1.5)^2$ slug-ft², $\omega_{\text{final}} = 20/2$ rad/sec, and $\omega_{\text{initial}} = 10/2$ rad/sec. Then solve equations (*1*), (*2*) and (*3*) simultaneously to obtain $t = 1.91$ sec.

17. Determine the weight of B necessary to cause the 100 lb block A to attain a speed of 12 ft/sec after moving from rest for 4 sec. Assume the drum rotates in frictionless bearings. Refer to Fig. 19-17.

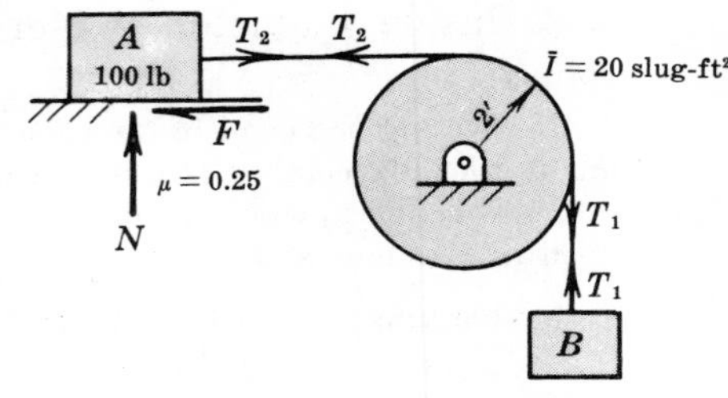

Fig. 19-17

Solution:

Apply the equation for linear impulse-momentum to the 100 lb block and B. Apply the angular impulse-angular momentum equation to the drum.

By inspection, $N = 100$ lb and $F = 25$ lb. The equations for the 100 lb block, the weight B, and the drum are respectively:

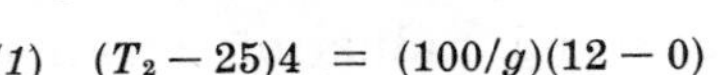

$$(1)\quad (T_2 - 25)4 = (100/g)(12 - 0)$$
$$(2)\quad (W_B - T_1)4 = (W_B/g)(12 - 0)$$
$$(3)\quad [T_1(2) - T_2(2)]4 = \bar{I}\omega = 20(12/2)$$

Solve simultaneously to obtain $W_B = 54.3$ lb.

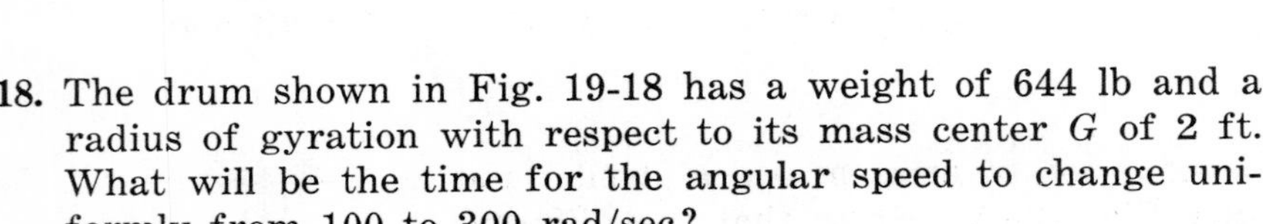

18. The drum shown in Fig. 19-18 has a weight of 644 lb and a radius of gyration with respect to its mass center G of 2 ft. What will be the time for the angular speed to change uniformly from 100 to 300 rad/sec?

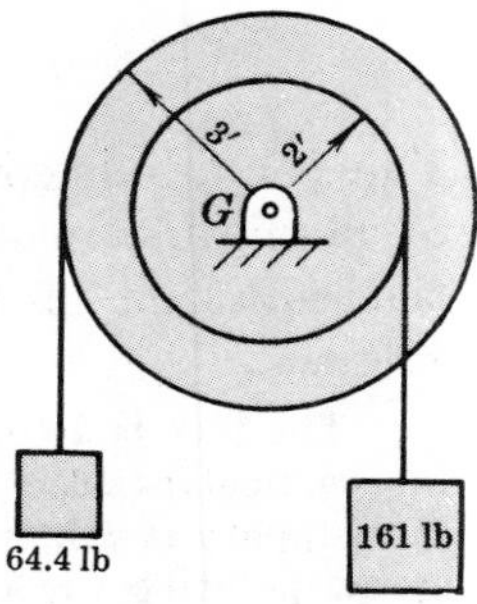

Fig. 19-18

Solution:

Use the mass center G as the point about which moments are to be taken. The linear impulse of the 161 lb weight has a moment about G of (2 ft)(161 lb)(t sec) = $322t$ lb-sec-ft. Likewise, the moment of the linear impulse of the 64.4 lb weight is $(-3)(64.4)(t) = -193t$ lb-sec-ft. The algebraic sum is $+129t$ lb-sec-ft. The angular impulse is equal to the change in the angular momentum of the system with reference to G. For the drum this change is

$$+\bar{I}(\omega_2 - \omega_1) = m\bar{k}^2(\omega_2 - \omega_1) = (644/32.2)(2)^2(300 - 100) = +16{,}000 \text{ lb-sec-ft}$$

The linear momentum of each weight must be multiplied by its arm to determine the moment of linear momentum (angular momentum) about G. The 161 lb moves down, changing its speed from 2(100) to 2(300) ft/sec. The moment of the change in its linear momentum about G is positive (clockwise is assumed as the positive direction) and equals $+5[2(300) - 2(100)]2 = +4000$ lb-sec-ft. A similar expression is shown below for the 64.4 lb weight. Then

$$\Sigma(\text{Ang Imp})_G = \Delta H_G$$
$$129t = +16{,}000 + 4000 + (64.4/32.2)[3(300) - 3(100)]3 = 23{,}600$$

from which the time to change the angular speed is $t = 183$ sec.

19. Refer to Fig. 19-19 below. A cylinder of radius r, mass m, and moment of inertia $\bar{I}$ (about the center) rolls from rest down a plane inclined θ degrees with the horizontal. What is the speed of its center t sec from rest?

Solution:

Draw a free body diagram showing all external forces acting on the cylinder. Do not make the mistake of assuming the friction force F is equal to the product of coefficient of friction and the normal force N.

Since the cylinder is in plane motion, the equations of impulse-momentum for this type of motion are

(1) $\Sigma(\text{Imp})_x = \Delta G_x$ and (2) $\Sigma(\text{Ang Imp})_O = \Delta H_O$

Note that F is the only external force with a moment about the center O. Equations (1) and (2) become:

(3) $[(W \sin\theta - F)\text{ lb}](t\text{ sec}) = (m\text{ slugs})(\bar{v} - 0)$

(4) $[(F\text{ lb})(r\text{ ft})](t\text{ sec}) = (\bar{I}\text{ slug-ft}^2)(\omega - 0)$

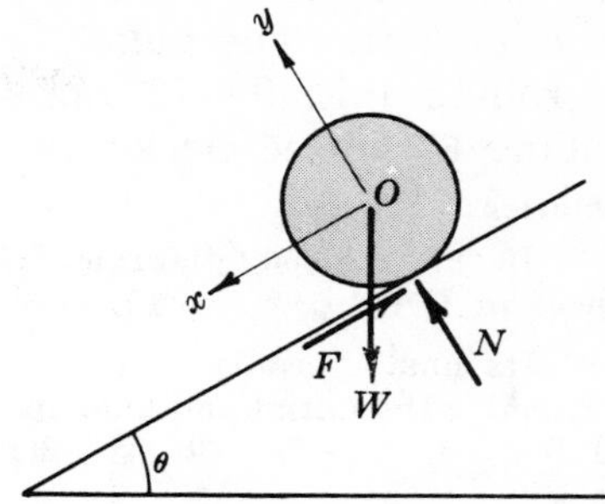

Fig. 19-19

In these equations the friction force F and the speed $\bar{v}$ are unknown. ($\omega = \bar{v}/r$, since no slipping is assumed.) In (3), $\bar{v}$ has units of lb-sec/slug = ft/sec; and in (4), ω has units of lb-ft-sec/slug-ft^2 = rad/sec.

From (4), $F = \dfrac{\bar{I}\omega}{rt}$. Substitute this into (3), with $\omega = \dfrac{\bar{v}}{r}$, to obtain $\bar{v} = \dfrac{W \sin\theta}{m + \bar{I}/r^2}t$.

20. A solid homogeneous cylinder 3 ft in diameter weighing 300 lb is rolled up a 20° incline by a force of 250 lb applied parallel to the plane. Assuming no slipping, determine the speed of the cylinder after 6 sec if the initial speed is zero. Refer to Fig. 19-20 below.

Solution:

Draw the free body diagram of the cylinder, assuming that the frictional force F acts up the plane. The impulse-momentum equations of plane motion are

(1) $\Sigma(\text{Imp})_{||} = \Delta G_{||}$ or $(250 + F - 300 \sin 20°)6 = \dfrac{300}{g}\bar{v}$

(2) $\Sigma(\text{Ang Imp})_G = \bar{I}\omega$ or $\left[250\left(\tfrac{3}{2}\right) - F\left(\tfrac{3}{2}\right)\right]6 = \dfrac{1}{2}\dfrac{300}{g}\left(\dfrac{3}{2}\right)^2 \dfrac{\bar{v}}{3/2}$

from which the speed $\bar{v}$ after 6 sec = 171 ft/sec, parallel to the plane.

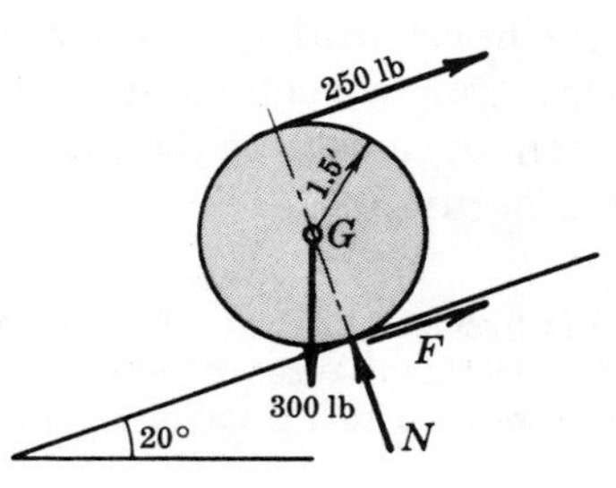

Fig. 19-20

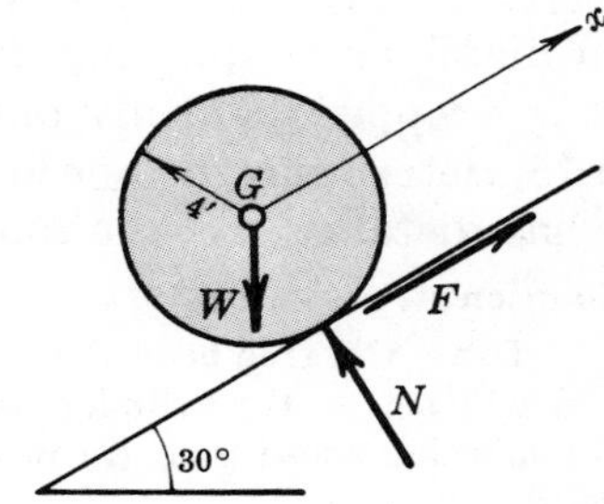

Fig. 19-21

21. Refer to Fig. 19-21 above. A solid homogeneous cylinder starts up a 30° inclined plane with a linear speed of 20 ft/sec. If it rolls freely to rest, how long does it take to reach its highest point?

Solution:

The initial speed $\bar{v}$ of the mass center G is 20 ft/sec. The initial angular velocity $\omega = 20/4 = 5$ rad/sec. The final speed is zero. The free body diagram shows the forces acting on the cylinder.

Apply the two impulse-momentum equations of plane motion.

(1) $\Sigma(\text{Imp})_x = \Delta G_x$ or $(F - W \sin 30°)t = W/g(0 - 20)$

(2) $\Sigma(\text{Ang Imp})_G = \Delta H_G$ or $(-4F)t = \tfrac{1}{2}(W/g)(4^2)(0 - 5)$

from which $t = 1.86$ sec.

22. In Fig. 19-22 below, the wheel weighs 161 lb and has a radius of gyration with respect to G of 2 ft. The pulley is weightless and runs in frictionless bearings. The wheel is rolling initially 10 rad/sec counterclockwise. How long will it take until it is rolling 6 rad/sec clockwise?

Solution:

In the free body diagram, friction is assumed to act to the left. By kinematic relations the initial speed of G is $v_G = -3(10) = -30$ ft/sec, i.e. to the left.

Its final speed is $+3(6) = +18$ ft/sec, to the right. To find the speeds of the 60 lb weight, determine the initial and final speeds of point B on the wheel: $\mathbf{v}_B = \mathbf{v}_{B/G} + \mathbf{v}_G$. Hence the initial speed of B is $v_B = -2\omega - 30 = -2(10) - 30 = -50$ ft/sec. The final speed is by similar reasoning, +30 ft/sec.

Write equations (*1*) and (*2*) for the wheel, and equation (*3*) for the 60 lb weight:

$$(1)\quad \Sigma(\text{Imp})_h = \Delta G_h \quad\text{or}\quad (T-F)t = (161/32.2)[+18-(-30)]$$

$$(2)\quad \Sigma(\text{Ang Imp})_G = \Delta H_G \quad\text{or}\quad (2T+3F)t = (161/32.2)(2)^2[+6-(-10)]$$

$$(3)\quad \Sigma(\text{Imp})_v = \Delta G_v \quad\text{or}\quad (60-T)t = (60/32.2)[+30-(-50)]$$

Simplifying: (1′) $Tt - Ft = 240$, (2′) $\frac{2}{3}Tt + Ft = 106.7$, (3′) $60t - Tt = 149.1$.

Add (1′) and (2′) to obtain $Tt = 208$. Substitute this into (3′) to find $t = 5.95$ sec.

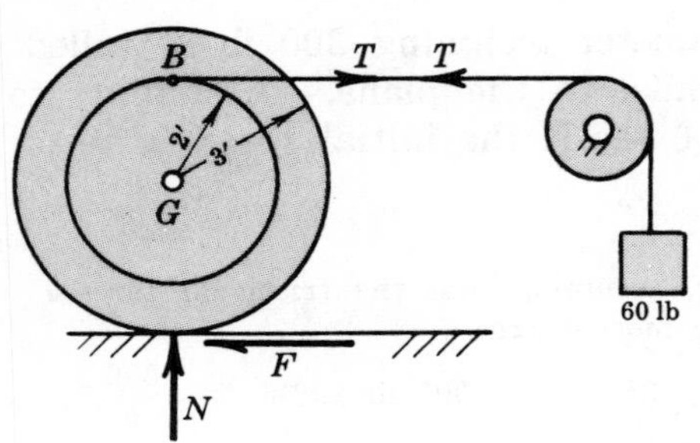

Fig. 19-22

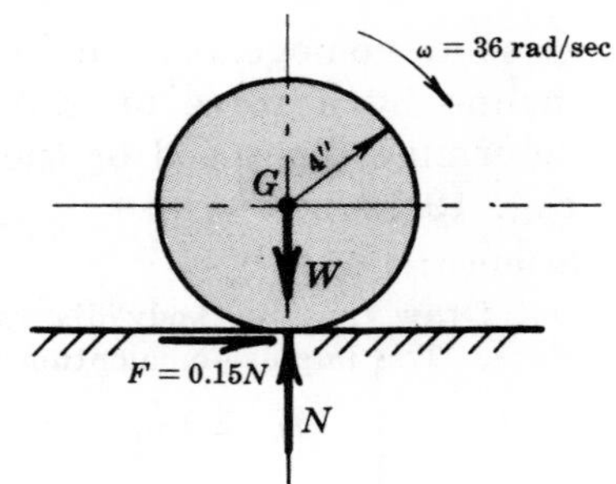

Fig. 19-23

23. Refer to Fig. 19-23 above. A solid homogeneous cylinder weighing 96.6 lb and 8 in. in diameter is spinning 36 rad/sec clockwise about its horizontal geometric axis when it is dropped suddenly onto a horizontal plane. Assuming a coefficient of friction of 0.15, determine the speed $\bar{v}$ of the mass center G when pure rolling begins. Through what distance has the mass center moved before this occurs?

Solution:

Draw the free body diagram when contact with the plane is first established. The same force system will act on the cylinder until pure rolling is attained. The frictional force F acts (1) to decrease the angular speed and (2) to increase the mass center speed from zero to its value $\bar{v} = r\omega$ when pure rolling obtains.

Here $N = W = 96.6$ lb; $F = 0.15\,N = 14.5$ lb; $m = 3$ slugs; $\bar{I} = \frac{1}{2}mr^2 = \frac{1}{2}(3)(\frac{1}{3})^2 = \frac{1}{6}$ slug-ft².

The impulse-momentum equations are

$$(1)\quad \Sigma(\text{Imp})_h = \Delta G_h \quad\text{or}\quad Ft = m(\bar{v}-0)$$

$$(2)\quad \Sigma(\text{Ang Imp})_G = \Delta H_G \quad\text{or}\quad -Frt = \bar{I}(\omega - 36)$$

In the second equation, assume the clockwise direction is positive. The equations become $14.5t = 3\bar{v}$ and $-14.5(\frac{1}{3})t = \frac{1}{6}(\omega - 36)$, where $\omega = \bar{v}/r = 3\bar{v}$. The solution is thus $\bar{v} = 4.0$ ft/sec, $t = 0.83$ sec. Then $s = \frac{1}{2}(\bar{v}+0)t = 1.66$ ft.

24. A stream of water 2 inches in diameter, moving 80 ft/sec horizontally, strikes a flat vertical plate as shown in Fig. 19-24 below. After striking, the water moves parallel to the plate. What is the force exerted on the plate by the stream of water?

Solution:

Consider all the particles of water in time interval Δt. The total mass m of these particles is

$$m = Av(\Delta t)\delta = \tfrac{1}{4}\pi(2/12)^2(80)(\Delta t)(62.4/32.2) = 3.39(\Delta t) \text{ slugs}$$

where A = area of stream in ft², v = speed of stream in ft/sec, δ = density of water in slugs/ft³.

Let P = force of plate on the mass m of water. Then

$$\Sigma F_x(\Delta t) = \Delta G_x = m(v_x'' - v_x')$$

or

$$P(\Delta t) = 3.39(\Delta t)(0 - 80)$$

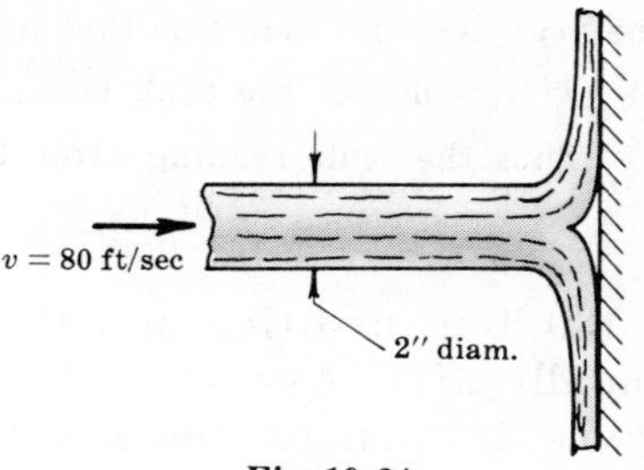

Fig. 19-24

from which $P = -271$ lb. The force of the water on the plate is $+271$ lb, i.e. to the right.

25. Rework Problem 24, but assume the plate is moving to the right with speed 20 ft/sec.

Solution:

Again consider the mass of all particles of water in time Δt. The speed of the water relative to the plate is $v = (80 - 20) = 60$ ft/sec. Then $m = Av(\Delta t)\delta = 2.54(\Delta t)$, and using the linear impulse-momentum equation in the x direction we obtain

$$P(\Delta t) = 2.54(\Delta t)(20 - 80)$$

where the final speed of water is that of the plate. Solving, the force of the plate on the water is $P = -153$ lb. Of course the force of the water on the plate is $+153$ lb, i.e. to the right.

26. A stream of water with cross sectional area 3 in² and moving 30 ft/sec horizontally, strikes a fixed blade curved as shown in Fig. 19-25. Assuming the speed of the water relative to the blade is constant (no friction is considered), determine the horizontal and vertical components of the force of the blade on the stream of water.

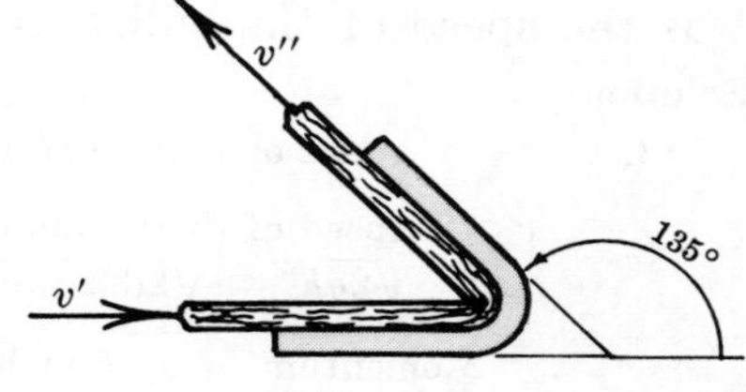

Fig. 19-25

Solution:

Note that the final velocity $\mathbf{v}''$ has the same magnitude v as the initial velocity $\mathbf{v}'$ but a different direction. The mass m of all particles of water in time interval Δt is

$$m = Av(\Delta t)\delta = (3/144)(30)(\Delta t)(62.4/32.2) = 1.21(\Delta t)$$

Using the linear impulse-momentum equation for the x and y directions,

$$\Sigma F_x(\Delta t) = \Delta G_x = m(v_x'' - v_x') \quad \text{or} \quad P_x(\Delta t) = 1.21(\Delta t)(-30\cos 45° - 30)$$

$$\Sigma F_y(\Delta t) = \Delta G_y = m(v_y'' - v_y') \quad \text{or} \quad P_y(\Delta t) = 1.21(\Delta t)(+30\sin 45° - 0)$$

where P_x and P_y are respectively the x and y components of the blade force. Solving, $P_x = -61.9$ lb (to the left on the water) and $P_y = +25.7$ lb (up on the water).

27. A tank weighing 150 lb rests on platform scales. A vertical jet empties water into the tank with a speed of 20 ft/sec. Its cross sectional area is 0.20 square inches. What will be the scale reading 1 minute later?

Solution:

The water from the jet exerts a continuous force F on the bottom of the tank and thus to the scale. From the impulse-momentum equation, considering the water at any time t as a free body, we may write $\Sigma(\text{Imp})_v = \Delta G_v$ or $Ft = m(0 - 20)$, where $m = Avt\delta$. Then

$$Ft \;=\; (0.20/144)(20)(t)(62.5/32.2)(0-20) \qquad \text{or} \qquad F = -1.1 \text{ lb}$$

The negative sign indicates that an upward force (from the scale) is required to stop the water.

At $t = 60$ sec the tank holds $(0.20/144)(20)(60)(62.5) = 104$ lb of water.

Thus the scale reading after 1 minute will be $(1.1 + 104 + 150) = 255$ lb.

28. A 150 lb man sitting in a 150 lb iceboat fires a gun discharging a 2 ounce bullet horizontally aft. Assuming the muzzle speed to be 1600 ft/sec and that there is negligible friction on the ice, with what speed will the boat move after the shot is fired?

Solution:

Since the impulse on the entire system is zero, the total momentum remains constant. This is true because the action on the bullet is equal to the reaction on the man, gun, and boat. The initial momentum is zero. Hence the momentum after the shot is fired must also be zero. Assume the bullet speed to be positive. The equation is then

$$\frac{W_{\text{boat}} + W_{\text{man}}}{g}v + \frac{W_{\text{bullet}}}{g}1600 \;=\; 0 \qquad \text{or} \qquad \frac{150+150}{g}v + \frac{2/16}{g}1600 \;=\; 0$$

and $v = -0.667$ ft/sec. The minus sign indicates that the boat and man move in the opposite direction from the bullet.

29. A bullet weighing 2 oz was fired horizontally into a 100 lb sand bag suspended on a rope 3 ft long as shown in Fig. 19-26. It was calculated from the observed angle θ that the bag with the bullet imbedded in it swung to a height of 0.097 ft. What was the speed of the bullet as it entered the bag?

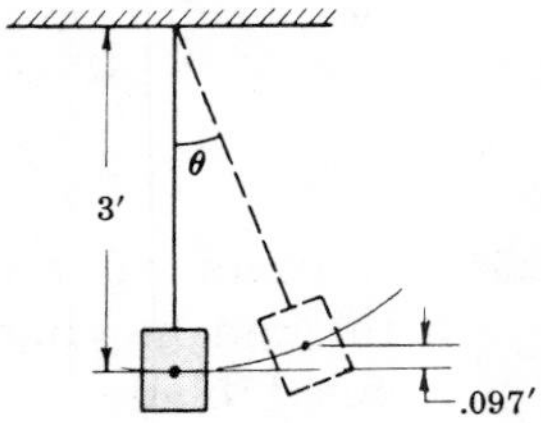

Fig. 19-26

Solution:

Let $v_1 \;=\;$ speed of bullet before impact,

$v_2 \;=\;$ speed of (bag + bullet) after impact

$\quad = \sqrt{2gh} \;=\; \sqrt{2(32.2)(0.097)} \;=\; 2.5$ ft/sec.

Momentum of system before impact $=$ momentum of system after impact

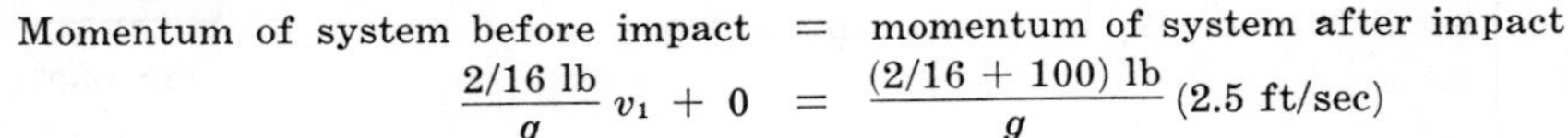

$$\frac{2/16 \text{ lb}}{g}v_1 + 0 \;=\; \frac{(2/16 + 100)\text{ lb}}{g}(2.5 \text{ ft/sec})$$

from which $v_1 = 2000$ ft/sec.

30. As another example of conservation of linear momentum, consider the recoil of guns.

Initially the projectile is at rest in the gun. The charge explodes, pushing the projectile from the gun and at the same time pushing back with the same force on the gun. Since no external forces act during this explosion, the sum of the momenta of the projectile forward and the gun backward must be the same as the initial momentum of the system, i.e. zero. Thus

$$m_p v_p + m_g v_g \;=\; 0$$

where $m_p, m_g \;=\;$ masses of projectile and gun respectively,

$v_p, v_g \;=\;$ speeds of projectile and gun immediately after the explosion.

Solving for speed of recoil, $v_g = -(m_p/m_g)v_p$. The minus sign indicates that the gun moves in the opposite direction to the projectile.

The greater the mass of the gun, the less will be the speed of recoil. Hence the energy to be absorbed by the recoil spring or other devices will be correspondingly less. Of course, the weight is limited by the need for mobility.

Both the projectile and the gun possess kinetic energy as recoil starts. The projectile, however, possesses the greater amount since kinetic energy depends not only on the mass but also on the square of the speed.

31. A bullet weighing 2 oz and moving with a speed of 1600 ft/sec strikes a block weighing 10 lb and moving in the same direction with a speed of 100 ft/sec. What is the resultant speed v of the bullet and the block assuming the bullet to be imbedded in the block?

Solution:

$$\text{Momentum of system before impact} \;=\; \text{momentum of system after impact}$$

$$\frac{2/16\text{ lb}}{g}(1600\text{ ft/sec}) + \frac{10\text{ lb}}{g}(100\text{ ft/sec}) \;=\; \frac{(2/16+10)\text{ lb}}{g}v, \qquad v = 119\text{ ft/sec}$$

32. A spring normally 6 inches long is connected to the two weights shown in Fig. 19-27 and compressed 2 inches. If the system is released on a smooth horizontal plane, what will be the speed of each block when the spring is again its normal length? The spring constant is 12 lb/in.

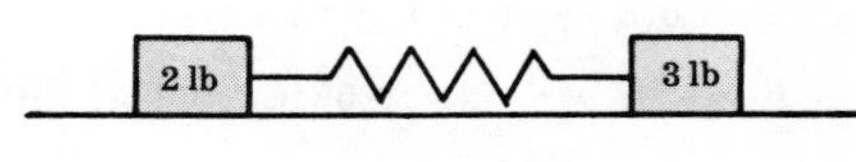

Fig. 19-27

Solution:

Since the same spring force acts on the two weights, but in opposite directions, the total linear impulse on the system of both weights is zero. Hence the momentum of the two weights is constant, i.e. zero, and $(2/g)v_2 + (3/g)v_3 = 0$, where the subscripts refer to the 2 and 3 lb weights. From this equation, $v_2 = -(3/2)v_3$ at all times.

Another equation involving v_2 and v_3 is necessary. The work done by the spring in expanding to its original length (by virtue of its potential energy) is equal to the change in kinetic energy of both weights. As the spring expands a distance x from its compressed position, its compressive force $=$ $12(2-x)$ lb.

Hence the total work done $= \int_0^2 12(2-x)dx = 24$ in-lb or 2 ft-lb. Equating this to the final kinetic energy (initial is zero), $2 = \frac{1}{2}(2/g)v_2^2 + \frac{1}{2}(3/g)v_3^2$. Solve simultaneously with the previous equation to obtain $v_3 = 4.14$ ft/sec (to the right) and $v_2 = -6.22$ ft/sec (to the left).

33. A bob of weight W travels in a circular path of radius R on a smooth horizontal plane under the restraint of a string which passes through a hole O in the plane as shown in Fig. 19-28. If the angular speed of the bob is ω_1 rad/sec when the radius is R, what will be the angular speed if the string is pulled from underneath until the radius of the path is $R/2$? Find the ratio of the final tension in the string to its initial value.

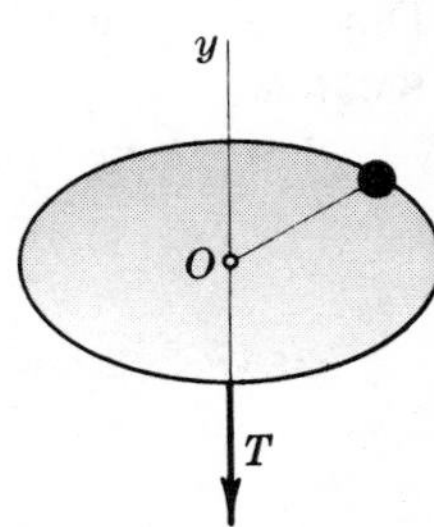

Fig. 19-28

Solution:

Note that no forces have moments about the y-axis. Hence the angular impulse about that axis is zero, meaning that there can be no change in angular momentum.

The initial angular momentum is $I_O\omega_1$ or $(W/g)R^2\omega_1$. In the new position, when the radius is $R/2$ and the angular speed is ω_2, the angular momentum is $(W/g)(R/2)^2\omega_2$. Equating the two and solving, $\omega_2 = 4\omega_1$.

The initial tension in the cord $= (W/g)R\omega_1^2$. The final tension $= (W/g)(R/2)(4\omega_1)^2 = 8(W/g)R\omega_1^2$. Then the final tension is 8 times its initial value.

34. The two small spheres, each weighing 2.5 lb, are connected by a light string as shown in Fig. 19-29 below. The horizontal platform is rotating under no external moments at 36 rad/sec when the string breaks. Assuming no friction between the spheres and the groove in which they ride, determine the angular speed of the system when the spheres hit the outer stops. The moment of inertia I for the disk is 0.4 slug-ft².

Solution:

The initial moment of inertia I_i is that of the disk and the two spheres at distance 3 inches from the center. Hence

$$I_i = 0.4 + 2(2.5/32.2)(3/12)^2 = 0.41 \text{ slug-ft}^2$$

The initial angular momentum is $I_i\omega_i$ or 0.41(36). The final moment of inertia I_f is

$$I_f = 0.4 + 2(2.5/32.2)(11/12)^2 = 0.531 \text{ slug-ft}^2$$

Since no external moments act on the system, there is conservation of angular momentum; then

$$I_f\omega_f = I_i\omega_i, \qquad 0.531\omega_f = 0.41(36), \qquad \text{and} \qquad \omega_f = 27.8 \text{ rad/sec}$$

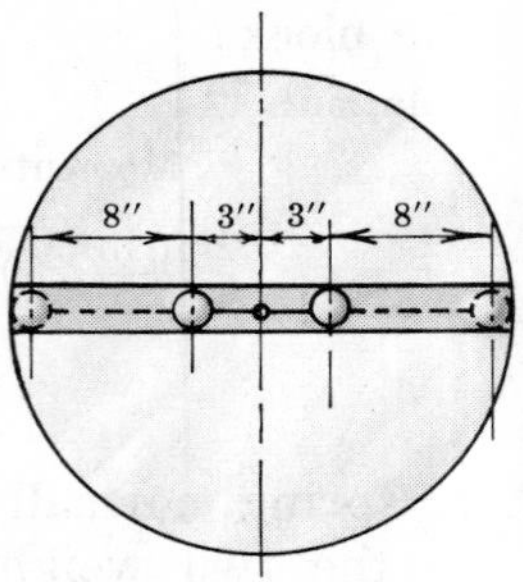

Fig. 19-29

35. Solve the following impact equations (*1*) and (*2*) for unknowns v_1 and v_2.

$$(1)\quad e = \frac{v_2 - v_1}{u_1 - u_2} \qquad (2)\quad m_1u_1 + m_2u_2 = m_1v_1 + m_2v_2$$

where e = coefficient of restitution
u_1, u_2 = speeds of bodies 1 and 2 respectively before impact
v_1, v_2 = speeds of bodies 1 and 2 respectively after impact
m_1, m_2 = masses of bodies 1 and 2 respectively.

Solution:

Multiply (*1*) by $(u_1 - u_2)m_1$ to obtain: (*3*) $em_1(u_1 - u_2) = m_1v_2 - m_1v_1$.

Add (*2*) and (*3*) to get: $m_1u_1 + m_2u_2 + em_1(u_1 - u_2) = (m_2 + m_1)v_2$.

Then $v_2 = \dfrac{m_1u_1(1+e) + u_2(m_2 - em_1)}{m_2 + m_1}$. Similarly $v_1 = \dfrac{m_2u_2(1+e) + u_1(m_1 - em_2)}{m_2 + m_1}$.

36. Discuss impact for purely elastic bodies ($e = 1$).

Solution:

For elastic impact substitute $e = 1$ in the equations for v_2 and v_1 derived in Problem 35.

$$(1)\quad v_1 = \frac{2m_2u_2 + u_1(m_1 - m_2)}{m_2 + m_1} \qquad (2)\quad v_2 = \frac{2m_1u_1 + u_2(m_2 - m_1)}{m_2 + m_1}$$

Of special interest is the case when $m_1 = m_2 = m$. Then the equations become

$$(1')\quad v_1 = \frac{2mu_2 + 0}{m + m} = u_2 \qquad (2')\quad v_2 = \frac{2mu_1 + 0}{m + m} = u_1$$

The above equations explain what takes place when a moving penny strikes a stationary penny on a smooth surface. The final speed of the moving penny will be the initial speed of the other penny (in this case zero), while the final speed of the formerly stationary penny will be the initial speed of the moving penny. In other words, the moving penny stops and the stationary one assumes its speed. The same thing is true of a straight row of pennies. The first one (farthest away from the moving penny) moves away while the others remain stationary.

37. What happens during inelastic impact?

Solution:

This is the case when one body absorbs the other or clings to it. Common sense indicates that they have a common final speed. Note that if $e = 0$ is substituted into the equations for v_1 and v_2 in Problem 35 this is actually so.

$$v_1 = \frac{m_2u_2 + m_1u_1}{m_2 + m_1}, \qquad v_2 = \frac{m_1u_1 + m_2u_2}{m_2 + m_1}, \qquad \text{or} \qquad v_1 = v_2$$

38. Two equal billiard balls meet centrally with speeds of 6 ft/sec and −8 ft/sec. What will be their final speeds after impact if the coefficient of restitution is assumed to be 0.8?

Solution:

Let subscript 1 refer to the 6 ft/sec ball and subscript 2 to the −8 ft/sec ball. The masses are equal $(m_1 = m_2 = m)$.

Hence $u_1 = 6$ ft/sec and $u_2 = -8$ ft/sec. The problem is to determine v_1 and v_2. Work directly from the following fundamental equations rather than applying the results of Problem 35.

$$(1)\quad e = \frac{v_2 - v_1}{u_1 - u_2}, \qquad 0.8 = \frac{v_2 - v_1}{6-(-8)}, \qquad \text{or} \qquad v_2 - v_1 = 11.2$$

$$(2)\quad m_1u_1 + m_2u_2 = m_1v_1 + m_2v_2, \qquad m(+6) + m(-8) = mv_1 + mv_2, \qquad \text{or} \qquad v_1 + v_2 = -2$$

Solve the two equations simultaneously to get $v_1 = -6.6$ ft/sec, $v_2 = 4.6$ ft/sec.

39. A ball rebounds vertically from a horizontal floor to a height of 64 ft. On the next rebound it reaches a height of 46 ft. What is the coefficient of restitution between the ball and the floor?

Solution:

Let subscript 1 refer to the ball. The initial and final speeds of the floor, u_2 and v_2, are zero since it is assumed the floor remains stationary during impact.

The second time the ball hits the floor it has fallen from a height of 64 ft. Its speed u_1 is

$$u_1 = \sqrt{2gh} = \sqrt{2(32.2\text{ ft/sec}^2)(64\text{ ft})} = 64.2\text{ ft/sec}$$

By similar reasoning the speed v_1 of rebound may be found. The ball with speed v_1 rises to a height of 46 ft. Hence its starting speed v_1 is $v_1 = \sqrt{2(32.2)(46)} = 54.5$ ft/sec.

Now apply the following equation of impact, where downward speeds are chosen positive:

$$e = \frac{v_2 - v_1}{u_1 - u_2} = \frac{0-(-54.5)}{64.2 - 0} = 0.85$$

The value of e could be determined by considering the square roots of the successive heights to which the ball bounced. This is true since the value of e in this case is actually the ratio of the speeds which depend on the square roots of the heights. $e = \sqrt{46/64} = 0.85$.

40. The 2 lb ball moving horizontally as shown in Fig. 19-30(a) with a speed of 12 ft/sec hits the bottom of the rigid slender 5 lb bar which pivots about its top. If the coefficient of restitution is 0.7, determine the angular speed of the bar and the linear speed of the ball just after impact.

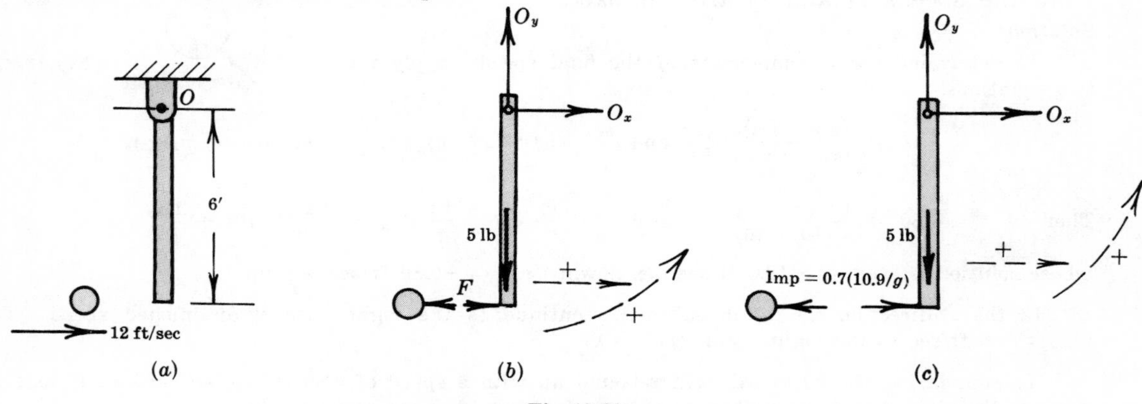

Fig. 19-30

Solution:

During the first phase of the impact, from time zero to t_1, the linear impulse acting between the ball and bar will cause the sphere to slow down and the lower end of the bar to speed up to a common speed u. The free body diagrams for this first phase are shown in Fig. 19-30(*b*) above. The linear impulse of the interacting force F causes the change in linear momentum of the ball as indicated in equation (*1*), where to the right is considered positive.

$$-\int_0^{t_1} F\,dt \quad = \quad \Delta G \quad = \quad (2/g)(u-12) \tag{1}$$

This linear impulse has a moment about the pivot point O of the bar, and causes a change in the angular momentum of the bar as indicated in equation (*2*) where counterclockwise is considered positive.

$$+\int_0^{t_1} F(6)\,dt \quad = \quad \Delta H_O \quad = \quad I_O(\omega - 0) \tag{2}$$

where $I_O = \frac{1}{3}mL^2 = \frac{1}{3}(5/g)(6^2) = 60/g$ slug-ft², $\omega = u/6$ rad/sec. Putting these values into equation (*2*) and dividing by 6, we have

$$+\int_0^{t_1} F\,dt \quad = \quad 5u/3g \tag{3}$$

Adding (*1*) and (*3*), we find $u = 6.55$ ft/sec. Thus at the end of this first phase (compression phase) the ball and the lower end of the bar have a common speed of 6.55 ft/sec to the right. The bar then has an angular speed $\omega = 6.55/6 = 1.09$ rad/sec counterclockwise.

The linear impulse acting during this phase can be calculated from (*1*) as

$$\int_0^{t_1} F\,dt \quad = \quad -(2/g)(6.55-12) \quad = \quad 10.9/g \text{ lb-sec}$$

This value can be checked by substituting $u = 6.55$ ft/sec into (*3*).

In the second phase (restitution phase) the linear impulse acting is only 0.7 of that acting in the compression phase, as shown in Fig. 19-30(*c*) above. Proceeding as before, using v as the speed of the ball immediately after impact and ω' as the angular speed of the bar immediately after impact, the equations for the ball and bar are respectively

$$(4) \quad -0.7(10.9/g) \; = \; (2/g)(v-6.55) \qquad\qquad (5) \quad +0.7(10.9/g)(6) \; = \; (60/g)(\omega' - 1.09)$$

from which $v = 2.72$ ft/sec to the right, and $\omega' = 7.73$ rad/sec counterclockwise.

41. Refer to the adjacent Fig. 19-31. The 20 lb ball moving down at 10 ft/sec strikes the 12 lb ball moving as shown at $6\sqrt{2}$ ft/sec. The coefficient of restitution $e = 0.80$. Find the speeds v_1 and v_2 after impact.

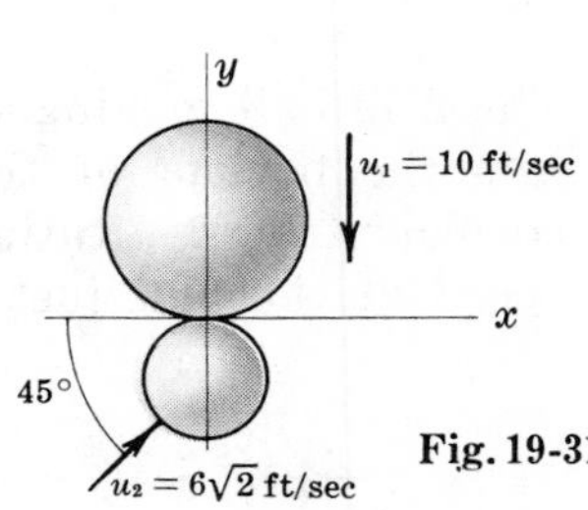

Fig. 19-31

Solution:

To determine the y components of the final speeds, apply the two equations:

$$e \; = \; \frac{(v_2)_y - (v_1)_y}{(u_1)_y - (u_2)_y} \qquad \text{and} \qquad m_1(v_1)_y + m_2(v_2)_y \; = \; m_1(u_1)_y + m_2(u_2)_y$$

Then $$0.80 \; = \; \frac{(v_2)_y - (v_1)_y}{(-10)-(6)} \qquad \text{and} \qquad \frac{20}{g}(v_1)_y + \frac{12}{g}(v_2)_y \; = \; \frac{20}{g}(-10) + \frac{12}{g}(+6)$$

whose solution is: $(v_2)_y = -12.0$ ft/sec, i.e. down; $(v_1)_y = +0.80$ ft/sec, i.e. up.

In the x direction the 12 lb ball will continue to the right with undiminished speed. Thus $(v_2)_x = +6$ ft/sec to the right, and $(v_1)_x = 0$.

To summarize, the 20 lb ball will rebound up with a speed of 0.80 ft/sec, and the 12 lb ball will move to the right and down with components of 6 and 12 ft/sec respectively.

42. Refer to Fig. 19-32. A body at rest is free to move on a smooth table top. It is struck by an impact force F at a distance d from its mass center G. Investigate the motion and locate the instantaneous center of the body in terms of d and its radius of gyration about its mass center.

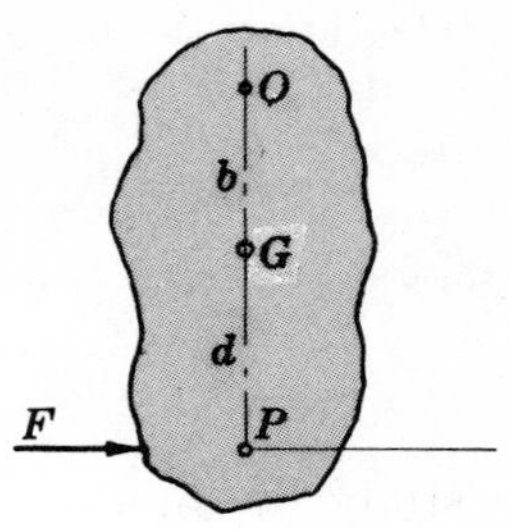

Fig. 19-32

Solution:

Employing the impulse-momentum equations for plane motion,

$$(1)\quad \Sigma(\text{Imp})_h = \Delta G_h \qquad \text{or} \qquad \int F\,dt = (W/g)(\bar{v} - 0)$$

$$(2)\quad \Sigma(\text{Ang Imp})_G = \Delta H_G \qquad \text{or} \qquad \int Fd\,dt = \bar{I}(\omega - 0)$$

Since the distance d is constant it may be removed from under the integral sign. Then $\bar{v} = (g/W)\int F\,dt$ and $\omega = (d/\bar{I})\int F\,dt$. But $\bar{I} = (W/g)\bar{k}^2$; hence $\omega = (gd/W\bar{k}^2)\int F\,dt$, where $\bar{k}$ is the radius of gyration of the body with respect to the mass center.

To locate the instantaneous center – a point about which the body tends to pivot only – select any point O at a distance b from the mass center on a line perpendicular to the action line of the impact force F, as shown in the figure. The velocity of O is $\mathbf{v}_O = \mathbf{v}_{O/G} + \bar{\mathbf{v}}$. But $\bar{v}$ has been found in terms of the linear impulse. Also, $v_{O/G} = b\omega$ where ω has just been defined in terms of the linear impulse. Note that for the object shown, $\bar{v}$ is to the right (+) and ω is counterclockwise. Hence $v_{O/G}$ is to the left (–).

Substituting in the formula, $\quad v_O = -b\omega + \bar{v} = -\dfrac{bdg}{W\bar{k}^2}\displaystyle\int F\,dt + \dfrac{g}{W}\int F\,dt.$

To make O an instantaneous center, v_O must be zero. Set the above expression equal to zero and solve to obtain $b = \bar{k}^2/d$.

This may be viewed slightly differently. Suppose the distance d was chosen originally as $\bar{k}^2/b$. Then point O will remain at rest. When this occurs the force F is said to be applied through the *center of percussion P*.

43. Apply the results of the preceding problem to find the height h above the plane at which a solid cylinder of radius R should be struck by a horizontal impact force F in order to have no sliding at the point of contact O. See Fig. 19-33.

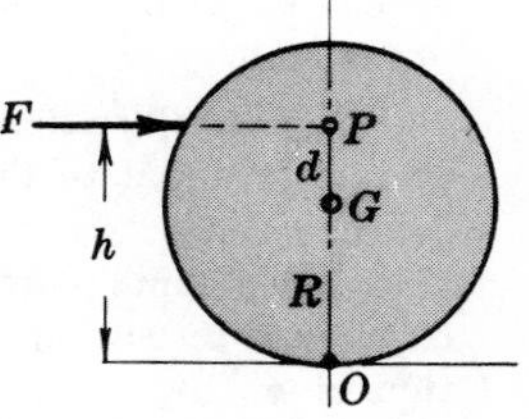

Fig. 19-33

Solution:

According to the theory, F should be struck through the center of percussion P in order that the point of contact O be the instantaneous center.

Hence $h = d + R = \bar{k}^2/R + R$. But for a cylinder, $\bar{k}^2 = \bar{I}/m = \frac{1}{2}mR^2/m = \frac{1}{2}R^2$. Then $h = \frac{1}{2}R + R = \frac{3}{2}R$.

44. A rocket and its fuel has an initial mass m_0. Fuel is burned at a constant rate, $dm/dt = C$. The gases exhaust at a constant speed relative to the rocket. Neglecting air resistance, find the speed of the rocket at time t.

Solution:

Let $\quad m$ = mass of rocket and remaining fuel at time t, i.e. $m = m_0 - Ct$
$\quad v$ = speed of rocket, u = speed of gases.

We will assume vertical motion with up being positive. The only external force to the system at time t is the weight mg which is negative. Our equation is

$$\Sigma F = m\frac{dv}{dt} - \frac{dm}{dt}(v - u)$$

where $(v - u)$ is the relative speed of the rocket to the gases and is a constant K. Then

$$-mg = m\frac{dv}{dt} - CK, \qquad -(m_0 - Ct)g = (m_0 - Ct)\frac{dv}{dt} - CK, \qquad \text{or} \qquad \frac{dv}{dt} = -g + \frac{CK}{m_0 - Ct}$$

Integrating the last equation,

$$\int_0^v dv = \int_0^t -g\,dt + \int_0^t \frac{CK}{m_0 - Ct}\,dt$$

or $$v = -gt + [-K\ln(m_0 - Ct)]_0^t = -gt - K\ln(m_0 - Ct) + K\ln m_0 = K\ln\frac{m_0}{m_0 - Ct} - gt.$$

45. Study the motion of the gyroscope shown schematically in Fig. 19-34(a) as a spinning wheel and rotor which is attached to point O in such a way that it is free to turn in any direction about O.

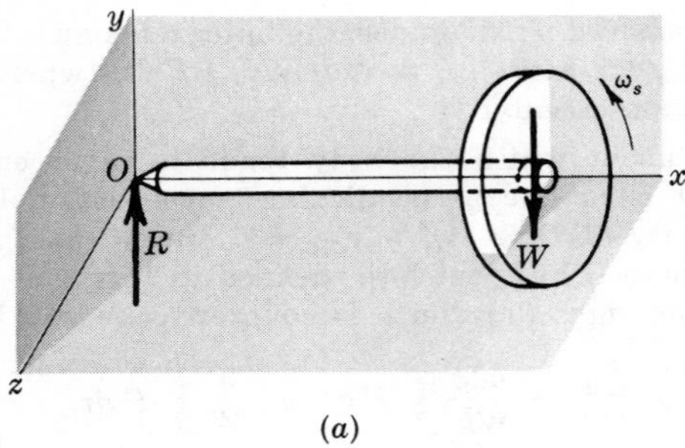

(a) (b)

Fig. 19-34

Analysis:

Assume that the wheel is spinning with a large angular velocity $\boldsymbol{\omega}_s$ about the x-axis which is called the spin axis. Also assume that the system is released; the external forces acting on it are the weight W of the wheel (neglect weight of rotor) and the upward reaction R at O. The weight W exerts a torque on the system about the z-axis which is called the moment axis. It is known by actual experiment that the wheel does not fall but turns about the y-axis which is called the axis of precession.

The explanation for this follows. Before the system is released, the angular momentum of the system is equal to the product of its moment of inertia and its angular velocity $\boldsymbol{\omega}_s$, i.e. $\bar{I}\boldsymbol{\omega}_s$. This can be represented by a vector **OB** directed along the x-axis to the right. (According to the right hand rule, the thumb points to the right when the fingers are curled in the direction of $\boldsymbol{\omega}_s$. This is also expressed by saying that a right handed screw would advance to the right when turned in the direction of $\boldsymbol{\omega}_s$.)

Immediately upon release there is an external torque **M** acting on the system. It acts clockwise about the z-axis when viewed from the positive end of that axis, and its magnitude is equal to the product of the weight and the perpendicular distance from the z-axis to the center of mass. For a short time interval dt, the impulse of this torque is **M**dt. Its units are the same as the units of angular momentum and it may be represented according to the right hand rule by a vector along or parallel to the z-axis and acting in the negative direction. Fig. 19-34(b) illustrates this vector **BC**. But this angular impulse **M**dt is the change in the angular momentum of the system. The new angular momentum is the vector sum of **OB** and **BC**. This line **OC** is in the xz-plane but at an angle $d\phi$ with the original x-axis. Note that the new axis of spin is along OC. Using scalar quantities and substituting $d\phi$ for $\tan d\phi$, there results $d\phi = \dfrac{M\,dt}{\bar{I}\omega_s}$ or $M = \bar{I}\omega_s\dfrac{d\phi}{dt}$.

But $d\phi/dt$ is the angular speed of the spin axis about the precession axis, say ω_p.

The equation of the gyroscope is $M = \bar{I}\omega_s\omega_p$. Note that the spin axis remains in the xz-plane, provided ω_s remains large.

46. A rotor weighing 4000 lb and with a radius of gyration 3.00 ft is mounted with its geometric axis along the fore and aft line of a ship that is turning 1 rpm counterclockwise as viewed from above. The rotor turns 300 rpm counterclockwise when observed from the rear of the ship. Assuming the center to center distance between bearings to be 3.50 ft, determine the front and rear bearing reactions on the rotor.

Solution:

Draw a free body diagram of the rotor with the x-axis as the fore and aft line of the ship. See Fig. 19-35. The front and rear reactions are R_F and R_R respectively. According to the right hand rule, the angular momentum about the x-axis ($\bar{I}\boldsymbol{\omega}_s$) is drawn to the left along the x-axis. To precess in the given direction about the y-axis, the rotor must be acted upon by an angular impulse ($\mathbf{M}\,dt$) as shown. According to the right hand rule, this moment (caused by the reactions) must be counterclockwise when viewed from the positive end of the z-axis. The magnitude of this moment is $M = (3.50/2)(R_F - R_R)$.

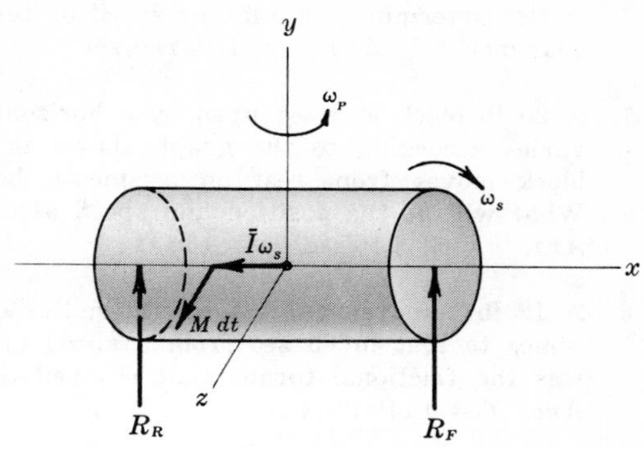

Fig. 19-35

The values given are: $\omega_s = 2\pi(300/60) = 31.4$ rad/sec; $\omega_p = 2\pi(1/60) = 0.105$ rad/sec; $\bar{I} = m\bar{k}^2 = (4000)/(32.2)(3.00)^2 = 1118$ slug-ft^2.

Then $M = \bar{I}\omega_s\omega_p = 3690$ lb-ft, and $(R_F - R_R) = M(2/3.50) = 2110$ lb.

A vertical sum yields $R_F + R_R = 4000$ lb. The solution is $R_F = 3055$ lb and $R_R = 945$ lb.

Supplementary Problems

47. In Problem 6, show that the selection of the top point as the moment center yields $2PR = \frac{d}{dt}(4mR^2\omega)$ which does not agree with the results obtained in that problem. [It should not agree because the top point does not fit the conditions imposed on point O in equation (*9*).]

48. A 100 lb weight falls freely from rest for 4 sec. Find its momentum at that time.
Ans. 400 lb-sec

49. A pile driver hammer weighing 1000 lb is dropped from a height of 25 ft. If the time required to stop the pile driver is 1/20 sec, determine the average force acting during that time.
Ans. $F_{av} = 25{,}000$ lb

50. A body is thrown upward with an initial speed of 50 ft/sec. Find the time required for the body to attain a speed of 20 ft/sec downward. *Ans.* 2.18 sec

51. A 50 lb body is projected up a 30° plane with initial speed 20 ft/sec. If the coefficient of friction is 0.25, determine the time required for the body to have an upward speed of 10 ft/sec.
Ans. 0.433 sec

52. The magnitude of a force applied tangentially to the rim of a 4 ft diameter pulley varies according to the law $F = 0.03t$. Determine the angular impulse of the force about the axis of rotation for the interval 0 to 35 sec. Hint: Use the integral definition of angular impulse. *Ans.* 36.8 lb-sec-ft

53. A 300 lb disk rotating 1000 rpm has a diameter of 3 ft. Find its angular momentum.
Ans. 1100 lb-sec-ft

54. A disk of weight 24 lb and with a radius of gyration equal to 1.9 ft is subjected to a torque $M = 2t$ lb-ft. Determine the angular speed of the disk 2 sec after it started from rest. Assume frictionless bearings. *Ans.* $\omega = 1.49$ rad/sec

55. A 20 lb block is acted upon by a horizontal force P which varies according to the graph shown in Fig. 19-36. The block moves from rest on a smooth horizontal surface. What will be the position and speed after 8 sec?
Ans. $v = 33.8$ ft/sec, $s = 134$ ft

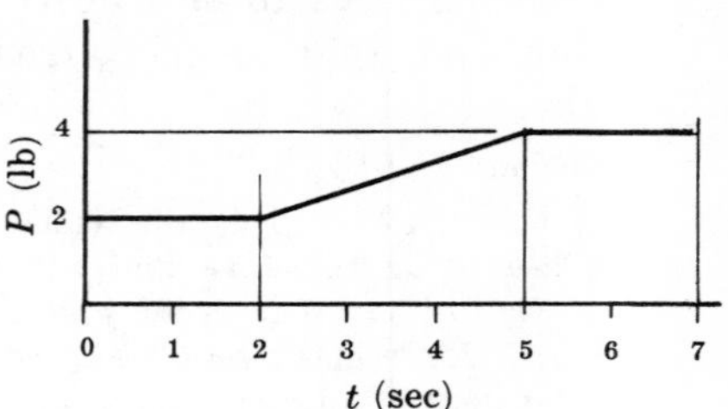

Fig. 19-36

56. A 12 lb homogeneous rotor with radius of gyration 1.2 ft comes to rest in 85 sec from a speed of 180 rpm. What was the frictional torque that stopped the rotor?
Ans. $T = 0.119$ lb-ft

57. The two weights shown in Fig. 19-37 below are connected by a light inextensible cord that passes over a homogeneous cylinder of weight 8 lb and diameter 4 ft. How long will it take the 14 lb weight to move from rest to a speed of 6 ft/sec? *Ans.* $t = 1.30$ sec

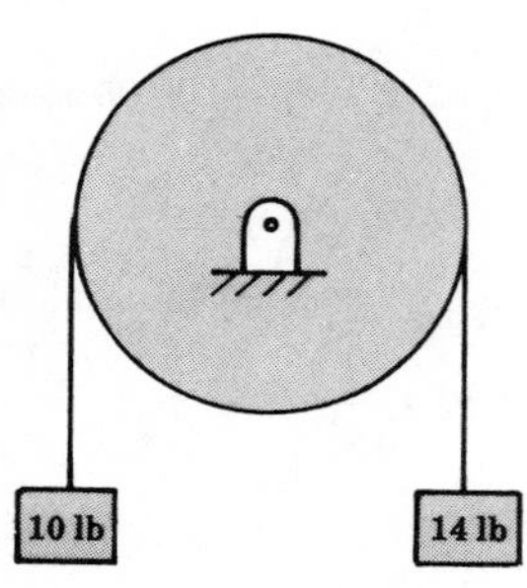

Fig. 19-37

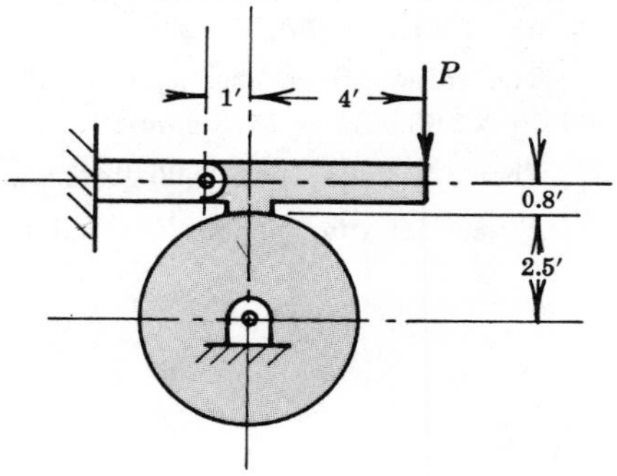

Fig. 19-38

58. A rotor weighing 230 lb has a radius of gyration 2.42 ft. The rotor is rotating 120 rpm counterclockwise when the force P is applied to the brake as shown in Fig. 19-38 above. What is the magnitude of P if the rotor stops in 92 sec? The coefficient of friction between the brake and rotor is 0.4. *Ans.* $P = 0.78$ lb

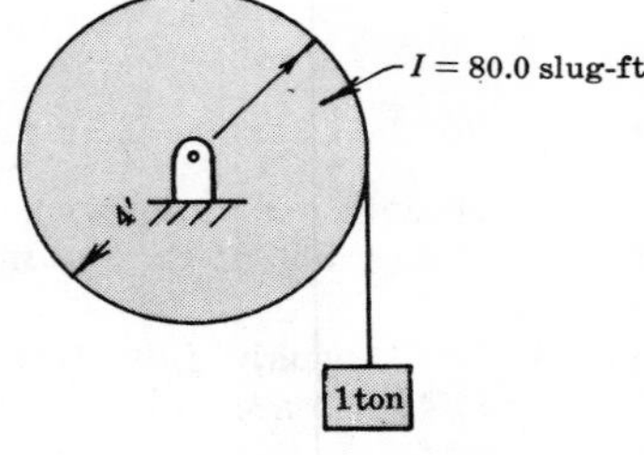

Fig. 19-39

59. The drum rotating 20 rpm is lifting a one ton cage connected to it by a light inextensible cable as shown in Fig. 19-39. If power is cut off, how long will it take the cage to come to rest? Assume frictionless bearings. *Ans.* $t = 0.172$ sec

60. Two homogeneous disks of weights W_1 and W_2 and radii r_1 and r_2 are free to turn in frictionless bearings in the fixed vertical frame shown in Fig. 19-40. The upper disk W_1 is turning ω rad/sec clockwise when it is dropped onto the lower disk W_2. Assuming the coefficient of friction μ is constant, determine the time until slipping stops. What are the angular velocities at that time?

Ans. $t = \dfrac{r_1\omega}{2\mu g(1 + W_1/W_2)}$,

$\omega_1 = \dfrac{\omega}{1 + W_2/W_1}$ clockwise,

$\omega_2 = \dfrac{r_1}{r_2}\left(\dfrac{\omega}{1 + W_2/W_1}\right)$ counterclockwise

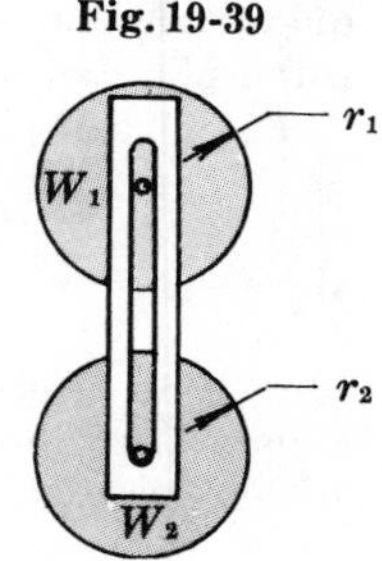

Fig. 19-40

61. Water issues at the rate of 60 ft/sec from a 4 inch diameter nozzle. Determine the reaction of the nozzle against its supports. *Ans.* 610 lb

62. A jet of water 2 inches in diameter exerts a force of 300 lb on a flat vane perpendicular to the stream. What is the nozzle speed of the jet? *Ans.* 84.2 ft/sec

63. A jet of water issues from a 1 inch diameter nozzle at 100 ft/sec. Determine the total force against a circularly curved vane where the direction of the jet changes 45°. Assume no friction.
Ans. 81.2 lb

64. The force on a curved vane that changes the direction of a jet of water by 45° is 450 lb. If the nozzle speed of the jet is 80 ft/sec, calculate the diameter of the nozzle. *Ans.* 2.95 in.

65. Refer to Fig. 19-41. A stream of water 4 in^2 in cross section and moving horizontally with a speed of 100 ft/sec splits into two equal parts against the fixed vane. Assuming no friction between the water and the blade, determine the force of the water on the blade.
Ans. $P_x = 919$ lb to right

66. In the preceding problem, assume the stream splits so that two-thirds of the water moves to the top portion of the blade and one-third to the lower portion of the blade. Determine the reaction of the water on the blade.
Ans. $P_x = 919$ lb to right, $P_y = 127$ lb up

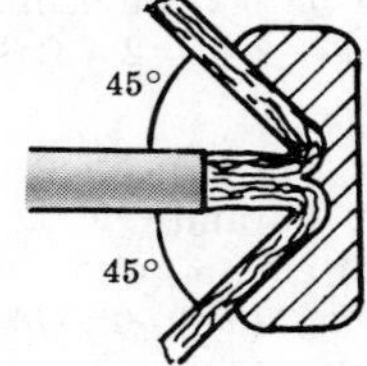

Fig. 19-41

67. A nozzle with cross sectional area 2.63 in^2 discharges a horizontal stream of water with a speed of 80 ft/sec against a fixed vertical plate as shown in Fig. 19-42 below. What is the force against the plate? *Ans.* 226 lb to right

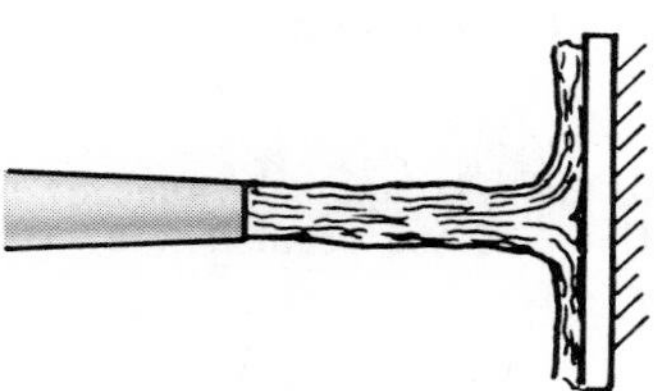

Fig. 19-42

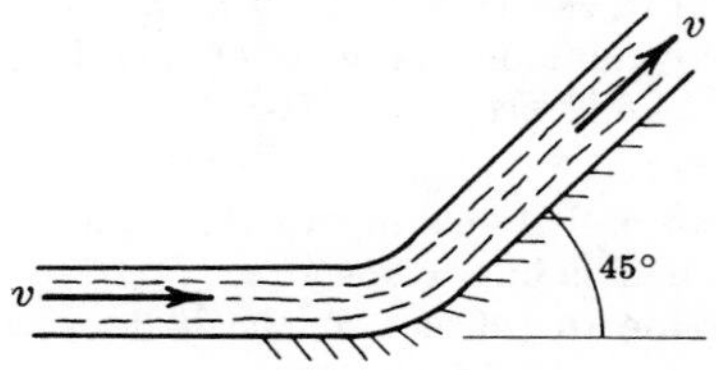

Fig. 19-43

68. A stream of water with cross sectional area 3 in^2 and moving 30 ft/sec strikes a stationary curved blade as shown in Fig. 19-43 above. Assuming that friction is negligible, find the horizontal and vertical components of the force of the blade on the stream.
Ans. $P_x = 10.7$ lb to left, $P_y = 25.7$ lb up

69. Two sliding disks A and B are mounted on the same shaft. Disk A is 3 ft in diameter, 2 inches thick and weighs 100 lb, and is at rest. Disk B is 3 ft in diameter, 4 inches thick and weighs 200 lb, and is rotating clockwise 600 rpm. If the disks are engaged so as to rotate together, what is their common angular speed? *Ans.* 400 rpm

70. In Problem 69, what is the percent loss in kinetic energy of the system because the disks are coupled together? *Ans.* 33.3%

71. A 70 lb boy is standing in a 100 lb boat that is initially at rest. If the boy jumps horizontally with a speed of 6 ft/sec relative to the boat, determine the speed of the boat. *Ans.* 2.47 ft/sec

72. After a flood, a goat weighing 40 lb finds itself adrift on one end of a 50 lb log 6 ft long. As the other end touches shore the goat maneuvers to that end. When it gets there, how far is it from shore? Assume the log is at right angles to the shore and the water is almost calm after the storm.
Ans. 2.67 ft

73. A 0.2 lb bullet is fired from a 14 lb rifle with a muzzle speed of 1000 ft/sec. What is the speed of the rifle recoil? *Ans.* 14.3 ft/sec

74. A 100,000 lb gun fires a projectile that weighs 1000 lb. If the recoil apparatus exerts a constant force of 90,000 lb and the gun moves back 7.8 inches, calculate the muzzle speed of the projectile.
Ans. 613 ft/sec

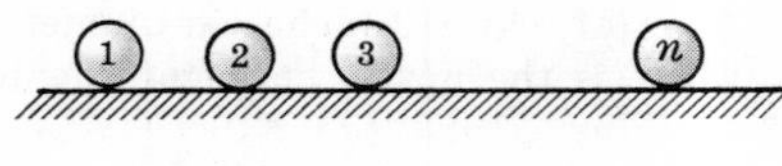

Fig. 19-44

75. In Fig. 19-44, a series of n identical balls is shown on a smooth horizontal surface. If number 1 moves horizontally with speed u into number 2 which in turn collides with number 3, etc., and if the coefficient of restitution for each impact is e, determine the speed of the nth ball.
Ans. $v_n = (1+e)^{n-1}u/2^{n-1}$

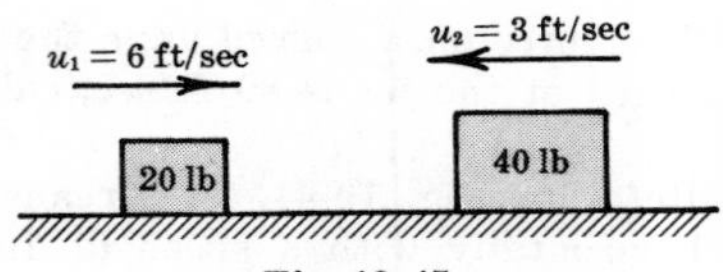

Fig. 19-45

76. In Fig. 19-45, the 20 lb block is moving to the right with a speed of 6 ft/sec. The 40 lb block is moving to the left with a speed of 3 ft/sec. If the coefficient of restitution is assumed to be 0.4, determine the speeds immediately after impact.
Ans. $v_{20} = 2.4$ ft/sec to left, $v_{40} = 1.2$ ft/sec to right

77. A 4 lb ball and a 3 lb ball move on a smooth horizontal plane along a straight line path with speeds of +6 and −8 ft/sec respectively. Determine their speeds after impact if the impact is (*a*) inelastic, (*b*) elastic, and (*c*) such that the coefficient of restitution is 0.5.
Ans. (*a*) 0, 0; (*b*) −6, +8; (*c*) −3, +4 ft/sec

78. Block A weighing 20 lb and moving 12 ft/sec horizontally to the right meets block B weighing 16 lb and moving 8 ft/sec horizontally to the left. If the coefficient of restitution is $e = 0.7$, determine the speeds of A and B immediately after impact.
Ans. $v_A = 3.11$ ft/sec to left, $v_B = 10.9$ ft/sec to right

79. A 2000 lb car traveling 30 mph overtakes a 1500 lb car traveling 15 mph in the same direction. What is their common speed after coupling? What is the loss in kinetic energy?
Ans. 34.6 ft/sec, 6300 ft-lb

80. A 30 lb ball A moving to the right with a speed of 30 ft/sec strikes squarely another ball B weighing 10 lb which has a speed of 7 ft/sec in the opposite direction. If the coefficient of restitution is 0.6, determine the speeds of the balls after impact.
Ans. $v_A = 15.2$ ft/sec right, $v_B = 37.4$ ft/sec right

81. A ball falls freely from rest through a distance of 100 ft. If it strikes the ground and bounces to a height of 64 ft, what is the coefficient of restitution between the ball and the ground? *Ans.* $e = 0.80$

82. A glass ball is dropped onto a smooth horizontal floor from which it bounces to a height of 9 ft. On the second bounce it attains a height of 6 ft. What is the coefficient of restitution between the glass and the floor? *Ans.* $e = 0.82$

83. A 4 lb weight falls 0.5 ft onto a 2 lb platform mounted on springs whose combined $k = 50$ lb/ft. If the impact is fully plastic ($e = 0$), determine the maximum distance the platform moves down from its position before impact. See Fig. 19-46 below. *Ans.* 3.90 in.

84. In Fig. 19-47 below, the weight W is moving with speed u when the string becomes taut. What will be the speed of weight w if the coefficient of restitution is e. The weights are on a smooth horizontal plane. *Ans.* $v = Wu(1+e)/(W+w)$

85. A mass m_1 moving with speed u strikes a stationary mass m_2 hanging on a string of length L ft as shown in Fig. 19-48 below. If the coefficient of restitution is e, determine (*a*) the speed of each mass immediately after the impact and (*b*) the height h to which mass m_2 will rise.

Ans. $v_1 = u\dfrac{m_1 - em_2}{m_1+m_2}$, $v_2 = u\dfrac{m_1(1+e)}{m_1+m_2}$, $h = \dfrac{u^2m_1^2(1+e)^2}{(m_1+m_2)^2\,2g}$

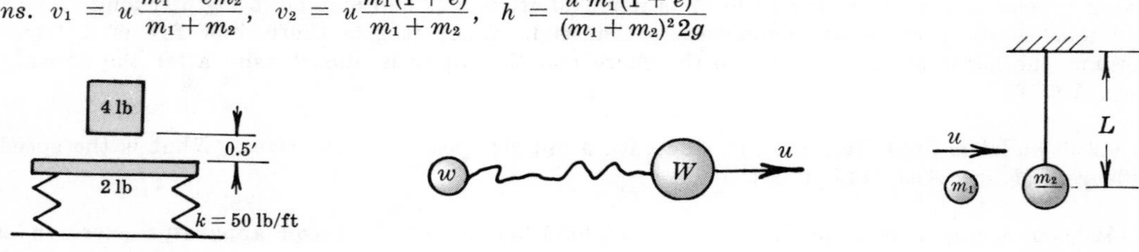

Fig. 19-46 Fig. 19-47 Fig. 19-48

86. A 2 lb ball traverses the frictionless tube shown in Fig. 19-49 below, falling a vertical distance of 3 ft. It then strikes a 2 lb ball hanging on a 3 ft cord. Determine the height to which the hanging ball will rise (*a*) if the collision is perfectly elastic, (*b*) if the coefficient of restitution is 0.7.
Ans. (*a*) 3 ft, (*b*) 2.17 ft

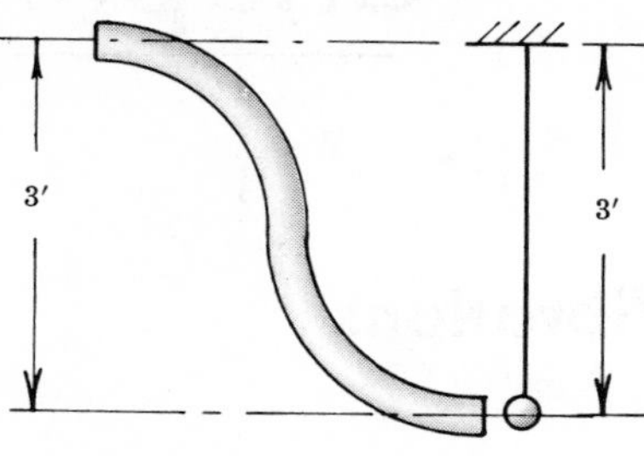

Fig. 19-49

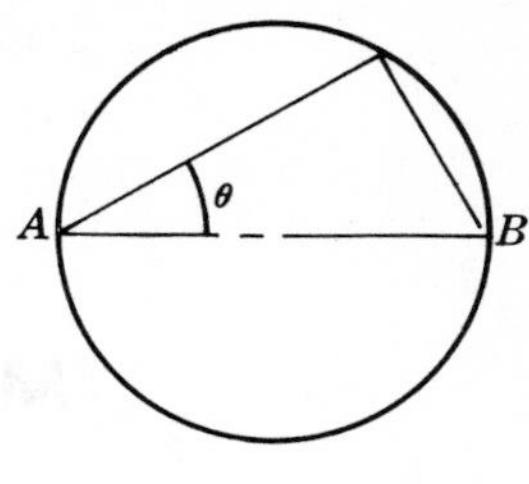

Fig. 19-50

87. Refer to Fig. 19-50 above. A ball thrown from position A against a smooth vertical circular wall rebounds and hits position B at the other end of the diameter through A. Show that the coefficient of restitution is equal to the square of the tangent of angle θ. Extraneous information: ball weighs 2 ounces, diameter of the circle is 10 ft, and speed of the ball is 28 ft/sec.

88. The billiard ball moving 12 ft/sec strikes a smooth horizontal plane at an angle of 35° as shown in Fig. 19-51 below. If the coefficient of restitution is 0.6, what is the velocity with which the ball rebounds? *Ans.* $v = 10.7$ ft/sec, $\theta = 22.8°$

89. In Fig. 19-52 below, a large drum is filled with liquid weighing 50 lb/ft³. The drum and liquid weigh 200 lb and are at rest on a horizontal sheet of ice for which the coefficient of friction is 0.05. If a 4 inch plug 3 ft below the surface of the liquid is suddenly removed, will the drum move? *Ans.* Yes

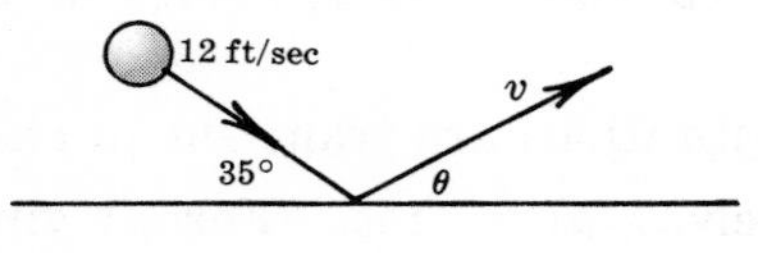

Fig. 19-51

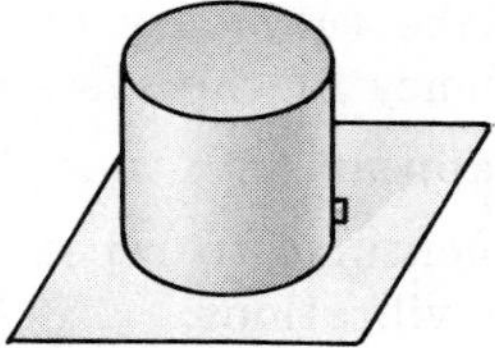
Fig. 19-52

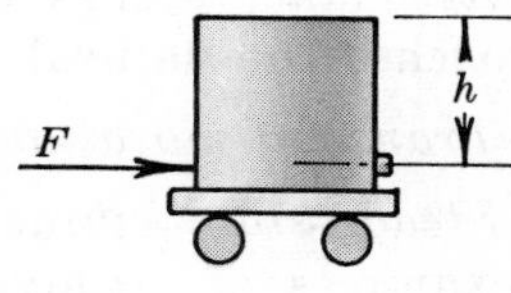

Fig. 19-53

90. A tank of liquid weighing δ lb/ft³ rests on a cart. If h is the height of the fluid above the orifice, what horizontal force F is necessary to hold the tank at rest when the fluid starts issuing from the nozzle? The cross sectional area of the orifice is A ft². See Fig. 19-53 above. *Ans.* $F = 2Ah\delta$

91. A rocket weighs 6000 lb empty. If it is projected vertically up from the earth with a fuel load of 15,000 lb, calculate the initial acceleration. Assume the gases exhaust at 6000 ft/sec relative to the rocket and that the initial weight rate of fuel burning is 300 lb/sec. *Ans.* $a = 53.5$ ft/sec²

92. The rotor of a gyroscope is a homogeneous cylinder 4 inches in diameter and weighing 6 ounces. It is mounted horizontally midway between bearings 6 inches apart. The rotor is turning 9000 rpm in a clockwise direction when viewed from the rear. The assembly is turning 2 rad/sec about a vertical axis in a clockwise direction when viewed from above. What are the bearing reactions on the rotor shaft? *Ans.* $R_F = 12.8$ oz up, $R_R = 6.8$ oz down

93. Refer to Fig. 19-54. A solid wheel 6 in. in diameter and 2 in. thick rotates 6000 rpm. Neglect the mass of the shaft attached to it. Assume a weight density of 480 lb/ft³. Determine the speed of precession.
Ans. $\omega_p = 1.10$ rad/sec clockwise about the y-axis when viewed from above.

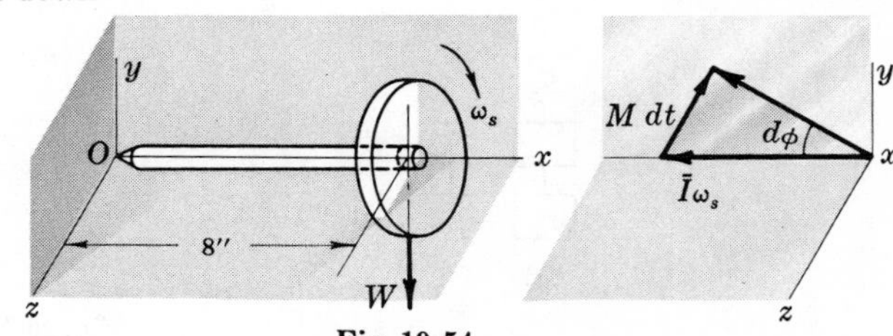

Fig. 19-54

Chapter 20

Mechanical Vibrations

DEFINITIONS

Mechanical vibration of a system possessing masses and elasticity is motion which repeats itself in a definite time interval.

The period is the time for the vibration to repeat itself.

A cycle is each repetition of the entire motion completed during the period.

The frequency is the number of cycles in a unit of time.

Free vibrations occur in a system acted upon by forces within the system – such as weights of the component parts, springs or other elastic components.

Natural frequency is the frequency of a system undergoing free vibrations.

Forced vibrations occur in a system acted upon by periodic external disturbing forces.

Resonance occurs when the frequency of the forced vibrations coincides or at least approaches the natural frequency of the system.

Transient vibrations disappear with time. Free vibrations are transient in character.

Steady-state vibrations continue to repeat themselves with time. Forced vibrations are examples of steady-state vibrations.

DEGREES of FREEDOM

The degrees of freedom of a system depend on the number of variables (coordinates) needed to describe its motion.

For example, in Fig. 20-1(a) the motion of the mass on a spring which is assumed to vibrate only in a vertical line can be described with one coordinate and thus possesses a single degree of freedom. A bar supported by the two springs in Fig. 20-1(b) needs two variables as shown and therefore possesses two degrees of freedom.

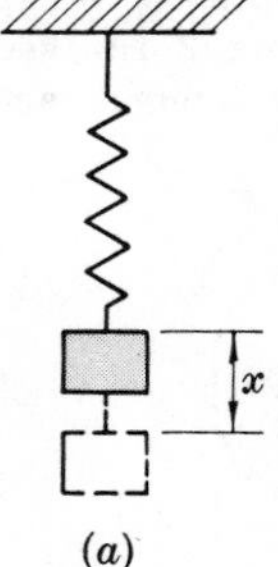

(a)

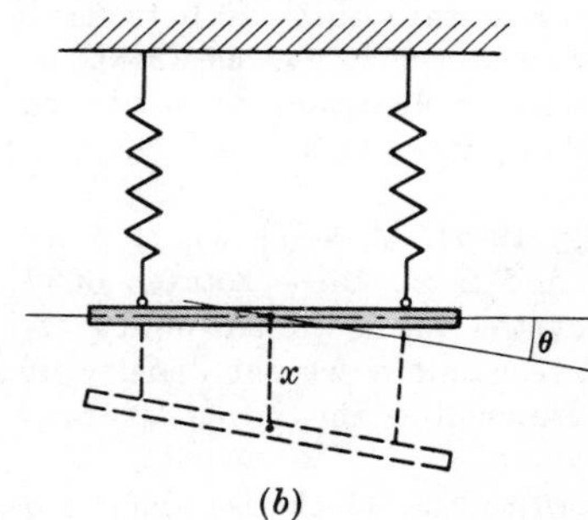

(b)

Fig. 20-1

SIMPLE HARMONIC MOTION

Simple harmonic motion can be represented by a sine or cosine function of time. Thus $x = X \sin \omega t$ is an equation of simple harmonic motion. It could represent the projection on a diameter of a vector X inches long as the tip of the vector X rotates on a circular path with a constant angular velocity ω rad/sec. For this motion

x = length of the projection in inches on a diameter

X = length of the rotating vector in inches

ω = circular frequency in rad/sec

$\tau = 2\pi/\omega$ = period in sec

$f = \omega/2\pi$ = frequency in cycles/sec.

COMPLICATED SYSTEMS

A complicated system is analyzed by replacing it with an equivalent system of masses, springs, and damping devices. The differential equations of this idealized system, when solved, will approximate the desired results. Engineering judgment will suggest proper modifications to fit the actual system.

The problems which follow illustrate free vibrations with and without damping and forced vibrations with and without damping. Only viscous damping (where the damping force is proportional to the velocity of the body) is considered. However, it is well to point out two other types of damping: (1) Coulomb damping which is independent of the velocity and arises because of sliding of the body on dry surfaces (its force is thus proportional to the normal force between the body and the surface on which it slides), and (2) solid damping which occurs as internal friction within the material of the body itself (its force is independent of the frequency and proportional to the maximum stress induced in the body itself).

Solved Problems

FREE VIBRATIONS — LINEAR

1. A weight W hangs on a vertical spring whose spring constant or modulus is k lb/in. Assuming the weight of the spring may be neglected, study the motion of the weight if it is released at a distance x_0 below the equilibrium position with an initial velocity v_0 downward.

Analysis:

In Fig. 20-2, various positions of the weight W are shown. The distances are exaggerated for clarity. The value X is the amplitude of the motion. Of course, the weight will also rise to a height X above the equilibrium position.

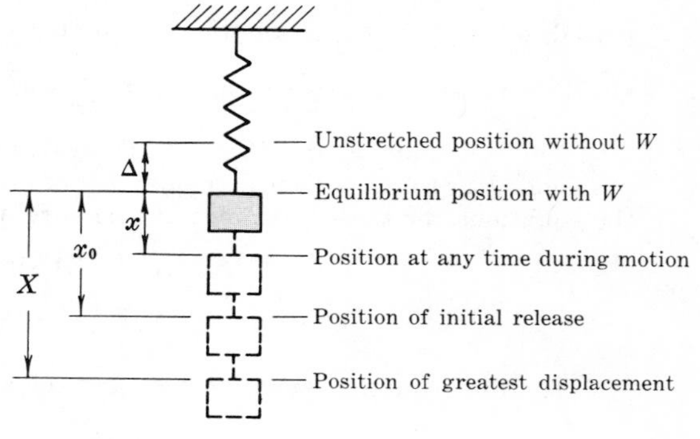

Fig. 20-2

The tension in the spring is equal to the product of the spring modulus k and the distance the spring is stretched or compressed from its unstretched position (without W). Draw free body diagrams of the weight in its equilibrium position and at the position x below equilibrium. Note that the position x of the body at any time is expressed from its position of static equilibrium.

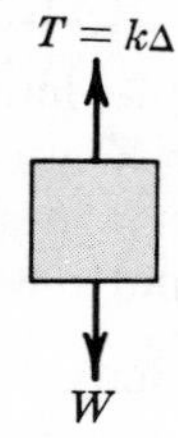

Equilibrium position

Fig. 20-3

In Fig. 20-3 no acceleration is shown since the system is in equilibrium. Hence $T = k\Delta = W$. Note that consequently $\Delta = W/k$.

In Fig. 20-4, assume displacements below the equilibrium position are positive. Since the direction of the acceleration is unknown, assume it to be positive. A negative sign would then mean it is directed up. Apply Newton's Laws to this free body diagram and obtain

$$\Sigma F_v = ma \qquad \text{or} \qquad W - T = (W/g)a$$

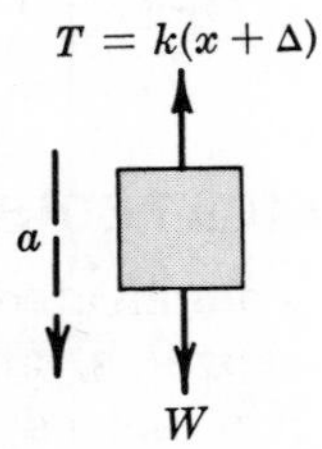

Position below equilibrium

Fig. 20-4

Substitute $T = k(\Delta + x)$ and $W = k\Delta$ to get $k\Delta - k\Delta - kx = (W/g)a$. Since $a = d^2x/dt^2$, the equation of motion becomes $\dfrac{d^2x}{dt^2} + \dfrac{kg}{W}x = 0$. It should be apparent that whenever x is positive (below the equilibrium position) the acceleration (which is then negative) is directed up. This means that as the weight moves down to its lowest position, the acceleration is up or the weight is decelerating. Just after reaching the bottom and starting up, the displacement is still positive and the acceleration is still directed oppositely, i.e. up. Therefore the weight will accelerate up to the equilibrium position. Above this position, the displacement is negative and the acceleration is directed down to the equilibrium position. Hence up to its top point of travel, the weight decelerates and between the top point and the position of equilibrium it accelerates. The acceleration is always directed toward the equilibrium position.

Assume that the solution of the above second order differential equation has the form

$$x = A \sin \omega t + B \cos \omega t$$

where ω is the circular frequency in radians/sec.

To determine whether or not this is a solution, take the second derivative with respect to time (d^2x/dt^2) and substitute it into the differential equation. Note that

$$\frac{dx}{dt} = A\omega \cos \omega t - B\omega \sin \omega t \qquad \text{and} \qquad \frac{d^2x}{dt^2} = -A\omega^2 \sin \omega t - B\omega^2 \cos \omega t = -\omega^2 x$$

Substitute the value of d^2x/dt^2 into the equation of motion to obtain $-\omega^2 x + (kg/W)x = 0$. Hence ω must equal $\sqrt{kg/W}$ if the assumed value of x is to be a solution. Then the solution is thus far

$$x = A \sin \sqrt{kg/W}\, t + B \cos \sqrt{kg/W}\, t$$

Note that a cycle of motion will be completed at intervals of 2π radians, i.e. when $\sqrt{kg/W}\,\tau = 2\pi$ where τ is the period or time for one cycle. Hence $\tau = 2\pi\sqrt{W/kg}$. The frequency is the inverse of the period, or $f = \dfrac{1}{\tau} = \dfrac{1}{2\pi}\sqrt{kg/W}$. As pointed out previously, $\Delta = W/k$; hence the above formulas may also be written $\tau = 2\pi\sqrt{\Delta/g}$ and $f = \dfrac{1}{2\pi}\sqrt{g/\Delta}$.

The constants A and B should be evaluated on the basis of the boundary conditions given in the problem. Here it is assumed that at $t = 0$, $x = x_0$ and $v = v_0$. Substitute these values of x and v into the equations for these variables but be sure also to substitute the time $t = 0$.

$$x_0 = A \sin(\omega \cdot 0) + B \cos(\omega \cdot 0)$$
$$v_0 = A\omega \cos(\omega \cdot 0) - B\omega \sin(\omega \cdot 0)$$

The first equation yields $B = x_0$, and the second equation gives $A = v_0/\omega$.

Hence the solution is: $\quad x = \dfrac{v_0}{\omega} \sin \omega t + x_0 \cos \omega t.$

The solution may be written in another form: $x = X\cos(\omega t - \phi)$, where X = amplitude = $\sqrt{(v_0/\omega)^2 + (x_0)^2}$ and ϕ = phase angle = $\tan^{-1}\frac{v_0}{x_0\omega}$. Note that $\omega = \sqrt{kg/W}$.

This problem is the most commonly used example of free vibrations which theoretically would continue indefinitely after the weight is set in motion. Since only one variable has been used to describe the motion, the system possesses one degree of freedom. Many additional problems can now be solved by reducing them to this type. In other words, replace the actual elastic suspension by an equivalent spring attached to the vibrating body.

2. A motor weighing 50 lb is mounted on four springs each with a spring constant k = 10 lb/in. Find the frequency and period of the motor as it vibrates.

Solution:

Each spring carries 12.5 lb if the weight is uniformly distributed. Hence the frequency in cycles per second is

$$f = \frac{1}{2\pi}\sqrt{\frac{kg}{W}} = \frac{1}{2\pi \text{ rad/cycle}}\sqrt{\frac{(10 \text{ lb/in})(386 \text{ in/sec}^2)}{12.5 \text{ lb}}} = 2.80 \text{ cps}$$

Note that the acceleration of gravity g must be in inches/sec² in order that the equation be dimensionally correct. The period $\tau = 1/f = 0.36$ sec.

3. Solve Problem 1 using the conservation of energy theorem.

Solution:

This theorem states that the sum of the potential energy V and kinetic energy T of the system is a constant provided the system is conservative (no friction or damping is assumed at this point in the discussion).

At any distance x below the equilibrium position, the spring tension is $W + kx$. Hence the potential energy V_s of the spring is equal numerically to the work done by this force in stretching the spring:

$$V_s = \int_0^x (W + kx)\,dx = Wx + \tfrac{1}{2}kx^2$$

During this same displacement x, the weight has lost potential energy of the amount Wx. Then the total potential energy of the system is $\frac{1}{2}kx^2$.

At a distance x below equilibrium, the kinetic energy T of the weight which is moving with a velocity dx/dt is $T = \frac{1}{2}(W/g)(dx/dt)^2$. Since the spring is assumed to be weightless, its kinetic energy is zero. Therefore the kinetic energy of the system is that of the weight only.

The conservation of energy theorem states that $T + V$ = constant, or that $\frac{d}{dt}(T+V) = 0$. Then

$$\frac{d}{dt}\left[\frac{1}{2}\frac{W}{g}\left(\frac{dx}{dt}\right)^2 + \frac{1}{2}kx^2\right] = 0 \quad \text{or} \quad \frac{1}{2}\frac{W}{g}2\left(\frac{dx}{dt}\right)\frac{d^2x}{dt^2} + \frac{1}{2}k\,2x\frac{dx}{dt} = 0$$

This reduces to the same differential equation obtained in Problem 1: $\frac{d^2x}{dt^2} + \frac{kg}{W}x = 0$.

Hence the frequency is $\sqrt{kg/W}$ rad/sec or $\frac{1}{2\pi}\sqrt{kg/W}$ cycles/sec.

Note that in a linear differential equation of the type shown above with the second order term having a coefficient 1, the circular frequency ω in rad/sec equals the square root of the coefficient of the x term. ($\omega = 2\pi f$, where ω is in rad/sec and f is in cycles/sec.)

4. A small weight W is fastened to a vertical wire which is under tension T as shown in Fig. 20-5. What will be the natural frequency of vibration of the weight if it is displaced laterally a slight distance and then released?

Solution:

Assume the weight W at some time during the motion is at a distance x to the right of equilibrium.

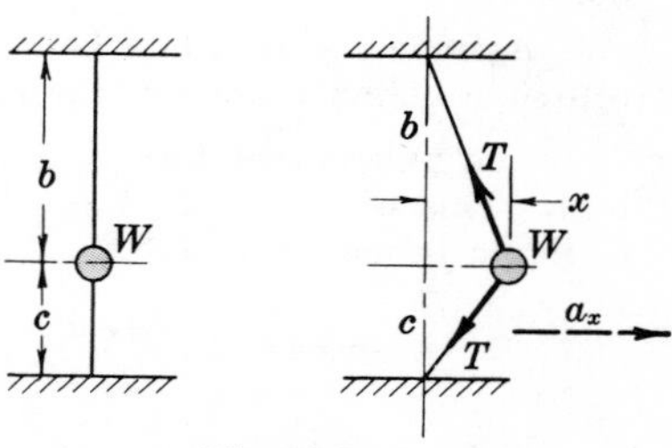

Fig. 20-5

In the horizontal direction this weight is acted upon by the components of the two tensions T shown. For *small* displacements, these x components of the tensions will be Tx/b and Tx/c both acting to the left or negative, if we assume x to the right is positive. Hence the differential equation of motion is

$$-T\frac{x}{c} - T\frac{x}{b} = \frac{W}{g}a_x = \frac{W}{g}\frac{d^2x}{dt^2} \qquad \text{or} \qquad \frac{d^2x}{dt^2} + \frac{g}{W}\left(\frac{T}{c} + \frac{T}{b}\right)x = 0$$

Note that this differential equation is entirely similar, except for the coefficient of the x term, to the spring equation. Hence, as indicated in Problem 3, the frequency is

$$f = \frac{1}{2\pi}\sqrt{\frac{T}{W}\left(\frac{b+c}{bc}\right)g} \text{ cycles/sec}$$

5. Refer to Fig. 20-6 below. The cylinder is weighted to float as shown in the figure. If the cross section of the cylinder is A square inches and the weight W lb, what will be the frequency of oscillation if the cylinder is depressed somewhat and released? The density of the liquid is δ. Neglect the damping effects of the liquid as well as the inertia effects of the moving liquid.

Solution:

When the cylinder is a distance x below its equilibrium position it is buoyed up by a force equal to the weight of the displaced liquid. This is exactly analogous to the spring Problem 1. Using Newton's laws, the unbalanced force, which is up for a downward displacement, is equated to the product of the mass and its acceleration. Call a downward displacement positive. Then the differential equation of motion is

$$-\delta A x = \frac{W}{g}\frac{d^2x}{dt^2} \qquad \text{or} \qquad \frac{d^2x}{dt^2} + \frac{\delta A g}{W}x = 0$$

Thus the frequency is $f = \dfrac{1}{2\pi}\sqrt{\dfrac{\delta A g}{W}}$ cycles/sec.

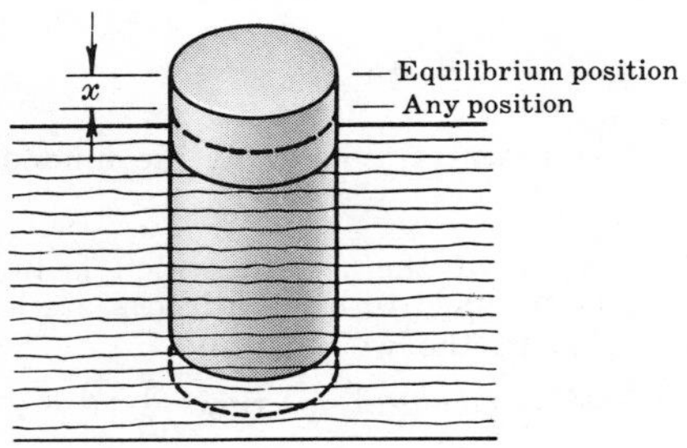

Fig. 20-6

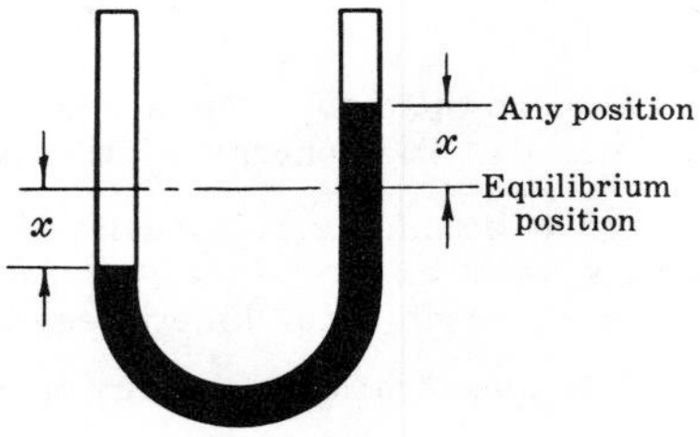

Fig. 20-7

6. Refer to Fig. 20-7 above. Liquid of density δ and total length l is used in a manometer shown in the figure. A sudden increase in pressure on one side forces the liquid down. Upon release of the pressure, the liquid oscillates. Neglecting any frictional damping, what will be the frequency of vibration?

Solution:

Assume the liquid to be a distance x below the equilibrium position in the left column and, of course, a distance x above the equilibrium position in the right column.

The unbalanced force tending to restore equilibrium is the weight of a column of the liquid $2x$ high. This weight is $2xA\delta$, where A is the area of the cross section of the liquid. The total weight of liquid in motion is $lA\delta$. Using Newton's law,

$$-2xA\delta = \frac{lA\delta}{g}\frac{d^2x}{dt^2} \qquad \text{or} \qquad \frac{d^2x}{dt^2} + \frac{2g}{l}x = 0, \qquad \text{and } f = \frac{1}{2\pi}\sqrt{2g/l} \text{ cps}$$

Note that f is independent of the density of the liquid and of the cross section area.

7. Determine the natural undamped frequency of the system shown in Fig. 20-8. The bar is assumed weightless and the lower spring is attached to the bar midway between the points of attachment of the upper springs. The spring constants are as shown.

Solution:

The total static displacement Δ of the weight W is equal to the stretch x_3 of spring k_3 and the average of x_1 and x_2 which are the extensions of springs k_1 and k_2 respecitvely.

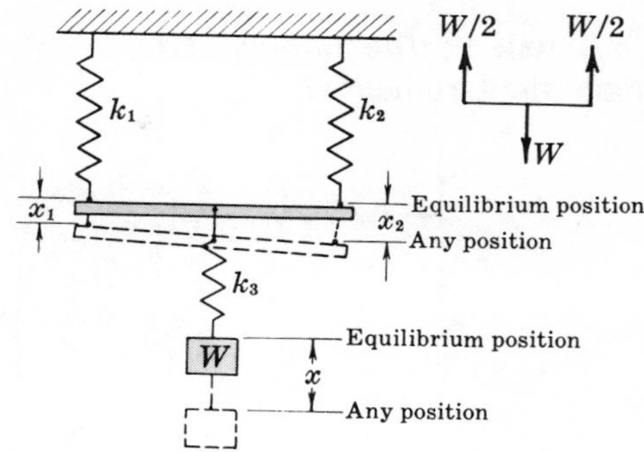

Fig. 20-8

This problem may be solved by substituting this value of Δ into the equation $f = (1/2\pi)\sqrt{g/\Delta}$ (refer to Problem 1). Note that the entire weight W is transferred through spring k_3 whereas $W/2$ is transferred through each spring k_1 and k_2 to the support. The latter fact can be ascertained by referring to the free body diagram of the bar shown to the right of the figure. Hence

$$x_3 = \frac{W}{k_3}, \quad x_2 = \frac{W/2}{k_2}, \quad x_1 = \frac{W/2}{k_1} \quad \text{and} \quad \Delta = x_3 + \tfrac{1}{2}(x_2 + x_1) = W\left(\frac{4k_1k_2 + k_1k_3 + k_2k_3}{4k_1k_2k_3}\right)$$

Substituting,

$$f = \frac{1}{2\pi}\sqrt{\frac{g4k_1k_2k_3}{W(4k_1k_2 + k_1k_3 + k_2k_3)}}$$

8. In Fig. 20-9 below, the weight W is suspended by means of spring k_2 from the end of a rigid weightless beam which is of length l and attached to the frame at its left end. It is also supported in a horizontal position by a spring k_1 attached to the frame as shown. What is the natural frequency f of the system?

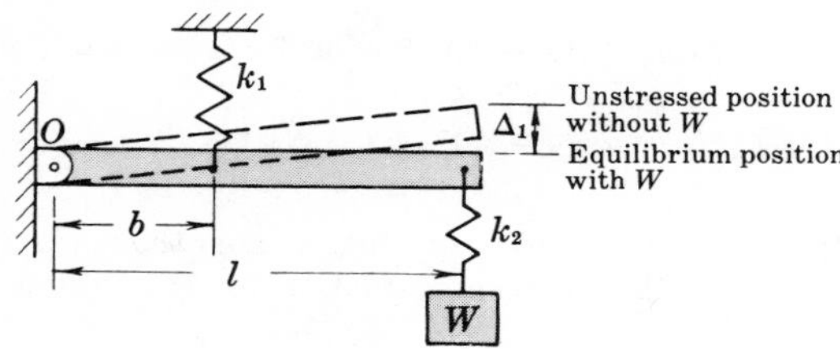

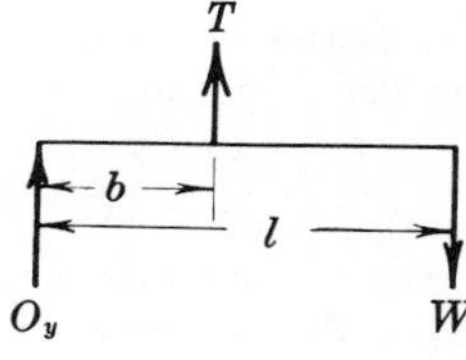

Fig. 20-9

Solution:

The horizontal position is the state of equilibrium. Determine the static displacement Δ_1 of the end of the beam from its original position when no load is impressed. In the free body diagram of the beam in the equilibrium position, note that the tension in spring k_1 must be $T = (l/b)W$ to hold the beam in equilibrium. (This can be seen by taking moments about O.) Hence the spring k_1 elongates under this tension an amount Wl/bk_1. If a point on the beam a distance b from the left end is displaced this amount Wl/bk_1, the right end is displaced (by similar triangles) an amount $\Delta_1 = \frac{l}{b}\frac{Wl}{bk_1} = \frac{W}{k_1}\left(\frac{l}{b}\right)^2$.

The total static displacement Δ of W is then equal to the sum of Δ_1 and the elongation of spring k_2 in transmitting the weight W to the beam. Then

$$\Delta = \frac{W}{k_2} + \left(\frac{l}{b}\right)^2\frac{W}{k_1} = W\left(\frac{k_1 + (l/b)^2k_2}{k_1k_2}\right)$$

and

$$f = \frac{1}{2\pi}\sqrt{g/\Delta} = \frac{1}{2\pi}\sqrt{\frac{gk_1k_2}{W[k_1 + (l/b)^2k_2]}} \text{ cps}$$

FREE VIBRATIONS — ANGULAR

9. What will be the natural frequency of the system shown in Fig. 20-10(a) below for small displacements?

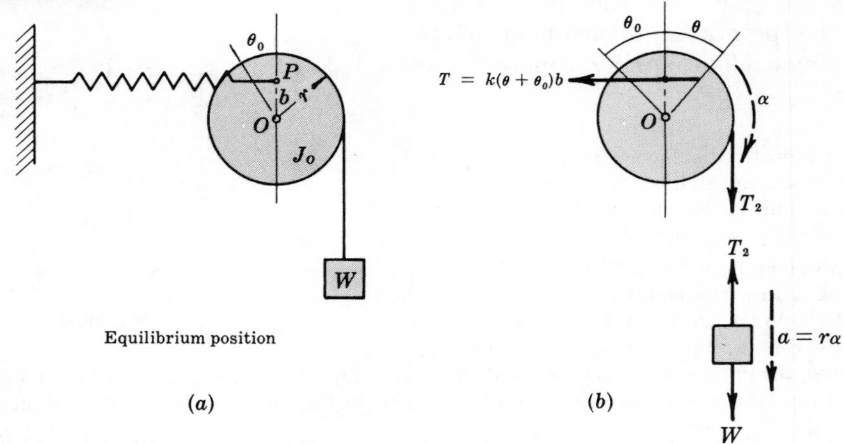

Fig. 20-10

Solution:

In the equilibrium position shown in Fig. 20-10(a), the spring has been stretched a distance equal to $b\theta_0$. In any other position during the motion such as that shown in Fig. 20-10(b), the cylinder is displaced through an additional angle θ and the spring is stretched a total distance of $b(\theta + \theta_0)$. Note that the angular displacements are assumed small. Also assume clockwise angular displacements are positive. The acceleration of the weight W is $a = r\alpha$.

The equations of motion for the weight and the cylinder are respectively

$$(1) \quad \Sigma F = (W/g)a \quad \text{or} \quad W - T_2 = (W/g)r\alpha$$

$$(2) \quad \Sigma M_O = J_O\alpha \quad \text{or} \quad T_2 r - bk(\theta + \theta_0)b = J_O\alpha$$

where J_O is the centroidal polar moment of inertia. Substitute the value of T_2 from equation (1) into equation (2) to obtain

$$\left(W - \frac{W}{g}r\alpha\right)r - b^2k\theta - b^2k\theta_0 = J_O\alpha$$

But in the equilibrium position with P directly above O, $\Sigma M_O = 0$; this means that $(k\theta_0 b)b = Wr$. (Note that $T_2 = W$ when the system is in equilibrium.) The equation of motion for the cylinder becomes

$$\left(J_O + \frac{W}{g}r^2\right)\frac{d^2\theta}{dt^2} + b^2k\theta = 0, \qquad \text{from which} \qquad f = \frac{1}{2\pi}\sqrt{\frac{b^2k}{J_O + (W/g)r^2}} \text{ cps}$$

10. A disk with moment of inertia J_O is rigidly attached to a slender shaft (or wire) of torsional stiffness K. K is the number of in-lb necessary to twist the shaft through one radian. What will be the frequency of oscillations if the shaft is twisted through a small angle and then released?

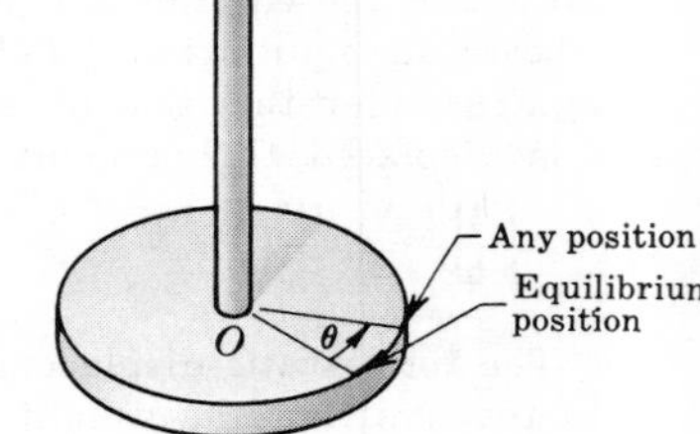

Fig. 20-11

Solution:

When the shaft is twisted through any angle θ as shown in Fig. 20-11, there is a restoring torque $= K\theta$ exerted by the twisted shaft. Hence the equation of motion becomes $-K\theta = J_O\alpha$. Rewritten, this is

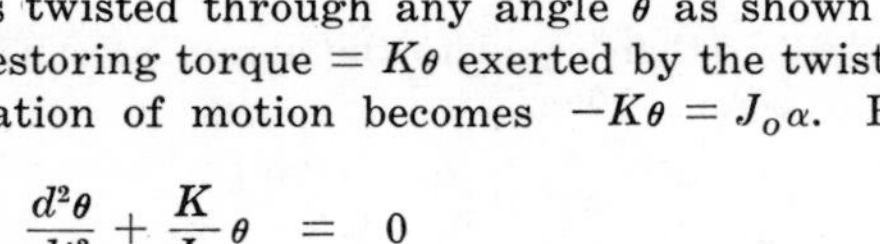

$$\frac{d^2\theta}{dt^2} + \frac{K}{J_O}\theta = 0$$

Thus

$$f = \frac{1}{2\pi}\sqrt{K/J_O} \text{ cps}$$

11. A steel disk 4 inches in diameter and 1 inch thick is rigidly attached to a steel wire 1/16 inch in diameter and 2 feet long. What is the natural frequency of the system?

Solution:

According to the theory developed in Strength of Materials, the angle of twist for the wire is $\theta = \dfrac{Tl}{(\pi d^4/32)G}$ where T is the torque in lb-in, l is the length in inches, G is the shear modulus of elasticity (12×10^6 lb/in² for steel), d is the diameter in inches.

Hence the torsional constant $K = \dfrac{T}{\theta} = \dfrac{\pi d^4 G}{32l} = \dfrac{\pi(1/16)^4(12 \cdot 10^6)}{32(24)} = 0.748$ lb-in/rad.

Moment of inertia of disk is $J_O = \dfrac{1}{2}mr^2 = \dfrac{1}{2}\left(\dfrac{\pi(4)^2(1)(0.286)}{4(386)}\right)2^2 = 0.0186$ lb-sec²-in.

Note that the density of steel was used as 0.286 lb/in³ and that g must be 386 in/sec² in order to make the equation dimensionally correct.

From the preceding problem, the frequency $f = \dfrac{1}{2\pi}\sqrt{\dfrac{K}{J_O}} = \dfrac{1}{2\pi}\sqrt{\dfrac{0.748}{0.0186}} = 1.01$ cps.

12. Discuss the motion of two heavy masses with moments of inertia J_1 and J_2 and connected by a shaft of small diameter d, as shown in Fig. 20-12.

Analysis:

If one mass is held and the other rotated and then both released, the system will oscillate. Since no external torques are assumed to act on the system, the angular momentum of the system is conserved. Thus $J_1\omega_1 + J_2\omega_2 = 0$; hence $\omega_2 = -(J_1/J_2)\omega_1$.

Note that ω_1 and ω_2 represent the varying angular velocities of J_1 and J_2 during a complete cycle for each mass.

Since the previous equation indicates that the masses always are rotating in opposite directions, there is a section of the shaft that is always at rest. This nodal section can be used to study the motion of the masses, since each can be treated as a torsional pendulum (see Problems 10 and 11).

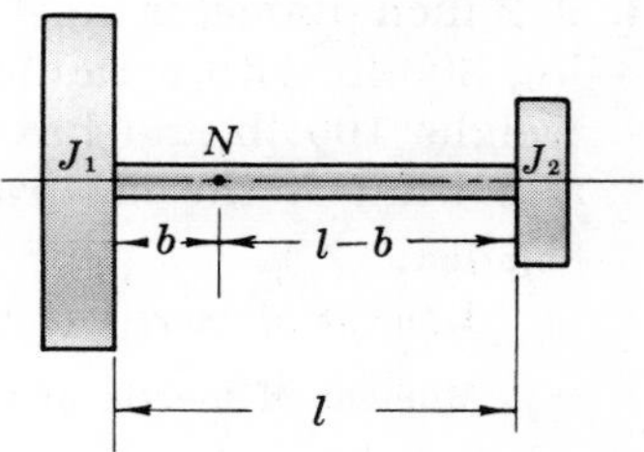

Fig. 20-12

The time for one mass to complete a cycle must equal the time for the other to complete a cycle. If they differed, one mass would eventually rotate in the same direction as the other. But the above equation indicates that the masses always rotate in opposite directions. Since the periods are equal, the number of cycles per second for each must be equal. Thus

$$f = \frac{1}{2\pi}\sqrt{\frac{K_1}{J_1}} = \frac{1}{2\pi}\sqrt{\frac{K_2}{J_2}}$$

where K_1 and K_2 are the torsional spring constants of the parts of the shaft from the nodal section to each end. Hence those constants are related by $K_1/K_2 = J_1/J_2$.

For a cylindrical shaft $K = \pi d^4 G/32l$, where d is its diameter, G is the shearing modulus of elasticity and l its length. Assume the nodal section is a distance b from the J_1 mass; then

$$K_1 = \frac{\pi d^4 G}{32b} \quad \text{and} \quad K_2 = \frac{\pi d^4 G}{32(l-b)}$$

Hence $\dfrac{K_1}{K_2} = \dfrac{l-b}{b} = \dfrac{J_1}{J_2}$ from which $b = \dfrac{J_2 l}{J_1 + J_2}$. This locates the nodal section.

For the left portion, $f = \dfrac{1}{2\pi}\sqrt{\dfrac{K_1}{J_1}}$ where $K_1 = \dfrac{\pi d^4 G}{32b} = \dfrac{\pi d^4 G(J_1 + J_2)}{32lJ_2}$. Then

$$f = \frac{1}{2\pi}\sqrt{\frac{\pi d^4 G(J_1 + J_2)}{32lJ_1J_2}}$$

Since the polar moment of inertia of the area of a circle is $J = \pi d^4/32$, the above expression may be written $f = \frac{1}{2\pi}\sqrt{\frac{JG(J_1+J_2)}{J_1 J_2 l}}$. Note that J_1, J_2 refer to the mass polar moments of inertia of the cylinders and J refers to the area polar moment of inertia of a cross section of the shaft.

13. An engine has a flywheel weighing 500 lb on each end of a steel shaft. Assuming each flywheel has a radius of gyration r of 16 inches, and that the equivalent shaft between them is 26 inches with a diameter of 2.60 inches, determine the natural frequency of torsional oscillation.

Solution:

Moment of inertia for each flywheel is $J_f = (W/g)r^2 = (500/386)(16)^2 = 332$ lb-sec²-in.

Moment of inertia for the area of cross section of shaft is $J = \pi(2.60)^4/32 = 4.48$ in⁴.

Using the formula of the preceding Problem 12, we have

$$f = \frac{1}{2\pi}\sqrt{\frac{JG(J_1+J_2)}{lJ_1J_2}} = \frac{1}{2\pi}\sqrt{\frac{4.48(12\times10^6)(332+332)}{26(332)(332)}} = 17.8 \text{ cps}$$

14. A 2 inch diameter steel shaft 15 inches long is attached at one end to a flywheel weighing 300 lb with a radius of gyration of 6 inches and at the other end to a rotor which weighs 100 lb and has a radius of gyration of 4 inches. Where is the nodal section and what is the natural frequency of torsional oscillation?

Solution:

Moment of inertia of the flywheel is $J_f = (W/g)r^2 = (300/386)(6)^2 = 28.0$ lb-sec²-in.

Moment of inertia of the rotor is $J_r = (W/g)r^2 = (100/386)(4)^2 = 4.15$ lb-sec²-in.

Moment of inertia of the shaft cross section is $J = \pi d^4/32 = \pi(2)^4/32 = 1.57$ in⁴.

From Problem 12, the distance of the nodal section from the flywheel is $b = \frac{J_r l}{J_f + J_r} = 1.94$ in.

The frequency $f = \frac{1}{2\pi}\sqrt{\frac{JG(J_f+J_r)}{lJ_fJ_r}} = \frac{1}{2\pi}\sqrt{\frac{1.57(12\times10^6)(28.0+4.15)}{15(4.15)(28.0)}} = 94.1$ cps.

15. What is the natural frequency of a weight W attached to the end of the cantilever beam of length l shown in Fig. 20-13? Neglect the weight of the beam.

Solution:

This is an example of a weight which vibrates because of the elasticity of the beam rather than that of a spring. To solve for the frequency $f = \frac{1}{2\pi}\sqrt{g/\Delta}$, it is necessary to determine the static deflection Δ of the weight W.

By the use of theory developed in Strength of Materials the static deflection of a cantilever beam due to a concentrated load W on the free end is $\Delta = Wl^3/3EI$, where E is the tensile (or compressive) modulus of elasticity and I is the moment of inertia of the cross sectional area about the neutral axis. (For a circular cross section, the neutral axis is a horizontal diameter.) Hence

$$f = \frac{1}{2\pi}\sqrt{\frac{3gEI}{Wl^3}} \text{ cps}$$

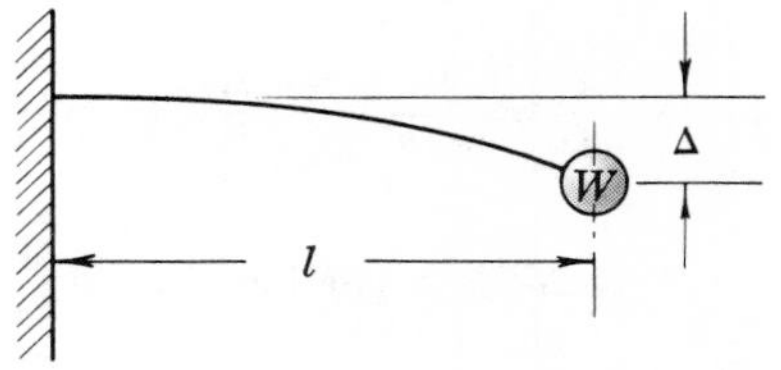

Fig. 20-13

16. A reed-type tachometer is composed of small cantilever beams with weights attached to their free ends. If the vibration frequency of a disturbing force corresponds to the natural vibration frequency of one of the reeds, it will vibrate. Since each reed is calibrated, it is possible to determine the frequency of the disturbance immediately. What size weight W should be placed on the free end of a spring steel reed 0.05 inch thick, 0.20 inch wide, and 4.00 inches long so that its natural frequency is 50 cps? See Fig. 20-14.

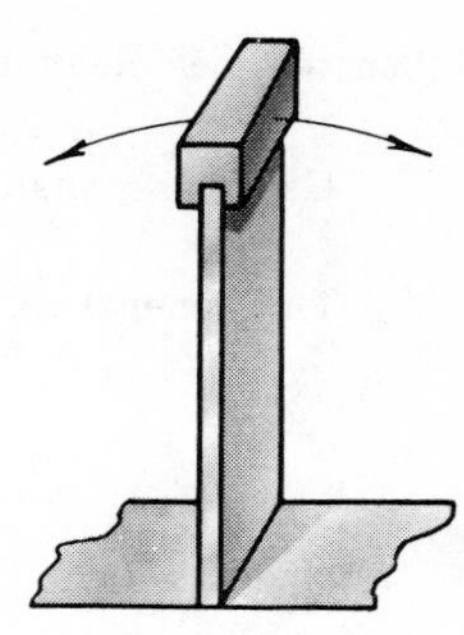

Fig. 20-14

Solution:

By Problem 15, $f = \dfrac{1}{2\pi}\sqrt{\dfrac{3gEI}{Wl^3}}$ where $f = 50$ cps, $g = 386$ in/sec², $E = 30 \times 10^6$ lb/in², $l = 4.00$ in., and $I = \frac{1}{12}bh^3 = \frac{1}{12}(0.20)(0.05)^3 = 2.08 \times 10^{-6}$ in⁴. Substituting values in the above equation, we find $W = 0.011$ lb.

Special Note: There are two possible values of the moment of inertia $[\frac{1}{12}(0.20)(0.05)^3$ or $\frac{1}{12}(0.05)(0.20)^3]$ of the rectangular cross sectional area. The reed will vibrate in the direction shown in the figure, necessitating the use of the smaller value of the moment of inertia. This yields the lower of two possible frequencies.

FREE VIBRATIONS with VISCOUS DAMPING

17. A weight W suspended from a spring whose modulus is k lb/in is subjected to viscous damping represented in Fig. 20-15 below as occurring by means of a dashpot. This damping force is proportional to the velocity, i.e. $F = c(dx/dt)$ where c is the damping constant in lb-sec/in. Discuss the motion as the damping coefficient c varies.

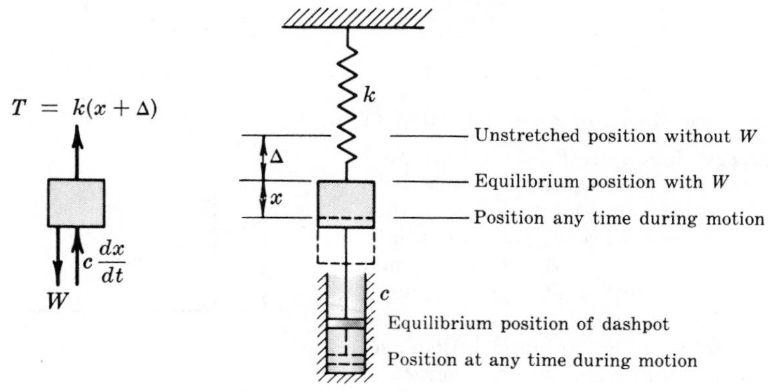

Fig. 20-15

Analysis:

Fig. 20-15 indicates the essential data. The free body diagram to the left illustrates all the forces acting on the body when displaced a distance x below equilibrium and traveling down.

Note as before that in the equilibrium position $k\Delta = W$. Note also that the damping force $c(dx/dt)$ opposes motion. The equation of motion $[\Sigma F = (W/g)a$ with an assumed position down$]$ becomes

$$W - k(x+\Delta) - c\frac{dx}{dt} = \frac{W}{g}\frac{d^2x}{dt^2} \qquad \text{or} \qquad \frac{d^2x}{dt^2} + \frac{cg}{W}\frac{dx}{dt} + \frac{kg}{W}x = 0$$

Assume a solution of this differential equation in the form $x = A\,e^{st}$, where A and s are constants $\neq 0$. Substitute this value into the equation (noting that $dx/dt = A\,s\,e^{st}$ and $d^2x/dt^2 = A\,s^2\,e^{st}$) to obtain

$$As^2e^{st} + A\frac{cg}{W}se^{st} + A\frac{kg}{W}e^{st} = 0 \qquad \text{or} \qquad \left(s^2 + \frac{cg}{W}s + \frac{kg}{W}\right)e^{st} = 0$$

The desired solution must be such that the above equations be zero. Since e^{st} cannot be zero, then its coefficient must be zero, i.e. $s^2 + \frac{cg}{W}s + \frac{kg}{W} = 0$.

Using the quadratic formula, the two solutions of s are $s = \frac{-cg}{2W} \pm \sqrt{\left(\frac{cg}{2W}\right)^2 - \frac{kg}{W}}$.

The general solution is of the form

$$x = Ae^{\left[\frac{-cg}{2W} + \sqrt{\left(\frac{cg}{2W}\right)^2 - \frac{kg}{W}}\right]t} + Be^{\left[\frac{-cg}{2W} - \sqrt{\left(\frac{cg}{2W}\right)^2 - \frac{kg}{W}}\right]t}$$

The radical may be real, imaginary, or zero depending on the magnitude of the damping coefficient c. The value of c which makes the radical zero is called the critical damping coefficient c_c. Its value is obtained by equating the radicand to zero and is

$$c_c = \frac{2W}{g}\sqrt{\frac{kg}{W}} = \frac{2W}{g}\omega_n$$

Note that ω_n is the undamped natural frequency of the system.

The ratio of the damping coefficient c in any system to the critical damping coefficient c_c is called the damping factor d. Its use simplifies the analysis of the problem.

Multiply $\frac{cg}{2W}$ by $\frac{c_c}{c_c}$ and substitute $d = \frac{c}{c_c}$ and $c_c = \frac{2W}{g}\omega_n$ to get

$$\frac{cg}{2W} = \frac{c}{c_c}\frac{g}{2W}c_c = d\frac{g}{2W}\frac{2W}{g}\omega_n = d\omega_n$$

The solution for x obtained above may be written

$$x = Ae^{[-d\omega_n + \sqrt{d^2\omega_n^2 - \omega_n^2}]t} + Be^{[-d\omega_n - \sqrt{d^2\omega_n^2 - \omega_n^2}]t}$$

$$= Ae^{(-d + \sqrt{d^2 - 1})\omega_n t} + Be^{(-d - \sqrt{d^2 - 1})\omega_n t}$$

Three cases arise, depending on the value of d.

Case A. Large damping ($d > 1$) means that the radical is real and less than d. Hence both exponents are negative. The value of x is then equal to the sum of two decreasing exponentials. When $t = 0$, $x = Ae^0 + Be^0 = A + B$. The plot of either part of the solution of such aperiodic motion indicates the frictional resistance is so large that the weight, after its initial displacement, creeps back to equilibrium without vibrating. See Fig. 20-16. Since there is no period to the motion, it is called aperiodic.

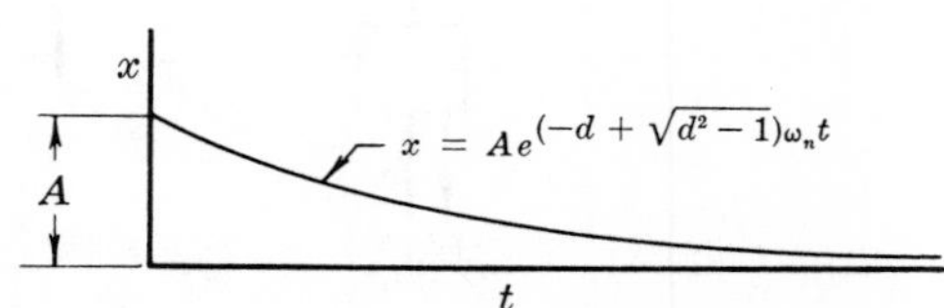

Fig. 20-16

Case B. Light damping ($d < 1$) means that the radical is imaginary. Using $i = \sqrt{-1}$, the solution may be rewritten

$$x = Ae^{[-d + i\sqrt{1 - d^2}]\omega_n t} + Be^{[-d - i\sqrt{1 - d^2}]\omega_n t}$$

$$= e^{-d\omega_n t}[Ae^{i\sqrt{1 - d^2}\,\omega_n t} + Be^{-i\sqrt{1 - d^2}\,\omega_n t}]$$

The term in brackets may be expressed in terms of a sine or cosine function. When this is done, we obtain $x = Xe^{-d\omega_n t}\sin(\sqrt{1 - d^2}\,\omega_n t + \phi)$ where $X\sin\phi$ = displacement at $t = 0$ and ϕ = phase angle. Note that ω_n, d, X and ϕ are all constants. A plot of this solution shows the sine curve with its height continuously decreasing because it is multiplied by the factor $e^{-d\omega_n t}$ which decays with time. See Fig. 20-17 below.

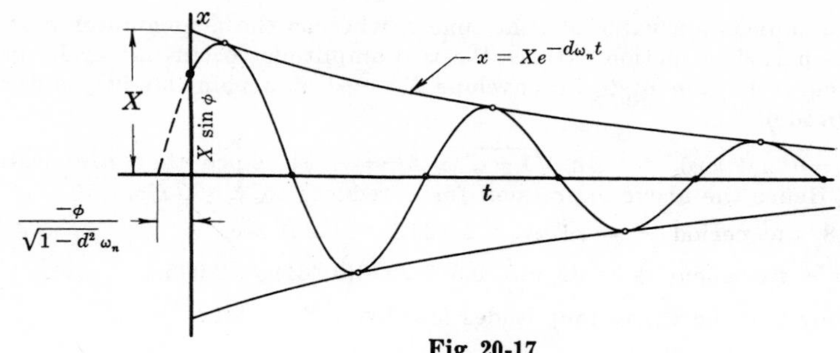

Fig. 20-17

Case C. Critical damping ($d = 1$) means that the solution must be written

$$x = (A + Bt)e^{-\omega_n t}$$

The t term is inserted with B because otherwise only one of the two solutions would be found. This method is developed in a Differential Equations course.

The graph here is similar to *Case A*.

The motion is aperiodic, but the time of return to equilibrium is a minimum when the damping is critical.

18. In Fig. 20-18, the weight W is suspended from a spring whose constant is 20 lb/in and is connected to a dashpot providing viscous damping. The damping force is 10 lb when the velocity of the dashpot plunger is 20 in/sec. The weight of W and the plunger is 12 lb. What will be the frequency of the damped vibrations?

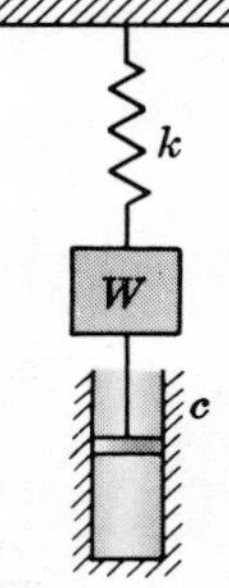

Fig. 20-18

Solution:

The damping coefficient $c = \dfrac{10 \text{ lb}}{20 \text{ in/sec}} = 0.5$ lb-sec/in.

The natural frequency of the undamped system (let us say the oil has been removed from the dashpot) is

$$\omega_n = \sqrt{kg/W} = \sqrt{20(386)/12} = 25.4 \text{ rad/sec}$$

The critical damping coefficient according to Problem 17 is

$$c_c = \frac{2W}{g}\omega_n = \frac{2(12)}{386}(25.4) = 1.58 \text{ lb-sec/in}$$

The damping factor $d = c/c_c = 0.5/1.58 = 0.316$, which is less than 1. According to Problem 17, this is light damping and oscillations will be present. The solution is of the form

$$x = Xe^{-d\omega_n t}\sin(\sqrt{1-d^2}\,\omega_n t + \phi)$$

The frequency of the damped vibration ω_d is the coefficient of the time t in the sine term.

$$\omega_d = \sqrt{1-d^2}\,\omega_n = \sqrt{1-(0.316)^2}\,(25.4) = 24.1 \text{ rad/sec}$$

Note that the period of the damped vibration is $2\pi/24.1 = 0.26$ sec, and the period of the undamped system is $2\pi/25.4 = 0.25$ sec.

19. In Problem 18, determine the rate of decay of the oscillations.

Solution:

This is conveniently expressed by introducing a new term, logarithmic decrement δ which is the natural logarithm of the ratio of any two successive amplitudes one cycle apart.

$$\delta = \ln\frac{x_1}{x_2} = \ln\frac{Xe^{-d\omega_n t}\sin(\sqrt{1-d^2}\,\omega_n t + \phi)}{Xe^{-d\omega_n(t+\tau)}\sin[\sqrt{1-d^2}\,\omega_n(t+\tau) + \phi]}$$

The numerator indicates a value of x at time t, whereas the denominator gives the value at time $t+\tau$ where τ is the period of motion. Hence the two amplitudes occur one cycle apart (we neglect the fact that the sine curve is tangent to its envelope $Xe^{-d\omega_n t}$ at a point slightly different from the point of maximum amplitude).

Now $\sin(\sqrt{1-d^2}\,\omega_n t+\phi) = \sin[\sqrt{1-d^2}\,\omega_n(t+\tau)+\phi]$ since they are evaluated one cycle or 2π radians apart. Hence the above expression for δ reduces to $\delta = d\omega_n\tau$.

In Problem 18, the period $\tau = 2\pi/\omega_d = 2\pi/24.1 = 0.261$ sec.

The logarithmic decrement $\delta = d\omega_n\tau = 0.316(25.4)(0.261) = 2.095$.

The ratio of any two successive amplitudes is $e^{\delta} = e^{2.095} = 8.12$.

20. Set up the differential equation of motion for the system shown in Fig. 20-19 below. Determine the natural frequency of the damped oscillations.

Solution:

In linear vibration we expressed the distance a body moves from equilibrium as a function of time. In rotation such as this we are interested in expressing the angular displacement of a body in terms of time.

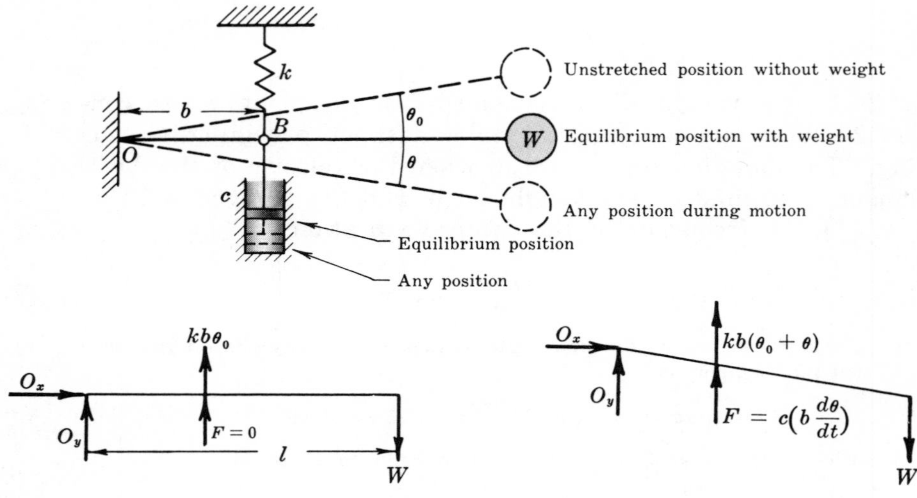

Fig. 20-19

The free body diagram for any phase during the motion is shown separately. Note that if the arm is moving down, the damping force opposes motion and acts up. It is equal to the product of the damping coefficient c and the velocity of the plunger in the dashpot. This is the velocity of the point B on the rod which is a distance b from O and hence has a linear velocity equal to the product of b and the angular velocity $d\theta/dt$ of the rod.

The spring force is the product of k and the total linear displacement of the spring. This displacement is that of point B and hence equals $b(\theta_0+\theta)$. The force is $kb(\theta_0+\theta)$.

The equilibrium free body diagram is also shown. Take moments about O to show $kb^2\theta_0 = Wl$. This information will simplify the differential equation of motion.

A summation of moments about O for the free body diagram for any phase is equated to $I_O\alpha$. But for a concentrated weight W, $I_O = (W/g)l^2$ and $\alpha = d^2\theta/dt^2$. Therefore the equation of motion $\Sigma M_O = I_O\alpha$ becomes

$$+Wl - kb^2(\theta_0+\theta) - cb^2\frac{d\theta}{dt} = \frac{W}{g}l^2\frac{d^2\theta}{dt^2}$$

But $Wl = kb^2\theta_0$. Then, simplifying,

$$\frac{d^2\theta}{dt^2} + \frac{cb^2g}{Wl^2}\frac{d\theta}{dt} + \frac{kb^2g}{Wl^2}\theta = 0$$

To solve, let $\theta = e^{st}$. Then the equation becomes (if e^{st} is a solution)

$$s^2e^{st} + \frac{cb^2g}{Wl^2}se^{st} + \frac{kb^2g}{Wl^2}e^{st} = 0 \qquad \text{or} \qquad s = \frac{-cb^2g}{2Wl^2} \pm \frac{1}{2}\sqrt{\frac{c^2b^4g^2}{W^2l^4} - \frac{4kb^2g}{Wl^2}}$$

Critical damping occurs when the radicand is zero. Hence $c_c = 2\frac{l}{b}\sqrt{\frac{Wk}{g}}$.

If vibrations occur, the radicand will be negative and the solution will be of the form

$$\theta = Ce^{-(cb^2g/Wl^2)t}\sin\left\{\sqrt{-\frac{c^2b^4g^2}{4W^2l^4} + \frac{kb^2g}{Wl^2}}\,t + \phi\right\}$$

where C and ϕ are determined by the conditions of the problem. Compare this solution with that of Problem 17, *Case B*.

The frequency ω_d of the damped vibration is the coefficient of the time t in the sine term, i.e.,

$$\omega_d = \sqrt{-\frac{c^2b^4g^2}{4W^2l^4} + \frac{kb^2g}{Wl^2}} = \frac{b}{l}\sqrt{\frac{kg}{W} - \left(\frac{cbg}{2Wl}\right)^2} \text{ rad/sec}$$

21. Determine the natural frequency of the oscillations of the system shown in Fig. 20-20 below.

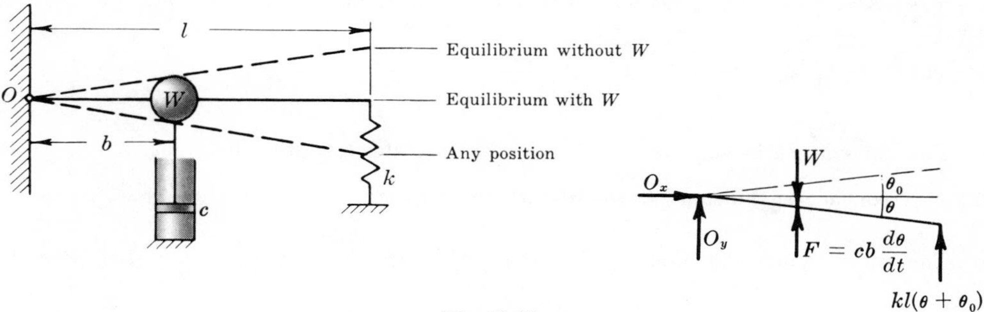

Fig. 20-20

Solution:

As shown in the previous Problem 20, the equation of motion may be found by referring to a free body diagram for any phase other than equilibrium. Thus the equation is

$$\Sigma M_O = I_O\alpha = I_O\frac{d^2\theta}{dt^2} \qquad \text{or} \qquad +Wb - cb^2\frac{d\theta}{dt} - kl^2(\theta + \theta_0) = \frac{W}{g}b^2\frac{d^2\theta}{dt^2}$$

But in the equilibrium position $Wb - kl^2\theta_0 = 0$. Then the equation of motion is

$$\frac{d^2\theta}{dt^2} + \frac{cg}{W}\frac{d\theta}{dt} + \frac{kl^2g}{Wb^2}\theta = 0$$

Again let $\theta = e^{st}$ to get $s^2 + \frac{cg}{W}s + \frac{kl^2g}{Wb^2} = 0$. From this, $s = \frac{-cg}{2W} \pm \sqrt{\frac{c^2g^2}{4W^2} - \frac{kl^2g}{Wb^2}}$.

Remembering that the radicand must be negative for oscillations to occur, the frequency will be

$$\omega_d = \sqrt{\frac{kl^2g}{Wb^2} - \frac{c^2g^2}{4W^2}} \text{ rad/sec}$$

FORCED VIBRATIONS (without DAMPING)

22. In Fig. 20-21 below, a weight W is suspended on a spring whose modulus is k lb/in. The weight is subjected to a periodic disturbing force $F\cos\omega t$. Discuss the motion.

Analysis:

The differential equation now has an additional term when compared with the equation of free vibrations.

$$-kx + F\cos\omega t \quad = \quad \frac{W}{g}\frac{d^2x}{dt^2}$$

or

$$\frac{d^2x}{dt^2} + \frac{kgx}{W} \quad = \quad \frac{Fg}{W}\cos\omega t$$

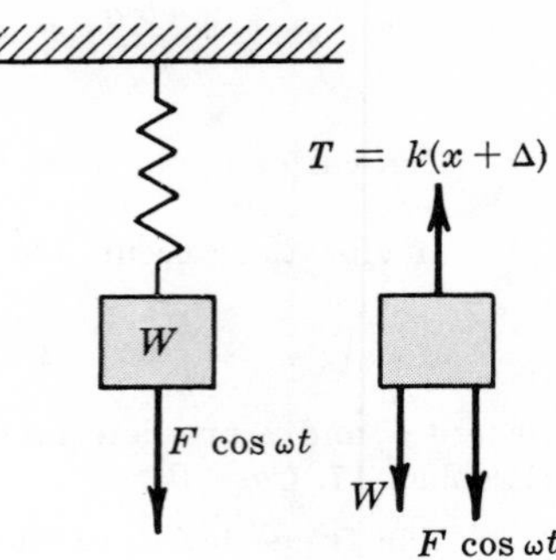

Fig. 20-21

According to the theory of Differential Equations, the solution of this equation consists of the sum of two parts: (1) the solution previously determined in Problem 1 for the equation when the right side is set equal to zero (the transient part), and (2) a solution to make $\frac{d^2x}{dt^2} + \frac{kg}{W}x = \frac{Fg}{W}\cos\omega t$.

Assume that the second solution, which is called the steady-state solution, is of the form $x = X\cos\omega t$. Then $dx/dt = -X\omega\sin\omega t$ and $d^2x/dt^2 = -X\omega^2\cos\omega t$.

Substituting, $-X\omega^2\cos\omega t + \frac{kg}{W}X\cos\omega t = \frac{Fg}{W}\cos\omega t$. Hence X must be $\frac{F/k}{1 - \omega^2 W/gk}$.

Let Δ_F be the deflection which the force F would impart to the spring if acting on it statically, i.e. $\Delta_F = F/k$. Also note that $\omega_n^2 = kg/W$ where ω_n is the natural frequency when the disturbing force is absent.

Then X may be written $\frac{\Delta_F}{1 - (\omega/\omega_n)^2}$.

For convenience in analysis let $\frac{\omega}{\omega_n} = r$; then the steady-state solution is $x = \frac{\Delta_F}{1 - r^2}\cos\omega t$. Note that its frequency is the same as the disturbing frequency.

The entire solution is then $x \quad = \quad A\sin\sqrt{\frac{kg}{W}}\,t + B\cos\sqrt{\frac{kg}{W}}\,t + \frac{1}{1 - r^2}\Delta_F\cos\omega t$.

The first two terms, representing the free vibrations, are transient in character because some damping is always present to cause those vibrations to decay. Hence study only the solution $x = \Delta_F\frac{1}{1 - r^2}\cos\omega t$. Its maximum value, which occurs when $\cos\omega t = 1$, is $\Delta_F/(1 - r^2)$ and is called the amplitude. The ratio of the amplitude of the steady-state solution to the static deflection Δ_F which F would cause is called the magnification factor. Its value is $\frac{\Delta_F/(1 - r^2)}{\Delta_F} = \frac{1}{1 - r^2}$. Since $r = \frac{\omega}{\omega_n} = \frac{f}{f_n}$, this can be written $\frac{1}{1 - (f/f_n)^2}$. Its value can be positive or negative depending on whether f is less than f_n or not. When $f = f_n$, resonance occurs and the amplitude is theoretically infinite. Actually, the damping which is always present holds the amplitude to a finite amount.

A plot of magnification factor versus the frequency ratio r is shown in Fig. 20-22.

The negative value when $r > 1$ indicates that the force F is directed one way while the displacement x is opposite.

Note that when $r = \sqrt{2}$, the magnification factor is $\frac{1}{1 - (\sqrt{2})^2} = -1$. This means that if the ratio r is made greater than $\sqrt{2}$, the magnitude of the magnification factor will be less than 1. Thus the disturbing force under these conditions will cause less movement than if it were applied statically.

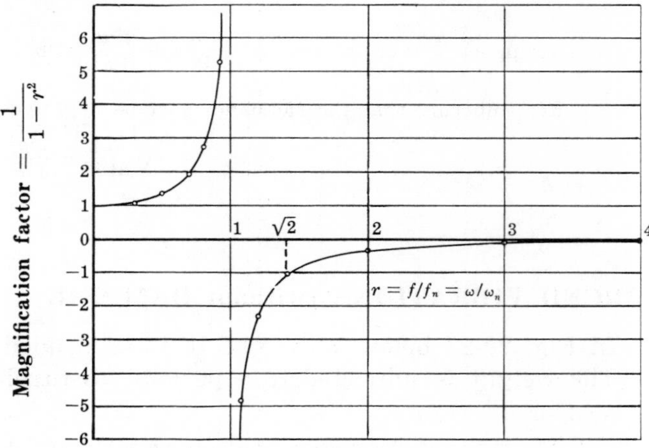

Fig. 20-22

23. A disturbing force of 2 lb acts harmonically on a 12 lb weight suspended on a spring whose modulus is 36 lb/in. What will be the amplitude of the block if the disturbing frequency is (*a*) 1 cps, (*b*) 5.40 cps, (*c*) 50 cps?

Solution:

The natural frequency of the system is $f_n = \frac{1}{2\pi}\sqrt{\frac{kg}{W}} = \frac{1}{2\pi}\sqrt{\frac{36(386)}{12}} = 5.42$ cps.

The deflection which the disturbing force would give to the spring if applied statically is $\Delta_F = 2/36 = 0.056$ inch.

(*a*) The frequency ratio $r = f/f_n = 1/5.42 = 0.1845$; hence the amplitude will be $\Delta_F/(1-r^2) = 0.056/(1-0.0341) = 0.058$ inch.

(*b*) The ratio $r = 5.40/5.42$; hence the amplitude will be 7 inches.

(*c*) The ratio $r = 50/5.42$; hence the amplitude will be -0.0007 inch. Note that in this case the amplitude is opposite to the direction in which the force is exerted but is negligible in magnitude.

24. A refrigerator unit weighing 60 lb is supported on three springs, each with a modulus of k lb/in. The unit operates at 600 rpm. What should be the value of k if one-twelfth of the disturbing force of the unit is to be transmitted to the supporting box?

Solution:

Assume that the disturbing force transmitted is proportional to the amplitude of the motion of the unit. This is logical because the supporting springs transmit forces proportional to their deformation which equals the amplitude of the motion of the unit.

From Problem 22, the ratio of the amplitude of the steady-state motion to the static deflection which the disturbing force would cause (in this case $-1/12$) is equal to $1/(1-r^2)$. Note that the ratio is negative because the natural frequency of the springs must be less than the disturbing frequency for a reduction to occur, and thus according to the graph in Problem 22, the magnification factor is below the line.

Hence in this problem $-1/12 = 1/(1-r^2)$, from which $r^2 = 13$, $r = f/f_n = \sqrt{13}$, and $f_n = (600 \div 60)/\sqrt{13} = 2.77$ cps. Using the weight W on one spring, f_n is equated to $\frac{1}{2\pi}\sqrt{\frac{kg}{W}}$. Then $2.77 = \frac{1}{2\pi}\sqrt{\frac{k(386)}{20}}$, from which $k = 15.7$ lb/in.

FORCED VIBRATIONS with VISCOUS DAMPING

25. In Fig. 20-23 a weight W is suspended from a spring whose modulus is k lb/in. It is also connected to a dashpot which provides viscous damping. Discuss the motion if the weight is subjected to a harmonic disturbing force $F_0 \cos \omega t$.

Solution:

The free body diagram to the left shows the forces acting on the weight. Assuming down is positive, the equation of motion is

$$\Sigma F = \frac{W}{g}a$$

or

$$W + F_0 \cos \omega t - k(x+\Delta) - c\frac{dx}{dt} = \frac{W}{g}\frac{d^2x}{dt^2}$$

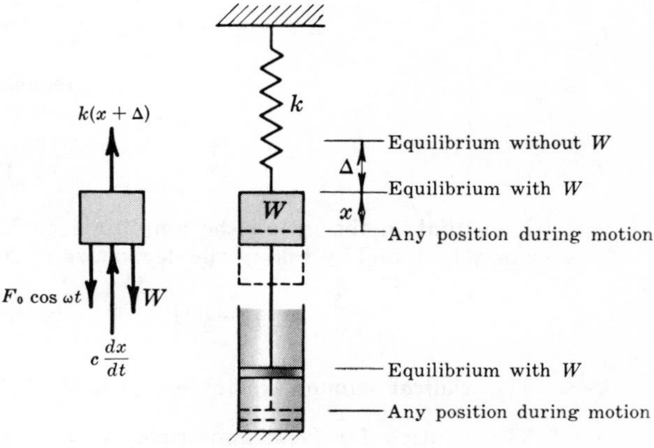

Fig. 20-23

As in previous problems, the free body diagram for the equilibrium position shows $W - k\Delta = 0$. Thus the equation becomes

$$\frac{d^2x}{dt^2} + \frac{cg}{W}\frac{dx}{dt} + \frac{kg}{W}x = \frac{g}{W}F_0 \cos \omega t$$

The transient solution will be neglected as in Problem 22. The steady-state solution is

$$x = \frac{F_0 \cos(\omega t - \phi)}{\sqrt{(k - W\omega^2/g)^2 + (c\omega)^2}} \qquad \text{with} \qquad \tan\phi = \frac{c\omega/k}{1 - W\omega^2/gk}$$

Since usually only the amplitude X of the motion is considered, the equation for the amplitude may be written

$$X = \frac{F_0}{\sqrt{(k - W\omega^2/g)^2 + (c\omega)^2}} = \frac{F_0/k}{\sqrt{(1 - W\omega^2/kg)^2 + (c\omega/k)^2}}$$

This is further simplified by noting that F_0/k is the static deflection ΔF_0 which the disturbing force would give the spring. Also, $W/kg = 1/\omega_n^2$ where ω_n is the natural undamped frequency of the system (rad/sec). If d is the ratio of the given damping coefficient c to the critical damping coefficient c_c, the last term in the radicand may be expressed $\dfrac{c\omega}{k} = \dfrac{c}{c_c}\dfrac{c_c\omega}{k} = d\dfrac{c_c\omega}{k}$. But the critical damping coefficient (Problem 17) is $c_c = (2W/g)\omega_n$. Then the last term of the radicand is $d\left(2\dfrac{W}{g}\omega_n\right)\dfrac{\omega}{k} = d\left(2\dfrac{\omega_n\omega}{\omega_n^2}\right) = 2rd$, where $r = \dfrac{\omega}{\omega_n}$. Thus

$$\frac{X}{\Delta_{F_0}} = \frac{1}{\sqrt{(1 - r^2)^2 + (2rd)^2}} \qquad \text{and} \qquad \tan\phi = \frac{2rd}{1 - r^2}$$

The ratio X/Δ_{F_0} is called the magnification factor.

In Fig. 20-24, a graph of the magnification factor against the frequency ratio shows the peaking which occurs near $r = \omega/\omega_n = 1$ and the influence which damping exerts on the heights of these peaks.

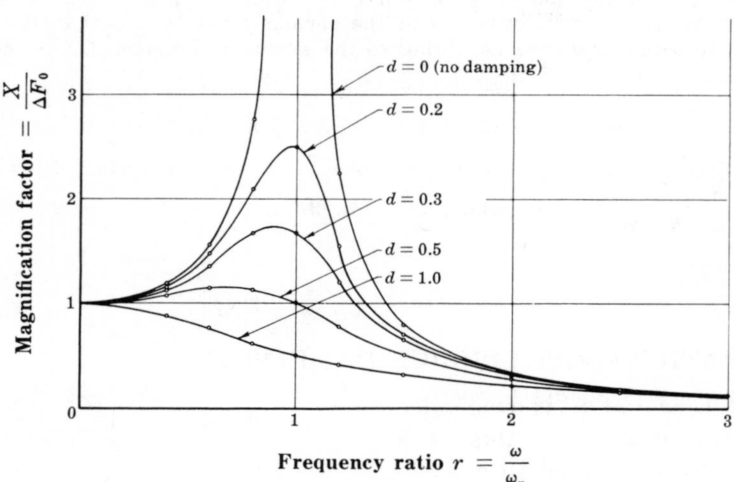

Fig. 20-24

Note that in the graph the amplitude peaks are at a value slightly less than $r = 1$. The exact value may be found by taking the derivative of X/Δ_{F_0} with respect to r and equating the result to zero.

$$\frac{d}{dr}\left(\frac{X}{\Delta_{F_0}}\right) = -\tfrac{1}{2}[(1 - r^2)^2 + (2rd)^2]^{-3/2}\,[2(1 - r^2)(-2r) + 2(2rd)2d] = 0$$

The radical cannot equal zero; hence $2(1 - r^2)(-2r) + 2(2rd)2d = 0$ or $r(r^2 + 2d^2 - 1) = 0$.

The solution for peak amplitude is $r = \sqrt{-2d^2 + 1}$.

In Fig. 20-25 below, a plot is also shown of the phase angle ϕ for various damping coefficients.

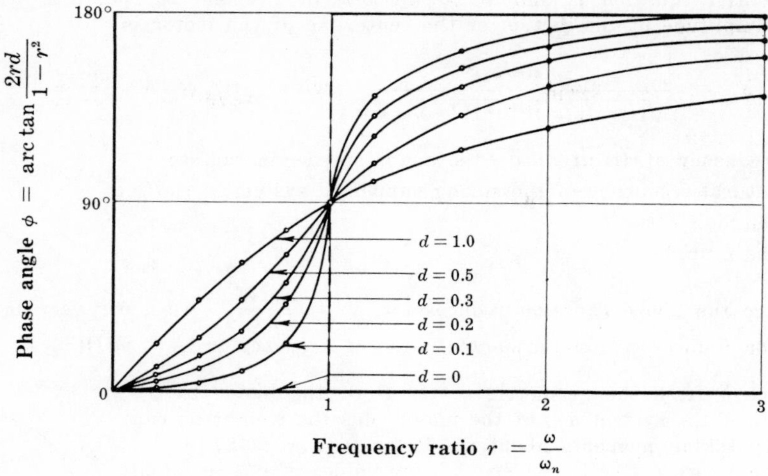

Fig. 20-25

26. A small weight w is attached with an eccentricity e to the flywheel of a motor mounted on springs as shown in Fig. 20-26. The modulus of each of the springs is k lb/in. The dashpot introduces viscous damping of an amount c lb-sec/in. If the total weight of the motor and small weight is W, study the motion of the system under the action of the disturbing force caused by the eccentrically mounted weight w. (This problem illustrates the effect of either reciprocating or rotating unbalance.)

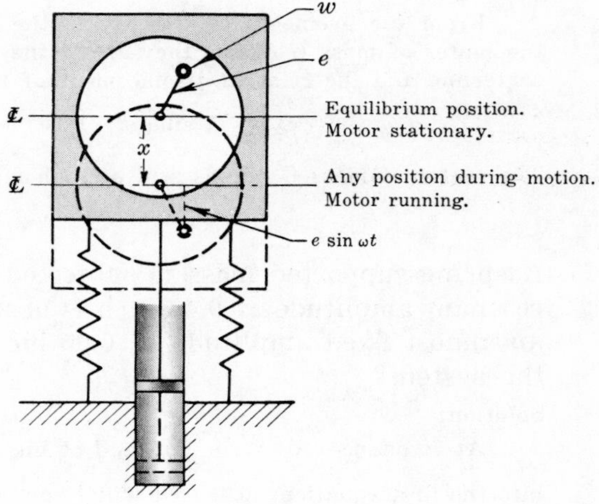

Fig. 20-26

Analysis:

In the position shown, the weight w is below the centerline by an amount $e \sin \omega t$. Hence, if the positive direction is assumed as down, the absolute displacement x_1 of the weight w is the sum of its vertical displacement $(e \sin \omega t)$ relative to the centerline and the absolute displacement x of the centerline, i.e. $x_1 = x + e \sin \omega t$.

Let F = unbalanced force exerted by motor on weight w to impart to it the acceleration d^2x_1/dt^2. Then

$$F = \frac{w}{g}\frac{d^2x_1}{dt^2} = \frac{w}{g}\frac{d^2(x + e \sin \omega t)}{dt^2} = \frac{w}{g}\frac{d^2x}{dt^2} - \frac{w}{g} e\,\omega^2 \sin \omega t$$

The equation of motion of the motor without the small weight w is

$$-F - kx - c\frac{dx}{dt} = \frac{W-w}{g}\frac{d^2x}{dt^2}$$

Note that F is used with a negative sign because this unbalanced force of the weight w on the motor is opposite in direction to that of the motor on w. Substitute the value of F obtained above into the motor equation and simplify to obtain

$$\frac{d^2x}{dt^2} + \frac{cg}{W}\frac{dx}{dt} + \frac{kg}{W}x = \frac{w}{W} e\,\omega^2 \sin \omega t$$

This differential equation is similar to the one in Problem 25, provided F_0 is replaced by $(w/W)e\omega^2$. The amplitude of the motion of the centerline of the motor is

$$X = \frac{\frac{w}{W}e(\omega/\omega_n)^2}{\sqrt{[1-(\omega/\omega_n)^2]^2 + (2d\,\omega/\omega_n)^2}} \qquad \text{with} \qquad \tan\phi = \frac{2d\,\omega/\omega_n}{1-(\omega/\omega_n)^2}$$

where ω = frequency of disturbance — the motor speed — in rad/sec
ω_n = natural frequency of the spring supported system in rad/sec
d = damping ratio
ϕ = phase angle.

At resonance the above equation reduces to $X = \dfrac{we/W}{2d}$. Also, for very large values of $\dfrac{\omega}{\omega_n}$ the same equation reduces within reasonable limits of accuracy to $X = we/W$.

The distance b from the geometric center O of the motor to the center of mass G of the system, i.e. of the motor plus the weight w, can be determined by taking moments about O. Refer to Fig. 20-27.

Thus $bW = ew$ or $b = we/W$ which at large values of ω/ω_n is equal to X. Hence at high motor speeds (ω/ω_n very large) the magnitude of the displacement X of the centerline is equal to the magnitude of b. However, at large values of ω/ω_n the value of ϕ as deduced from the equation $\tan\phi = \dfrac{2d\,\omega/\omega_n}{1-(\omega/\omega_n)^2}$ approaches 180°. Hence the displacement X equals b but is 180° out of phase.

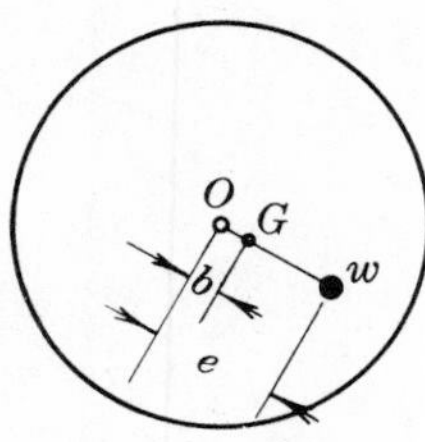

Fig. 20-27

From the geometry of the figure, the absolute displacement x_G of the center of mass G equals the sum of the absolute displacement of the centerline and the relative displacement of the center of mass to the centerline, or

$$x_G = X\sin(\omega t - 180°) + b\sin\omega t = -b\sin\omega t + b\sin\omega t = 0$$

The center of mass G stands still at high motor speeds.

27. A spring-supported mass is subjected to a disturbing force of varying frequency. A resonant amplitude of 0.46 inch is observed. Also, at very high disturbing frequencies an almost fixed amplitude of 0.05 inch is observed. What is the damping ratio d of the system?

Solution:

At resonance $X = \dfrac{we/W}{2d}$, and at high frequencies $X = \dfrac{we}{W}$. Thus $\dfrac{we}{W} = 0.05$. Substituting into the first equation, $0.46 = 0.05/2d$ or $d = 0.054$.

CRITICAL SPEEDS of SHAFTS

28. A disk of weight W mounted on a vertical shaft which is rotating ω rad/sec, has its center of mass G a distance e from its geometric center S, the point through which the shaft passes (see Fig. 20-28). Determine the critical (or whipping) speeds of the shaft, i.e. those speeds at which the shaft tends to vibrate excessively in a transverse direction. The disk moves horizontally under this condition.

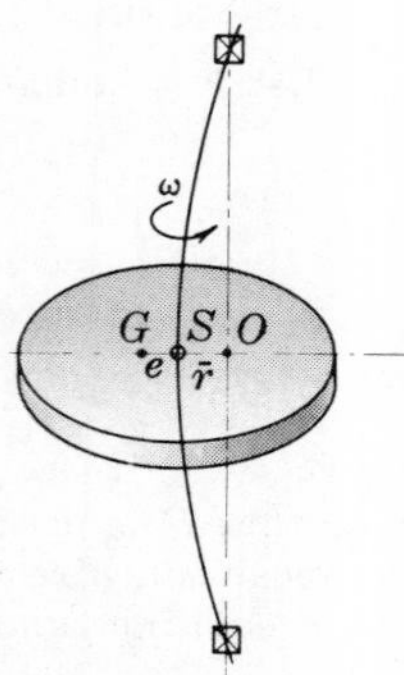

Fig. 20-28

Solution:

Let $\bar{r}$ be the distance OS that the shaft center moves from the line of bearing centers. Let e be the eccentricity of the mass, i.e. SG. If the shaft has an equivalent spring constant k (where k is the force required to deflect the shaft unit distance from its equilibrium position), the restoring pull of the shaft in the stressed position shown is $k\bar{r}$. The normal acceleration toward O of the mass is $(\bar{r}+e)\omega^2$. The normal force exerted on the mass to produce this acceleration is $(W/g)(\bar{r}+e)\omega^2$. Since the force $k\bar{r}$ supplies this normal force, write $k\bar{r} = (W/g)(\bar{r}+e)\omega^2$. But kg/W equals the square of the natural frequency ω_n of the system when the shaft is not rotating. (To determine ω_n experimentally, merely pull the disk horizontally and release it and observe the vibrations, i.e. $\omega_n = 2\pi f$.) Hence the equation may be rewritten $\omega_n^2\bar{r} = \bar{r}\omega^2 + e\omega^2$. Then

$$\bar{r} = \frac{e(\omega/\omega_n)^2}{1-(\omega/\omega_n)^2}$$

The amount of whip is a maximum when $\omega = \omega_n$. Thus if the speed of rotation of the shaft ω equals the natural frequency ω_n of the system, whipping may occur. Naturally, damping is always present and limits the amplitude.

If $\omega/\omega_n < 1$, $\bar{r}$ is of the same sign as e, i.e. OS is in the same direction as SG. This means that S is between O and G, and the rotation takes place with G on the outside as shown in the figure. On the other hand, if $\omega/\omega_n > 1$, G is between O and S during the rotation.

If ω/ω_n is very much greater than 1, then $\bar{r}$ approaches e in magnitude and is opposite in sign. Thus for large speeds, $OS = -SG$; and the system rotates with G at O, in the centerline of the bearings.

29. The rotor of a machine weighs 50 lb and is keyed to the center of a 2.0 inch diameter steel shaft 20 inches between bearing centers. Determine (1) the critical speed, (2) the amplitude of vibration of the rotor which is turning 4000 rpm, assuming an eccentricity e of 0.002 inch, and (3) the force transmitted to the bearings.

Solution:

(1) To find the critical speed, it is necessary first to determine the equivalent spring modulus of the shaft with a concentrated load W at the center. According to problems previously worked, $k = W/\Delta$ where for a simple beam such as this the static deflection $\Delta = Wl^3/48EI$ with l in inches, E = modulus of elasticity (for steel, $E = 30\times10^6$ psi), and I = moment of inertia of the circular cross sectional area about a diameter, i.e. $\pi d^4/64$. Then

$$k = \frac{48EI}{l^3} = \frac{48(30\times10^6)}{(20)^3}\frac{\pi(2)^4}{64} = 141{,}500 \text{ lb/in}$$

Neglecting weight of shaft, the natural frequency in rpm with a weight W at the midpoint is

$$\omega_n = \frac{60}{2\pi}\sqrt{\frac{kg}{W}} = \frac{60}{2\pi}\sqrt{\frac{141{,}500(386)}{50}} = 9980 \text{ rpm}$$

(2) From Problem 28, $\bar{r} = \dfrac{e(\omega/\omega_n)^2}{1-(\omega/\omega_n)^2}$ where $\left(\dfrac{\omega}{\omega_n}\right)^2 = \left(\dfrac{4000}{9980}\right)^2 = 0.161$.

Then $\bar{r} = \dfrac{0.002(0.161)}{1-0.161} = 0.00038$ inch, amplitude of vibration of rotor.

(3) The force transmitted to both bearings is

$$\frac{W}{g}(\bar{r}+e)\omega^2 = \frac{50}{386}(0.00038+0.002)\left(\frac{4000\times2\pi}{60}\right)^2 = 53.8 \text{ lb}$$

This can also be obtained from $k\bar{r} = 141{,}500(0.00038) = 53.8$ lb.

AMPLITUDE of VIBRATION MEASURING INSTRUMENTS

30. In Fig. 20-29 below, the weight W is mounted on springs whose combined modulus is k lb/in. The dashpot provides a damping factor c lb-sec/in. The supporting frame is subjected to a vibration $x = X\sin\omega t$. (1) Determine the motion of W. (2) Discuss the action of a vibrometer. (3) Apply the theory to explain how an accelerometer functions.

Solution:

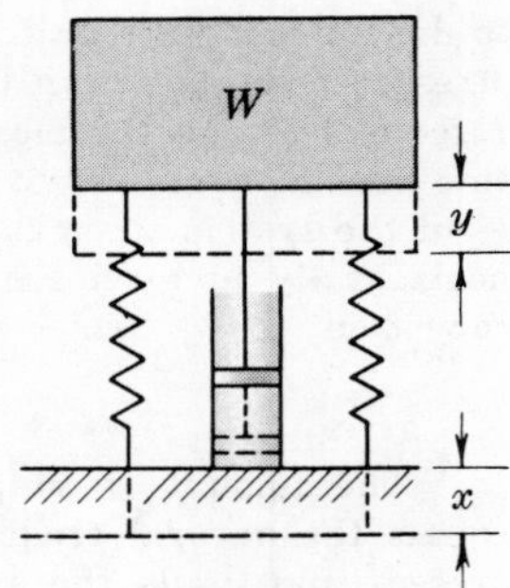

Fig. 20-29

Let the frame be displaced a distance x from equilibrium. Then at the same time let the weight W have an absolute displacement y from its equilibrium position. The relative displacement of the weight to the frame is then $y - x$ which can be represented by z.

The spring force in this position is proportional to $z + W/k$ and the damping force is proportional to dz/dt. The equation of motion for the weight W acted on by these two unbalanced forces and gravity is

$$W - k\left(z + \frac{W}{k}\right) - c\frac{dz}{dt} = \frac{W}{g}\frac{d^2y}{dt^2} \quad \text{or} \quad -kz - c\frac{dz}{dt} = \frac{W}{g}\frac{d^2y}{dt^2}$$

But $y = z + x$ and $\dfrac{d^2y}{dt^2} = \dfrac{d^2z}{dt^2} + \dfrac{d^2x}{dt^2}$. Also, $\dfrac{d^2x}{dt^2} = \dfrac{d^2}{dt^2}(X \sin \omega t) = -X\omega^2 \sin \omega t$.

Substituting, the equation can be written

$$\frac{d^2z}{dt^2} + \frac{cg}{W}\frac{dz}{dt} + \frac{kg}{W}z = -\frac{d^2x}{dt^2} = X\omega^2 \sin \omega t$$

As indicated in Problem 26, the amplitude Z of the relative motion is

$$Z = \frac{X(\omega/\omega_n)^2}{\sqrt{[1 - (\omega/\omega_n)^2]^2 + (2d\,\omega/\omega_n)^2}}$$

A vibrometer is a weight W suspended on soft springs (k is small). Hence its natural frequency $\omega_n = \sqrt{kg/W}$ will also be small. If the vibrometer is attached to a vibrating frame of any kind, the ratio ω/ω_n will be large. Then

$$Z = \frac{X}{\sqrt{[1/(\omega/\omega_n)^2 - 1]^2 + [2d/(\omega/\omega_n)]^2}} = X \text{ approximately}$$

This indicates that the magnitude of Z is equal to the magnitude of X. However, they are 180° out of phase because ϕ is approximately 180° when ω/ω_n is very large. (From Problem 26, $\tan \phi = \dfrac{2d(\omega/\omega_n)}{1 - (\omega/\omega_n)^2}$.) Hence actually $Z = -X$. The absolute displacement of the weight is $Z + X = 0$. Thus the weight W will be almost stationary. A pencil attached to the frame could be used to mark on paper attached to the weight the displacement of the frame relative to the almost stationary weight.

An accelerometer is similar but the springs are hard (k is large). Hence its natural frequency $\omega_n = \sqrt{kg/W}$ will also be large. If the accelerometer is attached to a vibrating framework, the ratio ω/ω_n will be small. Then

$$Z \cong \frac{X(\omega/\omega_n)^2}{\sqrt{(1 - 0^2)^2 + (0)^2}} \cong \frac{X\omega^2}{\omega_n^2}$$

This means that the relative displacement Z is approximately equal to the product of the absolute displacement X and ω^2 divided by the constant ω_n^2.

But $X\omega^2$ is the amplitude of the acceleration of the framework which is vibrating with harmonic motion $x = X \sin \omega t$. Hence the relative displacement Z is a measure of the acceleration to which the framework is being subjected — important information for its proper design.

Supplementary Problems

31. A weight of 10 lb vibrates with harmonic motion $x = X \sin \omega t$. If the amplitude X is 4 inches and the weight makes 1750 vibrations per minute, find the maximum acceleration of the weight in ft/sec^2. *Ans.* $a = 11{,}200$ ft/sec^2

32. A cylinder oscillates about a fixed axis with a frequency of 10 cpm. If the motion is harmonic with an amplitude of 0.10 radian, find the maximum acceleration in rad/sec^2. *Ans.* $\alpha = 0.11$ rad/sec^2

33. An instrument weighing 4.4 lb is fastened to four rubber mounts each of which is rated at 0.125 inch deflection per pound loading. What will be the natural frequency of vibration in cps? *Ans.* $f = 8.44$ cps

34. A maple log weighing 50 lb/ft^3 is 5 in. in diameter and 5 ft long. If while it is floating vertically in the water it is displaced downward from its equilibrium position, what will be the period of oscillation? *Ans.* $\tau = 2.21$ sec

35. Determine the natural frequency of vertical vibration of a horizontal simple beam of length l to which a weight W is fastened at the midpoint. Neglect the weight of the beam. Note that the deflection of W is $Wl^3/48EI$. *Ans.* $f = (2/\pi)\sqrt{3EIg/Wl^3}$ cps

36. A weight of 4 oz is fastened to the midpoint of a 6 inch long vertical wire in which the tension is 3.2 lb. What will be the period of vibration of the weight if it is displaced laterally and then released? *Ans.* $\tau = 0.11$ sec

37. A simple pendulum consists of a small bob of weight W tied to the end of a string of length l. Show that for small oscillations the natural frequency is $(1/2\pi)\sqrt{g/l}$ cps.

38. Refer to Fig. 20-30 below. Determine the natural frequency of the system composed of a weight W lb suspended by a spring, whose constant is k lb/in, from the end of a weightless cantilever beam of length l. Hint: A 1 lb force applied at W will cause it to deflect a total distance of $\left(\frac{1}{k} + \frac{1}{3EI/l^3}\right)$ in. *Ans.* $f = \frac{1}{2\pi}\sqrt{\frac{3EIkg}{W(3EI + kl^3)}}$ cps

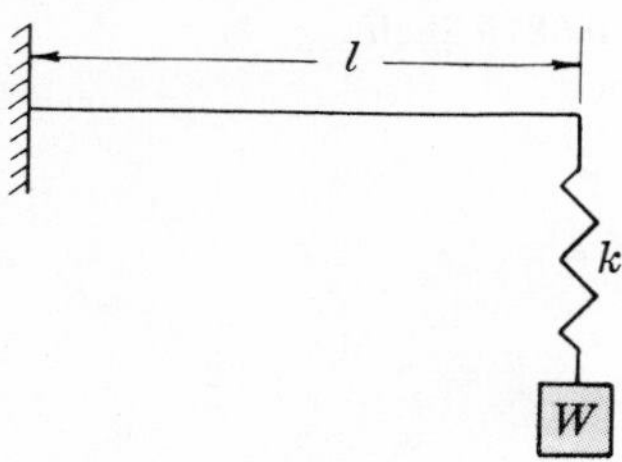

Fig. 20-30

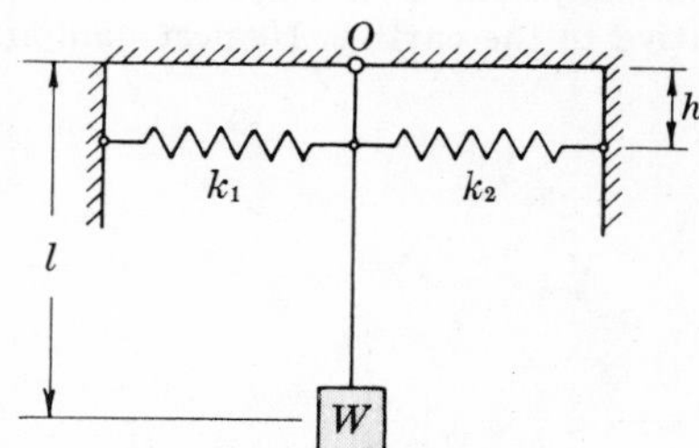

Fig. 20-31

39. Refer to Fig. 20-31 above. Determine the natural frequency in cps of the pendulum system shown in the figure. The springs have respectively constants k_1 and k_2 lb/in and are unstretched in the equilibrium position. Neglect the weight of the stiff rod. Hint: The pendulum in its displaced position is acted on by W and the sum of the forces in the springs.

Ans. $f = \frac{1}{2\pi}\sqrt{\frac{Wl + (k_1 + k_2)h^2}{Wl^2}g}$ cps

40. In the adjacent Fig. 20-32, the weight $W = 12$ lb. The spring modulus $k = 30$ lb/in. Neglecting the weight of the bell crank, determine the frequency of the system in cps. *Ans.* $f = 1.65$ cps

41. A steel disk 5 inches in diameter and 1/8 in. thick is rigidly attached to a steel wire 1/32 in. in diameter and 20 in. long. What is the natural frequency of this torsional pendulum? *Ans.* 0.50 cps

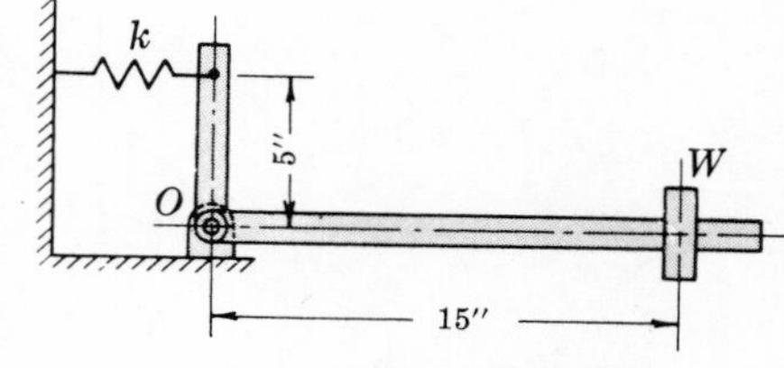

Fig. 20-32

42. An engine has a flywheel of 150 lb at each end of a steel shaft that is 2 inches in diameter. Assuming that the equivalent length of shaft between the flywheels is 2 ft, determine the natural frequency of torsional oscillation in cps. The radius of gyration for each flywheel is 8.8 inches and $G = 12 \times 10^6$ psi. *Ans.* 36.3 cps

43. A weight $W = 6$ lb is attached to a spring with constant $k = 14$ lb/in. Determine the critical damping coefficient. *Ans.* $c_c = 0.933$ lb-sec/in

44. A vibrating system consists of a weight $W = 10$ lb, a spring with constant $k = 20$ lb/in, and a dashpot with damping constant $c = 0.55$ lb-sec/in. Determine: (*a*) the damping factor, d; (*b*) the damped natural frequency, ω_d; (*c*) the logarithmic decrement, δ; (*d*) the ratio of any two successive amplitudes.
Ans. (*a*) $d = 0.382$, (*b*) $\omega_d = 25.7$ rad/sec, (*c*) $\delta = 2.60$, (*d*) ratio $= 13.5$

45. A machine weighing 200 lb is supported on three springs each with a constant $k = 60$ lb/in. A harmonic disturbing force of 5 lb acts on the machine. Determine (*a*) the resonant frequency and (*b*) the maximum distance the machine moves from equilibrium if the disturbing frequency is 200 cpm. Assume negligible damping. *Ans.* (*a*) $f = 2.97$ cps, (*b*) 0.106 in.

46. A spring supported mass is subjected to a harmonic disturbing force of varying frequency. A resonant amplitude of 0.82 inch is observed. At very high disturbing frequencies, an almost fixed amplitude of 0.07 inch is observed. What is the damping ratio d of the system? *Ans.* $d = 0.043$

47. A solid thin disk with weight $W = 6$ lb is attached to the center of a $\frac{3}{8}$ inch steel shaft. If the distance between bearing centers is 30 inches, what is the critical speed of the shaft? Neglect the weight of the shaft. *Ans.* 552 rpm

48. In Problem 47, the eccentricity of the disk is 0.0005 inch. What will be the amplitude of vibration of the disk at a shaft speed of 1750 rpm? What vibratory unbalance will be transmitted to the bearings?
Ans. 0.00055 in., 0.029 lb

49. A vibrometer has a natural frequency of 0.40 cps. Assuming negligible damping, what is the lower frequency limit of the instrument for 3% accuracy? Hint: Since the vibrometer is operated above its natural frequency, $Z/X = -1.03$ for a 3% error. *Ans.* 2.34 cps

50. A weight $W = 16.1$ lb is suspended in a box by a vertical spring whose constant $k = 22$ lb/in. If the box is subjected to a vibration $x = 0.08 \sin \pi t$, determine the absolute amplitude y of the weight (relative to the earth). Neglect damping effects. *Ans.* 0.0815 inch

Chapter 21

Special Topics: Beams and Virtual Work

Part A – Beams

A beam is a structural member which has a length considerably larger than its cross sectional dimensions and which carries loads usually perpendicular to the axis of the beam (thus the loads are at right angles to the length). The loads may be distributed over a very small distance along the beam, in which case they are called *concentrated,* or they may be distributed over a measurable distance, in which case they are called *distributed.*

TYPES of BEAMS

(*a*) Simple – supports are at the ends. See Fig. 21-1(*a*)

(*b*) Cantilever – one end is mounted in a wall and the other end is free (this is the only type considered here). See Fig. 21-1(*b*)

(*c*) Overhanging – at least one support is not at the end. See Fig. 21-1(*c*)

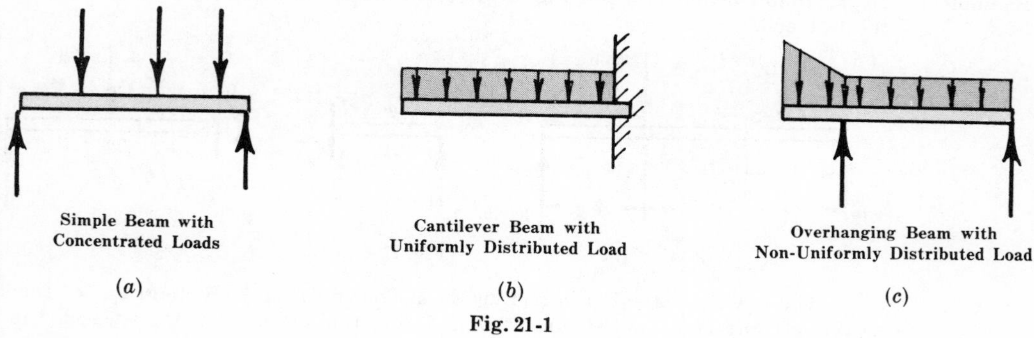
Simple Beam with Concentrated Loads (*a*)
Cantilever Beam with Uniformly Distributed Load (*b*)
Overhanging Beam with Non-Uniformly Distributed Load (*c*)

Fig. 21-1

SHEAR and MOMENT

Shear and moment at a cross section *C-D* in a beam are best visualized by dividing the beam into two parts *A* and *B* (see Fig. 21-2) to the left and right respectively of *C-D*. A free body diagram of part *A* must show all the external forces acting on *A* as well as the forces which *B* exerts on *A* to hold it in equilibrium.

The equilibrating forces which the right portion exerts on the free body diagram are (*a*) a vertical force *S* and (*b*) a set of horizontal distributed forces which because they have a zero summation are represented only by their moment *M*.

Shear at the section *C-D* is the vertical force *S* obtained by equating the summation of all the vertical forces to zero.

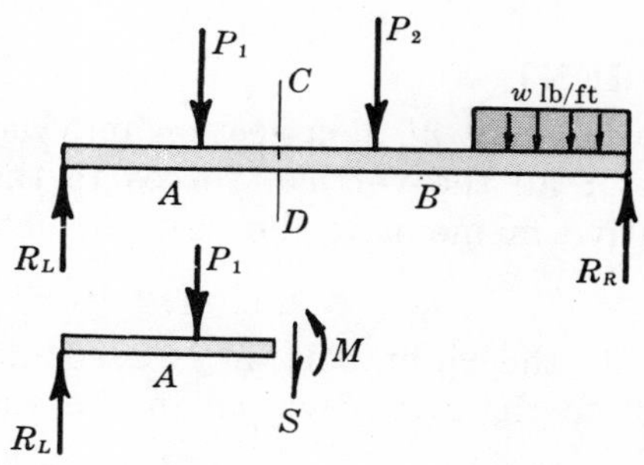

Fig. 21-2

Moment M at the section *C-D* is obtained by equating to zero the sum of the moments of all forces about a point in the cross section. In the types of problems considered here, any point in the cross section can be used as a moment center.

SIGN CONVENTION

In beam theory it is customary to call shear S positive if it acts down on the left portion A (if the right part B is used as a free body diagram then positive shear S must act up). Bending moment M at a section is positive if it acts counterclockwise on the left portion A (if the right part B is used as a free body diagram then positive moment M must act clockwise). Fig. 21-3 below illustrates these points.

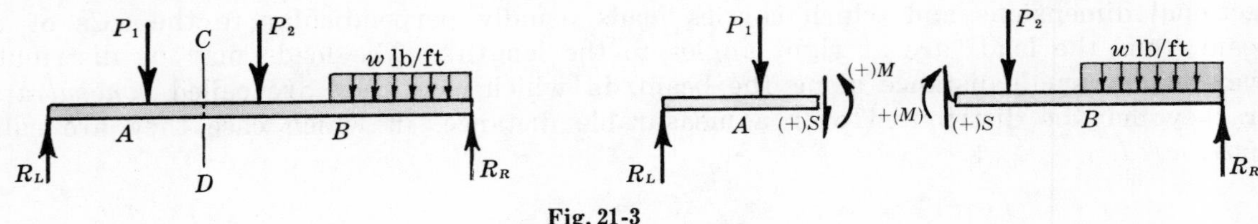

Fig. 21-3

SHEAR

Shear S may now be thought of as the sum of all the vertical forces to the left of the section where an up force causes positive shear (if the sum to the right is used, a down force causes positive shear).

Example 1: In the simple beam shown in Fig. 21-4 below, determine shear S at section C-D 2 ft from the left end.

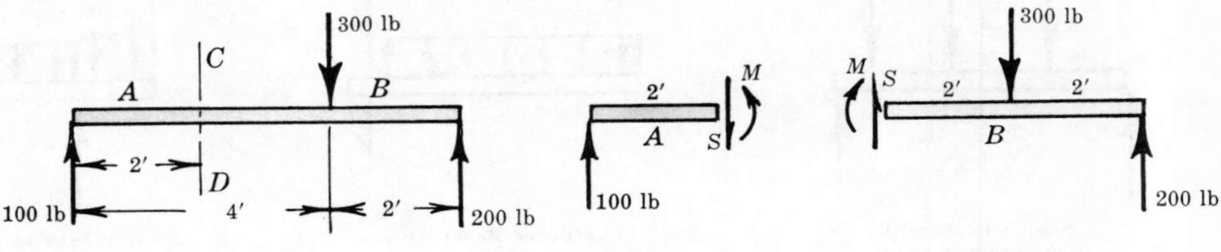

Fig. 21-4

The reactions are 100 lb and 200 lb as shown. Free body diagrams of the sections to the left and right of C-D are shown. Using the left part A, the shear is the sum of the forces to the left or

$$S = +100 \text{ lb}$$

If the right part B is used, the shear is the sum of the forces to the right of the section (but now a down force is positive); hence

$$S = +300 - 200 = +100 \text{ lb}$$

MOMENT

Moment M at a section may be evaluated as the sum of the moments about the section of all the vertical forces to the left. An up force to the left of the section causes positive moment at the section; hence in the figure above

$$M = +100(2) = +200 \text{ lb-ft}$$

If the right portion is used, an up force again contributes positive moment (in contrast to the sign reversal in the shear); hence for the right part B,

$$M = -300(2) + 200(4) = +200 \text{ lb-ft}$$

Shear and moment vary with the location of the section C-D. If x is the distance of C-D from the left end of the beam, shear and moment equations may then be expressed in terms of x. Refer particularly to Solved Problem 1.

SHEAR and MOMENT DIAGRAMS

Shear and moment diagrams present a graphical picture of the variation of S and M across the beam. Summations of forces and moments of forces to the left of section (or to the right, of course, as indicated before) can be used to plot S and M directly in the less difficult problems.

Example 2: Draw shear and moment diagrams for Example 1. It is necessary to use two free body diagrams as shown in Fig. 21-5 below.

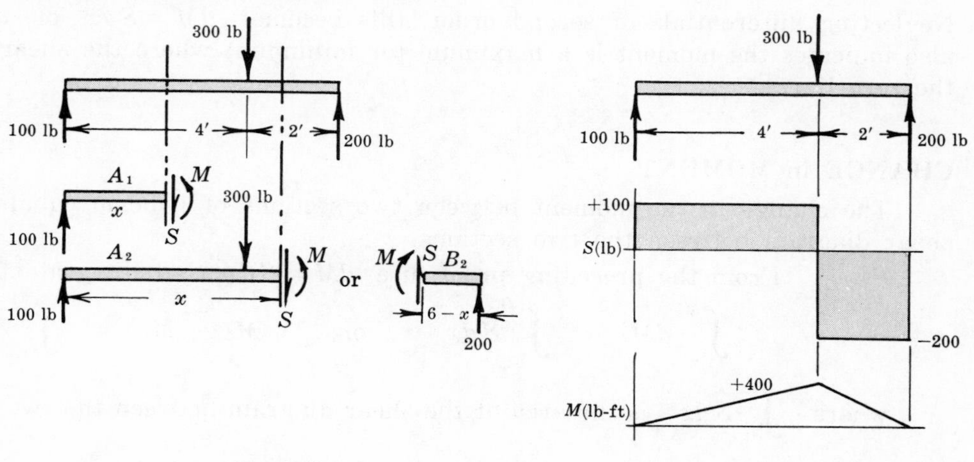

Fig. 21-5 **Fig. 21-6**

The free body diagram A_1 shows that for $0 < x < 4$ ft, $S = +100$ lb and $M = +100x$ lb-ft. The free body diagram A_2 shows that for $4 < x < 6$ ft, $S = -200$ lb and $M = +100x - 300(x-4) = -200x + 1200$ lb-ft. If the free body diagram B_2 is used, then $M = +200(6-x) = -200x + 1200$ lb-ft (be sure to use x as the distance from the left end). The value of M at $x = 4$ ft is $M = +400$ lb-ft.

The above information may now be plotted as shear and moment diagrams (see Fig. 21-6 above).

SLOPE of the SHEAR DIAGRAM

Slope of the shear diagram at any section along the beam is the negative of the unit load at that point.

Proof: Fig. 21-7 shows a portion dx of a beam. The unit load is approximately w lb/ft on this short piece. The shear S and moment M at the left are assumed positive. The shear and moment at the right are assumed to increase to $S + dS$ and $M + dM$ respectively.

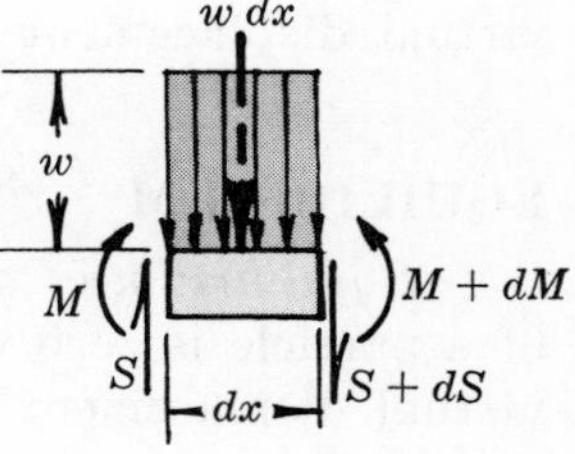

Fig. 21-7

Then $\quad \Sigma F_v \;=\; 0 \;=\; +S - w\,dx - S - dS$

from which $dS = -w\,dx$ or $dS/dx = -w$.

CHANGE in SHEAR

The change in shear between two sections of a beam carrying a distributed load equals the negative of the area of the load diagram between the two sections.

Proof: From the preceding proof, use $dS = -w\,dx$ and integrate from x_1 to x_2. Thus

$$\int_{S_1}^{S_2} dS \;=\; \int_{x_1}^{x_2} -w\,dx \qquad \text{or} \qquad S_2 - S_1 \;=\; -\int_{x_1}^{x_2} +w\,dx$$

where $\displaystyle\int_{x_1}^{x_2} +w\,dx$ is the area of the load diagram between the two sections.

SLOPE of the MOMENT DIAGRAM

Slope of the moment diagram at any section along the beam is the value of the shear at that section.

Proof: Using the diagram of the preceding proofs, equate the sum of the moments about the right end to zero and obtain

$$-M \;-\; S\,dx \;+\; w\,dx\,(dx/2) \;+\; M \;+\; dM \;=\; 0$$

Neglecting differentials of second order, this becomes $dM = S\,dx$ or $dM/dx = S$. This also indicates the moment is a maximum (or minimum) where the shear diagram crosses the zero line.

CHANGE in MOMENT

The change in the moment between two sections of a beam equals the area of the shear diagram between the two sections.

Proof: From the preceding proof, use $dM = S\,dx$ and integrate from x_1 to x_2.

$$\int_{M_1}^{M_2} dM \;=\; \int_{x_1}^{x_2} S\,dx \qquad \text{or} \qquad M_2 - M_1 \;=\; \int_{x_1}^{x_2} S\,dx$$

where $\int_{x_1}^{x_2} S\,dx$ is the area of the shear diagram between the two sections.

Refer to Problems 1, 2, and 3.

Part B – Virtual Work

VIRTUAL DISPLACEMENT and VIRTUAL WORK

A virtual displacement δs *of a particle* is any arbitrary infinitesimal change in the position of the particle consistent with the constraints imposed on the particle. This displacement can be imagined; it does not have to take place.

Virtual work δU is the work done by all the forces acting on the particle during a virtual displacement δs.

EQUILIBRIUM

Equilibrium of a Particle. The necessary and sufficient condition for the equilibrium of a particle is zero virtual work done by all the forces acting on the particle during any virtual displacement δs.

Equilibrium of a Rigid Body. The necessary and sufficient condition for the equilibrium of a rigid body is zero virtual work done by all the external forces acting on the body during any virtual displacement consistent with the constraints imposed on the body.

Equilibrium of a Connected System of Rigid Bodies is defined as above for rigid bodies. Keep in mind that for a virtual displacement consistent with constraints no work is done by internal forces, by reactions at smooth pins, by forces normal to the direction of motion. The external forces which do work (including friction if present) are called active or applied forces.

Equilibrium of a System exists if the potential energy V has a stationary value. Thus, if V is a function of one independent variable such as x, then $dV/dx = 0$ will yield the equilibrium value(s) of x.

STABLE EQUILIBRIUM

Stable equilibrium occurs if the potential energy V is a minimum. If in Fig. 21-8(a) below the bead is placed at the bottom of the frictionless wire bent as a circle, intuition indicates this is a position of stable equilibrium with the potential energy of the bead a minimum because any disturbance will be followed by a return to the bottom position. Using the x-axis as a standard (datum), the potential energy of the bead any place below the x-axis is

$$V = -Wy = -W\sqrt{a^2 - x^2}$$

Set dV/dx equal to zero to find the equilibrium position: $\dfrac{dV}{dx} = +\dfrac{Wx}{\sqrt{a^2 - x^2}} = 0$

Hence the solution needed here is $x = 0$ (the bead is then at the bottom). In determining the type of equilibrium, it is necessary to evaluate d^2V/dx^2 at the equilibrium position. Thus

$$\frac{d^2V}{dx^2} = +\frac{W}{\sqrt{a^2 - x^2}} + \frac{Wx^2}{(a^2 - x^2)^{3/2}}$$

and at $x = 0$, $d^2V/dx^2 = +W/a$ (positive) showing stable equilibrium.

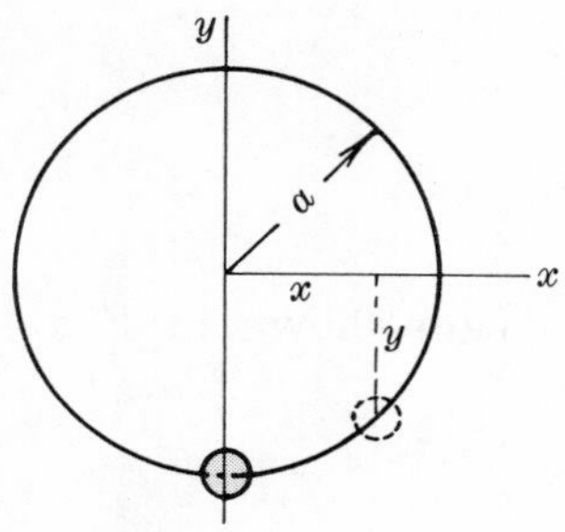

(a) Stable Equilibrium

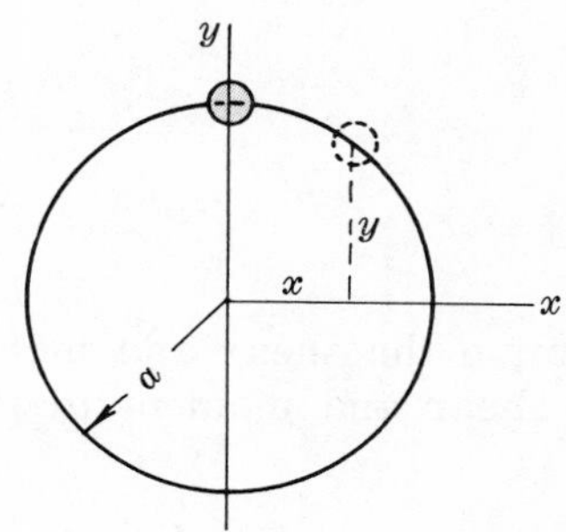

(b) Unstable Equilibrium

Fig. 21-8

UNSTABLE EQUILIBRIUM

Unstable equilibrium occurs if the potential energy V is a maximum. If in Fig. 21-8(b) above the bead is placed at the top of the wire, intuition indicates this is a position of unstable equilibrium with the potential energy of the bead a maximum. Using the x-axis as a standard (datum), the potential energy of the bead any place above the x-axis is

$$V = +Wy = +W\sqrt{a^2 - x^2}$$

Set dV/dx equal to zero to find the equilibrium position: $\dfrac{dV}{dx} = -\dfrac{Wx}{\sqrt{a^2 - x^2}} = 0.$

Hence the solution needed here is $x = 0$ (the bead is then at the top). Note also,

$$\frac{d^2V}{dx^2} = -\frac{W}{\sqrt{a^2 - x^2}} - \frac{Wx^2}{(a^2 - x^2)^{3/2}}$$

and at $x = 0$, $d^2V/dx^2 = -W/a$ (negative).

NEUTRAL EQUILIBRIUM

Neutral equilibrium exists if the system remains in any position in which it is placed. For example, the bead can be located any place on a horizontal wire and will remain there.

SUMMARY

To determine the value(s) of the variable(s) for which a system is in equilibrium, express the potential energy V of the system as a function of the variable(s). In the above discussion x was the variable. Then set $dV/dx = 0$ to determine the value(s) of x for equilibrium. Evaluate d^2V/dx^2 to ascertain the type of equilibrium.

If $d^2V/dx^2 > 0$, equilibrium is stable

If $d^2V/dx^2 < 0$, equilibrium is unstable

If $d^2V/dx^2 = 0$, equilibrium is neutral

Solved Problems

1. Determine the shear and moment equations for the beam shown in Fig. 21-9 below. Draw shear and moment diagrams.

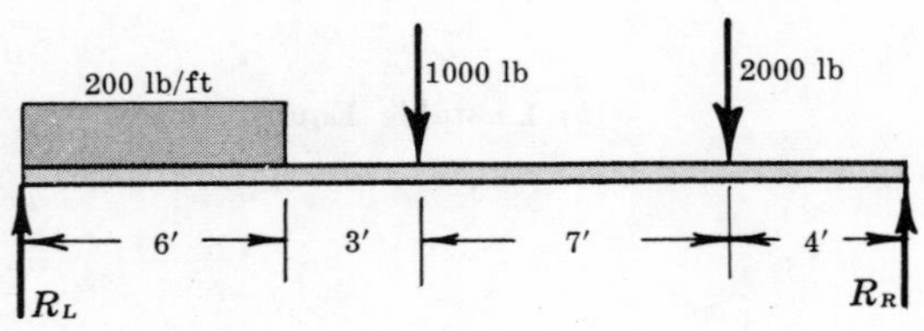

Fig. 21-9

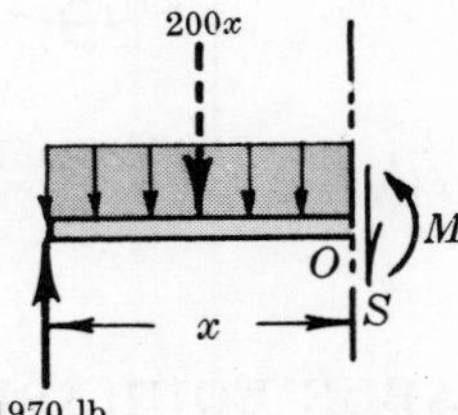

Fig. 21-10

Solution:

First determine the right reaction R_R by equating the sum of the moments of all external forces about the left end to zero. The 200 lb/ft load may be replaced (this should only be done to determine the reactions) by a 200(6) or 1200 lb load at its midpoint.

$$\Sigma M_{R_L} = 0 = -1200(3) - 1000(9) - 2000(16) + 20R_R, \qquad R_R = 2230 \text{ lb}$$

Similarly,

$$\Sigma M_{R_R} = 0 = -20R_L + 1200(17) + 1000(11) + 2000(4), \qquad R_L = 1970 \text{ lb}$$

In the interval $0 < x < 6$ the free body diagram shown in Fig. 21-10 above holds.

A vertical summation of all forces acting on the free body diagram yields

$$+1970 - 200x - S = 0 \qquad \text{or} \qquad S = 1970 - 200x \text{ lb}$$

This shows that S is the sum of the forces to the left of the section.

Next sum moments about a point O in the cross section at x.

$$-1970x + 200(x/2) + M = 0 \qquad \text{or} \qquad M = 1970x - 100x^2 \text{ lb-ft}$$

This shows that the moment at the section is the sum of the moments of all vertical forces to the left of the section.

Free body diagrams are shown in Fig. 21-11 below for various sections along the beam.

$6' < x < 9'$

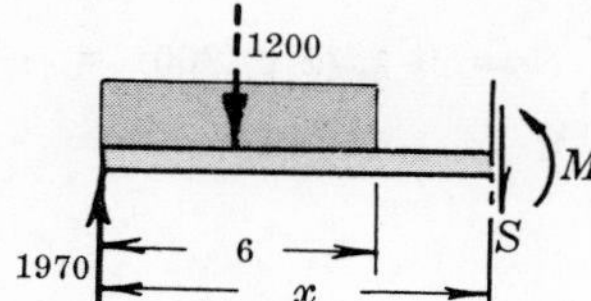

$$S = 1970 - 1200 = 770 \text{ lb}$$

$$M = 1970x - 1200(x-3) = 770x + 3600 \text{ lb-ft}$$

$9' < x < 16'$

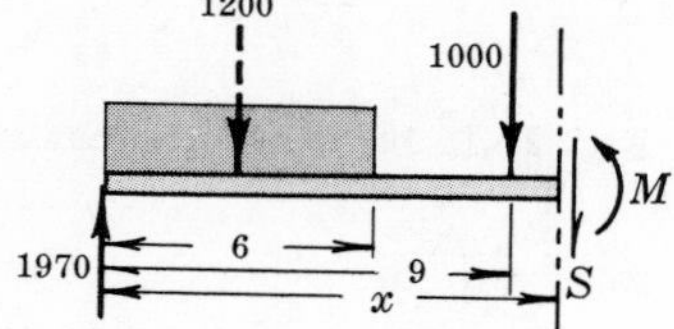

$$S = 1970 - 1200 - 1000 = -230 \text{ lb}$$

$$M = 1970x - 1200(x-3) - 1000(x-9)$$
$$= -230x + 12{,}600 \text{ lb-ft}$$

$16' < x < 20'$

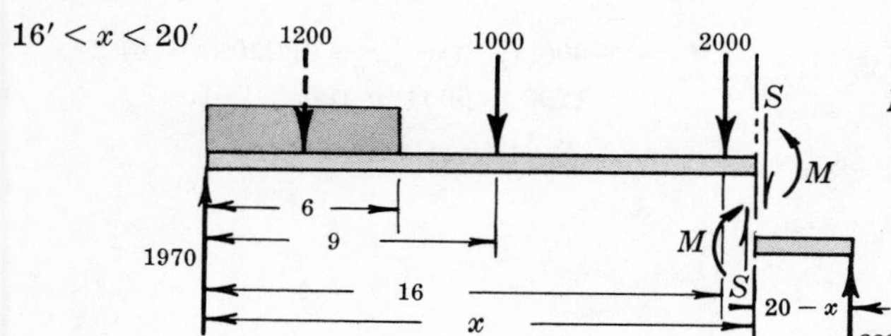

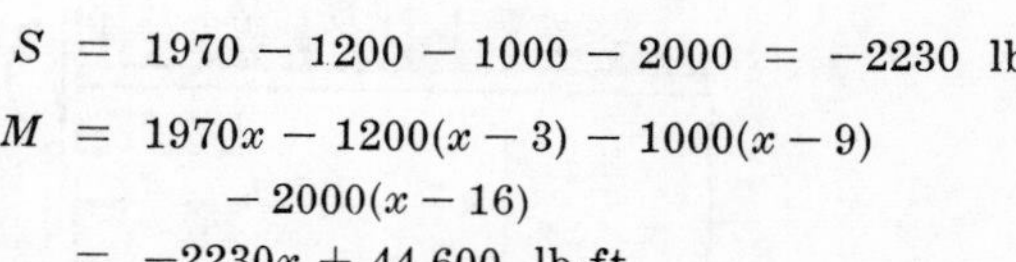

$$S = 1970 - 1200 - 1000 - 2000 = -2230 \text{ lb}$$

$$M = 1970x - 1200(x-3) - 1000(x-9)$$
$$- 2000(x-16)$$
$$= -2230x + 44{,}600 \text{ lb-ft}$$

Or from right portion,

$$M = +2230(20-x)$$
$$= -2230x + 44{,}600 \text{ lb-ft}$$

Fig. 21-11

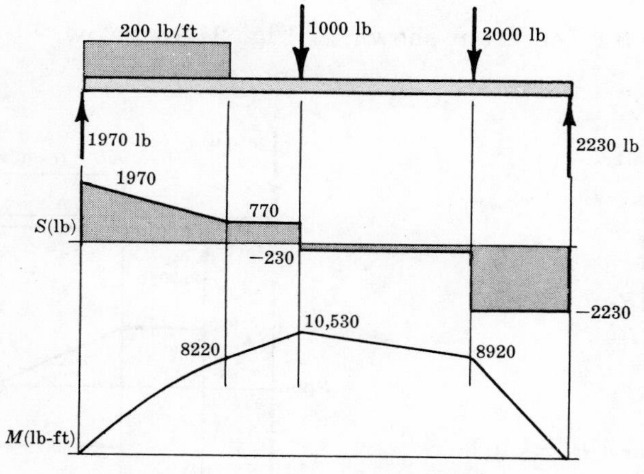

Fig. 21-12

Shear and moment diagrams are shown in Fig. 21-12.

2. The cantilever beam supports the triangular load and the uniformly distributed load as shown in Fig. 21-13. Derive the shear and moment equations.

Solution:

Two free body diagrams are needed because the loading changes at 6′.

In the interval $0 < x < 6'$, the diagram (see Fig. 21-14 below) and equations are as follows. The height of the load at x using similar triangles is $(x/6)(200)$ lb/ft. The total load is then $\frac{1}{2}x(x/6)(200)$ lb located at the point $x/3$ to the left of the section.

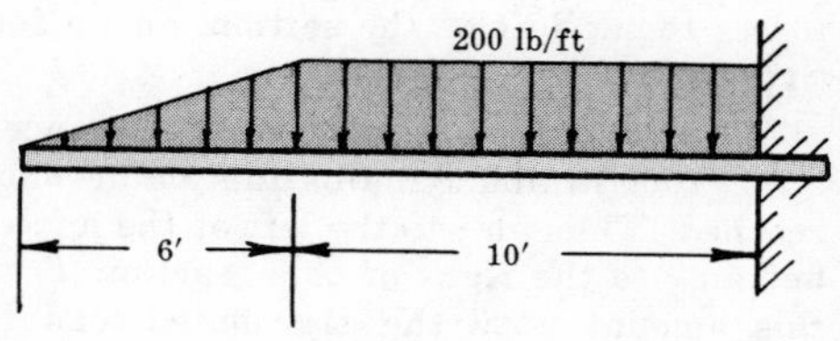

Fig. 21-13

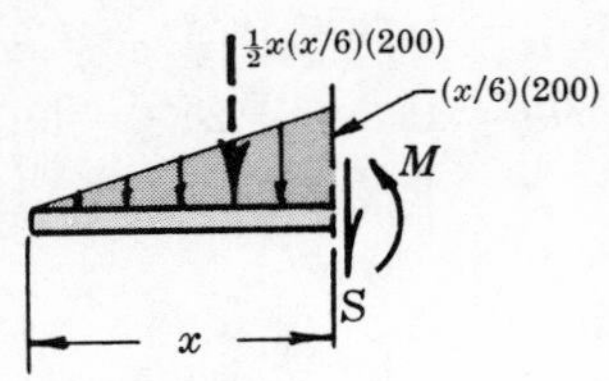

Fig. 21-14

$$S = -\tfrac{1}{2}x(x/6)(200) = -(50/3)x^2 \text{ lb}$$
$$M = -(50/3)x^2(x/3) = -(50/9)x^3 \text{ lb-ft}$$

In the interval $6' < x < 16'$, the diagram (see Fig. 21-15 below) and equations becomes

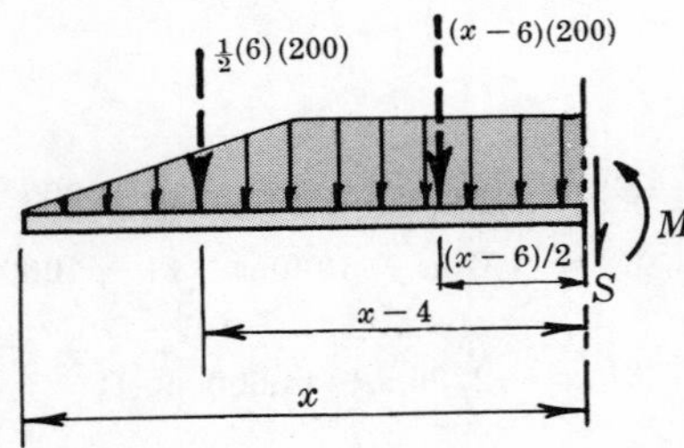

Fig. 21-15

$$S = -600 - (x-6)(200) = +600 - 200x \text{ lb}$$
$$M = -600(x-4) - (x-6)(200)(x-6)/2$$
$$= -1200 + 600x - 100x^2 \text{ lb-ft}$$

3. Draw shear and moment diagrams for the beam shown in Fig. 21-16 below.

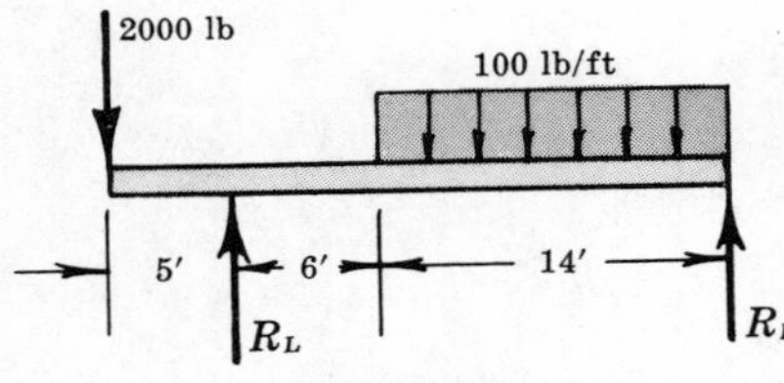

Fig. 21-16

Solution:

Summing moments about R_L,

$$+2000(5) - 1400(13) + 20R_R = 0 \quad \text{or} \quad R_R = 410 \text{ lb}$$

Summing moments about R_R,

$$+2000(25) - 20R_L + 1400(7) = 0 \quad \text{or} \quad R_L = 2990 \text{ lb}$$

In drawing the shear diagram (see Fig. 21-17) make use of the fact that the shear at any section is the sum of the forces to the left of the section, an up force contributing positive shear.

For a section a very small distance (ϵ) in from the left, $S = -2000$ lb and remains this value until the left reaction is reached. Thus an ϵ to the left of the left reaction, $S = -2000$ lb; but an ϵ to the right of this reaction, $S = +990$ lb. It remains this amount until the distributed load is reached. Then the shear decreases (load is down) at the rate of 100 lb/ft. It reaches zero value at 9.9 ft (990/100). Its value an ϵ to the left of the right end is -410 lb. The right reaction thus closes the shear diagram.

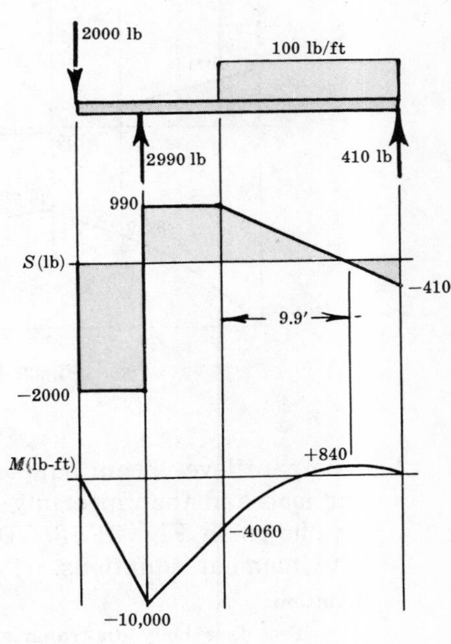

Fig. 21-17

To draw the moment diagram (see Fig. 21-17), first determine the moment at the left end (really an ϵ to the right of the left end). This value is -2000ϵ lb-ft or zero at the left end. From $x = 0$ to $x = 5'$, the shear is negative (constant value); hence the slope of the moment diagram is a constant negative value. The moment just to the left of the left reaction equals the moment at the left end (zero) plus the area of the shear diagram from the left end to the left reaction [negative 5(2000) = −10,000 lb-ft].

The shear is a constant positive value from $x = 5'$ to $x = 11'$; hence the slope of the moment diagram is positive. The change in moment is the area of the shear diagram from $x = 5'$ to $x = 11'$, or $+6(990) = +5940$. Thus the moment changes from −10,000 to $(-10{,}000 + 5940) = -4060$ lb-ft.

The shear area is positive over the next 9.9′ at which point the moment is a maximum (where $S = 0$). The increase in moment from $x = 11'$ to $x = 20.9'$ is the area of the triangular shear diagram, i.e. $\frac{1}{2}(990)(9.9) = 4900$ lb-ft.

Thus M at $x = 20.9'$ is $(-4060 + 4900) = +840$ lb-ft. Since the shear is positive but decreasing in this interval $11' < x < 20.9'$, the moment diagram has a positive but decreasing slope up to the value of 840 lb-ft.

To complete the moment diagram note that the last shear area is negative but the magnitude of S is increasing. Hence the moment curve has a negative slope that becomes more negative. Also the moment curve drops in value an amount equal to the last shear area, which is $\frac{1}{2}(4.1)(410) = 840$ lb-ft. This means the moment at the right end is zero, as it should be.

4. A homogeneous ladder of weight W and length l is held in equilibrium by a horizontal force P as shown in Fig. 21-18. Using virtual work methods, express P in terms of W.

Solution:

Assuming x is positive to the right, the virtual work done by P for an increase δx is $-P\,\delta x$, since P is directed to the left.

Assuming y is positive up, the virtual work done by W for an increase δy is $-W\,\delta y$, since W is directed down.

The total virtual work δU is zero; hence

$$\delta U \quad = \quad -P\,\delta x \;-\; W\,\delta y \quad = \quad 0 \qquad (a)$$

From the figure, $x = l \sin\theta$ and $y = \frac{1}{2} l \cos\theta$. Then

$$\delta x = l(\cos\theta)\,\delta\theta \quad \text{and} \quad \delta y = -\tfrac{1}{2} l(\sin\theta)\,\delta\theta.$$

Fig. 21-18

Substituting these values into equation (a), there results $P = \frac{1}{2} W \tan\theta$.

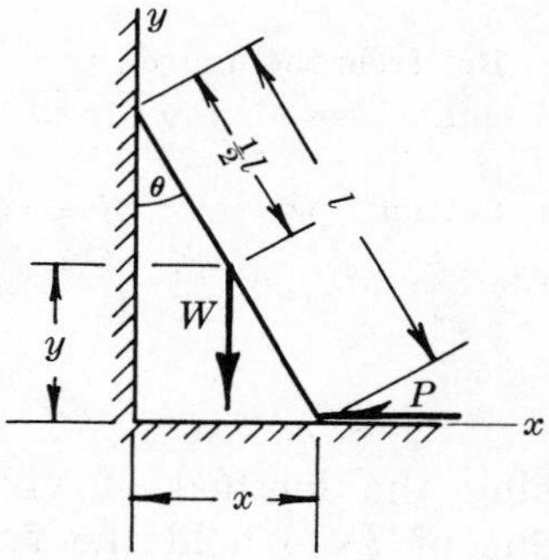

5. In the toggle device shown in Fig. 21-19, express the relationship between forces F and P in terms of angle θ.

Solution:

Assuming x positive to the right, the virtual work done by F for an increase δx is $-F\,\delta x$, because F is directed to the left. Assuming y positive up, the virtual work done by P for an increase δy is $-P\,\delta y$, because P is directed down. The total virtual work δU is zero; hence

$$\delta U \quad = \quad -F\,\delta x \;-\; P\,\delta y \quad = \quad 0 \qquad (a)$$

From the figure, $x = 2l\cos\theta$ and $y = l \sin\theta$. Then

$$\delta x = -2l(\sin\theta)\,\delta\theta \quad \text{and} \quad \delta y = l(\cos\theta)\,\delta\theta$$

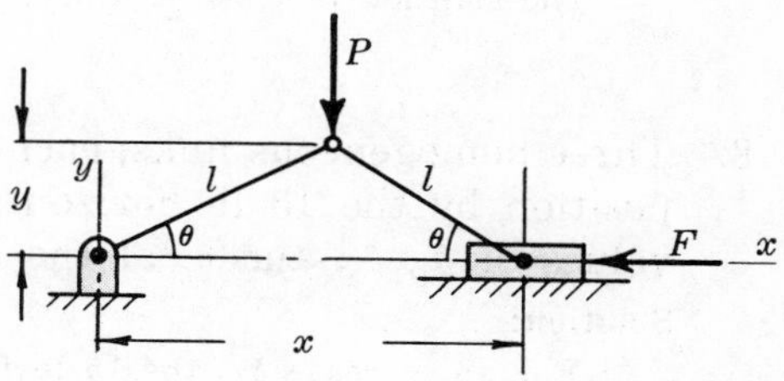

Fig. 21-19

Substituting these values into equation (a), there results $P = 2F\tan\theta$.

6. Using the method of virtual work, determine the relationship between the moment M applied to the crank R and the force F applied to the crosshead in the slider crank mechanism shown in Fig. 21-20 below.

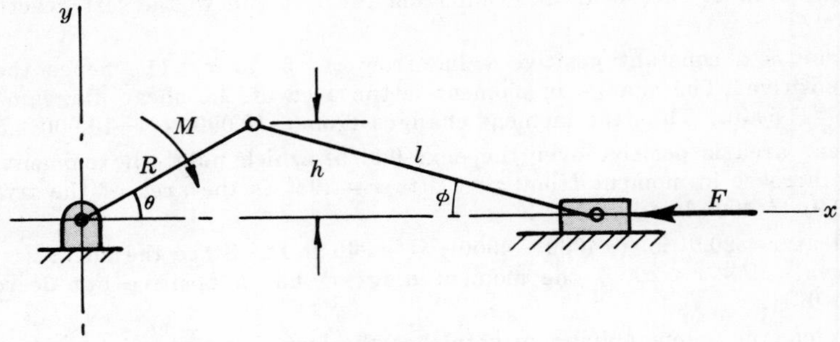

Fig. 21-20

Solution:

Assuming θ positive counterclockwise, the virtual work done by M for an increase $\delta\theta$ is $-M\,\delta\theta$. Assuming x positive to the right, the virtual work done by F for an increase δx is $-F\,\delta x$. The total virtual work δU is zero; hence

$$\delta U = -M\,\delta\theta - F\,\delta x = 0 \qquad (a)$$

But from the figure, $x = R\cos\theta + l\cos\phi$. To express ϕ in terms of θ, use $h = R\sin\theta = l\sin\phi$ and obtain $\cos\phi = \sqrt{1-\sin^2\phi} = \sqrt{1-(R/l)^2\sin^2\theta}$.

Putting $\delta x = -R(\sin\theta)\,\delta\theta - (R^2/l)\dfrac{\sin\theta\cos\theta}{\sqrt{1-(R/l)^2\sin^2\theta}}\,\delta\theta$ into (a),

$$M = FR(\sin\theta)\left(1 + \frac{R\cos\theta}{l\sqrt{1-(R/l)^2\sin^2\theta}}\right)$$

7. Using the method of virtual work, determine the value of F to hold the frame in equilibrium under the action of force P. Each link is of length $2a$, and $\theta = 45°$.

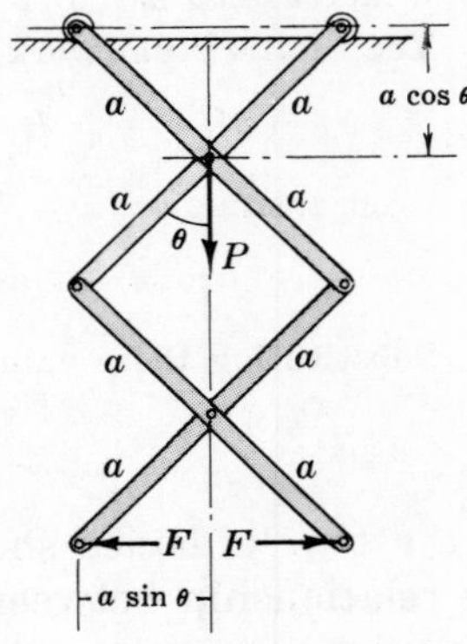

Fig. 21-21

Solution:

Let θ be any angle in the analysis. Later the actual value 45° will be used. If there is an increase $\delta\theta$, P will rise doing negative work. The magnitude of the differential rise is $|\delta(a\cos\theta)| = a\sin\theta\,\delta\theta$. At the same time the F forces will do positive work. The magnitude of the differential movement of either F is $\delta(a\sin\theta) = a\cos\theta\,\delta\theta$.

Thus $\delta U = -P(a\sin\theta\,\delta\theta) + 2F(a\cos\theta\,\delta\theta) = 0$.

The solution is $F = \frac{1}{2}P\tan\theta$; at $\theta = 45°$, $F = 0.5P$.

8. Three homogeneous links, each weighing 10 lb and 6 ft long, are held in the equilibrium position by the 15 lb horizontal force as shown in Fig. 21-22 below. Determine the values of θ_1, θ_2, and θ_3 at the equilibrium position.

Solution:

For an increase δx, the 15 lb force will do positive work. For an increase δy_1, δy_2, and δy_3, the 10 lb weights will do positive work. Hence

$$\delta U = +15\,\delta x + 10\,\delta y_1 + 10\,\delta y_2 + 10\,\delta y_3 = 0$$

From the figure, $x = 6\sin\theta_1 + 6\sin\theta_2 + 6\sin\theta_3$,

$$y_1 = 3\cos\theta_1, \quad y_2 = 6\cos\theta_1 + 3\cos\theta_2, \quad \text{and} \quad y_3 = 6\cos\theta_1 + 6\cos\theta_2 + 3\cos\theta_3$$

Hence

$$\begin{aligned}\delta x &= 6\cos\theta_1\,\delta\theta_1 + 6\cos\theta_2\,\delta\theta_2 + 6\cos\theta_3\,\delta\theta_3\\ \delta y_1 &= -3\sin\theta_1\,\delta\theta_1,\\ \delta y_2 &= -6\sin\theta_1\,\delta\theta_1 - 3\sin\theta_2\,\delta\theta_2,\\ \delta y_3 &= -6\sin\theta_1\,\delta\theta_1 - 6\sin\theta_2\,\delta\theta_2 - 3\sin\theta_3\,\delta\theta_3\end{aligned}$$

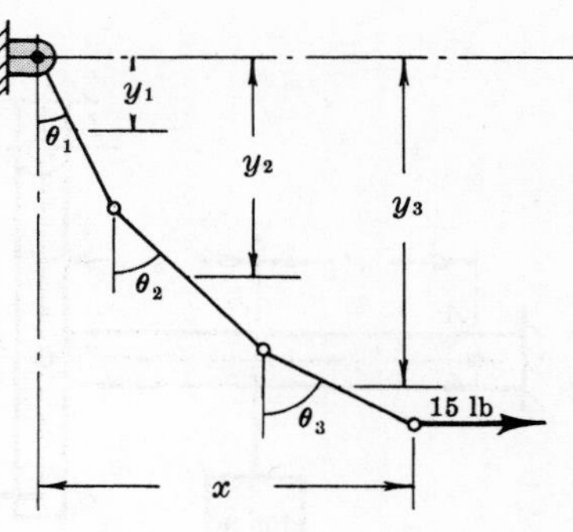

Fig. 21-22

Substituting,

$$\begin{aligned}\delta U = &+15[6\cos\theta_1\,\delta\theta_1 + 6\cos\theta_2\,\delta\theta_2 + 6\cos\theta_3\,\delta\theta_3]\\ &-10[3\sin\theta_1\,\delta\theta_1] - 10[6\sin\theta_1\,\delta\theta_1 + 3\sin\theta_2\,\delta\theta_2]\\ &-10[6\sin\theta_1\,\delta\theta_1 + 6\sin\theta_2\,\delta\theta_2 + 3\sin\theta_3\,\delta\theta_3] = 0\end{aligned}$$

To solve for θ_1, allow only $\delta\theta_1$ to exist. This leads to

$$(90\cos\theta_1 - 30\sin\theta_1 - 60\sin\theta_1 - 60\sin\theta_1)\,\delta\theta_1 = 0$$

from which $\tan\theta_1 = 90/150$ or $\theta_1 = 30°58'$.

To solve for θ_2, allow only $\delta\theta_2$ to exist, and obtain

$$(90\cos\theta_2 - 30\sin\theta_2 - 60\sin\theta_2)\,\delta\theta_2 = 0 \qquad \text{from which} \qquad \theta = 45°$$

Similar reasoning yields $(90\cos\theta_3 - 30\sin\theta_3)\,\delta\theta_3 = 0$ from which $\theta_3 = 71°33'$.

9. Using the method of virtual work, determine the force in the top member of the truss shown in Fig. 21-23(a) below. The triangles are equilateral.

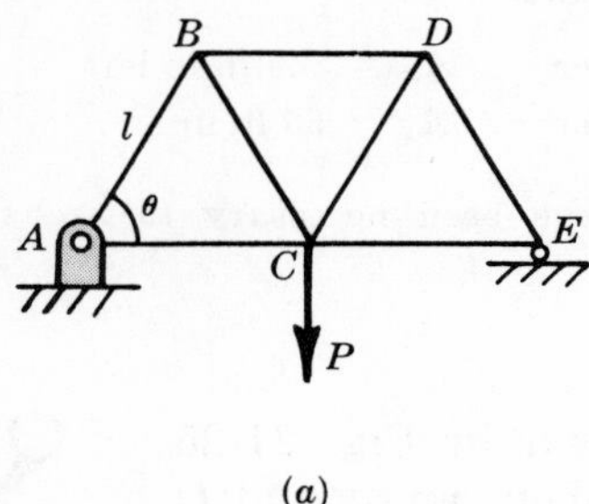

(a)

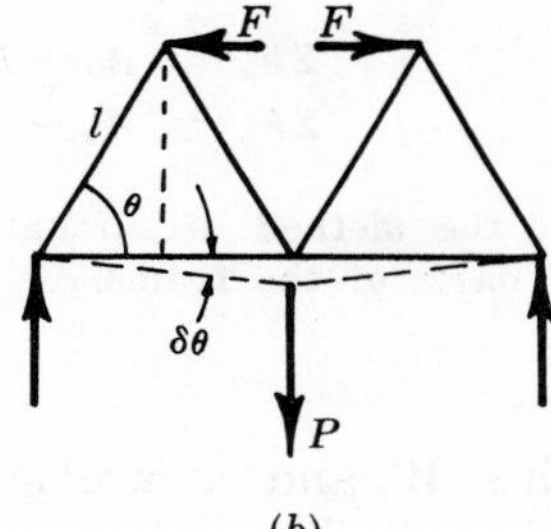

(b)

Fig. 21-23

Solution:

Fig. 21-23(b) above shows two forces F replacing the top member. The point of application of the left force F is $l\cos\theta$ away from the left reaction. If P falls a small amount when θ increases $\delta\theta$, P will do positive work $+Pl\,\delta\theta$ (assuming the infinitesimal drop is the differential arc of a circle of radius l). The left F will oppose the increase in θ and so do negative work of an amount $-F|\delta(l\cos\theta)| = -Fl\sin\theta\,\delta\theta$. The other F does an equal amount of work. Hence

$$\delta U = +Pl\,\delta\theta - 2Fl\sin\theta\,\delta\theta = 0 \qquad \text{or} \qquad F = P/(2\sin\theta)$$

Thus when $\theta = 60°$, $F = 0.577P$.

10. Using the principles of virtual work, determine the components of the pin reactions at A and B in Fig. 21-24(a) below. Neglect friction at all pins. The force at E is horizontal.

Solution:

A free body diagram in Fig. 21-24(b) below shows the pin B removed and the frame given a virtual angular displacement $\delta\theta$ about pin A. The virtual displacement of B to B' can be thought of as a displacement δy down and a displacement δx to the right. For the small angular displacement, all points on the member BDE move down an amount δy; hence point C will undergo one-half the displacement of D. Since the bar BDE rotates through the angle $\delta\theta$, point E undergoes a horizontal displacement which is two-thirds that of point B ($\frac{2}{3}\delta x$). The pin reactions at A do no work. The virtual work

$$\delta U = +100(\tfrac{1}{2}\,\delta y) - B_y\,\delta y + B_x\,\delta x - 200(\tfrac{2}{3}\,\delta x) = 0$$

or

$$\delta U = (50 - B_y)\,\delta y + (B_x - 133)\,\delta x = 0$$

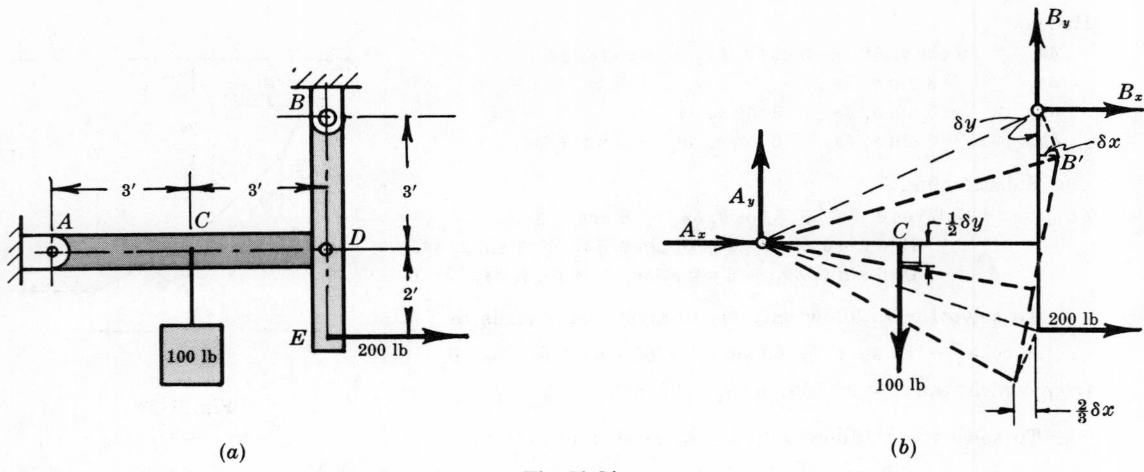

Fig. 21-24

Now since δx and δy are completely independent and arbitrary virtual displacements, δU will be zero if and only if

$$50 - B_y = 0 \qquad \text{and} \qquad B_x - 133 = 0$$

From these, $B_y = 50$ lb up and $B_x = 133$ lb to right.

The equations of equilibrium can be applied as follows:

$$\Sigma F_x = A_x + B_x + 200 = 0 \qquad \text{or} \qquad A_x = 333 \text{ lb to left}$$
$$\Sigma F_y = A_y - 100 + B_y = 0 \qquad \text{or} \qquad A_y = 50 \text{ lb up}$$

Without the method of virtual work it would have been necessary to use several free body diagrams of parts of the frame.

11. Two weights W and w are supported as shown in Fig. 21-25 by a bar of negligible weight which is pivoted about an axis at O perpendicular to the plane of the paper. Discuss equilibrium.

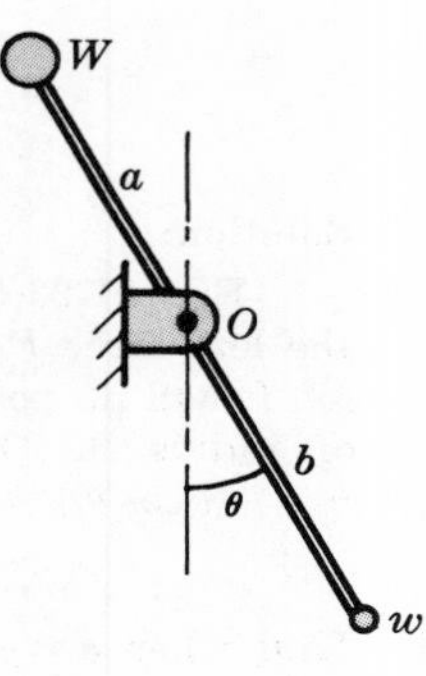

Fig. 21-25

Solution:

If $\theta = 0°$ is selected as the standard or reference configuration, then at any angle θ, W has lost potential energy $Wa(1 - \cos\theta)$ and w has gained $wb(1 - \cos\theta)$. The total potential energy V of the system at angle θ is

$$V = -Wa(1 - \cos\theta) + wb(1 - \cos\theta)$$

To study equilibrium, set $dV/d\theta = 0$:

$$dV/d\theta = -Wa \sin\theta + wb \sin\theta = 0$$

This is satisfied when $\sin\theta = 0 (\theta = 0°)$ or when $wb = Wa$ (θ may then be any angle).

To discuss stability, find $d^2V/d\theta^2 = -Wa\cos\theta + wb\cos\theta$.

Next evaluate $d^2V/d\theta^2$ for the two conditions of equilibrium. At $\theta = 0°$, $d^2V/d\theta^2 = -Wa + wb$. For stable equilibrium, the value must be positive; this obtains when $wb > Wa$.

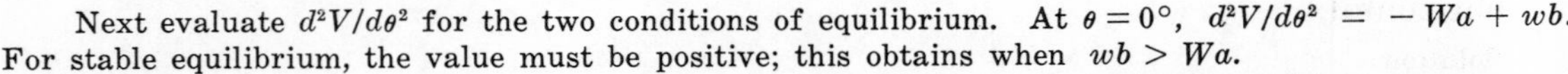

The other state of equilibrium occurs when $wb = Wa$, i.e. when $d^2V/d\theta^2 = 0$; this means neutral equilibrium.

To summarize, if $wb > Wa$ the system will be in stable equilibrium when $\theta = 0°$. If $wb = Wa$ the system will stay in any position in which it is placed.

12. A uniform bar of length l and weight W is held in equilibrium by a spring with constant k. When $\theta = 0°$, the spring is not stressed. Discuss equilibrium.

Solution:

Refer to Fig. 21-26. Using $\theta = 0°$ as the standard or reference configuration, the bar in position θ has lost potential energy equal to $\frac{1}{2}Wl(1-\cos\theta)$ and the spring has gained an amount $\frac{1}{2}kx^2$. From the geometry, $x^2 = a^2 + l^2 - 2al\cos\theta$. Hence the potential energy V for the system at any angle θ is

$$V = -\tfrac{1}{2}Wl(1-\cos\theta) + \tfrac{1}{2}k(a^2 + l^2 - 2al\cos\theta)$$

and
$$dV/d\theta = -\tfrac{1}{2}Wl\sin\theta + kal\sin\theta$$

which will be zero for $\sin\theta = 0 (\theta = 0°)$ or for $k = \frac{1}{2}W/a$.

Next find $d^2V/d\theta^2 = -\frac{1}{2}Wl\cos\theta + kal\cos\theta$. At $\theta = 0°$, $d^2V/d\theta^2 = -\frac{1}{2}Wl + kal$; this will be positive if $k > \frac{1}{2}W/a$. At any other angle, $d^2V/d\theta^2 = 0$ if $k = \frac{1}{2}W/a$.

In conclusion, if $k = \frac{1}{2}W/a$ the system will remain in any position in which it is placed.

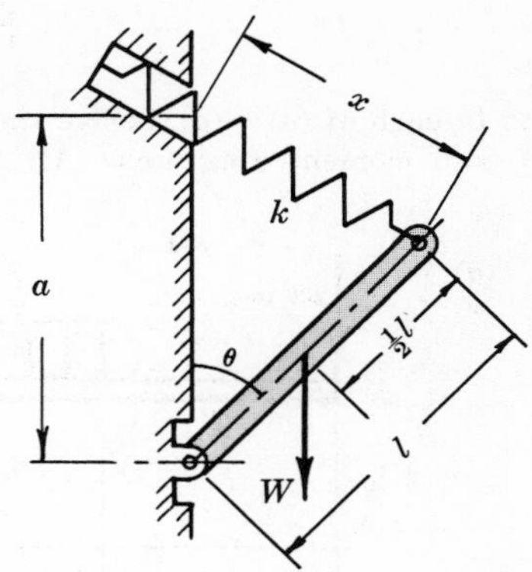

Fig. 21-26

13. In Fig. 21-27, the two identical gears turn about frictionless pivots. A weightless 2 ft bar rigidly attached to the gear holds the 20 lb weight. The other gear is attached to a vertical spring with constant $k = 12$ lb/in. Determine the angle(s) θ for equilibrium.

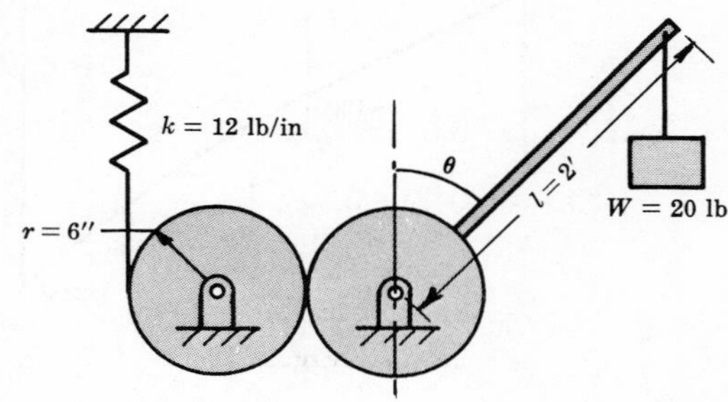

Fig. 21-27

Solution:

If $\theta = 0°$ is used as the standard or reference configuration and the spring is unstressed, then for any other angle θ the weight loses potential energy in amount $Wl(1-\cos\theta)$. The spring gains an amount $\frac{1}{2}k(r\theta)^2$. The total potential energy of the system is

$$V = -Wl(1-\cos\theta) + \tfrac{1}{2}kr^2\theta^2$$

Then
$$dV/d\theta = -Wl\sin\theta + kr^2\theta, \qquad d^2V/d\theta^2 = -Wl\cos\theta + kr^2$$

The first derivative will be zero if $\theta = 0°$ or if $\sin\theta = (kr^2/Wl)\theta$.

At $\theta = 0°$, $d^2V/d\theta^2 = -Wl + kr^2 = -20(2) + (144)(\frac{1}{2})^2 = -4$ (not stable).

The other angle for equilibrium is determined from $\sin\theta = \dfrac{kr^2}{Wl}\theta = \dfrac{144(\frac{1}{2})^2}{20(2)}\theta = 0.9\theta$ which by trial and error is satisfied by $\theta = 44.1°$. Then $d^2V/d\theta^2 = -20(2)\cos 44.1° + 36 = +7.2$, which indicates stable equilibrium.

14. A uniform rope of weight W is placed on a sphere of radius r as shown in Fig. 21-28. Find the tension T in the rope when it is in a horizontal plane at a vertical distance b below the top. Use the method of virtual work.

Solution:

The rope is at a height y above the xz-plane. Its length is $l = 2\pi x = 2\pi\sqrt{r^2 - y^2}$.

If the rope is given a virtual displacement δy down, the virtual work done is

$$\delta U = +W\,\delta y + T\,\delta l = 0$$

Substituting $\delta l = \dfrac{2\pi(\frac{1}{2})(-2y\,\delta y)}{(r^2-y^2)^{1/2}} = \dfrac{-2\pi y\,\delta y}{(2br-b^2)^{1/2}}$, we obtain

$$T = \frac{W(2br-b^2)^{1/2}}{2\pi(r-b)}$$

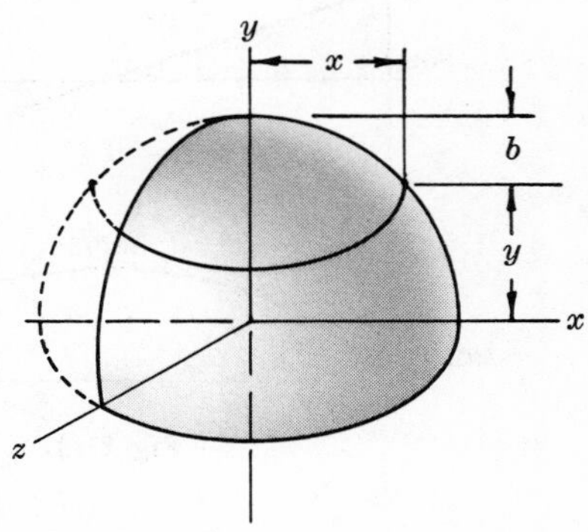

Fig. 21-28

Supplementary Problems

15. In each of (a) – (g), derive the shear and moment equations for the beam shown. Also verify the shear and moment diagrams. All x distances are measured from the left end of each beam.

(a)

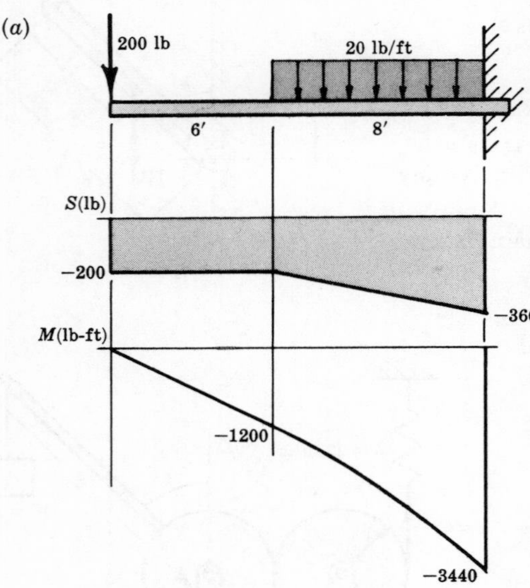

Fig. 21-29

Ans. $0 < x < 6'$ $\begin{cases} S = -200 \text{ lb} \\ M = -200x \text{ lb-ft} \end{cases}$

$6' < x < 14'$ $\begin{cases} S = -80 - 20x \text{ lb} \\ M = -360 - 80x - 10x^2 \text{ lb-ft} \end{cases}$

(b)

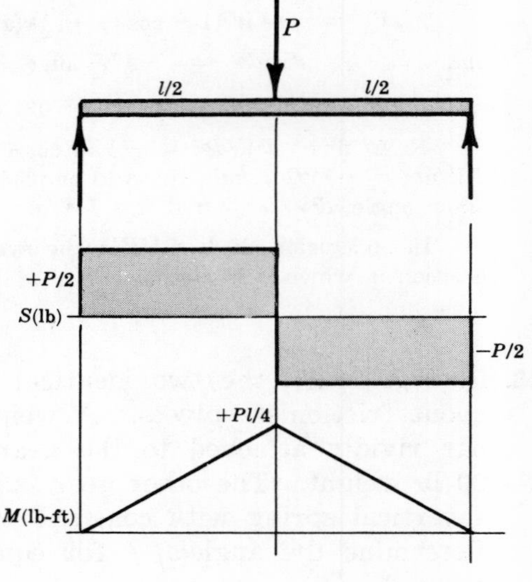

Fig. 21-30

Ans. $0 < x < l/2$ $\begin{cases} S = +P/2 \text{ lb} \\ M = +Px/2 \text{ lb-ft} \end{cases}$

$l/2 < x < l$ $\begin{cases} S = -P/2 \text{ lb} \\ M = Pl/2 - Px/2 \text{ lb-ft} \end{cases}$

(c)

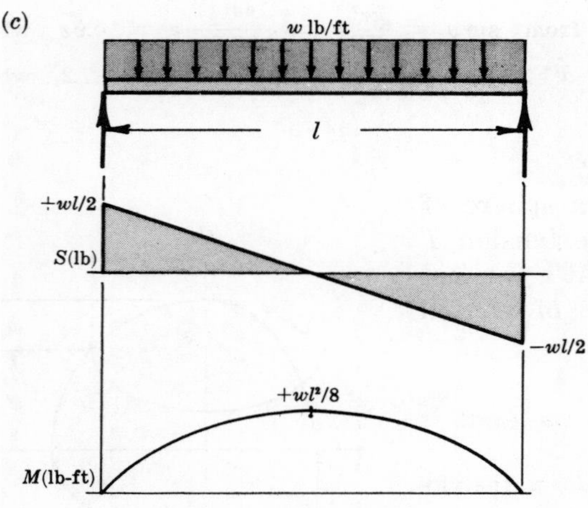

Fig. 21-31

Ans. $0 < x < l$ $\begin{cases} S = +wl/2 - wx \text{ lb} \\ M = +wlx/2 - wx^2/2 \text{ lb-ft} \end{cases}$

(d)

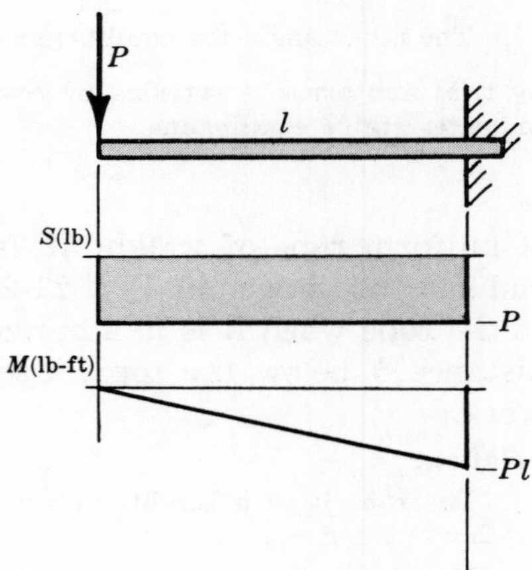

Fig. 21-32

Ans. $0 < x < l$ $\begin{cases} S = -P \text{ lb} \\ M = -Px \text{ lb-ft} \end{cases}$

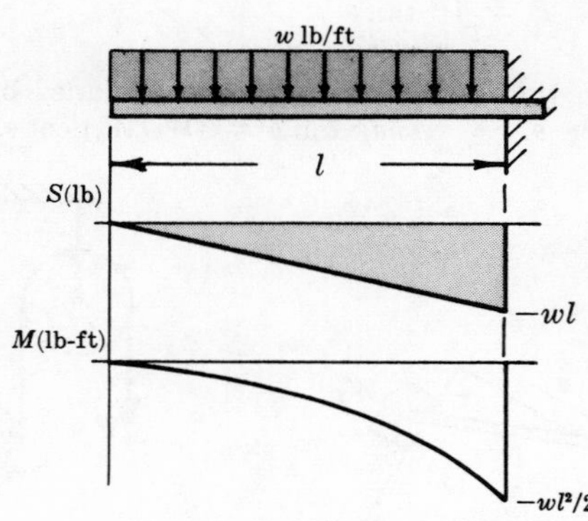

Fig. 21-33

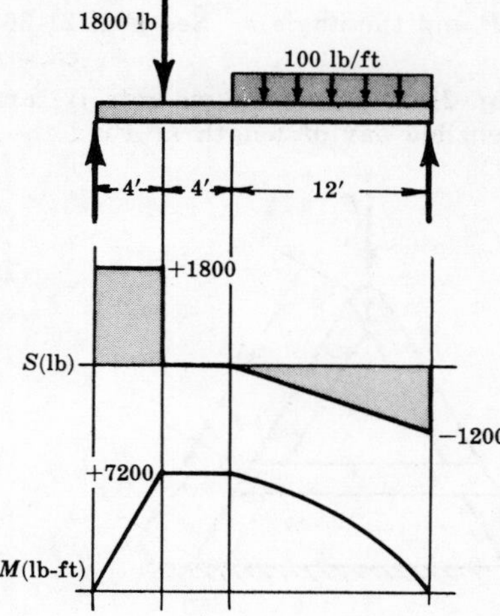

Fig. 21-34

Ans. $0 < x < l \begin{cases} S = -wx \text{ lb} \\ M = -wx^2/2 \text{ lb-ft} \end{cases}$

Ans. $0 < x < 4' \begin{cases} S = +1800 \text{ lb} \\ M = +1800x \text{ lb-ft} \end{cases}$

$4' < x < 8' \begin{cases} S = 0 \\ M = 7200 \text{ lb-ft} \end{cases}$

$8' < x < 20' \begin{cases} S = +800 - 100x \text{ lb} \\ M = +4000 + 800x - 50x^2 \text{ lb-ft} \end{cases}$

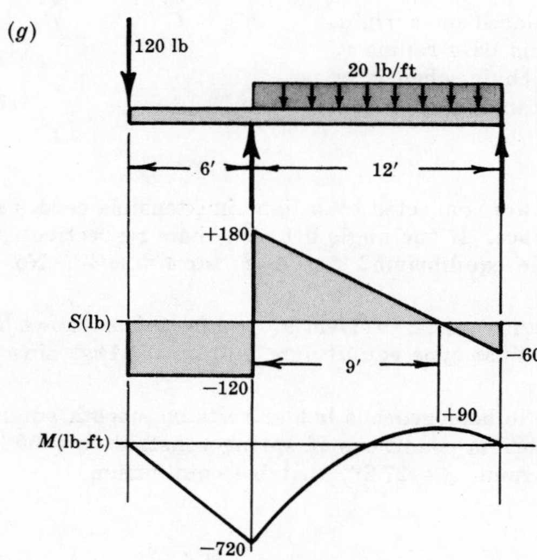

Fig. 21-35

Ans. $0 < x < 6' \begin{cases} S = -120 \text{ lb} \\ M = -120x \text{ lb-ft} \end{cases}$

$6' < x < 18' \begin{cases} S = +300 - 20x \text{ lb} \\ M = -2160 + 300x - 10x^2 \text{ lb-ft.} \end{cases}$

16. Using the method of virtual work, find the force T in the horizontal cross member in terms of the load P and the angle θ. See Fig. 21-36 below. *Ans.* $T = \frac{3}{4}P \tan\theta$

17. In Fig. 21-37 below, the weights W_1 and W_2 are held on the frictionless orthogonal planes by a rigid inextensible bar of length l. Find the equilibrium angle θ. *Ans.* $\tan\theta = (W_2/W_1)\cot\theta_1$

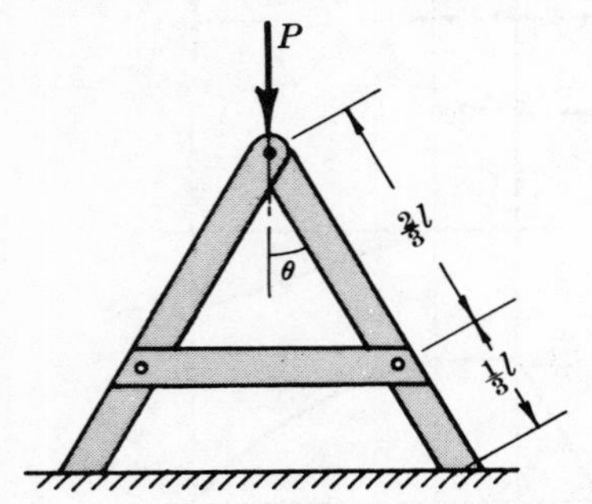

Fig. 21-36

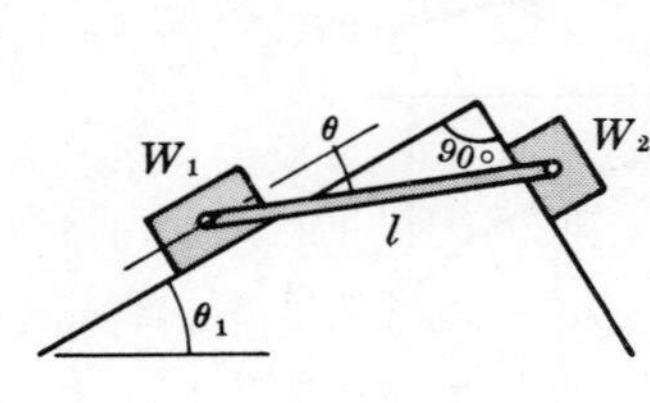

Fig. 21-37

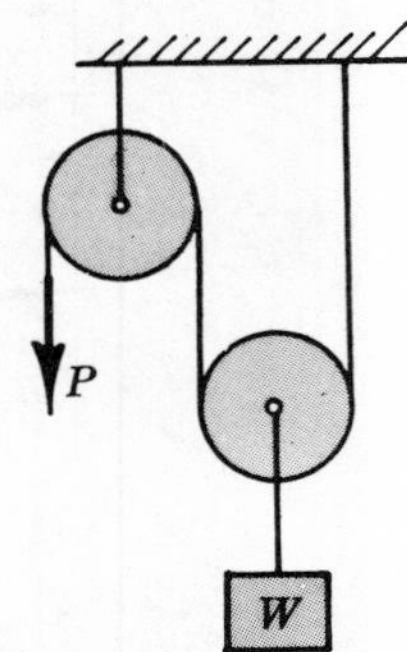

Fig. 21-38

18. Refer to Fig. 21-38 above. Using the method of virtual work find the force P to hold the weight W in equilibrium. *Ans.* $P = \frac{1}{2}W$

19. In Fig. 21-39, the bars AB, AC, HK and KI are a ft long. The other bars are $2a$ ft long. The bars are connected by frictionless pins. Using the method of virtual work determine the relationship between P and Q.
Ans. $Q = 4P$

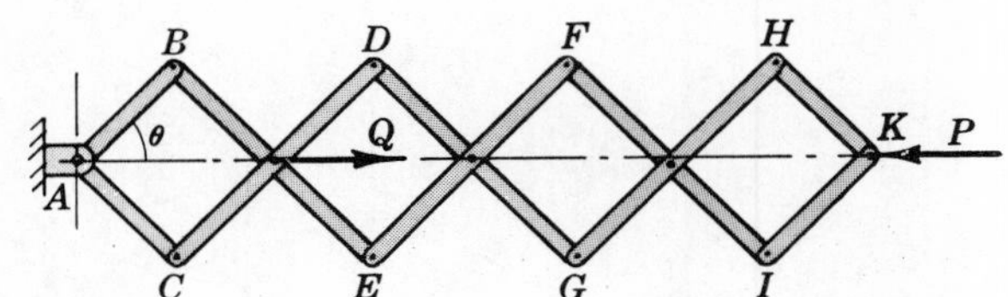

Fig. 21-39

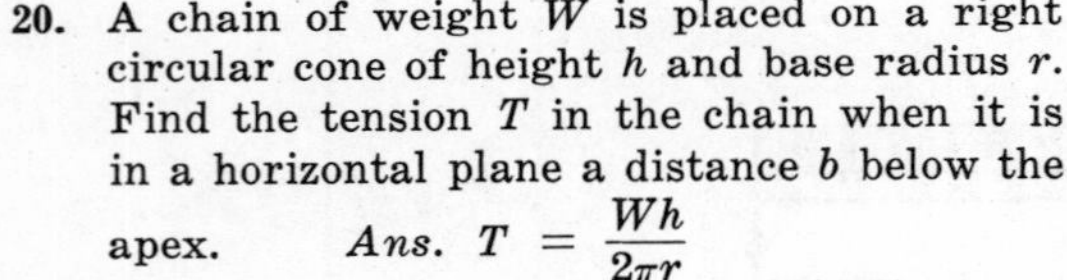

20. A chain of weight W is placed on a right circular cone of height h and base radius r. Find the tension T in the chain when it is in a horizontal plane a distance b below the apex. *Ans.* $T = \dfrac{Wh}{2\pi r}$

21. The two weights w and W are connected by a light inextensible cord, as shown in Fig. 21-40 below, on the smooth cylindrical surface. If the angle between their respective radii is 90°, determine the angle of equilibrium. Is it stable equilibrium? *Ans.* $\tan\theta = w/W$. No.

22. Suppose the weights of the preceding problem are connected as shown in Fig. 21-41 below. Determine the angle of equilibrium. What type equilibrium is it? *Ans.* $\sin\theta = w/W$. Unstable

23. In Fig. 21-42 below, the 40 lb homogeneous ladder rests on smooth surfaces. The spring is unstretched when $\theta = 0°$. Study equilibrium conditions if spring constant $k = 50$ lb/ft.
Ans. $\theta = 0°$, stable equilibrium; $\theta = 27.2°$, unstable equilibrium

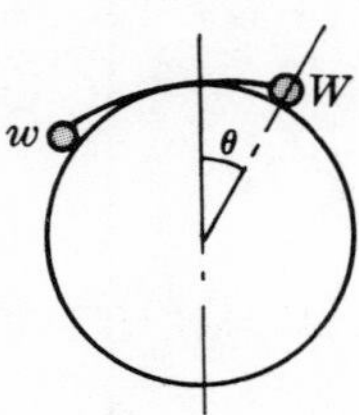

Fig. 21-40

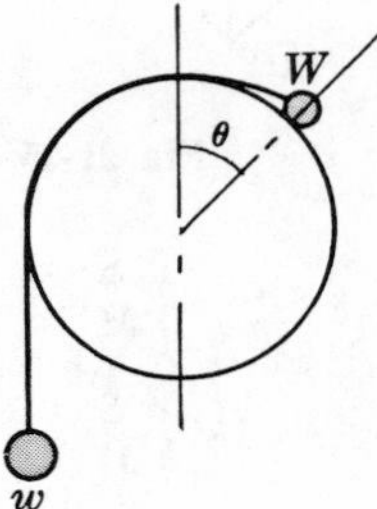

Fig. 21-41

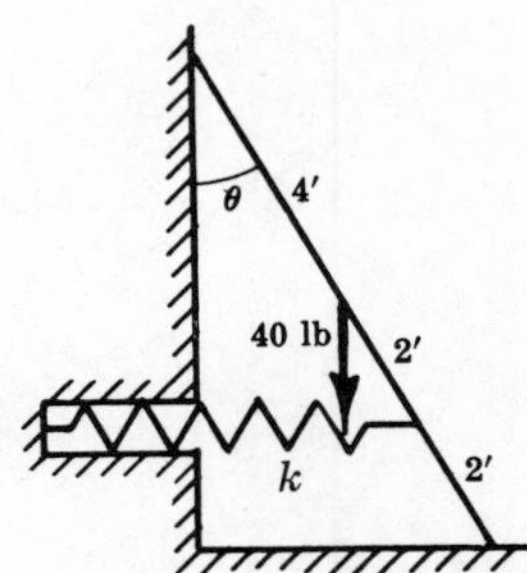

Fig. 21-42

INDEX